MOLECULAR AND CELLULAR ASPECTS OF MUSCLE CONTRACTION

ADVANCES IN EXPERIMENTAL MEDICINE AND BIOLOGY

A Continuation Order Plan is available for this series. A continuation order will bring delivery of each new volume immediately upon publication. Volumes are billed only upon actual shipment. For further information please contact the publisher.

MOLECULAR AND CELLULAR ASPECTS OF MUSCLE CONTRACTION

Edited by

Haruo Sugi
Teikyo University
Tokyo, Japan

Supported by the Fujihara Foundation of Science

Springer Science+Business Media, LLC

Library of Congress Cataloging-in-Publication Data

Molecular and cellular aspects of muscle contraction/edited by Haruo Sugi.
    p. ; cm. (Advances in experimental medicine and biology; v. 538)
    "Supported by the Fujihara Foundation of Science."
    Proceedings of a seminar.
    Includes bibliographical references and index.
    ISBN 978-1-4613-4764-4
      1. Muscle contraction—Congresses. 2. Muscle cells—Congresses. 3. Muscles—Molecular aspects—Congresses. I. Sugi, Haruo, 1933– II. Fujihara Kagaku Zaidan. III. Series.

QP312.M757 2003
573.7'5—dc22

          2003060728

ISBN 978-1-4613-4764-4   ISBN 978-1-4419-9029-7 (eBook)
DOI 10.1007/978-1-4419-9029-7

Proceedings of the Muscle Symposium, which was supported by a grant from the Fujihara Foundation of Science, held as the Fourth Fujihara Seminar, October 28–November 1, 2002, at Hakone, Japan.

ISSN 0065-2598

ISBN 978-1-4613-4764-4

©2003 Springer Science+Business Media New York
Originally published by Kluwer Academic / Plenum Publishers, New York 2003
Softcover reprint of the hardcover 1st edition 2003

http://www.wkap.nl/

10 9 8 7 6 5 4 3 2 1

A C.I.P. record for this book is available from the Library of Congress

PREFACE

This volume presents the proceedings of a muscle symposium, which was supported by the grant from the Fujihara Foundation of Science to be held as the Fourth Fujihara Seminar on October 28 - November 1, 2002, at Hakone, Japan. The Fujihara Seminar covers all fields of natural science, while only one proposal is granted every year. It is therefore a great honor for me to be able to organize this meeting.

Before this symposium, I have organized muscle symposia five times, and published the proceedings: " Cross-bridge Mechanism in Muscle Contraction (University of Tokyo Press, 1978), "Contractile Mechanisms in Muscle " (Plenum, 1984); "Molecular Mechanisms of Muscle Contraction" (Plenum, 1988); "Mechanism of Myofilament Sliding in Muscle contraction" (Plenum, 1993); "Mechanisms of Work Production and Work Absorption in Muscle" (Plenum, 1998). As with these proceedings, this volume contains records of discussions made not only after each presentation but also during the periods of General Discussion, in order that general readers may properly evaluate each presentation and the up-to-date situation of this research field.

It was my great pleasure to have Dr. Hugh Huxley, a principal discoverer of the sliding filament mechanism in muscle contraction, in this meeting. On my request, Dr. Huxley kindly gave a special lecture on his monumental discovery of myofilament-lattice structure by X-ray diffraction of living skeletal muscle. I hope general readers to learn how a breakthrough in a specific research field can be achieved.

When a Cold Spring Harbor Symposium, entitled "The Mechanism of Muscle Contraction" was held in 1971 (Cold Spring Harb. Symp. Quant. Biol. 37, 1972), many investigators felt that the mechanism of muscle contraction was almost solved. At that time, however, I felt reluctance against such optimism, and this was my motivation to organize several symposia in the past. At present, everybody participating in this meeting feels that the mechanism of muscle contraction still remains to be a mystery, clearly indicating that Nature is indeed much wiser than human being.

Finally, I hope that this volume will stimulate young investigators and lead them to achieve breakthroughs in this research field.

Haruo Sugi

ACKNOWLEDGEMENT

The editor would like to express sincere thanks to the Fujihara Foundation of Science for generous financial support, which made this meeting possible.

The editor also owes a debt of gratitude to Ms. Yuka Suzuki and Ms. Ibuki Shirakawa at Teikyo University for their enormous efforts in preparing the discussion records in this volume, and Drs. Teizo Tsuchiya, Takenori Yamada, Shigeru Chaen, Seiryo Sugiura, and Kaoru Katoh for compiling the indices.

CONTENTS

I. MOLECULAR FACTORS INFLUENCING CARDIAC AND SMOOTH MUSCLE CONTRACTION

II. MOLECULAR MECHANISM OF ACTIN-MYOSIN INTERACTION

III. MOLECULAR BASIS FOR REGULATORY MECHANISM OF MUSCLE CONTRACTION

IV. STRUCTURAL CHANGES DURING CONTRACTION

V. NONMUSCLE MOTILE SYSTEMS

VI. EXCITATION-CONTRACTION COUPLING

VII. MUSCLE MECHANICS

VIII. MUSCLE FATIGUE AND ENERGETICS

I. MOLECULAR FACTORS INFLUENCING CARDIAC AND SMOOTH MUSCLE CONTRACTION

MYOSIN LIGHT CHAIN COMPOSITION IN NON-FAILING DONOR AND END-STAGE FAILING HUMAN VENTRICULAR MYOCARDIUM

J. van der Velden[1], Z. Papp[2], N.M. Boontje[1], R. Zaremba[1], J.W. de Jong[3], P.M.L. Janssen[4], G. Hasenfuss[4], and G.J.M. Stienen[1*]

1. INTRODUCTION

In the heart changes occur in the composition and phosphorylation status of different contractile proteins during development and under pathological conditions[1]. Such changes may have important consequences for myocardial pump function. In rat myocardium changes in myosin heavy chain (MHC) composition during development and cardiac disease are correlated with changes in myocardial ATPase activity[2, 3] and in Ca^{2+}-responsiveness of force[4]. However, species differences in myocardial contractile protein composition make it difficult to extrapolate these findings to the human heart. Indeed, in several studies it has been proposed that the concepts derived from studies in rodents are not applicable to large mammals[5, 6].

It has been suggested that in contrast to alterations in MHC composition in rodent hearts, the diseased human myocardium is characterized by alterations in myosin light chain 1 composition[5, 7]. Indeed, expression of atrial light chain 1 (ALC-1) has been reported in failing human ventricles[7], while the opposite shift was observed in hypertrophied human atria[8]. Moreover, replacement of atrial light chain 2 (ALC-2) by ventricular light chain 2 (VLC-2) has been reported in hypertrophied human atria[8], while degradation of VLC-2 has been observed in the ventricles of patients with idiopathic dilated cardiomyopathy[9].

Apart from changes in isoform expression, changes may occur in the phosphorylation status of myosin light chains (MLCs). The phosphorylation status is determined by the action of several protein kinases and protein phosphatases, which may be altered under

* [1]Laboratory for Physiology, Institute for Cardiovascular Research (ICaR-VU), VU University Medical Center, Amsterdam, the Netherlands. [2]UDMHSC Department of Cardiology, Debrecen, Hungary. [3]Thorax Center, Erasmus University Rotterdam, the Netherlands. [4]Department of Cardiology and Pneumology, Georg-August-University Göttingen, Göttingen, Germany.

pathological conditions. Two isoforms of the VLC-2 are known (VLC-2 and VLC-2*)[10], which may both be phosphorylated by Ca^{2+}/calmodulin-dependent myosin light chain kinase (MLCK)[11] and protein kinase C (PKC)[12] and dephosphorylated by light chain phosphatase[13]. Recently, Arrell et al.[14] have shown that in addition to VLC-2, the ventricular light chain 1 (VLC-1) isoform may be phosphorylated, possibly by PKC. In human heart failure increased activity has been reported for PKC[15] and for Ca^{2+}/calmodulin-dependent protein kinase (CaM-kinase), which may phosphorylate and activate MLCK[16]. In addition, type 1 phosphatase activity was increased in failing human myocardium[17]. Recently a decreased VLC-2 phosphorylation was observed in end-stage failing human hearts[18, 19].

The present chapter summarizes our efforts to identify the functional relevance of changes in isoform composition and phosphorylation status of MLCs in ventricular myocardium from end-stage failing human hearts. In addition, we show to what extent these changes contribute to the increased myocardial Ca^{2+}-responsiveness of force development previously observed in end-stage human heart failure[7, 18, 20]. Furthermore, since basal VLC-2 phosphorylation is decreased in end-stage human heart failure, the effect of VLC-2 dephosphorylation by Protein Phosphatase-1 on isometric force development was studied.

2. PROTEIN ANALYSIS

Left ventricular biopsies were taken during heart transplantation surgery from 14 explanted end-stage failing (New York Heart Association class IV) hearts and from 6 nonfailing donor hearts. Heart failure resulted from ischemic (n=9), dilated (n=4) or unknown (n=1) cardiomyopathy. Biopsies were transferred in cardioplegic solution and upon arrival in the laboratory, stored in liquid nitrogen. Samples were obtained after informed consent and with approval of the local Ethical Committees. One part of the biopsy was used for protein analysis and the other part was used for myocyte isolation[21].

Two dimensional gel electrophoresis (2D-PAGE) as described previously[18, 19] was performed to determine the isoform expression and degree of phosphorylation of ventricular MLCs. Samples were treated with trichloroacetic acid to fixate the phosphorylation status of contractile proteins[22]. The ratio between VLC-1 and VLC-2, re-expression of ALC-1 and phosphorylation of VLC-1 was investigated using a mini-protean II system (Bio-Rad, Hercules, CA, USA) as described previously by Morano et al[7, 22]. Isoelectric focusing was performed in glass capillary tubes (length 7 cm, diameter 1 mm) with a pH gradient of 4.0-6.5 (Amersham Pharmacia Biotech AB, Uppsala, Sweden). Capillary gels contained 6.8% total acrylamide (acrylamide to bis-acrylamide ratio 17.5:1). Comparable samples (300 µg dry weight) were loaded in each tube. In the second dimension the proteins were separated based on molecular weight in a slab gel using an acrylamide to bis-acrylamide ratio of 37.5:1 (total acrylamide 13.5%; pH 8.8).

To investigate phosphorylation of VLC-2, trichloroacetic acid treated samples (600 µg dry weight) were loaded on immobiline strips with a pH gradient of 4.5 to 5.5 (Amersham Pharmacia Biotech, Uppsala, Sweden)[18, 19]. In the second dimension proteins were separated by SDS-PAGE[23]. All gels were stained with Coomassie blue, scanned and analyzed using Image Quant (Molecular Dynamics).

To investigate Protein Phosphatase-1 (PP-1) specificity a suspension of Triton-skinned donor cardiomyocytes was incubated in 1 ml of relaxing solution with and without PP-1 (0.5 U/ml, lot no. 16757; Upstate Biotechnology) for 60 minutes at room temperature. Subsequently, TCA-treated cells (75 µg dry weight) were analyzed by 2D-PAGE. These gels were silver-stained to enhance resolution.

3. MECHANICAL MEASUREMENTS ON ISOLATED MYOCYTES

Cardiomyocytes were mechanically isolated and mounted in the experimental set-up as described previously[21]. Before mechanical isolation, tissue was defrosted in relaxing solution (pH 7.0; in mmol/l: free Mg^{2+} 1, KCl 145, EGTA 2, ATP 4, imidazole 10). During the isolation the tissue was kept on ice. Isolated myocytes were immersed for 5 minutes in relaxing solution containing 1% Triton X-100. Triton removes soluble and membrane-bound kinases and phosphatases and thereby arrests the phosphorylation status of myofibrillar proteins. To remove Triton, cells were washed twice in relaxing solution. Thereafter, a single myocyte was attached between a force transducer and a piezoelectric motor (Figure 1).

Isometric force measurements were performed at 15°C and a sarcomere length, measured in relaxing solution, of 2.2 µm. The composition of relaxing and activating solutions used during force measurements was calculated as described by Fabiato[24]. The pCa, i.e. $-\log_{10}[Ca^{2+}]$, of the relaxing and activating solution (pH 7.1) were, respectively, 9 and 4.5. Solutions with intermediate free $[Ca^{2+}]$ were obtained by mixing of the activating and relaxing solutions. After the first control activation at saturating (maximal) $[Ca^{2+}]$ (pCa=4.5), resting sarcomere length was readjusted to 2.2 µm, if necessary. The second control measurement was used to calculate maximal isometric tension (i.e. force divided by cross-sectional area). The next 4 to 5 measurements were carried out at submaximal $[Ca^{2+}]$ followed by a control measurement. Force values obtained in solutions with submaximal $[Ca^{2+}]$ were normalized to the interpolated control values. In a separate series of force measurements, myocytes from 3 donor and 6 failing hearts were incubated in relaxing solution containing 0.5 U/ml PP-1 and 6 mM dithiothreitol (DTT) for 60 minutes at 20°C after the initial force-pCa series. Thereafter, the force-pCa series was repeated.

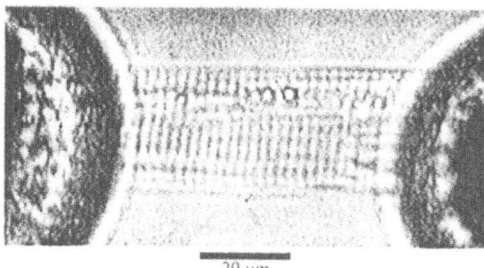

20 µm

Figure 1. Single cardiomyocyte from a failing heart in relaxing solution glued between a force transducer and a piezoelectric motor.

Force-pCa relations were fit to a modified Hill equation:

$$F(Ca^{2+})/F_0 = (1-k) \cdot [Ca^{2+}]^{nH}/(Ca_{50}^{nH} + [Ca^{2+}]^{nH}) + k$$

F is steady-state force. F_0 denotes this force at saturating $[Ca^{2+}]$, nH reflects the steepness of the relationship, and Ca_{50} (or pCa_{50}) represents the midpoint of the relation. Force-pCa relations obtained after PP-1 contained a small component approximated by an offset, k.

Values are given as means ± SEM of n experiments. n is number of hearts if not stated otherwise. Mean values for donor and failing samples were compared using an unpaired Student *t*-tests. Paired Student *t*-tests were used when comparing maximal force and Ca^{2+}-sensitivity of single cardiomyocytes before and after PP-1 treatment. A two tailed P-value of less than 0.05 was considered significant.

4. PROTEIN COMPOSITION

4.1. Myosin Light Chain Isoform

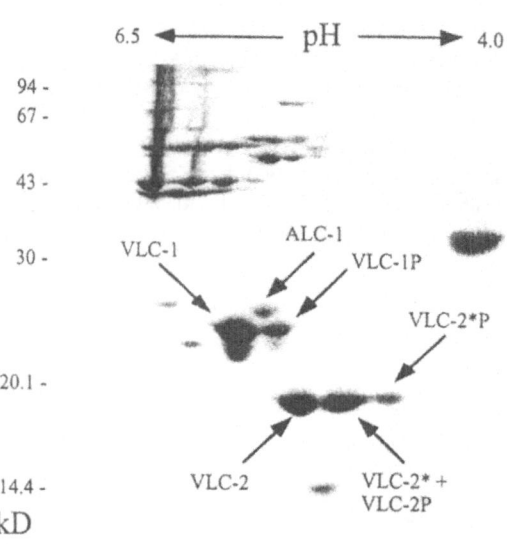

Figure 2. Mini 2D-gel to illustrate the expression of atrial light chain 1 (ALC-1) in a failing ventricular sample. Horizontally: isoelectric point, vertically: molecular weight. VLC-1, ventricular light chain 1; VLC-1P, phosphorylated VLC-1; VLC-2, ventricular light chain 2 composed of two unphosphorylated isoforms (VLC-2 and VLC-2*), which are both partly phosphorylated (respectively, VLC-2P and VLC-2*P).

Figure 2 shows a mini 2D-gel (pH gradient: 4.0-6.5) from failing ventricular tissue expressing a small amount of ALC-1. Comparable amounts of ALC-1 were observed in two donor samples and in six other end-stage failing ventricles, ranging from 0.3 to 17.9% of total myosin light chain 1. The average values obtained corresponded to 1.9±1.4% in donor and to 2.4±1.4% in failing ventricles. To assess possible VLC-2 degradation the ratio of total VLC-1 (including ALC-1) to VLC-2 was analyzed. No significant difference was found in this ratio between donor (0.92±0.13) and failing (1.19±0.26) hearts.

4.2. Phosphorylation of Myosin Light Chains

VLC-2 in the human heart consists of two isoforms (VLC-2 and VLC-2*), which may both be phosphorylated (VLC-2P and VLC-2*P). Figure 3 illustrates that the degree of VLC-2 phosphorylation was lower in failing than in donor tissue. Phosphorylation of the VLC-2 isoform was significantly decreased in failing compared with donor ventricles from 32.1 to 19.5 % (Table 1). The percentage of phosphorylated and unphosphorylated VLC-2* isoform did not differ between failing and donor myocardium.

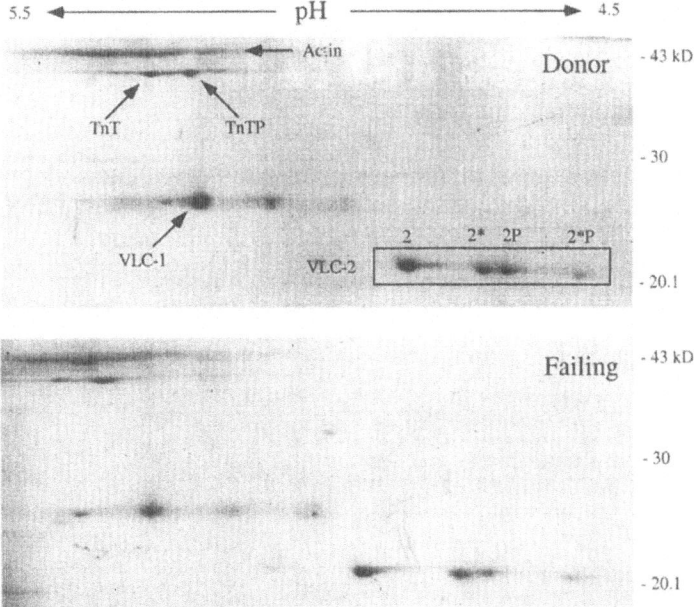

Figure 3. 2D-gels to illustrate ventricular light chain 2 composition in donor and failing ventricular tissue. VLC-2 was less phosphorylated in the failing heart. TnT and TnTP, unphosphorylated and mono-phosphorylated troponin T; VLC-1, ventricular light chain 1; VLC-2, ventricular light chain 2 composed of two isoforms (2 and 2*), which are both partly phosphorylated (2P and 2*P, respectively).

Table 1. Phosphorylation of ventricular light chain 2

Protein	Donor (n=5)	Heart failure (n=14)
VLC-2	38.1±3.9	51.2±2.1*
VLC-2*	20.6±1.2	22.5±1.2
VLC-2P	32.1±2.6	19.5±1.6*
VLC-2*P	9.2±2.4	6.8±0.8

Values are given as percentage of total VLC-2. *$P<0.05$, donor versus failing. n is number of hearts.

Recently it was shown that in addition to VLC-2, VLC-1 may be phosphorylated[14]. On our mini 2D-gels (Figure 2) two spots were evident at the level of VLC-1, which were both recognized by a specific antibody directed against VLC-1 in Western immunoblotting (not shown). Therefore, these two VLC-1 spots may represent unphosphorylated and phosphorylated VLC-1 (respectively, VLC-1 and VLC-1P in Figure 2). The amount of phosphorylated VLC-1 as a percentage of total myosin light chain 1 did not differ between donor (11.6±2.6%) and failing (11.9±1.2%) hearts.

5. FORCE DEVELOPMENT IN CARDIAC MYOCYTES

Maximal isometric tension amounted to 32.4±4.5 and 30.1±3.0 kN/m^2 in donor (n=6; 47 myocytes) and failing (n=14; 59 myocytes) hearts, respectively. Passive tension did not differ between donor (1.7±0.2 kN/m^2) and failing (2.1±0.2 kN/m^2) myocardium. In Figure 4 recordings of isometric force development obtained at saturating (pCa 4.5) and at sub-maximal [Ca^{2+}] (pCa 5.6) are shown.

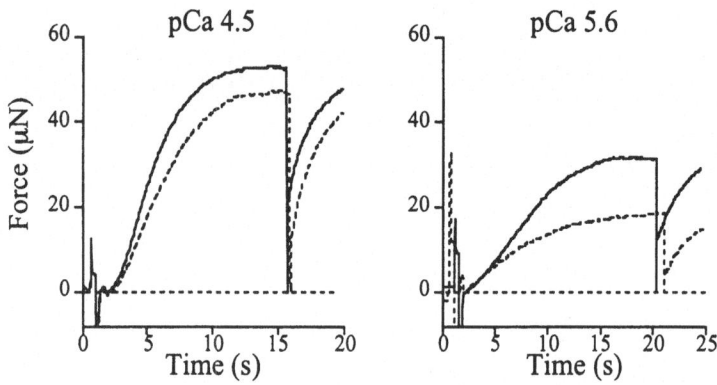

Figure 4. Recordings of isometric force development during maximal (pCa 4.5) and submaximal (pCa 6.5) activation before (continuous recording) and after (dashed recording) PP-1 incubation in a failing myocyte. The abrupt changes in force mark the transitions of the preparation through the interface between solution and air. The dashed horizontal lines indicate the passive force level, determined from the slack test (near the end of each recording).

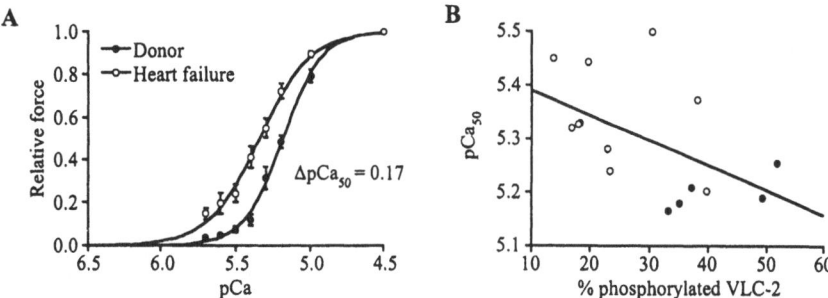

Figure 5. A. Ca^{2+}-sensitivity of the contractile apparatus was significantly higher in failing than in donor hearts. **B.** A negative correlation was present between Ca^{2+}-sensitivity of the contractile apparatus (pCa_{50}) and percentage VLC-2 phosphorylation: $pCa_{50}=(5.44\pm0.07)-(0.005\pm0.002)\cdot\%$ phosphorylated VLC-2 (R=-0.53, P<0.05).

The average force-pCa relationships obtained in donor (n=6) and failing (n=10) hearts are shown in Figure 5A. It can be noted that Ca^{2+}-responsiveness was substantially increased in failing hearts (pCa_{50} 5.35±0.03) compared to donor myocardium (pCa_{50} 5.18±0.02)(P<0.05). The steepness of the force-pCa curves, nH, did not differ significantly between donor (3.74±0.18) and failing (3.28±0.17) myocardium.

No significant correlation was present between Ca^{2+}-sensitivity of the contractile apparatus (pCa_{50}) and percentage of ALC-1 (not shown), while an inverse correlation was observed between pCa_{50} and percentage of phosphorylated VLC-2 (Figure 5B; correlation coefficient R=-0.53; P<0.05).

6. EFFECT OF PP-1 ON CALCIUM SENSITIVITY

To investigate PP-1 specificity a suspension of donor myocytes was incubated without (control) and with PP-1 and analyzed on silver-stained 2D-gels (Figure 6A). The corresponding densitometric scans shown in Figure 6B indicate that VLC-2 was completely dephosphorylated by incubation of myocytes in PP-1 containing (0.5 U/ml; 60 minutes) relaxing solution (dashed line), while incubation in relaxing solution without PP-1 did not alter VLC-2 phosphorylation (continuous line). Since silver-staining is linear only in a narrow concentration range, these gels cannot be used for detailed quantitative analysis, but nevertheless it can be concluded that after PP-1 treatment only negligible quantities of the phosphorylated VLC-2 isoforms remain. Phosphorylation of TnT and VLC-1 was preserved after PP-1 treatment (data not shown).

Figure 4 illustrates that maximal force (pCa 4.5) decreased by approximately 10% after PP-1. This decline was attributable to the duration of the incubation period as it was also found in control experiments described below. However, PP-1 significantly decreased Ca^{2+}-sensitivity of force in donor (n=3; 13 myocytes) and failing (n=6; 15 myocytes) hearts (Figure 7). The shift in pCa_{50} after PP-1 treatment was significantly larger in failing (0.20 pCa units) than in donor (0.10 pCa units) cardiomyocytes. Thus PP-1 caused a marked reduction in the difference in Ca^{2+}-responsiveness between donor and failing hearts to 0.08 pCa units (P<0.05).

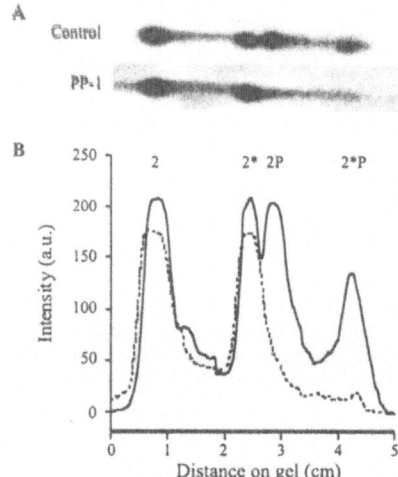

Figure 6. A. VLC-2 region from silver-stained 2D-gels of donor myocytes treated without (Control) and with (PP-1). See abbreviations Figure 3. Panel **B** shows corresponding densitometric scans of VLC-2 composition in control myocytes (continuous line) and PP-1 treated myocytes (dashed line). PP-1 specifically dephosphorylated VLC-2, while TnT and VLC-1 phosphorylation were preserved (not shown).

At low $[Ca^{2+}]$ (pCa>6) relative force values were somewhat elevated after PP-1 in both donor and failing myocytes. This could be due to limitations in the determination of the passive force levels in the myocytes and/or changes therein after PP-1 treatment. The pCa_{50} and nH values before and after PP-1 are summarized in Table 2. The nH values indicate that PP-1 did not alter the steepness of the force-pCa relationships.

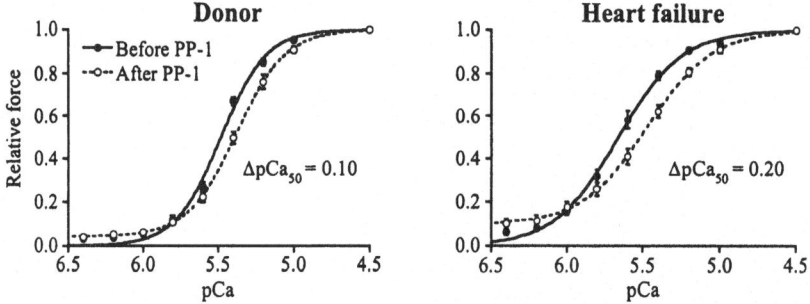

Figure 7. Ca^{2+}-sensitivity of force before and after PP-1 in donor (13 myocytes) and failing (15 myocytes) myocardium. Force was measured at different $[Ca^{2+}]$ before and after PP-1 in the same myocyte. Force at submaximal $[Ca^{2+}]$ was normalized to the control force at saturating $[Ca^{2+}]$.

Table 2. pCa_{50} and nH values before and after PP-1

		Donor (n=13)	Heart failure (n=15)
Before PP-1	pCa_{50}	5.48±0.01	5.66±0.02†
	nH	3.06±0.17	2.38±0.15†
After PP-1	pCa_{50}	5.38±0.02*	5.46±0.03*†
	nH	2.95±0.19	2.29±0.12†

Myocytes from 3 donor and 6 failing hearts were treated with PP-1. Force was measured at different [Ca^{2+}] before and after PP-1 in the same myocyte. *P<0.05, before versus after PP-1. †P<0.05, donor versus heart failure; n, denote the number of cardiomyocytes.

Control experiments were performed in 2 donor cardiomyocytes to investigate if the decrease in Ca^{2+}-responsiveness could be attributed to PP-1 or was due to time-dependent alterations during the incubation period. These cells were incubated in relaxing solution for 60 minutes at 20°C without addition of PP-1 (time-control). These experiments clearly indicated that no alterations occurred in the midpoint (before: 5.44±0.02; after: 5.45±0.01) and steepness (before: 2.71±0.29; after 2.76±0.13) of the force-pCa relationship after incubation in relaxing solution.

To assess whether the PP-1 induced shift in Ca^{2+}-sensitivity was maximal, 2 failing cardiomyocytes were incubated for 60 minutes with a ten-fold higher dose of PP-1 (5 U/ml). The decrease in pCa_{50} (0.21 pCa units) was similar to the decrease after incubation with 0.5 U/ml PP-1.

Troponin I might also be dephosphorylated by PP-1. TnI dephosphorylation would increase Ca^{2+}-sensitivity of force and could thereby potentially interfere with the effect of VLC-2 dephosphorylation on Ca^{2+}-responsiveness. To investigate if TnI dephosphorylation occurred as a side-effect of PP-1 treatment, myocytes were incubated with protein kinase A (PKA), which is known to phosphorylate TnI and to decrease Ca^{2+}-sensitivity of force subsequent to the PP-1 treatment. In a previous study[25], we observed a small non-significant decrease in Ca^{2+}-responsiveness of force after PKA in donor cardiomyocytes (ΔpCa_{50}=0.02), but a significant decrease in failing (ΔpCa_{50}=0.24) cells [25]. Hence, if dephosphorylation of TnI would occur by PP-1 the effect of PKA on Ca^{2+}-sensitivity of force would be enlarged. However, after PP-1 treatment, PKA did not significantly alter Ca^{2+}-responsiveness of force in donor cardiomyocytes (n=4), while the decrease in Ca^{2+}-responsiveness after PKA in failing cardiomyocytes (n=4) was smaller (ΔpCa_{50}=0.10) than in the absence of PP-1[25]. From these data we consider it highly unlikely that PP-1 caused dephosphorylation of TnI. These experiments also showed that the difference in Ca^{2+}-responsiveness between donor and failing myocytes was completely abolished by the combined action of PP-1 and PKA. This suggests that the difference observed between Ca^{2+}-responsiveness of force in donor and end-stage failing human hearts can be fully attributed to the combination of alterations in both VLC-2 and TnI phosphorylation.

7. Ca^{2+}-SENSITIVITY IN END-STAGE FAILING HEARTS

7.1. Myosin Heavy Chains versus Myosin Light Chains

In accordance with previous studies, Ca^{2+}-sensitivity of the contractile apparatus was increased in end-stage failing human hearts compared to non-failing donor hearts[7, 18, 20]

Alternative explanations have been given for the origin of the increased Ca^{2+}-responsiveness in human heart failure. A study by Metzger et al.[3] indicated a role for the MHC composition in Ca^{2+}-sensitivity of the contractile apparatus. A shift from α- to β-MHC in hypothyroid rats was associated with a decrease in Ca^{2+}-responsiveness. Moreover, recently it was demonstrated that a minor shift in MHC composition (12%) was sufficient to alter power output in rat cardiomyocytes by 50%[26]. Although previous studies have reported predominance of β-MHC in human ventricles, in a recent study[27] some α-MHC protein (7%) was found in all non-failing human ventricles, while α-MHC could only be detected in one out of ten failing ventricles. Thus, analogous to rodent hearts, a shift in MHC composition with human heart failure might be responsible for Ca^{2+}-sensitivity changes, albeit that the direction of the shift in rodents would be opposite to the shift observed in human heart failure. However, previous analysis of the MHC composition in our samples showed small amounts of α-MHC protein only in one donor and one failing sample. Therefore, we consider it unlikely that a shift in MHC composition is the cause of the difference in Ca^{2+}-responsiveness observed between non-failing and failing human hearts. This notion was confirmed by the observation that PP-1 and PKA treatment completely abolished the difference in Ca^{2+}-responsiveness between donor and failing hearts.

Previous studies indicated a role for myosin light chains in human heart failure[5, 7-9] According to Morano et al.[7] the increased Ca^{2+}-sensitivity of the contractile apparatus is due to expression of atrial light chain 1 (ALC-1) in the left ventricle. A positive correlation between ALC-1 expression and Ca^{2+}-sensitivity of the contractile apparatus was found, though ALC-1 was not present in all failing tissues[7]. Margossian et al. have suggested that a decrease in MgATPase activity at high calcium concentrations is partly due to proteolysis of VLC-2 in idiopathic dilated cardiomyopathy[9]. In our group of donors and patients with heart failure, expression of ALC-1 was small and did not correlate with pCa_{50}. Moreover, the ratio between MLC-1 and MLC-2 did not differ between donor and failing myocardium, indicating that the stoichiometry of MLC-1 versus MLC-2 was preserved in heart failure. These results suggest that the observed increase in Ca^{2+}-responsiveness is not due to degradation or re-expression of MLCs in the failing human ventricle. However, they do not completely rule out the possibility that ALC-1 affects Ca^{2+}-sensitivity of force development, bur they reveal that high ALC-1 expression is not a general feature in human heart failure.

7.2. Phosphorylation Status of Ventricular Light Chains

Apart from the changes in isoform composition alterations may occur in the phosphorylation status of MLCs due to altered activities of kinases[15, 16] and phosphatases[17]. Phosphorylation of VLC-2 was decreased in failing human hearts, though no difference was observed in phosphorylation levels of VLC-1 and the less abundant VLC-2* isoform. An increase in protein phosphatase 1 activity has been found in failing human hearts[17], which may account for the decreased VLC-2 phosphorylation levels observed in the failing human samples. It is noteworthy that, *in vivo*, PKA phosphorylates and thereby activates phosphatase inhibitor 1[28], which is known to be a potent inhibitor of PP-1. Thus, β-adrenergic receptor desensitization observed in human heart failure might result in a decrease in VLC-2 phosphorylation.

A significant inverse correlation was found between pCa_{50} and percentage of phosphorylated VLC-2, that is, VLC-2 phosphorylation is associated with a decrease in Ca^{2+}-responsiveness. However, Morano[13] has shown that phosphorylation of VLC-2 causes an increase in Ca^{2+}-responsiveness. An explanation for our paradoxical findings could be as follows. In addition to thick filament proteins, the phosphorylation status of thin filament proteins may be altered. Previously, we observed a decrease in TnI phosphorylation in failing human myocardium, while TnT phosphorylation was unaltered[19]. Therefore, the Ca^{2+}-desensitizing effect of decreased basal VLC-2 phosphorylation may be overruled by the effect of decreased TnI phosphorylation, which results in an increased Ca^{2+}-sensitivity of the contractile apparatus in end-stage failing human hearts.

7.3. Effect of VLC-2 Dephosphorylation

Surprisingly, although the basal level of VLC-2 phosphorylation was decreased in failing myocardium, the response to VLC-2 dephosphorylation was enhanced in human heart failure. After PP-1 the difference in Ca^{2+}-responsiveness between failing and donor myocardium was diminished, while it was completely abolished after subsequent treatment with PKA. This suggests that the difference in Ca^{2+}-responsiveness does not reflect intrinsic differences in contractile protein isoform composition, but rather differences in endogenous phosphorylation status of thin and thick filament proteins. The enhanced response to VLC-2 dephosphorylation in failing myocytes might also result from differences in TnI phosphorylation levels between donor and failing hearts discussed above. Since in non-failing donor hearts Ca^{2+}-responsiveness is lower compared to failing hearts, the desensitizing effect of PP-1 may already be saturated, while failing hearts may be more susceptible to VLC-2 dephosphorylation due to the enhanced Ca^{2+}-sensitivity of force.

An alternative explanation for the difference in responsiveness to VLC-2 dephosphorylation between donor and failing hearts could be differences in the phosphorylation level of myosin binding protein-C (MyBP-C). It has been suggested that phosphorylation of MyBP-C performs a permissive role in contraction of the heart muscle[29, 30]. Phosphorylation of MyBP-C would alleviate the axial movement of the myosin head and would thereby facilitate the actin-myosin interaction[29]. Thus, a difference in MyBP-C phosphorylation between donor and failing hearts could alter the response to VLC-2 (de)phosphorylation. An increased activity of Ca^{2+}/calmodulin-dependent protein kinase, which may phosphorylate MyBP-C[31], has been found in idiopathic dilated cardiomyopathy[16]. Knowledge on the phosphorylation status of MyBP-C in healthy and failing myocardium is of great interest and warrants further study, particular in human tissue.

The decrease in Ca^{2+}-sensitivity of force after dephosphorylation of VLC-2 is in agreement with the results obtained after phosphorylation of MLC-2 in cardiac and skeletal muscle[13, 32]. The effect of MLC-2 dephosphorylation on force development in human cardiomyocytes may be explained by a decrease in the apparent rate of crossbridge attachment (f_{app}). It has been proposed that upon VLC-2 dephosphorylation myosin heads move toward the backbone of the thick filament away from the thin filament, diminishing the probability of actin to myosin binding[33]. Hence, the decrease in crossbridge attachment may originate from a structural change in the thick filament.

8. SUMMARY

The increased Ca^{2+}-responsiveness in end-stage human heart failure cannot be attributed to contractile protein isoform changes, but rather is the complex resultant of changes in degree of phosphorylation of VLC-2 and TnI. Despite the decreased basal level of VLC-2 phosphorylation the response to VLC-2 dephosphorylation is enhanced in failing myocytes, which might result from differences in endogenous phosphorylation of thin and thick filament proteins between donor and failing hearts. Taken together decreased VLC-2 phosphorylation in end-stage human heart failure might represent a compensatory process leading to an improvement of myocardial contractility by opposing the detrimental effects of increased Ca^{2+}-responsiveness of force and impaired Ca^{2+}-handling on diastolic function.

9. ACKNOWLEDGMENTS

Support by the Netherlands Heart Foundation (grant 99.155) is gratefully acknowledged.

10. REFERENCES

1. B. Swynghedauw. Developmental and functional adaptation of contractile proteins in cardiac and skeletal muscles. Physiol. Rev. 66, 710-771 (1968).
2. J.J. Mercadier, A.M. Lompre, C. Wisnewsky, J.L. Samuel, J. Bercovici, B. Swynghedauw, and K. Schwartz. Myosin isoenzymic changes in several models of rat cardiac hypertrophy. Circ. Res. 49, 525-532 (1981).
3. L. Gorza, P. Pauletto, A.C. Pessina, S. Sartore, and S. Schiaffino. Isomyosin distribution in normal and pressure-overloaded rat ventricular myocardium. Circ. Res. 49, 1003-1009 (1981).
4. J.M. Metzger, P.A. Wahr, D.E. Michele, F. Albayya, and M.V. Westfall. Effects of myosin heavy chain isoform switching on Ca^{2+}-activated tension development in single adult cardiac myocytes. Circ. Res. 84, 1310-1317 (1999).
5. O. Ritter, H.P. Luther, H. Haase, L.G. Baltas, G. Baumann, H.D. Schulte, and I. Morano. Expression of atrial myosin light chains but not α-myosin heavy chains is correlated in vivo with increased ventricular function in patients with hypertrophic obstructive cardiomyopathy. J. Mol. Med. 77, 677-685 (1999).
6. S.J. Kim, R.K. Kudej, A. Yatani, Y.K. Kim, G. Takagi, R. Honda, D.A. Colantonio, J.E. van Eyk, D.E. Vatner, R.L. Rasmusson, and S.F. Vatner. A novel mechanism for myocardial stunning involving impaired Ca^{2+}-handling. Circ. Res. 89, 831-837 (2001).
7. I. Morano, K. Hädicke, H. Haase, M. Böhm, E. Erdmann, and M.C. Schaub. Changes in essential myosin light chain isoform expression provide a molecular basis for isometric tension regulation in the failing human heart. J. Mol. Cell. Cardiol. 29, 1177-1187 (1997).
8. P. Cummins. Transitions in human atrial and ventricular myosin light-chain isoenzymes in response to cardiac-pressure-overload-induced hypertrophy. Biochem. J. 205, 195-204 (1982).
9. S.S. Margossian, P.A.W. Anderson, P.D. Chantler, M. Deziel, P.K. Umeda, H. Patel, W.F. Stafford, P. Norton, A. Malhotra, F. Yang, J.B. Caulfield, and H.S. Slayter. Calcium regulation in the human myocardium affected by dilated cardiomyopathy: A structural basis for impaired Ca^{2+}-sensitivity. Mol. Cell. Biochem. 194, 301-313 (1999).
10. I. Morano, M. Wankerl, M. Böhm, E. Erdman, and J.C. Rüegg. Myosin P-light chain isoenzymes in the human heart: evidence for diphosphorylation of the atrial P-light chain isoform. Basic Res. Cardiol. 84, 298-305 (1989).
11. N. Frearson, and S.V. Perry. Phosphorylation of the light-chain components of myosin from cardiac and red skeletal muscles. Biochem. J. 151, 99-107 (1975).
12. R.C. Venema, R.L. Raynor, T.A. Noland, and J.F. Kuo. Role of protein kinase C in the phosphorylation of cardiac myosin light chain 2. Biochem. J. 294, 401-406 (1993).

13. I. Morano. Effects of different expression and posttranslational modifications of myosin light chains on contractility of skinned human cardiac fibers. Basic Res. Cardiol. 87, 129-141 (1992).

14. D.K. Arrell, I. Neverova, H. Fraser, E. Marban, and J.E. van Eyk. Proteomic analysis of pharmacologically preconditioned cardiomyocytes reveals novel phosphorylation of myosin light chain 1. Circ. Res. 89, 480-487 (2001).

15. N. Bowling, R.A. Walsh, G. Song, T. Estridge, G.E. Sandusky, R.L. Fouts, K. Mintze, T. Pickard, R. Roden, M.R. Bristow, H.N. Sabbah, J.L. Mizrahi, G. Gromo, G.L. King, and C.J. Vlahos. Increased protein kinase C activity and expression of Ca^{2+}-sensitive isoforms in the failing human heart. Circulation 99, 384-391 (1999).

16. U. Kirchhefer, W. Schmitz, H. Scholz, and J. Neumann. Activity of cAMP-dependent protein kinase and Ca^{2+}/calmodulin-dependent protein kinase in failing and nonfailing human hearts. Cardiovasc. Res. 42, 254-261 (1999).

17. J. Neumann, T. Eschenhagen, L.R. Jones, B. Linck, W. Schmitz, H. Scholz, and N. Zimmermann. Increased expression of cardiac phosphatases in patients with end-stage heart failure. J. Mol. Cell. Cardiol. 29, 265-272 (1997).

18. J. van der Velden, L.J. Klein, R. Zaremba, N.M. Boontje, M.A.J.M. Huybregts, W. Stooker, J. Witkop, L. Eijsman, J.W. de Jong, C.A. Visser, F.C. Visser, and G.J.M. Stienen. Effects of calcium, inorganic phosphate and pH on isometric force in single skinned cardiomyocytes from donor and failing human hearts. Circulation 104, 1140-1146 (2001).

19. J. van der Velden, Z. Papp, R. Zaremba, N.M. Boontje, J.W. de Jong, V.J. Owen, P.B.J. Burton, P. Goldmann, K. Jaquet, and G.J.M. Stienen. Increased Ca^{2+}-sensitivity of the contractile apparatus in end-stage human heart failure results from altered phosphorylation of contractile proteins. Cardiovasc. Res. (in press).

20. M.R. Wolff, S.H. Buck, S.W. Stoker, M.L. Greaser, and R.M. Mentzer RM. Myofibrillar calcium sensitivity of isometric tension is increased in human dilated cardiomyopathies. J. Clin. Invest. 98, 167-176 (1996).

21. J. van der Velden, L.J. Klein, M. van der Bijl, M.A.J.M. Huybregts, W. Stooker, J. Witkop, L. Eijsman, C.A. Visser, F.C. Visser, and G.J.M. Stienen. Force production in mechanically isolated cardiac myocytes from human ventricular muscle tissue. Cardiovasc. Res. 38, 414-423 (1998).

22. I. Morano, H. Arndt, C. Gärtner, and J.C. Rüegg. Skinned fibers of human atrium and ventricle: Myosin isoenzymes and contractility. Circ. Res. 62, 632-639 (1988).

23. J. van der Velden, L.J. Klein, M. van der Bijl, M.A.J.M. Huybregts, W. Stooker, J. Witkop, L. Eijsman, C.A. Visser, F.C. Visser, and G.J.M. Stienen. Isometric tension development and its calcium sensitivity in skinned myocyte-sized preparations from different regions of the human heart. Cardiovasc. Res. 42, 706-719 (1999).

24. A. Fabiato. Myoplasmic free calcium concentration reached during the twitch of an intact isolated cardiac cell and during calcium-induced release of calcium from the sarcoplasmic reticulum of a skinned cardiac cell from the adult rat or rabbit ventricle. J. Gen. Physiol. 78, 457-497 (1981).

25. J. van der Velden, J.W. de Jong, V.J. Owen, P.B.J. Burton, and G.J.M. Stienen. Effect of protein kinase A on calcium sensitivity of force and its sarcomere length dependence in human cardiomyocytes. Cardiovasc. Res. 46, 487-495 (2000).

26. T.J. Herron, and K.S. McDonald. Small amounts of α-myosin heavy chain isoform expression significantly increase power output of rat cardiac myocyte fragments. Circ. Res. 90, 1150-1152 (2002).

27. S. Miyata, W. Minobe, M.R. Bristow, and L.A. Leinwand. Myosin heavy chain isoform expression in the failing and nonfailing human heart. Circ. Res. 86, 386-390 (2000).

28. J. Neumann, R.C. Gupta, W. Schmitz, H. Scholz, A.C. Nairn, and A.M. Watanabe. Evidence for isoproterenol-induced phosphorylation of phosphatase inhibitor-1 in the intact heart. Circ. Res. 69, 1450-1457 (1991).

29. S. Winegrad. Cardiac myosin binding protein C. Circ. Res. 84, 1117-1126 (1999).

30. G. McClellan, I. Kulikovskaya, and S. Winegrad. Changes in cardiac contractility related to calcium-mediated changes in phosphorylation of myosin-binding protein C. Biophys. J. 81, 1083-1092 (2001).

31. K.K. Schlender, and L.J. Bean. Phosphorylation of chicken cardiac C-protein by calcium/calmodulin-dependent protein kinase II. J. Biol. Chem. 266, 2811-2817 (1991).

32. I. Morano, H. Arndt, C. Bächle-Stolz, and J.C. Rüegg. Further studies on the effects of myosin P-light chain phosphorylation on contractile properties of skinned cardiac fibers. Basic Res. Cardiol. 81, 611-619 (1986).

33. R.J. Levine, R.W. Kensler, Z. Yang, J.T. Stull, and H.L. Sweeney. Myosin light chain phosphorylation affects the structure of rabbit skeletal muscle thick filaments. Biophys. J. 71, 898-907 (1996).

STUDY OF THE VERTEBRATE MHC MULTIGENE FAMILY DURING HEART DEVELOPMENT

Rumiko Matsuoka*

1. INTRODUCTION

In vertebrates such as birds and mammals, looping and septation generate the four-chambered heart. The origin and lineage relationships of cardiac cell types, endocardial endothelia, ventricular myocytes, and atrial myocytes, constitute the tubular heart when it begins rhythmic contraction. Each cardiac cell type is established by lineage diversification of embryonic cells which arise from one of three distinct origins, cardiogenic mesoderm, neural crest, or proepicardium. The structural and functional diversity of the striated muscle is reflected by the presence of a variety of myosin isoforms and is composed of two heavy chains and four light chains. A family of MHC genes encodes myosin heavy chains (MHCs). The vertebrate MHC multigene family has been subdivided into fast skeletal muscle and cardiac/slow skeletal muscle subfamilies (Stedman et al., 1990). The type of MHC expressed in a muscle cell defines the specific type of muscle fiber, significantly affects its contractile properties, and serves as an excellent marker for differentiated cardiac and skeletal muscle (Masaki and Yoshizaki, 1974; Moore et al., 1992; Nguyen et al., 1982, Nudel et al., 1980; Reiser et al., 1988; Robbins et al., 1986: Sartore et al., 1978). The expression of each MHC isoform is regulated in tissue-specific and developmental stage-specific ways (Bader et al., 1982; Bandman et al., 1982; Evans et al., 1988; Lompre et al., 1984). At present, 9 distinct chick sarcomeric MHC isoforms have been described at the protein level in the skeletal muscle, atria, ventricles, and conduction system (de Jong et al., 1988; Bandman and Rosser, 2000; Evans et al., 1988; Gonzalez-Sanchez and Bader, 1984, 1985; Zhang et al., 1986). To date, at both the gene and protein levels, atrial (Oana et al., 1995, 1998; Yutzey et al., 1994) and ventricular (Bisaha and Bader, 1991; Machida et al., 2000a; Stewart et al., 1991) MHCs have been characterized well.

Formation of the cardiac conduction system is a critical step during normal heart

---

*Department of Pediatric Cardiology, Division of Genomic Medicine, Institute of Advanced Biomedical Engineering and Science, Graduate School of Medicine, Tokyo Women's Medical University, 8-1 Kawada-cho, Shinjuku-ku, Tokyo 162-8666, Japan

Molecular and Cellular Aspects of Muscle Contraction
Edited by H. Sugi, Kluwer Academic/Plenum Publishers, 2003

development. Immunohistochemical evidence has suggested that slow skeletal MHCs are present in the developing and adult conduction system (Gonzalez-Sanchez and Bader, 1985; de Groot et al., 1987; Sanders et al., 1986). However, it was not known, which of these isoforms was (were) expressed in the conduction system. Chick and quail embryos and neonates have been used as model systems for studying heart organogenesis in vertebrates.

In this report, I reviewed regulation of the expression of 9 kinds of each MHC isoform in tissue-specific and developmental stage-specific ways, and demonstrated the unique expression pattern of the 5 MHCs, which is distinct from atrial, ventricular and developmental patterns of skeletal MHCs in the conduction systems.

2. CHARACTERIZATION OF CHICK CARDIAC MHCS

In mammals, the complexity and diversity of MHC have been extensively studied in multiple species, including rat, mouse, and human (Weiss et al., 1999). Comparative studies of the mammalian MHC gene family have provided unique insights into the evolutionary relationships among the divergent MHC genes. The mammalian heart expresses α and β cardiac/slow skeletal MHCs (Mahdavi et al., 1982). The α cardiac MHC is predominantly expressed in the adult atria and the β cardiac/slow skeletal type is found mainly in the adult ventricle and adult slow skeletal muscle (Lompre et al., 1984). The amino acid sequences of the α and β /slow MHCs in mammals show a high homology (99%) and are well conserved (Matsuoka et al., 1991). In contrast, homology among chick cardiac MHCs is considerably less. The type of MHC expressed in a muscle cell defines the specific type of muscle fiber, significantly affects its contractile properties, and serves as an excellent marker for differentiated cardiac and skeletal muscle. The expression of each MHC isoform is regulated in tissue-specific and developmental stage-specific ways. Chick and quail embryos and neonates have been used as model systems for studying heart organogenesis in vertebrates.

To better understand the molecular mechanisms responsible for the cardiac chamber-specific expression of MHCs during heart development, we constructed, λ gt10 expression cDNA library, using embryonic day 15 (E15,H.H. stage 41 chick embryo) chick ventricle poly(A)$^+$ RNA. Then, as shown in Figure1, we isolated and characterized cDNAs corresponding to different chick MHCs, including atrial MHC /skeletal MHC 3 (Oana et al., 1995,1998), ventricular MHC (Machida S. et al., 2000a), neonatal fast skeletal MHC (Machida et al., 2000b) and slow skeletal MHC 2 (Machida et al., 2002). Subsequently, in chicken, the full coding regions of 6 of the 9 chick MHC cDNA have been sequenced by Molina (1987) for embryonic skeletal MHC, by Chao and Bandman (1997) for adult chicken pectoralis, and our group. We found that these four MHCs and embryonic skeletal MHCs mRNA were expressed in heart and chamber-specific ways during development, based on our Northern blot analysis, S1-nuclease mapping analysis, *in situ* hybridization and immunohistochemistry data. We also found that the chick neonatal fast skeletal MHC is expressed in the heart conduction cells during the late embryonic to newly hatched stages but not in adults (Machida et al., 2000b). In the avian slow skeletal muscle, three different slow MHC isoforms have been identified, which, according to their appearance during development, have been called slow skeletal MHC 3 or atrial HCM because of its specific expression in chick atrial muscle, slow skeletal MHC 1, and slow skeletal MHC 2 isoforms. Interestingly, slow skeletal MHC 2 was restricted to the subendocardial clusters of cells and around the blood vessels of late

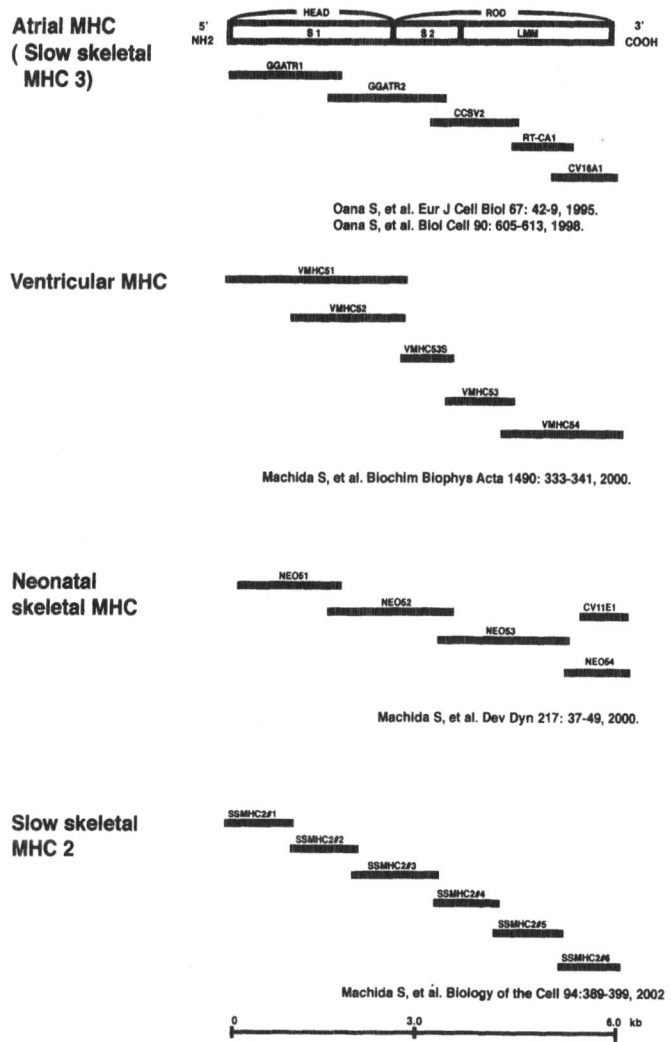

Figure 1. Relationship of the 4 chick myosin heavy chain cDNA clones (atrial MHC/slow skeletal MHC3, ventricular MHC, neonatal MHC and slow skeletal MHC) against the domain map of myosin heavy chain. The MHC domains shown are S1 head, S2 portion of the rod and the light meromyosin (LMM) portion of the rod.

embryonic and adult hearts, and not expressed in the myocardium throughout the life of the chicken (Machida et al., 2002). Therefore, we suggested that the previously uncharacterized slow tonic MHC localized in the conduction system is slow skeletal MHC 2 (Machida et al., 2002). To investigate the relationships between slow skeletal MHC 2 and other MHC isoforms, we compared the amino acid sequence data of these 4 (atrial MHC /skeletal MHC 3, ventricular MHC, neonatal fast skeletal MHC and slow skeletal MHC 2) and embryonic skeletal MHC (Fig. 2). The amino acid sequences of the α and β cardiac/slow skeletal MHCs in mammals show a high homology (99%) and are conserved well (Matsuoka et al., 1991). In contrast, homology among chick cardiac MHCs is considerably less. Slow skeletal MHC 2 was 81.8% identical to chick ventricular MHC and 75.6% identical to chick atrial MHC. The primary sequence of these MHC types allowed us to determine the structural features of this isoform and its phylogenetic position within the sarcomeric MHC multigene family.

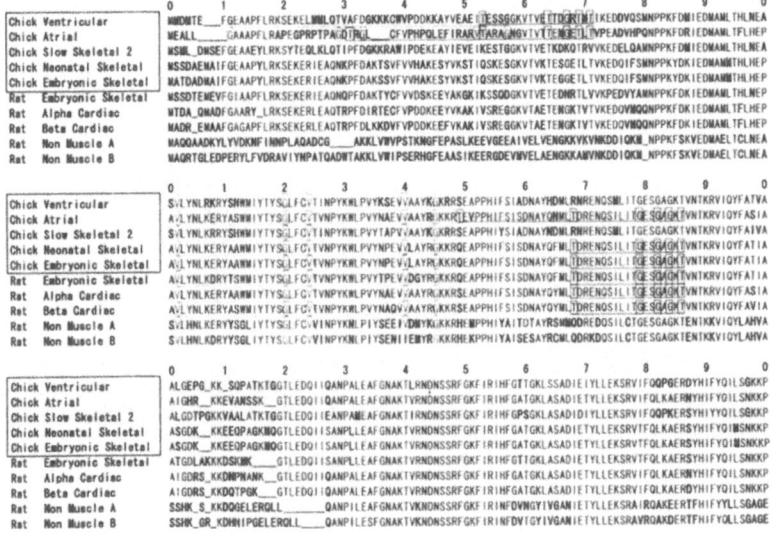

Figure 2.   Comparison of amino acid sequencing of 5 chick myosin heavy chain (MHC)s: atrial/ slow skeletal MHC 3, ventricular, slow skeletal MHC 2, neonatal skeletal and embryonic skeletal.

3. EXPRESSION OF MHCS IN THE DEVELOPING CONDUCTION SYSTEM

Formation of the cardiac conduction system is a critical step during normal heart development. Purkinje fibers are distinguished from heart muscle cells by a distinct localization pattern of myofibrillar proteins (Mikawa and Fischman, 1996; Mikawa, 1999; Moorman et al., 1998; Schiaffino, 1997).   Gonzalez-Sanchez and Bader (1985) reported that the differentiated conduction cells of chick heart expressed a slow tonic type of skeletal MHC, which was not present in the myocardium. Using S1-nuclease mapping analyses, we have earlier shown that atrial MHC/slow skeletal MHC 3 is expressed in the developing but not in adult ventricle (Oana et al., 1998). However, immunohistochemical studies have suggested that atrial MHC is present in the adult conduction system of the ventricle (de Groot et al, 1987; Sanders et al, 1986). We reinvestigated using *in situ* hybridization if the atrial MHC gene was indeed expressed in the adult conduction system of the ventricle. Including this result, our data showed that atrial MHC/slow skeletal MHC 3 and neonatal skeletal MHC mRNA were expressed strongly in the atrium, and also weakly in the ventricle during the late embryonic and neonatal stages (Fig. 3A and 3C). Furthermore, in the ventricle, the expression of atrial MHC/slow skeletal MHC 3 (Fig. 3D, 3F), slow skeletal MHC 2 (Fig. 3E and 3G) and neonatal skeletal MHC mRNA (Fig. 3H) was observed in the subendocardium structures. To confirm specific

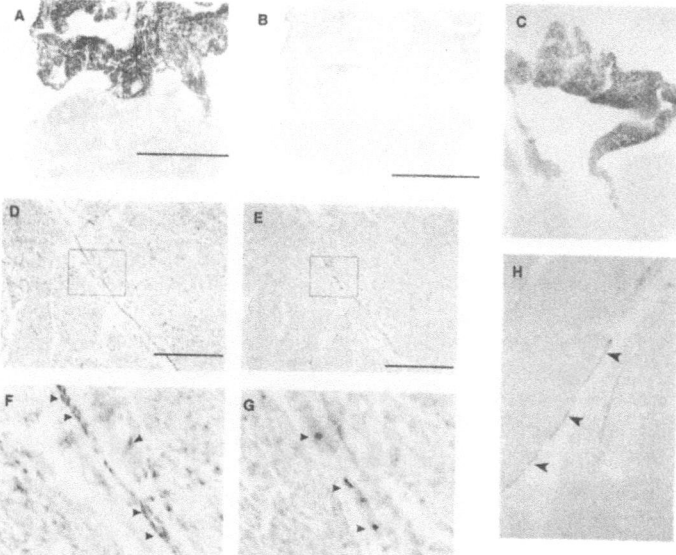

Figure 3. *In situ* hybridization analysis for atrial MHC or slow skeletal MHC 2 mRNAs on frozen sections of late embryonic and adult chick heart, and neonatal skeletal MHC mRNAs on frozen sections of neonatal 0 day chick heart. Sections are from ED 19.5 (A, and B), neonatal day 0 (C and H), and adult (F and G) hearts. They were incubated with the digoxigenin-labeled atrial MHC mRNA (A, D and F) or the probe of slow skeletal MHC 2 (D, E and G) or the probe of neonatal skeletal MHC (C and H). Specific hybridization signals were not detected in the myocardium (B). In the ventricle, signals were detected in subendocardial structures (D-G, arrowheads). Atrial MHC mRNA was expressed strongly in the atrium (A) and was also expressed in the subendocardial structures (F, arrowheads) of the ventricular myocardium. Slow skeletal MHC 2 mRNA was not expressed in the atrium (B) but was expressed in the subendocardial structures (G, arrowheads) of the ventricular myocardium. Neonatal MHC mRNA was expressed strongly in the atrium (C) and was also expressed in the subendocardial structures (H, arrowheads) of the ventricular myocardium.

Bars: 1000 μ m (A and B), 400 μ m (D and E).

The boxed areas in D and E are shown in higher magnification in F and G.

expression of various MHC isoforms in the conduction cells, we performed immunohistochemical staining using 5 monoclonal antibodies of embryonic skeletal MHC, neonatal skeletal MHC, atrial MHC/slow skeletal MHC 3 and slow skeletal MHC 2 (NA8), ventricular MHC (HN11) and slow skeletal MHC 2 (NA3) (Fig. 4 and 5).

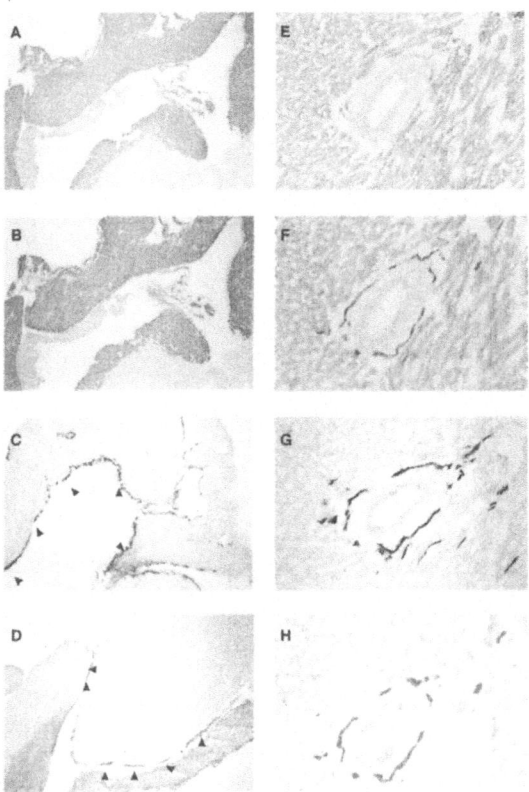

Figure 4.   Immunohistochemical staining of frozen sections of atrium.   Sections are from ED 18 (A-D), and adult (E-H) hearts.   They were incubated with 4 monoclonal antibodies of embryonic skeletal MHC, NA8 (atrial MHC/slow skeletal MHC 3 and slow skeletal MHC 2), HN11 (ventricular MHC) and NA3 (slow skeletal MHC 2) followed by a horseradish peroxidase-linked secondary antibody. Embryonic skeletal MHC and NA8 stained strongly the atrial myocardium (A and B), but HN11- and NA3-positive cells were only observed in the subendocardial structures (C and D) at the late embryos (ED18). In the adult atrium, embryonic skeletal MHC- (E), NA8- (F), HN11- (G) and NA3-positive cells (H) were detectable within the subendocardial structures and around the blood vessels .

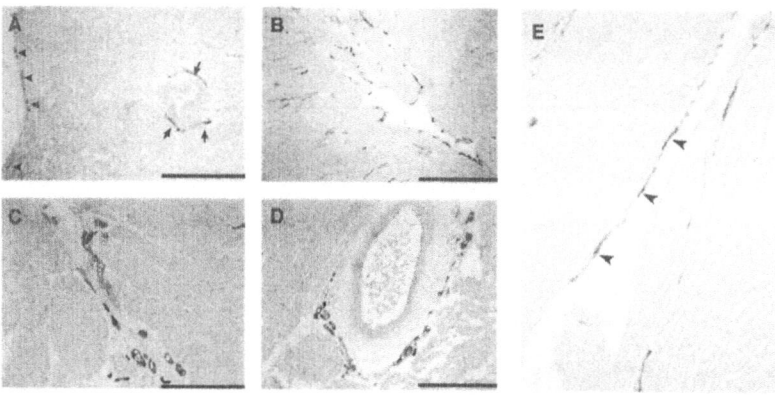

Figure 5. Immunohistochemical staining of frozen sections of ventricles at various developmental stages with the NA8 antibody. Sections are from ED 15 (A), ED 18 (B), ND0 (E), and adult (C and D) hearts. They were incubated with NA3 or neonatal skeletal antibody and then with a horseradish peroxidase-linked secondary antibody. Although the myocardium was not stained with NA3 at early embryonic stages, some intensely NA3-positive cells began to appear as subendocardial clusters and around the blood vessels in the ventricular myocardium in later developmental stages (ED15). At ED 18, many NA3-positive cells appeared in the same location (B). In the adult, intensely NA3-positive cells were detectable within the subendocardial structures (C) and around the blood vessels of the ventricular myocardium (D). At neonatal day 0, many of the neonatal skeletal MHC positive cells were located in the subendocardial layer. Bars: 400 μm (A and B), 200 μm (C and D).

However, immunohistochemical positive-staining cells were not clearly found in the proximal components of the conduction system (atrioventricular node and His bundle), but the subendocardial conduction fiber-like structures and the cells around blood vessels in the developing and adult atrium (Fig. 4). Embryonic skeletal MHC and NA8 stained strongly the atrial myocardium (A and B), but HN11- and NA3-positive cells were only observed in the subendocardial structures (C and D) at the late embryos (ED18). Although the myocardium was not stained with NA3 at early embryonic stages, some intensely NA3-positive cells began to appear as subendocardial clusters and around the blood vessels in the atrial myocardium in later developmental stages (ED15). No reactivity was found with NA7, which reacts specifically with slow skeletal MHC1, suggesting that slow skeletal MHC1 is not expressed in the developing and adult heart.

Previous immunohistochemical studies have suggested that the NA3 (monoclonal antibody for slow skeletal MHC2)- and NA8 (monoclonal antibodies for slow skeletal MHC2 and atrial MHC)-positive cells are those of the conduction system (Gonzalez-Sanchez and Bader, 1985). To verify that the subendocardial cells that stained with anti-neonatal skeletal MHC, NA8, NA3 and reagents are indeed the conduction tissue cells, we performed immunofluorescents double-staining of heart sections using antibodies to neonatal skeletal (Fig. 6A), NA8 (Fig. 6B) or NA3 (Fig. 6C), and desmin (Fig. 6D-F). Desmin is the intermediate filament type present in all types of muscle cells. It is

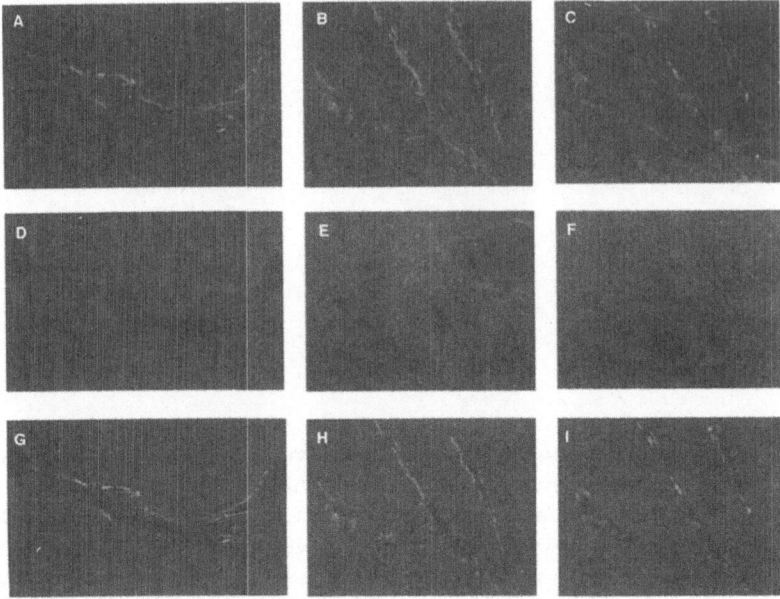

Figure 6. Double-immunofluorescence staining of frozen chick ventricle sections at neonatal day 7 with antibodies against slow skeletal MHCs and desmin. Frozen sections were incubated with neonatal skeletal MHC (A) or NA8 (B) or NA3 (C) and anti-desmin (D, E and F) followed by an FITC-conjugated (A-C) or a Texas Red-conjugated (D-F) secondary antibody. The neonatal skeletal MHC or NA8- or NA3-positive cells were located in the subendocardium of the ventricle (A and B). Anti-desmin stained uniformly the ventricular myocytes but also the same subendocardial structures (D-F). The FITC and Texas Red images were superimposed (G-I), showing that the NA3- and NA8-positive cells contained a high level of desmin.

particularly abundant in Purkinje cells, especially in birds and larger mammals (Filogamo et al., 1990; Machida et al., 2000a, 2002, Thornell et al., 1985) and is a good marker for these cells. Figure 6A shows the staining patterns of a day 7 neonate chick atrium with the neonatal skeletal antibody, and Figure 6B and 6C show the staining patterns of a day 7 neonate chick ventricle with the two MHC antibodies. The entire neonatal skeletal antibody (Fig. 6A), NA8 (Fig. 6B) and NA3 (Fig. 6C) labeled only a population of cells located in the subendocardium. Anti-desmin stained the same ventricle sections diffusely, but more importantly, it strongly labeled the structures present in the subendocardium (Fig. 6D-F). Both the appearance and anti-desmin expression of the subendocardial structures indicate that they are conduction fibers. When the staining patterns by anti-MHC (green) and anti-desmin (red) were superimposed, these subendocardial structures became yellow (Fig. 6G-I), indicating co-localization of slow skeletal MHC and desmin. These results show that the linear structures stained with the anti-MHC antibodies are conduction fibers. Based on their histological appearance, we suggest that these structures are the conduction system of the ventricle. Therefore, interestingly, in addition to the expected skeletal MHC expression in the chick cardiac muscle, during the late embryonic to newly hatched stages, these 5 MHCs (embryonic skeletal MHC, neonatal fast skeletal MHC, atrial MHC /skeletal MHC 3, ventricular MHC, and slow skeletal MHC 2) were expressed in the subendocardial clusters of the cells, and around the blood vessels within the atrium and ventricle (Fig.7). In adult the 4 MHCs (atrial MHC /skeletal MHC 3, ventricular MHC, slow skeletal MHC 2 and embryonic skeletal MHC) were also expressed in the subendocardial clusters of cells, and around the blood vessels within the atrium and ventricle (Fig.7). Based on various morphological criteria as well as rich desmin expression, we believe that the subendocardial clusters of cells are Purkinje cells.

4. The physiological significance of MHCs in the developing conduction system

Heart contraction is coordinated by conduction of electrical excitation through specialized tissues of the cardiac conduction system. The cardiac conduction system develops as a separate functional entry from the contracting working myocardium. Although the physiological significance of the expression of these 5 MHCs in Purkinje cells remains unclear, these MHC isoforms are excellent molecular markers for terminally differentiated Purkinje cells. It is well known that Purkinje fibers are distinguished from heart muscle cells by a distinct localization pattern of myofibrillar

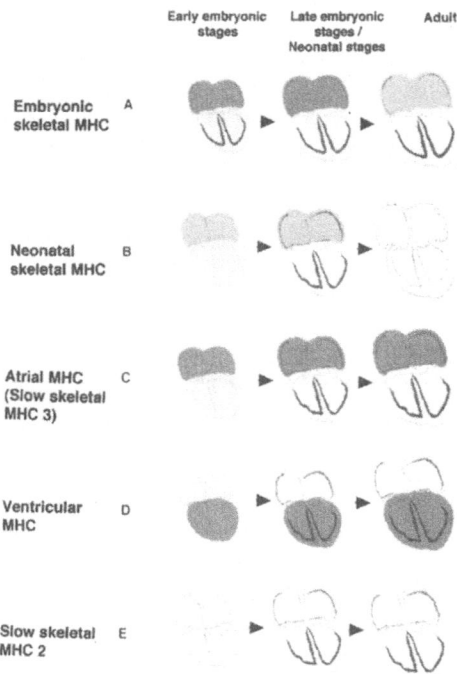

Figure 7. In summary of localization of myosin heavy chain in chick heart.

proteins (Mikawa and Fischman, 1996; Moorman et al., 1998; Schiaffino, 1997). Gonzalez-Sanchez and Bader (1985) reported that the differentiated conduction cells of chick heart expressed a slow tonic type of skeletal MHC, which was not present in the myocardium.   Immunohistochemical analyses have provided evidence that atrial MHC is present in the adult conduction system of the ventricle (de Groot et al., 1987; Sanders et al., 1986). In the avian slow skeletal muscle, three different slow MHC isoforms have been identified, which, according to their appearance during development, have been called slow skeletal MHC 3 (atrial MHC), slow skeletal MHC 1, and slow skeletal MHC 2 isoforms (Page et al., 1992). The expression of slow skeletal MHC 2 was detected beginning from the late embryonic stage to adult. This MHC isoform appeared to be expressed exclusively in Purkinje cells, as its presence was not detected in the myocardium. Based on our *in situ* hybridization and immunohistochemistry data, we concluded that the previously uncharacterized slow tonic MHC localized in the conduction system (Gonzalez-Sanchez and Bader, 1985) is slow skeletal MHC 2 (Machida et al., 2002). This MHC isoform is abundantly expressed in adult slow skeletal muscles (Page et al., 1992). DiMario and Stockdale (1997) have shown that when muscle fibers derived from myoblasts of the slow muscle are cocultured with the neural tube, they express slow skeletal MHC 2, while muscle fibers formed from myoblasts of the fast muscle origin continue to express only fast MHC. They demonstrated that slow skeletal MHC 2 gene expression in vitro was regulated by a combination of cell lineage (intrinsic mechanisms) and innervation (extrinsic mechanisms). Interestingly, slow skeletal MHC 2 expression was delayed or absent in denervated avian slow muscle in vivo (Lefeuvre et al., 1996). Kirby et al. (1980) has reported that during sympathetic innervation to the chick heart, nerves are first seen (in the heart) at ED 10. Autonomic innervation of the conduction system in the developing chick heart may trigger the expression of slow skeletal MHC proteins in this system. Our study showed that atrial MHC/slow skeletal MHC3 was present in the early embryonic myocardium and then in ventricular Purkinje cells. Gourdie et al. (1995), using the retroviral cell tagging procedure, demonstrated that peripheral elements of the chick cardiac conduction tissue are derived directly from the embryonic myocardium. Based on the expression pattern of atrial MHC, we suggest a possibility that the cells expressing this MHC in the primary ventricular myocardium differentiate into conduction cells. It is interesting to note that sympathetic innervation in developing chick heart correlates with or slightly precedes the appearance of the neonatal skeletal or NA8- or NA3-positive cells (Fig. 6A, 6B and 6C).

Purkinje cells of the adult heart express not only two of the slow skeletal MHCs (atrial MHC and slow skeletal MHC 2) but also ventricular MHC (Fig. 4B-D and 4F-H). In addition, chick embryonic skeletal MHC was weakly localized in adult Purkinje cells (Fig. 4A and 4E). Our previous study demonstrated that chick neonatal fast skeletal MHC appeared during early cardiogenesis and then was localized in the Purkinje fibers at the late embryonic and newly hatched stages, but not in the adult stage (Machida et al., 2000a). Thus these data indicate that the chick ventricle initially co-expresses ventricular MHC, atrial MHC and developmental types (embryonic and neonatal type) of skeletal MHCs but that later, the atrial and developmenal types of skeletal MHCs are completely down regulated. However, this regulation does not seem to occur in the ventricular conduction tissue (Purkinje myocytes).

Alyonycheva et al. (1997) demonstrated that skeletal myosin-binding protein-H, which is present in skeletal muscle but not in myocardium, was expressed in the late embryonic to adult chicken conduction system. The bHLH family of myogenic transcription factors, such as myoD, Myf-5, myogenin, and MRF-4 regulates expression of many of the skeletal muscle-specific proteins. However, it is generally thought that cardiac muscle cells do not express any of the members of the myoD gene family. Recently, Takebayashi-Suzuki et al. (2001) reported that myoD is expressed, although only weakly, in the chick differentiated Purkinje cells. Thus, it appears that a skeletal muscle cell gene expression program is activated in differentiated conduction cells (Moorman et al., 1998). In terms of their protein expression, differentiated Purkinje cells have characteristics of both cardiac and skeletal muscle cells. For example, while expressing myoD, Purkinje cells also expressed ventricular MHC (Fig. 4C and 4G). In fact, this MHC isoform was expressed at a higher level in Purkinje cells than in ventricular myocytes. Nkx2.5/Csx gene is a cardiac specific homeobox gene, and it has been shown to be activated in chick differentiated Purkinje cells. Similar to the expression levels of ventricular MHC in Purkinje cells and ventricular myocytes, the level of Nkx2.5/Csx mRNA is significantly higher in the conduction cells than in the cardiomyocytes of the ventricle (Takebayashi-Suzuki et al., 2001). A question raised why some areas of the embryonic heart develop into the nodal direction. The mechanisms involved in this down-regulation in the ventricular myocytes and in its absence in the Purkinje cells during development are unknown.

Habets et al. (2002) reported that, intriguingly, these areas express the transcriptional repressor, Tbx2, and its co-operation with Nkx2.5/Csx leads to repression of the transcriptional activity of the gene encoding Atrial natriuretic factor (*Anf*), which is a highly specific marker for the developing atrial and ventricular working myocardium.

In summary, we have demonstrated a unique expression pattern of the 5 MHCs, distinct from atrial, ventricular and developmental types of skeletal MHCs in the conduction systems. The slow skeletal MHC 2, in particular, was exclusively expressed in differentiated conduction cells during development. It is essential to define the molecular mechanisms for the regulation of the unique expression pattern of the 5 MHCs, which is distinct from that of the atrial, ventricular and developmental types of skeletal MHCs in the conduction systems. Identification of the transcription factors of these MHC genes which may interact with the transcriptional repressor Tbx2 in co-operation with Nkx2.5/Csx, thus repressing the transcriptional activity of the gene encoding *Anf*, is one of the urgent tasks we must address.

ACKNOWLEDGEMENT

I wish to thank Drs. Machida, Oana, Nakajima, Noda and Takagaki for investigating or support of these series, and Ms. Eriko Hiratsuka, Mr. Yoshiyuki Furutani Ms. Michiko Furutani and Ms. Michiko Tamai for technical support on *in situ* hybridization and sequence analyses.

REFERENCES

Alyonycheva T, Cohen-Gould L, Siewert C, Fischman DA and Mikawa T (1997). Skeletal muscle-specific myosin binding protein-H is expressed in Purkinje fibers of the cardiac conduction system. Circ. Res. 80, 665-672

Bader D, Masaki T and Fischman DA (1982). Immunochemical analysis of myosin heavy chain during avian myogenesis in vivo and in vitro. J. Cell Biol. 95, 763-770

Bandman E, Matsuda R and Strohman RC (1982). Developmental appearance of myosin heavy and light chain isoforms in vivo and in vitro in chicken skeletal muscle. Dev. Biol. 93, 508-518

Bandman E and Rosser BW (2000). Evolutionary significance of myosin heavy chain heterogeneity in birds. Microsc. Res. Tech. 50, 473-491

Bisaha JG and Bader D (1991). Identification and characterization of a ventricular-specific avian myosin heavy chain, VMHC1: expression in differentiating cardiac and skeletal muscle. Dev. Biol. 148, 355-364

Bourke DL, Wylie SR, Theon A and Bandman E (1995). Myosin heavy chain expression following myoblast transfer into regenerating chicken muscle. Basic Appl. Myol. 5, 43-56

Chan-Thomas PS, Thompson R, Robert B, Yacoub MH and Barton PJR (1993). Expression of homeobox genes Msx-1 (Hox-7) and Msx-2 (Hox-8) during cardiac development in the chick. Dev. Dyn. 197, 203-216

Chao TH and Bandman E (1997). Cloning nucleotide sequence and characterization of a full-length cDNA encoding the myosin heavy chain from adult chicken pectoralis major muscle. Gene 199, 265-270

Chen Q, Moore LA, Wick M and Bandman E (1997). Identification of a genomic locus containing three slow myosin heavy chain genes in the chicken. Biochim. Biophys. Acta 1353, 148-156

Chomczynski P and Sacchi N (1987). Single-step method of RNA isolation by acid guanidinium thiocyanate-phenol-chloroform extraction. Anal. Biochem. 162, 156-159

de Groot IJM, Sanders E, Visser SD, Lamers WH, de Jong F, Los JA and Moorman AFA (1987). Isomyosin expression in developing chicken atria: a marker for the development of conductive tissue. Anat. Embryol. 176, 515-523

de Jong F, de Groot IJM, Geerts WJC, Wessels A, Peschar AW, Lamers WH, and Moorman AFM (1988). Immunohistochemical evidence for two differentially expressed atrial myosin heavy chain isoforms during avian cardiogenesis. In: Sarcomeric and Non-sarcomeric Muscles: Basic and Applied Research Prospects for the 90s (Cazzazo U, editor) Unipress Padova, Padova, 299-304

DiMario JX and Stockdale FE (1997). Both myoblast lineage and innervation determine fiber type and are required for expression of the slow myosin heavy chain 2 gene. Dev. Biol. 188, 167-180

Evans D, Miller JB and Stockdale FE (1988). Developmental patterns of expression and coexpression of myosin heavy chains in atrial and ventricles of the avian heart. Dev. Biol. 127, 376-383

Filogamo G, Corvetti G and Daneo LS (1990). Differentiation of cardiac conducting cells from the neural crest. J. Autono. Nerv. System 30, S55-S58

Fishman MC and Olson EN (1997). Parsing the heart: genetic modules for organ assembly. Cell 91, 153-156

Gonzalez-Sanchez A and Bader D (1984). Immunochemical analysis of myosin heavy chains in the developing chicken heart. Dev. Biol. 103, 151-158

Gonzalez-Sanchez A and Bader D (1985). Characterization of a myosin heavy chain in the conductive system of the adult heart and developing chicken heart. J. Cell Biol. 100, 270-275

Gourdie RG, Miwa T, Thompson RP and Mikawa T (1995). Terminal diversification of the myocyte lineage generates Purkinje fibers of the cardiac conduction system. Development 121, 1423-1431

Habets PEMH, Moorman AFM, Clout DEW, van Roon MA, Lingbeek M, van Lohuizen M, Campione M and Christoffels VM (2002). Cooperative action of TBx2 and Nkx2.5 inhibits ANF expression in the atrioventricular canal: implications for caerdiac chamber formation.Genes. & Dev. 16,1234-1246

Kirby ML, McKenzie JW and Weidman TA (1980). Developing innervation of the chick heart: a histofluorescence and light microscopic study. Anat. Rec. 196, 333-340

Lefeuvre B, Crossin F, Fontaine-Perus J, Bandman E and Gardahaut MF (1996). Innervation regulates myosin heavy chain isoform expression in developing skeletal muscle fibers. Mech. Dev. 58, 115-127

Lompre AM, Nadal-Ginard B and Mahdavi V (1984). Expression of the cardiac ventricular a- and b-myosin heavy chain genes is developmentally and hormonally regulated. J. Biol. Chem. 259, 6437-6446

Machida S, Matsuoka R, Noda S, Hiratsuka E, Takagaki Y, Oana S, Furutani Y, Nakajima H, Takao A and Momma K (2000a). Evidence for the expression of neonatal skeletal myosin heavy chain in primary myocardium and cardiac conduction tissue in the developing chick heart. Dev.Dyn. 217, 37-49

Machida S, Noda S, Furutani Y, Takao A, Momma K and Matsuoka R (2000b). Complete sequence and characterization of chick ventricular myosin heavy chain in the developing atria. Biochim. Biophys. Acta 1490, 333-341

Machida S, Noda S, Takao A, Nakazawa M and Matsuoka R (2002). Expression of slow skeletal myosin heavy chain 2 gene in Purkinje fiber cells in chick heart. Biol. Cell 94, 389-399

Mahdavi V, Periasamy M and Nadal-Ginard B (1982). Molecular characterization of two myosin heavy chain genes expressed in the adult heart. Nature 297, 659-664

Masaki T and Yoshizaki C (1974). Differentiation of myosin in chick embryos. J. Biochem. 76, 123-131

Matsuoka R, Beisel, KW , Furutani M, Arai S and Takao A (1991). Complete sequence of human cardiac Δ- and β-myosin heavy chain gene and amino acid comparison to other myosin based on structural and functional differences. Am. Med. Genet. 41, 537-547

Mikawa T and Fischman DA (1996). The polyclonal origin of myocyte lineages. Annu. Rev. Physiol. 58, 509-521

Molina IM, Kropp KE, Gulick J and Robbins J (1987). The sequence of an embryonic myosin heavy chain gene and isolation of its corresponding cDNA. J. Biol. Chem. 262, 6489-6493

Moore LA, Arrizubieta MJ, Tidyman WE, Herman LA and Bandman E (1992). Analysis of the chicken fast myosin heavy chain family. Localization of isoform-specific antibody epitopes and regions of divergence. J. Mol. Biol. 225, 1143-1151

Moorman AFM, de Jong F, Denyn MMFJ and Lamers WH (1998). Development of the cardiac conduction system. Circ. Res. 82, 629-644

Nguyen HT, Gubits RM, Wydro RM and Nadal-Ginard B (1982). Sarcomeric myosin heavy chain is coded by a highly conserved multigene family. Proc. Natl. Acad. Sci. U.S.A. 79, 5230-5234

Nudel U, Katcoff D, Carmon Y, Zevin-Sonkin D, Levi Z, Shaul Y, Shani M and Yaffe D (1980). Identification of recombinant phages containing sequences from different rat myosin heavy chain genes. Nucleic Acids Res. 8, 2133-2146

Oana S, Matsuoka R, Nakajima H, Hiratsuka E, Furutani Y, Takao A and Momma K (1995). Molecular characterization of a novel atrial-specific myosin heavy-chain in the chick embryo. Eur. J. Cell Biol. 67, 42-49

Oana S, Machida S, Hiratsuka E, Furutani Y, Momma K, Takao A and Matsuoka R (1998). The complete sequence and expression patterns of the atrial myosin heavy chain in the developing chick. Biol. Cell 90, 605-613

Olson EN and Klein WH (1994). bHLH factors in muscle development: dead lines and commitments, what to leave in and what to leave out. Genes Dev. 8, 1-8

Page S, Miller JB, DiMario JX, Hager EJ, Moser A and Stockdale FE (1992). Developmentally regulated expression of three slow isoforms of myosin heavy chain: diversity among the first fibers to form in avian muscle. Dev. Biol. 154, 118-128

Reiser PJ, Greaser ML and Moss RL (1988). Myosin heavy chain composition of single cells from avian slow skeletal muscle is strongly correlated with velocity of shortening during development. Dev. Biol. 129, 400-407

Robbins J, Horan T, Gulik J and Kroop K (1986). The chicken myosin heavy chain family. J. Biol. Chem. 261, 6606-6612

Sanders E, de Groot IJM, Geerts WJ, de Jong F, van Horssen AA, Los JA and Moorman AFA (1986). The local expression of adult chicken heart myosins during development. II. Ventricular conducting tissue. Anat. Embryol. 174, 187-193

Sartore S, Pierobon-Bormioli S and Schiaffino S (1978). Immunohistochemical evidence for myosin polymorphism in the chicken heart. Nature 274, 82-83

Schiaffino S and Reggiani C (1996). Molecular diversity of myofibrillar proteins: gene regulation and functional significance. Physiol. Rev. 76, 371-423

Schiaffino S (1997). Protean patterns of gene expression in the heart conduction system. Circ. Res. 80, 749-750

Stedman HH, Eller M, Jullian EH, Fertels SH, Sarkar S, Sylvester JE, Kelly AM and Rubinstein NA (1990). The human embryonic myosin heavy chain. Complete primary structure reveals evolutionary relationships with other developmental isoforms. J. Biol. Chem. 265, 3568-3576

Stewart AFR, Camoretti-Mercado B, Perlman D, Gupta M, Jakovcic S and Zak R (1991). Structural and phylogenetic analysis of the chicken ventricular myosin heavy chain rod. J. Mol. Evol. 33, 357-366

Takebayashi-Suzuki K, Pauliks LB, Eltsefon Y and Mikawa T (2001). Purkinje fibers of the avian heart express a myogenic transcription factor program distinct from cardiac and skeletal muscle. Dev. Biol. 234, 390-401

Thornell LE, Eriksson A, Johansson B, Kjorell U, Franke WW, Virtanen I and Lehto VP (1985). Intermediate filament and associated proteins in heart purkinje fibers: A membrane-myofibril anchored cytoskeletal system. Ann. N.Y. Acad. Sci. 455, 213-240

Weiss A, Schiaffino S and Leinwand LA (1999). Comparative sequence analysis of the complete human sarcomeric myosin heavy chain family: implications for functional diversity. J. Mol. Biol. 290, 61-75

Yutzey KE, Rhee JT and Bader D (1994). Expression of the atrial-specific myosin heavy chain AMHC1 and the establishment of anteroposterior polarity in the developing chicken heart. Development 120, 871-883

Zhang Y, Shafiq SA and Bader D (1986). Detection of a ventricular-specific myosin heavy chain in adult and developing chicken heart. J. Cell Biol. 102, 1480-1484

MYOSIN-BINDING PROTEIN C (MYBP-C) IN CARDIAC MUSCLE AND CONTRACTILITY

Saul Winegrad

1. INTRODUCTION

Myosin binding protein C (MyBP-C) was first discovered in skeletal muscle by Offer et al, (1973), and subsequent work from several laboratories (Koretz 1979, Craig and Offer1976, Davies 1988, Seiler et al 1996, Sebillon et al 2001) has shown that it can play an important role in the formation of normal thick filaments in skeletal muscle. In its absence, isolated myosin does not form thick filaments with uniform thickness, uniform length or helically ordered myosin heads (Koretz 1079, Rhee et al 1994, Lin et al 1994). Surprisingly, cardiac muscle from mice with MyBP-C knocked out form normally appearing sarcomeres and myofibrils (Harris et al 2002), raising the possibility that some form of compensation has occurred or formation of thick filaments differs between cardiac and skeletal muscle.

Two to four (most likely 3) molecules are distributed every 43 nm within the C zone of the sarcomere (Craig and Offer 1976) and are oriented perpendicularly to the axis of the thick filament. Three isoforms of MyBP-C, two skeletal and one cardiac, exist. All three isoforms have two myosin-binding sites, one in C10 at the C terminal and the other near the N terminal. The two skeletal isoforms, not normally expressed in the heart, have 10 modules (labeled C1-C10) that resemble either fibronectin or immunoglobulin domains (Gilbert et al 1996, Schwartz et al 1995, Gautel et al 1995). The cardiac isoform is found only in cardiac muscle.

There are other differences between cardiac and skeletal isoforms. The cardiac isoform has an extra module at the N terminal (C0), 28 additional residues in C5 and 3 phosphorylation sites in a sequence connecting C1 to C2 (the MyBP-C motif) (Gautel et al 1995). Bound to the thick filament in heart muscle there is a Ca –calmodulin regulated kinase that is specific for these sites (Schlender and Bean 1991, Hartzell and Glass 1984).

The recognition that about 50% of the mutations responsible for familial hypertrophic cardiomyopathy occur in MyBP-C has fostered considerable interest in its function and the

Saul Winegrad, Department of Physiology, School of Medicine, University of Pennsylvania, Philadelphia, PA 19072 changes in structure and function produced by its phosphorylation (Bonne 1995, Watkins 1996, Flavigny et al 1999). Although phosphorylation of the protein in reconstituted contractile systems does not seem to have a major effect on actomyosin ATPase activity, the degree of phosphorylation of MyBP-C in intact heart muscle is changed in the same direction as force development by the 5 most common modulators of contractility (alpha and beta adrenergic agonists, cholinergic transmitters, Ca^{++}, endothelin; McClellan et al 1994, Winegrad 1999). Phosphorylation is associated with a change in structure of thick filament (Weisberg and Winegrad 1996, 1998) and in the probability of interaction between isolated thick and thin filaments (Levine et al 2001). These results have led to the hypothesis that the probability of forming weak interactions between myosin in thick filaments and actin in thin filaments can be regulated by the degree of phosphorylation of MyBP-C (McClellan et al 2001, Levine et al 2001).

Recently Kunst et al (2000) prepared fragments of cardiac MyBP-C and examined their effects on the contractile properties of skeletal muscle, where phosphorylation of endogenous MyBP-C does not occur. Of particular interest were the results with fragments C1 to C2 (C1C2), which contained the phosphorylation sites in the MyBP-C motif, and C0C1, which did not contain the phosphorylation sites. Unphosphorylated C1C2 was bound to the skinned slow skeletal fibers (in which the isoform of myosin is very similar to that in cardiac muscle) and reduced maximum Ca activated force (Fmax) by about 50%. The binding changed some of the physical properties of fibers such as stiffness of cross bridges. Phosphorylated C1C2 did not bind to the skinned fibers and did not alter Fmax. C0C1 did not alter Fmax. From these biochemical data, Kunst et al (2000) proposed that the properties of the cross bridges could be modified by binding of C1C2 in MyBP-C to myosin near its hinge region (Gruen and Gautel 1999, Gruen et al 1999).

2. REVERSIBLE CHANGE IN CONTRACTILITY WITH EXTRACTION

Mutations of MyBP-C gene leading to the absence of normal MyBP-C or almost complete absence of any MyBP-C can result in impairment of contractility, hypertrophy, dilation and even cardiac failure (Watkins et al 1995; Carrier et al 1997; Rottbauer et al 1997), and knockout of the cardiac MyBP-C gene results in hypertrophy and abnormal function of the heart (Harris et al 2002). Extraction of 60-70% of MyBP-C from a single, mechanically disrupted myocardial cell over an hour, though it increases Ca sensitivity, has no significant effect on Fmax (Hofmann et al 1991), suggesting that all effects of the decrease in MyBP-C are unlikely to occur concurrently with its extraction from the thick filament. We have followed the structure and performance of skinned heart for 4 hours after the onset of extraction of MyBP-C and found changes that were not apparent after one hour. Our data suggest that processes affecting contractility were initiated during 1 hour of extraction of MyBP-C, but were not detectable by measurement of Fmax until they had evolved further. Of considerable importance is the fact that these effects on Fmax and Ca sensitivity can be reversed by incubating the tissue in relaxing solution containing rat cardiac MyBP-C. These two sets of results following extraction of MyBP-C suggest that MyBP-C does not alter force generation by a direct effect on the force generating interactions. More likely, MyBP-C causes a change in the entire thick filament that in some way influences the availability of the myosin population for interaction with actin.

3. STRUCTURE OF FILAMENTS

Native thick filaments isolated from rat ventricle exhibit one of three different structures as defined by the thickness of the filament and the degree of order of the myosin heads. In one structure thick filaments have an average diameter outside the bare zone of 29.8 $\pm$ 0.12 nm. The myosin heads lie along the backbone of the filament and produce a repeating structure with a periodicity of about 43 nm visible as a mildly convex contour on the surface of the filament. In the optical diffraction pattern produced by filaments in the micrographs there are reflections on the meridian as well as on layer lines characteristic for helically arranged myosin heads in the filament. This structure has been called the tight structure because it has the narrowest diameter of ordered thick filaments. In the second structure myosin heads are extended from the thick filaments at different angles to the backbone of the filament and are highly disordered. The myosin heads arise from the surface of the filament at about 43 nm intervals, but they extend at different angles from the backbone of the filament. Because of this variability in angle, the mass of the myosin heads is not distributed at uniformly repeating intervals. This produces an image with visibly disordered myosin heads and an absence of reflections along the 43 nm layer line. Therefore this filament is said to have a disordered structure. The filaments with the third structure are thicker, having an average diameter of 34.2$\pm$0.8 nm. The myosin heads were ordered and not obviously extended. The high degree of regularity in the position of the myosin heads results in strong reflections along the 43 nm layer line in optical diffraction pattern. Although there is insufficient resolution to determine whether the increase in diameter is solely due to looser packing of the filament backbone or to the position of the myosin heads, the filament appears to be less tightly packed than in either of the other structures. Therefore these filaments are said to have a loose structure. In 8 control preparations the average distribution of the three structures was 58$\pm$5% tight, 12$\pm$4% disordered and 30$\pm$4% loose structure.

4. INFLUENCE OF CA ON THICK FILAMENT STRUCTURE

Because a Ca-calmodulin activated kinase can phosphorylate two thick filament proteins (MyBP-C and RLC), we examined the effect of different extracellular concentrations of Ca on the relative distribution of the three structures of the thick filament. Trabeculae were isolated from rat right ventricle and allowed to recover from the dissection for 20-30 min in normal Krebs' solution without stimulation. The reason for maintaining quiescence was to separate the effects of phosphorylation of MyBP-C from those due to phosphorylation of RLC. In the absence of stimulation and with the maintenance of intracellular Ca below the threshold for activation of contraction, RLC is less than 5% phosphorylated (McClellan et al,

2001). After the period of recovery, quiescent trabeculae were treated in one of three different ways. They were soaked in Krebs' solution containing 2.5 mM or 1.25 mM Ca for 120 min and then half of the trabeculae soaked in low Ca were bathed in Krebs' solution containing 7.5 mM Ca for an additional 10 min.

The structure of the majority of the isolated thick filaments, judged by the diameter of the filament and its optical diffraction pattern, was different for each of the three protocols. The structure of the vast majority of filaments separated from cardiac muscle that had been exposed to 2.5 mM Ca throughout the experiment had the tight structure. They were generally ordered, with myosin heads arranged along the filament backbone. These filaments produced diffraction patterns that displayed myosin layer-lines. A bare zone, free of myosin heads, occupied the central 0.18-0.20 μm, and each intact filament (ca. 1.6 μm long) displayed tapered ends.

Most of the thick filaments separated from quiescent cardiac muscle that had been soaked in 1.25 mM Ca for 120 minutes were disordered. The myosin heads extended from the backbone at various angles and to varying distances, producing disorder of the normally helical arrangement on the surface of relaxed filaments. Myosin layer lines were largely absent from diffraction patterns obtained from images of these filaments and when present were very weak. The location, appearance and extent of the bare zones were the same, regardless of the concentration of Ca in the Krebs' solution. An additional 10 minutes exposure to 7.5 mM Ca following the 120 min in 1.25 mM calcium had a dramatic effect on the structure of the thick filaments. The myosin heads once more appeared to be well-ordered, even more so than in filaments soaked for 120 minutes in solution with 2.5 mM calcium. The periodicity produced by the ordered heads of a helical array was more pronounced than before. Diffraction patterns obtained from images of these filaments had strong reflections along myosin layer-lines especially the 43 nm layer line. The filaments from muscle bundles soaked in 7.5 mM Ca had significantly greater diameters along their myosin head-bearing limbs than did any of the other filaments. The diameter, position and length of the bare zones, however, remained unchanged. These filaments were indistinguishable from the filaments with loose structure described above.

5. RELATION BETWEEN FILAMENT STRUCTURE AND PHOSPHORYLATION OF MYBP-C

In view of the apparent importance of MyBP-C in the formation of thick filaments and the effect of PKA induced phosphorylation of MyBP-C on the structure of the thick filament, the frequencies of given thick filament structures at different degrees of phosphorylation of MyBP-C were compared with electron microscopy and optical diffraction. Two types of experiments provided the data: measurements of phosphorylation and determination of structure carried out on the same set of trabeculae and measurements on trabeculae from different sets that had been treated in the same fashion. Exposure to PKA or to different concentrations of extracellular Ca and to 0.1 μM isoproterenol were used to produce the different structures and levels of phosphorylation. There is an excellent direct correlation between the relative amount of unphosphorylated, monophosphorylated and di- plus tri-phosphorylated MyBP-C and the relative amount of, respectively, disordered, tight and loose structures.   There is also an excellent inverse correlation between the relative amount of disordered and loose structures and the relative amount of, respectively, di plus tri-phosphorylated and unphosphorylated MyBP-C.   As phosphorylation increases, the fraction of thick filaments with disordered structure falls exponentially and the amount of loose structure rises exponentially.   The relative amount of the tight structure rises, reaches a maximum and then falls.

In the absence of phosphorylation of MyBP-C, the thick filament has a disordered structure.   Addition of the first phosphate produces order and the tight structure.   A further increase in order of myosin heads and a looser packing of myosin follows the addition of a second and third phosphate to each MyBP-C.   At the level of our resolution we have not been able to detect any significant difference in the structure of the thick filament produced by the increase in phosphorylation from two to three.   These changes in structure with phosphorylation are not due to changes in the phosphorylation of RLC because there is less than a 5% change in its phosphorylation associated with the changes in structure (McClellan et al, 2001).

6. INTERACTIONS BETWEEN THICK AND THIN FILAMENTS

Isolated thick and thin filaments were prepared in a solution containing 5 mM Mg ATP without Ca to prevent both formation of rigor links between the two types of filaments and activation of force-generating cycling of myosin heads.   There was no significant difference in either number of filaments per square micron or ratio of thin to thick filaments between control and the PKA treated preparations.   Thin and thick filaments were counted if at least a half a length (0.5 and 0.8 μm for respectively thin and thick filaments) was visible in the micrograph.   This result indicated that the type of structure of the thick filament did not influence the concentration of thin filaments in a preparation.

When isolated thick and thin filaments were parallel and lay near each other, extension of a portion of the thick filament to the thin filament was sometimes seen.   In most cases a distinct periodicity of about 40 nm existed between the connecting structures near the surface of the thick filament, suggesting that the structures might be myosin heads extended from the thick to the thin filaments. Where these connecting structures were present a reflection in the optical diffraction pattern on the 36-38 nm layer line was often seen in addition to the 43 nm layer line.   Thick-thin filament pairs never produced this reflection in the absence of the periodic structures connecting thick to thin filaments.   The presence of this reflection with an actin periodicity indicates that the thick filament extensions have a site-specific interaction with actin in the thin filament.   The lack of extension on the side of the thick filament away from the thin filament is additional support for this interpretation.   The probability of multiple interactions between thick and thin filaments is significantly enhanced by phosphorylation of MyBP-C and the formation of the loose, more ordered structure. The presence of parallel nearby thick and thin filaments without the repeated periodic structures, seen most commonly with the tightly structured thick filaments, is probably due to an occasional interaction as might occur with lower probability of interaction between filaments.

At the time of their staining, thick and thin filaments were incubated in a solution containing 2.5 mM MgATP, 0 Ca and 1.0 mM EGTA to prevent the formation of rigor links or actively cycling cross bridges. The ionic strength was kept near physiological to permit the formation of weak attachments of cross bridges to thin filaments. This weakly bound state is non-force producing and exists before cross bridges with bound ADP and Pi release Pi to form a strongly bound state that leads to generation of force and movement. In view of the conditions under which the interactions between cross bridges and thin filaments were observed, it is likely that the interactions consisted of weak, non-force generating bonds, similar to what Matsubara observed during diastole in rhythmically contracting mammalian heart (Matsubara et al 1979).

Among the PKA treated preparations 92% of thick filaments parallel to nearby thin filaments had mass extending to the surface of the thin filaments with 43 nm periodicity. Of the thick filaments from untreated preparations, only 47% of the thick filaments parallel to and near thin filaments in the untreated preparations had mass extending to the thin filaments with a 43 nm periodicity. This difference is highly significant (P<0.01).

The frequency of interactions between thick filaments and thin filaments was significantly greater when the thick filaments showed greater helical order regardless of whether the degree of order was induced by Ca or PKA This was the case after the muscle had been soaked in 2.5 mM Ca, or in 7.5 mM Ca after a period in 1.25 mM Ca. Since ordered myosin heads in the tightly structured filaments were more likely to have interactions with thin filaments than extended disordered myosin heads the distance of the myosin heads from the axis of the thick filaments appears to be less important than the degree of helical order of the heads

7. EFFECT ON KINETICS OF CONTRACTION

Formation of the weakly bound state may be a rate limiting step in the cycling of cross bridges in cardiac muscle or simply determine which myosin heads will enter the force generating cycle. An increase in the concentration of weakly bound cross bridges would increase the number of force generating cross bridges when the muscle is activated by Ca. Such a mechanism would explain the relationship between phosphorylation of MyBP-C and contractility when contractility is changed by stimulation with alpha or beta adrenergic agonists, endothelin, cholinergic stimulation or Ca.

In reconstituted systems of actin and myosin lacking the steric constraints of an intact filament lattice, the weakly bound state should form more easily and be less dependent on phosphorylation of MyBP-C. In intact cardiac muscle on the other hand, enhancement of maximum Ca activated force from increased formation of weak bonds between actin and myosin would be expected with phosphorylation of MyBP-C.

In the preparations of thick and thin filaments treated with activated PKA, the inhibitory subunit of troponin (TNI) should also be phosphorylated. This will not affect the structure of the thick filaments, but it could modify the interaction of myosin heads with thin filaments if phosphorylation of TNI has other effects besides its decrease in the affinity of TNC for Ca. .Decreased affinity normally diminishes the likelihood of myosin head binding to actin. In this study, EGTA was present in the solutions, and therefore Ca binding should not have been a factor. When thick filament structure was changed by varying extracellular Ca concentration, TNI phosphorylation should not change. Although phosphorylation of MyBP-C changes thick filament structure, it may complement changes in the thin filament from phosphorylation of troponin and thereby lead to modulation of the actin-myosin interaction.

Ca regulated changes in thick filament structure that modulate the maximum level of force production of contraction fit well with changes in contractile activity that occur with Ca induced alterations in excitation-contraction coupling. Increase in cytoplasmic Ca contraction not only leads to greater Ca binding of the ion by TNC and greater thin filament activation. The increase in cytoplasmic Ca also alters thick filament structure and may produce a more rapid rate of cross bridge attachment and greater force production as a result of an enhanced duty cycle. Lowering cytoplasmic Ca moves cross bridges away from the thin filament decreasing the probability of formation of the weakly binding state.

8. FUNCTIONAL IMPLICATIONS OF THE C5 MODULE OF MYBP-C

Most mutations of cMyBP-C produce truncations that eliminate the C terminus of the protein, which contains an important binding site for the rod portion of myosin (Starr and Offer 1978, Okagaki et al 1993). However with either of two mutations in C5, the LMM binding site is present in the abnormal protein, but the heart still becomes myopathic. This suggests that C5 *per se* may have an important function in cardiac muscle.

We have examined the role of C5 in the contractile process by bathing skinned cardiac trabecula with fragments of MyBP-C containing C5 to interfere with the normal interactions of the C5 in endogenous MyBP-C. C2C5 fragments produced a marked reduction in force generated in pCa 5.0, the concentration that normally produces Fmax in our skinned trabecula preparation. Fmax occurred at pCa 4.5 and was lower than control by a small but significant amount. C2C4 fragments at the same or 3 times higher concentration did not change force. These data implicate C5 as a key region in the modification of contractility by MyBP-C. The change in force with C2C5 completely reversed with removal of the fragment. The time course for the decline of force as well as the washout and the restoration of force were very similar.

The effect of C5 on contractility was characterized more completely with a fragment consisting only of C5. Consistent with the results using C2C4 and C2C5, force at pCa 5.0 was markedly reduced by C5, and its removal of C5 completely or nearly completely reversed the decline in force. Gruen and Gautel (1999) found that C2C5 does not bind to myosin in solution, but Moolman-Smook et al (2002) have recently shown that C5 binds to C8 and that the strength of binding is reduced in the two mutations of C5 that cause cardiomyopathy without truncation of MyBP-C. They also have produced evidence that C7 and C10 interact. The effect of added C5 or C2C5 on contractility was most likely due to displacement of endogenous C5 from its binding to C8. Since MyBP-C is not extracted from the trabeculae and the effect on contractility is completely reversed with removal of C5 or C2C5, release of MyBP-C from the myofilaments cannot be the cause of the decline in contractility. Apparently the binding of the C terminus (C8-C10) to LMM remains to prevent loss of MyBP-C from the skinned fibers.

Based on these observations Moolman-Smook et al (2002) have proposed a model with 3 molecules of MyBP-C forming a collar around a thick filament every 43 nm in the C zone of the sarcomere. The 3 molecules overlap in a parallel arrangement. They propose flexibility in the region of C5. Their proposal of a collar formed around the rod portion of the thick filament by 3 MyBP-C molecules every 43 nm is similar to one suggested by Winegrad based on thickness measurements of thick filaments (1999) although the nature of the interaction is different. The former is based on interactions between MyBP-C molecules and the latter to interactions between MyBP-C and myosin.

Our data provide support for the functional importance of C5 interaction. When C5 is added to the bathing medium to interfere with the normal interactions of endogenous C5, Ca sensitivity is substantially decreased and Fmax declines. The myofibril is not irreversibly altered since removal of the C5 from the bathing medium restores the original level of contractility. The relatively rapid response of the trabeculae to addition and removal of C5 from the bathing medium is consistent with relatively moderate affinity of C5 for C8 found by Flashman et al. (2002)

9. SUMMARY

Both the MyBP-C motif between C1 and C2 and the C5 module are important regions for implementing the effect of MyBP-C on myosin and on contractility but in different ways. C5 may determine the folding of MyBP-C and the manner in which MyBP-C interacts with myosin. In spite of its apparent importance this interaction does not appear to be physiologically regulated. Its alteration by mutation however can have a major effect on contractility. On the other hand, the effect of the motif is regulated by phosphorylation and appears to be an important part of a physiological mechanism(s) for modulating contractility.

Thick filaments isolated from cardiac muscle exist in one of three different structures (Levine et al 2001). Different degrees of phosphorylation of MyBP-C can produce transitions among the three structures. The combination of the binding results of Flashman et al (2002) with the data of McClellan et al (2001) suggests that the C5 interaction with C8 is critical in maintaining the normal structure of thick filaments and the normal function of the force generators in the filaments. The cardiac-specific sequence in C5 and its normal interaction with another part of the same or a different MyBP-C may be required for the correct 3 dimensional shape of the three MyBP-C molecules at each locus in the C zone and the normal structure of the thick filament. The normal interactions may be necessary to allow transitions in binding and filament structure that are associated with phosphorylation of the MyBP-C motif.

10. REFERENCES

Bonne, G., Carrier, L., Bercovici, J., Cruard, C., Richard, P., Hainque, B., Gautel,M., Labiet, S., James, M., Beckman, J., Weissenbach, J., Vosberg, HP., Fiszman, M., Komajda, M., Schwartz, K. 1995 Cardiac myosin binding protein C gene splice acceptor site mutation is associated with familial hypertrophic cardiomyopathy. Nature Genet. 11: 438-440.

Carrier, L., Bonne, G., Bahrend, E., Yu, B., Richard, P., Niel, F., Hainque, B., Cruard, C., Gary, F., Labiet, S., Bouhour, J-B.,Dubourg, O., Desnos, M., Hagege, A., Trent, Komajda, M, Fiszman, M., Schwartz, K. Organization and sequence of human cardiac myosin binding protein C gene (MyBP-C3) and identification of the mutations predicted to produce truncated proteins in familial hypertrophic cardiomyopathy. Circ. Res. 80: 427-434, 1997.

Craig, R.,and Offer, G. 1976. The location of C-protein in rabbit skeletal muscle. Proc. R. Soc. (Lond) 192:451-461

Davies, J. 1988. Interaction of C-protein with pH 8.0 synthetic thick filaments prepared from the myosin of vertebrate skeletal muscle. J. Musc. Res. Cell Motil. 9: 174-183

Flavigny,J, Souchet. M, Sebillon P, Berrebi-Bertrand I, Hainque B, Mallet A, Bril A, Schwartz K, And Carrier L. 1999. COOH-terminal truncated cardiac myosin binding protein C mutants resulting from familial hypertrophic cardiomyopathy mutations exhibit altered expression and/or incorporation into fetal rat cardiomyocytes. J. Mol. Biol. 294: 443-456.

Gautel,M., Zuffardi,O., Freiberg,A.,and Labeit,S. 1995. Phosphorylation switches specific for the cardiac isoforms of myosin binding protein C: a modulator of cardiac contraction. EMBO J. 14: 1952-1960.

Gilbert, R., Kelly, M.G.,Mikawa, T.,and Fischman,D.A. 1996. The carboxyl terminus of myosin binding protein-C (MyBP-C), C-protein) specifies incorporation into the A-band of striated muscle. J. Cell Sc. 109: 101-111.

Gruen, M.,and Gautel, M.. 1999 Mutations in beta-myosin S2 that cause familial hypertrophic cardiomyopathy (FHC) abolish the interaction with the regulatory domain of myosin binding protein-C. , J.Mol. Biol. 286, 933-949.

Gruen, M, Prinz, H, and Gautel, M. 1999 cAPK-phosphorylation controls the interaction of the regulatory domain of cardiac myosin binding protein C with myosin-S2 in an on-off fashion. FEBS Letters. 453(3):254-9.

Harris, SP., Bartley, CR., Hacker, TA,. McDonald, KS., Douglas, PS,. Greaser, ML,. Powers, PA., Moss, RL. 2002; Hypertrophic cardiomyopathy in cardiac myosin binding protein-C knockout mice. Circulation Research. 90:594-601.

Hartzell,C.,and Glass,D. 1984. Phosphorylation of purified cardiac muscle protein by purified cAMP-dependent and endogenous Ca-calmodulin-dependent protein kinases. J. Biol. Chem 259: 15587-15596.

Hofmann, PA, Hartzell, HC and Moss, RL. 1991 Alterations in Ca sensitive tension due to partial extraction of C-protein from rat skinned cardiac myocytes and rabbit skeletal muscle fibers. J. Gen. Physiol. 97: 1141-1163.

Koretz, J. F. 1979. Effects of C-protein on synthetic myosin filament structure. Biophys. J. 27: 433-446.

Kunst, G. ,Kress, K.R,. Gruen, M, Uttenweiler, D. , Gautel, M. ,and Fink, R.H.A.. 2000. MyBP-C (C-Protein)- a phosphorylation dependent force regulator in muscle that controls the attachment of myosin heads by its interaction with myosin – S2. Circ. Res 86: 51-58.

Levine, R.J.C., Weisberg A. , Kulikovskaya, I., McClellan, G. and Winegrad, S. 2001. Multiple structures of thick filaments in resting cardiac muscle and their influence on cross bridge interactions. Biophys. J. 81: 1070-1082.

Lin, Z., Lu, M.H.,Schultheiss, T., Choi, J., Holtzer, S., DiLuulo, C., Fischman, D.A.,and Holtzer, H. 1994. Sequential appearance of muscle-specific proteins in myoblasts as a function of time after cell division: evidence for a conserved myoblast differentiation program in skeletal muscle. Cell Motil. Cytoskel. 29: 1-19.

Matsubara, I., Yagi,N., Endoh, M. 1979 Movement of myosin heads during a heart beat. Nature 278: 474-476.

McClellan,G., Weisberg, A., and Winegrad, S. 1994. CAMP can raise or lower cardiac actomyosin ATPase activity depending on alpha adrenergic activity. Amer. J. Physiol. 267: H431-442

McClellan,G., Kulikovskaya, I. and Winegrad,S. 2001 Structural and functional responses of the contractile proteins to changes in calcium concentration in the heart. Biophys. J. 81: 1083-1092

Moolman-Smook,J., Flashman,E., de Lange, W., Li,Z., Corfield,V., Redwood,C., Wathins,H. 2002. Idnentification of novel interactions between domains of myosin binding protein C that are modulated by hypertrophic cardiomyopathy missense mutations Circ. Res. 91: 704-711.

Offer, G., Moos,C.,and Starr, R. 1973. A new protein of the thick filaments of vertebrate skeletal myofibrils. Extraction. purification and characterization. J. Mol. Biol.; 74: 653-676.

Okagaki,T., Weber,F.E., Fischman,D.A., Vaughan,K.T., Mikawa,T.,and Reinach,F.C. 1993. The major myosin binding domain of skeletal muscle MyBP-C (C protein) resides in the COOH-terminal, immunoglobulin C2 repeat. J. Cell Biol. 123: 619-626.

Rhee, D. Sanger, J.M.,and Sanger, J.W. 1994. The premyofibril: evidence for its role in myofibrillogenesis. Cell Motil. Cytoskel. 28: 1-24.

Rottbauer, W., Gautel, M., Zehelein, J., Labeit, S., Franz, W.M., Fischer, C., Vollrath, B., Mall, G., Dietz, R., Kubler, W., Katus, H.A. Novel splice donor site mutation in the cardiac myosin -binding protein C gene in familail hypertrophic cardiomyopathy. Charaterization of cardiac transcript and protein. J. Clin. Invest. 100: 475-482.1997

Sebillon, P, Bonne, G, Flavigny, J, Venin,S, Rouche, A, Fiszman,M, Vikstrom,K, Leinwand,L, Carrier,L, Schwartz,K. 2001 COOH-terminal truncated human cardiac MyBP-C alters myosin filament organization. Comp. Rendus Acad. Sc. 324: 251-260.

Schlender,K.,and Bean,L. 1991. Phosphorylation of chicken cardiac C protein by calcium calmodulin-dependent protein kinase II. J. Biol. Chem. 266: 2811-2817

Schwartz, K., Carrier, L., Guicheney, P., Komajda, M. 1995 The molecular basis of cardiomyopathies. Circulation91: 1336-1347.

Seiler,S.H., Fischman,D.A., Leinwand,L.A. 1996 Modulation of myosin filament organization by C protein family members. Molec. Biol. Cell 7: 113-127.

Starr,R. and Offer, G. 1978 The interaction of C-protein with heavy meromyosin and subfragment-2. Biochem. J. 171: 813-816.

Watkins, H, Conner, D, Thierfelder, L, Jarcho, JA, MacCrea, C, McKenna, WJ, Maron, BJ, Seidman, JG, Seidman CE. 1995; Mutations in the cardiac myosin binding protein C on chromosome 11 cause familial hypertrophic cardiomyopathy. Nat. Genetics 1: 433-438.

Weisberg,A.,and Winegrad, S. 1996. Alteration in myosin cross bridges by phosphorylation of myosin-binding protein C in cardiac muscle. Proc. Nat. Acad. Sc (USA) 93: 8999-9003.

Weisberg, A.,and Winegrad, S. 1998. Relation between cross bridge structure and actomyosin ATPase activity in rat heart. Circ. Res. 1998; 83: 60-72

Winegrad, S. 1999. Cardiac myosin binding protein C. Circ. Res. 84: 1117-1126.

DISCUSSION

Huxley: What do you think may be the mechanism by which disorder of myosin heads affects tension development?

Winegrad: The most attractive working hypothesis at the present time is that there is an optimal alignment of thick and thin filaments relative to each other, that produces the largest percentage of myosin heads that can bind to actin and generate force simultaneously. Binding of a distensable cMyBPC between filaments could modulate this interaction, and therefore the number of force generators acting simultaneously. Another possibility is that change in the tightness of packing of myosin rods in the thick filament resulting from change in phosphorylation of cMyBPC can modify the movement of the myosin heads. In any case, the mechanism must be cooperative because force can be changed by as much as 70% but there is only 1 molecule of C protein for 7 molecules of myosin. Therefore a change in the C zone, where C protein is located, can change the properties of force generators outside the C zone.

Huxley: Does C-protein bind to a particular region of the titin molecule, i.e. along its length?

Winegrad: C protein binds with a titin in the C zone with 43nm periodicity. The titin binding site is in the C8-C10 region of C protein. The location of the binding site in titin is very close to the myosin binding site in titin. Both C protein and titin bind to myosin with the same periodicity.

Ranatunga: Have effects of C protein been examined in rigor muscle fibers?

Winegrad: The only study that I know is that of Gautel, who showed that exposure of skinned slow skeletal fiber to cardiac C1C2 fragment of C protein changes the stiffness of muscle in rigor.

Solaro: As you know, in a transgenic model without cTnI phosphorylation, there was no effect of C-protein phosphorylation on force. Do you think that the thin filament binding of C-protein may be important here?

Winegrad: In view of the possible role of C protein binding to actin and its modulation by phosphorylation of C protein, this possibility is one that we have been giving considerable attention. It is possible that the phosphorylation of both proteins is involved in the regulation.

Pollack: How do you think the significance of the fact that C-protein is distributed not in the entire thick filament, but only in a limited region?

Winegrad: This is a very important point because the modulation of contractility by C protein moves at least 60-70% of force generators. Some cooperative or quasi cooperative mechanism must be moved. Perhaps imposing a specific steric relation between myosin heads and actin within the C zone can have on effect outside the C zone.

ter Keurs: Is MyBPC long enough to bind both to titin and to myosin as well as to actin?

Winegrad: Yes, the proline rich regions in C0C1 and C5 should produce flexibility. The similarity of C0C1 proline rich region to the PEVK region of titin would suggest that it could extend considerably and provide the length that is necessary.

THE ENDOCARDIAL ENDOTHELIAL Na$^+$/K$^+$ ATPase AND CARDIAC CONTRACTION

Paul Fransen, Jan Hendrickx, and Gilles De Keulenaer[*]

1. INTRODUCTION

The cavitary sides of the ventricles are covered by the endocardium, which consists of an innermost thin layer of endothelial cells (1-3 μm, the endocardial endothelium, EE) and a subjacent layer of interstitial tissue. Although the EE was initially considered as a passive monolayer of cells at the border between the circulating blood and the underlying cardiomyocytes, current insights now suggest an obligatory role of the EE in the regulation of normal cardiac function (for review, see Brutsaert, 2003). Since Brutsaert and co-workers in 1988 for the first time showed that selective removal of the EE layer from the myocardium in papillary muscles from cat hearts altered mechanical performance of the muscle, this observation was confirmed in cardiac muscle preparations from different species (Smith et al., 1991, Wang and Morgan, 1992) and also in the intact animal in-vivo (Gillebert et al., 1992; De Hert et al., 1993). Coronary endothelial cells have been described to have similar inotropic activity (Li et al., 1993). How the presence of an endothelial monolayer covering a thick layer of cardiomyocytes influences the mechanical performance of these myocytes, is presently under investigation. In their initial paper, Brutsaert et al. (1988) stated: " As for the mechanisms involved ..., we can only speculate at present. We might think of at least three possible ways by which the endocardium could affect myocardial performance: by the release of a chemical substance or messenger, by an electrochemical barrier, or by both."

Today, we know that EE cells respond to humoral and mechanical stimuli by releasing biologically active substances, which modulate myocardial contractility (stimulus-secretion-contraction coupling) (Figure 1). This cross-talk mechanism between endothelium and muscle cells has been extensively studied in blood vessels, especially following the observation of Furchgott and Zawadski (1980) about the obligatory role of the vascular endothelium in acetylcholine-induced vasorelaxation. Bioassay studies demonstrated that endothelial cells released a labile, diffusable non-prostanoid factor identified as nitric oxide (Rubanyi and Vanhoutte, 1986; Furchgott and Vanhoutte, 1989).

[*] Paul Fransen, Jan Hendrickx and Gilles De Keulenaer, Department of Pharmacology, University of Antwerp, Groenenborgerlaan, 171, B-2020 Antwerpen, Belgium. E-mail: paul.fransen@ua.ac.be

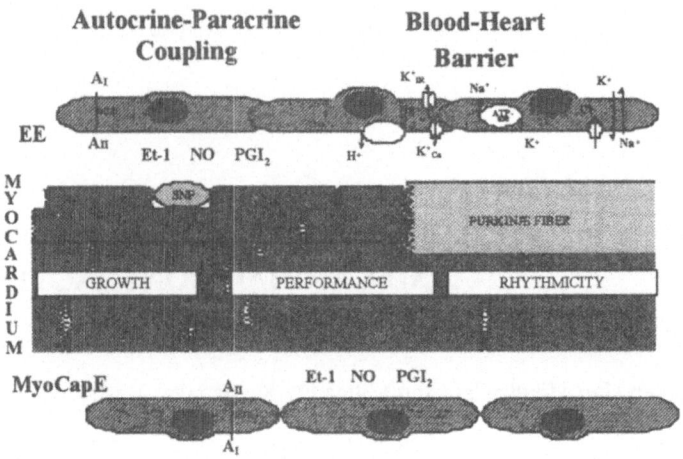

Figure 1. Growth, performance and rhythmicity of the myocardium, the subneural plexus (SNP) and Purkinje fibers can be affected by EE (EE) and endothelium of myocardial capillaries (MyoCapE) via two pathways: autocrine-paracrine coupling (left) and/or blood-heart barrier (right). A_I and A_{II}: angiotensin I and II, ACE: angiotensin converting enzyme, Et-1: endothelin I, NO: nitic oxide, PGI_2: prostacyclin, K^+_{IR}: inwardly rectifying K^+ channel, K^+_{Ca}: Ca^{2+} activated K^+ channel

Besides NO and other relaxing factors such as endothelium-derived hyperpolarizing factor (EDHF, Chen and Suzuki, 1990), vascular endothelial cells also release contracting factors among which arachidonic acid metabolites and endothelin-1.

In EE cells, release of prostacyclin (PGI_2) and prostaglandin E_2 (PGE_2) (Mebazaa et al., 1993a), endothelin-1 mRNA expression, endothelin-release (Mebazaa et al., 1993b) and NO-release (Balligand et al., 1993) has been demonstrated and inotropic activity of these factors have been characterized. Besides these factors, a number of unidentified factors have been suggested to count for the positive inotropic action of coronary and EE cell superfusate on myocyte contractility (Smith et al., 1991; Shah et al., 1994). Although the release of endothelium-derived factors such as ET-1, NO, prostaglandins and others are important in the modulation of myocardial performance, the contribution of these messengers in the contractile response following selective EE damage seems to be rather insignificant.

Indeed, the addition to intact rabbit papillary muscles of BQ-123, L-NMMA and indomethacin to antagonize the tonic effects of ET-1, EDRF and prostaglandins, did not result in the typical negative inotropic and twitch abbreviating response observed after selective EE damage, suggesting that other, presumably non-paracrine effects could play a role (De Keulenaer et al., 1995).

According to the second cross-talk mechanism (Figure 1), the EE may establish an active barrier between circulating blood and cardiomyocytes, analogous to the blood-brain endothelial barrier between circulating blood and the brain interstitial fluid (Brutsaert, 1989; Brutsaert et al., 1998; Brutsaert, 2003). Cardiomyocytes make up to 80% of the mass of the myocardium, but their number accounts for less than 40% of all cardiac cells

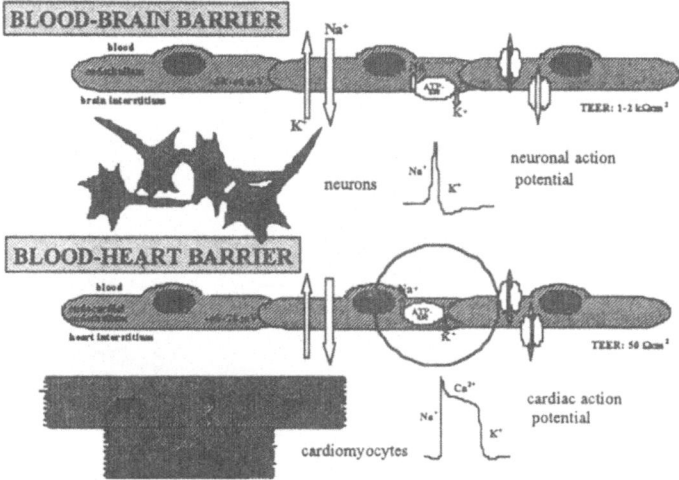

Figure 2. In parallel with the blood-brain barrier, the blood-heart barrier might control the ionic composition surrounding the cardiomyocytes, which are extremely sensitive to fluctuations of Ca^{2+}, K$^+$, Na$^+$, and Cl$^-$. The maintenance of a homogeneous ionic composition of the extracellular fluids surrounding the cardiomyocytes is, therefore, a prerequisite for normal function of the heart and its disturbance may lead to electrical and mechanical abnormalities. TEER: transendothelial electrical resistance

and the surface area of non-myocytes may exceed the myocytic surface area. Therefore, non-myocytes such as EE cells may have profound effects on the regulation of contraction of cardiomyocytes. It is well characterized, for example, that subendocardial myocardium better survives coronary artery occlusion because of diffusion from the ventricular lumen via the EE (Reimer and Jennings, 1979). Nonetheless, because of specific phenotypic properties of EE cells, only a limited number of substances may be able to pass from the blood into the excitable tissue of the heart and vice versa. In parallel with the best studied endothelial barrier, i.e. the blood-brain barrier (BBB), the endocardial barrier might protect the heart from toxic substances in the blood and from ionic fluctuations in the blood that would disturb the complex electrical and mechanical activity of the heart (Figure 2). These activities are for example extremely sensitive to fluctuations in the extracellular concentration of ions (Ca^{2+}, K$^+$, Na$^+$, and Cl$^-$). The maintenance of a homogeneous ionic composition of the extracellular fluids surrounding the cardiomyocytes is, therefore, a prerequisite for normal function of the heart and its disturbance may lead to electrical and mechanical abnormalities.

In papillary muscles, a thin muscle bundle enveloped by EE cells, the negative inotropic and contraction shortening effect of EE removal can, for example, be easily reversed by a decrease of the extracellular Cl$^-$ concentration and can be mimicked by a slight decrease of extracellular K$^+$ (Figure 3). Furthermore, assuming a role of the EE as a blood-heart barrier and, as a consequence, assuming a different ionic constitution of the plasma and the myocytal interstitium, removal or dysfunction of the barrier may have dramatic implications for virtually all the functions of the heart. Is there additional experimental evidence that supports this concept?

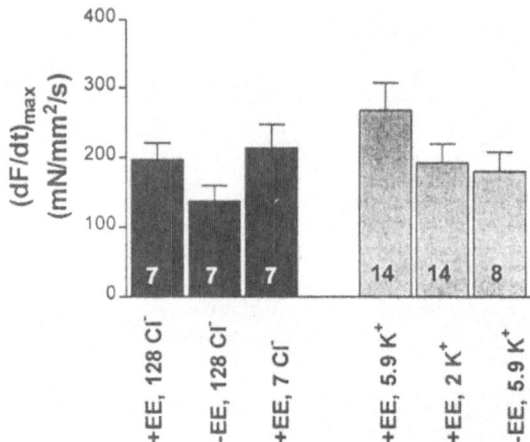

Figure 3. Maximal rate of force development ($(dF/dt)_{max}$ of isometric contractions of cat (black bars) and rabbit (white bars) papillary muscle in control conditions (with intact EE, +EE, 128 mM external Cl⁻ and 5.9 mM external K⁺), following removal of EE (-EE) and following decrease of Cl⁻ (7 mM Cl⁻) or decrease of K⁺ (2 mM K⁺). Reduction of Cl⁻ reverses the negative inotropic effect of EE removal, whereas reduction of K⁺ mimics the negative inotropic effect of EE removal.

2. POLARITY OF ENDOCARDIAL ENDOTHELIUM

One of the most prominent features of a cell barrier is the asymmetrical distribution of transport systems between luminal and basolateral side of the cell. This has been extensively studied in BBB endothelial cells and in endothelial cells of the cornea (corneal endothelial barrier, CEB). In both endothelial barrier models, the asymmetrical occurrence of ion pumps, ion channels and ion co-transporters leads to specific transendothelial ion transport and different ionic constitutions of the blood and brain interstitial milieu (cerebrospinal fluid, CSF) for the BBB and of the aquous humor and stroma for the CEB.

The asymmetric distribution of ions between blood and CSF suggests active transport of K^+, Ca^{2+}, HCO_3^- and amino acids from the CSF to the blood and Mg^{2+} and Cl⁻ in the reverse direction. The Na^+/K^+ ATPase has been shown to be strictly localized at the basal membrane (Betz et al., 1980,1983) and to be one of the driving forces of the directed transport of Na^+ and K^+. An ameloride-sensitive non-selective cationic channel (probably important in transendothelial Na^+ transport) is present in the luminal membrane (Betz, 1983, Vigne et al., 1989). Hoyer et al. (1991) and Popp et al. (1992) found inwardly rectifying K^+ channels to be mainly expressed in the antiluminal membrane, as well as a stretch-activated non-selective but Ca^{2+} permeable cationic channel. A furosemide-sensitive $Na^+/K^+/2Cl^-$ co-transporter was localized at the luminal membrane (Betz, 1983). The cartoon presented in Figure 4 is probably far from complete, but at least suggests that the BBB endothelial monolayer might regulate unidirectional flux of ions from the blood to the brain interstitium and vice versa, thereby controlling the ionic environment and complex electrical activity of the neurons.

The corneal endothelium has been extensively studied with respect to directed HCO_3^- and water transport. Although this barrier has a much lower transendothelial resistance than the BBB (±30 Ω cm^2 versus 1000-2000 Ω cm^2), the asymmetrical distribution of ion transport systems between luminal and abluminal cell membrane leads to different ionic compositions of aquous humor (apical side) and stroma (basolateral side). Also here, the basolateral localisation of the Na$^+$/K$^+$ ATPase (α_1 isoform) contributes to the barrier properties of the endothelium (Guggenheim & Hodson, 1994). Furthermore, the endothelium has been studied extensively also for the asymmetrical distribution of pH regulating transport systems (Bonanno and Giasson, 1992; Zhang et al., 2002; Sun et al., 2000; Bonanno et al., 1999).

In EE cells, the asymmetrical distribution of ion channels has been studied with the single channel patch-clamp technique. In the luminal membrane of EE cells from bullfrog atrium, Ito et al. (1993) described a class of outwardly rectifying non-selective cationic channels and two types of ohmic channels of which the larger conductance channel was also non-selective for K$^+$ and Na$^+$. A more detailed study of ion channels in luminal and abluminal membranes of endothelial cells of guine-pig atria was performed by the same group (Table 1, Manabe et al., 1995). Non-selective cation channels and the inwardly rectifying K$^+$ channel were apparent at the luminal side of the cells, whereas large conductance Ca^{2+} activated K$^+$ channels were observed at the abluminal membrane.

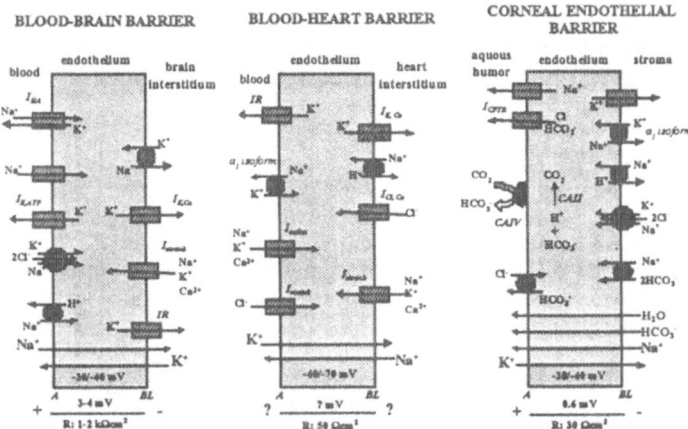

Figure 4. Comparison of the asymmetrical distribution of ion channels, ion pumps and ion transporters in the endothelial cells of blood-brain barrier, blood-heart barrier and corneal endothelial barrier. I_{HA}: hyperpolarisation-activated current, $I_{K_{ATP}}$: ATP dependent K$^+$ current, $I_{K,Ca}$: Ca^{2+} activated K$^+$ current, $I_{stretch}$: stretch activated current, IR: inwardly rectifying K$^+$ current, $I_{Cl,Ca}$: Ca^{2+} activated Cl$^-$ current, I_{cation}: non-selective cation current, I_{CFTR}: cystic fibrosis transmembrane conductance regulator current, CAII and IV: carbonic anhydrase II and IV, R: transmembrane electrical resistance, A: apical, BL: basolateral

Table 1. Number of observations for different channel types in guinea-pig atrial EE cells (modified after Manabe et al., 1995)

channels	conductance (pS)	Appearance (%) luminal	Appearance (%) abluminal
NSC	36.3 +1.4	11.6 (43/371)	2.6 (4/154)
NSC	11.2 + 0.5	4.9 (18/371)	0.6 (1/154)
Cl	408.6 + 24.4	8.6 (32/371)	13.6 (21/154)
K_{Ca}	33.5 + 3.8	5.7 (21/371)	3.9 (6/154)
K_{Ca}	210.7 + 8.0	1.9 (7/371)	4.5 (7/154)
K_{IR}	31.7 + 2.2	1.6 (6/371)	0 (0/154)
unc		11.1 (41/371)	3.9 (6/154)
total		45.3 (168/371)	29.2 (45/154)

Numbers in parentheses indicate actual number of experiments. Overlap of multiple channel openings of a given type of channels was neglected in calculating the appearance rate. Unc: unclassified channels, NSC: non-selective cation channel, K_{Ca}: Ca^{2+}-dependent K^+ channel; K_{IR}: inwardly rectifying K^+ channel. The slope conductance was measured with a K^+ gradient of 150/5.4 mM.

In porcine EE (right atrium), Hoyer et al. (1994) demonstrated at the luminal side a Ca^{2+} influx through stretch-activated channels, which activated maxi K^+ channels at the same side. Taking together these observations, a cartoon similar to the BBB and CEB cartoon can be drawn (Figure 4). Although a direct transendothelial unidirectional flux of ions from blood to the myocardial interstitium or vice versa and a different ionic composition of the blood and myocardial interstitium has not yet been demonstrated, the above observations strongly suggest that the EE in the heart may be able to control and regulate such a flux, thereby controlling part of the complex mechanical and electrical activity of cardiomyocytes. Furthermore, the transendothelial electrical resistance (TEER) of about 50 Ωcm^2 was lower than in BBB endothelium, but was higher than in CEB (Brutsaert et al., 1998). Because asymmetrical localization of the Na^+/K^+ ATPase in BBB and CBE is at the basis of directed transport of ions across the endothelia, we wondered whether the Na^+/K^+ ATPase in EE cells was also asymmetrically distributed.

3. CARDIAC Na^+/K^+ ATPase

Although the Na^+/K^+ ATPase has been studied extensively in cardiac muscle, its presence and function in cardiac endothelial cells was unknown. Na^+/K^+ ATPase consists of two subunits: the α (ouabain-sensitive) and β subunit. In the heart, expression of three α isoforms (α_1, α_2 and α_3) has been demonstrated. In EE, electrophysiological measurements supported the presence of Na^+/K^+ ATPase in cardiac endothelium (Daut et al., 1988; Laskey et al., 1990; Fransen et al., 1995). In cultured EE cells, it was found that the resting membrane potential at 2.7 mM external K^+ suddenly dropped to values around –20 mV with an increase in outwardly directed currents. This was attributed to inhibition of an active sodium pump in these cells (similar results were obtained with removal of external K^+ and with addition of 100 μM ouabain) (Fransen et al., 1995) and was suggestive for the presence of the α_1 isoform of the Na^+/K^+ ATPase in EE. The α_1 and α_2 isoforms have been described to show different sensitivity to external K^+ (Gao et al., 1995): the apparent dissociation constants for the α_1 and α_2 isoforms are 3.7 and 0.5 mM, respectively. With molecular (mRNA and protein), electrophysiological, morphological and microfluorimetric techniques, subsequent direct evidence was found for the presence of α_1, but not α_2 isoforms in EE of rat and rabbit. Furthermore, in these

EE cells, the α_1 isoform was distributed asymmetrically and was present at the luminal membrane, where it co-localized with platelet endothelial cell adhesion molecule (PECAM) (Figure 5).

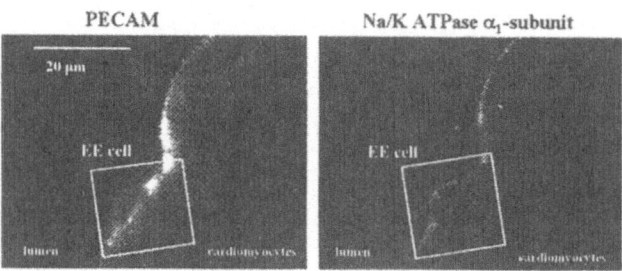

Figure 5. Double immunostaining of cryostat sections of rat heart for PECAM (left) and α_1 isoform of the Na$^+$/K$^+$ ATPase.

In cardiomyocytes, the different isoforms have been described to display different localisation and function with respect to myocardial contraction (Gao et al., 1995; Juhaszova & Blaustein, 1997; Gao et al., 1999, James et al., 1999; Pfeiffer et al., 1999). This was confirmed in our laboratory in rat and rabbit cardiac muscle (Figure 6, Fransen et al., 2001).

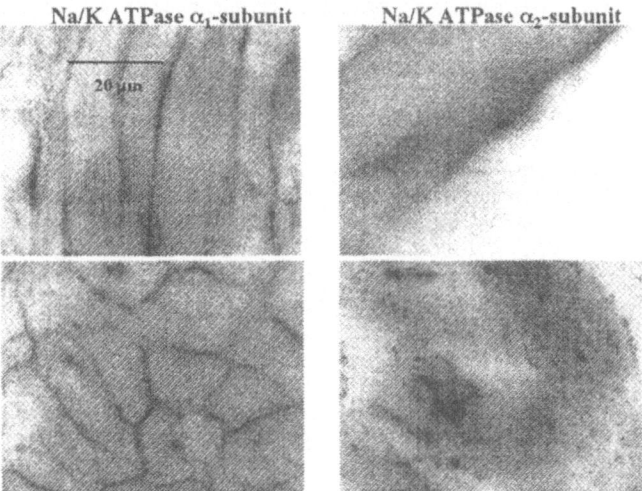

Figure 6. Immunostaining of cryostat sections through left ventricle of rabbit hearts. Sections were stained for α_1 (left) and α_2 (right) isoforms of the Na$^+$/K$^+$ ATPase.

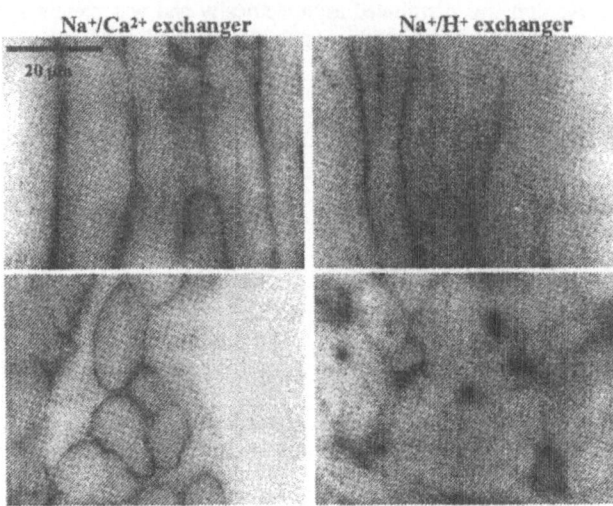

Figure 7. Immunostaining of cryostat sections through left ventricle of rabbit hearts. Sections were stained for Na^+/Ca^{2+} exchanger (left) and Na^+/H^+ exchanger (right).

In the rabbit heart, the α_1 subunit isoform was present mainly at the myocytal cell borders and in intermyocytal blood vessels, whereas the α_2 isoform was evident at the myocytal T-tubules. We investigated also the localization of the Na^+/Ca^{2+} (NCE) and Na^+/H^+ (NHE) exchanger, which were both present at the cell borders (NCE>NHE) and T-tubules. NHE was further extensively present in the intermyocytal capillaries (Figure 7). These observations suggested different functions for the Na^+/K^+ ATPase α subunit isoforms with α_1 regulating global Na^+, K^+ and Ca^{2+} following action potentials and α_2 regulating local Na^+, K^+ and Ca^{2+} at the excitation-contraction site: the T-tubules (Blaustein and Lederer, 1999; Gao et al., 2002). Both isoforms may be influenced by pH via the activity of co-localized NHE.

The functional significance of all these observations was assessed by measuring isotonic and isometric contractions of rabbit right ventricular isolated papillary muscles with intact EE (+EE muscles) and with EE removed (–EE muscles). EE cells display the α_1 isoform of the Na^+/K^+ ATPase only. So, all factors that affect the α_1 subunit isoform will have endothelium-dependent effects on cardiac performance. In initial experiments, in which we studied the effects of higher external K^+ variations on $(dF/dt)_{max}$ in +EE and –EE muscles, we found that increasing external K^+ from 5 to 12.5 mM caused no effect (+EE, n = 6) or only a slight positive inotropic effect (±5%, -EE, n = 7), suggesting that isolated intact rabbit papillary muscles are very tolerant to "high" external K^+ (Figure 8). They are, however, not tolerant to low external K^+. This was not very surprising because in guinea-pig cardiomyocytes, $[K^+]_o$ activated α_1 isoform of Na^+/K^+ ATPase half-maximally at 3.7 mM $[K^+]_o$ (Gao et al., 1995). At reduced external K^+, biphasic inotropic effects were observed: reducing external K^+ to 2.5 mM caused endothelium-dependent negative inotropy, whereas further decrease of external K^+ caused endothelium-independent positive inotropy (Figure 8).

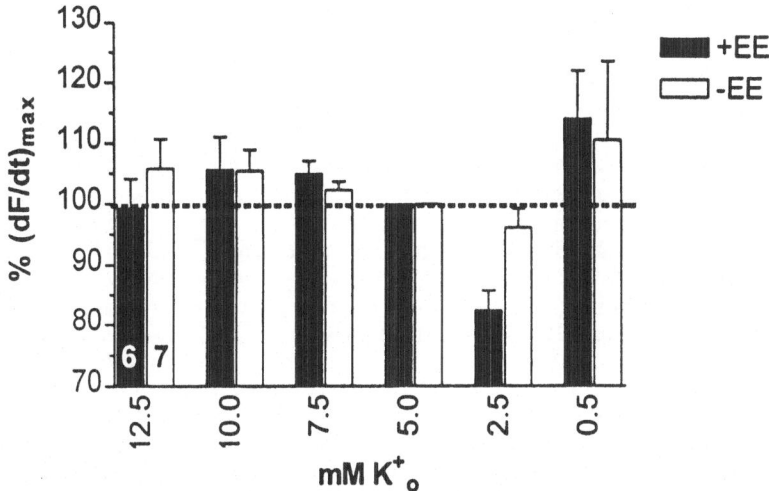

Figure 8. Relative maximal rate of force development of isometric contractions of isolated papillary muscles of rabbit right ventricle as a function of the external K^+ concentration. $(dF/dt)_{max}$ of isometric contactions of muscles with intact EE (+EE, n = 6) and with EE removed (-EE, n = 7) were normalized against the values at 5.0 mM external K^+.

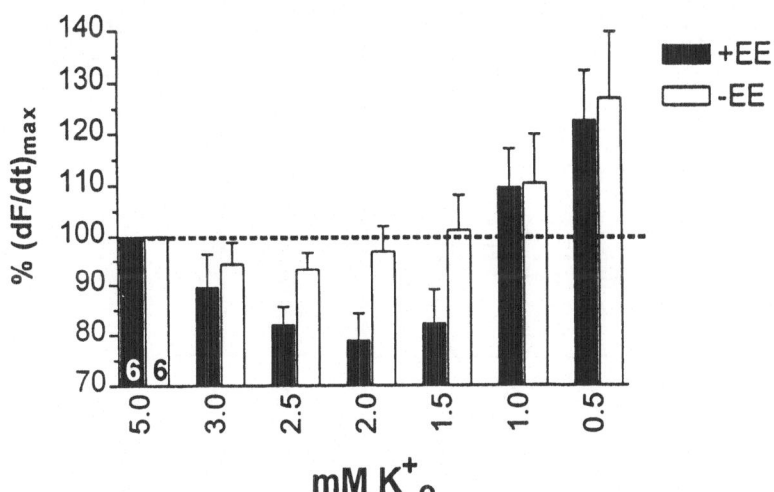

Figure 9. Relative maximal rate of force development of isometric contractions of isolated papillary muscles of rabbit right ventricle as a function of the external K^+ concentration. $(dF/dt)_{max}$ of isometric contractions of muscles with intact EE (+EE, n = 6) and with EE removed (-EE, n = 6) were normalized against the values at 5.0 mM external K^+.

Because the K^+ sensitivity of sodium pump inhibition is highest around the apparent K_D values of 3.5 mM K^+ for α_1 and 0.5 mM K^+ for α_2, experiments were repeated within a narrower K^+ concentration range (Figure 9). Similar results were obtained: a negative inotropic effect at K^+ of 3 down to 2 mM, turning to positive inotropy at K^+ below 2 mM. The negative inotropic effect was dependent on the presence of an intact EE, whereas the positive inotropic effect seemed endothelium-independent.

Release of endothelium-derived substances is dependent on endothelial Ca^{2+}, which itself is dependent on the membrane potential and external K^+. Therefore, it was tested whether inhibition of NO, PGI_2 and endothelin (angiotensin) pathways in rabbit papillary muscles could be involved in the negative inotropic effect of moderate external K^+ reduction. The negative inotropic effect at 2.5 mM external K^+ (-23.4 ± 2.7%, n = 6) was not affected by NO-inhibition (-26.2 ± 7.5% in the presence of L-NMMA, 50 μM, n=6), prostacyclin-inhibition (-21.5 ± 3.4% in the presence of indomethacin, 10 μM, n=6), endothelin-inhibition (-25.1 ± 4.4% in the presence of BQ-123, 1 μM, n = 6), or ACE-inhibiton (-25.0 ± 2.9% in the presence of captopril, 1 μM, n = 6). If the negative inotropic effect of 2.5 mM external K^+ is not due to altered release of inotropic substances, it is probably related to altered ion transport via inhibition of the endothelial or muscle Na^+/K^+ ATPase across the EE. As a result, it can be suggested that the EE acts to keep the interstitial K^+ concentration above the lower limit of 3 mM by a directed transport of K^+ from the plasma to the myocytes and that the EE can be considered as a sensor for plasma K^+, which at low concentrations is arrhythmogenic.

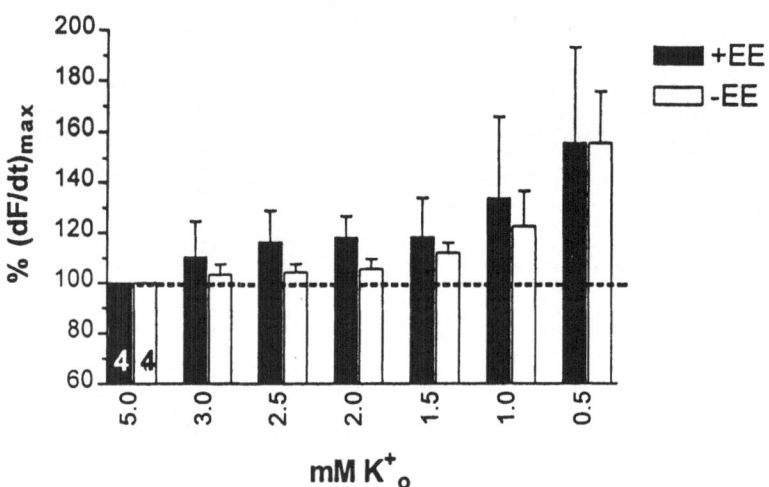

Figure 10. Relative maximal rate of force development of isometric contractions of isolated papillary muscles of rabbit right ventricle incubated with 50 μM DMA as a function of the external K^+ concentration. $(dF/dt)_{max}$ of isometric contractions of muscles with intact EE (+EE, n = 4) and with EE removed (-EE, n = 4) were normalized against the values at 5.0 mM external K^+.

At 0.5 mM external K⁺, the preparation became somewhat unstable, which we believe is due to Ca^{2+} overload, because of α_1 and α_2 inhibition and increase of internal Na⁺. Remarkable is that this Ca^{2+} overload, which can be seen as an increase of passive tension at 1 and 0.5 mM external K⁺ ($+41.3\pm5.6$ %) is significantly ($p=0.005$) reduced in the presence of dimethylameloride (DMA, $+10.3\pm4.9\%$). Increase of passive tension is a typical phenomenon of acidification (Orchard and Kentish, 1990) and might occur due to Na^+_i overload via pump inhibition and Na⁺/H⁺ exchange activation. With respect to this, it can be mentioned that cariporide (NHE-1 inhibitor) has cardioprotective effects during ischemia/reperfusion and especially improved diastolic function (Portman et al., 2001).

Furthermore, in the presence of DMA, the negative inotropic effect of reduction of external K⁺ was completely reversed (Figure 10) and only positive inotropic effects of K⁺ reduction were observed. These results can be explained as follows. By blocking the endothelial sodium pump (external K⁺ to 2.5 or 2 mM in +EE muscles) an interstitial accumulation of protons in the subjacent myocytal interstitium is induced via activation of the endothelial NHE, leading to negative inotropy. By blocking the exchanger with DMA before decrease of external K⁺ and inhibition of the endothelial sodium pump, this acidification is avoided (Figure 11, Fransen et al., 2001)

At present, the different acidifying and alkalinising transporters in cultured EE cells and in isolated rabbit papillary muscles are under investigion together with the role of the sodium pump herein.

5. CONCLUSIONS

The EE is more than a passive monolayer of closely apposed cells covering the cardiac cavities. Besides release of endothelium-derived factors, the present chapter indicates that the EE can function as a blood-heart barrier, thereby controlling the ionic environment of the underlying cardiomyocytes. There is a striking similarity with the endothelia of the BBB and CEB. Not only ion channels but also the Na⁺/K⁺ ATPase, which is at the basis of the barrier properties of BBB and CEB (basolateral) with directed transendothelial transport of ions, are asymmetrically distributed between abluminal and luminal side of the EE cell membranes. Although localized at the basolateral side of the membrane in BBB and CEB endothelia, Na⁺/K⁺ ATPase was localized in the luminal membrane in EE cells. Furthermore, the EE Na⁺/K⁺ ATPase is the α_1 isoform, similarly as in CEB and different epithelia, suggesting that its function is to regulate global transendothelial transport of ions from blood to myocardial interstitium and vice versa. Inhibition of the endothelial α_1 isoform has negative inotropic effects on cardiac performance, which are probably related to pH changes. Therefore, it is suggested that the EE is able to modulate the ionic constitution of myocardial interstitium with directed transport of K⁺ and H⁺ from blood to cardiomyocytes and of Na⁺ from myocytes to the blood (Figure 11). With respect to K⁺, it seemed that muscles were highly tolerant to high external K⁺, but very sensitive to low external K⁺. With respect to H⁺, it is worthwhile mentioning that Muller-Borer et al. (1998) observed an interstitial pH of 7.3 for a bathing pH of 7.45 in perfused rabbit papillary muscles and that, apparently, the external and internal pH for endocardial cardiomyocytes was lower than for myocytes 200 μm below the endocardial surface. The authors suggested that EE cells might be involved. As cardiac Na⁺/K⁺ ATPase has been ascribed a key role in pathophysiological conditions such as hypoxia-reoxygenation, ischemia-reperfusion, primary hypertension, hypertrophy and heart failure, one may wonder to what extent the asymmetrical distribution of the α_1

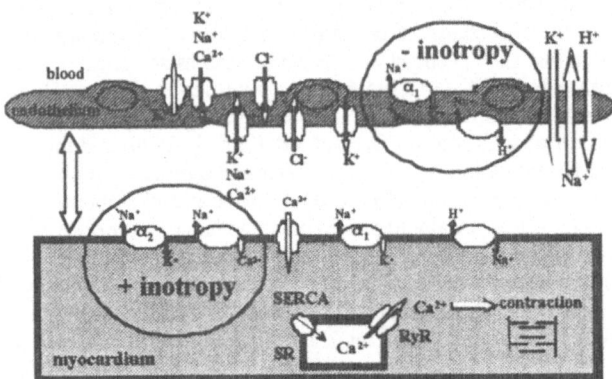

Figure 11. The asymmetrical distribution of ion channels and the luminal position of the α_1 subunit of the Na$^+$/K$^+$ ATPase are at the basis of a putative transendothelial transport of Na$^+$, K$^+$ and H$^+$. Inhibition of the α_1 isoform leads to negative inotropy, whereas inhibition of the muscle α_2 isoform causes positive inotropy. SERCA: smooth endoplasmic reticulum Ca^{2+} pump, RyR: ryanodine receptor

isoform of the Na$^+$/K$^+$ ATPase α subunit in EE cells and the relation with pH regulating ion transporters and exchangers are involved.

5. REFERENCES

Balligand, J. L., Kelly, R. A., Marsden, P. A., Smith, T. W., and Michel, T., 1993, Control of cardiac muscle cell function by an endogenous nitric oxide signaling system. *Proc. Natl. Acad. Sci. USA* **90**:347-351.

Betz, A. L., Firth, J. A., and Goldstein, G. W., 1980, Polarity of the blood-brain barrier: distribution of enzymes between the luminal and the antiluminal membranes of brain capillary endothelial cells. *Brain Res.* **192**:17-28.

Betz, A. L., 1983, Sodium transport from blood to brain: inhibition by furosemide and amiloride. *J. Neurochem.* **41**:1158-1164.

Blaustein, M. P., and Lederer, W. J., 1999, Sodium/calcium exchange: its physiological implications. *Physiol. Rev.* **79**:763-854.

Bonanno, J. A., and Giasson, C., 1992, Intracellular pH regulation in fresh and cultured bovine corneal endothelium. II. Na$^+$:HCO$_3^-$ cotransport and Cl$^-$/HCO$_3^-$ exchange. *Inv. Ophtalmol. Vis. Sci.* **33**:3068-3079.

Bonanno, J. A., Guan, Y., Jelamskii, S., and Kang, X.J., 1999, Apical and basolateral CO$_2$-HCO$_3^-$ permeability in cultured bovine corneal endothelial cells. *Am. J. Physiol.* **277**:C545-C553.

Brutsaert, D. L., 2003, Cardiac endothelial-myocardial signalling. Its indispensable role in cardiac growth, contractile performance and rhythmicity. *Physiol. Rev.* in press.

Brutsaert, D. L., 1989, The endocardium. *Annu. Rev. Physiol.* 51:263-273.

Brutsaert, D. L., Meulemans, A. L., Sipido, K. R., and Sys, S. U., 1988, Effects of damaging the endocardial surface on the mechanical performance of isolated papillary muscle. *Circ. Res.* **62**:358-366.

Brutsaert, D. L., Fransen, P., Andries, L. J., De Keulenaer, G. W., and Sys, S. U., 1998, Cardiac endothelium and myocardial function. *Cardiovasc. Res.* **38**:281-290.

Chen, G. F., and Suzuki, H., 1990, Calcium dependency of the endothelium-dependent hyperpolarization in smooth muscle cells of the rabbit carotid artery. *J. Physiol.* **421**:521-534.

Daut, J., Mehrke, G., Nees, S., and Newman, W. H., 1988, Passive electrical properties and electrogenic sodium transport of cultured guinea-pig coronary endothelial cells. *J. Physiol.* **402**:237-254.

De Hert, S. G., Gillebert, T. C., and Brutsaert, D. L., 1993, Alteration of left ventricular endocardial function by intracavitary high-power ultrasound interacts with volume, inotropic state, and alpha-1 adrenergic stimulation. *Circulation 87*: 1275-1285.

De Keulenaer, G. W., Fransen, P., Brutsaert, D. L., and Sys, S. U., 1995, Decreased myocardial contractility after damage to endocardial endothelium is not merely caused by loss of endothelin production. *Cardiovasc. Res.* **30**:646-647.

Fransen, P. F., Demolder, M. J. M., and Brutsaert, D. L., 1995, Whole cell membrane currents in cultured pig endocardial endothelial cells. *Am. J. Physiol.* **268**:H2036-H2047.

Fransen, P., Hendrickx, J., Brutsaert, D. L., and Sys, S. U., 2001, Distribution and role of Na$^+$/K$^+$ ATPase in endocardial endothelium. *Cardiovasc. Res.* **52**:487-499.

Furchgott, R. F., and Vanhoutte, P. M., 1988, Endothelium-derived relaxing and contracting factors. *FASEB J.* **3**:2007-2018.

Furchgott, R. F., and Zawadski, J. V., 1980, The obligatory role of endothelial cells in the relaxation of arterial smooth muscle by acetylcholine. *Nature* **288**:373-376.

Gao, J., Mathias, R. T., Cohen, I. S., and Baldo, G. J., 1995, Two functionally different Na/K pumps in cardiac ventricular myocytes. *J. Gen. Physiol.* **106**:995-1030.

Gao, J., Wymore, R., Wymore, R. T., Wang, Y., McKinnon, D., Dixon, J. E., Mathias, R. T., Cohen, I. S., and Baldo, G. J., 1999, Isoform-specific regulation of the sodium pump by α- and β-adrenergic agonists in the guinea-pig ventricle. *J. Physiol.* **516**:377-383.

Gao, J., Wymore, R. S., Wang, Y., Gaudette, G. R., Krukenkamp, I. B., Cohen, I. S., and Mathias, R. T., 2002, Isoform-specific stimulation of cardiac Na/K pumps by nanomolar concentrations of glycosides. *J. Gen. Physiol.* **119**: 297-312.

Gillebert, T. C., De Hert, S. G., Andries, L. J., Jageneau, A. H., and Brutsaert, D. L., 1992, Intracavitary ultrasound impairs left ventricular performance: presumed role of endocardial endothelium. *Am J Physiol.* **263**:H857-H865.

Guggenheim, J. A., and Hodson, S. A., 1994, Localization of Na$^+$/K$^+$ ATPase in the bovine corneal endothelium. *Biochim. Biophys. Acta* **1189**:127-134.

Hoyer , J., Popp, R. Meyer, J., Galla, H.-J., and Gogelein, H., 1991, Angiotensin II, vasopressin and GTP$_\gamma$-s inhibit inward-rectifying K+ channels in porcine cerebral capillary endothelial cells. *J. Membr. Biol.* **123**:55-62.

Hoyer, J., Distler, A., Haase, W., and Gogelein, H., 1994, Ca^{2+} influx through stretch-activated cation channels activates maxi K$^+$ channels in porcine endocardial endothelium. *Proc. Natl. Acad. Sci. USA* **91**:2367-2371.

Ito, H., Matsuda, H. and Noma, A., 1993, Ion channels in the luminal membrane of endothelial cells of the bull-frog heart. *Jap. J. Physiol.* **43**:191-206.

James, P. F., Grupp, I. L., Grupp, G., Woo, A., Askew, G. R., Croyle, M. L., Walsh, R. A., and Lingrel J.B., 1999, Identification of a specific role for the Na,K-ATPase α_2 isoform as a regulator of calcium in the heart. *Mol. Cell* **3**:555-563.

Juhaszova, M., and Blaustein, M. P., 1997, Na$^+$ pump low and high ouabain affinity α subunit isoforms are differently distributed in cells. *Proc. Natl. Acad. Sci. USA* **94**:1800-1805.

Laskey, R. E., Adams, D. J., Johns, A., Rubanyi, G. M., and van Breemen, C., 1990, Membrane potential and Na$^+$-K$^+$ activity modulate resting and bradykinin-stimulated changes in cytosolic free calcium in cultured endothelial cells from bovine atria. *J. Biol. Chem.* **265**:2613-2619.

Li, K., Rouleau, J. L., Andries, L. J., and Brutsaert, D.L., 1993, Effect of dysfunctional vascular endothelium on myocardial performance in isolated papillary muscles. *Circ. Res.* **72**:768-777.

Manabe, K., Ito, H., Matsuda, H., Noma, A., and Shibata, Y., 1995, Classification of ion channels in the luminal and abluminal membranes of guinea-pig endocardial endothelial cells. *J. Physiol.* **484**:41-52.

Mebazaa, A., Martin, L. D., Robotham, J. L., Maeda, K., Gabrielson, E. W., and Wetzel, R. C., 1993, Right and left ventricular cultured endocardial endothelium produces prostacyclin and PGE2. *J. Mol. Cell. Cardiol.* **25**:245-248.

Mebazaa, A., Mayoux, E., Maeda, K., Martin, L. D., Lakatta, E. G., Robotham, J. l., and Shah, A. M., 1993, Paracrine effects of endocardial endothelial cells on myocyte contraction mediated via endothelin. *Am. J. Physiol.* **265**:H1841-H1846.

Muller-Borer, B. J., Yang, H., Marzouk, S. A. M., Lemasters, J. J., and Cascio, W. E., 1998, pH$_i$ and pH$_o$ at different depths in perfused myocardium measured by confocal fluorescence microscopy. *Am. J. Physiol.* **275**:H1937-H1947.

Orchard, C. H., and Kentish, J. C., 1990, Effects of changes of pH on the contractile function of cardiac muscle. *Am. J. Physiol.* **258**, C967-C981

Pfeiffer, R., Beron, J., and Verrey, F., 1999, Regulation of Na$^+$ pump function by aldosterone is α-subunit isoform specific. *J. Physiol.* **516**:647-655.

Popp, R., Hoyer, J., Meyer, J., Galla, H.-J., and Gogelein, H., 1992, Stretch-activated non-selective cation channels in the antiluminal membrane of porcine cerebral capillaries. *J. Physiol.* **454**:435-449.

Portman, M. A., Panos, A. L., Xiao, Y., Anderson, D. L., and Ning, X, 2001, HOE-642 (cariporide) alters pH_i and diastolic function after ischemia during reperfusion in pig hearts in situ. *Am. J. Physiol.* **280**:H830-H834.

Reimer, K. A., and Jennings, R. B., 1979, The changing anatomic reference base of evolving myocardial infarction. Underestimation of myocardial collateral blood flow and overestimation of experimental anatomic infarct size due to tissue edema, hemorrhage and acute inflammation. *Circ.* **60**:866-876.

Rubanyi, G. M., and Vanhoutte, P. M., 1986, Oxygen-derived free radicals, endothelium, and responsiveness of vascular smooth muscle. *Am. J. Physiol.* **250**:H815-H821.

Shah, A. M., Mebazaa, A., Wetzel, R. C., and Lakatta, E. G., 1994, Novel cardiac myofilament desensitizing factor released by endocardial and vascular endothelial cells. *Circ.* **89**:2492-2497.

Smith, J. A., Shah, A. M., and Lewis, M. J., 1991, Factors released from endocardium of the ferret and pig modulate myocardial contraction. *J. Physiol.* **439**:1-14.

Sun, B., Vaughan-Jones, R. D., and Kambayashi, J.-I., 2000, Two distinct HCO_3^- dependent H^+ efflux pathways in human vascular endothelial cells. *Am. J. Physiol.* **277**:H28-H32.

Vigne, P., Champigny, G., Marsault, R., Barbry, P, Frelin, C., and Lazdunski, M., 1989, A new type of amiloride-sensitive cationic channel in endothelial cells of brain microvessels. *J. Biol. Chem.* **264**:7663-7668.

Wang, J., and Morgan, J. P., 1992, Endocardial endothelium modulates myofilament Ca^{2+} responsiveness in aequorin-loaded ferret myocardium. *Circ. Res.* **70**:754-760.

Zhang, Y., Xie, Q., Sun, X. C., and Bonanno, J. A., 2002, Enhancement of HCO_3^- permeability across the apical membrane of bovine corneal endothelium by multiple signaling pathways. *Invest. Ophthalmol. Vis. Sci.* **43**:1146-1153.

DISCUSSION

ter Keurs: What is the EC50 of ouabain for $\alpha 1$ & $\alpha 2$ Na^+/K^+ pumps in the heart?

Fransen: In our experimental setting, in which we wanted to block the endothelial $\alpha 1$–subunit isoform of the pump, ouabain could not be used because it has a higher affinity for the $\alpha 2$-isoform than for the $\alpha 1$-isoform. In rat, the sensitivity of the $\alpha 2$-isoform is even 100 to 1,000 times higher than that of the $\alpha 1$–isoform. In rabbit, the different sensitivities of $\alpha 1$ and $\alpha 2$ for ouabain are less clear. In our experiments in rabbit papillary muscles, the EC50 value of ouabain was very high, being 50-100μM.

Gonzalez-Serratos: What degree of acidification is produced by blocking the endothelial Na^+/K^+ pump?

Fransen: We do not know. At present, we are able to measure only intracellular myocytal pH_i, but not interstitial extracellular pH_o, which could directly demonstrate acidification because of endothelial Na^+/K^+ pump block.

KLF5/BTEB2, A Krüppel-like Zinc-finger Type Transcription Factor, Mediates Both Smooth Muscle Cell Activation And Cardiac Hypertrophy

Ryozo Nagai, Takayuki Shindo, Ichiro Manabe, Toru Suzuki, Masahiko Kurabayashi

1. ABSTRACT

Cardiac and vascular biology need to be approached interactively because they share many common biological features as seen in activation of the local renin-angiotensin system, angiogenesis, and extracellular matrix production. We previously reported KLF5/BTEB2, a Krüppel-like zinc-finger type transcription factor, to activate various gene promoters that are activated in phenotypically modulated smooth muscle cells, such as a nonmuscle type myosin heavy chain gene SMemb, plasminogen activator inhibitor-1 (PAI-1), iNOS, PDGF-A, Egr-1 and VEGF receptors at least in vitro. KLF5/BTEB2 mRNA levels are downregulated with vascular development but upregulated in neointima that is produced in response to vascular injury. Mitogenic stimulation activates KLF5/BTEB2 gene expression through MEK1 and Egr-1. Chromatin immunoprecipitation assay showed KLF5/BTEB2 to be induced and to bind the promoter of the PDGF-A gene in response to angiotensin II stimulation. In order to define the role of KLF5/BTEB2 in cardiovascular remodeling, we targeted the KLF5/BTEB2 gene in mice. Homozygous mice resulted in early embryonic lethality whereas heterozygous mice were apparently normal. However, in response to external stress, arteries of heterozygotes exhibited diminished levels of smooth muscle and adventitial cell activation. Furthermore, cardiac fibrosis and hypertrophy induced by continuous angiotensin II infusion. We also found that RARa binds KLF5/BTEB2, and that Am80, a potent synthetic RAR agonist, inhibits angiotensin II-induced cardiac hypertrophy. These results indicate that KLF5/BTEB2 is an essential transcription factor that causes not only smooth muscle phenotypic modulation but also cardiac hypertrophy and fibrosis.

Ryozo Nagai, Takayuki Shindo, Ichiro Manabe, Toru Suzuki, Department of Cardiovascular Medicine, University of Tokyo Graduate School of Medicine, Bunkyo-ku, Tokyo 113-8655, Japan, Masahiko Kurabayashi, The Second Department of Internal Medicine, Gunma University School of Medicine, Maebashi, Gunma 371-8510, Japan

Molecular and Cellular Aspects of Muscle Contraction
Edited by H. Sugi, Kluwer Academic/Plenum Publishers, 2003

2. INTRODUCTION

Metabolic and/or mechanical stresses induce structural remodeling of the heart and blood vessels, such as cellular hypertrophy and hyperplasia and interstitial fibrosis, which constitute the pathogenesis of heart failure and atherosclerosis [1]. Locally expressed growth factors play key roles in these processes; however, little is known about the transcriptional regulatory mechanisms underlying them. We recently identified and characterized KLF5/BTEB2 as a transcription factor for SMemb/NMHC-B, which is a molecular marker of phenotypically modulated smooth muscle cells and fibroblasts [2-5] KLF5 belongs to the Krüppel-like transcription factor family, which has diverse functions for cell differentiation and embryonic development [6]. Normally, KLF5 is abundantly expressed in developing blood vessels, but is downregulated in adult vessels [6] Its expression, however, is strongly upregulated in activated vascular smooth muscle cells in athero- and arteriosclerosis and myofibroblasts in hypertrophied heart[5]. In the present study, to elucidate the role of KLF5 in the cardiovascular remodeling, we generated KLF5 knockout mice and analyzed the *in vivo* function of KLF5 [7]

3. Method

3.1 Generation of KLF5 knockout mice

Knockout mice were generated as described previously [8]. A plasmid targeting vector was constructed to replace the 3.3 kb fragment encompassing exons 2-3 of KLF5 with the neomycin resistance gene, after which it was linearized and introduced into 129/Sv-derived SM-1 embryonic stem cells by electroporation. Male chimeras were crossbred with C57BL/6 females.

3.2 Vascular injury by cuff placement

Cuff placement surgery was carried out on 10 to 12-week-old male mice according to a method described previously [9]. A polyethylene tube was opened longitudinally, loosely placed around the artery and then closed with sutures. After 28 days, the mice were sacrificed with an overdose of anesthetic and perfused first with PBS and then with 10% neutral buffered formalin at 100 mmHg. The sections were stained with Elastica van Gieson or Masson-Trichrome stain.

3.3 Angiotensin II infusion

An incision was made in the midscapular region under sterile conditions, and osmotic minipumps (Alzet model 2002, Alza Corp, Mountain View, California) containing angiotensin II (Wako, Osaka, Japan) dissolved in 0.15 M NaCl and 1 mM acetic acid were implanted. Thereafter, angiotensin II was delivered for 14 days at a rate of 3.2 mg/kg/day.

3.4 Cell culture

Primary rat cardiac fibroblasts were obtained from newborn rats and grown to confluence in DMEM supplemented with 10% FBS. Before angiotensin II stimulation, cells were placed in serum-free, defined medium for 4 days.

3.5 Quantitative reverse transcriptase PCR

The methods for reverse transcription of RNA and quantitative relative PCR have been described elsewhere[10]. For quantitation of transcript levels within tissues, real-time PCR was performed using LightCycler (Roche) and QuantiTect SYBR green PCR kits (Qiagen).

3.6 Chromatin immunoprecipitation assays

Quiescent cardiac fibroblasts were treated with or without 1 _ M angiotensin II and then fixed in 1% formaldehyde. The fixed chromatin samples were subjected to immunoprecipitation as described previously, with minor modifications. Anti-KLF5 antibody was raised against recombinant KLF5 protein, and its specificity was thoroughly examined by Western analyses. Protein G (Roche) was used to preclear samples and for immunoprecipitation instead of the protein A used in the original protocol33. Samples of purified reverse-crosslinked immunoprecipitated DNA were subjected to PCR analyses. The sequences of the PCR primers were:
PDGF-A promoter, 5'- GCACTGAAGGGTGGGCAAGCTCGAGGGAGG-3' and 5'-CGGGCCGGGGACCCGCACCTCGGAAGCGCT-3';
PDGF-A 3'-end (the region in the last exon of the PDGF-A gene near the 3'-end of the coding sequence), 5'-GTGACATTCCTGAACATACTATGTATGGTG-3' and 5'-GTCTCTCCGAGTGCTACAGTACTTGCTTTG-3'.

3.7 Co-immunoprecipitation assay

The lysates of HeLa S3 cells were centrifuged at 18,000 g for 10 min, and the protein concentration in the extracts was adjusted to 1 mg/ml. One mg of anti-RAR antibody (Santa Cruz Biotechnology, Santa Cruz, California) or 1 mg of control rat IgG was bound to 10 ml of protein G sepharose, after which the antibody-bound protein G sepharose was stirred with 1 ml of HeLa S3 extract for 6 h. The resultant immunoprecipitate was washed ten times with RIPA buffer (50 mM Tris-HCl, pH 7.9, 150 mM NaCl, 1% NP-40, 0.5% sodium deoxycholate, 0.1% SDS, 1 mM phenylmethylsulfonyl fluoride, 0.5 mg/ml leupeptin and 1 mg/ml pepstatin A) and then subjected to SDS-PAGE and immunoblotted with anti-KLF5 antibody.

4. RESULTS

4.1 Homozygote lethality in KLF5 knockout mice

The number of pups was significantly reduced when heterozygotes (KLF5+/-) were intercrossed. Analysis of the embryos from timed KLF5+/- intercrosses showed that KLF5-/- homozygotes died prior to E 8.5.

4.2 Reduced responses to vascular injury and tumor implantation in KLF5 +/- mice

KLF5+/- mice survived until adulthood and were apparently normal and fertile, though expression of KLF5 was reduced to about half that in wild-type mice. In the femoral arteries injured with a polyethylene tube cuff, we found that in KLF5+/+ mice a thick layer of granulation tissue, containing numerous microvessels and inflammatory cells and a large amount of extracellular matrix, developed around the cuff (Fig. 1). The

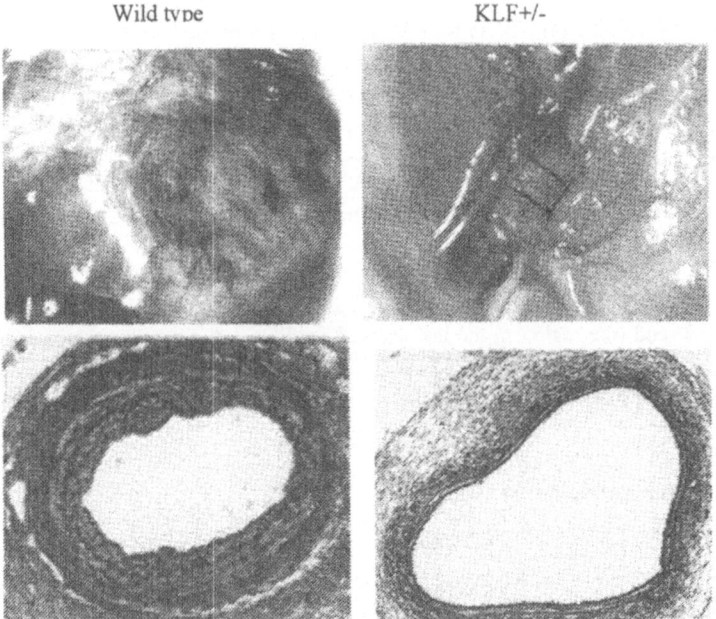

Figure 1. Vascular injury model entailing placement of a cuff around the femoral artery in KLF5 +/+ (wild-type) and KLF5 +/- mice. Asterisks in c-f indicate the lumen of the femoral artery. In wild-type mice, a thick layer of highly vascular granulation tissue surrounded the polyethylene cuff, making it barely visible, whereas KLF5 +/- mice exhibited markedly less granulation tissue, less angiogenesis and less severe fibrosis. Cuff placement also caused intimal hyperplasia in wild-type mice, but not in KLF+/-.

KLF5+/- mice, by contrast, developed markedly less granulation tissue and showed less angiogenesis (Fig. 1b, d). Moreover, within the cuffs, the arteries of the heterozygotes were thin-walled and dilated (Fig. 1e), which was in striking contrast to the wild-type animals.

4.3 Reduced angiotensin II-induced cardiac hypertrophy and fibrosis in KLF5 +/- mice

The reduced vascular remodeling observed in KLF5+/- mice prompted us to examine the role of KLF5 in responses elicited by angiotensin II, a potent growth factor known to play a central role in both cardiac hypertrophy and vascular remodeling. Following a continuous, 14-day infusion of angiotensin II, the hearts of wild-type mice were found to be significantly heavier than those of heterozygous mice (Fig. 2).

Wild type   KLF+/-

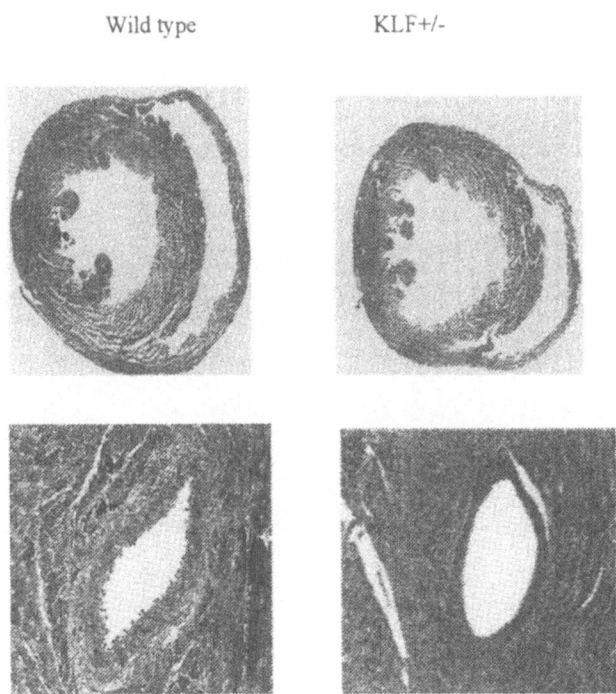

Figure 2. Cross-sections of heart after 14 days of angiotensin II infusion. Substantial perivascular and interstitial fibrosis were observed in wild-type mice, but such changes were less severe or almost undetectable in KLF5 +/- mice.

4.4 Angiotensin II-induced expression of the KLF5 and PDGF-A genes

We next examined the molecular mechanisms by which KLF5 controls cardiovascular remodeling. Because PDGF-A is a well-known growth factor involved in mesenchymal cell activation, angiogenesis and tissue remodeling [11], we analyzed angiotensin II-stimulated upregulation of KLF5 and PDGF-A in cultured cardiac fibroblasts. Upregulation of KLF5 was detected within 2 h after the start of angiotensin II stimulation, was sustained for > 4 h, and was followed by upregulation of PDGF-A (Fig. 3). Furthermore, overexpression of KLF5 markedly increased PDGF-A promoter activity in transiently transfected cells.

To further confirm that KLF5 controls the PDGF-A gene promoter in response to angiotensin II, we carried out a series of chromatin immunoprecipitation (ChIP) assays using anti-KLF5 antibody. When cultured cardiac fibroblasts treated with 1 _ M angiotensin II for 3.5 h and untreated cells were formalin-fixed and subjected to ChIP analysis, angiotensin II was found to clearly increase the binding of KLF5 to the promoter region of the PDGF-A gene. On the other hand, KLF5 did not bind to either the last exon of the PDGF-A gene, which does not contain KLF5 binding sites, or to the ß-globin gene promoter, which is silent in fibroblasts (data not shown).

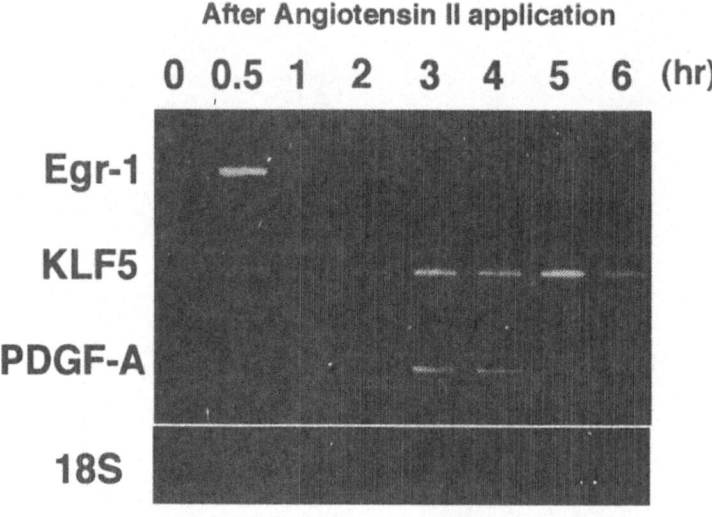

Figure 3. RT-PCR analyses of KLF5, Egr-1 and PDGF-A expression in angiotensin II-stimulated cardiac fibroblasts. Upregulation of Egr-1 and KLF5 was respectively detected beginning 0.5 h and 2 h after angiotensin II application; expression of Egr-1 persisted for only 2 h, while that of KLF5 was sustained for > 4 h and was followed by upregulation of PDGF-A expression.

4.5 Effects of retinoid derivatives on KLF5-dependent PDGF-A transcription and cardiovascular remodeling

We next screened a number of compounds for their ability to modulate KLF5 activity using a PDGF-A promoter reporter construct cotransfected with the KLF5 expression vector. We found that Am80, a synthetic RAR agonist [4, 12], activated PDGF-A promoter activity in cells overexpressing KLF5 and RARa but did not significantly affect PDGF-A promoter activity if either KLF5 or RARa was not overexpressed. Furthermore, we found that KLF5 coimmunoprecipitates with RAR , indicating direct physical interaction between KLF5 to RAR.

Finally we administered Am80 to wild-type mice with cuffed femoral arteries and continuous angiotensin II treatment. We found that Am80 reduced development of granulation tissue and the neointima in wild-type mice until they approximated those seen in KLF5 +/- mice and suppressed angiotensin II-induced cardiac hypertrophy (Fig. 4), suggesting that Am80 modulates KLF5 function both *in vitro* and *in vivo*.

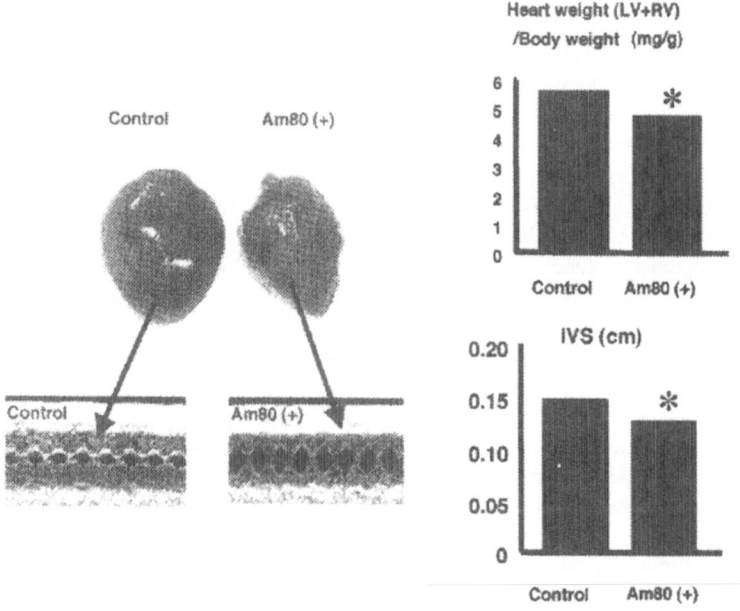

Figure 4. Am80 reduced angiotensinII-induced cardiac hypertrophy and fibrosis. Echocardiographic analysis comparing heart weight (left ventricle + right ventricle)/body weight ratios (l) and intraventricular septal (IVS) thickness in wild-type mice after 14 days of angiotensin II infusion with or without AM80 administration (5mg/kg/day given orally for 1 week before surgery and for 2 weeks after it). Am80 suppressed angiotensin II-induced cardiac hypertrophy.

5. DISCUSSION

Cardiovascular remodeling is a complex process involving activation of mesenchymal cells, production of extracellular matrix, and angiogenesis, all of which contribute to the pathogenesis of atherosclerosis and heart failure. Following initial activation of mesenchymal cells (e.g., smooth muscle cells and fibroblasts) by an external stress, several immediate early response genes are activated during the acute phase of the stress response; remodeling processes are then sustained over longer periods of time via autocrine/paracrine loops involving a number of humoral factors[13]. Cardiac hypertrophy is also developed mainly by the individual cell hypertrophy of myocytes in response to increased external workload. At the same time, however, various kinds of qualitative and quantitative modulations in cardiac myocytes, non-myocytes, and extracellular matrix are observed during the progression of LVH [14, 15].

We originally cloned KLF5 cDNA from a smooth muscle cell cDNA library as a transcription factor for SMemb/NMHC-B6, which is markedly induced in proliferating smooth muscle cell and cardiac myofibroblasts in pressure-overloaded heart. KLF5 furthermore activates many varieties of gene promoters such as iNOS, PAI-1 and Flt-1, all of which are known to be upregulated during cardiovascular remodeling. As we expected, neointimal formation was markedly attenuated in KLF5+/- mice, which also showed attenuated interstitial and perivascular fibrosis in cardiac hypertrophy and reduced adventitial thickening in cuff-injured arteries. It is suggested from these data that KLF5 plays an essetial role not only in vascular smooth muscle cells, but also in activated fibroblasts.

From genetic studies using gene targeting technique, it has been demonstrated that PDGF-A is required for lung alveolar morphogenesis via inducing alveolar myofibroblasts[16], and PDGF-B/PDGF receptor b signaling is required for kidney glomerular morphogenesis possibly via inducing glomerular myofibroblasts [17], suggesting that PDGF may also play a role in the induction of myofibroblasts. Because both PDGF (unpublished data) and endothelin-I are up-regulated in pressure-overloaded hearts, it is possible that multiple growth factors are induced in concert in myofibroblast via KLF5 during development of cardiac hypertrophy. Given that KLF5 is induced following activation of immediate early response genes (e.g., Egr-1)[18] and that it, in turn, controls expression of various growth factors (e.g., PDGF-A and TGFß), KLF5 may be in a position to mediate between the acute response to an external stress and tissue remodeling in general.

In summary, the results of the present study provide strong evidence that KLF5 is a crucial determinant of the cellular response to cardiovascular injury, playing a key role in mediating tissue remodeling. The present study provides a clear basis for the development of drugs modulating KLF5 function to control cardiovascular remodeling and angiogenesis.

REFERENCES

1. P. Libby. Changing concept of atherogenesis. J Intern Med. 247:349-358 (2000).

2. Kuro-o, R. Nagai, H. Tsuchimochi, H. Katoh, Y. Yazaki, A. Ohkubo, et al. Developmentally regulated expression of vascular smooth muscle myosin heavy chain isoforms. J Biol Chem. 264:18272-18275. (1989).

3. I. Manabe, M. Kurabayashi, Y. Shimomura, M. Kuro-o, N. Watanabe, M. Watanabe, et al. Isolation of the embryonic form of smooth muscle myosin heavy chain (SMemb/NMHC-B) gene and characterization of its 5'-flanking region. Biochem Biophys Res Commun. 239:598-605 (1997).

4. N. Watanabe, M. Kurabayashi, Y. Shimomura, K. Kawai-Kowase, Y. Hoshino, I. Manabe, et al. BTEB2, a Kruppel-like transcription factor, regulates expression of the SMemb/Nonmuscle myosin heavy chain B (SMemb/NMHC-B) gene. Circ Res. 85:182-191 (1999).

5. I. Shiojima, M. Aikawa, J. Suzuki, Y. Yazaki, R. Nagai. Embryonic smooth muscle myosin heavy chain SMemb is expressed in pressure-overloaded cardiac fibroblasts. Jpn Heart J. 40:803-818 (1999).

6. J.J. Bieker. Kruppel-like factors: three fingers in many pies. J Biol Chem. 276: 34355-34358 (2001).

7. T. Shindo, I. Manabe, Y. Fukushima, K. Tobe, K. Aizawa, S. Miyamoto, et al. Kruppel-like zinc-finger transcription factor KLF5/BTEB2 is a target for angiotensin II signaling and an essential regulator of cardiovascular remodeling. Nature Med. 8:856-863 (2002).

8. T. Shindo, H. Kurihara, K. Kuno, H. Yokoyama, T. Wada, Y. Kurihara, et al. ADAMTS-1: a metalloproteinase-disintegrin essential for normal growth, fertility, and organ morphology and function. J Clin Invest. 105:1345-1352. (2000).

9. M. Moroi, L. Zhang, T. Yasuda, R. Virmani, H.K. Gold, M.C. Fishman, et al. Interaction of genetic deficiency of endothelial nitric oxide, gender, and pregnancy in vascular response to injury in mice. J Clin Invest. 101:1225-1232 (1998).

10. I. Manabe, G.K. Owens. Recruitment of serum response factor and hyperacetylation of histones at smooth muscle-specific regulatory regions during differentiation of a novel P19-derived in vitro smooth muscle differentiation system. Circ Res. 88:1127-1134 (2001).

11. B.C. Berk. Vascular smooth muscle growth: autocrine growth mechanisms. Physiol Rev. 81:999-1030 (2001).

12. L. Eyrolles, H. Kagechika, E. Kawachi, H. Fukasawa, T. Iijima, Y. Matsushima, et al. Retinobenzoic acids. 6. Retinoid antagonists with a heterocyclic ring. J Med Chem. 37:1508-1517 (1994).

13. S.L. Friedman. Molecular regulation of hepatic fibrosis, an integrated cellular response to tissue injury. J Biol Chem. 275:2247-2250 (2000).

14. K.T. Weber, J.S. Janicki, S.G. Shroff, R. Pick, R.M. Chen, R.I. Bashey. Collagen remodeling of the pressure-overloaded, hypertrophied nonhuman primate myocardium. Circ Res. 62:757-765 (1998).

15. K.O. Leslie, D.J. Taatjes, J. Schwarz, M. von Turkovich, R.B. Low. Cardiac myofibroblasts express alpha smooth muscle actin during right ventricular pressure overload in the rabbit. Am J Pathol. 139:207-216 (1991).

16. H. Bostrom, K. Willetts, M. Pekny, P. Levén, P. Lindahl, H. Hedstrand, et al. PDGF-A signaling is a critical event in lung alveolar myofibroblast development and alveogenesis. Cell. 85:863-873 (1996).

17. P. Levén, M. Pekny, S. Gebre-Medhin, B. Swolin, E. Larsson, C. Betsholtz. Mice deficient for PDGF-B show renal, cardiovascular, and hematological abnormalities. Genes Dev. 8:1875-1887 (1994).

18. K. Kawai-Kowase, M. Kurabayashi, Y. Hoshino, Y. Ohyama, R. Nagai. Transcriptional activation of the zinc finger transcriptional factor BTEB2 gene by Egr-1 through mitogen-activated protein kinase pathways in vascular smooth muscle cells. Circ Res. 85:787-795 (1999).

DISCUSSION

Sweeney: Was TNF-α induction in response to Ang 2 blocked by either giving Am 80 or in the KLF 5/BTEB 2 mouse?

Nagai: TNF-α levels were not examined. But when beterozygotes were treated with LPS, they showed similar response as wild type.

Gonzalez-Serratos: Have you studied the hearts with SM80, where hypertrophy does not change the Ca^{2+} handling or the contractile properties?

Nagai: No, we have not. We have only examined echocardiography, whose finding was clearly improved.

Matsuoka: Is there any kind of congenital heart disease in these knockout animals? KLF5/BTEB2 +/- mouse is there any up-regulation of NMHC-A protein?

Nagai: Homozygotes are early embryoric lecthal, so we do not know cardiovascular abnormality, at least heterozygotes do not show any cardiovascular defect.

REGULATION OF THE RHO SIGNALING PATHWAY BY EXCITATORY AGONISTS IN VASCULAR SMOOTH MUSCLE

Yoh Takuwa

1. INTRODUCTION

It has been established in smooth muscle that binding of excitatory agonists to specific cell surface receptors, most of which belong to a heptahelical type of G protein-coupled receptors, leads to the activation of phospholipase C and the opening of Ca^{2+} channels on the plasma membrane.[1, 2] Phospholipase C activation generates the two second messengers, inositol-1, 4, 5-trisphosphate and 1, 2-diacylglycerol. Inositol-1, 4, 5-trisphosphate stimulates Ca^{2+} release from an intracellular Ca^{2+} store, which, together with stimulated Ca^{2+} influx across the plasma membrane, brings about an increase in the intracellular free Ca^{2+} concentration ($[Ca^{2+}]i$). An increase in the $[Ca^{2+}]i$ activates the Ca^{2+}, calmodulin-dependent enzyme myosin light chain kinase (MLCK), which phosphorylates the 20 kDa myosin light chain (MLC), leading to the initiation of a contractile response.[3] An increase in the membrane content of 1, 2-diacylglycerol leads to activation of protein kinase C, which activates a contractile mechanism independently of or synergistically with an increase in the $[Ca^{2+}]i$.[1, 2]

It was initially thought that physiological agonists causes an increase in the level of phosphorylated MLC in vascular smooth muscle primarily through inducing an increase in the $[Ca^{2+}]i$.[3] However, subsequent investigations showed that receptor activation by excitatory agonists in vascular smooth muscle upregulates the Ca^{2+} sensitivity of MLC phosphporylation and, hence, contraction[4]; several laboratories showed that receptor agonists induce larger increases in the extents of MLC phosphorylation and contraction than depolarization with KCl at a given level of the $[Ca^{2+}]i$.[5-7] Thus, agonists somehow enhance the Ca^{2+} sensitivity of MLC phosphorylation and contraction. The development of membrane-permeabilized smooth muscle preparation contributed to our understanding the mechanisms underlying agonist-induced enhancement of contraction.[8] In smooth muscle permeabilized with staphylococcal α-toxin and β-escin, Kitazawa et al.[10] demonstrated that excitatory receptor agonists including phenylephrine and thromboxane A2 potentiated contraction at a constant $[Ca^{2+}]i$.[9] This potentiation of contraction was accompanied by an increase in the level of phosphorylated MLC. These actions of

*Yoh Takuwa, Department of Physiology, Kanazawa University Graduate School of Medicine, 13-1 Takara-machi, Kanazawa, Ishikawa, Japan 920-8640.

agonists required GTP, and were mimicked and inhibited by the non-hydrolyzable guanine nucleotides, GTPγS and GDPβS, respectively. These observations suggested that agonists potentiated MLC phosphorylation and contraction in a G protein-dependent manner. However, it was not known how a particular G protein is involved in agonist-induced potentiation of MLC phosphorylation and contraction.

2. DOWNREGULATION OF MYOSIN PHOSPHATASE IN CALCIUM SENSITIZATION

Agonist- or GTPγS-induced Ca^{2+} sensitization of MLC phosphorylation in smooth muscle could be due to enhancement of MLC phosphorylating enzyme (myosin kinase) activity or inhibition of protein phosphatase activity toward MLC. Kitazawa et al.[10] demonstrated that GTPγS or phenylephrine decreased the rate of dephosphorylation of phosphorylated MLC in permeabilized vascular smooth muscle, suggesting that the stimulation inhibited protein phosphatase activity toward MLC.   They also showed that GTPγS did not enhance thio-phosphorylation of MLC. This latter observation indicates that GTPγS does not enhance myosin kinase activity, because thio-phosphorylated MLC is a very poor substrate for phosphatase. In accordance with this report, Kubota et al.[11] showed that GTPγS induced inhibition of the protein phosphatase activity toward both MLC and heavy meromyosin, but not stimulation of myosin kinase activity, in homogenates of tracheal smooth muscle. It is also known that when the phosphatase activity in permeabilized smooth muscle is inhibited by the addition of phosphatase inhibitors such as calyculin A and microcystin LR, Ca^{2+}-induced contraction is enhanced. These results suggest that excitatory agonists and GTPγS alter the Ca^{2+} sensitivity of MLC phosphorylation through inhibition of phosphatase activity toward myosin, but not through stimulation of myosin kinase activity, by a G protein-dependent mechanism.

We directly demonstrated that GTPγS stimulation of vascular smooth muscle is indeed accompanied by a decrease in the *in vitro* activity of immunoprecipitated cellular myosin phosphatase (immunoprecipitation-phosphatase assay).[12] Myosin phosphatase consists of three subunits of the 38 kDa catalytic subunit (PP1δ), 110 kDa myosin targeting regulatory subunit (MYPT1/MBS), and 21 kDa regulatory subunit (M21). We immunoprecipitated myosin phosphatase from permeabilized SMCs stimulated with GTPγS, by using anti-110 kDa MYPT1antibody. The anti-MYPT1 immunoprecipitate was verified to comprise the three subunits as analysed by Western blotting, and showed phosphatase activity toward MLC. The anti-MYPT1 immunoprecipitate from GTPγS-stimulated vascular smooth muscle cells displayed a 50 % decrease in the phosphatase activity toward MLC, as compared to non-stimulated control smooth muscle cells.

3. INVOLVEMENT OF THE SMALL G PROTEIN RHO IN CALCIUM SENSITIZATION

Available evidence[8, 9, 11] suggested that a G protein mediates Ca^{2+}-sensitization of MLC phosphorylation. A G protein that mediates Ca^{2+}-sensitization could be a heterotrimeric G protein or a small molecular weight G protein. Hirata et al.[15] first reported that the low-molecular-weight G protein Rho is involved in GTPγS enhancement of Ca^{2+}-induced contraction in permeabilized vascular smooth muscle. They observed that GTPγS increased the Ca^{2+} sensitivity of contraction and that the GTPγS-induced Ca^{2+} sensitization of contraction was abolished by each of Staphylococcal exoenzyme epidermal cell differentiation inhibitor (EDIN) and Clostrium botulinum exotoxin C3, both of which specifically ADP-ribosylates and inactivates Rho. Several other groups[16-18] showed that C3 toxin inhibited agonist-induced contraction as well as GTPγS-induced contraction. Thus, these observations suggested that Rho is involved in Ca^{2+} sensitization of contraction. However, how Rho is involved in Ca^{2+} sensitization of contraction was still unknown.

We demonstrated for the first time that pretreatment of permeabilized vascular smooth muscle cells with C3 completely inhibited GTPγS enhancement of Ca^{2+}-induced MLC phosphorylation.[19] This observation indicated that Rho mediates potentiation of MLC phosphorylation and thus is involved in Ca^{2+} sensitization of contraction. We observed that C3 pretreatment also completely abolished inhibition by GTPγS of MLC dephosphorylation. We further showed that pretreatment of permeabilized smooth muscle cells with C3 toxin abolished GTPγS-induced inhibition of myosin phosphatase by immunoprecipation-phosphatase assay.[12] Thus, Rho is a critical G protein that mediates inhibition of myosin phosphatase, resulting in potentiation of MLC phosphorylation.[2, 19-21]

4. MOLECULAR MECHANISMS OF RHO-MEDIATED CALCIUM SENSITIZATION

Recent studies[22-24] identified a number of Rho target molecules that mediate many of Rho actions. These include the serine/threonine kinase Rho kinase/ROCK/ROK, protein kinase N (PKN) and Citron kinase, and the non-kinase proteins mDia, Rhotekin and Rhophilin. Kimura et al.[25] demonstrated that Rho kinase, but not PKN, phosphorylated the 110 kDa MYPT1 subunit of purified myosin phosphatase holoenzyme in vitro. Importantly, Rho kinase-catalyzed phosphorylation of MBS decreased phosphatase activity. Feng et al.[26] identified Thr^{695} of MYPT1 (numbering of chicken M133 isoform) as an inhibitory phosphorylation site. We found that the Rho kinase inhibitor HA1077 abolished GTPγS-induced Ca^{2+} sensitization of MLC phosphorylation in permeabilized SMCs, but not MLC phosphorylation induced by Ca^{2+} alone.[12, 21] We also observed that a Rho kinase inhibitor protein, the pleckstrin homology region of Rho kinase, similarly inhibited GTPγS-induced sensitization of MLC phosphorylation. Moreover, we found that HA1077 abolished GTPγS-induced inhibition of myosin phosphatase by the immunoprecipation-phosphatase assay. These inhibitory actions of HA1077 were accompanied by inhibition of GTPγS-induced MYPT1 phosphorylation. These observations provided evidence that Rho kinase mediates GTPγS-induced, Rho-dependent downregulation of myosin phosphatase activity most likely through phosphorylation of MBS *in vivo* in smooth muscle. HA1077 potently inhibits agonist-induced smooth muscle contraction with a decrease in MLC phosphorylation. These observations, together with previous findings[16-18] that C3 toxin inhibits agonist-induced contraction, suggest that Rho- and Rho kinase-mediated inhibition of myosin phosphatase operates as a mechanism of receptor agonist-induced sensitization of MLC phosphorylation and contraction.[12, 19-21]

Thr[695] of MYPT1, the inhibitory phosphorylation site, was also shown to be phosphorylated *in vitro* by other kinases including ZIP/MYPT1-kinase[27] and integrin-linked kinase (ILK)[28]. However, it is not known whether these kinases are involved in regulation of smooth muscle myosin phosphatase activity *in vivo*. It is not known either whether Rho is somehow linked to regulation of activities of these kinases.

CPI-17 is an endogenous smooth muscle-specific inhibitor phosphoprotein for myosin phosphatase.[29] CPI-17 is abundantly expressed particularly in vascular smooth muscle. The phosphatase-inhibitory activity of CPI-17 is very low in its non-phosphorylated state and becomes 1000-fold stimulated when phosphorylated at Thr[38] *in vitro*.[29] Excitatory receptor agonists were demonstrated to increase Thr[38] phosphorylation of CPI-17 in smooth muscle,[30] suggesting that CPI-17 participates in Ca^{2+}-sensitization. CPI-17 was shown to be phosphorylated *in vitro* by several protein kinases including protein kinase C[29], Rho kinase[31], PKN[32], ZIP/MYPT1-kinase[33] and ILK[34]. Protein kinase C inhibitors reduced agonist-induced phosphorylation of CPI-17 at Thr[38] with inhibition of contraction.[35] Rho kinase inhibitors were also shown to partially inhibit agonist-induced phosphporylation of CPI-17. [35] These observations suggest that CPI-17 participates in protein kinase C-and/or Rho kinase-mediated inhibition of myosin phosphatase.

5. REGULATION OF RHO BY EXCITATORY AGONISTS

Like Ras, Rho cycles between a GDP-bound inactive state and a GTP-bound active state.[24, 36] Despite accumulation of the observations showing the importance of the Rho-signaling pathway in the regulation of smooth muscle contraction, no direct effects of excitatory agonists on Rho activity in smooth muscle have been determined. We recently developoed a new biochemical assay which employs a recombinant RhoA-binding protein, Rhotekin, for determining amounts of an active form of RhoA, GTP-bound RhoA (GTP-Rho), in smooth muscle tissues.[37] We found in rabbit aortic smooth muscle that the thromboxane A_2 mimetic, U46619, which induced a sustained contractile response, induced a sustained rise in the amount of GTP-Rho in a dose-dependent manner with an EC_{50} value similar to that for the contractile response.[37] U46619-induced Rho activation was thromboxane A_2 receptor-mediated and reversible. Other agonists including noradrenalin, serotonin, histamine and endothelin-1 also stimulated Rho, albeit to lesser extents than U46619. In contrast, angiotensin II and phorbol 12, 13-dibutyrate failed to increase GTP-Rho. The tyrosine kinase inhibitor genistein substantially inhibited Rho activation by these agonists, except for endothelin-1. Thus, the magnitude and mode of agonist-induced Rho activation does not appear to be uniform among agonists.

The cycling of Rho between the GTP-bound active and GDP-bound inactive states is under the tight regulation by the two major groups of proteins, guanine nucleotide exchange factors (GEFs) and GTPase-activating proteins (GAPs).[24, 36] Previous studies largely on non-muscle cells demonstrated that stimulation of GPCRs with receptor agonists including lysophosphatidic acid, endothelin-1 and thrombin induced Rho activation, which was mediated through the receptor coupling to the $G_{12/13}$ family of the heterotrimeric G proteins.[38] Indeed, direct physical and functional interaction of $G_{12/13}$

with a group of structurally related GEFs acting on Rho (RhoGEF) was demonstrated.[39] More recent studies[40, 41] showed that G_q also had the ability to mediate stimulation of Rho, but the signaling pathway that mediates G_q–mediated Rho stimulation is still incompletely understood although the $G\alpha_q$-RhoGEFs interaction of the more limited extents was observed. In cultured VSM cells, the expression of activated forms of $G\alpha_{12}$ and $G\alpha_{13}$, but not $G\alpha_q$, was shown to induce a contraction that was inhibited by C3 toxin and a Rho kinase inhibitor, whereas the expression of dominant negative forms of $G\alpha_{12}$ and $G\alpha_{13}$ inhibited receptor agonist-induced, C3 toxin- and Rho kinase inhibitor-sensitive contraction.[42] Thus, these observations indicated that $G_{12/13}$ mediates Rho- and Rho kinase-dependent contraction in receptor agonist-stimulated cultured VSM cells. However, our observations described above suggested that more than a single mechanism for activating Rho might be present in smooth muscle.[37]

We recently found novel, Ca^{2+}-dependent activation of Rho in vascular smooth muscle. High KCl-induced membrane depolarization as well as noradrenalin stimulation

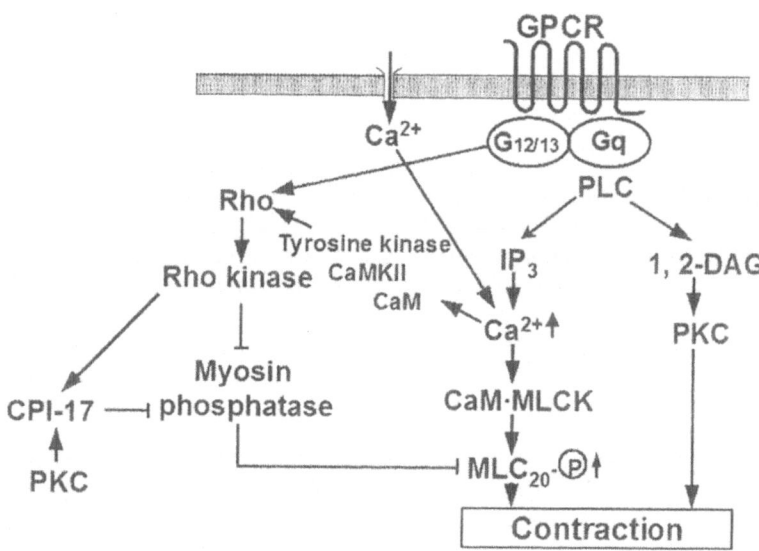

Figure 1. Signaling mechanisms for vascular smooth muscle contraction. Vasoconstrictors act on G protein-coupled receptors(GPCRs) to stimulate phospholipase C(PLC) via G_q and Rho via both $G_{12/13}$ and G_q. The G_q-mediated Rho activation involves calmodulin(CaM), calmodulin kinase II (CAMKII) and a tyrosine kinase. PLC stimulation leads to Ca^{2+}-mobilization, resulting in the activation of the calmodulin-dependent enzyme myosin light chain kinase (MLCK). Rho inhibits myosin phosphatase via Rho kinase, resulting in potentiation of Ca^{2+}-dependent MLC phosphorylation. 1, 2-diacylglycerol (1, 2-DAG) produced via the action of PLC activates protein kinase C, which contributes to contraction independently of or synergistically with Ca^{2+}. The myosin phosphatase inhibitor protein CPI-17 is involved in inhibition of myosin phosphatase activity.

induced similar extents of time-dependent, sustained increases in GTP-Rho. Consistent with this, the Rho kinase inhibitors HA 1077 and Y 27632 inhibited both contraction and MLC phosphorylation induced by KCl as well as noradrenalin, with similar dose-response relationships. Either removal of extracellular Ca^{2+} or addition of a dihydropyridine Ca^{2+} channel antagonist totally abolished KCl-induced Rho stimulation and contraction. A calmodulin antagonist, Ca^{2+}/calmodulin-dependent protein kinase II (CaMKII) inhibitors and also tyrosine kinase inhibitors inhibited KCl-induced Rho activation and contraction. The combination of extracellular Ca^{2+} removal and depletion of the intracellular Ca^{2+} store greatly reduced noradrenalin- and the thromboxane A2 analogue-induced Rho stimulation and contraction. These observations indicated the existence of thus far unrecognized Ca^{2+}-dependent Rho stimulation mechanism, which involves Ca^{2+}/calmodulin, CaMKII and a tyrosine kinase, in smooth muscle. Our data also suggested that the Ca^{2+}-dependent mechanism is involved in excitatory receptor agonist-induced Rho activation. Most likely, this Ca^{2+}-dependent mechanism for activating Rho appears to operate in cooperation with the $G_{12/13}$-mediated mechanism. Thus, multitudes of heterotrimeric G protein-mediated signaling pathways regulate Rho activity in VSM, contributing to the Ca^{2+}-sensitivity regulation of MLC phosphorylation and contraction (Figure 1).

6. CONCLUSION

Since we reported in 1995 that agonist-induced Ca^{2+} sensitization of smooth muscle MLC phosphorylation and contraction involves Rho-dependent inhibition of myosin phosphatase[19], great progress has been made in understanding the mechanisms of Rho-dependent Ca^{2+} sensitization. Excitatory receptor agonists, most of which act on the heptahelical GPCRs, stimulate Rho through multitudes of mechanisms involving $G_{12/13}$ and G_q–Ca^{2+}. The latter mechanism has thus far been not unrecognized in non-muscle and muscle cells and may be unique to smooth muscle. Activation of Rho likely stimulates Rho kinase, which phosphorylates the regulatory subunit MYPT1 of myosin phosphatase. Rho kinase may also phosphporylates to activate the myosin phosphatase inhibitor phosphoprotein CPI-17. Recent studies demonstrated the possibilities that several other serine/threonine kinases may also be involved in phosphorylation of both MYPT1 and CPI-17 and thus regulation of myosin phosphatase activity. Consistent with the roles for Rho kinase in agonist-induced contraction, Rho kinase inhibitors reduce agonist-induced contraction. These observations support the current model for the signaling of agonist-induced contraction, in which activation of receptors by excitatory agonists triggers Ca^{2+} mobilization and consequent activation of MLCK via G_q, as well as Rho-Rho kinase-dependent inhibition of myosin phosphatase via $G_{12/13}$ and G_q. The activation of MLCK and inhibition of myosin phosphatase operate in concert to result in efficient phosphorylation of MLC and contraction. Protein kinase C was recently shown to be involved in inhibition of myosin phosphatase through phosphorylation of CPI-17. However, a number of previous studies also suggested that protein kinase C contributes to contraction in a manner independently of MLC phosphorylation. Thus, the protein kinase C branch of the G_q-signaling may regulate both MLC phosphorylation-dependent and -independent mechanisms of contraction.

ACKNOWLEDGMENTS

I am very grateful to my colleagues, Sotaro Sakurada, Noriko Takuwa, Naotoshi Sugimoto, Hiromitru Nagumo and Masakuni Noda. This work was supported by grants from the Japan Society for the Promotion of Science Research for the Future Program,and Hoh-Ansha Foundation. A portion of this study is carried out as a part of "Ground-based Research Announcement for Space Utilization" promoted by Japan Space Forum.

REFERENCES

1. H. Rasmussen, Y. Takuwa, and S. Park, Protein kinase C in the regulation of smooth muscle contraction, *FASEB J.* 1, 177-185 (1995)

2. Y. Takuwa, Regulation of vascular smooth muscle contraction. The roles of Ca2+, protein kinase C and myosin light chain phosphatase, *Jpn. Heart J.* 37, 793-813 (1996)

3. K. E. Kamm, and J. T. Stull, The function of myosin and myosin light chain kinase phosphorylation in smooth muscle, *Annu. Rev. Pharmacol. Toxicol,* 25, 593-620 (1985)

4. P. Somlyo, and A. V. Somlyo. Signal transduction by G-proteins, Rho-kinase and protein phosphatase to smooth muscle and non-muscle myosin II, *J Physiol(Lond)* 522, 177-185 (2000)

5. J. P. Morgan, and K. G. Morgan, Stimulus-specific patterns of intracellular calcium levels in smooth muscle of ferret portal vein, *J. Physiol. (London),* 351, 155-167 (1984)

6. M. Rembold, and R. A. Murphy, Myoplasmic [Ca2+]i determines myosin phosphorylation in agonist-stimulated swine arterial smooth muscle, *Cir. Res.* 63, 593-603 (1988)

7. D-C. Tang, J. T. Stull, Y. Kubota, and K. E. Kamm, Regulation of the Ca2+ dependent of smooth muscle contraction, *J. Biol. Chem.* 267, 11839-11845 (1992)

8. T. Kitazawa, S. Kobayashi, K. Horiuti, A. V. Somlyo, and A. P. Somlyo, Receptor-coupled, permeabilized smooth muscle. Role of the phosphatidylinositol cascade, G proteins, and modulation of the contractile response to Ca^{2+}, *J. Biol. Chem.* 264, 5339-5342 (1989)

9. T. Kitazawa, B. D. Gaylinn, G. H. Denney, and A. P. Somlyo, G-protein-mediated Ca^{2+} sensitization of smooth muscle contraction through myosin light chain phosphorylation, *J. Biol. Chem.* 266, 1708-1715 (1991)

10. T. Kitazawa, M. Masuo, and A. P. Somlyo, G Protein-mediated inhibition of myosin light-chain phosphatase in vascular smooth muscle, *Proc. Natl. Acad. Sci. USA* 88, 9307-9310 (1991)

11. Y. Kubota, M. Nomura, K. E. Kamm, M. C. Mumby, and J. T. Stull, GTPrS-dependent regulation of smooth muscle contractile elements, *Am. J. Physiol.* 262, C405-C410 (1992)

12. H. Nagumo, Y. Sasaki, Y. Ono, H. Okamoto, M. Seto, and Y. Takuwa, Rho kinase inhibitor HA1077 prevents Rho-mediated myosin phosphatase inhibition in smooth muscle cells. *Am. J. Physiol.* 278, C57-C65 (2000)

13. Alessi, L. K. Macdougall, M. M. Sola, M. Ikebe, and P. Cohen, The control of protein phosphatase-1 by targeting subunits. The major myosin phosphatase in avian smooth muscle is a novel form of protein phosphatase-1. *Eur. J. Biochem.* 210, 1023-1035 (1992)

14. .H. Shimizu, M. Ito, M. Miyahara, K. Ichikawa, S. Okubo, T. Konishi, M. Naka, T. Tanaka, K. Hirano, and D. J. Hartshorne, Characterization of the myosin-binding subunit of smooth muscle myosin phosphatase, *J. Biol. Chem.* 269, 30407-30411(1994)

15. K. Hirata, A. Kikuchi, S. Sasaki, S. Kuroda, K. Kaibuchi, Y. Matsuura, H. Seki, K. Saida, and Y. Takai, Involvement of rho p21 in the GTPγS-enhanced calcium ion sensitivity of smooth muscle contraction, *J. Biol. Chem.* 267, 8719-8722 (1992)

16. Fujita, T. Takeuchi, H. Nakajima, H. Nishio, and F. Hata, Involvement of heterotrimeric GTP-binding protein and Rho protein, but not protein kinase C, in agonist-induced Ca2+ sensitization of skinned muscle of guinea pig vas deferens. *J. Pharmacol. Exp. Ther.* 274,555-561(1995)

17. N. Kokubu, M. Satoh, and I. Takayanagi, Involvement of botulinum C3-sensitive GTP-binding proteins in α1-adrenoceptor subtypes Ca^{2+}-sensitization, *Eur. J. Pharmacol.* 290, 19-27 (1995)

18. A. Lucius, A. Anders, A. Steusloff, M. Troschka, F. Hofman, K. Aktories, and G. Pfitzer, Clostridium difficile toxin B inhibits carbachol-induce force and myosin light chain phosphorylation in guinea-pig smooth muscle: role of Rho proteins, *J. Physiol.(Lond)* 506, 83-93 (1998)

19. M. Noda, C. Yasuda -Fukazawa, K. Moriishi, T. Kato, T. Okuda, K. Kurokawa, and Y. Takuwa, Involvement of Rho in GTPγS-induced enhancement of phosphorylation of 20 kDa myosin light chain in vascular smooth muscle cells: inhibition of phosphatase activity, *FEBS Lett.* 367, 246-250 (1995)

20. Y. Takuwa, M. Noda, C. Yasuda, M. Kumada, and K. Kurokawa, 1995, Regulation of Ca^{2+}-dependent pholsphoaryltaion of 20kDa myosin light chain by the small molecular weight G protein Rho p21 in vascular smooth muscle cells, in: Regulation of the contractile cycle insmooth muscle edited by T. Nakano, D. J. Hartshorne.Tokyo: Springer, pp. 103-110.

21. Y. Takuwa, 1999, Regulation of smooth muscle myosin phosphatase and contraction by Rho and Rho kinase, in Molecular mechanisms of smooth muscle contraction, edited by K. Kohama and Y. Sasaki. Georgetown. R. G. Landes. pp. 47-58.

22. S, Narumiya, T. Ishizaki, and N. Watanabe, Rho effectoors and reorganization of actin cytoskeleton, *FEBS. Lett.* 410, 68-72 (1997)

23. L. Lim, E. Manser, T. Leung, and C. Hall, Regulation of phosphorylation pathways by p21 GTPases. The p21 Ras-related Rho subfamily and its role in phosphorylation signalling pathways, *Eur. J. Biochem.* 242, 171-185 (1996)

24. L. Van Aelst, and C. D. Souza-Schorey, Rho GTPases and signaling networks,   *Genes Dev.* 11, 2295-2322 (1997)

25. K. Kimura, M. Ito, M. Amano, K. Chihara, Y. Fukata, M. Nakafuku, B. Yamamori, J. Feng, T. Nakano, K. Okawa, A. Iwamatsu, and K. Kaibuchi, Regulation of myosin phosphatase by Rho and Rho-associated kinase (Rho-kinase), *Science* 273, 245-248 (1996)

26. J. Feng , M. Ito, K. Ichikawa, M. Nishikawa, D. J. Hartshorne, and T. Nakano,   Inhibitory phosphorylation site for Rho-associated kinase on smooth muscle myosin phosphatase, *J. Biol. Chem.* 274, 37385-37390 (1999)

27. J. A. MacDonald, M. A. Borman, A. Muranyi, A. V. Somlyo, D. J. Hartshorne, and T. A. J. Haystead, Identification of the endogenous smooth muscle myosin phosphatase-associated kinase, *Proc. Natl. Acad. Sci. USA* 98, 2419-2424 (2001)

28. Muranyi, J. A. MacDonald, J. T. Deng, D. P. Wilson, T. A. J. Haystead, M. P. Walsh, F. Erdodi, E. Kiss, Y. Wu, and D. J. Hartshorne, Phosphorylation of the myosin phosphatase target subunit by integrin-linked kinase, *Biochem. J.* 366, 211-216 (2002)

29. M. Eto, T. Ohmori, M. Suzuki, K. Furuya, and F. Morita, A noved protein phosphatase1 inhibitory protein potentiated by protein kinase C isolated from porcine aorta media and characterization. *J. Biochem.* 118, 1104-1107 (1995)

30. T. Kitazawa, M. Eto, T. P. Woodsome, and D. L. Brautigan, Agonist trigger G protein-mediated activation of the CPI-17 inhibitor phopsphoprotein of myosin light chain phosphatase to enhance cvascular smooth muscle contractility, *J. Biol. Chem.* 275, 9897-9900 (2000)

31. M. Koyama, M. Ito, J. Feng, T. Seko, K. Shiraki, K. Takase, D. J. Hartshorne, and T. Nakano, Phosphorylation of CPI-17, an inhibitory phosphoprotein of smooth muscle myosin phosphatase by Rho kinase, *FEBS Lett.* 475, 197-200 (2000)

32.   T. Hamaguchi, M. Ito, J. Feng, T. Seko, M> Koyama, H. Machida, K. Takase, M. Amano, K. Kaibuchi, D. J. Hartshorne, and T. Nakano, Phosphorylation of CPI-17, an inhibitory phosphoprotein of smooth muscle myosin phosphatase by protein kinase N, *Biochem. Biophys. Res. Commun.* 274, 825-830 (2000)

33.   J. A. MacDonald, M. Eto, M. A. Borman, D. L. Brautigan, and T. A. J. Haystead, Dual Ser and Thr phosphorylation of CPI-17, an inhibitor of myosin phosphatase, by MYPT-associated kinase, *FEBS Lett.* 493, 91-94 (2001)

34.   J. T. Deng, C. Sutherland, D. L. Brautigan, M. Eto, and M. P. Walsh, Phosphorylation of the myosin phosphatase inhibitors, CPI-17 and PHI-1, by integrin-linked kinase, Biochem. J. 367, 517-524 (2002)

35.   M. Eto, T. Kitazawa, M. Yazawa, H.. Mukai, Y. Ono, and D. L. Brautigan, Histamine-induced vasoconstriction involves phosphorylation of a specific inhibitor protein for myosin phosphatase by protein kinase C alpha and delta isoforms, *J. Biol. Chem.* 276, 29072-29078 (2001)

36.   L. Bishop, and A. Hall, Rho GTPase and their effector proteins, *Biochem. J.* 348, 241-255 (2000)

37.   S. Sakurada, H. Okamoto, N. Takuwa, N. Sugimoto, and Y. Takuwa, Rho activation in excitatory agonist-stimulated vascular smooth muscle, *Am. J. Physiol.* 281, C571-C578 (2001)

38.   Gohla, S. Offermanns, T. M. Wilkie, and G. Schultz, Differential involvement of $G\alpha_{12}$ and $G\alpha_{13}$ in receptor-mediated stress fiber formation, *J. Biol. Chem.* 274, 17901-17907 (1999)

39.   S. Fukumoto, H. Chikumi, and J. S. Salvo, RGS-containing RhoGEFs: the missing link between transforming G proteins and Rho, *Oncogene* 20, 1661-1668 (2001)

40.   M. A. Booden, D. P. Siderovski, and C. J. Der, Leukemia-associated Rho guanine nucleotide exchange factor promotes $G\alpha q$-coupled activation of RhoA, *Mol. Cell. Biol.* 22, 4053-4061 (2002)

41.   H. Chikumi, J. Vazquez-Prado, J-M. Servitja, H. Miyazaki, and J. S. Gutkind, Potent activation of RhoA by $G\alpha q$ and $G\alpha q$-coupled receptors, *J. Biol. Chem.* 277, 27130-27134 (2002)

42.   Gohla, G. Schlutz, and S. Offermanns, Role for $G_{12/13}$ in agonist-induced vascular smooth muscle cell contraction, *Circ. Res.* 87, 221-227 (2000)

DISCUSSION

Nishimura: There are several Rho family G proteins such as Rho A, B and C. When you talk about Rho activation, do you mean that all Rho family is activated?

Takuwa: We have only determined the activity of Rho A.

Nishimura: Actually you don't know yet whether Rho B and Rho C are activated or not.

Takuwa: No, we don't.

Morano: What is the link between KCl depolarization with subsequent Ca^{2+}-calmodulin stimulation and Rho-activation?

Takuwa: Our observation that the CAMK II inhibitors inhibit Rho activation suggests that phosphorylation of a Rho-GEF or other proteins that regulate Rho-GEF activity is likely a link between Ca^{2+} signal and Rho stimulation.

HUMAN HEART FAILURE: DILATED VERSUS FAMILIAL HYPERTROPHIC CARDIOMYOPATHY[1]

Norman R. Alpert and David M. Warshaw*

1. INTRODUCTION

Heart failure is an increasing public health problem with high morbidity and mortality (Cowie, 1997). The survival rates closely correlate with the severity of the heart disease with 50% of the patients classified as NYHA IV surviving only one year (Calif et al 1997; SOLVD Investigators, 1991). The survivability of heart failure patients correlates with the ejection fraction (Cohn et al 1997). The deficit in ventricular performance in failing hearts is directly related to a power deficit of the contractile apparatus. Power is the rate of doing work and at the cellular level is a mechanical expression of the myocyte's force velocity relationship (power = force X velocity). The ability of the myocyte to develop force and velocity depends on the characteristics of the molecular motor myosin interacting with actin with the obligatory hydrolysis of ATP and the resulting power stroke. In this review we will describe and compare the mechanical and kinetic characteristics of the mechano-enzyme myosin from heart failure patients with dilated cardiomypathy and familial hypertrophic cardiomyopathy (Alpert et al 2002 Palmiter et al 2000, Hasenfuss et al 1992). The studies on heart tissue from patients with dilated cardiomyopathy take advantage of myothermal techniques for assessing the cross-bridge characteristics. In experiments involving tissues from patients with familial hypertrophic cardiomyopathy we use laser trap techniques for evaluating single molecule mechanics.

2. DILATED CARDIOMYOPATHY (DCM)

2.1 Protection of Human Left Ventricular Strips During Transport and Dissection

We were confronted with two problems in preparing strips of human heart muscle for myothermal analysis. First the heart tissue had to be protected during transport from

---

*Norman R. Alpert, David M. Warshaw, University of Vermont, Burlington, VT 05405
[1] Supported in part by USPHS Grants HL66157, HL55641, HL59408.

Molecular and Cellular Aspects of Muscle Contraction
Edited by H. Sugi, Kluwer Academic/Plenum Publishers, 2003

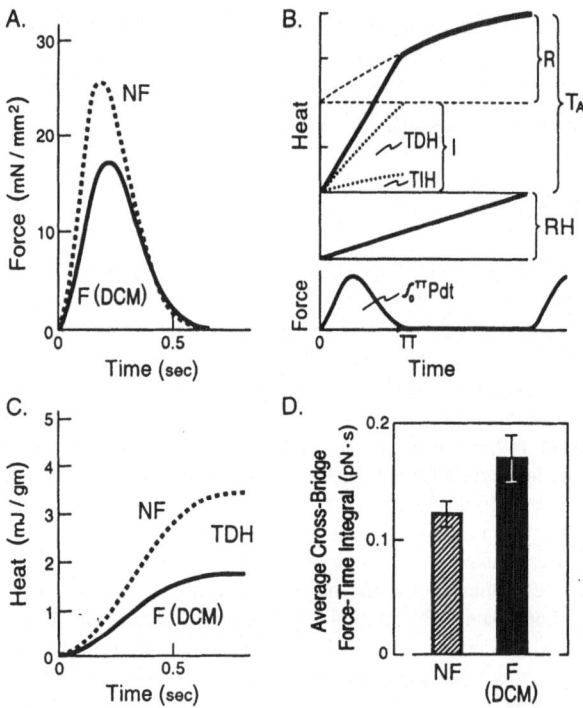

Figure 1. A) The time course of isometric force development for the non failing (NF) and dilated cardiomyopathic failing (F-DCM) hearts. Note the peak isometric force, rate of force development and rate of relaxation are lower in the F-DCM preparations. B) The time course of heat production relative to the isometric twitch. The isometric myogram is seen at the bottom of the figure. The heat production is divided into resting heat (RH), total activity related heat (T_A), recovery heart (R), initial heat (I), tension dependent heat (TDH) and tension independent heat (TIH). See text for a detailed description. C) The time course of tension dependent heat liberation (TDH) for the non failing (NF) and failing (F-DCM) heart muscle strips. D) The average cross-bridge force time integral for the non failing (NF) and failing (F-DCM) heart strips that form the basis for the muscle force time integral.

the surgical suites and transplant centers. In addition, strips had to be thin enough so that oxygenation remained adequate at 37°C with stimulation frequencies up to 180 beats per minute.   The use of a 2.3-butanedione monoxime protective solution (BDM) (Krebs-Ringer, 30 mM BDM, 11.2 mM glucose. 2.5 mM Ca^{2+}, 10 IU insulin per liter) permitted protected transport and dissection of viable, excitable strips that are < 30 mm^2 and produce high tension (Mulieri et al 1989).

2.2 Patient Population and Strip Mechanics

Control muscle strips were prepared from left ventricular epicardial dissections obtained during coronary artery by-pass surgery from seven patients with normal left ventricular mechanical performance (Mulieri et al 1992, Hasenfuss et al 1992). Strips from DCM hearts were obtained at the time of transplant from six patients (NYHA IV). Isometric twitch tension in the non failing compared with DCM failing heart strips was reduced from 25.9 ± 3.9 to 13.9 ± 2.0 mN/mm^2 (P<0.02)(Fig. 1A). The maximum rate of tension development and relaxation were reduced from 193 ± 26 and 148 ± 23 to 95 ± 11 and 80 ± 13 mN/mm^2 s (p<0.02)(Fig. 1A).

2.3 Thermo-mechanical investigation of human heart muscle strips.

Ultrasensitive vacuum deposited bismuth/antimony thin film thermopiles in conjunction with a capacitance force transducer were used to evaluate heat production and mechanics in strips of parallel coursing human heart muscle fibers (Mulieri et al 1977). The time course of mechanics along with energy consumption provides a window for viewing the extent and rate of cellular and molecular mechanical phenomenon along with the underlying chemical reactions.   Under steady state conditions in a repetitively stimulated muscle strip all mechanical and chemical conditions at the start of each contraction are identical (pH, ATP, ADP, CrP, and Pi). Since contraction and relaxation are energy consuming processes this implies the recovery phenomena (also energy requiring) take place throughout the contraction, relaxation and diastolic phases (Balaban 1990). In an activated muscle, 1) intracellular calcium is released, 2) myosin combines with actin, rotates developing force and/or motion and then detaches with the cross-bride cycles repeating themselves,   3)   calcium is removed from the cytosol resulting in relaxation. Contraction, relaxation and calcium cycling are energy consuming processes powered by the energy contained in the terminal phosphate of ATP during hydrolysis (ATP → ADP + Pi). ATP is then resynthesized from ADP by the closely coupled creatine transphosphorylase .reaction   (CrP + ADP→ ATP + Cr). Mitochondrial oxidative phosporylation then recharges the system. Under these conditions the muscle and thermopile can be viewed as a thermodynamically closed system. For a closed system, the enthalpy change (ΔH) is equal to the heat liberated (q) and the work done (w)(Eq 1). Under isometric conditions where no work is done the heat liberated is equal to the enthalpy change (Eq 2). The enthalpy change is equal to the enthalpy changes in all of the energy requiring reactions coupled to the contraction relaxation cycle. Accordingly, if n_i is the number of moles involved in the ith reactions and ΔH_i is the enthalpy change in the ith reaction then the total energy exchange is described in Eq 3.

$$q + w = -\Delta H \qquad\qquad (1)$$

$$q \quad = -\Delta H \qquad\qquad (2)$$

$$-\Delta H = \sum_{0}^{ith} (n)_i (-\Delta H_i) \qquad\qquad (3)$$

Since we know the primary source of energy in the muscle contraction relaxation cycle is ATP hydrolysis and since under isometric conditions no work is done, we can use the heat production to provide an index of the extent and rate of ATP hydrolysis.. The muscle strip at rest, in contact with the thermopile system, liberates heat at a steady rate (Fig. 1B, RH). When the muscles is activated, heat is liberated at an initial rapid and then a secondary slower rate (Fig. 1B, T_A) This represents the total activity related heat (cross-bridge cycling, calcium cycling and recovery processes). The secondary slower rate of heat production is liberated at a mono exponentially decreasing rate and thus can be extrapolated back to zero time. This represents the heat production associated with mitochondrial oxidative recovery processes (Fig. 1B, R). Subtracting the recovery heat (R) from the total heat (T_A) leaves the initial heat (Fig. 1B, I) The initial heat (I) can be partitioned into the tension dependent (TDH) (cross-bridge cycling) and tension independent (TIH)(calcium cycling) heats by incubating the muscle in low concentrations of BDM (<5 mM) to eliminate tension and leave only the tension independent heat. This latter quantity is subtracted from the initial heat to give the TDH(Fig. 1B)..

2.4 Determination of the Average Cross-bridge Force-time Integral

The TDH is a reflection of the number of molecules of ATP hydrolyzed by the cycling cross-bridges and thus the number of cross-bridge cycles that occur during the duration of the isometric twitch. The TDH for the non-failing and DCM failing tissues was 3.39 ± 0.59 and 1.34 ± 0.22 mJ/g (Fig. 1C). Dividing the TDH produced during contraction and relaxation by the enthalpy change per molecule of ATP hydrolyzed (56 pico nano Joules) provides a calculation of the number of cross-bridge cycles that occur in the muscle ($XbrCyc_{muscle}$) during contraction and relaxation (Eq 4), The cross-bridge cycles per half sarcomere ($XbrCyc_{hs}$) is obtained by dividing the number of half sarcomeres ($\#hs_{muscle}$) per muscle into the cross-bridge cycles per muscle ($XbrCyc_{muscle}$). The $\#hs_{muscle}$ is calculated by the ratio (l_{max},the muscle length)/(hs, the half sarcomere length), The muscle force time integral (FTI_{muscle}) for the contraction relaxation cycle is equal to the number of cross-bridge cycles that occur in a half sarcomere ($XbrCyc_{hs}$) multiplied by the average cross-bridge force time integral (FTI_{XBr}) or the cross-bridge force-time integral (FTI_{XBr}) is equal to the ratio of the muscle force-time integral (FTI_{muscle}) to the number of crossbridge cycles per half sarcomere ($XBrCyc_{hs}$) (Eq 5) The average cross-bridge force time integral for the DCM failing hearts was 21 % greater than the controls (Fig 1D). This is surprising in view of the substantial reduction in peak isometric force and is attributable to an accompanying decrease in activation (Alpert et al 2002, Hasenfuss et al 1992). The increase in the average cross-bridge force-time integral in the failing heart can result from an increase in unitary force (F_{uni}) or attachment time (t_{on})[$FTI_{XBr} = (F_{uni}) \times (t_{on})$]. In a recent study on failing hearts using sinusoidal analysis it was found that the attachment time was increased (Mulieri 2002).

$$XbrCyc_{muscle} = TDH/56 \; pnJ \qquad (4)$$

$$FTI_{XBr} = (FTI_{muscle})/ (XbrCyc_{hs}) \qquad (5)$$

2.5 The Development of Dilated Cardiomyopathy

At the present there is no completely sastisfactory explanation for the development of dilated carrdiomyupathy. An attractive hypothesis was suggested by Olsen et al (Olsen et al 2002, Olson et al 1998) in which they suggest that depression in force transmission leads ultimately to dilated cardiomyopathy. In the studies reported above we have shown a depression in force and in velocity of shortening. Accordingly, the power output in the DCM failing heart is markedly decreased. If decreased force transmission produces dilated cardiomyopathy then the decreased power output in the DCM failing hearts may be the cause of the dilation.

2.6. Generalizations

The increase in cross-bridge force time integral and the associated increase in economy of isometric contraction raises the question as to whether all forms of failure exhibit these qualitative changes. To answer that question we examined cross-bridge mechanics in Familial Hypertrophic Cardiomyopathy.

3. FAMILIAL HYPRTROPHIC CARDIOMYOPATHY

3.1 Background

Familial hypertrophic cardiomyopathy is a disease of the sarcomere with occurrence of about 1 in 500 (Maron 1995). This inherited heart disease is autosomally dominant and linked to mutations in genes of 10 different sarcomeric proteins presented in order of prevalence (β-myosin heavy chain [35%], cardiac troponin T [15%], cardiac myosin binding protein C [15%], α-tropomyosin [<5%], the essential and regulatory light chains, cardiac troponin I, α-actin, titin). Most of the population with the disease are free of symptoms until the advent of sudden death (in the young), left and right ventricular hypertrophy, and failure. In addition to the hypertrophy the heart is characterized by myocyte disarray and fibrosis (Spirito et al 1997).

3.2 R403Q and L908V Mutations

The motor and lever arm domains of the mysosin molecule conissts of an upper and lower 50 kD region, a 25 and 20 kD region (Fig. 2). These contain the nucleotide and actin binding sites. The connection between the lever arm and the 25/20 kD region funcitons as a converter resulting in the movement of the lever arm. The lever arm is stabilized by the associated regulatory and essential light chains. Amino acid 403 is located in the vicinity of the actin binding region with the mutation involving an arginine to gluatamine substitution (Fig. 2). Amino acid 908 is loacated about 15 nm from the motor domain in the S2 region about 10 heptides from the lever arm with the mutation consisting of a leucine to valine substitution (Fig. 2)(Rayment et al 1993, Rayment et al 1995). Patients with R403Q compared with L908V exhibits earlier onset, greater penetrance, greater hypertrophy and results in a higher incidence of failure and sudden death (Fananapazir et al 1994).

3.3 Molecular Mechanics of the Mutant Myosins

Velocity and force are the two parameters that determine myocardial power output. The velocity potential for the mutant myosin molecules was determined by means of the *in vitro* motility assay which assesses the ability of the myosin molecules to propel actin filaments (Palmiter et al 2000). It was surprising that in our assay system the mutant myosin propelled actin faster than did the control myosin molecules (control , 1.39 ± 0.03 um/s; R403Q, 1.87 ± 0.07 um/s; L908V, 1.76 ± 0.15 um/s [for control versus 403 or 908, $p < 0.05$]).

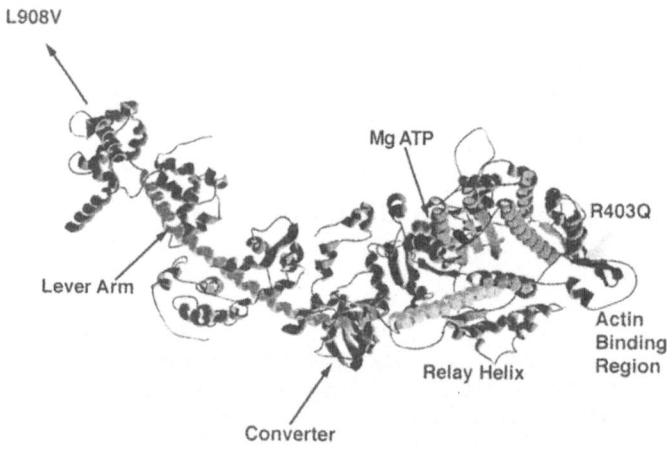

Figure 2. The location of the cardiac 403 and 908 mutations based on the previously described crystal structure of myosin (Rayment et al 1993). Note the actin and nucleotide binding sites, the converter region, the lever arm and the relay helix.

3.4 Partitioning the Velocity Into Primary Components

The primary components of the velocity measurements made by means of the *in vitro* motility assay were evaluated using the laser trap system. This system allows the partitioning of velocity into unitary displacement (D_{uni}) and attachment time (t_{on}) with actin filament velocity being proportional to the ratio D_{uni}/t_{on} (Palmiter et al 2000). The laser trap system consisted of a three bead configuration where two of the beads were sequestered one in each of the two energy wells while the third bead was used as platform for attaching single myosin molecules (Fig. 3). The beads are coated with NEM myosin in order to attach them to a TRITC-phalloidin labeled actin filament. The beads are moved so as to stretch the actin filament with pre-tension force of about 4 pN. Then the bead-actin-bead system is lowered to make contact with the single myosin molecule previously positioned on the bead platform. The force on the actin filament is then lowered to < 0.02 pN allowing for the recording of virtually unloaded isotonic interaction of the single myosin molecule and the actin filament (Fig. 3, top). An actual data trace is seen in Fig 3, bottom where there is a clear indication of the unitary displacement and attachment times. Note that there is considerable noise in the system due to Brownian motion so that the signal to noise ratio is about one. We deal with this difficulty by using the mean variance analysis technique for separating the true signal from the noise (Patlak 1993, Guilford et al, 1997).

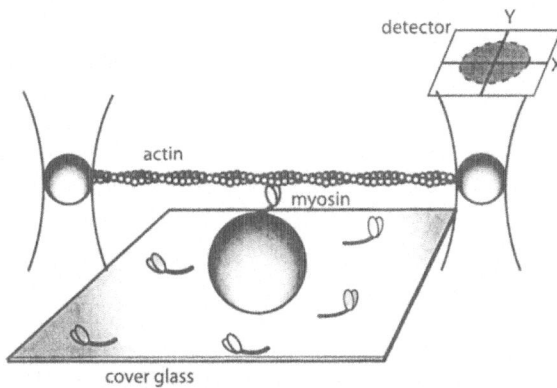

Figure 3. The three bead laser trap system with feedback control Above is a diagram of the system with beads attached to the actin filament and the myosin molecules on the pedestal. Below is an actual record of displacement and associated attachment time.

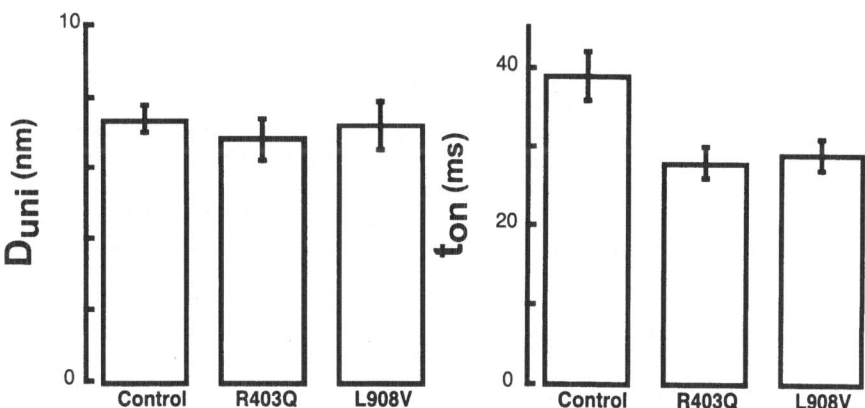

Figure 4. Unitary diaplacement (D_{uni}) and attachment time (t_{on}) for the control, R403Q and L908V myosin molecules.

The results of these experiments indicate that there is no difference in unitary displacement between the control myosin and the R403Q or the L908V myosin mutations (Fig. 4). In contrast, the attachment times for R403Q and L908V are significantly lower (p < 0.05) than that for the control myosin (Fig. 4). Therefore the increase in velocity of produced by the R403Q and L908V mutations results from kinetic differences (t_{on}; Control 39 $\pm$ 3 msec, R403Q 28 $\pm$ 2 msec, L908V 29 $\pm$ 2 msec, p < 0.05) probably ADP release (Palmiter et al 1999) and not from mechanical difference (D_{uni}; Control 7.4 $\pm$ 0.4 nm, R403Q 6.8 $\pm$ 0.6 nm, L908V 7.2 $\pm$ 0.7 nm, ns)(Palmiter et al. 2000).

3.5 The Average Force

The average force is the product of the unitary force (F_{uni}) and the duty cycle (DC, i.e., the percent of the total cycle time spent in the attached strongly bound state. The force production of the mutant myosins compared with control was assessed using the mixture technique based on the interaction model where actin filament velocity is measured with mixtures of cardiac myosin (Control, R403Q, L908V) and the faster chicken skeletal myosin (Harris et al 1994). Using this assay, the average force for the mutant myosins was found not to differ from that of the control myosin (Palmiter et al 2000).

3.6 Speculation on the mechanism involved in the increased actin filament velocity found with R403Q and L908V

The increase in actin filament velocity in the *in vitro* motility assay for R403Q and L908V has been shown to result from the decrease in the on time (t_{on}) and not from unitary displacement (D_{uni}). The mechanism by which the myosin mutations produce these changes in unknown. R403Q is located in the motor domain of the molecule in the vicinity of the actin binding site (Fig. 2). The arginine to glutamine subtitution results in a change from a positively charged side chain to an uncharged polar side chain (Fig. 5). This shift in charge may effect the binding of myosin to actin and thus the kinetics of the interaction. The 908 amino acid is located about 10 heptides into the S2 alpha helical coiled coil region. This is about 15 nm from the actin and nucleotide regions of the myosin motor domain (Fig. 2). The leucine to valine mutation (L908V) involves two hydrophobic amino acids (Fig. 5). The smaller size of the valine might reduce the stability of the S2 alpha helical coiled coil. How this might effect the motor domain remains unknown. We can speculate that the change in the coiled coil gives more flexibility to the head region of the myosin molecule and thus changes how it interacts with actin. Alternatively, the mutation may change the compliance of the S2 region and through the relay helix force sensor (Fig 2) might effect the myosin kinetics. The speculation about the mechanism by which the mutated myosin alters kinetics requires definitive experiments to rule out the alternate hypotheses.

3.7 The development of hypertrophy, myocyte disarray and fibrosis

The phenotypic characterization of familial hypertrophic cardiomyopathy involves the development of asymmetric hypertrophy, myocyte disarray and fibrosis. How does a mutant motor that moves actin faster lead to these phenotypic changes. Since an individual has both a normal and mutant allele, every muscle cell should have both the

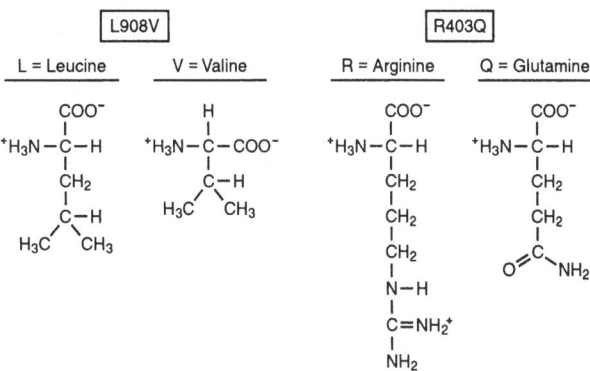

Figure 5. The lysine to valine mutation (L908V) and the arginine to glutamine mutation (R403Q).

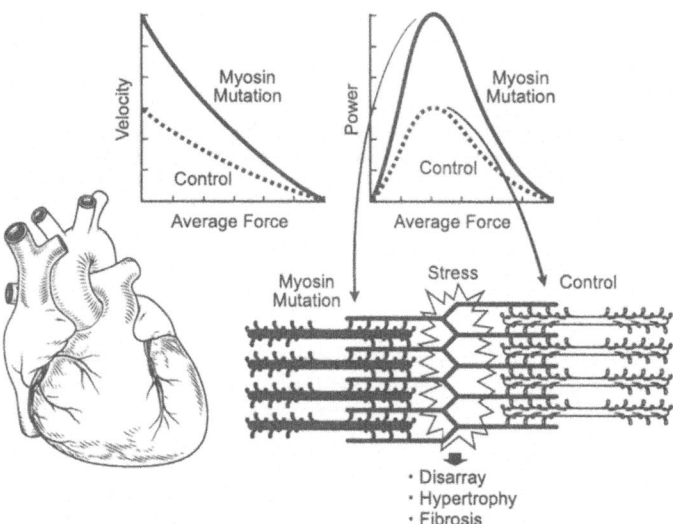

Figure 6. The force velocity and power output for the control versus the mutant myosins. This suggests a linear arrangement where the differential power results in internal stress leading to the hypertrophy, alterations in the extracellular matrix and myofiber disarray.

slower (wild type) and faster (mutant) myosin present. We know that the average force for the wild type and mutant myosins is the same. Accordingly the power output of the mutant myosin (force X velocity) is greater (Fig. 6). The tug of war between the two myosin species having different mechanical characteristics can lead to internal mechanical stress. We hypothesise that this internal stress (Fig 6) results in intracellular signals that lead to hypertrophy, reorganization of the intracellular matrix and fibrosis.

4. REFERENCES

Alpert N.R., Mulieri L.A. and Warshaw D. The failing human heart. *Cardiovasc Res* **54**: 1-10, 2002.

Balaban R.S. Regulation of oxidative phosphorylation in the mammalian cell. *Am J Physiol* 1990; **258**: C377-C389.

Cakuf R.M., Adams K.F., McKenna W.J., et al. A randomized controlled trial of epoprostenol therapy for severe congestive heart failure: the Flolan International Randomized Survival Trial (FIRST) *Am Heart J* 1997; **134**: 44-54.

Cohn J.N., Zieche S., Smith R., et al. Effet of the calcium antagonist flodipine as a supplementary vasodilator therapy in patients with chronic heart failure treated with enalapril: V-HeFT III Vasodilator heart failure trial (V-HEFT) study group. *Circ* 1997; **96**(3): 856-863.

Cowie M.R., Monsteaad A., Wood D.A. et al. The Epidemiology of heart failure. *Eur. Heart J.* 1997; **18**: 208-225

Fananapazir L., Epstein N.D. Genotype-phenotypecorelations in hypertrophic cardiomyopathy. *Circ.* 1994; **89**: 22-32.

Guilford W.H., Dupuis D.E., Kennedy G., et al. Smooth muscle and skeletal muscle myosins produce similar unitary forces and displacements in the laser trap. *Biophys J* 1997; **72**: 1006-1021.

Harris D.E., Work S.S., Wright R.K., Alpert N.R., Warshaw D.M. Smooth, cardiac and skeletal muscle myosin force and motion generation assesed by cross-bridge mechanical interactions in vitro. *J Muscle Res Cell Motil* 1994; **15**: 11-19.

Hasenfuss G., Mulieri L.A., Leavitt B.J., Allen P.D., Haeberle J.R. and Alpert N.R. Alteration of contractile function and excitation-contraction coupling in dilated cardiomyopathy. *Circ Res* **70**: 1225-1232, 1992.

Maron B.J., Gordin J.M., Flack J.M., Gidding S.S., Kurosaki T.T., Bild D.E. Prevalence of hypertrophic cardiomyopathy in a general population of you adults. Echocardiographic analysis of 4111 subjects in the CARDIA study. Coronary artery risk development in young adults. *Circ* 1995; **92**: 785-789.

Mulieri L.A., Barnes W., Leavitt B.J., Ittleman F.P., LeWinter M.M., Alpert N.R., Maughan D.W. Alterations of myocardial dynamic stiffness implicating abnormal crossbridge function in human mitral regurgitation heart failure. *Circ Res* 2002; **90**: 66-72.

Mulieri L.A., Hasenfuss G., Ittleman F., Blanchard E.M., Alpert N.R. Protection of human left ventricular myocardium from cutting injury with 2,e-buteanedione monoxime. *Circ Res* 1989; **65**: 1441-1444.

Mulieri L.A., Leavitt B.J., Hasenfuss G., Allen P.D., Alpert N.R. Contraction frequency dependence of twitch and diastolic tension in hum dilated cardiomyapathy (tension frequency relationship in cardiomyopathy. *Basic Res Cardiol* 1992; **87** (Suppl 1): 199-312.

Mulieri L.A., Luhr G., Trefry J., Alpert N.R. Metal film thermopiles for use with rabbit right ventricular papillary muscles. *Am J Physiol* 1977; **233**: C146-C156.

Olsen T.M., Illenberger S., Kishimoto N.Y., Huttelmaier S., Keating M.T., Jockusch B.M. Metavinculin mutations alter actin interaction in dilated cardiomyopathy. *Circ* 2002; **105**: 431-437.

Olsen T.M., Michels V.V., Thibodeau, Tai Y.-S., Keating MT.Actin mutations in dilated cardiomyopathy, a heritable form of heart failure. *Science* 1998; **280**: 750-752.

Palmiter K.A., Tyska M.J., Dupuis D.E., Alpert N.R., Warshaw D.M. Kinetic differences determined at the single molecule level account for the functional diversity of cardiac V_1 and V_3 isoforms. *J. Physiol* 1999; **519**: 669-678.

Palmiter K.A., Tyska M.J., Haeberle J.R., Alpert N.R., Fananapazir L., Warshaw D.M.. R403Q and L908V mutant β-cardiacmyosin from patients with familial hypertrophic cardiomyopathy exhibit enhanced mechanical performance at the single molecule level. *J Muscle Res and Cell Motility* 2000; **21**: 609-620.

Patlak J.B. Measuring kdinetics of complex single ion channel data using mean-variance histograms. *Biophys. J* 1993; **65**: 29-42.

Rayment I., Holden H.M., Seller J.R., Fananapazir L., Epstein N.D. Structural interpretation of mutation in the β-cardiac myosin that have been implicated in familial hypertrophic cardiomyopathy. *Proc Natl Acad Sci USA* 1995; **92:** 3864-3868.

Rayment I., Rypnieski W.R., Schmidt-Base K., et al Three-dimensional structure of myosin subfragment 1: a molecular motor. *Science* 1993; **261:** 50-58.

Spirito P., Seidman C.E., McKenna W.J., Maron B.J. The management of hypertrophic cardiomyopathy. *New Eng J Med* 1997; **336:** 775-785.

DISCUSSION

Huxley: Have you actually measured directly the ADP off rate in the FHC mutants, or is it an implication from other considerations?

Alpert: No, we have not measured the ADP off rate in the FHC preparations. We are speculating about this based on cardiac V1/V3 studies.

Winegrad: What was the status of patients with double mutations in the myosin?

Alpert: There are so few patients with a double mutation that it is premature to talk about the phenotype. There is a large variance in phenotypic expression resulting from a point mutation in a sarcomeric protein. In addition these are gender and racial differences.

Cecchi: Your model for explaining the disarray assumes that the mutated myosin is distributed in a very non-uniform way, is this the case?

Alpert: It is not necessary for our model that the mutated myosin distributes in a non-uniform way.

ter Keurs: Severe pressure overload for the heart such as in pulmonary stenosis also leads to myofibrillar disarray in the absence of myosin mutations. What mechanism would you propose in this scenario?

Alpert: I don't know. However I speculate that, in the case of pulmonary stenosis, there is an increase in sympathetic activity as well as the production of intracellular hypertrophic stimuli; this could result in disarray and hypertrophy.

Morano: Do you believe that the hypercontractile state is related to myosin or to the Law of La Place?

Aplert: I think the hypercontractile state results from the mutated myosin (super myosin), the hypertrophy and the Law of La Place.

Maughan: Relating to Cecchi's question, wouldn't the "tug of war" between mutant and wild-type myosin more likely occur between adjacent myosin heads, and if so, can you speculate how that stress point leads to the kind of myocyte disarray?

Alpert: The internal stress may lead to the production of enzymes that remodel the extracellular matrix as well as the production of hypertrophic growth factors.

Sugi: You use the term of "step size." Is it equivalent to so-called "powerstroke" or so called "interaction distance"? I think your "step size" is very close to "powerstroke." Is that right?

Alpert: That's right. But remember the step size here is measured under very lightly loaded conditions, so in reality there is a step but little power. None the less it is equivalent to the powerstroke.

REGULATION OF SMOOTH MUSCLE CONTRACTION BY CALCIUM, MONOMERIC GTPASES OF THE RHO SUBFAMILY AND THEIR EFFECTOR KINASES

G. Pfitzer, A. Wirth, C. Lucius, D. Brkic-Koric, E. Manser, P. de Lanerolle, and A. Arner[*]

1. INTRODUCTION

A key event in the activation of smooth muscle contraction ist the phosphorylation of the regulatory light chains of myosin (r-MLC) at Ser^{19} which is predominantly catalyzed by the Ca^{2+} and calmodulin dependent myosin light chain kinase, MLCK (Gallager et al., 1997). Recently, it has been suggested that, in addition, r-MLC phosphorylation and contraction may be induced in a Ca^{2+}-independent manner by several protein kinases, such as Rho associated kinase, ZIP kinase, and integrin linked kinase, ILK (reviewed in Ganitkevich et al, 2002). These kinases may, hence, be involved in Ca^{2+}-independent contractions (Kureishi et al., 1999) leading to an increased Ca^{2+}-sensitivity of r-MLC phosphorylation and contraction. Although the physiological role of these kinases is far from clear, they may be of importance during the maintained phase of a contraction when intracellular $[Ca^{2+}]$ has returned to near resting values (Himpens and Somlyo, 1988; Lucius et al., 1998). One goal of this study, therefore, was to test whether Ca^{2+} is required

[*] G. Pfitzer (corresponding author), A. Wirth, D. Brkic-Koric, Department of Vegetative Physiologie, University of Cologne, Robert-Koch Str. 39, 50931 Koeln, Germany; phone: +492214786950, fax: +492214783538, e-mail: Gabriele.Pfitzer@uni-koeln.de; C. Lucius, Insitute of Physiology, Humboldt-University of Berlin; E. Manser, Glaxo-IMCB Group, Institute of Molecular and Cell Biology, National University of Singapore, Kent Ridge, Singapore 0511; P. de Lanerolle, Dept. of Physiology and Biophysics, University of Illinois, Chicago, Il 60612, USA; A. Arner,Dept. of Physiologcial Sciences, Lund University, Solvegatan 19, Lund, Sweden.

Molecular and Cellular Aspects of Muscle Contraction
Edited by H. Sugi, Kluwer Academic/Plenum Publishers, 2003

for tension maintenance of an agonist induced contraction using the membrane permeant form of the caged Ca-chelator, diazo2.

Dephosphorylation of r-MLC is catalyzed by a type 1 phosphatase (MLCP), a constitutively active enzyme that consists of a regulatory, targeting subunit (MYPT1), which binds the phosphatase to its substrate myosin, further a catalytic subunit and a 20 kDa subunit of unknown function (Hartshorne et al., 1998 for review). MLCP activity is inhibited in response to agonists and the poorly hydrolyzed GTP analogue, GTPγS (Somlyo and Somlyo 2000 for review). Inhibition of MLCP may be mediated by an endogenous inhibitory phosphopetide, CPI-17, which initially has been shown to be a substrate of PKC (Kitazawa et al., 1999) and later of integrin-linked kinase (Deng et al., 2002), and of Rho-associated kinase, ROK (Koyama et al. 2000), an effector of the monomeric GTPase, Rho (Matsui et al., 1996). *In vitro*, MLCP activity is further inhibited by phosphorylation of its regulatory subunit, MYPT1 by Rho-associated kinase (Kimura et al., 1996), and a cop-purifying kinase, that is related to ZIP-like kinase (MacDonald et al., 2001).

The importance of activation of Rho/ROK signaling cascade for tension maintenance was shown by membrane permeant toxins that covalently modify Rho thereby inactivating it and by the relatively specific ROK inhibitor, Y27632 (Pfitzer 2001 for review). Incubation of ileal smooth muscle with toxin B induced an inhibition of the tonic phase of the contraction which was associated with inhibition of phosphorylation of r-MLC (Lucius et al., 1998). Unfortunately the effects of toxin B are not entirely specific for RhoA because this toxin also inhibits other members of the Rho family, namely Rac and cdc42 (Just et al., 1995). Effectors of these GTPases are the p21 activated protein kinases (Manser et al., 1994). In non-muscle cells, these kinases phosphorylate and inhibit MLCK (Sanders et al., 1999) and r-MLC (Chew et al. 1998) and in skinned smooth muscle, caldesmon, desmin and r-MLC (van Eyk et al., 1998). Phosphorylation of r-MLC by PAK2 induced a contraction of endothelial cells (Chew et al., 1998). In triton skinned smooth muscle, PAK3 induced a Ca^{2+}-independent contraction which was associated with phosphorylation of caldesmon and preceded the increase in phosphorylation of r-MLC (van Eyk et al., 1998). Thus, it is conceivable that the inhibition of force observed in the presence of toxin B is due to inhibition of Rac and inactivation of PAK rather than to inhibition of the Rho/ROK pathway. Another goal of this study, therefore, was to further analyze the mechanism of inhibition of force by toxin B.

2. EXPERIMENTAL

2.1. Tissue Preparation

Guinea pigs (Dunkin Harley) or rabbits of either sex were anaesthetised with halothan and sacrificed by exsanguination. The smooth muscle (longitudinal smooth muscle from the ileum from the guinea pig, and carotid artery from the rabbit) were rapidly removed and placed in oxygenated PSS of the following composition (in mM) NaCl 118, KCl 5, Na_2HPO_4 1.2, $MgCl_2$ 1.2, $CaCl_2$ 1.6, HEPES 24, glucose 10, pH 7.4. All experiments were carried out at room temperature (21-24 °C) unless stated otherwise.

2.2. Flash photolysis of diazo2

Intact longitudinal ileal smooth muscle strips (≈0.2 mm wide and 5 mm long) were loaded with 10 μM diazo2/AM (Molecular Probes, Eugene, OR) in PSS containing 1% DMSO and 0.02% Pluronic F-127 (BASF from Molecular Probes) for 2-3 h. Thereafter, the muscle strips were mounted between a fixed hook and the extended arm of an AME force transducer and held in 0.5 ml perspex baths. Before photolysis the muscle was transfered to a 100 μl bath. The muscle was illuminated, when in solution, through a quartz window in the cuvette. Flash photolysis was performed after a control contraction/relaxation cycle, at the second peak of the carbachol-induced contraction using a high-pressure xenon flash lamp system giving light flashes with a duration of <5ms (Arner et al., 1998).

2.3. RhoA translocation

Separation of particulate and cytosolic fraction was performed according to Gong et al., (1997). In brief, ileal smooth muscle was dissected in Ca^{2+} free PSS containing 10^{-4} M atropine to avoid translocation of Rho during preparation. The tissues were then allowed to equilibrate in PSS containing Ca^{2+} for 30 min followed by stimulation with 100 μM carbachol. At the desired time points, the strips were shock frozen, homogenized in ice-cold homogenization buffer (in mM: Tris-HCl 10, pH 7.5, $MgCl_2$ 5, EDTA 2, sucrose 250, dithiothreitol 1, leupeptin 1, aprotinin 20 μg/ml), centrifuged at 100,000 g for 30 min with an air fuge at 4°C (Beckmann) and the supernatant was collected as the cytosolic fraction. Pellets were resuspended and membrane proteins were extracted by incubation for 30 min in homogenization buffer containing 1% Triton X-100 and 1% sodium cholate. The extract was centrifuged at 800 g for 10 min and the supernatant was collected as the detergent soluble particulate fraction. Proteins of the cytosolic and the detergent soluble fraction were separated by SDS PAGE and RhoA was detected by Western blot as described (Pfitzer et al., 2001). In separate experiments we verified that the detergent-insoluble particulate fraction did not contain RhoA immunoreactivity.

2.4. Permeabilized smooth muscle

For determination of the rate of dephosphorylation, the longitudinal ileal smooth muscle was permeabilized with α-toxin (1000 U/ml) in relaxing solution for 30 min after mounting in the myograph (Steusloff et al., 1995). Relaxing solution consisted of (in mM) imidazol 20, EGTA 10, Mg-acetate 10, NaN_3 5, ATP 7.5, creatine phosphate 10, pH 7.0 at room temperature, ionic strength was adjusted to 150 mM with methane sulfonate. The contraction solution contained in addition 10 mM $CaCl_2$. Rigor solution had the same composition as the relaxing solution exept that it did not contain ATP and creatine phosphate and Mg-acetate was reduced to 2.5. mM, pH was 7.0 at 16°C. For heavy permeabilization with Triton X 100, the carotid arteries were incubated in 1 % Triton X-100 for 4 h on ice as described (Pfitzer et al., 2001). The skinned fiber bundles were stored at – 20°C in relaxing solution (see below) containing 50 % (v/v) glycerol and used within one week. Ring segments were mounted in relaxing solution containing: imidazole 20, EGTA 4, $MgCl_2$ 10, ATP 7.5, DTT 2, NaN_3 1, creatine phosphate 10, leupeptin 1 μM, calmodulin 0.5 or 1 μM, creatine kinase 140 U/ml. The pH was adjusted to 6.7 at room temperature with KOH. The contracting solution contained in addition 4 mM $CaCl_2$. Alterations in the free Ca^{2+}-concentration were obtained as in Pfitzer et al., (2001).

2.5. Protein preparations and statistical evaluation

The GST-PAK1 with an activating mutation (T422E) was cloned, expressed and purified as described previously (Manser et al., 1997). Values are shown as means ± S.E.M., n is the number of observations. Difference of responsiveness among groups was analysed by ANOVA followed by the Newman-Keuls test. Student's t test was used when appropriate. A P value < 0.05 was considered to indicate significant differences.

3. RESULTS AND DISCUSSION

3.1. The tonic phase of the contraction depends on elevated intracellular $[Ca^{2+}]$

Activation of the longitudinal smooth muscle from guinea pig ileum by carbachol (100 μM) typically induces a biphasic contraction (fig. 1A) whereby an inital rapid increase in tension is followed by a partial relaxation and a second slower transient increase in force which reaches its maximum about 50 to 60 s after addition of carbachol and amounts to 50-70% of the first maximal force (F_{max}). Force then slowly declines to 12% of F_{max} (Lucius et al., 1998). We have previously shown that the initial rapid phase of the contraction is preceded by an increase in intracellular $[Ca^{2+}]$ which peaks at about 3 s after onset of stimulation and then declines to low but still suprabasal levels (336 ±28 nM as compared to resting $[Ca^{2+}]$ of 182 ±32 nM, n=6; Lucius et al., 1998). In other words, the second peak of the contraction occurs in the absence of a concommitant increase in intracellular $[Ca^{2+}]$ (cf. also Himpens and Somlyo, 1988).

To test whether this phase is Ca^{2+}-dependent we loaded intact longitudinal smooth muscle strips with the membrane permeable acetoxymethyl ester form of the "caged" Ca^{2+} chelator, diazo2. The Ca^{2+}-affinity of the native form of diazo2 is low ($K_d \approx 2.2$ μM), but after an intense UV light flash it is converted to a high-affinity Ca^{2+}-binding form ($K_d \approx 0.07$ μM) and can therefore be used to rapidly lower the free $[Ca^{2+}]$ (Adams et al., 1989). Loading the strips with diazo2-AM had no effect on the force transient, i.e. the times to develop maximal force (peak 1) and the delayed contraction (peak 2) and the half time of relaxation from peak 2 were similar to the corresponding values during contractions in nontreated muscles. Illumination of the muscles at peak 2 of the contraction, however, induced a rapid nearly complete relaxation ($\approx$80-90% of peak 1) with a half time of 4.9 ± 0.3 s (fig. 1b). Relaxation was biphasic (fig. 1c), a quasi linear initital phase with a duration of 3.6 ±0.4 s preceded an exponential relaxation with a rate constant of 0.234 ±0.016 s^{-1} (n=6). In nontreated muscles, illumination did not affect force when performed at peak 2 showing that the light pulse itself did not influence contraction or relaxation processes.

The events determining the rate of relaxation are complex in smooth muscle reflected in the biphasic nature of the time course of relaxtion. Such a biphasic relaxation has been reported in permeabilized rabbit femoral artery, a tonic smooth muscle (Khromov et al., 1995). To the best of our knowledge this is the first report of a biphasic

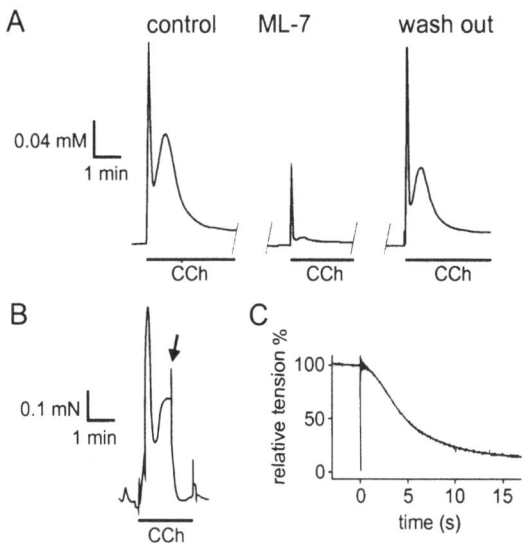

Figure 1. Pannel A: typical force tracing of a carbachol-induced contraction in ileal longitudinal smooth muscle from the guinea pig (control), in the presence of the MLCK inhibitor, ML-7 (1 μM); pannel B: relaxation induced by flash photolysis of ileal smooth muscle at arrow which has been loaded with the membrane permeant form of diazo2 ; pannel C : expanded time course of flash induced the relaxation. Note the biphasic nature of the relaxation (results are representative of 6 individual experiments).

relaxation in a phasic smooth muscle. The fall in free [Ca^{2+}] is rapid ($k \geq 2000$ s^{-1}, Adams et al., 1989) but the dissociation of calmodulin from MLCK (3.5 s^{-1}, Kasturi et al., 1993) and dephosphorylation of r-MLC may contribute to the linear phase. Yet, we cannot say with certainty that the Ca^{2+}-dependency is solely due to activation of MLCK as it has also been suggested that Ca^{2+} is required for activation of Rho (c.f.Takuwa, this volume). Still, activation of MLCK is required because the second phase of the carbachol induced contraction is inhibited by the MLCK inhibitor, ML-7 (Fig. 1a). Thus, the time course of relaxation appears to reflect complex processes, involving inactivation of MLCK and desinhibition of MLCP, dephosphorylation of r-MLC and continuously cycling crossbridges, that remain phosphorylated after photolysis of diazo-2. The second phase could be well fitted by a single exponential rate constant being about 3 fold faster than in the permeabilized phasic bladder smooth muscle (Khromov et al. 1995) which may reflect the properties of intact smooth muscle.

It is of interest to note that relaxation was not complete. This may be due to the low Ca^{2+} buffering capacity of diazo2. Alternatively it may reflect the involvement of Ca^{2+} independent myosin light chain kinases (Ganitkevich et al., 2002 for review). In this context it is also interesting that toxin B does not induce complete inhibition of force (Lucius et al., 1998) suggesting that the kinase is not ROK. In any case, the contribution of a Ca^{2+}-independent contraction to force in intact ileal smooth muscle appears to be minor.

3.2. Carbachol induces translocation of RhoA to the particulate fraction

The second peak of the contraction is inhibited by toxin B, which inactivates monomeric GTPases of the Rho family (Otto et al., 1996) and the relatively specific inhibitor of Rho-kinase, Y27632 (Sward et al., 2000) suggesting that activation of this intracellular signalling cascade is also necessary for the delayed phase of the contraction. In line with this we find a redistribution of RhoA immunoreactivity from the cytosolic to the particulate, i.e. membrane fraction reaching a plateau 1 min after addition of carbachol (fig. 2). Translocation of Rho from the cytosolic pool to the particulate fraction has been taken as a measure of activation of Rho (Gong et al., 1997). Interestingly, the RhoA remained in the particulate fraction even when tension had decreased to its final low steady state value. This raises the question as to the mechanism of relaxation in the continued presence of the agonist. Inactivation of MLCK appears to be unlikely because the concentration of free $[Ca^{2+}]$ does not decrease to resting values (Lucius et al., 1998). On the other hand, RhoA may be transformed into a membrane bound, inactive form and/or its effectors may be downregulated as has been suggested by Gong et al. (1997).

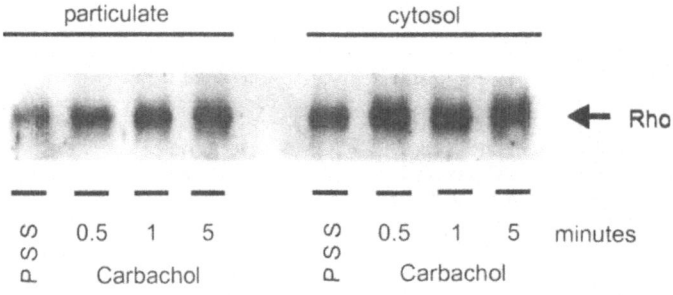

Figure 2. Time course of translocation of RhoA during carbachol-induced contraction. Note, that after 5 min of stimulation with carbachol, force had declined to about 12% of F_{max}.

3.3. Inhibition of force by toxin B is associated with desinhibition of MLCP

To determine, whether incubation of ileal smooth muscle with toxin B leads to an increased activity of MLCP which is to be expected if Rho inhibits MLCP activity we examined the rate of relaxation and dephosphorylation under conditions were MLCK activity is fully inhibited. For this we first incubated the ileal muscle strips with toxin B and then permeabilized them with α-toxin. In such treated preparations neither carbachol nor GTPγS induced a Ca^{2+}-sensitization as described before (Otto et al., 1996). Following a maximal activation of these preparations (pCa 4.35), MLCK activity was rapidly inactivated by transfering the muscle strips into a rigor solution (pCa >8) in the presence of the MLCK inhibitor, ML-9 (100 μM, fig. 3). After a lag period of 30 s, force declined in the toxin B treated strips significantly faster than in the control strips ($t_{1/2}$: 0.97 ±0.05 min (control), 0.062 ±0.05 min (toxin B). In the control strips steady state phosphorylation at pCa 4.35 was significantly higher (43 ±5%, n=4) than in the toxin B treated strips (14 ±2 %, n=4). These values were not different from those determined 15 s after transfer into the rigor solution while after 30 s, phosphorylation in the toxin B treated strips had declined by 60% compared to only 40% in the control strips. The respective rate constants of dephosphorylation were 0.078 ± 0.02 s^{-1} in toxin B treated and 0.036 ±0.001 in control strips. Thus, incubation of ileal smooth muscle with toxin B, i.e. inactivation of Rho leads to a desinhibition of MLCP.

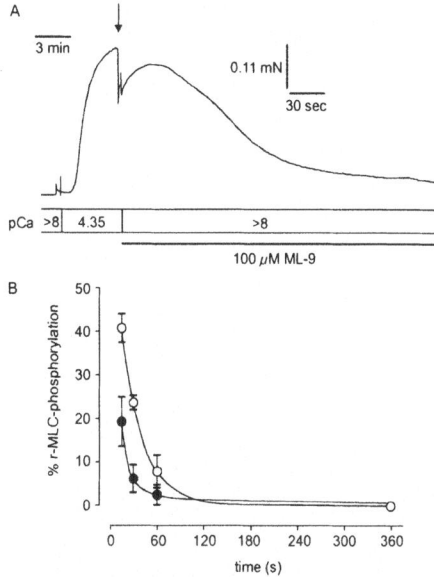

Figure 3. A: Experimental protocol to determine the time course of r-MLC dephosphorylation in α-toxin permeabilized ileal smooth muscle that had been treated with toxin B or dialysis buffer before permeabilization. Following maximal activation (pCa 4.35) the fibers were transfered into Ca^{2+}-free rigor solution containing the MLCK inhibitor, ML-9 (100 μM) and shock frozen at the indicated times after transfer to the rigor solution. Arrow : change in chart recorder speed. B: Time course of dephosphorylation of r-MLC. Opend symbols, control fibers, closed symbols, toxin B treated fibers (n=3-5). The rate of dephosphorylation could be well fitted by a single exponential with a rate constant of 0.078 ±0.02 s^{-1} for toxin B treated fibers and 0.036 ±0.001 s^{-1} for control fibers.

It was suggested that Rho activates ROK which in turn phosphorylates MYPT1 thereby inhibiting MLCP (Kimura et al., 1996). In line with this, the non-specific ROK inhibitor, HA1077 also increased the rate of dephosphorylation (Sward et al., 2000). This hypothesis was recently questioned because stimulation of smooth muscle with agonists does not increase phosphate incorporation into the ROK phosphorylation sites of MYPT1 (Kitazawa et al., 2002). Furthermore, thiophosphorylation induced inhibition of MPYT1 was not inhibited by inhibitors of the Rho/ROK pathway (Pfitzer et al., 2001) supporting the possibility that ROK does not directly phosphorylate MYPT1 (MacDonald et al., 2001).

The inhibitory effect of toxin B on force could also be due to inhibition of Rac and or cdc42 and its effector PAK. We therefore, determined whether a recombinant PAK1 (PAK1*) would lead to an activation of contraction. For these experiments, we used triton skinned arterial rather than ileal smooth muscle, because it was shown that certain agonists activate Rac and its downstream effector, PAK1 in cultured arterial smooth muscle cells (Schmitz et al., 1998). Following a control submaximal (pCa 6.01) and maximal activation (pCa 4.35), the preparations were incubated with PAK1* or buffer and 10 nM okadaic acid for 60 min. Control experiments showed that okadaic acid at this low concentration, which inhibits only type 2A phosphatases (Takai et al., 1989) does not increase r-MLC phosphorylation and has no effect on force. In contrast to a recent report (van Eyk et al., 1998), PAK1* did not induce a Ca^{2+}-independent contraction or r-MLC phosphorylation (data not shown). Rather, the subsequent submaximal contraction elicited at pCa 6.01 was significantly inhibited compared to the buffer control (fig. 4). This was associated with an inhibition of r-MLC phosphorylation. The extent of inhibition is, however, smaller than in the ileal smooth muscle where we have previously shown that PAK1* phosphorylates and inactivates MLCK (Wirth et al., 2002). It remains to be determined whether this is due to simultaneously inhibiting myosin light chain phosphatase (Takizawa et al., 2002).

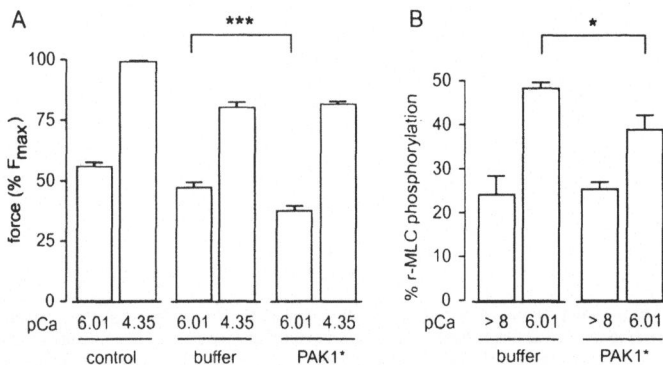

Figure 4. Inhibition of submaximal Ca^{2+} (pCa 6.01) activated force (A) and r-MLC phosphorylation (B) by recombinant PAK1 in triton skinned rabbit carotid arteries. Force was normalized to the maximal activation (pCa 4.35, 1 μM calmodulin) before incubation with PAK1. Values are mean ±SEM for n= 9-10 (force measurements) and n=7-13 (phosphorylation) determinations. * and *** denote a P-value of <0.05 and <0.001 respectively.

4. CONCLUSIONS

The major finding of this study is that the quasi tonic phase of the carbachol-induced contraction in longitudinal guinea pig smooth muscle depends both on Ca^{2+} and the activation of the monomeric GTPase Rho. In contrast, the effector PAK1 of the monomeric GTPase Rac leads to an inhibition of force which is opposite to what would be expected if the effect of toxin B were due to inhibition of Rac (fig. 4). Rather, our results suggest that activation of Rho and Rac may have opposing actions on smooth muscle contraction similar to what has been previously shown to be the case in non muscle cells. Further studies are required to test whether Ca^{2+} has a dual action, namely activating MLCK as well as Rho and the precise mechanism of inhibition of MLCP by the Rho/ROK pathway.

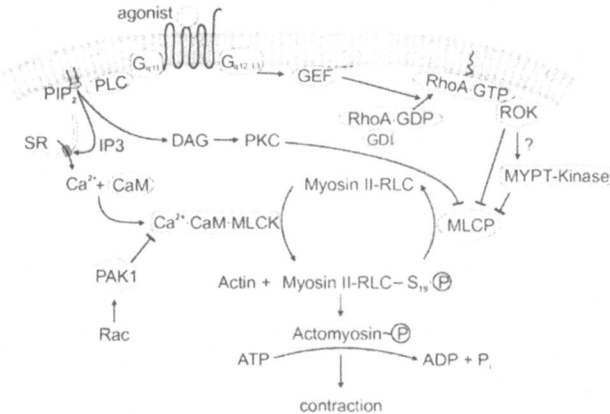

Figure 5. Proposed mechanism of regulation of smooth muscle contraction via Ca^{2+}-dependent activation of MLCK, inhibition of MLCP via the Rho/ROK pathway and inhibition of MLCK via the Rac/PAK pathway. Contractile agonists act via receptors that are coupled to heterotrimeric G proteins. $G_{\alpha q 11}$ activates phospholipase C (PLC) which hydrolyses phosphatidylinositol 4,5 bisphosphate (PIP_2) thereby generating inositol 1,4,5-trisphosphate (IP_3) which releases Ca^{2+} from the sarcoplasmatic reticulum. Ca^{2+} then binds to calmodulin (CaM) and activates myosin light chain kinase (MLCK) which phosphorylates the regulatory light chains of myosin II and initiates crossbridge cycling. $G_{q12/13}$ activates the monomeric GTPase, Rho, by activating a guanosine exchange protein (GEF) which triggers the exchange of GDP by GTP. The inactive, GDP bound form of RhoA is bound to a guanosine exchange inhibitor (GDI) and located in the cytosol. In the active, GTP-bound form, RhoA is bound to the membrane by its geranylgeranyl moiety. RhoA-GTP activates ROK which either directly or indirectly (e.g. via activation of MYPT-kinase identified as ZIP-like kinase, MacDonald et al., 2001) phosphorylates the regulatory subunit of MLCP, MYPT1. Activation of the monomeric GTPase Rac in turn phosphorylates and inhibits MCLK. Not shown are PKC-dependent mechanisms of inhibition of MLCP involving phosphorylation of the inhibitory peptide,CPI 17 (c.f. Introduction).

5. ACKNOWLEDGMENTS

Toxin B was generously provided by K. Aktories, Freiburg, Germany. The excellent techical assistance of D. Metzler, and R. Kemkes is gratefully acknowledged. The authors thank Dr. P. Scherer, University of Cologne for many stimulating discussions. This work was supported by grants from the DFG, DAAD and the Medical Faculty of the University of Cologne (Koeln Fortune).

6. REFERENCES

Adams, S.R., Kao, J.P.Y., and Tsien, R.Y., 1989. Biologically useful chelators that take up Ca^{2+} upon illumination, *J. Am. Chem. Soc.* **111**:7957.

Arner, A., Malmqvist, U., and Rigler, R., 1998. Calcium transients and the effect of a photolytically released calcium chelator during electrically induced contractions in rabbit rectococcygeus smooth muscle, *Biophys. J.* **75**:1895.

Chew, T.L., Masaracchia, R.A., Goeckeler, Z.M., and Wysolmerski, R.B., 1998, Phosphorylation of non-muscle myosin II regulatory light chain by p21-activated kinase (gamma-PAK).*J. Muscle Res. Cell Motil.* **19**:839.

Deng, J.T., Sutherland, C., Brautigan, D.L., Eto, M. and Walsh, M.P., 2002, Phosphorylation of the myosin phosphatase inhibitors, CPI-17 and PHI-1, by integrin-linked kinase, *Biochem. J.* **367**:517:

Gallagher, P.J., Herring, B.P., and Stull, J.T., 1997. Myosin light chain kinases, *J. Muscle Res. Cell Motil.* **18**:1.

Ganitkevich, V., Hasse, V., and Pfitzer, G., 2002, Ca^{2+}-dependent and Ca^{2+}-independent regulation of smooth muscle contraction.*J. Muscle Res. Cell Motil.* **23**:47.

Gong, M.C., Fujihara, H., Somlyo, A.V., and Somlyo, A.P., Translocation of rhoA associated with Ca2+ sensitization of smooth muscle, *J. Biol. Chem.***272** :10704

Hartshorne, D.J., Ito, M., and Erdödi, F. 1998. Myosin light chain phosphatase: subunit composition, interactions and regulation, *J. Muscle Res. Cell Motil.* **19**:325

Himpens, B., and Somlyo, A.P. 1988. Free-calcium and force transients during depolarization and pharmacomechanical coupling in guinea-pig smooth muscle, *J. Physiol.* **395**:507.

Just, I., Selzer, J., Wilm. M., von Eichel-Streiber, C., Mann, M., and Aktories, K., 1995. Glucosylation of Rho proteins by Clostridium difficile toxin B, *Nature* **375**:500.

Kasturi, R., Vasulka, C., and Johnson, J.D., 1993, Ca2+, caldesmon, and myosin light chain kinase exchange with calmodulin, *J. Biol. Chem.* **268**:7958.

Khromov, A., Somlyo, A.V., Trentham, D.R., Zimmermann, B., and Somlyo, A.P., 1995. The role of MgADP in force maintenance by dephosphorylated cross-bridges in smooth muscle: a flash photolysis study, *Biophys. J.* **69**:2611.

Kimura, K., Ito, M., Amano, M., Chihara, K., Fukata, Y., Nakafuku, M., Yamamori, B., Feng, J., Nakano, T., Okawa, K., Iwamatsu, A., and Kaibuchi, K., 1996, Regulation of myosin phosphatase by Rho and Rho-associated kinase (Rho-kinase), *Science* **273**:245.

Kitazawa, T., Khalequzzaman, M.D., Woodsome, T.P., and Eto, M., 2002, Evaluation of signaling pathways for Ca^{2+}-sensitization in smooth muscle, *Biophys. J.* **82**:421a.

Kitazawa, T., Takizawa, N., Ikebe, M., and Eto, M. 1999. Reconstitution of protein kinase C-induced contractile Ca2+ sensitization in triton X-100-demembranated rabbit arterial smooth muscle, *J. Physiol.* **520**:139.

Koyama, M., Ito, M., Feng, J., Seko, T., Shiraki, K., Takase K., Hartshorne, D.J., and Nakano, T., 2000. Phosphorylation of CPI-17, an inhibitory phosphoprotein of smooth muscle myosin phosphatase, by Rho-kinas, *FEBS Lett.* **475**:197.

Kureishi Y., Ito, M., Feng, J., Okinaka, T., Isaka, N., and Nakano, T. 1999, Regulation of Ca2+-independent smooth muscle contraction by alternative staurosporine-sensitive kinase, *Eur. J. Pharmacol.* **376**:315

Lucius, C., Arner, A., Steusloff, A., Troschka, M., Hofmann, F., Aktories, K., and Pfitzer, G. 1998. Clostridium difficile toxin B inhibits carbachol-induced force and myosin light chain phosphorylation in guinea-pig smooth muscle: role of Rho proteins, *J. Physiol.* **506**: 83.

MacDonald, J.A., Borman, M.A., Muranyi, A., Somlyo, A.V., Hartshorne, D.J., and Haystead, T.A., 2001, Identification of the endogenous smooth muscle myosin phosphatase-associated kinase, *Proc. Natl. Acad. Sci. USA* **98**:2419.

Manser, E., Leung, T., Salihuddin, H., Zhao, Z., and Lim, L., 1994, A brain serine/threonine protein kinase activated by Cdc42 and Rac1, *Nature* **367**:40.

Manser, E., Huang, H.Y., Loo, T.H., Chen, X.Q., Dong, J.M., Leung, T., Lim, L.,1997, Expression of constitutively active alpha-PAK reveals effects of the kinase on actin and focal complexes, *Mol. Cell Biol.* **17**:1129.

Matsui, T., Amano, M., Yamamoto, T., Chihara, K., Nakafuku, M., Ito, M., Nakano, T., Okawa, K., Iwamatsu, A., and Kaibuchi, K., 1996. Rho-associated kinase, a novel serine/threonine kinase, as a putative target for small GTP binding protein Rho, *EMBO J.* **15**:2208.

Otto, B., Steusloff, A., Just, I., Aktories, K., and Pfitzer, G., 1996, Role of Rho proteins in carbachol-induced contractions in intact and permeabilized guinea-pig intestinal smooth muscle, *J. Physiol.* **496**:317.

Pfitzer G., 2001, Invited review: regulation of myosin phosphorylation in smooth muscle, *J. Appl. Physiol.* **91**:497.

Pfitzer, G., Sonntag-Bensch, D., and Brkic-Koric, D., 2001, Thiophosphorylation-induced Ca(2+) sensitization of guinea-pig ileum contractility is not mediated by Rho-associated kinase, *J. Physiol.* **533**:651.

Sanders, L.C., Matsumura, F., Bokoch, G.M. and de Lanerolle, P. 1999, Inhibition of myosin light chain kinase by p21-activated kinase, *Science,* **283**:2083.

Schmitz, U., Ishida, T., Ishida, M., Surapisitchat, J., Hasham, M.I., Pelech, S., and Berk, B.C. 1998, Angiotensin II stimulates p21-activated kinase in vascular smooth muscle cells: role in activation of JNK, *Circ. Res.* **82**:1272.

Somlyo, A.P., and Somlyo, A.V. 2000. Signal transduction by G-proteins, rho-kinase and protein phosphatase to smooth muscle and non-muscle myosin II, *J. Physiol.* **522**:177.

Steusloff, A., Paul, E., Semenchuk, L .A., Di Salvo, J., and Pfitzer, G., 1995, Modulation of Ca2+ sensitivity in smooth muscle by genistein and protein tyrosine phosphorylation, *Arch. Biochem. Biophys.* **320** : 236.

Swärd, K., Dreja, K., Susnjar, M., Hellstrand, P., Hartshorne, D.J., and Walsh, M.P., 2000, Inhibition of Rho-associated kinase blocks agonist-induced Ca2+ sensitization of myosin phosphorylation and force in guinea-pig ileum, *J. Physiol.* **522**:33.

Takai, A., Troschka, M., Mieskes, G., and Somlyo, A.V., 1989, Protein phosphatase composition in the smooth muscle of guinea-pig ileum studied with okadaic acid and inhibitor 2, *Biochem. J.* **262**:617.

Takizawa, N., Koga, Y., and Ikebe, M., 2002, Phosphorylation of CPI17 and myosin binding subunit of type 1 protein phosphatase by p21-activated kinase, *Biochem. Biophys. Res .Commun.* **297**:773

Van Eyk, J.E., Arrell, D.K., Foster, D.B., Strauss, J.D., Heinonen, T.Y.K., Furmaniak- Kazmierczak, E., Côté, G.P., and Mak, A.S., 1998, Different molecular mechanisms for Rho family GTPase-dependent, Ca²⁺-independent contraction of smooth muscle, *J. Biol. Chem.* **273**:23433.

Wirth, A., Schroeter, M., Manser, E., der Lanerolle, P. and Pfitzer, G., 2002 Inhibition of smooth muscle myosin light chain kinase (MLCK) activity and contraction by P21 activated protein kinase1 (PAK1), *Pflügers Arch.* **443**:193

DISCUSSION

Takuwa: Did you examine how Rac is regulated by receptor agonists?

Pfizer: No, we have not yet studied whether Rac is activated in our system. We are in the process of doing so. However, in cultured smooth muscle cells, it was shown that certain agonists can activate Rac.

II. MOLECULAR MECHANISM OF ACTIN-MYOSIN INTERACTION

BIO-NANOMUSCLE PROJECT: CONTRACTILE PROPERTIES OF SINGLE ACTIN FILAMENTS IN AN A-BAND MOTILITY ASSAY SYSTEM

Madoka Suzuki, Hideaki Fujita, and Shin'ichi Ishiwata[*]

1. ABSTRACT

We have developed a new microscopic technique to measure the force generated on a single actin filament (FA) in the A-band in which the intact lattice structure composed of myosin thick filaments is maintained; we call this newly developed system "*Bio-nanomuscle* (or an A-band motility assay system)". The A-bands were prepared by selective removal of thin filaments from rabbit skeletal glycerinated myofibrils under optical microscope with the use of gelsolin (a severing and barbed (B)-end capping protein of FA) that was prepared from bovine serum. A polystyrene bead of 1 μm in diameter attached to the B-end of FA (through a gelsolin molecule attached to the surface of the bead) was trapped and manipulated with optical tweezers. The displacement of the bead up to 200 nm (corresponding to the force of ~ 40 pN) was determined by phase-contrast image analysis. At the initial stage of this study, the overlapping length of an FA with the A-band was determined from the fluorescence image of FA labeled with rhodamine-phalloidin (Rh-Ph) and the phase-contrast image, but we later improved the method of determination by moving the sample stage stepwise using the piezo actuator. The average force per overlap was subsequently estimated and the histogram was fitted with two Gaussian distributions. Each peak is supposed to correspond to the force developed by FA interacting outside or inside the A-band, and the peak value of the latter was estimated to be 140 pN/μm. From this value, the average force developed per each cross-bridge (CB; a two-headed myosin molecule) was determined to be 1.3 pN.

[*] Department of Physics, School of Science and Engineering, Waseda University,
3-4-1 Okubo, Shinjuku-ku, Tokyo 169-8555, Japan
ishiwata@mn.waseda.ac.jp

2. INTRODUCTION

The contractile system of muscle is composed of two kinds of myofilaments: a thick filament mainly composed of myosin and a thin filament composed of actin and associated regulatory proteins, including tropomyosin and troponin. These two kinds of filaments are hexagonally packed, and slide past one another when muscle contracts[1, 2]. Muscle fibers[3, 4] or myofibrils[5-7] have been an appropriate preparation for studying the properties of the interaction between FA and myosin, and they have shown many insights into the mechanism and physiological roles of both cardiac and skeletal muscle contraction. Over the last decade, an *in vitro* motility assay has been widely used for studying the sliding movement and force generation on a single FA to explain the molecular mechanism of muscle contraction. However, these studies have been performed only at low ionic strength (usually lower than 50 mM) because myosin molecules easily detach from FA at higher ionic strength (even at a physiological one). Furthermore, myosin molecules adhere to the glass surface two-dimensionally in random fashion, and some of the molecules are denatured.

Our aim is to establish a new experimental system that bridges the gap between the muscle contractile system and the *in vitro* motility assay system. We expect that the bio-nanomuscle (A-band motility assay) system reported here may reveal new aspects of the mechanism of muscle contraction and its regulation.

3. MATERIALS AND METHODS

3.1. Myofibrils and Proteins

Myofibrils were prepared by homogenizing rabbit psoas muscle fibers glycerinated in 50% (v/v) glycerol containing 0.5 mM $NaHCO_3$, 5 mM EGTA, and 1 mM leupeptin for more than 3 weeks at $-20\ °C$ as described previously[8]. Glycerol in the suspension of myofibrils was removed by centrifugation at $3000 \times g$ in a rigor solution (60 mM KCl, 5 mM $MgCl_2$, 10 mM 3-(N-morpholino)propanesulfonic acid (MOPS) (pH 7.0) and 1 mM EGTA) at $4\ °C$.

Actin was prepared from rabbit skeletal white muscle according to a standard procedure[9]. Bovine plasma gelsolin was prepared according to the method of Kurokawa et al.[10]; during this preparation, the incubation time of gelsolin in Ca^{2+}-free buffer was kept as short as possible as the severing activity was lowered by incubation in the Ca^{2+}-free buffer.

3.2. Preparation of an A-band Motility Assay System from the Myofibrils on the Coverslip

About 100 μl of the suspension of myofibrils in the rigor solution was placed on one side of the coverslip and the rigor solution was exchanged with a gelsolin solution A (60 mM KCl, 4 mM $MgCl_2$, 20 mM MOPS (pH7.0), 2 mM EGTA, 1.9 mM $CaCl_2$, 1.5 mM NaN_3, 2 mM leupeptin and 0.3 mg/ml gelsolin) with a pipette, taking care not to suck up myofibrils (gelsolin treatment A). After a 30-min gelsolin treatment, the gelsolin soluion A was exchanged to gelsolin solution B (gelsolin solution A plus 20 mM BDM and 1 mM ATP) with a pipette (gelsolin treatment B). After a 20-min gelsolin treatment B, the myofibrils were washed first with relaxing solution (120 mM KCl, 4 mM MgCl$_2$, 20 mM MOPS (pH7.0), 4 mM EGTA, 4 mM ATP and 10 mM dithiothreitol (DTT)), then with the rigor solution containing 10mM DTT. The suspension of myofibrils after gelsolin treatment A and B was then moved to the opposite side of the coverslip using a pipette, where gelsolin was not present on the glass surface. Myofibrils without thin filaments were treated with 2 volumes of the rigor solution containing 10 mM DTT and 0.5% (v/v) TritonX-100 and washed several times with the rigor solution containing 10 mM DTT to remove TritonX-100. All procedures were carried out at 0 °C.

After the flow cell was washed with an assay buffer (AB) (100 mM KCl, 2 mM Mg^{2+}, 2 mM MgATP, 25 mM Imidazole-HCl (pH7.4), 1 mM EGTA, 0.5 mg/ml bovine serum albumin, 10 mM DTT, 4.5 mg/ml D(+)-Glucose, 50 units/ml Glucose Oxidase, 50 units/ml Catalase, 15 mM Creatine Phosphate and 150 units/ml Creatine Phosphokinase), bead-tailed FAs in AB-buffer were applied and both edges of the flow cell were sealed with non-fluorescent nail polish. Bead-tailed FAs were prepared as previously reported[11, 12]. A-bands shorter than 1.6 µm in length or that had ends which looked dark in a phase-contrast image were not used for our assays as they were considered to have contracted or partly dissociated during gelsolin treatment. All the experiments were done at 27-29 °C.

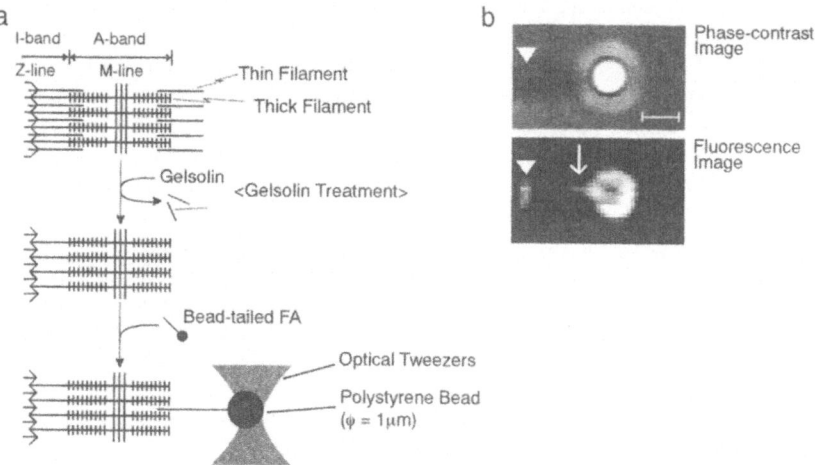

Figure 1. a, Schematic diagram illustrating the experimental procedure for preparing the A-band motility assay system in a myofibril. b, Phase-contrast image and fluorescence image during tension generation. Both images were integrated for one second (30 frames) while FA (indicated by an arrow) was interacting with the A-band. Arrowheads indicate the Z-line. Scale bar, 1 µm.

4. RESULTS AND DISCUSSION

Figure 1 illustrates the experimental procedure for preparing A-band motility assay system from skeletal myofibrils. Myofibrils were treated with gelsolin, and the thin filaments were selectively removed from the filament lattice. Since myofibrils are usually cut at the I-band by homogenization, some of them had no Z-lines at their ends, although the others did. The Z-lines in myofibrils were observed in phase-contrast microscope and even in fluorescence microscope after the bead-tailed FAs were applied because the Z-lines remaining were easily labeled with free Rh-Ph, and thus were rendered fluorescent. When the bead-tailed FA was attached to the Z-line remaining at the end of the A-band, the FA was stuck to the Z-line strongly without generating tension, so that it could not be detached by pulling with optical tweezers. There were some myofibrils that did not have the Z-lines at their ends but instead had dark ends of high density that were distinguishable in a phase-contrast image. It was difficult for FA to attach to the ends of these A-bands. We consider this dense edge is attributable to connectin/titin not being removed successfully by homogenization. Some of the A-bands were shorter than 1.6 μm and the both ends looked dense under the phase-contrast microscope. Also, some of the A-bands were spreading thin. All such "unfavorable" myofibrils were not used for the following motility assay.

The bead-tailed FA was kept near the A-band until the FA began to interact with it. Before the FA began to interact with the A-band, the bead stayed at the trap center and showed rotational Brownian motion. As soon as the FA was attached to the edge of the A-band, tension was generated continuously some of the time, and transiently at other times. At the beginning of this study, the length of an overlap of FA with the A-band was determined by comparing the fluorescence image of FA with phase-contrast images of the bead and the edge of the A-band. As it was often difficult, however, to determine the position of the edge of the A-band (even when the edge can be identified, the resolution is low), we altered the method as follows: after the steady tension was obtained, that is, after the tension generated by CBs in the A-band and the load imposed by the optical tweezers were balanced, the sample stage was moved stepwise by moving a piezo actuator. The tension and the load were balanced again at a new position of higher or lower level of tension depending on the direction of the movement of the sample stage. The degree of the change in overlap was calculated from the displacement of the sample stage and that of the bead due to the movement of the stage (Fig. 2). The difference in tension before and after the movement of the stage was calculated from the displacement of the bead and the stiffness of the optical tweezers (0.12 pN/nm in Fig. 2). Tension per unit length overlap (pN/μm) thus estimated from the microscopic image analysis and from the displacement of the stage was summarized as a histogram in Fig. 3.

The histogram in Fig. 3 was fitted by two Gaussian distributions with peaks at 74 pN/μm and 140 pN/μm. We assume that these two values correspond to two situations for FA, that is, FA interacting outside or inside the A-band, respectively. It is thought that the FA interacting outside the A-band could interact with a smaller number of thick filaments than those inside the A-band because the former is thought to interact with the thick filaments two dimensionally arranged, whereas the latter is considered to be located at the trigonal position of the native thick filament lattice three dimensionally arranged. Thus, higher tension is expected to be generated inside the A-band than outside.

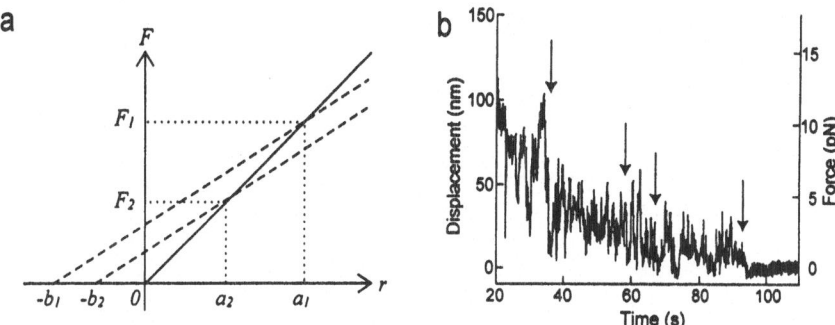

Figure 2. Method for estimating the force generated per unit length of overlap by stepwise moving the sample stage. a, Relationship between the load applied by the optical tweezers (solid line) and the force generated by the A-band (broken line) vs. displacement from the trap center (at the origin of the abscissa), r. At first, the two forces are balanced at $F = F_1 = k \cdot a_1$, where k is the stiffness of the optical tweezers and a_1 the average position of the bead. When the sample stage, or the A-band, is moved away from the trap center by $\Delta b = -b_2 - (-b_1) = b_1 - b_2$, the bead position must be shifted from a_1 to a_2 and the generated tension is decreased from F_1 to $F_2 = k \cdot a_2$. Therefore, the change of the overlap between the thick filaments and the FA, Δr, is $(a_1 - a_2) + \Delta b$. Thus, the tension generated per unit length of overlap can be calculated as $\Delta F/\Delta r = (F_1 - F_2) / \Delta r = k (a_1 - a_2) / \Delta r$. All the variables could be determined from the experiments. b, Time course of the displacement of the trapped bead interacting with the A-band. The stage was moved away from the trap center four times at the instances indicated by arrows. When FA was pulled out from the A-band, the bead went back to the trap center at the time shown by the fourth arrow.

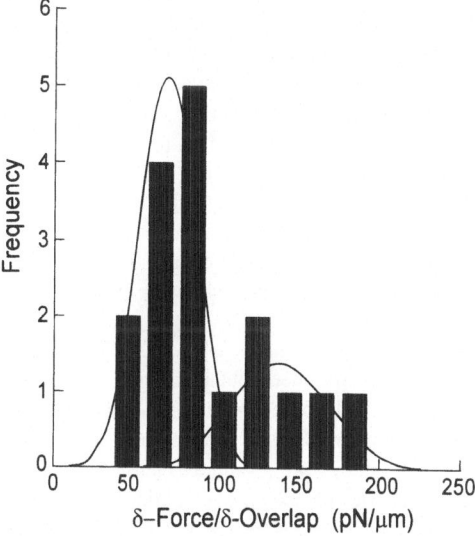

Figure 3. Histogram showing the tension generated on single FA per overlap (normalized to 1 μm) between FA and the A-band. The histogram was fitted by two Gaussian distributions, for which peak values were determined to be 74 and 140 pN/μm, respectively.

Next, we calculated the number of CBs (= myosin molecules) that can interact with a single FA inside the A-band. Since each myosin thick filament (1.6 μm long) consists of 300 myosin molecules, there are 150 CBs in half a sarcomere. As a single FA is surrounded by three thick filaments being located at the corner of the triangular lattice, one sixth (60°/360°) of the three thick filaments have a chance to interact with each FA. Consequently, we estimate that 150×3/6=75 CBs are available to each FA per half a sarcomere, in which the effective length of half an A-band is estimated to be 0.7 μm after the removal of the central bare zone of 0.2 μm. According to this estimation and the average value of the second peak in Fig. 3, we estimate that the average force over the ATPase cycle generated by each CB inside the A-band is 1.3 pN. If we remember that the number of CBs estimated by the way described above is the maximum number of CBs that can interact with each FA, it is safe to say that single CBs can generate a force of more than 1.3 pN.

In our motility assay system, the force is measured on the plane almost parallel to the glass surface, which is different from our previous *in vitro* motility assay system where the bead-tailed FA is positioned upward from the surface by optical tweezers[12-14]. The present system can also control the orientation of myosin molecules. They are arranged parallel to the long axis of a myofibril, so that this is considered to be the most suitable for the generation of force[15]. Ishijima et al.[16] estimated the force generated by myosin molecules oriented in single synthetic myosin rod cofilaments by the noise analysis of the generated force to be 2.1 ± 0.4 pN. The difference between these two values could be explained by the difference in the solvent conditions, especially different ionic strength (higher ionic strength in our experiments). The interaction between myosin and FA is generally weakened when the ionic strength increases. The *in vitro* motility assay has been limited to low ionic strength like 25 mM KCl because myosin molecules easily detach from FA at higher ionic strength unless FAs are suppressed by viscous polymer networks like methylcellulose added to the assay buffer. In our system, on the other hand, a FA is surrounded by the thick filaments and keeps the interaction even under higher ionic strength, e.g., higher than 100 mM KCl. It is also interesting that some FAs with large overlap kept interacting and generating force continuously even outside the A-band under the same conditions, whereas a large part of FAs only generated force transiently. This indicates that the number of myosin molecules interacting could be enough to generate weak but steady force even at 100 mM KCl. The array of thick filaments in our system that cannot be achieved in an *in vitro* assay made this phenomenon possible.

The value of the average force generated by each CB obtained in the present study was a little smaller than what has been obtained from the muscle contractile system. This may be due to us using only pure FAs and not reconstituted thin filaments. The association of regulatory proteins such as tropomyosin and troponin may elevate the tension[17]. In the present study, we examined tension generation under auxotonic condition instead of isometric or isotonic condition. The next step of our bio-nanomuscle project would be to measure tension generated under such physiological conditions. It will also be very interesting to examine the effects of the regulatory proteins under the well defined conditions. Using this approach, we aim to bridge the gap between the experimental systems and conditions mentioned above, and elucidate the mechanism of molecular motors in a protein assembly.

5. ACKNOWLEDGMENTS

We are grateful to Mark Chee of Duke University for his critical reading of the manuscript. This work was partly supported by Grants-in-Aid for Specially Promoted Research and for the Bio-venture Project from the Ministry of Education, Sports, Culture, Science and Technology of Japan.

6. REFERENCES

1. Huxley, A.F. & Niedergerke, R. *Nature* **173**, 971-973 (1954)
2. Huxley, H.E. & Hanson, J. *Nature* **173**, 973-976 (1954)
3. Goldman, Y.E. & Brenner, B. *Annu. Rev. Physiol.* **49**, 629-636 (1987)
4. Cooke, R. *CRC Rev. Biochem.* **21**, 53-118 (1986)
5. Anazawa T., Yasuda, K. & Ishiwata, S. *Biophys. J.* **61**, 1099-1108 (1992)
6. Yasuda, K., Shindo, Y. & Ishiwata, S. *Biophys. J.* **70**, 1823-1829 (1996)
7. Friedman, A.L. & Goldman, Y.E. *Biophys. J.* **71**, 2774-2785 (1996)
8. Ishiwata, S. & Funatsu, T. *J. Cell Biol.* **100**, 282-291 (1985)
9. Kondo, H. & Ishiwata, S. *J. Biochem.* **79**, 159-171 (1976)
10. Kurokawa, H., Fujii, W., Ohmi, K., Sakurai, T. & Nonomura, Y. *Biochem. Biophys. Res. Commun.* **168**, 451-457 (1990)
11. Suzuki, N., Miyata, H., Ishiwata, S. & Kinosita, K. Jr. *Biophys. J.* **70**, 401-408 (1996)
12. Nishizaka, T., Seo, R., Tadakuma, H., Kinosita, K. Jr. & Ishiwata, S. *Biophys. J.* **79**, 962-974 (2000)
13. Miyata, H., Hakozaki, H., Yoshikawa, H., Suzuki, N., Kinosita, K. Jr., Nishizaka, T. & Ishiwata, S. *J. Biochem.* **115**, 644-647 (1994)
14. Nishizaka, T., Miyata, H., Yoshikawa, H., Ishiwata, S. & Kinosita, K. Jr. *Nature* **377**, 251-254 (1995)
15. Tanaka, H., Ishijima, A., Honda, M., Saito, K. & Yanagida, T. *Biophys. J.* **75**, 1886-1894 (1998)
16. Ishijima, A., Kojima, H., Higuchi, H., Harada, Y., Funatsu, T. & Yanagida, T. *Biophys. J.* **70**, 383-400 (1996)
17. Fujita, H., Sasaki, D., Ishiwata, S. & Kawai, M. *Biophys. J.* **82**, 915-928 (2002)

DISCUSSION

Gonzalez-Serratos: As the number of cross-bridges attached to actin depends on the free Ca^{2+} concentration in muscle, what is the free Ca^{2+} concentration at which all the cross-bridges are attached to actin?

Ishiwata: We have not yet studied the effect of Ca^{2+} because we first focused on the tension generation in the nanomuscle system using pure actin filaments. It would be a next subject in our nanomuscle project.

Ranatunga: I just wondered as to what happens to the titin filaments in the A-band? Titin may interact with actin.

Ishiwata: We do not know about whether or not and how connectin/titin linkages remain at the edge of the A-band. Because the Z-band is removed, it is probably folded at the ends of thick filaments. It appears that the folded connectin/titin does not disturb the tension generation.

Sugi: The motion of an actin filament should be helicoidal when actin filament is made to interact with myosin at the outer edge of the myofibril. How do you think about this?

Ishiwata: The actin filament motion can be helicoidal to some extent even if an actin filament is made to interact inside the A-band.

Sugi: Yes, but the helicoidal motion should be most pronounced at the outer surface of the A-band.

Ishiwata: Well, if the helicoidal motion of an actin filament occurs, I believe that it is due to the torque produced by cross-bridges, so that the helicoidal (rotational) motion of an actin filament must occur equally in both regions.

Huxley: How did you estimate the tension per cross-bridge?

Ishiwata: Assuming that the number of myosin molecules (=cross-bridges) in each thick filament is 300, and that the actin filament is located within the triangular lattice of thick filaments, we estimate that 75 cross-bridges (c.b.) are available for interacting with the actin filament of $0.7\mu m$ long, so that the density of cross-bridges along the actin filament is 75c.b./700nm=0.1c.b./nm. Thus, the average tension generated by each cross-bridge is estimated to be (0.2pN/nm)/ (0.1c.b./nm)=2pN/c.b.

DEVELOPMENT OF AN APPARATUS TO CONTROL LOAD BY ELECTROMAGNET FOR A MOTILITY SYSTEM *IN VITRO*

Takashi Watari, Akira Ono, Yoshiki Ishii, Huang Zhenli, Shinichi Miyake and Teizo Tsuchiya*

1. ABSTRACT

We developed an electromagnet to perform quick changes in load in the motility system consisting of myosin molecules attached to a magnetizable bead and actin filaments. The electromagnet was combined with an inverted microscope and load could be quickly changed under optical observation. The magnetic field was generated by high electric current (6V, 0-125A) and the maximum field was 8,000 Oe. The maximum force exerted on a bead was 80pN at 2.5mm distance from a magnet. The change in force was 0.48 % at the distance of 5.0 mm from the magnet when a bead moved longitudinally for 30 μm. The time to change load was about 20 ms. The movements of a bead in water were recorded by video when step changes in magnetic field were applied and it was shown that a bead exactly followed the change in force. This apparatus is very much useful to analyze the transient changes in the movement of a bead, if the movement is relatively slow as in the interaction between actin and myosin from molluscan smooth muscle.

2. INTRODUCTION

The transient mechanical responses after quick change in load or in length were observed in the living single muscle fibers (Civan and Podolsky, 1966; Huxley and Simmons, 1971; Edman and Curtain, 2001) and Huxley and Simmons (1971) proposed the well-known hypothesis of cross-bridge rotation. Sugi and Tsuchiya (1981a, b) observed and analyzed the velocity transients in living single fibers that appeared when

Department of Biology, Faculty of Science, Kobe University, Nada-ku, Kobe 657-8501, Japan
*Correspondence should be sent to T.T., tsuchiya@kobe-u.ac.jp

Molecular and Cellular Aspects of Muscle Contraction
Edited by H. Sugi, Kluwer Academic/Plenum Publishers, 2003

the quick increase in load was applied during shortening. The above mechanical experiments, however, have all been performed on living single striated muscle fibres which had complex structures, e.g. myofibrils, sarcomeres, hexagonal lattice, Z-discs, elastic proteins and so on. In addition, the concomitant structural changes occur during contraction as "the staggering", therefore, the possibility that these structural changes might be related to the mechanical transients could not be ruled out. Edman and Tsuchiya (1996) showed that the strain of passive elements could be involved in isotonic velocity transients during force enhancement by stretch.

In recent years, the motility mechanisms in actin-myosin system have been often studied *in vitro* system but nobody has not observed the transient movements *in vitro* system as shown in single muscle fibers. Permanent magnet was utilized previously (Tregear et al., 1993) but the transient movements were not observed because of the difficulty in quick changes in the intensity of magnetic field. Therefore, we developed an electromagnetic apparatus to perform quick changes in load in the motility system in vitro to know whether the velocity transients could be observed or not. Preliminary results were reported at the XXXIV International Congress of Physiological Sciences (2001).

3. METHODS AND RESULTS

Fig. 1A shows the schematic drawing of a vertical view of the electromagnet and the inverted microscope and the photograph is shown in Fig. 2. The total apparatus was composed of a magnetic-field generator, an electric controller and an electric power supply (Kikusui, Yokohama; 6-120A). The direct current from the power supply was controlled in the range 0~120A, at 6V. A water-circulation tube to prevent an increase in temperature was wound together with the electric current coil. Continuous current for 15 s increased the temperature by 0.2 °C at the coil part. There was an air gap between the top of the magnet and an experimental bath, therefore, the increase in temperature of the solution in the bath was negligible. A sensor for temperature (Omron, Tokyo; E5CS) was installed in the magnet so that the increase in temperature could stop power supply. The intensity of magnetic field was measured by a gauss-meter (Toyo Technica, Tokyo; type 4048) and the maximum intensity at the top of the magnet was 8,000 Oe. The height of the magnet and the distance between the magnet and the bath could be adjusted by a micromanipulator so that the center of the magnet and the bath could be positioned in line and on the same horizontal plane. Fig. 1B shows a horizontal view of the magnet and the experimental bath. The three edges of a bath were covered with silicone rubber strips and one edge was coated with vaseline to prevent water from leaking and the upper part of the bath was covered with a cover glass. The distance between the preparation and the top of the magnet could be as short as 2.5 mm, but usually the measurements were made at the distance of 5-15 mm at various strengths of electric current.

Fig. 3 shows the relation between the intensity of magnetic field and distance. They were measured at five current intensities and in all cases, the intensity of magnetic field decreased with the increase in distance and the curve is steeper at shorter distance.

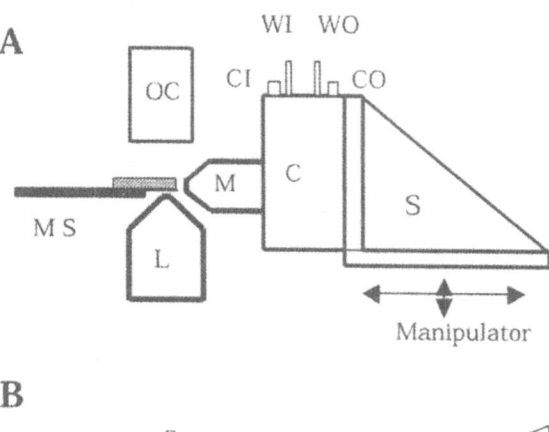

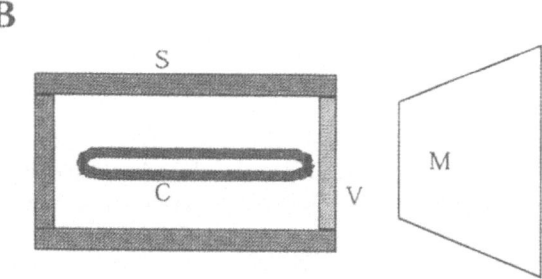

Fig. 1. Schematic drawing of the apparatus. **A**: an electromagnetic apparatus and an inverted microscope. C, coil; CI, current input; CO, current output; L, objective lens; M, magnet; MS, microscope stage; S, support of a magnet; OC, optical condenser; WI, water inlet; WO, water outlet. Note that the position of a magnet can be adjusted to be on the same horizontal plane and that there is an air gap between the top of a magnet and an experimental bath. **B**: an experimental bath and a magnet. C, an internodal cell preparation; M, top of a magnet; S, silicone rubber strips glued on three edges of a slide glass; V, Vaseline-coated part.

Fig. 2. The photograph of the magnet and the microscope. The observation stage of the inverted microscope was cut and the magnet was combined with the microscope. The top of the magnet and the experimental bath are on the same horizontal plane and the distance between them can be adjusted by a manipulator.

Fig. 4 shows the force exerted on a bead (diameter, 4.5μ m) by the magnet. The velocity of a bead in silicone oil of known viscosity was measured by video under a given intensity of magnetic field and the force exerted on the bead was calculated by Stoke's law, $F=6 \pi \eta rV$, where F is force, π is constant, η is viscosity, r is the radius of a bead and v is the velocity of a bead. The force on a bead decreased with the increased in distance and the curve is steeper at shorter distance. It was estimated how the force changed when a bead moved longitudinally for some distance. The change in force was 0.55 and 0.48 % at the distance of 2.5 and 5.0 mm respectively when a bead moved for 30 μ m. These changes are small enough in comparison with the applied step changes in force.

The force response time at the change in electric current was measured by a force transducer of rapid response (SensoNor, Oslo; AE801). A small piece of iron wire was attached to an arm of the transducer (natural oscillation frequency 4k Hz) and the time was typically 20 ms.

Fig. 5 shows the intensity of magnetic field along lines from the top of the magnet at different angles 0, 15, 30 ° as shown in the inset. The three curves almost coincide, suggesting that even if a cell preparation is at a slightly oblique angle, the intensity is almost same as at a right angle.

It was measured how a bead responded in water to the change in the intensity of magnetic field. The movements were recorded by video when step changes in force were applied as shown in the inset in Fig. 6. When the intensity was decreased from 242 to 0 Oe, a bead moving at a constant velocity suddenly stopped (Fig. 6A). A bead moving under 117 Oe increased the velocity quickly when the intensity was increased to 242 Oe (Fig. 6B). These results prove that a bead can exactly respond the change in the intensity of magnetic field.

4. DISCUSSIONS

In the analysis of the elemental process of actin-myosin interaction, the use of micro-beads is now very widespread (Sugiura et al., 1998; Nishizaka et al., 2000; Shin et al., 2000; Tanaka et al., 2002). In these experiments, beads have been used to support actin cables and the molecular interactive force or the length perturbation between actin and myosin has been measured using optical tweezers but it is impossible by these experiments to evaluate the previous physiologically important experiments that have contributed to the proposition of the basic ideas of cross-bridge movements (Hill, 1938; Civan and Podolsky, 1966; Huxley and Simmons, 1971; Edman, 1979). To address this problem, Ishii et al. (1997) and Tsuchiya et al. (1998) performed experiments using a centrifuge microscope and Tregear et al. (1993) tried to utilize magnet to control magnetizable beads. In the above two methods, however, the quick changes in load in the motility system were impossible.

The electromagnet developed in this experiment shows that it can exactly control the movement of a magnetizable bead and the application to a motility system *in vitro* seems to be very useful to know whether the transient movements can be observed or not in actin and myosin system without filament lattice structure. The problem in the present

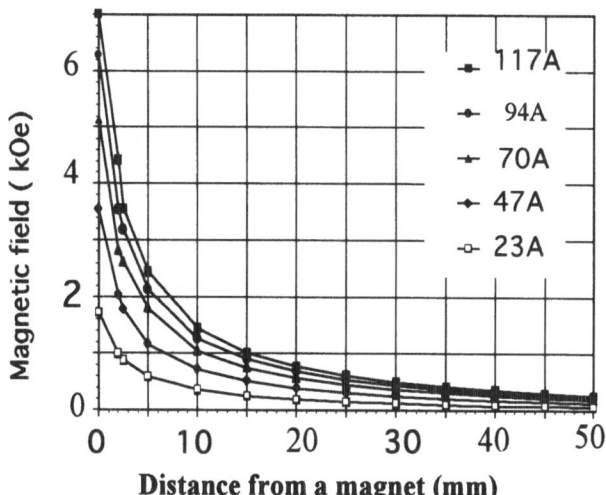

Fig₀ 3. The relation between the intensity of magnetic field and distance. The intensity of magnetic field was measured at five different current intensities as shown in the inset. The intensity decreases with the increase in distance and with the decrease in electric current.

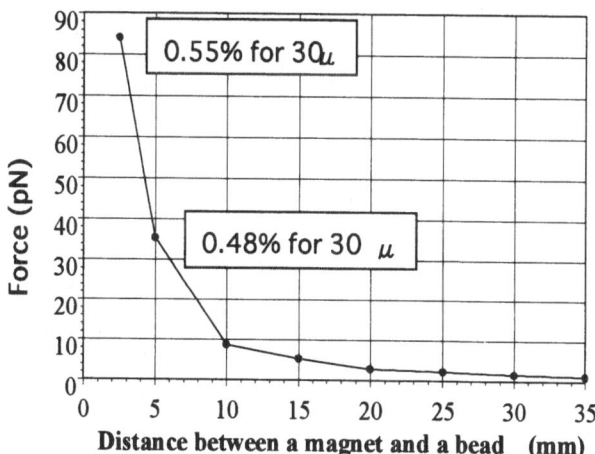

Fig. 4. The force acting on a bead (diameter, 4.5 μm) by the magnet. The velocity of a bead under magnetic field was measured in the silicone oil of known viscosity and the force was calculated by the Stalk's equation. At the distance of 2.5 and 5 mm from the magnet, the changes in force on a bead are 0.55 and 0.48 % respectively when a bead moves for 30 μm as shown in the inset.

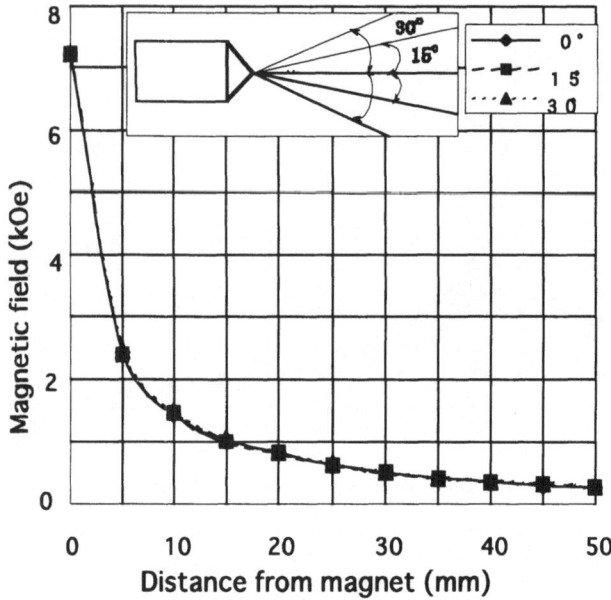

Fig. 5. The magnitude of magnetic field along the lines from the top of the magnet at different angles 0, 15 , 30°as shown in the inset. The three lines almost superimpose, suggesting that the magnitude of magnetic field in the lateral direction is nearly equal in the area examined.

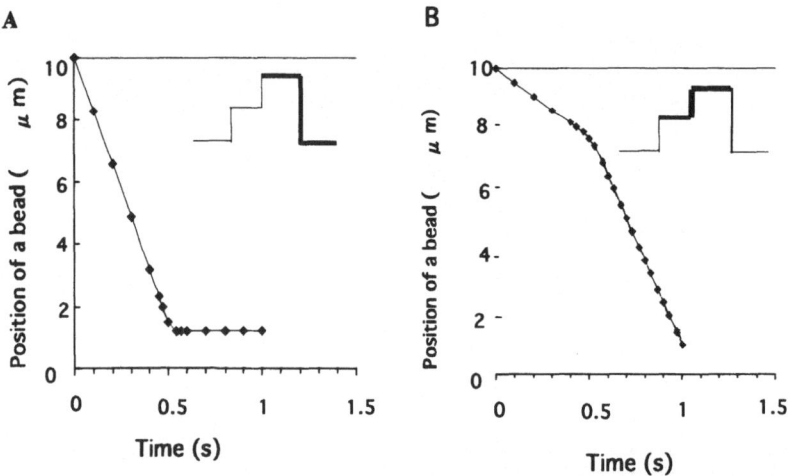

Fig. 6 The☐movement of a bead in water when the intensity of the magnetic filed was changed. The magnetic field was decreased from 242 to 0 Oe in **A** and increased from 117 to 242 Oe in **B**. The step changes in the intensity of magnetic field are shown by thick lines in the insets. Note that the movement of a bead exactly followed the change in the magnetic filed and that the velocity was very constant under the constant field. The time to change in force was less than 20 ms.

apparatus is that the force response is not quick enough for fast motility systems. However, in the system composed of myosin filaments from mollusks and actin cables in algae, the movement is very slow and the observation of transient movement may be possible, if any. The improvement of the magnet for quicker response is now in progress.

5. ACKNOWLEDGEMENTS

We should like to express our hearty thanks to Prof. H. Sugi, Teikyo University School of Medicine, for the generous use of his gauss-meter and also to Prof. T. Shimmen, Himeji Institute of Technology, for providing us with the green algae, *Chara corallina*.

6. REFERENCES

Civan, M. M. and Podolsky, R., 1966, Contraction kinetics of striated muscle fibres following quick changes in load. *J. Physiol.* **184**: 511-534.

Edman, K. A. P. ,1979, The velocity of unloaded shortening and its relation to sarcomere length and isometric force in vertebrate muscle fibres. *J. Physiol.* **291**: 143-159.

Edman, K. A. P. and Curtain, N. A. ,2001, Synchronal oscillations of length and stiffness during loaded shortening of frog muscle fibres. *J. Physiol.* **534**: 552-563.

Edman, K. A. P. and Tsuchiya T. ,1996, Strain of passive elements during force enhancement by stretch in frog muscle fibres. *J. Physiol.* **490**: 191-205.

Hill, A. V. ,1938, The heat of shortening and the dynamic constants of muscle. *Proc. Roy. Soc.* **B126**: 136-195.

Huxley, A. F. and Simmons, R. M. ,1971, Proposed mechanism of force generation in striated muscle. *Nature* **233**: 533-538.

Ishii, N., Tsuchiya, T. and Sugi, H.,1997, An in vitro motility assay system retaining the steady-state characteristics of muscle fibres under positive and negative load. *Biochim. Biophy Acta* **1319**: 155-162.

Nishizaka, T., Seo, R., Tadakura, H., Kinosita, K. Jr and Ishiwata, S. ,2000, characterization of single rigor bonds: load dependence of lifetime and mechanical properties. *Biophy. J.* **79**: 962-974.

Shin, W. M., Gryczynski, Z., Kakowicz, J. R. and Spudich, J. A. ,2000, A fret-based sensor reveals large ATP hydrolysis-induced conformational changes and three distinct states of the molecular motor myosin. *Cell* **102**: 683-694.

Sugi, H. and Tsuchiya, T. ,1981a, Isotonic velocity transients in frog muscle fibres following quick changes in load. *J. Physiol.* **319**: 219-238.

Sugi, H. and Tsuchiya, T. ,1981b, Enhancement of mechanical performance in frog muscle fibres after quick increases in load. *J. Physiol.* **319**: 239-252.

Sugiura, S., Kobayakawa, N., Fujita, H., Yamashita, H., Momomura, S., Chaen, S., Omata, M. and Sugi, H. ,1998, Comparison of unitary displacements and forces between 2 cardiac myosin isoforms by the optical trap technique; molecular basis for cardiac adaptation. *Circ. Res* **82**: 1029-1034.

Tanaka, H., Homma, K., Iwane-Hikikoshi, A., Katayama, E., Ikebe, R., Saito, J., Yanagida, T. and Ikebe, M. ,2002, The motor domain determines the large step of myosin V. *Nature* **415**: 192-195.

Tregear, R., Oiwa, K., Chaen, S. and Sugi, H.. ,1993, Relation between magnetically-applied force and velocity in beads coated with rabbit myosin, sliding on actin cables in *Nitellopsis* cells. *J. Mus. Res. Cell Motil.* **14**: 412-415.

Tsuchiya, T., Tanaka, H., Shirakawa, I., Karr, T. and Sugi, H. ,1998, Evidence of the essential role of myosin subfragment-2 in the ATP-dependent actin-myosin sliding in muscle contraction. *Jap. J. Physiol.* **48**: 383-387.

DISCUSSION

Saeki: How much was the remained magnetic field after electric current was off and did it affect the bead movement?

Tsuchiya: It was 10-20 Os amounting 0.25% of the maximum magnetic field. A bead was not made to move by the remained magnetic field.

Saeki: How do you explain the mechanism of the backward transients you observed?

Tsuchiya: Sugi and Tsuchiya (J. Physiol. 319: 219-238, 1981) observed back-and-force length oscillation of intact single fibers following a step increase in load, and put forward on hypothesis to explain this phenomenon. The hypothesis assumes a unique distribution of "f" and "g" (rate constants for making and breaking actin-myosin linkages), and in addition assumes that "g" changes with time after a step increase in load.

HIGH-RESOLUTION IMAGING OF MYOSIN MOTOR IN ACTION BY A HIGH-SPEED ATOMIC FORCE MICROSCOPE

Noriyuki Kodera, Tatsuya Kinoshita, Takahiro Ito, and Toshio Ando
Department of Physics, Kanazawa University, Kanazawa 920-1192, Japan

ABSTRACT

The atomic force microscope (AFM) is a powerful tool for imaging biological molecules on a substrate, in solution. However, there is no effective time axis with AFM; commercially available AFMs require minutes to capture an image, but many interesting biological processes occur at much higher rate. Hence, what we can observe using the AFM is limited to stationary molecules, or those moving very slowly. We sought to increase markedly the scan speed of the AFM, so that in the future it can be used to study the dynamic behaviour of biomolecules. For this purpose, we have developed various devices optimised for high-speed scanning. Combining these devices has produced an AFM that can capture a 100 x 100 pixel image within 80 ms, thus generating a movie consisting of many successive images of a sample in aqueous solution. This is demonstrated by imaging myosin V molecules moving on mica, in solution.

1. INTRODUCTION

Motor proteins are very sophisticated nano-machines. They produce force to pull cytoskeletal fibers, or move along these fibers. This mechanical function is coupled to ATP hydrolysis, and the chemical energy released by the hydrolysis is transduced to mechanical energy. To understand the mechanism by which motor proteins operate, we have to know (a) their physiological action, (b) their fine structures at atomic resolution, (c) the kinetics of ATP hydrolysis, (d) the structural dynamics in action, and we further have to know the relationship between these different aspects. The first aspect has been studied by optical microscopy, the second by x-ray crystallography, and the third has been studied by various transient techniques. The fourth aspect is the most difficult to study. In general, in life science it has been a dream to view the nanometer-scale dynamic behavior of individual biopolymers in solution. The structure of protein changes with time, and the function of protein appears during the structural changes. Hence, a technique, which can reveal temporal changes in the protein structure, has earnestly been desired in life science. Especially researchers who study biological molecular motors are very enthusiastic about viewing the nanometer-scale dynamic behavior of motor prtoteins. An atomic force microscope (AFM), which was invented 16 years ago (Binnig *et al.*, 1986), allows highly resolved imaging of protein in solution. Yet, it cannot be a candidate for the desired technique, because its scan speed is too slow to capture protein in motion. A high-speed AFM seems to be only the device to fulfill the capacity that life science has been our motivation for developing a high-speed AFM. We have optimized various devices involved in AFM for high-speed scanning, and have succeeded in producing an AFM that can capture a 100 x 100 pixel^2 image within 80 ms and therefore can generate a movie consisting of many successive images (80-ms intervals) of a sample in aqueous solution (Ando et al, 2001, 2002). This is demonstrated by imaging myosin V molecules moving on mica, in solution. The neck and head portions appear to move rigidly while the head/neck region and the neck/coiled-coil region are bending dramatically. At present, we are improving the performance of our AFM, and introducing a UV-irradiation system for producing ATP from caged-ATP. We hope that in the near future we can view nanometer-scale dynamic attitude of myosin V during it is moving along its track.

2. FACTORS LIMITING THE SCAN SPEED OF AFM

What factors of the AFM limit the scan speed? Here, we consider only the tapping mode of operation (Digital Instruments, Santa Barbara, CA). This is the mode most suitable for observing soft samples weakly attached to a substrate, in solution. In this mode the cantilever is oscillated at (or near) its resonance frequency. The oscillating tip briefly taps the surface of the sample at the bottom of each swing, resulting in a decrease in the oscillation amplitude. This decrease gives information of the sample height. The cantilever, therefore, has to oscillate for at least one cycle for each pixel of the image. To obtain an image consisting of NxN pixels using a cantilever having the resonance frequency of F_c, we require an imaging time (T), given by

$$T \geq 2nN^2 / F_c \qquad (1),$$

where n is the number of waves of oscillation required for measuring the oscillation amplitude. An imaging time of 80 ms for 100x100 pixels requires a resonance frequency higher than n x 250 kHz in water. Cantilevers with a higher resonance frequency tend to have a larger spring constant, undesirable for imaging soft samples. Therefore, to minimize the resonance frequency required for high-speed imaging, we need an RMS-DC converter that can output the amplitude voltage of the input sinusoidal signals as quickly as possible.

In addition to the resonance frequency of cantilevers, we have to consider another factor that limits the scan speed. Suppose that a sample on a substrate has a periodicity of λ, and the sample stage is moved horizontally with a velocity of Vs, the spatial frequency of $1/\lambda$ is converted to a temporal frequency of Vs/λ. The feedback system, that keeps the cantilever's oscillation amplitude constant, moves the sample stage up and down. The feedback bandwidth (F_b) should be wider than Vs/λ. Therefore, the scan speed is limited by the bandwidth as $Vs < \lambda F_b$. This scan speed determines the imaging time as

$$T = 2pN^2 / Vs \geq 2pN^2 / \lambda F_b \qquad (2),$$

where p is the pixel size. For example, if T=80 ms, N=100, p=2 nm, and λ=10 nm, then there is a required feedback bandwidth larger than 50 kHz. Various devices are involved in the feedback loop, as shown in Fig.1. It is not very difficult to achieve a high bandwidth for electronic devices. However, the scanner is the mechanical device most difficult to optimize for high-speed scanning. A well-known guiding principle for fabricating a mechanical device with a high resonance frequency is to make it with a small, compact, and

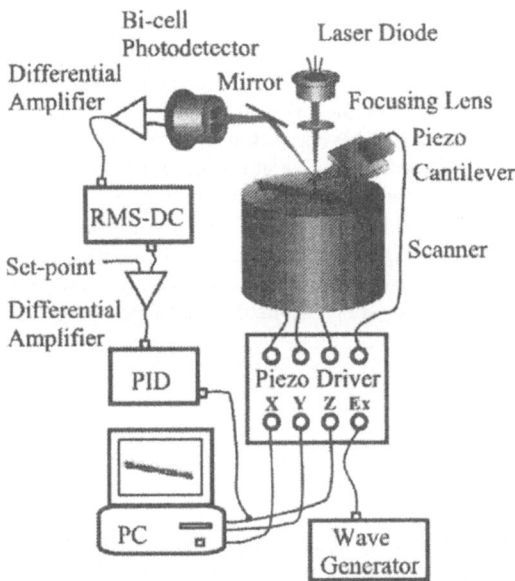

Fig.1   A schematic of a conventional AFM system for the tapping mode of operation.

light body. The sketches (Fig.2) show the conventional designs for the scanners that have been employed for the AFM. As long as the dimensions of these scanners are sufficiently large, the movements along the three axes do not interfere with each other. However, such large dimensions result in a low resonance frequency. We, therefore, require a different design for the high-speed scanner. Also required of the high-speed scanner is high rigidity against the impulsive forces produced by rapid movements of the piezo actuators. When

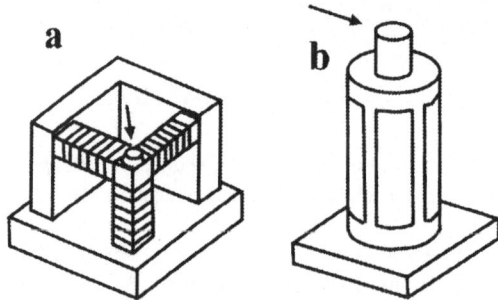

Fig.2   Sketches of AFM scanners with conventional designs. (a) a tripod type, (b) a cylindrical type. The arrows indicate the sample stages.

an object having the mass of 1 g is moved at 50 kHz with an amplitude of 10 nm, a peak impulsive force of 1.0 N is produced. The AFM scanner should not generate unwanted vibrations of even 1 nm against this impulsive force. Therefore, the required rigidity becomes 100 kg/μm. From simple calculations, it is evident that it is quite difficult to fabricate a mechanical device having such a high resonance frequency as well as high rigidity. Therefore, we need alternative guidelines for fabricating a high-speed scanner. Several laboratories throughout the world have been trying to develop a high-speed AFM (Sulcheck *et al.*, 2000; Viani *et al*, 1999). They must also have encountered the greatest difficulty when making a high-speed scanner. An alternative means of achieving a high-speed scanner is to use a cantilever with an integrated piezoelectric actuator such as zinc oxide. However, such integration inevitably results in a large spring constant of the cantilever.

3. NEW DEVICES

3.1.   Small Cantilevers

A high resonance frequency and a small spring constant are conflicting requirements for any mechanical device. This is evident from the following equations for a strip type cantilever.

$$F_c = 0.56 \frac{d}{L^2} \sqrt{\frac{E}{12\rho}} \ , \tag{3}$$

$$k = \frac{wd^3}{4L^3} E \ , \tag{4}$$

where k is the spring constant, d, L, and w are the thickness, the length and the width of the cantilever, and E and ρ are the Young's modulus and the density of material used for the cantilever, respectively. These conflicting requirements can be met only by using small dimensions. We fabricated small cantilevers from silicon nitride using micromachining techniques (Fig.3). They are 140 nm thick, 2 μm wide, and 9-11 μm long. The rear side of each cantilever is coated with gold of 20 nm thickness. The tips were grown by electron-beam deposition, with a growth rate of about 5 nm/s. The tip length was adjusted to about 1 μm. The radius of the tip end is 5-8 nm. The mechanical properties of the cantilevers were

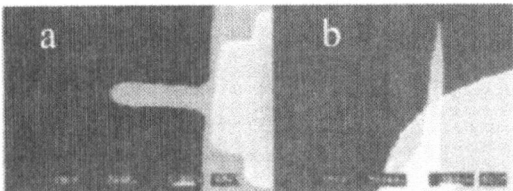

Fig.3 Electron micrographs of the small cantilever developed for our high-speed AFM. (a) the cantilever made from silicon nitride has no tip. (b) a tip was grown on the cantilever end by electron beam deposition.

tested by measuring the spectra of their thermal motions. The resonance frequencies are 1.3-1.8 MHz in air, and 450-650 kHz in water, and the spring constants are estimated to be 150-280 pN/nm. The highest resonance frequency in water, i.e., 650 kHz, can reduce the imaging time to 30 ms for 100x100 pixels.

3.2 RMS-DC Converter

Conventional RMS-DC converters require at least 5~6 waves for conversion. This requirement arises because the converter has to use a low-pass filter in order to separate the carrier (basic) wave from the amplitude-modulation wave. We designed a new converter that requires only a half wave for conversion (Fig.3 in Ando *et al.*, 2001). This converter is a type of peak-hold circuit. Two S/H circuits hold the peak and bottom voltages separately. The timing signals for this holding are made by the input sinusoidal signal itself. This guarantees stable and precise conversion even when the cantilever oscillation changes its frequency and phase. The difference between the two voltages held with the two S/Hs is output as the amplitude (not the RMS value) of the input sinusoidal signal. This new converter is satisfactory for an input sinusoidal signal of up to 1 MHz.

3.3 Optical Deflection Detection System

Since the cantilevers are very small, we required that the laser beam be focused onto the small cantilever as a small spot. Therefore, we could not use the optical deflection detection system that has widely been used in commercial AFMs. We designed an objective-lens type of deflection detection system (Fig.4). The incident laser beam is focused onto a small cantilever using an objective lens (CFI Plan Fluor ELWD 20xC, Nikon). The reflected beam is collected and collimated with the same objective lens. The incident and reflected beams are separated by a polarization beam splitter and a quarter-wavelength plate. The incident beam is entered into the objective lens at a slightly off-centered position to make the outgoing beam axis normal to the plane of the cantilever. The focused spot is 2-3 μm in diameter, sufficiently small for our small cantilevers. The optical lever magnification is about 2,000. This large magnification results from the short length of the cantilevers.

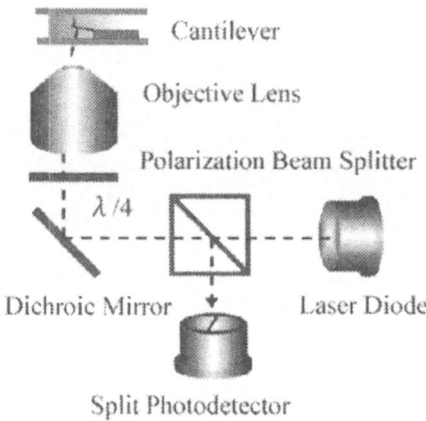

Fig.4   A schematic of the objective-lens type of optical deflection detection system.

3.4 Scanner

As mentioned above, it is quite difficult to fabricate a mechanical device with the high resonance frequency (>50 kHz) and high rigidity (> 100 kg/μ m) required of the high-speed scanner. We considered the following possibility: (1) We might somehow reduce the resonance amplitude even when the resonance occurs at low frequencies; (2) We might somehow counteract the impulsive forces produced by the quick movement of the piezo actuators. After making and testing a number of scanners with different designs, we reached the design illustrated in Figure 5. Stack-type piezoelectric actuators (AE0203D04, Tokin, Tokyo) are used in this scanner. They have a resonance frequency of 260 kHz in free oscillation, and their maximum displacement is 4.5 μ m. This scanner has a two-layered structure. One layer is for scanning in the y-direction, and the other layer is for scanning in the x- and z-directions. This structure guarantees little interference between the movements along the three axes. The z-scanner has two z-piezo actuators placed in opposite directions to one another. A sample stage is attached to one of the z-piezo actuators via a thin layer of vacuum grease. These actuators are displaced simultaneously in the same distance, but in counter directions, so that any impulsive forces produced are canceled out. The base plate (Base-2), to which the two z-actuators as well as an x-actuator are attached, is clamped in the z-direction by two flat surfaces (Base-1 and Plate-2) via steel ball bearings. This design allows smooth movement of the base (and hence, the sample stage) in the x-direction, and minimizes the vibrations of the base in the z-direction. When the z-piezo is displaced quickly, hydrodynamic force is generated as a reaction from the sample solution to the sample stage. To minimize this reactive force, a glass of circular-trapezoid shape with a small top surface of 1 mm diameter is used as the sample stage. The performance test of the z-scanner indicated that it was able to be driven stably up to about 60 kHz. Because the devices involved in the feedback loop, other than the scanner, have bandwidths much wider than 60 kHz, the feedback bandwidth is thus determined to be about 60 kHz.

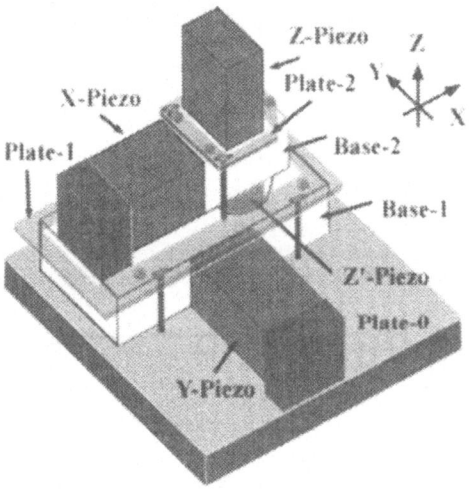

Fig.5 Scanner assembly. The piezo actuators are 5 mm long, 4 mm wide, and 2.7 mm thick. A sample stage is attached to the top of the z-piezo.

4. IMAGING

According to eqs. (1) and (2), the feedback bandwidth of 60 kHz as well as the high resonance frequencies of the small cantilevers can reduce the time for capturing an image with 100^2 2 nm pixels to 70 ms, as long as the apparent width of the sample is not too small. We examined whether imaging can really be carried out at (or near) the maximum rate predicted here. We chose a motor protein, myosin V as the first sample to be examined by our high-speed AFM, since it is the first unconventional myosin identified as a processive motor; it travels a great distance along an actin filament without detaching from actin (Sakamoto et al., 2000). We are aiming at viewing its nanometer-scale dynamic process in real time. We started with imaging myosin V alone directly attached to mica, in solution. In Fig.6 the images we first obtained are shown (270 nm scan range; 100^2 pixels; for 1.6 s (20 frames)). The scan rate was 1.25 kHz, corresponding to a tip speed of 0.68 mm/s, and the frame rate was 12.5/s. The myosin V molecule is attached to mica through one of the two heads, and the other moieties are free and moving. The typical Y shape is clearly seen. ATP is absent in this solution. Next, we show, in Fig.7,

AFM images of myosin V in a solution containing ATP. The angle of the long tail relative to the head/neck regions changed between the eighth and ninth frames among the successive 50 frames (see the second and third images in Fig.7). This quick change took place within 30 ms. After this change, the tail and the tail end are slowly moving. This marked contrast suggests that this quick orientational change may be driven by ATP.

Fig.6   Successive images, at 80 ms intervals, of myosin V weakly attached to mica surface, in buffer solution.

Fig.7   Successive images, at 80 ms intervals, of myosin V on mica in a solution containing ATP. The reconstructed movie and the other movies can be viewed at the web site (http://www.s.kanazawa-u.ac.jp/phys/ biophys/bmv_movie.htm).

5. DISCUSSION

In life science, it has been a dream to view the nanometer-scale dynamic behaviour of individual biopolymers in solution. The capacity to acquire successive images every 80 ms will allow a large expansion in the scope of biological processes that can be examined in real time. In the near future, we should be able to observe the behavior of processive motors such as kinesin and myosin V moving along their tracks, of molecular chaperones assisting a polypeptide chain to fold, or of a ribosome synthesizing a polypeptide. Such direct observations will provide insight into the mechanisms by which biomolecular machines operate. We think that the high-speed AFM has further potential in life science. If we can link dynamic images of a protein acquired by the high-speed AFM to its known atomic structure, we may be able to construct dynamic atomic models not obtainable by other techniques. How can we make such a link? Cryo-EM imaging may be a candidate that can mediate this linking process. Suppose cryo-EM images are obtained for protein molecules that were performing a function immediately before freezing; these molecules will be found with different conformations among these images. These conformations must occur on a single molecule of protein that dynamically changes its structure along the time axis. Therefore, the conformations found in the cryo-EM images can be aligned along the time axis, reflecting the dynamic AFM images. Then, we must deform the atomic structure so as to fit it to the conformations found in the cryo-EM images. In this way, we can construct dynamic atomic models that move and thus reflect the AFM movies. The static atomic structures of many proteins have been revealed by X-ray crystallography. However, the basic framework of structural biology has not been changed significantly since the first success by Professor Perutz in 1936. We hope that the high-speed AFM will enable a breakthrough in structural biology in the future.

REFERENCES

Ando, T., Kodera, N., Takai, E., Maruyama, D., Saito, K. & Toda, A. (2001) A high-speed atomic force microscope for studying biological macromolecules. *Porc. Natl. Acad. Sci. USA*, **98**, 12468-12472.

Ando, T., Kodera, N., Takai, E., Maruyama, D., Saito K. & Toda, A. A high-speed atomic force microscope for studying biological macromolecules in action. *Jpn. J. Appl. Phys.* **41**, 4851-4856.

Binnig, G., Quate, C.F. & Gerber, C. (1986) Atomic force microscope. *Phys. Rev. Lett.*, **56**, 930-933.

Sakamoto, T., Amitani, I., Yokota, E. & Ando, T. (2000) Direct observation of processive movement by individual myosin V molecules. *Biochem. Biophys. Res. Commun.* **272**, 586-590.

Sulcheck, T., Hsieh, R., Adams, J.D., Minne, S.C., Quate, C.F. & Adderton, D.M. (2000) High-speed atomic force microscopy in liquid. *Rev. Sci. Instrum.*, **71**, 2097-2099.

Viani, M.V., Schaffer, T.E., Paloczi, G.T., Pietrasanta, L.L., Smith, B.L. et. al. (1999) Fast imaging and fast force spectroscopy of single biopolymers with a new atomic force microscope designed for small cantilevers. *Rev. Sci. Instrum.*, **70**, 4300-4303.

TRANSLATION STEP SIZE MEASURED IN SINGLE SARCOMERES AND SINGLE FILAMENT PAIRS

Gerald H. Pollack[*], Xiumei Liu, Olga Yakovenko, Felix A. Blyakhman

1. INTRODUCTION

The quest to identify the size of the molecular contractile stroke has been paramount in the field of muscle contraction. The most common approach has been to employ the optical trap to study the interaction between single myosin molecules and actin filaments, and the results of these studies have been variable. Step size has ranged between ~4 nm and ~20 nm. Among the difficulties attendant with such measurements is high noise: peak-to-peak noise is not uncommonly on the order of 30 nm, which is many times the size of the putative step.

Another problem, perhaps more subtle, is that the myosin molecule lies in the unnatural environment. The molecule must be bonded in some way to the surface of a large microsphere. How it is physically bound, i.e., its orientation and immediate environment, and how this may affect the outcome, are unknown. The problem is potentially serious, for the goal is ultimately less one of identifying the "stroke" of the isolated molecule, but of identifying how any such molecular behavior translates to real muscle, so that comparison with large scale mechanical measurements can help reveal the underlying mechanism.

To this end, we have been pursuing measurements of step size on specimens more closely resembling those of intact muscle, yet on a near-molecular scale. We present the results of recent measurements made on single myosin and actin filaments sliding past one another, as well as measurements on single intact sarcomeres, where bundles of filaments slide past one another. The results of these experiments are virtually identical: step size is consistently an integer multiple of 2.7 nm, and this size paradigm is independent of the direction of sliding.

---

[*]Department of Bioengineering, University of Washington, Seattle, WA 98195,U.S.A.

Molecular and Cellular Aspects of Muscle Contraction
Edited by H. Sugi, Kluwer Academic/Plenum Publishers, 2003

2. METHODS

2.1 Filament Experiments

Experiments were based on silicon nitride nanolevers developed in this laboratory (Fauver et al., 1998). The lever deflects in response to a force, applied normal to its long axis at or near its tip. In essence, a thick filament was attached to a non-deflectable lever, while an actin filament was attached to a deflectable lever. The two filaments were brought together, and, in the presence of ATP and Ca^{2+}, they slid past one another.

Native thick filaments were isolated from the ABRM (Anterior Byssus Retractor Muscle) of the living blue mussel, *Mytilus*, as described (Sellers et al., 1991). Rhodamine-phalloidin-labeled F-actin extracted from rabbit skeletal muscles were prepared by the standard method (Pardee and Spudich, 1982), and were provided courtesy of A. M. Gordon and Charles Luo. The measurement apparatus was based on a Zeiss Axiovert 135 TV microscope system (Liu and Pollack, in press), used in either bright field, for measurements, and fluorescence and DIC, for mounting.

In the bright field mode, the flow cell was illuminated by an intensity-adjustable QTH (Quartz Tungsten Halogen) light source through a water-immersion condenser (Zeiss Achroplan, 63X/0.9W). The magnified images of the three levers, obtained using a 100X oil immersion objective lens and an intermediate lens (Zeiss Plan-Neofluar, 100X/1.30 oil), were projected onto a 1024-pixel photodiode array (Reticon RL1024K). The non-deflectable lever and each flexible lever cover ~180 and 45 pixels, respectively. The time course of lever position could be recorded continuously at a pixel sampling rate of 50 kHz, which yielded a temporal resolution of 22 ms/scan, given the array length of 1100 pixels.

Before attempting to manipulate filaments, the lever pairs' positions and light intensity were adjusted to optimize signal level and shape. The apparatus could then be switched to the fluorescence mode, whose light source was a 100 W mercury arc lamp directed through an optical fiber coupling. Fluorescence images of levers and filaments were monitored by a silicon-intensified camera (Sony XC-77 CCD).

The two lever sets were arranged in the flow cell with large (~10 mm) separation. Near one set, a small quantity (10 ml) of alpha-actinin was added, and near the other set a similarly small quantity (10 ml) of native thick filaments was pipetted. After ~5 minutes, the excess alpha-actinin and thick filaments were both washed away by a flow of buffer perpendicular to the lever shafts. This way, most of the thick filaments would attach at right angles to the stationary lever axis; also the cross contamination of two kinds of proteins would be minimized.

Orientation and density of attached thick filaments onto the non-deflectable stationary levers were examined carefully under DIC microscopy. Only thick filaments with a free segment projecting beyond the lever surface, and with axis at right angle to the lever, were acceptable for experimentation. Hence, myosin orientation was constrained in a natural direction relative to the actin filament. To promote visibility of thick filaments under fluorescence microscopy, F-actin fragments were bound to them, making them visibly fluorescent. This enabled simultaneous manipulation of F-actin and thick filaments.

Next, longer F-actin (5~10 μm) was added to the experimental chamber. A flow stream containing ATP and calcium, directed perpendicular to the lever shafts, was initiated immediately to exchange the chamber buffer. This flow also promoted the capture of single F-actin by the lever coated with alpha-actinin. When one end of a single F-actin was found attached to the tip of the deflectable lever, normal to the optical axis, it was manipulated into the vicinity of the target thick filament. Once the two filaments interacted, sliding was initiated. Then, the apparatus was immediately switched to the bright-field mode to facilitate displacement measurement.

2.2 Myofibril experiments

Isolated myofibrils were prepared from rabbit left ventricular trabecular muscles as described previously (Linke et al., 1994). Briefly, thin strips of muscle tissue were dissected for storage in rigor/glycerol solution (50/50 by volume) for a minimum of 5 days at -20°C. To obtain single myofibrils, glycerinated strips were minced, and the pieces were further skinned in a 4°C rigor solution containing 0.5% Triton X-100 for 30 minutes. After washing with fresh rigor solution, the tissue pieces were homogenized in a blender (Sorvall Omni Mixer) at low speed for 5-6 seconds in the same buffer.

To ensure that the results were not peculiar to any particular muscle type, we also examined invertebrate specimens. Specimens of bumblebee flight muscle were prepared as described previously (Blyakhman et al., 1999; Yang et al., 1998). Briefly, muscles were chemically skinned by soaking them alternately in solution A (1% v/w Triton X-100 in relaxing solution containing 10 mM leupeptin and 1mM DTT) and B (50% glycerol in relaxing solution) at 4° C. Muscles were kept in solution B at –20°C for immediate use and at -80°C for longer-term storage. Myofibrils were isolated by using fine needles to disaggregate specimens on a cover glass in a drop of rigor solution.

A drop of this "cardiac" or "flight-muscle" suspension was placed in the specimen chamber (volume approximately 300 µl), and myofibrils were allowed to settle and stick lightly to the chamber bottom where they could later be picked up by two microneedles. Excess rigor solution and extraneous tissue were washed out of the chamber via several rinses with relaxing solution. All experiments were performed at room temperature (20 – 22°C).

Single myofibrils were mounted in a specially constructed apparatus built around Zeiss Axiovert-35 microscope as previously described in detail (Blyakhman et al., 1999; Yang et al., 1998). Briefly, one end of a myofibril was attached either to the tip of a glass needle or in some instances to a nanofabricated tension transducer (Fauver et al., 1998). The other end was mounted on the moving glass tip of the piezoelectric motor, which could impose linear length changes on the specimen. These attachment fixtures were in turn mounted on hydraulic micromanipulators (Narishige) to facilitate positioning. The myofibril's striation pattern was projected onto a 1024-element photodiode array, which was scanned to produce a trace of intensity vs. position along the myofibril. Scans were made every 50 ms, a compromise between time resolution and integration time need to reduce noise.

Sarcomere length was calculated as the span between centroids of contiguous A-bands. To calculate the centroid we developed an algorithm based on the minimum average risk method originally developed by A. Kolomogorov (1931). For details, see Sokolov et al. (in press). The algorithm operates on repeated light intensity line scans of a sample, precisely quantifying movements of features of the sample between scans. The method is implemented by computing, in a limited region around the feature in question, the pointwise product of (a) the current intensity scan and (b) the first spatial derivative of the immediately previous scan. This is done by translating these scan regions relative to one other to find the optimal registration, determined as the minimum of their integrated product. The amount of shift required to achieve optimal registration is equal to the amount of feature shift.

As implemented, a 2-sarcomere-wide subsection of each successive digitized scan of a myofibril is selected to bracket a given A-band. The shift required to obtain the best fit, relative to the previous scan, is determined. From the relative shift between two A-bands, the change of sarcomere length can be computed. By repeating this computation for each successive scan, the time course of each A-band position, and thus of sarcomere length, can be obtained. Because the method is differential, high resolution is obtained. Through successive sarcomere-length computations, we could follow the time course of the length changes in single sarcomeres.

We first checked the striation quality by stretching and releasing the specimen several times in relaxing solution. If A-band and I-band widths failed to remain invariant after several rounds of stretch and release, the specimen was discarded. Myofibrils that passed this test were stretched slightly beyond rest length, to SL 2.4-2.6 μm (2.9-3.2 μm for flight muscle) to reveal distinct I-bands. Activation solution was then added, and a trapezoidal length change was imposed with the aid of the motor. Sometimes it was necessary to readjust needle positions in order to assure that the myofibril remained parallel to the array. Other times it was necessary to readjust the lighting conditions to assure that illumination remained uniform.

The stretch-release protocols were then carried out at moderate length (<2.6 μm for cardiac, and <3.3 μm for flight muscle). The protocol consisted of imposing a trapezoidal length change of 5-7%. Speed of stretch-release was generally 3-4 nm/sec/sarcomere. In total, 253 sarcomeres from 53 activated myofibrils, and 33 sarcomeres from nine relaxed myofibrils were analyzed in cardiac muscle, and 332 sarcomeres from 46 activated myofibrils were analyzed in bumblebee-flight muscle.

Analytical details follow largely along lines already presented (Blyakhman et al., 1999; Yang et al., 1998). To define a step it was necessary to define the pair of pauses surrounding the step. The pause was taken provisionally as a region of the trace whose estimated best-fit by eye was nominally parallel to the horizontal axis. The region had to contain a minimum of five consecutive sample points to qualify; most contained more. After assigning beginning and end points of the pause, a best-fit line was computed. This fit provided a guide for slight adjustment of the break points to yield pauses with slope closest to zero. Step size was then computed as the vertical span between midpoints of two successive best-fit line segments. The procedure was repeated for other pauses and steps.

To check for potential artifact, many controls have previously been carried out (Blyakhman et al., 1999; Yang et al., 1998). The effects of discreteness of the photodiode array were checked two ways: by using two magnifications and obtaining similar results and by replacing the photodiode array with a non-discrete sensor consisting of a scanning mirror and photomultiplier, and seeing comparable steps. Specimen-translation artifacts were checked by testing whether smoothly translating A-bands could produce steps. Analytical procedures were checked by testing whether each of two independent algorithms yielded similar step distributions. All tests proved negative for artifact.

In addition, controls were run to check for sub-nanometer detectability in the presence of noise (Blyakhman et al., 2001). Although baseline noise superimposed on a linear ramp could produce occasional step-like features, we found that any such steps had no consistent size or pattern. Further, by imposing a motor waveform that forced sarcomeres to step at a size slightly different from that of naturally occurring steps, we were able to distinguish step-size distributions with peaks separated by 0.4 nm or less. Thus, resolution was adequate to detect steps on a sub-nanometer scale.

Additional controls have been run in the present experiments. In the first control, the experimental conditions remained identical but the specimen was placed in relaxing solution instead of activating solution, to determine whether steps of similar size might show up. In a second control to check for resolution, we applied a stepped ramp to the myofibril and investigated the resulting sarcomere-length changes. Results of these controls were negative for artifact (Yakovenko et al., 2002).

3. RESULTS

3.1 Filaments

Figure 1 shows representative traces of filament translation vs. time. The smallest detectable step in trace 1 is 11 nm (see arrows), while larger steps appear as integer multiples of that value. These steps could be easily seen by eye because they were preceded and followed by long pauses. In trace 2a, step-size as small as 2.75 nm is also apparent. Also, clear steps ranging from ~5.5 nm and ~11.0 nm to nearly 16.5 nm are seen as well in trace 3a (arrows). Trace 3b demonstrates the same steps as that of 3a, although it was analyzed by a different (minimum-risk) algorithm, implying that the steps are unlikely to result from the artifacts of data analysis.

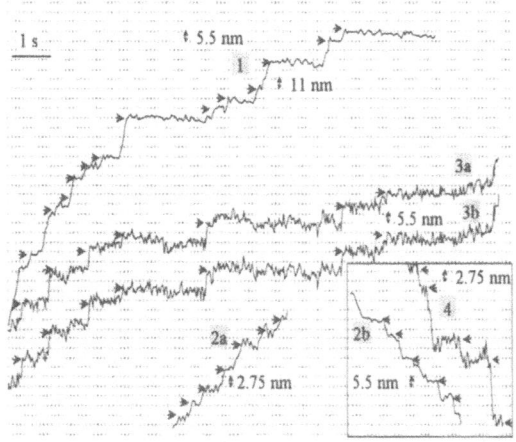

Figure 1. Stepwise interaction between single actin and thick filaments. Horizontal bar represents 1 sec, except for traces 1 and 2, where it is equal to 0.5 s. Traces in the box show backward steps while traces 1, 2a and 3 display mainly forward steps. Arrows indicate the positions of each pause.

Steps were not features of forward movement alone. Backward steps (corresponding to sarcomere lengthening) also appeared commonly, generally during a small amount of reverse sliding after tension had reached a maximum. Sliding was rapid at first, with short pauses (trace 2b in Fig. 1), and step size was uniformly ~11 nm. With the increase of backward motion, pauses became considerably longer and step size varied from 11 nm to ~44 nm (data not shown). Backward steps of 5.5 nm can also be seen in trace 4 of Fig. 1. Occasional backward steps also appeared during forward sliding.

Several possible artifacts were considered. One possibility is that the steps are generated by the discreteness of the photodiode array — as the lever sweeps across the array, a step is generated for each pixel that is traversed. However, each pixel corresponds to ~100 nm, which is far larger than the size of the steps. A second possibility is that the steps are generated by the software that converts the photodiode-array signal to a length trace. But as shown above, two independent algorithms generate virtually identical traces. Finally, random noise, when superimposed on a shortening or lengthening ramp, could create step-like signals. To avert this possibility, lever pairs were used instead of single levers, the two linked at their base, one attached to the actin filament, the other free. Any system noise should appear equally on both levers. Hence, we could compare each step measured on the deflectable lever with the behavior of the reference lever to determine whether the step arose from noise. Spurious signals were thereby eliminated from consideration.

To analyze step size, beginning and end points of each pause were identified, and an algorithm computed the best–fit straight line between those points (Blyakhman et al., 1999, Yakovenko et al., 2002). The algorithm then computed the vertical spacing between successive pauses, which gave step size. To qualify as a pause, the dwell time had to be at least 110 ms — in other words it had to include at least five consecutive data points. Sizes obtained from many steps were plotted as a continuous histogram.

A histogram of forward translation step size is shown as open circles in Fig. 2. The histogram includes 612 steps, at [ATP] of 10~100 μM. A histogram of lengthening or "backward" steps (n = 215), which were generally less distinct than the forward steps, is

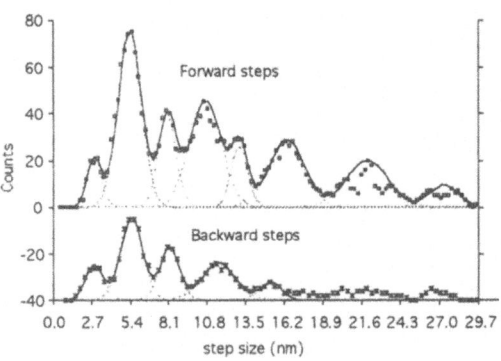

Figure 2. Continuous histogram of step-size distribution with bin width of 1 nm and increment of 00.1 nm. Tick marks on abscissa are spaced at 2.7 nm intervals.

plotted beneath in a similar way. In both cases, multiple peaks are seen at approximate integer multiples of 2.7 nm, such as 5.4 nm, 8.1 nm, 10.8 nm. The first peak, at 2.7 nm, may be less reliable than the others because the typical peak-to-peak noise level is of similar magnitude. However, peaks also appeared at integer multiples of 2.7 nm, such as 8.1 nm and 13.2 nm, implying that the 2.7-nm peak does not arise from noise. Also, the 2.7-nm peak is approximately as narrow as the others.

3.2 Myofibrils

Figure 3 shows the time course of sarcomere-length change in single sarcomeres of single isolated bumblebee myofibrils. The activated specimen was allowed to shorten, or was stretched by a motor-imposed ramp. Sarcomeres generally followed the ramp on a coarse scale. On finer scale, the pattern was staircase-like, with short pauses (arrows) interspersed between steps. The steps were typically 2 - 3 nanometers in size.

The results of analysis of these contractile steps are shown in Figure 4. Step size was computed in the following way. Beginning and end points of each pause were identified, and an algorithm computed the best-fit line between these points (see Methods). The algorithm then computed the vertical spacing between successive pauses, which gave the size of the step. Sizes obtained from many steps were plotted as a continuous histogram. For low velocity ramps (nominally 1 nm/sarc/sec) the histogram shows a single peak, whose maximum falls at 2.68 nm. For higher velocity ramps (nominally 8 nm/sarc/sec), the main peak fell at 2.67 nm, with a secondary peak at twice that value, and a hint of a peak at approximately three times the value. The inset shows lower-resolution data obtained earlier (Blyakhman, 1999), where multiple peaks likewise fell at approximately integer multiples of 2.7 nm.

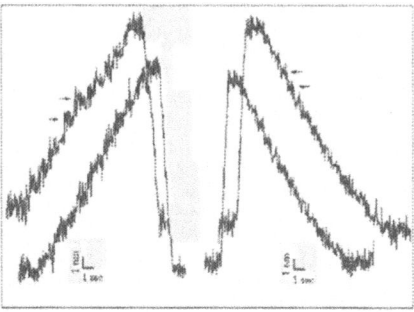

Figure 3. Time course of sarcomere-length change during motor-imposed trapezoid in single sarcomere of single bumblebee myofibrils. Left: slow lengthening ramp. Right: slow shortening ramp. Arrows denote representative pauses.

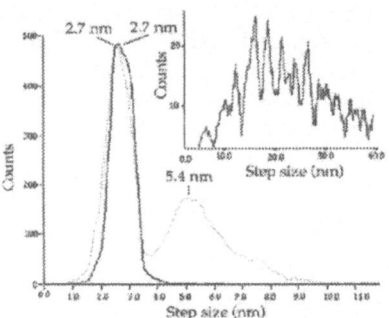

Figure 4. Histogram of shortening-step size in bumblebee flight muscle obtained by imposing shortening ramps. Low speed ramps shown dark (907 steps from 142 sarcomeres in 28 myofibrils); high speed ramps shown light (1011 steps from 190 sarcomeres in 40 myofibrils). Inset: Similar data obtained at lower resolution, from Blyakhman et al. (1999). In all histograms, peaks are situated at approximate integer multiples of 2.7 nm.

The results of parallel experiments carried out on cardiac myofibrils are shown in Figures 5 and 6. Figure 5 shows representative sarcomere-length-change traces. Ramp speed was nominally ~3 - 4 nm/sarcomere/sec, midway between the two values used for bumblebee samples. The traces are noisier than those of the bumblebee because the striation patterns are less regular. On the other hand, many of the pauses were of longer duration, and were therefore relatively more prominent.

Analytical results obtained from shortening steps are shown in Figure 6. They show a primary peak at 2.71 nm, indistinguishable from the bumblebee results. Several additional peaks are seen at approximate integer multiples of the primary value. (Little significance should be attributed to the relative amplitudes of the peaks, as these may change with loading conditions and with noise.) Hence, the cardiac histogram illustrates in a single figure, that the size of the shortening step is an integer multiple of 2.7 nm.

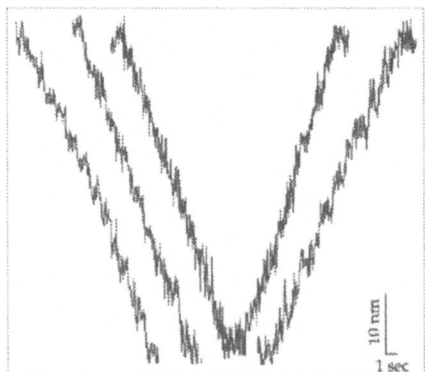

Figure 5. Representative traces of sarcomere-length change during motor-imposed trapezoid in rabbit cardiac muscle. First, second and fourth traces show segments of record; third trace shows entire record.

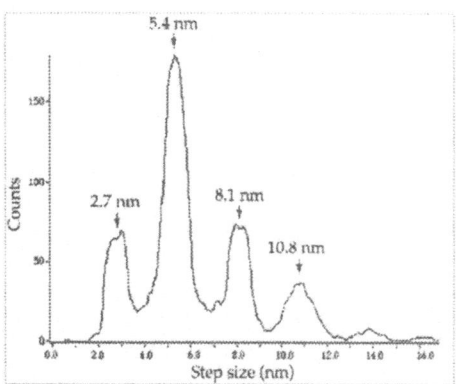

Figure 6. Histogram of shortening steps obtained from activated rabbit cardiac muscle. Histogram includes 493 steps from 75 sarcomeres in 25 myofibrils.

Lengthening steps were analyzed in a similar way. Pauses and steps were marginally less distinct than those observed during shortening (Figs. 3 and 5), but were clear enough to allow analysis. A series of histogram peaks was observed, each lying at an integer multiple of ~2.7 nm (Fig. 7). For comparison, bumblebee-muscle data are shown. A narrow peak is observed at 2.68 nm. The corresponding periodogram (not shown) had a peak situated at 2.65 – 2.70 nm.

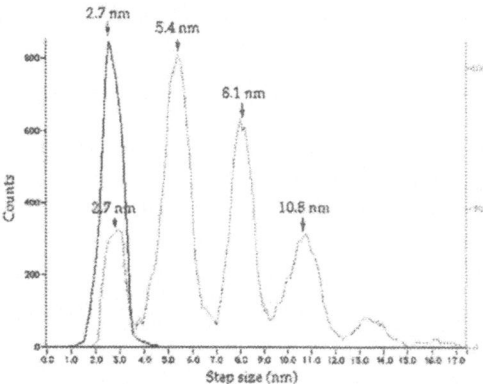

Figure 7. Histograms of lengthening steps in bumblebee flight muscle (light) and rabbit cardiac muscle (dark). Number of steps: bumblebee: 908; cardiac: 378.

4. DISCUSSION

The results show a high level of consistency among specimens. In both the single sarcomere and the single filament pair, step sizes are 2.7 nm and integer multiples thereof. These observations confirm and extend earlier observations of step size values in sarcomeres at integer multiples of 2.7 nm (Blyakhman et al., 1999), and establish that the previously implied quantum does indeed exist. Because the same quantal value is found in invertebrate and vertebrate specimens alike, and in two different experimental models, the value appears to be general. Thus, activated filaments slide in 2.7 nm steps or integer multiples thereof, whether in isolation or in the sarcomere.

Another significant finding is that the step is reversible in both experimental models: like the shortening steps, the lengthening steps are integer multiples of 2.7 nm. The fact that all results — shortening and lengthening — produced step sizes of 2.7 nm may seem suspiciously implicative of some type of systematic artifact; however, identical protocols carried out on unactivated myofibril specimens produced step-size distributions centered at 2.3 nm instead of 2.7 nm (Yakovenko et al., 2002). Thus, the possibility of systematic experimental artifact in the myofibril experiments yielding only 2.7-nm steps is ruled out. Further, the filament measurements are based on principles different from those of the myofibril, and yet the findings were similar. The fact that shortening and lengthening steps fall within the same paradigm implies reversibility of the contractile mechanism. Whatever process gives rise to contraction also appears to be involved in lengthening as well.

A third finding of interest is that contractile steps can be both small and large. The small step, or quantum, is 2.7 nm, but steps as large as five times that value were not infrequent. Such large steps could theoretically have arisen from inadvertently missed short pauses, but analysis of pause duration (Yakovenko et al., 2002) showed that this could be true only if the real pause-duration distribution were bimodal, with the "missed" peak separated from the observed peak by more than two times the SD of the observed peak, which seems unnatural. Hence, the contractile machinery apparently executes steps not only of 2.7 nm, but also of 5.4 nm and larger sizes as well. The capacity to generate such larger steps is also seen in various single molecule experiments (Kitamura et al., 1999; Yanagida et al., 2000; Molloy et al., 2000; Spudich et al., 2001). It implies a paradoxical combination of deterministic and stochastic components: a component that produce steps of size n x 2.7 nm, and a component that determines the value of n for each step, seemingly unpredictably, although no systematic attempt has been made to determine whether the size sequence might follow some subtle pattern. Interestingly, in the filament histogram and the sarcomere histogram, relative peak sizes were roughly similar.

A fourth finding of significance is that length changes occur with a high degree of cooperativity. The filament contains an appreciable number of parallel cross-bridges, and the myofibril contains many hundreds of parallel filaments. If length changes occurred randomly in each filament unit, the sarcomere shortening trace would be smooth, not stepwise. The consistently sharp transition between pauses and steps implies a high degree of cooperativity among parallel filaments.

The result of highest significance, perhaps, is the correspondence between dynamics and structure. Active shortening occurs in steps that are integer multiples of

2.7 nm. The 2.7-nm value is equal to the linear repeat of actin monomers along the thin filament. (Each of the two actin strands making up the filament has a 5.4 nm monomeric repeat, while the two strands are axially displaced by half that value, giving a linear repeat of 2.7 nm.) Correspondence between dynamics and structure is seen similarly in the microtubule-kinesin system, where the observed translation step is equal to the axial repeat of tubulin along the protofilament, 8 nm. Backwards steps are also integer multiples of 8 nm. Unless fortuitous, the striking parallelism between the two stepping paradigms implies a high level of mechanistic similarity: whatever mechanism governs kinesin-microtubule translation probably governs active translation in the myosin-actin system. The similarity of paradigms survives the difference in scale under which the respective measurements are carried out.

The results also show correspondence with at least some experiments in isolated actomyosin systems. The results reported by Kitamura et al. (1999) and Yanagida et al. (2000) using single myosin molecules translating along actin filaments showed consistent step sizes of approximately 5.4 nm. And, recent optical trap experiments from the Molloy laboratory (2000) show steps on the order of 5 – 6 nm. In the Guildford et al. (1997) experiments employing an optical trap, on the other hand, the step size was ~11 nm, almost exactly twice the Kitamura/Yanagida/Molloy value. Both these values were seen in our results — they show up as two and four times the more primary value of 2.7 nm. Whether such correspondence is more than coincidental awaits improvement in precision in the single molecule experiments.

The results also show correspondence with measurements from whole muscle fibers. When tetanically contracting fibers are abruptly released to shorten against a constant load, the shortening trace shows a series of oscillations that gradually damp out (Civan and Podolsky, 1966). That these oscillations may correspond to steps at the sarcomere level was first suggested by A. F. Huxley and group (Armstrong et al., 1966; Huxley, 1986), and the correspondence has been confirmed (Granzier et al., 1990). The sharp transient that occurs at the onset of the load clamp apparently synchronizes the steps, which show up as inflections in the fiber-length trace. The amplitudes of these inflections were recently measured by Edman and Curtin (2001), and were found to be 2.7 nm per half-sarcomere—the same as the step size measured here. Thus, the same fundamental step size is seen by two independent approaches.

What kind of mechanism fits these results? Steps that are integer multiples of the actin-monomer repeat are inevitable if each projecting myosin head is fixed in space and sticks transiently to one or another actin monomer as thin filaments translate past thick filaments. Step size should then be n x 2.7 nm, both for lengthening and for shortening. The larger steps ($n > 1$) would correspond to collectively "missed" opportunities for attachment.

Details of how the translational motion might be actively generated are open to speculation. One possibility is an inchworm-like mechanism, in which the actin filament crawls along the thick filament, each reptation producing an n x 2.7 nm step. A broad body of published evidence is consistent with this mechanism (Pollack, 2001). Whether this model proves adequate or not will be seen in the future. Nevertheless, the simple quantitative result reported here places a signature-like constraint on any proposed mechanism: The 2.7-nm paradigm is extremely robust, and would need to be explained outright by any proposed mechanism of contraction.

5. REFERENCES

Armstrong, C F., A. F. Huxley, and F. J. Julian. Oscillatory responses in frog skeletal muscle fibres. *J. Physiol.* 1966; 186: 26-27.

Blyakhman, F., Shklyar, T., and G. Pollack. 1999, Quantal length changes in single contracting sarcomeres. *J Muscle Res Cell Motil.* **20**: 529-538.

Blyakhman, T., Tourovskaya, A. and G. Pollack. 2001, Quantal Sarcomere-Length Changes in Relaxed Single Myofibrils. *Biophys J.* **81**: 1093-1100.

Civan, M. M. and R. J. Podolsky. Contraction kinetics of striated muscle fibres

Edman K.A. and N.A. Curtin. Synchronous oscillations of length and stiffness during loaded shortening of frog muscle fibres. *J. Physiol.* 2001; 534.2: 553-563.

Fauver, M., Dunaway. D., Lilienfeld, D., Craighead, H., and G. Pollack, 1998, Microfabricated cantilevers for measurement of subcellular and molecular forces. *IEEE Trans biomed eng.* **45**(7): 891-898.

following quick changes in load. J. Physiol. London 1966; 184: 511-534.

Granzier, H. L. M., Mattiazzi, A. and Pollack, G. H. Sarcomere dynamics during isotonic velocity transients in single frog muscle fibers. *Am. J. Physiol* 1990; 259: C266-278.

Guildford, W. H., Dupuis, D.E., Kennedy, G., Wu, J., Patlak, J. B., and D.M. Warshaw. Smooth muscle and skeletal muscle myosins produce similar unitary forces and displacements in the laser trap. *Biophys J.* 1997; 72(3): 1006-1021.

Huxley, A. F. Comments on "Quantal mechanism in cardiac contraction." *Circ. Res.* 1986; 59: 9-14.

Kitamura, K., Tokunaga, M., Iwane, A.H., and T. Yanagida. A single myosin head moves along an actin filament with regular steps of 5.3 nanometres. *Nature* 1999; 397: 129-134.

Kolmogoroff A. 1931, Uber die analytischen methoden in der Wahrscheinlichkeitsrechnung. *Math. Ann.* **104**:415-458.

Linke, W., Popov, V., and G. Pollack. 1994, Passive and active tension in single cardiac myofibrils. *Biophys J.* **67**: 782-792.

Liu, X. & Pollack, G. H., in press, Actin mechanics measured by nanofabricated cantilevers, *Biophys. J.*

Molloy, J., Kendrick-Jones, J., Viegel C., and R. Tregear. An unexpectedly large working stroke from chymotryptic fragments of myosin II. *FEBS Lett* 2000; 480: 293-297.

Molloy, J.E., Burns, J.E., Kendrick-Jones, J. Tregear, R.T. and D.C. White. Movement andforce produced by a single myosin head. *Nature* 1995; 378: 209-212.

Murphy, C., Rock, R., and J. Spudich. A myosin II mutation uncouples ATPase activity from motility and shortens step size. *Nature Cell Biol* 2001; 3: 311-315.

Pardee, J. D. & Spudich, J. A., 1982, Purification of muscle actin, *Methods Enzymol.* **85**:64-81.

Pollack, G.H. Cells, Gels and the Engines of Life. Ebner & Sons, Seattle, 2001.

Saito, K., Aoki, T., and T. Yanagida. Movement of single myosin filaments and myosin step size on an actin filament suspended in solution by a laser trap. *Biophys. J.* 1994; 66: 769-777.

Sellers, J. R., Han, Y. J., Kachar, B., 1991, The use of native thick filaments in in vitro motility assays, *J Cell Sci Suppl.* **14**:67-71.

Sokolov, S. (2002). IN PRESS

Svoboda, K., Schmidt, C., Schnapp, B., and S. Block. Direct observation of kinesin stepping by optical trapping interferometry. *Nature* 1993; 365: 721-727.

Trombitás, K. and G. Pollack. Actin filaments in honeybee-flight muscle move collectively. *Cell Motil Cytoskel* 1995; 32: 145-150.

Viegel C., Coluccio L., Jontes J., Sparrow J., Milligan R., and J. Molloy. The motor protein myosin-I produces its working stroke in two steps *Nature* 1999; 398: 530-533

Yakovenko, O., Blyakhman, F., Pollack, G. H., 2002, Fundamental step size in single cardiac and skeletal sarcomeres, *Am J Physiol Cell Physiol.* **283**: C735-C742.

Yanagida T, Esaki S, Iwane AH, Inoue Y, Ishijima A, Kitamura K, Tanaka H, and M. Tokunaga. Single-motor mechanics and models of the myosin motor. *Philos Trans R Soc Lond B Biol Sci* 2000; 355: 441-447.

Yang, P., Tameyasu, T., and G. Pollack, 1998, Stepwise dynamics of connecting filaments measured in single myofibrillar sarcomeres. *Biophys J.* 74: 1473-1483.

DISCUSSION

Curtin: In your 1st histogram of frequency vs step size, you commented that the stretch and shortening normally gave different frequencies for each step size. Was there a consistent difference between stretch and shortening (for example, small steps were more common during shortening than during stretch)?

Pollack: No, we have not detected a consistent difference between stretch and shortening, although we've not looked carefully. In the example I showed, the frequencies looked very similar for stretch and shortening at each step size, but this was not consistently the case.

Huxley: Could the factor which causes 2.3nm steps in passive muscle affect the sliding in active muscle and give slightly longer steps of 2.7nm?

Pollack: In the single filament preparation, there is no passive element; yet the same 2.7nm size paradigm is apparent. Thus, the passive elements do not seem to be involved in activated length changes.

Cecchi: Are the steps in the different sarcomeres synchronized?

Pollack: No, not in the single myofibril. All sarcomeres are in series, so if the steps were synchronized, the myofibril length change should be stepwise; but it was constrained to be a smooth, linear ramp. Therefore synchronized steps are impossible.

Ranatunga: You see a step size of 2.3nm in relaxed myofibril. What is that due to?

Pollack: I'm not sure. It may be titin, or it may be other structures. Measurements of length change in isolated titin molecules do show steps, but their size is much larger than those we measure.

Sugi: I am impressed by your long-lasting enthusiasm on the stepwise shortening phenomenon for over a quarter of a century. My question is how the stepsize changed according to the preparations used?

Pollack: Our earlier work showed integer multiples of 2.7nm steps, though the evidence was not convincing at that time. The 2.7nm steps were found by optical diffraction, high speed cinematography and phase-locked loop method, and were shown in my 1990 book to be reasonably well fit by a 2.7nm paradigm.

MOTOR FUNCTION OF UNCONVENTIONAL MYOSIN

Mitsuo Ikebe, Akira Inoue, So Nishikawa*, Kazuaki Homma, Hiroto Tanaka*, Atsuko Hikikoshi Iwane#, Eisaku Katayama**, Reiko Ikebe, and Toshio Yanagida#*

1. INTRODUCTION

Myosins are motor proteins that interact with actin filaments and convert energy from ATP hydrolysis into mechanical force. In addition to the well-characterized conventional, filament forming, two-headed myosin II of muscle and non-muscle cells, a number of myosin-like proteins have recently been discovered. Based upon their amino acid sequences, these newly found "myosins" do not seem to form myosin filaments, thus they are often called "unconventional" myosins. The discovery of these "myosin-like motor proteins" has fundamentally expanded the potential physiological importance of myosins in diverse biological processes such as chemotactic motility, endocytosis, exocytosis, phagocytosis, vesicular trafficking, secretion, *etc*. The myosins are classified based upon phylogenetic sequence comparisons of the motor domain (Cheney *et al.*, 1993; Goodson and Spudich, 1993; Mooseker and Cheney, 1995; Cope *et al.*, 1996; Titus, 1997; Hodge and Cope, 2000) and divided into at least 18 classes. In vertebrates, it has been shown that eleven classes of myosin, (including conventional filament forming myosin) are expressed. The N-terminal domains of these classes of unconventional myosins are relatively conserved and contain the primary force production machinery, whereas the C-terminal tail domains are highly divergent and are thought to function as targeting sites that bind to the cellular partner molecules. Between the motor and the diverse tail domains of myosin, there are neck regions that are composed of various numbers of light chain binding motifs (Mermall *et al.*, 1998).

One critical issue for the physiological relevance of each motor protein in diverse

Department of Physiology, University of Massachusetts Medical School, Worcester, Massachusetts 01655-0127, USA; * Single Molecule Process Project, ICORP, JST, 2-4-14, Mino, Osaka 562-0035; and #Graduate School of Frontier Biosciences, Osaka University, Suita, Osaka 565-0871; ** Division of Biomolecular Imaging, Institute of Medical University of Tokyo, Minato-ku, Tokyo 108-8639.

Correspondence to Mitsuo Ikebe: Department of Physiology, University of Massachusetts Medical School, 55 Lake Avenue North, Worcester, MA 01655. TEL: 508-856-1954   FAX: 508-856-4600

E-mail: mitsuo.ikebe@umassmed.edu

cellular motile systems is the processivity of each motor protein. It is known that conventional myosin is a typical non-processive motor, in which the motor protein dissociates from the track (actin) during each cross-bridge cycle. This non-processive property is favorable for filament forming conventional myosins. Since a large number of the motor proteins (myosin molecules) associate with a single actin filament, a prolonged binding with actin would interfere with the actin translocating activity of other myosins in the same filaments. Furthermore, the non-processive property is favorable for fast movement of actin since many myosins can interact with the same actin filaments without interfering each other. Recently it was found that myosin V is a processive motor, in which the motor molecule walks several steps on the track without dissociation from the track (actin filaments) (Mehta *et al.*, 1999; Sakamoto *et al.*, 2000; Walker *et al.*, 2000; Tanaka *et al.*, 2002; Veigel *et al.*, 2002). A motor protein having this property is suitable for cargo movement such as vesicular trafficking. Interestingly, it was found that the step size of myosin V is much larger than that of conventional myosin (Mehta *et al.*, 1999). According to the Lever-Arm model of the cross-bridge movement (Spudich, 1994), the step size is proportional to the length of the neck domain. Therefore, it was hypothesized that the extremely large step size of myosin V is due to its long neck domain (Lever-Arm) of myosin V (Mehta *et al.*, 1999). Subsequently, it was proposed that myosin V spans the helical repeat of actin filaments and walks along them with one head attached and the other head dissociated, based on electron microscopic observation of the shape of various myosin V molecules associating with actin (Walker *et al.*, 2000). However, our recent study has raised a question about this mechanism (Tanaka *et al.*, 2002). In the present study, we investigated the mode of movement of other unconventional myosins by multi-molecule and single molecule *in vitro* motility assay systems thus addressing the mechanism of "processive" movement of myosin.

Another critical issue for the physiological relevance of myosins is the directionality of the movement. In general, it is thought that actin filaments in cells protrude from the plasma membrane with the minus end at the tip of the filaments. It was originally thought that myosins are plus directed motors and that, myosin can only transport the cellular cargos from the inside of the cell to the cell surface. However, a recent finding of myosin VI as a myosin with reverse directionality (Wells *et al.*, 1999) has changed this view. The present study also addressed the directionality of myosin superfamily members.

2. PROCESSIVITY AND STEP SIZE

Processive movement of myosin VI having a short neck domain.

We studied whether myosin VI is a processive motor. While myosin VI is thought to be a two-headed myosin, it contains only one calmodulin light chain thus short neck domain. Based upon the hand-over-hand model of the processive movement of myosin V, myosin V's extremely long neck domain (lever-arm domain) is critical for the processive movement of myosin V. Myosin VI having a short neck domain is predicted to be a non-processive motor unless novel mechanism is operating for the movement of myosin VI. To address this question, we constructed chimeric myosin VI (GFP-M6HMMsRod) (Fig. 1A middle). It contained the motor domain, neck domain and entire coiled-coil of myosin VI. To stabilize the formation of the two-headed structure and visualize by fluorescence

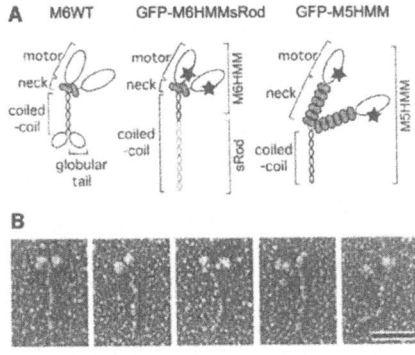

Figure 1. Structures of the myosin constructs. A, schematic representation of M6WT (left), GFP-M6HMMsRod (middle) and GFP-M5HMM (right). GFP moieties (★) were attached to the N-termini of M6HMMsRod and M5HMM. GFP and calmodulin/light chain are represented by stars and half tone, respectively. B, rotary-shadowed electron microscopic images of GFP-M6HMMsRods. The small globular domain protruded from the head domain is assumed to be a GFP moiety. Scale bar, 50 nm. (Nishikawa *et al.*, 2002, *Biochem. Biophys. Res. Commun.* **290**: 311-317, with permission)

microscopy, additional short coiled-coil domain (sRod) of myosin II that is too short to form filaments (Ikebe *et al.*, 2001) was attached to the C-terminal end of the coiled-coil domain of myosin VI. GFP (green fluorescent protein) was attached to the N-terminal ends of the construct to visualize the molecule under the fluorescence microscope. GFP-M6HMMsRod as well as wild type myosin VI (M6WT) was co-expressed with calmodulin in Sf9 insect cells. Electron microscopy revealed that GFP-M6HMMsRod had two heads connected with an extended rod as expected from the designed construct (Fig. 1B). GFP-M5HMM was also constructed for comparison (Fig. 1A right). The actin-activated ATPase activity (3 s^{-1}head^{-1}) of the GFP-M6HMMsRod was identical to M6HMM. The actin translocating velocity of GFP-M6HMMsRod (300 nm/s) in the surface gliding assay was identical to that of M6HMM and M6WT. Thus, N-terminal GFP moiety neither influenced the actin-activated ATPase activity nor the actin translocating velocity.

The movements of GFP-M6HMMsRods on actin filaments were directly observed using TIRFM (Funatsu *et al.*, 1995; Vale *et al.*, 1996; and Sakamoto *et al.*, 2000) (Fig. 2A). The binding to and the movement on actin filaments of GFP-M6HMMsRods were observed (Fig. 2C, left panels). The direction of movement was checked by polarity-marked actin filaments (Wells *et al.*, 1999; Homma *et al.*, 2001a). The direction of GFP-M6HMMsRod movement was consistent with the previous report for myosin VI subfragment-1 (Wells *et al.*, 1999) but opposite of that of GFP-M5HMM (Fig. 2C, right panels). To verify that moving fluorescent spots were indeed due to single GFP-M6HMMsRod molecules, the photobleaching characteristics and intensities of fluorescent spots were examined (Vale *et al.*, 1996). The fluorescent spots of surface-adsorbed GFP-M6HMMsRods disappeared primarily in a two-step process, as expected for photobleaching reactions of two GFP molecules bound to the GFP-M6HMMsRod (Fig. 2D). The fluorescence intensity measured just after binding to the glass surface, *i.e.*, before photobleaching (Fig. 2E, shaded bars) was approximately twice as large as that after the first photobleaching step (Fig. 2E, open bars). The fluorescence intensity of moving spots (Fig. 2F), which was measured just after encountering the actin filaments, was similar to that of the adsorbed ones before photobleaching. These results indicate that moving spots were due to single GFP-M6HMMsRod molecules.

The mean travel distance and duration attachment of single GFP-M6HMMsRod molecules were 240 nm and 0.44 s, respectively. The obtained travel distance (240 nm) is

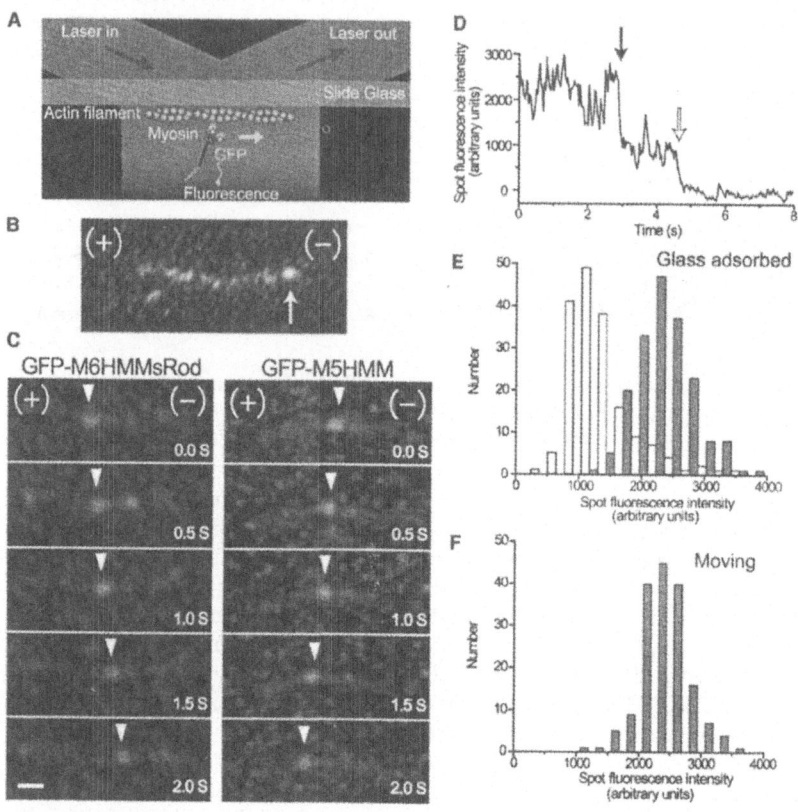

Figure 2. Processive movement of single molecules of myosin VI and V along actin filaments. A, schematic diagram used to observe movement of single myosin molecules by TIRFM. **B,** fluoresce image of a polarity-marked actin filament attached onto a glass surface via biotin-avidin system. A bright spot indicates the bright cap at the pointed end. **C,** sequential images of a single GFP-M6HMMsRod molecule (left panels) and a single GFP-M5HMM molecule (right panels) moving along an actin filament. Fluorescence images of a GFP-M6HMMsRod and an actin filament obtained by dual-view TIRFM were superimposed. (+) and (–) indicate the barbed and pointed ends of actin filaments, respectively. Arrowheads indicate the positions of myosin. Scale bar, 1 μm. **D,** a typical time course of photobleaching of a GFP-M6HMMsRod adsorbed onto a glass surface. Photobleaching took place by two steps (marked by arrows). **E,** histograms of fluorescence intensities of GFP-M6HMMsRods measured just after encountering with a glass surface, *i.e.*, before photobeaching (Shaded-bars) and after the first photobleaching step (Open-bars). **F,** a histogram of fluorescence intensities of GFP-M6HMMsRods moving along an actin filament. The fluorescence intensities of GFPM6HMMsRods were measured just after encountering with actin filaments, *i.e.*, before photobleaching. (Nishikawa *et al.*, 2002, *Biochem. Biophys. Res. Commun.* **290**: 311-317, with permission)

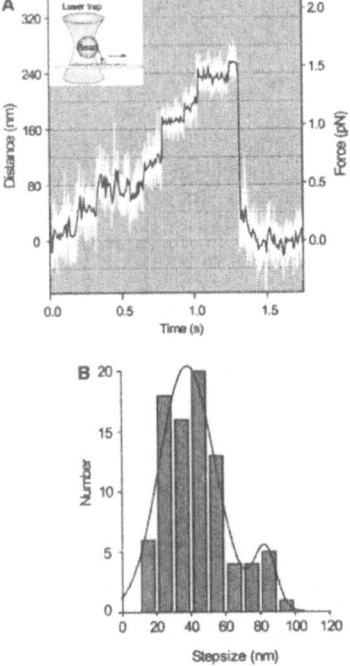

Figure 3. Optical trapping nanometry of bead-tagged M6WT moving along actin filaments. A, a typical record of the time course of motion of an optically trapped bead coated with M6WT along an actin filament. White dots, raw data. Black line, same data passed through a low-pass filter of 50 Hz bandwidth. Right ordinate indicates force calculated as (displacement) × (trap stiffness). Trap stiffness, 0.005–0.01 pN/nm. **B,** a histogram of the step size. Data were fit to two gaussians with peaks at 38 nm (SD = 16 nm) and at 82 nm (SD = 6 nm). A small gaussian at 82 nm may be due to two successive steps taking place rapidly within the temporal resolution (2 ms) of the measurement system. (Nishikawa *et al.*, 2002, *Biochem. Biophys. Res. Commun.* **290**: 311-317, with permission)

consistent with the recent enzyme kinetic analysis of myosin VI that the number of ATP molecules per encounter with actin is approximately 5 (De La Cruz *et al.*, 2001), if one ATP molecule is hydrolyzed per each 36 nm step (Rief *et al.*, 2000). About 60% of GFP-M6HMMsRods encountering with the actin filament moved more than 100 nm, that could be sufficiently determined by the computer image analysis. The mean velocity calculated from these values is 550 nm/s (= 240 nm/0.44 s), which is larger than the gliding velocity of actin filaments on GFP-M6HMMsRod- and M6WT-coated glass surfaces (~300 nm/s). A similar difference was observed for fluorescently-labeled wild type myosin V (Sakamoto *et al.*, 2000). This is probably due to the internal friction produced when many myosin molecules interact with an actin filament in the multi-molecule surface gliding assay.

A critical question is how myosin VI having a short lever-arm can move processively on actin filaments. To address this question, we attempted to determine the step size of myosin VI. We measured the movement of an optically-trapped bead coated with extremely low density thus only one molecule can interact with actin at a time (Mehta *et al.*, 1999). An optically-trapped M6WT-coated bead was brought into contact with an actin filament adsorbed on a glass surface. The movement of M6WT was determined by measuring the displacement of a bead with nanometer accuracy (Nishiyama *et al.*, 2001) (Fig. 3A, insertion). The displacement took place processively in a stepwise fashion, reached a plateau (the stall force ~1.5 pN) and fell suddenly to zero displacement (Fig. 3A). Figure 3B shows a histogram of step size. The average step size was ~38 nm. A small peak around 80 nm is probably due to two successive steps taking place rapidly within the temporal resolution (2 ms) of the measurement system. The fraction of moving beads was 0.25 at the molar ratio of molecules/beads of 1, which agreed that the number of molecules involved in the force generation is one. These results clearly demonstrate that myosin VI having only a short neck (lever-arm) moves processively with a large step size that is similar to myosin V. The results cannot be readily explained by the "lever-arm" hypothesis (Spudich, 1994). How does myosin VI move processively with a large step size? Does the "hand-over-hand" mechanism explain the processive movement of myosins? Further studies are required for answering these critical questions.

Processive movement of myosin IX, a single-headed myosin.

Class IX myosins are expressed in many tissues and cell types (Reinhard *et al.*, 1995; Chieregatti *et al.*, 1998; Wirth *et al.*, 1996). Several characteristic features of myosin IX have been revealed based upon its primary structure. There is an N-terminal extension that is similar to the Ras binding domain of Raf kinase and RalGEF (Kalhammer *et al.*, 1997). Within the motor domain, myosin IX contains a large insertion at the position of "loop 2", that may be involved in the interaction with actin (Schroder *et al.*, 1993) (Fig. 4A). Following the motor domain, there are several IQ motifs (four IQ motifs for myosin IXb) known as calmodulin/light chain binding motif. The tail domain contains a zinc binding region and a GAP (small G-protein GTPase activating protein) homology domain (Fig. 4A). Because of the presence of small G-protein binding domains, myosin IX has been thought to play a role in linking signal transduction to actin cytoskeleton (Sellers, 1999). There is no coiled-coil domain in myosin IX, that is present in the two-headed myosin to serve dimerization of myosin heavy chain and it is thought that myosin IX is a single-headed motor. Therefore, according to the "hand-over-hand" model, it is unlikely that myosin IX shows a processive movement.

We addressed whether myosin IX, a single headed-myosin, moves processively on actin filaments. We constructed myosin IX (M9BIQ4), containing the motor domain and the entire neck domain with four IQ motifs (Fig. 4A). A hexa-histidine (His) tag was introduced at the C-terminal end. M9BIQ4 was co-expressed with calmodulin in Sf9 cells because calmodulin has been identified to be light chains of myosin IX (Post *et al.*, 1998). The purified M9BIQ4 construct was composed of a high molecular mass band and a low molecular mass band and free from 200 kDa Sf9 conventional myosin and actin. The high molecular mass band (150 kDa) was consistent with the calculated molecular mass of M9BIQ4 heavy chain and was recognized by anti-His tag antibodies (Santa Cruz, CA) indicating that it is the M9BIQ4 heavy chain (Fig. 4B). The low molecular mass band showed a mobility shift with a change in Ca^{2+} that is characteristic of calmodulin (not shown), and also recognized by anti-calmodulin antibody (Fig. 4B). The stoichiometry of the bound calmodulin light chains was 4.1/heavy chain based upon densitometric analysis of the gel. Figure 4C shows the native gel electrophoresis of the purified M9BIQ4 and a full length myosin IXb. The mobility of M9BIQ4 was similar to that of myosin II S1 (one-headed form) but much larger than that of myosin II HMM (two-headed form). While the mobility of the full length myosin IXb was a little lower than that of M9BIQ4 because of its larger molecular mass, the mobility was consistent with the single headed structure. The result indicates that myosin IXb is single-headed in consistent with the primary structure of myosin IX having no coiled coil domain.

Figure 4D shows the binding of M9BIQ4 with actin in the presence of Mg^{2+}-ATP measured by actin co-sedimentation analysis. Nearly all M9BIQ4 was co-precipitated with actin in the presence of Mg^{2+}-ATP at very low actin concentration. M9BIQ4 was not precipitated without addition of actin (Fig. 4D). The result suggests that M9BIQ4 has a strong affinity for actin during the ATPase cycle. This is quite similar to myosin V, a processive motor, suggesting a processive nature of myosin IX. To further test whether myosin IX is a processive motor, we employed two approaches. First, we examined the dependence of the velocity of actin filament movement on the number of M9BIQ4 molecules. It has been shown that processive motor maintains a constant actin sliding

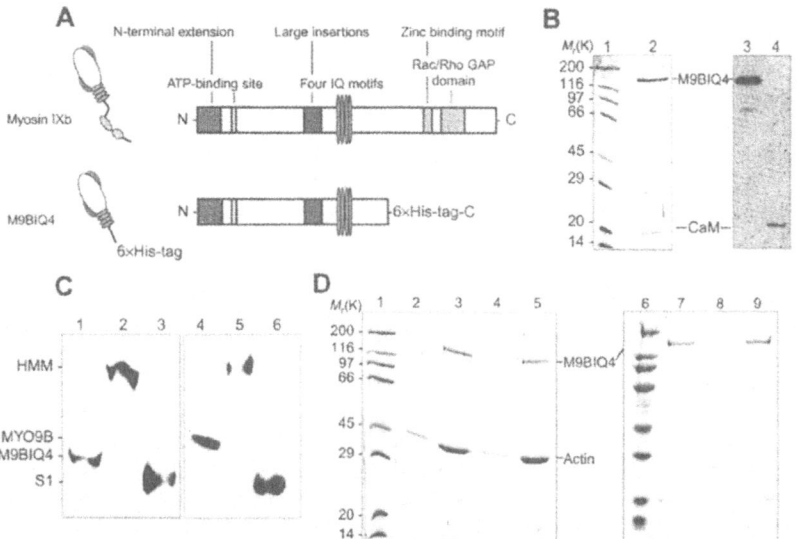

Figure 4. Structures of the myosin IX construct. A, schematic representation of myosin IXb and the M9BIQ4 mutant. M9BIQ4 contains a motor domain and a neck domain with four IQ motifs, each of which binds to calmodulin. **B**, SDS-PAGE analysis of purified M9BIQ4 protein demonstrates that it is composed of a high molecular mass component (heavy chain) and low molecular mass component (light chain). Lane 1, molecular mass standards; lane 2, purified M9BIQ4; lane 3, Western blot of M9BIQ4 with an anti-hexahistidine tag antibody; lane 4, Western blot of M9BIQ4 probed by anti-calmodulin antibody. Purified M9BIQ4 protein was transferred to PVDF membrane, then cut apart and incubated separately with anti-hexahistidine and anti-calmodulin antibodies. **C**, separation of purified myosin IXb on a native PAGE gel and analysis by Western blot. Lane 1, purified M9BIQ4; lane 2 and 5, smooth muscle myosin II HMM; lane 3 and 6, smooth muscle myosin II S1; lane 4, purified full-length myosin IXb. **D**, binding of M9BIQ4 to F-actin. The binding of M9BIQ4 was determined with an F-actin cosedimentation assay, in the presence and absence of ATP. Lane 1 and 6, molecular mass standards; lane 2, supernatant in the absence of ATP; lane 3, pellets in the absence of ATP; lane 4, supernatant in the presence of ATP; lane 5, pellets in the presence of ATP; lane 7, M9BIQ4 before centrifugation; lane 8, pellets of M9BIQ4 in the absence of actin and ATP; lane 9, supernatant of M9BIQ4 in the absence of actin and ATP. It should be noted that M9BIQ4 alone was not precipitated in the absence of F-actin. (Inoue *et al.*, 2002, *Nat. Cell Biol.* **4**: 302-306, with permission)

velocity independent of the surface motor density (Mehta *et al.*, 1999; Rock *et al.*, 2000), while the velocity of non-processive motors decreases with the reduction of the surface density of the motor molecules (Rock *et al.*, 2000). As shown in Fig. 5A, we measured the actin movement for a wide range of the M9BIQ4 surface densities. At high surface density (2×10^3 molecules/μm^2), M9BIQ4 supported continuous actin filament movement with the velocity of 0.08 ± 0.03 μm/sec. As we decreased the surface density of M9BIQ4, the actin filaments moved without decreasing the velocity. M9BIQ4 supported the movement as low as the surface density of 4 molecules/μm^2, which was similar to myosin V (Fig. 5A). In contrast, the velocity of actin filament movement decreased with decreasing the surface density of smooth muscle myosin II, a non-processive myosin, and myosin II could not support the actin filament movement at low surface density. At low surface densities of M9BIQ4, actin moved a limited distance presumably due to the presence of single M9BIQ4 in the area where the actin filament moved while the movement was more continuous at higher densities (Fig. 5B). Based upon a model determining the duty ratio (the fraction of the total cycle time that the myosin spends on actin) (Uyeda *et al.*, 1991), the duty ratio of M9BIQ4 was estimated to 1, suggesting a processive movement of M9BIQ4.

To further ascertain the processivity of M9BIQ4, we measured actin filament landing rate as a function of M9BIQ4 density. The landing assay tests the rate in which actin filaments land and move on the surface of a myosin coated coverslip. The landing rate decreases as the surface density of myosin molecules decreases. If only one myosin molecule is sufficient to support continuous movement, the landing rate shows pseudo first order dependence on myosin density (Mehta *et al.*, 1999; Rock *et al.*, 2000). For non-processive motors, a higher order dependence is obtained for this assay. When log (landing rate) is plotted against log (M9BIQ4) density, the slope of the obtained straight line provides the order of the landing process. As shown in Fig. 5C, the slope of the landing process of M9BIQ4 was 1. The result clearly indicates that myosin IX is a processive motor. At the very low surface density of M9BIQ4, the actin filaments were tethered to the surface by a single contact point (Fig. 6). During the movement, the actin filament remained connected at the in-focus single contact point, while the ends of the actin filament were out of focal plane, suggesting that the actin filament has interacted with only one molecule that is present at the in-focus contact point. As shown in Fig. 6, the actin filament tethered to the surface at a single point exhibiting nodal pivoting and moved its entire length through the single contact point. The actin filament finally diffused away when the end of the filament passed the contact point, presumably where M9BIQ4 exists. In the absence of myosin IX, while the filaments could be observed, they shifted from the focal plane within a few seconds by Brownian motion and the continuous motion of actin filament was not detected. All results obtained in this study indicate that myosin IX moves processively.

The mechanism by which myosin VI and myosin IX move processively.

How does myosin VI with a short neck domain processively move along the actin filament with large (38 nm) steps? It is proposed that myosin V with a long (23 nm) neck domain strides along the helical repeat of an actin filament by tilting the long neck domain of one head and leading the partner head to the neighbor helical pitch (36 nm)(Walker *et al.*, 2000), based on the lever-arm model (Spudich, 1994). However, myosin VI having a short neck domain is unlikely to span the actin helical pitch. One possibility is that the α-helix coiled-coil connecting the two heads unzips so that the heads can span the actin helical

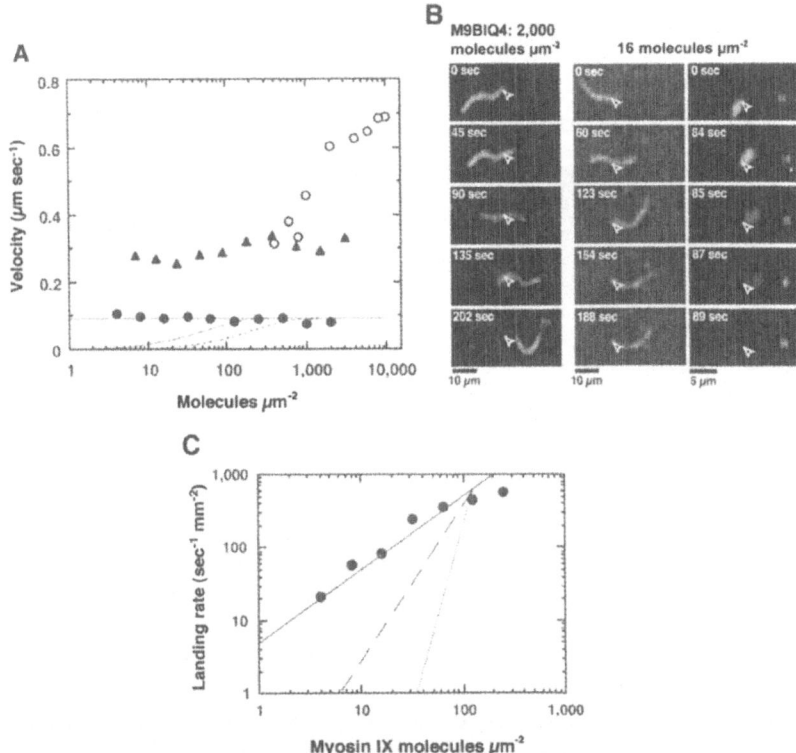

Figure 5. Processive movement of myosin IX. A, velocity of actin movement as a function of surface density of various myosins. Sliding actin filaments (7–20) were measured and the results are represented by means ± standard deviation. The solid black, broken and dotted lines represent the theoretical curves for a duty ratio of 1.0, 0.2 and 0.05, respectively. These curves were calculated with the equation, $V(\rho) = V_{max}[1-(1-f)^{\rho A}]$, where f is the duty ratio, ρ is the density and A is 0.08 μm^2. ●, M9BIQ4; ▲, myosin V HMM; O, smooth muscle myosin II. **B,** actin filament movement at high and low surface densities of M9BIQ4. Left panels show actin filament movement at a surface density of 2,000 molecules μm^{-2}. The middle and right panels show actin filament movement at a surface density of 16 molecules μm^{-2}. Arrowheads represent the terminus of actin filament at time zero. Times are indicated in each panel. For the long actin filaments, we did not observe the detachment of actin from the focal plane, even at low surface density (middle), because a buffer containing 0.5% methylcellulose was used to diminish the influence of Brownian motion. However, the detachment of the short actin filament, of 5 μm or less, could be observed (right) at low surface density, after the filament moved its entire length. At low surface density, the long actin filaments moved one actin filament length (middle, frames 1–3) and stopped (middle, frames 3–5). At higher surface density, actin filaments moved more than one actin filament length, presumably because of the presence of multiple myosin molecules (left). **C,** landing rate as a function of surface density of M9BIQ4. Solid, broken and dotted lines represent the theoretical curves of n = 1, 2, and 4, respectively. (Inoue *et al.*, 2002, *Nat. Cell Biol.* **4**: 302-306, with permission)

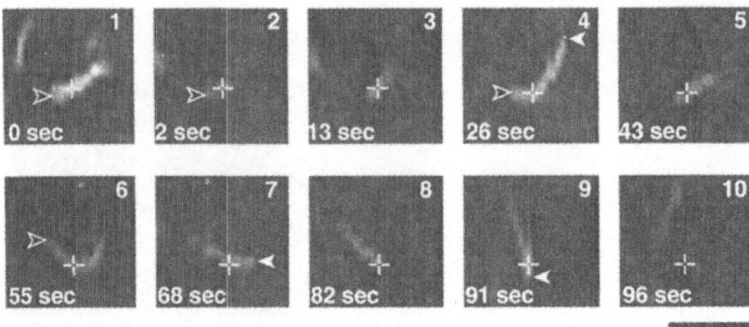

Figure 6. Actin filament movement with a very low surface density of myosin IX. Actin filament movement over a surface coated with 16 molecules of M9BIQ4 per μm². The contact point is indicated by the cross. Open and closed arrowheads represent the leading and trailing end of the filaments, respectively. The timecourse of actin filament movement is shown in the panels 1–10. The actin filament is attached at the in-focus contact point (presumably through an M9BIQ4 molecule). The ends of the filament are out of the focal plane (frames 2 and 3) and the images exhibit a diffusive rotation about the contact point. The actin filament movement started at frame 5. The actin filament moved through the contact point while exhibiting nodal pivoting (frames 6–9). When the end of the actin filament passed the contact point, the filament diffused away from the focal plane (frame 10). (Inoue *et al.*, 2002, *Nat. Cell Biol.* **4**: 302-306, with permission)

pitch. In rotary-shadowed electron micrographs of GFP-M6HMMsRods, some molecules appeared unzipping in the coiled-coil domain, though the majority of the molecules were not unzipped (Fig. 1B). The result suggests that there is flexibility in the coiled-coil of myosin VI. However, it is less likely that the possible unzipping enables myosin VI to step the helical pitch of actin. First, negatively stained images of electron micrographs did not show unzipping of myosin VI thus spanning the helical pitch of actin filaments. A GFP-M6HMMsRod molecule bound to the actin monomer with ~36 nm intervals corresponding to the actin helical pitch (Fig. 7A). Second, even if myosin VI coiled-coil is unzipped, such a flexible structure is unlikely to produce force and actually the 'lever-arm' model of the cross-bridge movement requires a rigid structure of 'lever-arm' domain.

Figure 7A shows a negatively-stained images of actin filament–GFP-M6HMMsRod complexes in the presence of 0.2 mM ATP. Interestingly, GFP-M6HMMsRod molecules bound to an actin filament with 35 ± 14 nm (mean $\pm$ SD, n = 407) intervals (Fig. 7B), coinciding with the actin helical pitch. This result suggests that GFP-M6HMMsRod molecules cooperatively bind to an actin filament, *i.e.*, when a myosin head binds to some position of an actin filament, the other head preferentially binds to the position ~36 nm (one helical pitch) apart from the pre-bound head. Recently, it was found by negative-stain electron microscopy that mutant myosin II subfragment-1 (one head), in which the 689th Gly is replaced by Val so that it forms a long-life actomyosin complex in the presence of ATP like myosin VI, periodically binds along one side of an actin filament at 36 nm intervals in the presence of ATP (E. Katayama and T. Q. P. Uyeda. Unpublished data). Therefore, it is plausible that the binding of a head of myosin VI evokes a 'hot spot' on the actin filament at 36 nm (actin helical pitch) apart from the initial binding site due to the

A

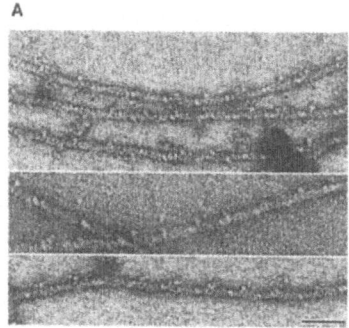

B

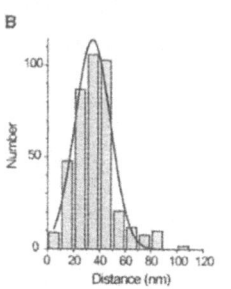

Figure 7. Binding of myosin VI to actin filaments during the movement. A, negatively-stained electron micrographs. GFP-M6HMMsRod interacted with actin filaments in the presence of 0.2 mM ATP. The buffer was rapidly replaced with a staining solution to instantaneously arrest the reaction (Katayama, 1989). Arrowheads indicate the positions of GFP-M6HMMsRod molecules bound to actin filaments. Scale bar, 50 nm. **B**, a histogram of the distance between the two adjacent GFP-M6HMMsRod molecules bound to actin filaments. A solid line is a gaussian fit to the data. The mean ± SD was 35 ± 14 nm (n = 407). (Nishikawa *et al.*, 2002, *Biochem. Biophys. Res. Commun.* **290**: 311-317, with permission)

conformational changes in an actin filament upon myosin binding. The partner head slides to the 'hot spot' thus producing the 36 nm step. While further evidence is required, the results suggest that the binding of myosin VI changes the actin filament structure. Such a change in actin filament structure would be involved in the processive movement of myosin VI and its large step size.

How can myosin IX, a single-headed myosin, move processively on actin filaments? It has been shown that myosin V, a two-headed myosin, moves processively along actin filaments with large steps (Mehta *et al.*, 1999). Walker *et al.* (2000) revealed by electron microscopic observation that myosin V can span the helical repeat of actin filaments. The two heads of myosin V on actin filament take polar conformation, in which one head is curved and the other straighter. This raised a hypothesis that the processive large steps of myosin V are produced by tilting the long neck domain of one head and leading the partner head to the neighbor helical pitch of an actin filament (Walker *et al.*, 2000). However, it is impossible for myosin IX to take this scenario for its processive movement since it only has one head. For microtubule based motors, KIF1A, a single headed kinesin family motor, shows a processive movement on microtubules (Okada and Hirokawa, 1999). It is proposed for KIF1A that electrostatic interaction between a Lys rich loop of KIF1A and Glu rich C-terminal end of tubulin (E-hook) interact with each other to prevent diffusion of KIF1A from microtubules (Okada and Hirokawa, 2000; Kikkawa *et al.*, 2001). It is plausible that a similar mechanism is operating for the processive movement of myosin IX on actin filaments. Understanding of the detailed mechanism of the processive movement of myosin IX requires further biophysical and biochemical studies.

3. DIRECTIONALITY

Directionality of the movement of the motor proteins is another important issue to evaluate the cellular function of the motor proteins. Among the myosin superfamily, it was found (Wells *et al.*, 1999) that class VI myosin moves towards the minus end of actin filaments in contrast to other characterized myosins, *i.e.*, myosin I, myosin II, and myosin V. It was proposed that the reverse directionality of myosin VI is due to the unique large insertion of myosin VI located between the motor domain and the neck domain, which changes the movement of the lever-arm of myosin to the opposite direction thus translocating actin backward. In other words, myosin VI is the only myosin that moves backward. This view was questioned by the recent finding and it was determined that the motor core domain is responsible for the directionality of myosin movement (Homma *et al.*, 2001a). Based upon this finding, it was predicted that there would be other myosin members that move backward. We examined the directionality of various unconventional myosins using the dual fluorescent labeled actin filaments. Among them, we found that myosin IX moves towards the minus end of actin filaments. Figure 8 shows the direction of the movement of myosin IXb (M9BIQ4). M9BIQ4 moved the dual fluorescence-labeled F-actin with the minus end at the rear of the movement. This means that myosin IX moved

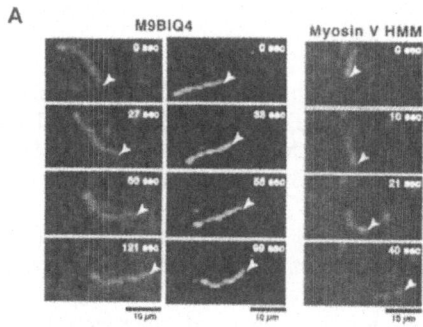

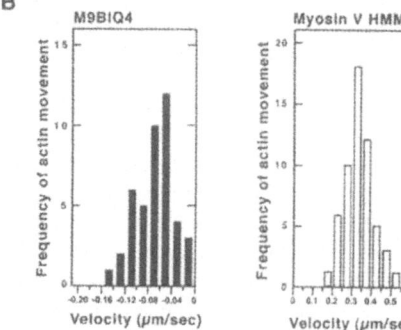

Figure 8. Direction of myosin IX movement. A, Movement of dual-labeled F-actin by M9BIQ4 or myosin V HMM. Times are indicated in each panel. The bright tip on the actin filament represents the minus end of the filament. The arrowheads indicate the leading part of the dual-labeled actin filaments at the front. Long actin filaments show clear orientation of the bright cap. Therefore, we observed the long actin filaments for the M9BIQ4 experiment. For 43 long actin filaments observed, the plus end of actin filaments lead the movement on M9BIQ4-coated glass surface. The result indicates that M9BIQ4 moves towards the minus end of F-actin. On the other hand, the minus end of the actin filaments lead the movement on myosin V HMM (a plus-end-directed myosin)-coated glass surface. B, Histogram of actin filament velocities. Movement towards the plus end of F-actin is defined as a positive value. Data was obtained with by a conventional *in vitro* motility assay with dual-fluorescence labeled F-actin. (Inoue *et al.*, 2002, *Nat. Cell Biol.* **4**: 302-306, with permission)

towards the minus end of actin filaments. The result agrees with the recent prediction by Homma *et al.* (2001a) that there are myosins moving backward in addition to class VI myosin and further supports that the motor domain but not the myosin VI unique insertion is responsible for the backward movement of myosin. On the other hand, myosin VII (not shown) and myosin X (Homma *et al.*, 2001b) moved towards the plus end of actin filaments. Recently, myosin XIV was found to be a plus directed motor (Herm-Gotz *et al.*, 2002). To date, the structural units that determine the directionality of myosin is not yet identified, but it is plausible that the difference in the structure down-stream of the "relay" element is involved in determining the directionality because the change in myosin head conformation during the ATPase cycle takes place in this region.

ACKNOWLEDGMENTS

We thank Dr. O. Sato (University of Massachusetts) for his help editing this manuscript.

REFERENCES

Cheney, R.E., Riley, M.A., and Mooseker, M.S., 1993, Phylogenetic analysis of the myosin superfamily, *Cell Motil. Cytoskelet.* **24**: 215-223.

Chieregatti, E., Gartner, A., Stofler, H.E., and Bahler, M., 1998, Myr7 is a novel myosin IX-rhoGAP expressed in rat brain, *J. Cell Sci.* **111**: 3597-3608.

Cope, M.J., Whisstock, J., Rayment, I., and Kendrick-Jones, J., 1996, Conservation within the myosin motor domain: Implications for structure and function. *Structure* **4**: 969-987.

De La Cruz, E.M., Ostap, E.M., and Sweeney, H.L., 2001, Kinetic mechanism and regulation of myosin VI, *J. Biol. Chem.* **276**: 32373-32381.

Funatsu, T., Harada, Y., Tokunaga, M., Saito, K. and Yanagida, T., 1995, Imaging of single fluorescent molecules and individual ATP turnovers by single myosin molecules in aqueous solution, *Nature* **374**: 555-559.

Goodson, H.V., and Spudich, J.A., 1993, Molecular evolution of the myosin family: relationships derived from comparisons of amino acid sequences, *Proc. Natl. Acad. Sci. U.S.A.* **90**: 659-663.

Herm-Gotz, A., Weiss, S., Stratmann, R., Fujita-Becker, S., Ruff, C., Meyhofer, E., Soldati, T., Manstein, D.J., Geeves, M.A., and Soldati, D., 2002, Toxoplasma gondii myosin A and its light chain: a fast, single-headed, plus-end-directed motor, *EMBO J.* **21**: 2149-2158.

Hodge, T., and Cope, M.J., 2000, A myosin family tree, *J. Cell Sci.* **113**: 3353-3354.

Homma, K., Yoshimura, M., Saito, J., Ikebe, R., and Ikebe, M., 2001a, The core of motor domain determines the direction of myosin movement, *Nature* **412**: 831-834.

Homma, K., Saito, J., Ikebe, R., and Ikebe, M., 2001b, Motor function and regulation of myosin X, *J. Biol. Chem.* **276**: 34348-34354.

Ikebe, M., Komatsu, S., Woodhead, J.L., Mabuchi, K., Ikebe, R., Saito, J., Craig, R., and Higashihara, M., 2001, The tip of the coiled-coil rod determines the filament formation of smooth muscle and nonmuscle myosin, *J. Biol. Chem.* **276**: 30293-30300.

Kalhammer, G., Bahler, M., Schmitz, F., Jockel, J., and Block, C., 1997, Ras binding domains: predicting function versus folding, *FEBS Lett.* **414**: 599-602.

Katayama, E., 1989, The effects of various nucleotides on the structure of actin-attached myosin subfragment-1 studied by quick-freeze deep-etch electron microscopy, *J. Biochem.* **106**: 751-770.

Kikkawa, M., Sablin, E.P., Okada, Y., Yajima, H., Fletterick, R.J., and Hirokawa, N., 2001, Switch-based mechanism of kinesin motors, *Nature* **411**: 439-445.

Mehta, A.D., Rock, R.S., Rief, M., Spudich, J.A., Mooseker, M.S., and Cheney, R.E, 1999, Myosin-V is a processive actin-based motor, *Nature* **400**: 590-593.

Mermall, V., McNally, J.G., and Miller, K.G., 1998, Transport of cytoplasmic particles catalysed by an unconventional myosin in living Drosophila embryos, *Nature* **369**: 560-562.

Mooseker, M.S., and Cheney, R.E., 1995, Unconventional myosins. *Annu. Rev. Cell Dev. Biol.* **11**: 633-675.

Nishiyama, M., Muto, E., Inoue, Y., Yanagida, T., and Higuchi. H., 2001, Substeps within the 8-nm step of the ATPase cycle of single kinesin molecules, *Nat. Cell Biol.* **3**: 425-428.

Okada, Y., and Hirokawa, N., 1999, A processive single-headed motor: kinesin superfamily protein KIF1A, *Science* **283**: 1152-1157.

Okada, Y., and Hirokawa, N., 2000, Mechanism of the single-headed processivity: diffusional anchoring between the K-loop of kinesin and the C-terminus of tubulin, *Proc. Natl. Acad. Sci. U.S.A.* **97**: 640-645.

Post, P.L., Bokoch, G.M., and Mooseker. M.S., 1998, Human myosin IXb is a mechanochemically active motor and a GAP for rho, *J. Cell Sci.* **111**: 941-950.

Reinhard, J., Scheel, A.A., Diekmann, D., Hall, A., Ruppert, C., and Bahler, M., 1995, A novel type of myosin implicated in signalling by rho family GTPases, *EMBO J.* **14**: 697-704.

Rief. M., Rock, R.S., Mehta, A.D., Mooseker, M.S., Cheney, R.E., and Spudich, J.A., 2000, Myosin-V stepping kinetics: A molecular model for processivity. *Proc. Natl. Acad. Sci. USA* **97**: 9482-9486.

Rock. R.S., Rief, M., Mehta, A.D., and Spudich, J.A., 2000, *In vitro* assays of processive myosin motors, *Methods* **22**: 373-381.

Sakamoto, T., Amitani, I., Yokota, E., and Ando, T., 2000, Direct observation of processive movement by individual myosin V molecules. *Biochem. Biophys. Res. Commun.* **272**: 586-590.

Schroder, R.R., Manstein, D.J., Jahn, W., Holden, H., Rayment, I., Holmes, K.C., and Spudich, J.A., 1993, Three-dimensional atomic model of F-actin decorated with *Dictyostelium* myosin S1, *Nature* **364**: 171-174.

Sellers, J.R., 1999, *Myosins (2nd ed.)*, Oxford University Press, New York.

Spudich, J.A., 1994, How molecular motors work, *Nature* **372**: 515-518.

Tanaka, H., Homma, K., Iwane, A., Ikebe, R., Saito, J., Katayama, E., Yanagida, T., and Ikebe, M., 2002, The motor domain determines the large step of myosin V, *Nature* **415**: 192-195.

Titus, M.A., 1997, Unconventional myosins: New frontiers in actin-based motors, *Trends Cell Biol.* **7**: 119-123.

Uyeda, T.Q.P., Warrick, H.M., Kron, S.J., and Spudich, J.A., 1991, Quantized velocities at low myosin densities in an *in vitro* motility assay, *Nature* **368**: 307-311.

Vale, R.D., Funatsu, T., Pierce, D.W., Romberg, L., Harada, Y., and Yanagida, T., 1996, Direct observation of single kinesin molecules moving along microtubules, *Nature* **380**: 451-453.

Veigel, C., Wang, F., Bartoo, M.L., Sellers, J.R., and Molloy, J.E., 2002, The gated gait of the processive molecular motor, myosin V, *Nat. Cell Biol.* **4**: 59-65.

Walker, M.L., Burgess, S.A., Sellers, J.R., Wang, F., Hammer, J.A., III, Trinick, J., and Knight, P.J., 2000, Two-headed binding of a processive myosin to F-actin, *Nature* **405**: 804-807.

Wells, A.L., Lin, A.W., Chen, L.Q., Safer, D., Cain, S.M., Hasson, T., Carragher, B.O., Milligan, R.A., and Sweeney, H.L., 1999, Myosin VI is an actin-based motor that moves backwards, *Nature* **401**: 505-508.

Wirth, J.A., Jensen, P.L., Post, W.M., Bement, M.S., and Mooseker, M.S., 1996, Human myosin IXb, an unconventional myosin with a chimerin-like rho/rac GTPase activating protein. *J. Cell Sci.* **109**: 653-661.

DISCUSSION

Katayama: Have you checked about the step size of myosin IX?

Ikebe: Now proceeding.

Yu: I noticed that, in your GFP-myosin VI pictures, there are multiple GFP-myosins on the same actin filament. How did you follow a particular one?

Ikebe: We followed the molecule in the movie.

SWITCH 1 OPENS ON STRONG BINDING TO ACTIN

Molecular and cellular aspects of muscle contraction

Kenneth C. Holmes and Rasmus R. Schröder*

1. INTRODUCTION

The 50K domain of the myosin cross bridge is split into two subdomains (upper and lower) by the major cleft that connects the actin binding site with the nucleotide binding site. A number of lines of experimental evidence now point to the major cleft closing on strong binding of the cross bridge to actin. In particular, our recent high resolution (14Å) cryo EFTEM images show that most of the 50K upper domain is involved in this movement. However, the switch 1 element of the nucleotide binding site is anchored in the 50K upper domain so that the strong binding to actin will lead to a 5-10Å movement of switch 1. Thus strong binding to actin moves switch 1. The movement of switch 1 opens the nucleotide binding site. This movement is quite distinct from the opening of switch 2, which is connected with the movement of the lever arm during the power stroke. The possibility that switch 1 moves (opens) on actin binding suggests new mechanisms for the initiation of the power stroke and for the release of ADP and phosphate during the cross bridge cycle.

2. GENERAL

At a Hakone Conference in 1986 we speculated on how the functionalities of a cross bridge, namely the power stroke, binding/releasing actin, and binding/releasing nucleotide, might be linked (Wray, Goody and Holmes 1988). In the mean time protein crystallography and electron microscopy have given us a rather precise view of the nano machines that carry out these functions. Our current problem is to explain these linkages in molecular terms.

---

* Max Planck Institute for Medical Research, Jahnstr. 29, 69120 Heidelberg, Germany

Molecular and Cellular Aspects of Muscle Contraction
Edited by H. Sugi, Kluwer Academic/Plenum Publishers, 2003

X-ray crystallography has yielded a very convincing structural basis for the power stroke that is consistent with a large body of experimental evidence (see review Geeves and Holmes 1999). Crystallographic studies have demonstrated that there are two major conformational states of the myosin cross bridge. These states differ in particular in the orientation of the long lever arm that is distal to actin in the actin-myosin complex. The rotation of this distal lever arm is responsible for the transport of actin filament past myosin cross bridges. On the other hand, the basis of actin binding/release and nucleotide binding/release and their strong inverse linkage has remained enigmatic. The reasons for this lie in the nature of the problem. Since the power stroke is essentially a property of the myosin cross-bridge alone, it can be studied by x-ray crystallography of isolated cross-bridges. Luckily, the two end states of the power stroke were identified quite quickly. However, to study the strong interaction with actin means dealing with the actin-myosin complex. Given the strange symmetry of decorated actin and its heterogeneity it not surprising that no crystals have been forthcoming. Therefore, some ten years ago, motivated by the need to get higher resolution structural data on decorated actin, we started a program of using energy filter cryoelectron microcopy to study the actin myosin cross-bridge complex (decorated actin). This research has now yielded a density map of decorated actin with sufficient resolution to show that the cleft in the 50K domain of myosin closes on binding to actin.

3. FITTING ATOMIC MODELS TO A 14Å RESOLUTION MAP OF DECORATED ACTIN

A 3D reconstruction has been obtained at 14Å resolution by the use of energy filter cryo-electron microscopy (Holmes et al. 2003), The use of an energy filter leads to a considerable increase in signal which in turn makes it possible to work closer to focus, thereby increasing the resolution. (Angert, Majorovits and Schröder 2000). However, this resolution is still well short of atomic. Our strategy therefore was to carry out a more quantitative version of the analysis first done in 1993 (Rayment et al. 1993, Schröder et al. 1993) namely, to fit atomic models of actin and the cross bridge derived from x-ray crystallography to the electron microscope density. We have carried out this fit by an automatic least squares procedure.

We used an improved f-actin structure (Holmes, unpublished) based on a four domain least squares refinement of a high resolution structure of g-actin (Dominguez et al. 1998) against x-ray fibre diffraction patterns of f-actin gels. (Holmes et al. 1990). For the cross-bridge we used the atomic coordinates of chicken S1 (Rayment et al. 1993) - 2MYS (the missing side chain coordinates were added for the light chains using Xplor - Lorenz, private communication). A refinement using 8 degrees of freedom (6 for the cross-bridge and 2 for the f-actin helix) yielded a good fit to the electron microscope density. However, it showed that the model was deficient in two ways: the 50K upper domain lay partly outside the density and the occupancy of the regulatory light chain in our preparations appeared to be about 30% so that adjustments to the effective density of the regulatory light chain needed to be made. In the subsequent refinement the upper 50K domain was allowed to refine as a separate object (see below) with 6 degreees of freedom. The resulting fit was excellent both for the actin and for the cross-bridge (density r-factor 22% with 14 degrees of freedom) and the stereochemistry of the actin-myosin interface was good. The parameters of the fit were robust: e.g. making adjustments to the degree of regulatory light chain occupancy produced very minor effects.

Allowing the 50K upper domain to refine separately produced a hinge-like movement of the upper 50K domain that brought the cardiomyopathy loop into contact with actin (Fig. 1). The area in contact with actin was 1000Å^2 when the original cross bridge coordinates were used and rose to 2000Å^2 when the upper 50K domain was allowed to refine separately. This appears to be the origin of strong binding.

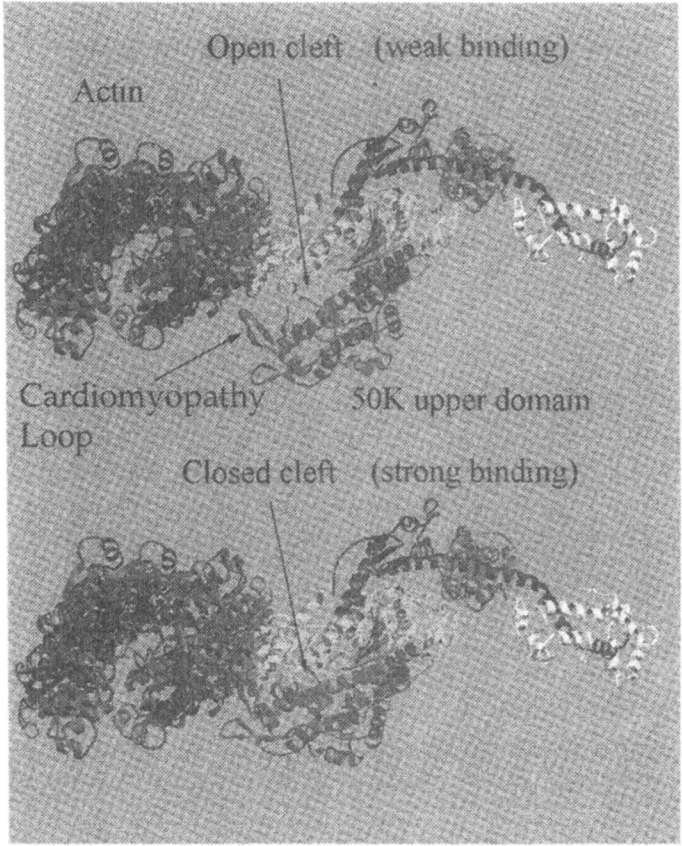

Figure 1. A view looking along the actin filament, which is to the left. The open actin binding cleft found in crystalline structures of the cross bridge (upper panel) closes on strong binding to actin (lower panel). The 50K upper domain rotates about 10° towards the actin filament.

Inspection of the structure leads to a clear demarkation of the 50K upper domain (Fig. 2) (all numbers in this section refer to chicken skeletal sequence). The 50K upper domain is linked to the body of the motor domain by three "loops", including the trypsin sensitive loops 1 and 2, some residues near 449 that are disordered in smooth muscle, and the highly conserved "strut" sequence (asp601, pro602) that has been mutated by

Sutoh (Sasaki, Ohkura and Sutoh 2000) to produce constitutively weak actin binding. In addition, there are two hinge points at 270 and 246 where the 50K upper domain connects onto the 7-stranded β-sheet of the motor domain. The switch 1 sequence, which forms part of the nucleotide binding site (see below), is an integral part of the 50K upper domain. It forms the end strand of a small 4 stranded β-sheet.

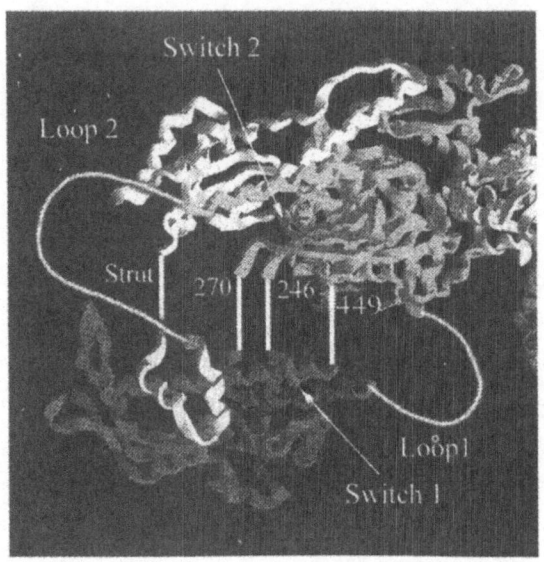

Figure 2. A diagram of part of the cross-bridge showing the 50K upper domain (grey). – same orientation as in Fig. 1. The 50K lower domain (actin binding) is shown in white and the body of the motor domain in light grey. (The correspondence between chicken skeletal and Dictyostelium is as follows: phe246=phe239; leu270=leu263; leu449=leu441).

4. STRONG BINDING MOVES SWITCH 1 AND OPENS THE NULEOTIDE BINDING POCKET

The myosin cross bridge is a P-loop protein with switch 1 and switch 2 elements similar to those of the G-proteins. The ATP is bound by the P-loop between the switch 1 and switch 2 elements. However, the switch 1 element of the nucleotide binding site is firmly anchored in the 50K upper domain. As a result of strong binding to actin, therefore, switch 1 moves away from the nucleotide binding site. Thus the strong binding to actin moves switch 1 and opens up the ATP binding site (Fig. 3). This movement is quite distinct from the opening of switch 2, which is caused by a 5° rotation

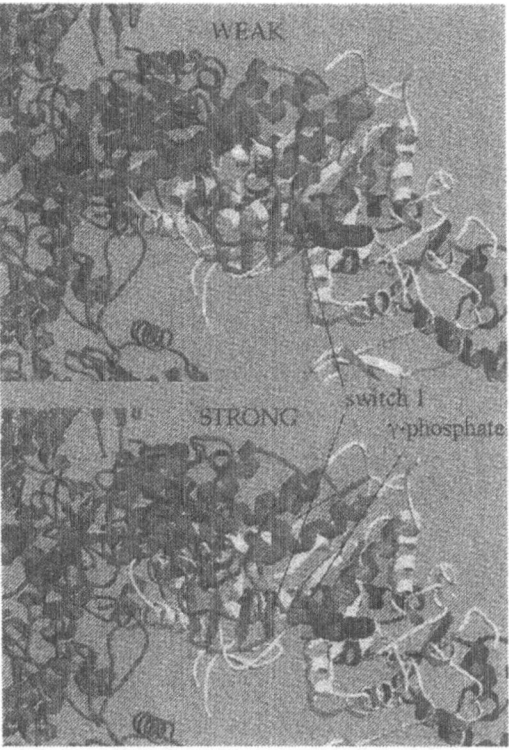

Figure 3. A view at 90° to Fig. 1 (actin filament on the left) showing the effect of strong binding. The 50K upper domain swings towards the actin filament and moves out switch 1, which is part of a small 4-stranded β-sheet. This breaks hydrogen-bonds made to the β and γ phosphates and turns the nucleotide binding tunnel into a nucleotide binding groove.

of the 50K lower domain with respect to the rest of the motor domain, which in turn leads to a 60° rotation of the lever arm. While the opening/closing of switch 2 has been observed a number of times by x-ray crystallography, the closing of the cleft has not yet been seen in isolated cross bridges. The movement of switch 1 breaks three hydrogen bonds to the polyphosphate moity and thereby weakens the nucleotide binding. Conversely, the binding of ATP to the active site forces the moving-in of switch 1 and closes the ATP binding site. This in turn leads to the opening of the cleft, which halves the area of the actin binding site and results in a major reduction of the binding affinity to actin. Thus the seesaw-like movement of the 50K upper domain provides a structural basis for the reciprocal relationship between actin affinity and ATP affinity.

Because loop 1 is particularly involved in binding to the γ-phosphate, the binding of ATP (rather than ADP) would appear to be necessary to pull in loop 1, which in turn opens the deep cleft and reduces the area of the cross-bridge in contact with actin. Conversely, the strong binding of actin opens the nucleotide binding site and reduces the coordination of the β and γ phosphates leading to release of nucleotide.

5. HOW DOES ACTIN BINDING INITIATE THE POWER STROKE?

To answer this question we need to know the structure of the pre-power stroke cross-bridge when strongly bound to actin. The pre-power stroke bound state is experimentally unavailable (at least in non-processive myosins), but one can model this state from the available structural data. To carry out this construction we assume that the actin-myosin interaction has constant geometry. On this assumption the 50K lower domain together with the 50K upper domain in its strong binding position make a stereospecific interaction with actin that is independent of the state of the cross bridge. Then we may model the pre-power stroke state bound to actin by using the 50K lower domain position found in the post-power stroke state when strongly bound to actin (rigor complex, Holmes et al. 2003) to determine the orientation of the motor domain in the pre-power stroke state. Further we assume that the 50K upper domain takes the same position in the pre- and post-power stroke states. Fortunately, pairs of structures are available for dictyosteleum in the pre- and post-power stroke states (Fisher et al. 1995; Smith and Rayment 1996a; Kull et al. 2003).

Mutational analysis has shown the importance of an invariant buried salt bridge that bridges switch 2 and switch 1. Dictyostelium myosin motor domains myosin with mutations E459R or R238E, that block salt-bridge formation, show defects in nucleotide-binding, and reduced rates of ATP hydrolysis. Inversion of the salt-bridge in double-mutant M765-IS eliminates most of the defects observed for the single mutants. In crystal structures the salt bridge is present in the closed form of the cross-bridge when it is not bound to actin (i.e. pre-power stroke state in crystals) but is broken in the open (post-power stroke state). It is broken because switch 2 swings away from the nucleotide binding site. It seems that the breaking of this salt bridge is one of the key events happening at the begining or during the power stroke. One of the consequences of the model described in 3. is to predict that the swinging in of the 50K upper domain to affect strong actin binding in the pre-power stroke state would also break this salt bridge (Fig. 4). This in turn suggests how the strong binding binding to actin could initiate the power stroke by breaking the salt bridge: switch 2 would no longer be tethered to switch 1 and could start to move out as the lever arm moves down.

A further effect of the moving of switch 1 is the opening up of the nucleotide binding site (Fig. 4) by breaking hydrogen bonds to the β- and γ-phosphate groups. This alone does not appear to be enough to cause phosphate release since the phosphate is not released until later in the power stroke. However, the moving of switch 2, which is strongly linked to the power stroke via the mechanism of the relay helix and the converter domain, breaks a strong main-chain amide H-bond to the γ-phosphate leaving it and the associated magnesium ion poorly coordinated. It appears therefore that the coincident moving out of both switch 1 and switch 2 is the final signal for product release.

A further effect of the moving of switch 1 is the opening up of the nucleotide binding site (Fig. 4) by breaking hydrogen bonds to the β- and γ-phosphate groups. This alone does not appear to be enough to cause phosphate release since the phosphate is not released until later in the power stroke. However, the moving of switch 2, which is strongly linked to the power stroke via the mechanism of the relay helix and the converter domain, breaks a strong main-chain amide H-bond to the γ-phosphate leaving

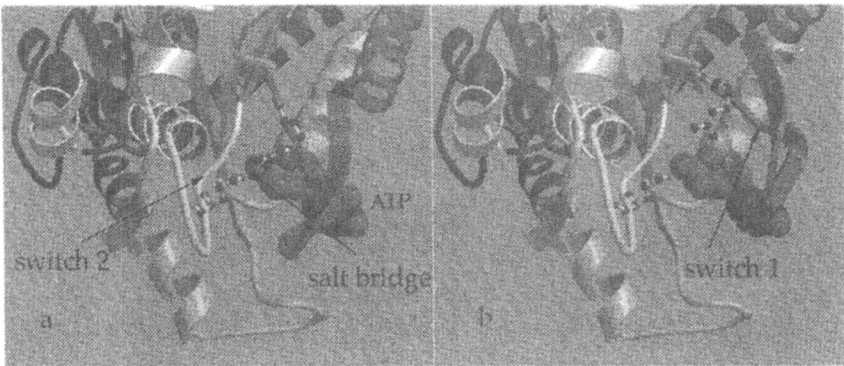

Figure 4. The two pictures show the environment of the ATP binding site. The P-loop and associated helix are behind the nucleotide. Switch 2 (lower 50K domain) is white and switch 1 (50K upper domain) is shown dark grey.
a) Cross-bridges in the pre-power stroke state but not attached to actin (Kull et al. 2003).
   The salt bridge between Glu459 and Arg238 can be seen.
b) A model of the pre-power stroke state when attached to actin. The effect of strong binding has been to move switch 1 up and out from the binding site. Note that the salt bridge is broken. This would allow switch 2 to move away from the _-phjosphate This movement in turn drives the power stroke.

it and the associated magnesium ion poorly coordinated. It appears therefore that the coincident moving out of both switch 1 and switch 2 is the final signal for product release.

6. ACKNOWLEDGEMENTS

The following computer programs have been used in constructing the figures: GRASP (Nicholls, Sharp and Honig 1991), Bobscript (Esnouf 1997), Molscript (Kraulis 1991), and Raster3D (Merritt and Bacon 1997).

7. REFERENCES

Angert, I., E. Majorovits and R. R. Schröder, 2000, Zero-loss image formation and modified contrast transfer theory in EFTEM, Ultramicroscopy. 81: 203-22.
Dominguez, R., Y. Freyzon, K. M. Trybus and C. Cohen, 1998, Crystal structure of a vertebrate smooth muscle myosin motor domain and its complex with the essential light chain: visualization of the pre-power stroke state, Cell. 94: 559-71.
Esnouf, R. M., 1997, An extensively modified version of MolScript that includes greatly enhanced coloring capabilities, J Mol Graph Model. 15: 132-4, 112-3.
Fisher, A. J., C. A. Smith, J. Thoden, R. Smith, K. Sutoh, H. M. Holden and I. Rayment, 1995, Structural studies of myosin:nucleotide complexes: a revised model for the molecular basis of muscle contraction, Biophysical Journal. 68: 27S-28S.

Geeves, M. A. and K. C. Holmes, 1999, Structural mechanism of muscle contraction, Ann. Rev. Biochemistry. 68: 687-727.

Holmes, K. C., F. J. Kull, W. Jahn, I. Angert and R. R. Schröder, 2003, High resolution cryo-electronmicroscopy of decorated actin shows that the actin bindng cleft is closed in the rigor state and that this opens switch 1, in preparation.

Holmes, K. C., D. Popp, W. Gebhard and W. Kabsch, 1990, Atomic model of the actin filament, Nature. 347: 44-9.

Kraulis, P. J., 1991, MOLSCRIPT: a program to produce both detailed and schematic plots of protein structures, Journal of Applied Crystallography. 24: 946-50.

Kull, J., I. Schlichting, A. Becker, D. Manstein and K. C. Holmes, 2003, The structure of dictyostelium myosin truncated at position 754 with bound ADP.BeF3, (in preparation).

Merritt, E. and D. Bacon, 1997, Raster 3D: Photorealistic Molecular Graphics, Methods in Enzzymology. 277: 505-524.

Nicholls, A., K. A. Sharp and B. Honig, 1991, GRASP: Graphical representation and analysis of structural properties, Proteins. 11: 281-296.

Rayment, I., H. M. Holden, M. Whittaker, C. B. Yohn, M. Lorenz, K. C. Holmes and R. A. Milligan, 1993, Structure of the actin-myosin complex and its implications for muscle contraction, Science. 261: 58-65.

Sasaki, N., R. Ohkura and K. Sutoh, 2000, Insertion or deletion of a single residue in the strut sequence of Dictyostelium myosin II abolishes strong binding to actin, Journal of Biological Chemistry. 275: 38705-38709.

Schröder, R. R., D. J. Manstein, W. Jahn, H. Holder, I. Rayment, K. C. Holmes and J. A. Spudich, 1993, Three-dimensional atomic model of F-actin decorated with Dictyostelium myosin S1, Nature. 364: 171-4.

Smith, C. A. and I. Rayment, 1996a, X-ray structure of the magnesium(ii).adp.vanadate complex of the dictyostelium-discoideum myosin motor domain to 1.9Å resolution, Biochemistry. 35: 5404-5417.

Wray, J. S., R. S. Goody and K. C. Holmes, 1988, Towards a molecular mechanism for the crossbridge cycle, Advances in Experimental Medicine & Biology. 226: 49-59.

DISCUSSION

Kushmerick: The essential step of the cross-bridge cycle chemistry is hydration of the nucleotide rather than hydrolysis of γ phosphate?

Holmes: One can only speculate. You obviously need hydrolysis to drive the process.

Katayama: Do you have any experimental evidence that the binding mode of pre-power stroke motor domain is the same as in rigor?

Holmes: No, but no evidence against it, either. Although low resolution is a problem, Peter Knight's work on myosin V shows that the motor domain binds with constant geometry.

Sugi: Can you prove that the lever arm motion of myosin molecules can generate torque strong enough to move high external loads?

Holmes: Myosin molecules have been shown to generate forces of piconewtons.

Sugi: But, I want to point out that the motility assay results do not necessarily mean that myofilament sliding is caused only by the lever arm mechanism.

Pollack: Ikebe showed that myosin IX, containing a single head, moves in 36nm steps along actin. How can that be reconciled with the lever arm mechanism that you suggest? Do different myosins then act by different mechanisms?

Holmes: Yes, different myosin apparently act by different mechanisms. However, the core sequences of the motor domains in all myosin are highly conserved, suggesting a common mechanism.

Yu: What is your suggested sequence of the movement of switches 1 and 2?

Holmes: You need closure of 1 and 2 for hydrolysis to occur. Once 2 opens, hydrolysis cannot happen. I would guess that the moving of switch 1 starts the power stroke. Switch 2 moves out as the power stroke proceeds.

Cecchi: Do you know what would be the position of the lever arm during isometric contraction?

Holmes: Somewhere in between the crystallographic positions.

Gene Transfer of Troponin I Isoforms, Mutants, and Chimeras

Margaret V. Westfall[1,2] and Joseph M. Metzger[1*]

1. SUMMARY

Thin filament proteins play an essential role in the regulation of myocardial pressure development. Within the thin filament of the sarcomere, troponin I (TnI) plays a key role in regulating the Ca^{2+} sensitivity of force. During myocardial development, there is a transition in TnI isoform expression from the slow skeletal isoform (ssTnI) in embryonic/fetal myocardium to the cardiac isoform (cTnI) expressed in adult hearts. Over a similar developmental time window, the calcium sensitivity of force development also decreases. Gene transfer of ssTnI, and chimeras derived from ssTnI and cTnI, into adult ventricular myocytes have provided insights into the isoform-specific domains of TnI responsible for differentially influencing myofilament Ca^{2+} sensitivity. Two separate isoform-specific regions, located in the carboxyl- and amino- portions of the protein, have been identified by comparing Ca^{2+}-activated isometric tension in myocytes expressing the TnI isoforms or chimeras. The carboxyl-portion of TnI also contributes to isoform-dependent differences in myofilament sensitivity to acidic pH, which ensues during several myocardial disease states. In contrast, the diminished Ca^{2+} sensitivity observed in response to β-adrenergic-mediated phosphorylation of cardiac TnI requires the amino-portion of the cardiac TnI isoform yet, does not depend on the presence of a specific isoform in the carboxyl-region of TnI. Recent studies with a mutation linked to hypertrophic cardiomyopathy have demonstrated that changes in protein charge also influence the ability of TnI isoforms to regulate myofilament Ca^{2+} sensitivity. Information gained from these, and future studies on more localized and specific changes in the amino acid sequence, may one day lead to the use of genetically engineered TnI for therapeutic manipulation of contractile function.

2. ROLE OF TROPONIN I IN MYOFILAMENT CONTRACTILE FUNCTION

---

[*] Departments of Physiology1 and Surgery, University of Michigan, 1301 E. Catherine St., Ann Arbor, MI 48109, phone: 734-763-0560, fax: 734-936-8813, e-mail:metzgerj@umich.edu

The ability of the heart to pump blood depends on the ability of individual cardiac myocytes to generate force during systole and subsequently relax during diastole. Thick and thin filaments organized into sarcomeres act as the force generators in these cells. Strong force-generating interactions between actin in the thin filament and myosin in the thick filament, are regulated by the thin filament proteins troponin and tropomyosin (Tm). Troponin is a heterotrimeric protein composed of a Ca^{2+} binding subunit, troponin C (TnC), an inhibitory subunit troponin I (TnI), and a Tm-binding subunit, troponin T (TnT). TnC binds Ca^{2+} as the cytosolic level of this ion increases during systole[1,2], and this initiates concerted conformational changes within TnI, TnT, Tm, and actin[1-3]. Based on the stearic hindrance model[4], these conformational changes collectively act to move Tm deeper into the actin groove[5], and allow the development of strong force-generating interactions between actin and the myosin cross-bridge. Further work to define the functional properties and domains within each of the thin filament proteins is needed to better understand cardiac force generation and relaxation. Improved understanding of functional domains within these proteins may one day reveal molecular targets for therapeutic treatment of cardiac dysfunction, a primary cause of morbidity and mortality in industrial nations.

Previously, knowledge about the contribution of TnI, and domains within TnI, to the regulation of contractile function has primarily come from biochemical/biophysical experiments with purified TnI[1,3]. Results from these studies demonstrated that TnI binds tightly to actin in the absence of Ca^{2+}. As Ca^{2+} levels rise, the interaction between TnI and actin weakens and TnI becomes more tightly bound to TnC. This toggling of TnI between actin and TnC in response to Ca^{2+} indicates that TnI works as part of a molecular switch within the thin filament. This approach also has been utilized to define functional domains within TnI[6]. In these studies, the inhibitory peptide (IP) region within TnI (residues 138-149 in rodent sequence) is the minimum sequence necessary to inhibit actomyosin interactions[1,6]. A secondary region (residues 150-182) further strengthens the inhibitory characteristics of TnI, although complete inhibition requires the presence of the full-length TnI[6].

3. DEVELOPMENTAL TRANSITION IN TROPONIN I EXPRESSION AND MYOFILAMENT FUNCTION

Insight into TnI structure-function also has come from studies on developing myocardium. The ssTnI isoform is expressed in embryonic myocardium, with a transition to the adult cardiac isoform observed during the late fetal or early neonatal period in all mammals[7] (see Fig. 1). This transition in TnI isoform expression is accompanied by decreasing myofilament sensitivity to Ca^{2+} [8], increased sensitivity to acidic pH[8], and increasing responsiveness to ••adrenergic-mediated enhancement of contractile function[9]. However, multiple myofilament proteins undergo isoform transitions over a similar period of time, making it difficult to assign a specific role for TnI. Recently, adenoviral-mediated gene transfer has distinguished the role of ssTnI from cTnI in adult cardiac myocytes[10]. This approach results in virtually complete replacement of endogenous cTnI with ssTnI, without changing the stoichiometry of the other contractile proteins. Force measurements made after permeabilization of these myocytes demonstrated that TnI isoform expression plays a key role in defining myofilament Ca^{2+} sensitivity[10], but does not significantly influence maximum force. Moreover, TnI isoforms significantly contribute to the pH sensitivity of the myofilaments, with decreased myofilament sensitivity to acidic pH observed in myocytes expressing ssTnI vs cTnI [10]. Myofilament pH sensitivity is important because acidosis often develops during pathophysiological states associated with myocardial ischemia[11,12], including diabetes, cardiomyopathies, and congestive heart failure. The decreased Ca^{2+} sensitivity observed with cTnI compared to ssTnI during acidosis likely contributes to reduced force generation in these dysfunctional states.

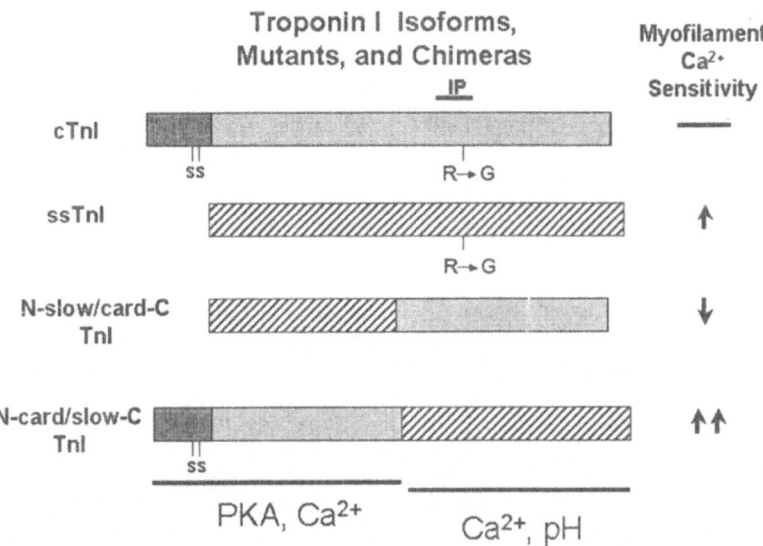

Figure 1. Alignment of troponin I isoform, mutant, and chimera proteins, and changes in myofilament Ca^{2+} sensitivity of tension observed in myocytes expressing these TnI proteins. Functional domains are shown below the aligned proteins. These domains were defined by measuring Ca2+-activated force in adult myocytes expressing these isoforms, mutants, and chimeras. Light gray indicates the cTnI isoform, while slow skeletal TnI (ssTnI) is shown in the hatched regions. The dark gray region within cTnI and N-card/slow-C TnI represents the 32 amino acid extension phosphorylated by protein kinase A (PKA). The mutant proteins studied are indicated below the wildtype isoform with an arrow between R and G. IP indicates the inhibitory peptide region.

The cardiac TnI isoform also contains a 32 amino acid extension that is a target for phosphorylation by protein kinase A (PKA), while ssTnI lacks this extension[9] (see Fig. 1). In experiments on myocytes expressing ssTnI after gene transfer, PKA did not phosphorylate this TnI isoform, and the decrease in myofilament sensitivity to Ca^{2+} observed in myocytes expressing cTnI was absent[13]. In subsequent experiments with transgenic mice expressing ssTnI, this maintenance of myofilament Ca^{2+} sensitivity in response to β-adrenergic activation resulted in partial loss of the accelerated myocyte relaxation observed in wildtype mice[14]. Collectively, results from gene transfer experiments demonstrate that TnI isoforms play a critical role in determining myofilament Ca^{2+} sensitivity at physiological pH, during the acidosis that often arises in response to myocardial ischemia, and in response to signal transduction via the PKA pathway.

4. FUNCTIONAL DOMAINS OF TROPONIN I DEFINED USING CHIMERAS OF TROPONIN I ISOFORMS

To further define isoform-specific functional domains, this gene transfer approach was used with TnI chimeras created from the ssTnI and cTnI isoforms (Fig. 1). Expression of the chimeras, N-card/slow-C TnI and N-slow/card-C TnI, was achieved in adult myocytes as was myofilament incorporation with virtually complete replacement of endogenous cTnI[15,16]. Two Ca^{2+} sensitivity domains were defined based on Ca^{2+}-activated force measurements[15,16]. Unexpectedly, the amino-terminal domain of ssTnI demonstrated reduced Ca^{2+} sensitivity compared to cTnI, while the carboxyl-region of ssTnI showed greater Ca^{2+} sensitivity than cTnI. The carboxyl-region of each isoform acts as the dominant domain in determining overall Ca^{2+} sensitivity. Thus, results obtained from studies with chimeras provide further evidence for the key contribution of TnI isoforms in determining myofilament Ca^{2+} sensitivity.

Insights into pH sensitivity and ··adrenergic-mediated influences of TnI were also obtained from studies with TnI chimeras. Force measurements made under acidic pH conditions indicated that the carboxyl-portion of TnI imparts pH sensitivity, with the cardiac isoform being more sensitive to acidic pH than ssTnI (Fig. 1). Phosphorylation experiments provided important information about the domains responsible for the TnI contribution to the PKA-mediated response to ··adrenergic activation[13]. The N-card/slow-C TnI chimera contains the 32 amino acid extension found in cTnI, and phosphorylation of this chimera resulted in a comparable decrease in myofilament Ca^{2+} sensitivity as endogenous cTnI. In contrast, N-slow/card-C TnI was not phosphorylated and as a result, Ca^{2+}-activated force did not change significantly in response to PKA. These results provide evidence that while the full-length TnI is likely required for the full effect of TnI phosphorylation on myofilament Ca^{2+} sensitivity, the TnI response to phosphorylation does not depend on isoform-specific sequences in the carboxyl-portion of TnI.

5. CHARGE CHANGE MUTATION IN TROPONIN I ISOFORMS LINKED TO CARDIAC DYSFUNCTION

Recently, gene transfer was used to investigate the effects of a single charge change mutation, associated with the human disease state hypertrophic cardiomyopathy (HCM)[17]. This mutation results in a switch from arginine to glycine at residue 146 in the rat sequence (aa 145 in human cTnI), and is part of the inhibitory peptide of cTnI. The R• G substitution in cTnI (Fig. 1) resulted in reduced replacement of endogenous cTnI, and a diminished ability to compete with cTnI for incorporation sites within the myofilament[18]. Functionally, increased myofilament Ca^{2+} sensitivity was observed in myocytes, with pH sensitivity that was comparable to that of endogenous cTnI. An analogous mutation in ssTnI (ssTnIR115G, Fig. 1) resulted in a comparable enhancement of Ca^{2+} sensitivity compared to cTnI, but did not differ from ssTnI. Interestingly, pH sensitivity in myocytes expressing ssTnIR115G more closely resembled cTnI and cTnIR145G than ssTnI. Unlike cTnIR145G, ssTnIR115G replaced virtually all of the endogenous cTnI. These results provide important evidence that charge differences within TnI isoforms are capable of conferring unique functional properties on the myofilament. Moreover, these findings have led us to postulate that charge differences between the cTnI and ssTnI isoforms that lie within the inhibitory peptide and secondary region are critically important in conferring isoform-specific influences on incorporation and function within the intact myofilaments of adult cardiac myocytes[19].

6. FUTURE DIRECTIONS

Adenoviral gene transfer provides a powerful tool for investigating the role played by thin filament proteins such as TnI in the adult cardiac myocyte. Using this approach, isoform-specific functions, as well as functional domains within TnI, have been identified in the adult myocyte. More refined domains and their influence on myocyte function can now be identified and investigated in the future. Studies on mutations linked to human disease states, such as HCM also have provided valuable insight into the overall function of TnI. Future studies with these mutations will not only enhance understanding of fundamental disease processes in the heart, but may also provide potential targets for gene therapy as more sophisticated vectors become available.

7. LITERATURE CITED

1. C.S. Farah, and F. C. Reinach. The troponin complex and regulation of muscle contraction. *FASEB J.* 9, 755-767 (1995).
2. A. M. Gordon, E. Homsher, and M. Regnier. Regulation of contraction in striated muscle. *Physiol. Rev.* 80, 853-924 (2000).
3. S.V. Perry. Troponin I: inhibitor or facilitator. *Mol. Cell Biochem.* 190, 9-32 (1999).
4. D. F. A. McKillop, and M. A. Geeves. Regulation of the interaction between actin and myosin subfragment a: Evidence for three states of the thin filament. *Biophys. J.* 65, 693-701 (1993).
5. P. Vibert, R. Craig, and W. Lehman. Steric-model for activation of muscle thin filaments. *J. Mol. Biol.* 266, 8-14 (1997).
6. B. Tripet, J. E. Van Eyk, and R. S. Hodges. Mapping of a second actin-tropomyosin and a second troponin C binding site within the C terminus of troponin I, and their importance in the Ca^{2+}-dependent regulation of muscle contraction. *J. Mol. Biol.* 271, 728-750 (1997).
7. L. Saggin, L. Gorza, S. Ausoni, S. Schiaffino. Troponin I switching in the developing heart. *J. Biol. Chem.* 264, 16299-16302 (1989).
8. R. J Solaro, J. A. Lee, J. C. Kentish, and D. G. Allen. Effects of acidosis on ventricular muscle from adult and neonatal rats. *Circ. Res.* 63, 779-787 (1988).
9. S. Bartel, I. Morano, H. D. Hunger, H. Katus, H. T. Pask, P. Karczewski, and E.-G. Krause. Cardiac troponin I and tension generation of skinned fibres in the developing rat heart. *J. Mol. Cell. Cardiol.* 26, 1123-1131 (1994).
10. M. V. Westfall, E. M. Rust, and J. M. Metzger. Slow skeletal troponin I gene transfer, expression, and myofilament incorporation enhances adult cardiac myocyte contractile function. *Proc. Nat. Acad. Sci.* 94,5444-5449 (1997).
11. J. P. Ebus G. J. Stienen, and G. Elzinga. Influence of phosphate and pH on myofibrillar ATPase activity and force in skinned cardiac trabeculae from rat. *J. Physiol. (London)* 476, 501-516 (1994).
12. C. H. Orchard and J. C. Kentish. Effects of changes of pH on the contractile function of cardiac muscle. *Am. J. Physiol.* 258, C967-C981 (1990).

13. M. V. Westfall, I. I. Turner, F. P. Albayya, and J. M. Metzger. Troponin I chimera analysis of the cardiac myofilament tension response to protein kinase A. *Am. J. Physiol.* 280, C324-C332 (2001).
14. R. C. Fentzke, S. H. Buck, J. R. Patel, H. Lin, B.M. Wolsaka, M. O. Stojanovic, A. F. Martin, R. J. Solaro, R. L. Moss, and J. M. Leiden, Impaired cardiomyocyte relaxation and diastolic function in transgenic mice expressing slow skeletal troponin I in the heart. J. Physiol. (London) 517, 143-157 (1999).
15. M. V. Westfall, F. Albayya, and J. M. Metzger. Functional analysis of troponin I regulatory domains in the intact myofilament of adult single cardiac myocytes. *J. Biol. Chem.* 274,22508-22516 (1999).
16. M. V. Westfall, F. P. Albayya, I. I. Turner, and J. M, Metzger. Chimera analysis of troponin I domains that influence Ca^{2-}-activated myofilament tension in adult cardiac myocytes. *Circ. Res.* 86:,470-477 (2000).
17. A. Kimura, H. Harada, J. E. Park, H. Nishi, M. Satoh, M. Takahashi, S. Hiroi, T. Sasaoka, N. Ohbuchi, T. Nakamura, T. Koyanagi, T.H. Hwang, J. A., Choo, K. S. Chung, A. Hasegawa, R. Nagai, O. Okasaki, H. Nakamura, M. Matsuzaki, T. Sakamoto, H. Toshima, Y. Koga, T. Imaizumi, T. Sasazuki. Mutations in the cardiac troponin I gene associated with hypertrophic cardiomyopathy. *Nat. Genet.* 16, 379-382 (1997).
18. M. V. Westfall, A. R. Borton, F. P. Albayya, and J. M. Metzger. Myofilament calcium sensitivity and cardiac disease: Insights from troponin I isoforms and mutants. *Circ. Res.* 91, 525-531 (2002).
19. M. Westfall, A. R. Borton, F. P. Albayya, S. Forfa, and J. M. Metzger. Targeted substitutions in charged residues within troponin I cause isoform-specific changes in the Ca^{2-}- and pH- sensitivities of tension in adult cardiac myocytes. *Biophys. J.* 82, 388a (2002).

DISCUSSION

ter Keurs: What is the feedback pathway that causes the cell to stop producing its own TnI when adenoviral-driven ss TnI protein is produced?

Metzger: Evidence from gene transfer and transgenic studies indicates that endogenous gene expression is not altered. Instead, competition appears to be post-transciptional, possibly at the level of ribosome.

Solaro: Is maximum tension also unaffected by acidic pH in the 132A mutant TnI?

Metzger: Our analysis so far has been focused on the shape and position of the tension-calcium relation. We also have the maximum tension data, but the statistics are not complete at this time.

Pfitzer: How many cells get expressed?

Metzger: I don't have a precise cell number count to give you. I can say that a major portion of the L V free wall is transduced from apex to base with this technique.

A HYPOTHESIS ABOUT MYOSIN CATALYSIS

Hirofumi Onishi, Takashi Ohki, Naoki Mozhizuki,
and Manuel F. Morales*

1. ABSTRACT

When ATP binds to the active site of myosin heads, Switch II undergoes a large conformational change and the cleft surrounding the bound γ-phosphate closes. In the closed state, Glu470 in Switch II comes together with Arg247 in Switch I to form a salt-bridge. Here, the functional significance of the two bridging residues was tested by using site-directed mutagenesis. We conclude from such tests that (a) the attractive force between Arg247 and the γ-phosphate of ATP moves the cleft to close, and (b) during hydrolysis, Glu470 is intimately involved in positioning the lytic water for the attack on the γ-phosphorus. We also speculate on how the salt-bridge between Arg247 and Glu470 is related to hydrolysis.

2. INTRODUCTION

Midway in the last century it became evident that "muscle contraction" results from a thermodynamic coupling between the conformational transitions of myosin molecules in muscle and the hydrolysis of ATP (Engelhardt and Ljubimowa, 1939; Szent-Gyorgyi A, 1945). Later, careful kinetic measurements of myosin ATPase set the movements in time, and rendered the overall process to be a sequence of transitions among several defined states (Kanazawa and Tonomura, 1965; Lymn and Taylor, 1970; Bagshaw and Trentham, 1974; Bagshaw et al., 1974). In the most recent advance, Rayment and collaborators have reported "still" pictures of the moving myosin molecules at atomic resolution, finally bringing us closer to the goal of explaining muscle contraction based on first principles (atoms, forces between them, and either classic or quantum laws of motion)(Fisher et al., 1995; Smith and Rayment, 1996b). Rayment's crystallographic "snapshots" could not be taken frequently enough during the time between states of

* Hirofumi Onishi, Takashi Ohki, and Naoki Mochizuki, Department of Structural Analysis, National Cardiovascular Center Research Institute, Suita, Osaka 565-8565. Manuel F. Morales, University of the Pacific, San Francisco, CA 94115

Molecular and Cellular Aspects of Muscle Contraction
Edited by H. Sugi, Kluwer Academic/Plenum Publishers, 2003

transition to generate movies of the various conformational changes coupled with hydrolysis.   By using unhydrolyzable nucleotide analogs (ADP.BeFx, ADP.AlF$_4^-$, ADP.VO$_4^-$, AMPPNP, ATPγS, ADP, etc.), however, they found stable "simulants" of the system in i-th and (i+1)st states.   By considering the initial and final structures, and having gross time-dependent measurements of the real system in action, initial guesses can be made of how the fine structure of the real system moves during the transition.   In this approach, Rayment and collaborators suggested that during the transition of M.ATP → M*.ATP, the two residues Arg247 and Glu470[a], once long-separated, come together to form a salt-bridge (see Fig. 1).

In the process of filling in important mechanistic details, site-directed mutation is a useful adjunct technique.   If the function, or a residue containing the function, is thought to be important, then (with certain provisos) substitution of a residue without the function serves to confirm or deny the proposed mechanism in the wild-type system.

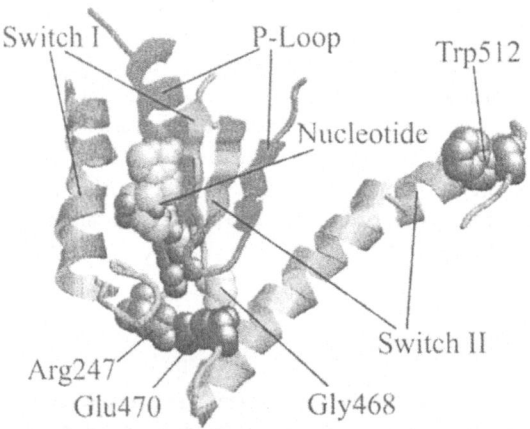

Fig. 1. The nucleotide-binding pocket in the transition state for hydrolysis.   The structure of the *Dictyostelium* myosin motor domain ligated at the active site with MgADP.vanadate has been solved by Smith and Rayment (Smith and Rayment, 1996b).   They have proposed that this structure mimics the transition state for the ATP hydrolysis.   As reported for many other nucleotide-binding proteins, the nucleotide-binding pocket is composed of three loops, viz., P-loop, Switch I, and Switch II.   Backbone atoms of the P-loop (168-202), the Switch I (218-257), and the Switch II (456-513) are shown as ribbon models.   Arg247, Gly468, Glu470, and Trp512 are shown as space-filled balls.   Trp512 is known to be the residue which is sensitive to close the cleft.   The cleft closure is also known to include a significant rotation in Gly468.   MgADP.AlF$_4^-$, simulating MgATP, is also shown as space-filled balls.   It should be noted that in this state, a salt-bridge is seen between Arg247 in Switch I and Glu470 in Switch II.

[a] The amino acid numeration in this paper corresponds to that of smooth muscle myosin (Yanagisawa et al., 1987).

Using this approach, we have recently investigated the significance of the two residues in ATP hydrolysis (Onishi et al., 2002). This paper centers on how we think the catalysis is achieved. For the convenience of the reader, however, we first summarize the background that leads us to our hypothesis.

3. EXPERIMENTAL BACKGROUND

We were able to study correlations between the presence of a residue and its functionality. By performing function-testing experiments, we concluded that both members of the salt-bridge pair had important, but distinct, functions (Onishi et al., 2002). For example, we found a very good correlation between the presence of Arg and the rate of mantATP (a fluorescent ATP analog) binding to the active site. We also found that there is a good correlation between the maximum increase in Trp fluorescence upon addition of ATP and the presence of Arg. Based on these results, we concluded that the attractive force between Arg247 and the γ-phosphate of the bound ATP probably helps to close the cleft and to initiate the "signal" that travels to Trp-512. Bearing

$$M + T \underset{k_{-1}}{\overset{k_{+1}}{\rightleftarrows}} M.T \underset{k_{-2}}{\overset{k_{+2}}{\rightleftarrows}} M^*.T \underset{k_{-3}}{\overset{k_{+3}}{\rightleftarrows}} M^{**}.DP_i \underset{k_{-4}}{\overset{k_{+4}}{\rightleftarrows}} M^*.D \underset{k_{-5}}{\overset{k_{+5}}{\rightleftarrows}} M + D$$

where M, T, D and P_i are myosin, ATP, ADP, and inorganic phosphate, respectively. k_{+i} and k_{-i} are forward and backward rate constants, respectively. * and ** refer to different conformations as detected by intrinsic protein fluorescence.

Scheme 1

directly on our present study, we found a strong correlation between the presence of Glu470 and the ability to catalyze ATPase. For example, the mutations at Glu470 resulted in a normal mantATP-binding rate and a normal ATP-induced increase in Trp fluorescence, but not a phosphate burst. These results allow us to assume that Glu at 470 plays a critical role especially in the transition from $M^*.T$ to $M^{**}.DP_i$ (step 3 in scheme I). In fact, neither R247A nor E470A exhibited a phosphate burst. In these cases, however, we assume that the reasons for failing to catalyse hydrolysis would be different. In the former, closure is poor and Glu470 does not move to the proper position, whereas in the latter, closure occurs, but Glu470 is lost.

These findings seemed to us, a strong indicator of a speculation that we earlier put forward (Onishi et al., 1997), and we set about to study the mechanism of the hydrolytic transition, $M^*.T \rightarrow M^{**}.DP_i$. In this transition, myosin in some way poises a water molecule (which we shall call the "lytic" water, w_1) to make the attack on the β-γ phosphate bond of the ATP bound at the active site. Although it has long been recognized that this hydrolysis is not *thermochemically* special in magnitude of free energy change, a great deal of interest is attached to the nature of this sub-process as compared to other catalytic devices. Moreover, it is puzzling that from Rayment's snapshot the residues suspected of participating in the sub-process seem to be too far away (>5 Å) from the vulnerable bond to be broken. An important antecedent to our proposal is the observation (Smith and Rayment, 1996b) that in the analogous image corresponding to M*.T there appears a second water molecule that we shall call w_2. Notable also is their suggestion that a structural "network" including water may confer "plasticity" to the catalytic mechanism.

4. MECHANISTIC DETAILS OF CATALYSIS

Our proposed myosin ATPase cycle is shown in Fig. 2. Step 1 corresponds to the formation of M.T by a collision between myosin and ATP (indicated by a small arrow in Fig. 2). In M.T, Arg247 indirectly interacts with the γ-phosphate of bound ATP via w_1. For this reason, nucleotide binding to the active site requires the presence of Arg247. As described above, the attractive force between Arg247 and the γ-phosphate of bound ATP is important for cleft closure (step 2) to occur in the conformational transition from M.T to M*.T. Following cleft closure, Glu470 moves toward Arg247 (indicated by a large arrow in Fig. 2) to form a salt-bridge.The movement of Glu470 results in the following two events. First, w_1 releases from the guanydyl group of Arg 247. Second, w_2 appears between the carboxyl group of Glu470 and the main-chain carbonyl oxygen of Gly468, and w_1 moves to the optimal position for nucleophilic attack on the γ-phosphate of bound ATP. Bound ATP is hydrolyzed by attack of w_1, and M**.DP$_i$ is formed (step 3). Following the bridge severance, the cleft re-opens (step 4) and the first product (orthophosphate) emerges from the backdoor. In the absence of actin, this sub-process is rate-limiting in the entire ATPase cycle. Finally, ADP release from the active site (step 5) allows the myosin to return the original conformation (M), which is ready to react with the next ATP.

Our speculated involvement of w_2 in myosin catalysis relates to catalysis in other systems. Structural comparisons suggested that myosin and G-proteins utilize a similar strategy for nucleotide hydrolysis (Smith and Rayment, 1996a). In many G-proteins, it is known that the δ-carboxyl oxygen of Gln in Switch II requires the movement of the lytic water to the in-line position on the γ-phosphorus (Coleman et al., 1994; Pai et al., 1990; Schweins et al., 1995; Sondek et al., 1994). As w_2 in myosin and Gln in G-proteins are placed in the corresponding positions, we can assume that in myosin catalysis, the hydrogen-bond network of Glu470-w_2-Ser246 may play the role that Gln plays in G-protein family. Recently, Okimoto et al. (Okimoto et al., 2001) have reported interesting theoretical studies of the hydrolysis mechnism of myosin. Their results suggested that the transition from M*.T to M**.DP$_i$ consists of a single elementary reaction in which a water molecule attacked the γ-phosphorus. However, their simulations did not assume the presence of w_2 in the γ-phosphate pocket. Perhaps, for this reason, they obtained an unusually large energy of activation. Theoretical studies are in progress to test the involvement of w_2.

5. POSSIBLE ROLES OF THE SALT-BRIDGE

As described above, the formation of the salt-bridge is important for positioning a lytic water molecule that acts in the hydrolysis. We also propose that the salt-bridge has a

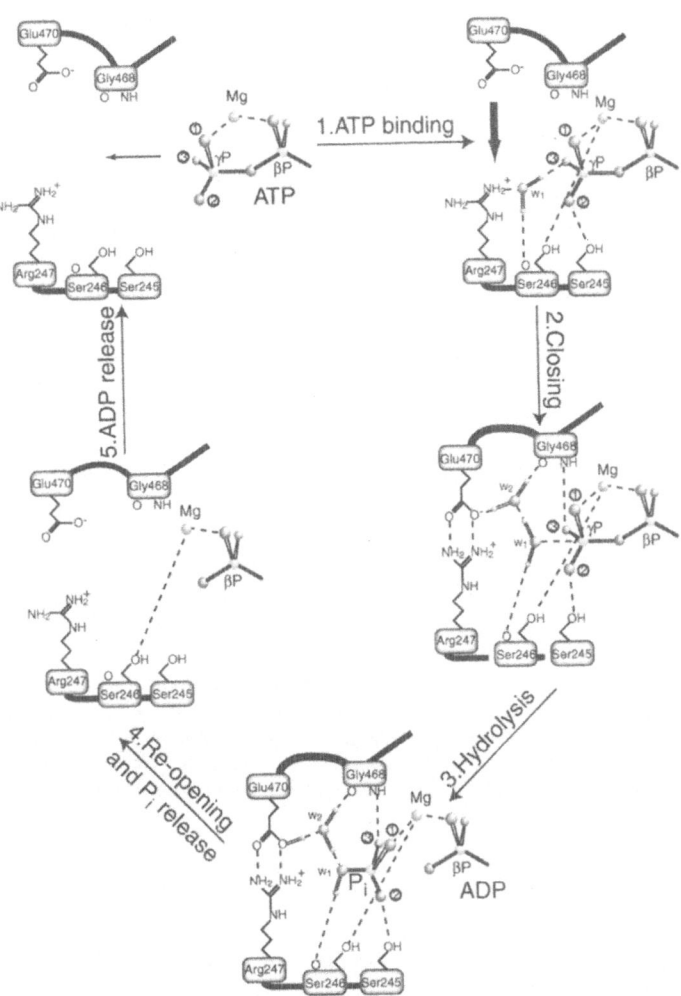

Fig. 2. A schematic drawing of the myosin ATPase cycle. ATP binding, closing of the cleft, hydrolysis, cleft re-opening plus Pi release, and ADP release correspond to steps in the 5-step reaction scheme described above (scheme I). In the prehydrolytic state, Arg247 interacts indirectly with γ-phosphate of the bound ATP via a water molecule (w_1). After closing the cleft, a second water (w_2) appears in the γ-phosphate pocket. The interaction between Glu470 and w_1 requires this w_2 water molecule. Bridging between Glu470 and Arg247 moves w_1 to its optimal position for attacking the γ-phosphate. Only those stereochemical interactions that participate in positioning w_1, w_2, and three oxygen atoms (①, ②, and ③) of the γ-phosphate are depicted. Covalent bonds are shown as solid lines and hydrogen bonds and ionic interactions as dashed lines. Phosphorus and magnesium are balls marked with P and Mg, respectively. Oxygen and hydrogen are depicted as large and small balls, respectively.

key role in holding orthophosphate in the γ-phosphate pocket, until the cleft re-opens (step 4 in Fig. 2).   Conformational changes between triphosphate and diphosphate states are similar for G-proteins and myosin (Smith and Rayment, 1996a).   In the cases of G-proteins, the flexible loops of Switch I and Switch II function as sensors for the γ-phosphate of bound GTP (Coleman et al., 1994; Pai et al., 1990).   The loops move towards each other when GTP is bound to the active site and move away when GTP is hydrolyzed.   In the case of myosin, ATP binding also results in a large conformational change of Switch II and the cleft closure (Fisher et al., 1995; Smith and Rayment, 1996b).   However, even after ATP is hydrolyzed, the products (orthophosphate and ADP) stay in the active site for a long time in the absence of actin (Bagshaw and Trentham, 1974; Bagshaw et al., 1974; Kanazawa and Tonomura, 1965; Lymn and Taylor, 1970).   In the metastable ternary complex, the structure of the molecule remains in the weakly-bound state for actin in readiness to generate the power stroke by its interaction with actin.   By this mechanism, myosin can store the energy of hydrolysis in the molecule, until its head encounters actin.

Elsewhere in the proceeding of this Symposium there is an interesting cryoEM study by Holmes and Schroader.   They report that actin binding to the myosin head causes changes of Switch I and Switch II structures.   Their observations would be consistent with events envisioned in our hypothesis.

6. REFERENCES

Bagshaw, C. R., Eccleston, J. F., Eckstein, F., Goody, R. S., Gutfreund, H., and Trentham, D. R. (1974). The magnesium ion-dependent adenosine triphosphatase of myosin. Two-step processes of adenosine triphosphate association and adenosine diphosphate dissociation. Biochem. J. *141*, 351-364.

Bagshaw, C. R. and Trentham, D. R. (1974). The characterization of myosin-product complexes and of product-release steps during the magnesium ion-dependent adenosine triphosphatase reaction. Biochem. J. *141*, 331-349.

Coleman, D. E., Berghuis, A. M., Lee, E., Linder, M. E., Gilman, A. G., and Sprang, S. R. (1994). Structures of active conformations of Gi alpha 1 and the mechanism of GTP hydrolysis. Science *265*, 1405-1412.

Engelhardt, W. A. and Ljubimowa, M. N. (1939). Myosine and adenosine triphosphatase. Nature *144*, 668-669.

Fisher, A. J., Smith, C. A., Thoden, J. B., Smith, R., Sutoh, K., Holden, H. M., and Rayment, I. (1995). X-ray structures of the myosin motor domain of Dictyostelium discoideum complexed with MgADP.BeFx and MgADP.AlF4-. Biochemistry *34*, 8960-8972.

Kanazawa, T. and Tonomura, Y. (1965). The pre-steady state of the myosin-adenosine triphosphate system. I. Initial rapid liberation of inorganic phosphate. J. Biochem. (Tokyo) *57*, 604-615.

Lymn, R. W. and Taylor, E. W. (1970). Transient state phosphate production in the hydrolysis of nucleotide triphosphates by myosin. Biochemistry *9*, 2975-2983.

Okimoto, N., Yamanaka, K., Ueno, J., Hata, M., Hoshino, T., and Tsuda, M. (2001). Theoretical studies of the ATP hydrolysis mechanism of myosin. Biophys. J. *81*, 2786-2794.

Onishi, H., Morales, M. F., Kojima, S., Katoh, K., and Fujiwara, K. (1997). Functional transitions in myosin: role of highly conserved Gly and Glu residues in the active site. Biochemistry *36*, 3767-3772.

Onishi, H., Ohki, T., Mochizuki, N., and Morales, M. F. (2002). Early stages of enrgy transduction by myosin: Roles of Arg in Switch I, of Glu in Switch II, and of the salt-bridge between them. Proc. Natl. Acad. Sci. U. S. A *99*, 15339-15344.

Pai, E. F., Krengel, U., Petsko, G. A., Goody, R. S., Kabsch, W., and Wittinghofer, A. (1990). Refined crystal structure of the triphosphate conformation of H-ras p21 at 1.35 A resolution: implications for the mechanism of GTP hydrolysis. EMBO J. *9*, 2351-2359.

Schweins, T., Geyer, M., Scheffzek, K., Warshel, A., Kalbitzer, H. R., and Wittinghofer, A. (1995). Substrate-assisted catalysis as a mechanism for GTP hydrolysis of p21ras and other GTP-binding proteins. Nat. Struct. Biol. *2*, 36-44.

Smith, C. A. and Rayment, I. (1996a). Active site comparisons highlight structural similarities between myosin and other P-loop proteins. Biophys. J. *70*, 1590-1602.

Smith, C. A. and Rayment, I. (1996b). X-ray structure of the magnesium(II).ADP.vanadate complex of the Dictyostelium discoideum myosin motor domain to 1.9 A resolution. Biochemistry *35*, 5404-5417.

Sondek, J., Lambright, D. G., Noel, J. P., Hamm, H. E., and Sigler, P. B. (1994). GTPase mechanism of Gproteins from the 1.7-A crystal structure of transducin alpha-GDP-AIF⁻₄. Nature *372*, 276-279.

Szent-Gyorgyi, A (1945). Chemistry of muscle contraction. (New York: Academic Press).

Yanagisawa, M., Hamada, Y., Katsuragawa, Y., Imamura, M., Mikawa, T., and Masaki, T. (1987). Complete primary structure of vertebrate smooth muscle myosin heavy chain deduced from its complementary DNA sequence. Implications on topography and function of myosin. J. Mol. Biol. *198*, 143-157.

DISCUSSION

Nishimura: When you inverted two amino acids at the position of 247 and 470, you assume that the salt bridge is again formed. Is this notion really proved by X-ray crystallography?

Onishi: We have to prove that our double mutant can form the salt-bridge. However, we have not completed the X-ray analysis of this mutant.

IN VITRO ACTOMYOSIN MOTILITY IN DEUTERIUM OXIDE

Shigeru Chaen, Naoto Yamamoto, Ibuki Shirakawa, and Haruo Sugi

1. ABSTRACT

Actin filament velocities in an in vitro motility assay system were measured both in heavy water (deuterium oxide, D_2O) and water (H_2O) to examine the effect of D_2O on the actomyosin interaction. The dependence of the sliding velocity on pD of the D_2O assay solution showed a broad pD optimum of around pD 8.5 which resembled the broad pH optimum (pH 8.5) of the H_2O assay solution, but the maximum velocity ($4.1 \pm 0.5\mu m/sec$, n=11) at pD 8.5 in D_2O was about 60% of that ($7.1 \pm 1.1\mu m/sec$, n=11) at pH 8.5 in H_2O. The K_m values of 95 and 80μM and V_{max} values of 3.2 and 5.1μm/sec for the D_2O and H_2O assay were obtained by fitting the ATP concentration dependence of the velocity (at pD and pH 7.5) to the Michaelis-Menten equation. The K_m value of actin-activated Mg-ATPase activity of myosin subfragment 1(S1) was decreased from 50μM[actin] in H_2O to 33μM[actin] in D_2O without any significant changes in V_{max} ($9.4 \; s^{-1}$ in D_2O and $9.3 \; s^{-1}$ in H_2O). The rate constants of ADP release from the acto-S1-ADP complex measured by the stopped flow method were $361 \pm 26 \; s^{-1}$ (n=27) in D_2O and $512 \pm 39 \; s^{-1}$ (n=27) in H_2O at 6 °C. These results suggest that the decrease in the in vitro actin-myosin sliding velocity in D_2O results from a slowing of the release of ADP from the actomyosin-ADP complex and the increase in the affinity of actin for myosin in the presence of ATP in D_2O.

2. INTRODUCTION

Sliding movement between actin and myosin filament in muscle contraction is generally thought to occur as the result of a cyclic association and dissociation of myosin heads with actin-binding sites involving ATP hydrolysis. In studying the interaction between actin and myosin, an in vitro motility assay system which visualizes the movement of fluorescently labeled actin filaments over a myosin-coated glass surface has been shown to be useful, and in vitro actin filament sliding has been studied extensively under various conditions, ionic strength, Mg-ATP concentration, temperature and pH, to examine the determinant of the velocity [1-7].

Shigeru Chaen, Department of Applied Physics, College of Humanities and Sciences, Nihon University, Tokyo 156-8550, Japan, Naoto Yamamoto, Ibuki Shirakawa, and Haruo Sugi, Department of Physiology, School of Medicine, Teikyo University, Tokyo 173-8605, Japan

Molecular and Cellular Aspects of Muscle Contraction
Edited by H. Sugi, Kluwer Academic/Plenum Publishers, 2003

In the present work, we have examined the effect of substituting D_2O for H_2O on the in vitro actin filament sliding velocities. Water is known to be involved in the myosin ATP hydrolysis step (MATP $\rightarrow$ MADPPi), in which the water attacks the γ-phosphorus atom of ATP [8], and besides the active role as the reactant of ATP hydrolysis, the hydration of motor proteins in determining the kinetics has also been studied [9,10]. Hotta and Morales [11] and Inoue et al. [12] reported that myosin ATPase activity in D_2O decreased to about 60% of that in H_2O. Experiments with skinned muscle fiber preparations have shown that at saturating calcium concentrations the muscle fiber generates about 20% more isometric force in D_2O than in H_2O [13]. It remains unclear, however, how D_2O affects the actomyosin interaction. We have demonstrated here that the actin filament velocity in the in vitro motility assay was about 40% slower in D_2O than in H_2O. The K_m value of the actin-activated Mg-ATPase activity of S1 was 33% smaller in D_2O than in H_2O without changing the V_{max} values in D_2O and H_2O, and the rate of ADP dissociation from the acto-S1-ADP complex measured by the stopped flow method was about 30% smaller in D_2O than in H_2O. Possible causes for the slower sliding velocity in D_2O are discussed.

3. MATERIALS AND METHODS

3.1 Materials

Myosin and actin were prepared from rabbit skeletal muscle by the method of Perry [14], and Spudich and Watt [15], respectively. Myosin was stored in 50% glycerol at – 20 °C. Heavy meromyosin (HMM) and S1 were prepared as described by Okamoto and Sekine [16]. D2O(99.9%) was from Aldrich (Milwaukee, WI, USA). ATP, bovine serum albumin, glucose oxidase and catalase were from Sigma (St. Louis, MO, USA). Other chemicals were of analytical grade.

3.2 In vitro motility assay

The sliding velocity of fluorescent actin filament was measured as described by Chaen et al. [17] with some modifications. After allowing the rhodamine-phalloidin-labeled actin filaments to bind to the HMM (80 µg/ml) over the nitrocellulose-coated glass coverslips in an ATP-free solution (25 mM KCl, 4 mM $MgCl_2$, 1 mM dithiothreitol, 0.04% NaN_3, and 25 mM imidazole (pH7.5)), assay solutions were introduced into the flow cell to start the movement of actin filaments. The temperature in the flow cell was adjusted by circulating temperature-controlled water into a brass block attached to the microscope stage and another block jacketing the objective. Experiments were carried out at 25 °C. The assay solutions contained 25 mM KCl, 4 mM $MgCl_2$, 1 mM dithiothreitol, 0.04% NaN_3, 4.5 mg/ml glucose, 0.22 mg/ml glucose oxidase, 0.036 mg/ml catalase, 2 mM ATP, and 25 mM imidazole (pH 7.5 and pD 7.5 for the H_2O and D_2O assay solution, respectively). The D_2O assay solution was prepared by dissolving the salts in 99.9% D2O, and the pD was adjusted by adding a D_2O solution of DCl or KOH until a pD with the relationship pD = pH reading + 0.4 was reached [18].

3.3 ATPase measurements

Actin-activated Mg-ATPase activity of S1 was measured in the H$_2$O and D$_2$O assay solutions using 0.25 M S1 at various concentrations of actin (2.5 – 30 μM) at 25 °C. D$_2$O content in the solution was about 98%; the protein sample in D$_2$O was prepared using S1 (12.5 μM) in H$_2$O, and F-actin (140 μM) in D$_2$O prepared by homogenizing the pelleted F-actin. The protein sample was incubated on ice for more than 12 h. Liberated inorganic phosphate was determined by the malachite green method [19].

3.4 Kinetic measurements

Measurement of ADP release from acto-S1 followed the procedure of White [20], using a stopped flow spectrometer (Applied Photophysics SX18MV, Leatherhead, UK). Since the acto-S1 dissociation by ATP in the presence of ADP is rate-limited by ADP release, the rate of ADP release was determined from the acto-S1 dissociation, which is detected in the decrease in light scattering (340 nm light scattered at 90° from the incident beam) when acto-S1-ADP (2 μM actin, 1.5 μM S1, and 50 μM ADP) was mixed with 2 mM ATP. Preparation of the protein sample in D$_2$O was similar to the above except for 2.5 μM S1, and 97% D$_2$O content. We performed the measurements at low temperature, 6 °C, since preliminary experiments at 20 °C showed a rate of ~1000 s-1, which near the limit of the stopped flow measurement, and the process has been reported to be temperature-dependent and the Arrhenius plot was linear from 0 °C to 25 °C [21].

4. RESULTS

4.1 Sliding velocities of actin filaments on HMM

The rate of an enzymatic reaction generally depends on hydrogen ion concentration. First we examined whether there were differences between the pH and pD dependence of the actin filament velocity. Fig.1 shows the dependence of the actin sliding velocity on pH and pD of the H$_2$O and D$_2$O assay solution, respectively. In the experiment, different pH(pD) buffer reagents were used: 25 mM imidazole at pH 6.0, 6.5, 7.0, 7.5 and pD 6.0, 6.5, 7.0, 7.5; 25 mM Tris at pH 8.0, 8.5 and pD 8.0, 8.5; 25 mM glycine at pH 9.0, 9.5 and pD 9.0, 9.5. The sliding movement showed a broad pD optimum of around pD 8.5 which resembled the broad pH optimum (pH 8.5) of the H$_2$O assay solution, but the maximum velocity (4.1 ± 0.5 μm/s, n=11) at pD 8.5 in D$_2$O was about 60% of that (7.1 ± 1.1 μm/s , n=11) at pH 8.5 in H$_2$O. These results indicate that the effect of deuterium ion concentration on the actin filament movement is similar to the case of hydrogen ions except for the slower velocities in D$_2$O. The dependence is somewhat different from prior studies on the in vitro motility assay [1,5] in which the dependence has a shoulder between pH 7.5 and 8.0. Although the optima were around pH and pD 8.5, the experiments described below were carried out at pH and pD 7.5, which have been used in most of the in vitro motility assay studies.

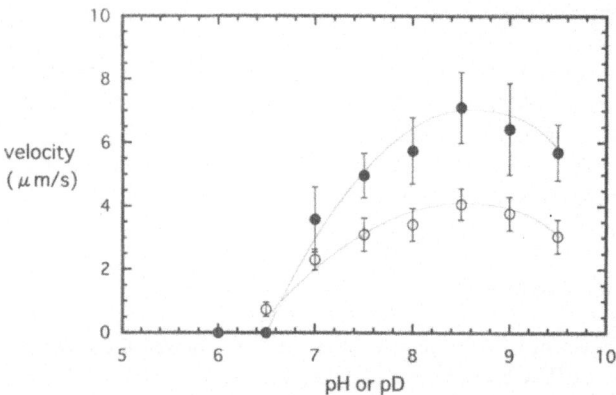

Figure 1. Dependence of the actin sliding velocity on pH (closed symbol) and pD (open symbol) of the H_2O and D_2O assay solution, respectively. Each data point represents the average velocity of 11 different filaments, and the vertical bar shows the standard deviation. The curves were drawn by eye.

The sliding velocity of actin filament in the in vitro motility assay has been shown to be dependent on the concentration of Mg-ATP [1,3,4,6]. Fig. 2 shows the effect of varying the Mg-ATP concentration from 20 µM to 2 mM on the actin filament velocity in H_2O and D_2O. The data were fitted with the Michaelis-Menten equation, yielding K_m values of 95 and 80 µM and V_{max} values of 3.2 and 5.1 µm/s for D_2O and H_2O assay, respectively. These results demonstrate that D_2O just scales down the Mg-ATP concentration dependence of the actin filament velocity in H_2O.

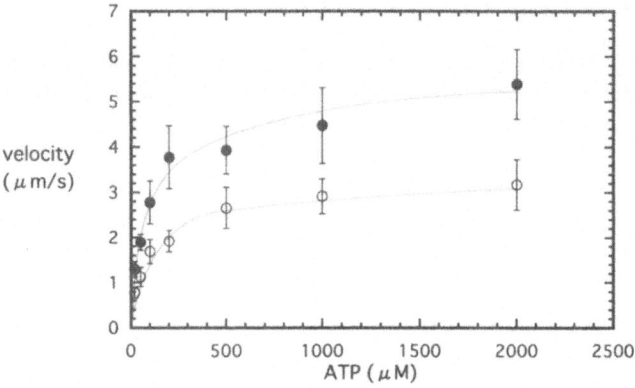

Figure 2. Dependence of the actin sliding velocity on Mg-ATP concentration in H_2O (closed symbols) and D_2O (open symbols). Each data point represents the average velocity of 11 different filaments, and the vertical bar shows the standard deviation. Curves are fit to the Michaelis-Menten equation, which gave K_m values of 80 and 95 µM and V_{max} of 5.1 and 3.2 µm/s for the H_2O and D_2O assay, respectively.

4.2 Actin-activated Mg-ATPase activity of S1

To examine whether D$_2$O affects the interaction between actin and S1 in the presence of ATP, the actin-activated Mg-ATPase activity of S1 was studied. The Mg-ATPase activities of S1 as a function of actin concentration both in D$_2$O and H$_2$O are shown in Fig. 3. Fits to the Michaelis-Menten equation show that the V$_{max}$ value in H$_2$O (9.3 s^{-1}) was scarely altered in D$_2$O (9.4 s^{-1}), while the K$_m$ value was decreased from 50 µM in H$_2$O to 33 µM in D$_2$O. These data indicate that D$_2$O affects the affinity of actin for S1 in the presence of ATP without changes in the maximum rate of the reaction.

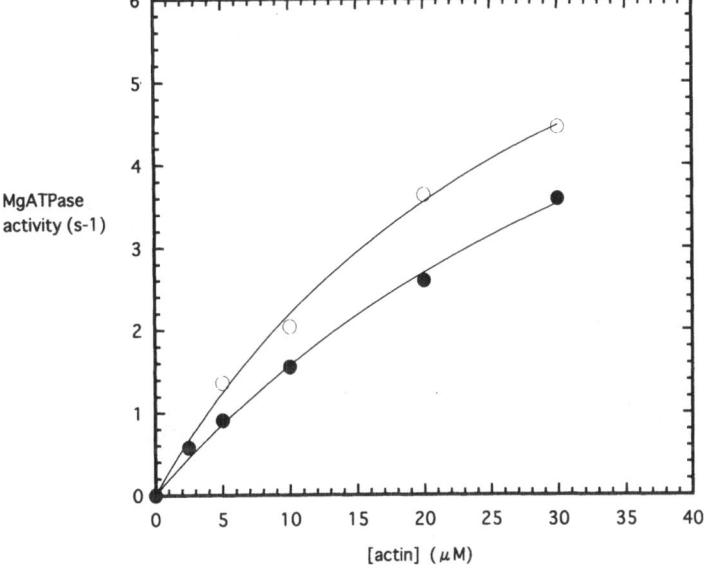

Figure 3. Mg-ATPase activity as a function of actin concentration in H$_2$O (closed symbol) and D$_2$O (open symbol). Curves are fit to the Michaelis-Menten equation, which gave K$_m$ values (actin concentration at which the Mg-ATPase activity is half of its maximum) of 50 µM and 33 µM and V$_{max}$ of 9.3 s^{-1} and 9.4 s^{-1} in the H$_2$O and D$_2$O assay solution, respectively.

4.3 Rate of ADP release from acto-S1

ADP release from the cross-bridge has been believed to limit unloaded shortening velocity [21]. Fig. 4 shows acto-S1 dissociation in the presence of ADP, the rate of which is the same as that of the ADP release, since the ADP release limits the acto-S1 dissociation [20]. The rate constant of ADP release was 30% slower in D$_2$O (361 ± 26 s^{-1} (n=27)) than in H$_2$O (512 ± 39 s^{-1} (n=27)) at 6 °C. The difference was significant (P<0.0001).

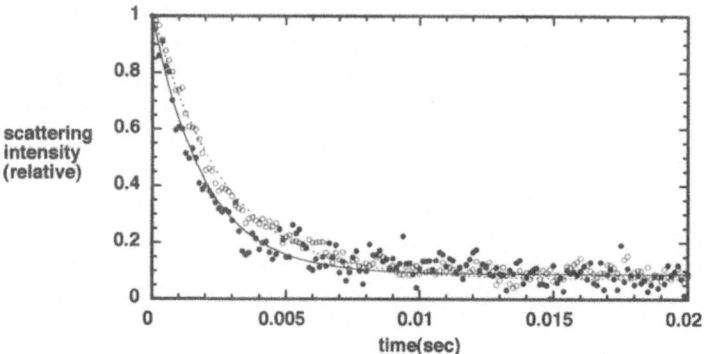

Figure 4. Changes in light scattering intensity on mixing acto-S1-ADP (2 μM actin, 1.5 μM S1, and 50 μM ADP) with 2 mM ATP measured by a stopped flow spectrometer. Solid and open symbols are the data points in H_2O and D_2O assay solution, respectively. Straight and dotted curves are the best fit to a single exponential equation, which gave a rate constant of 510 s^{-1} in H_2O and 377 s^{-1} in D_2O, respectively.

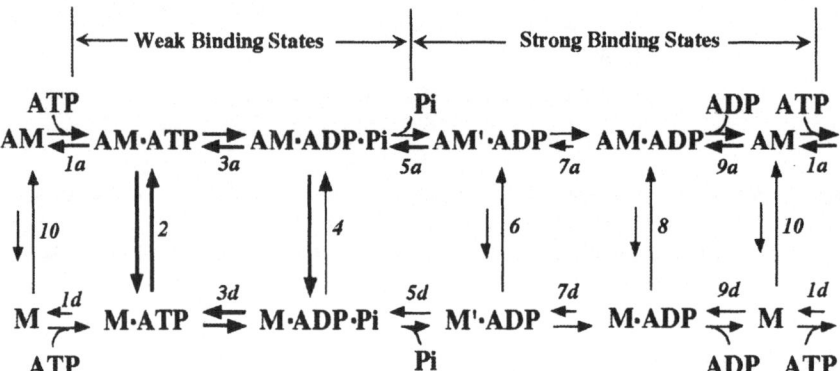

Figure 5. Kinetic scheme of the actomyosin ATPase, where A and M represent actin and myosin head, respectively.

5. DISCUSSION

5.1 Effects of D$_2$O on actomyosin kinetics

D$_2$O is known to have a higher viscosity, melting point, heat capacity and temperature of maximum density than H$_2$O [22]. These properties suggest that the degree of hydrogen bonding is higher in D$_2$O. Recent crystallographic data [23] have indicated that Mg-ADP and metallofluoride, a phosphate analogue, are held in the nucleotide-binding pocket of the myosin motor domain through hydrogen bonds. When the surrounding medium of the myosin head is changed from H$_2$O to D$_2$O, an exchange between hydrogen in the protein and deuterium of D$_2$O, and also an exchange between ordered water in the pocket and D$_2$O are expected to occur. Yamada et al. [24] showed that indole amide hydrogen of tryptophan residues in the myosin head incubated in D$_2$O were exchanged with deuterium at a rate of ~1s^{-1}. These data suggest that the exchange between hydrogen and deuterium in the nucleotide pocket would raise the degree of hydrogen bonding to hold ADP, which results in a slower rate of ADP release from the acto-S1-ADP complex (Fig. 5, step 9a).

5.2 Effects of D2O on the actin filament sliding velocity and its kinetic determinant

The frictional drag force against the sliding actin filament is calculated to be about 10 −14 N (the frictional drag force f = $\gamma \nu$, where ν =10 μm/s, and is the frictional coefficient, =2π η L/ln (2h/r) for =0.01 g/cm/s, the length of actin filament, L=1 μm, the height at which the filament is sliding above a surface h=10 nm, and the radius of the filament r = 5 nm), which is much smaller than the maximal force of one cross-bridge (10^{-12} N) [8,25]. So the slower velocity of actin filament in D$_2$O cannot be accounted fro by the 25% increase in the viscosity of D$_2$O compared with that of H$_2$O but by some rate determining step which decreased in the D$_2$O assay solution.

In 1985, Siemankowski et al. [21] suggested that the rate of ADP release can limit the unloaded shortening velocity of muscle, in contrast to the previous suggestion of Barany [26] that the rate of ATP hydrolysis by actomyosin limits the unloaded shortening velocity. Our in vitro data in D$_2$O are consistent with those of Siemankowski et al. that the actin filament sliding velocity does not correlate with Vmax of actin-activated Mg-ATPase activity of S1 but with the rate of ADP release from actomyosin. Namely, the decrease in actin filament velocity in D$_2$O is due to the higher degree of hydrogen bonding to hold the ADP in the pocket, which results in the slowing of the ADP release from actomyosin (Fig. 5, step 9a).

Another cause for the slower sliding velocity in D$_2$O may be from the Km value (the parameter of the affinity of actin for myosin in the presence of ATP) of the actin-activated Mg-ATPase activity of S1. The Km value for the actin-activated Mg-ATPase activity of S1 was 33 μM in D$_2$O and 50 μM in H$_2$O. Recently, Amitani et al. [27] have deduced that the in vitro sliding velocity is proportional to the value of (VmaxKm)$^{1/2}$ from their force-balance model, and shown that in vitro velocity data [28-34] were well fitted to this equation. In their model, the sliding velocity is determined by the balance between the force from active cross-bridges and the resistive drag force from weakly bound cross-bridges (shown as weak binding states in Fig. 5). The parameter of Km is included in the equation to express the time-resolved drag force on the assumption that the Km approximately equals the dissociation constant of weakly bound cross-bridges. Our in vitro and solution kinetic data fit to their equation, implying that D_2O might affect the equilibrium constant (step 2 and 4 in Fig. 5) of weakly bound cross-bridges so that the resistive drag force is increased.

The possible causes for the slower sliding velocity in D_2O are summarized as follows. (a) The decrease in the ADP release rate (step 9a in Fig. 5). (b) The increase in the affinity of F-actin for myosin at the weak binding state (step 2 and 4 in Fig. 5). Sweeny et al.[35] have shown that the actin sliding velocity was altered by the rate of ADP release but was not solely determined by it in their study on a surface loop (25/50 kDa loop), suggesting that another step in addition to the ADP release step may contribute to the rate of unloaded actin-based movement. Measurement of the stiffness-speed relationship of muscle fiber, which revealed the amount of attached cross-bridges in the relaxed state[36], would elucidate whether D_2O affects the equilibrium in the weak binding states in addition to the ADP release step.

REFERENCES

1. S.J. Kron, J.A. Spudich. Fluorescent actin filaments move on myosin fixed to a glass surface. Proc Natl Acad Sci U S A 83: 6272-6276 (1986).

2. Y. Harada, A. Noguchi, A. Kishino, T. Yanagida. Sliding movement of single actin filaments on one-headed myosin filaments. Nature 326:805-808 (1987).

3. S. Umemoto, J.R. Sellers. Characterization of in vitro motility assays using smooth muscle and cytoplasmic myosins. J Biol Chem 265:14864-14869 (1990).

4. D.M. Warshaw, J.M. Desrosiers, S.S. Work, K.M. Trybus. Effects of MgATP, MgADP, and Pi on actin movement by smooth muscle myosin. J Biol Chem. 266:24339-24343 (1991).

5. E .Homsher, F .Wang, J.R.Sellers. Factors affecting movement of F-actin filaments propelled by skeletal muscle heavy meromyosin. Am J Physiol. 262:C714-723(1992).

6. H .Yamashita, M.Sata, S. Sugiura, S.Momomura, T. Serizawa, M. Iizuka. ADP inhibits the sliding velocity of fluorescent actin filaments on cardiac and skeletal myosins. Circ Res. 74:1027-1033 (1994).

7. M.S. Kellermayer, G.H. Pollack. Rescue of in vitro actin motility halted at high ionic strength by reduction of ATP to submicromolar levels. Biochim Biophys Acta. 1277:107-114 (1996).

8. C.R. Bagshaw. Muscle Contraction. 2nd edn., Chapman and Hall, London (1993)

9. S. Highsmith, K. Duignan, R. Cooke, J. Cohen. Osmotic pressure probe of actin-myosin hydration changes during ATP hydrolysis.Biophys J. 70:2830-2837 (1996).

10. M. Suzuki, J.Shigematsu, Y. Fukunishi, Y. Harada, T. Yanagida, T.Kodama. Coupling of protein surface hydrophobicity change to ATP hydrolysis by myosin motor domain. Biophys J. 72:18-23 (1997).

11. K. Hotta, M.F. Morales. Myosin B nucleoside triphosphatase in deuterium oxide. J. Biol. Chem. 235:PC61-63 (1960).

12. A. Inoue, Y. Fukushima, Y. Tonomura. Effects of deuterium oxide on elementary steps in the ATPase reaction. Evidence for the similarity of key intermediates in contractile and transport ATPase. J Biochem (Tokyo). 78:1113-1121 (1975).

13. D.G. Allen, J.R. Blinks, R.E. Godt. Influence of deuterium oxide on calcium transients and myofibrillar responses of frog skeletal muscle. J Physiol. 354:225-251 (1984).

14. S.V. Perry. Myosin adenosinetriphosphatase. In Methods in Enzymology, vol. II, S.P. Colowick and N.O. Kaplan ed. Academic Press, New York   582-588 (1955).

15. J.A. Spudich, S.Watt. The regulation of rabbit skeletal muscle contraction. I. Biochemical studies of the interaction of the tropomyosin-troponin complex with actin and the proteolytic fragments of myosin.J Biol Chem. 246:4866-4871 (1971).

16. Y. Okamoto, T. Sekine. A streamlined method of subfragment one preparation from myosin.   J Biochem (Tokyo). 98:1143-1145 (1985).

17. S. Chaen, M. Nakaya, X.F. Guo, S. Watabe. Lower activation energy for sliding of F-actin on a less thermostable isoform of carp myosin.   J Biochem (Tokyo). 120:788-791 (1996).

18. P.K. Glasoe, F.A. Long. Use of glass electrodes to measure activities in deuterium oxide. J. Phys. Chem. 64:188-190 (1960).

19. T. Kodama, K. Fukui, K.Kometani. The initial phosphate burst in ATP hydrolysis by myosin and subfragment-1 as studied by a modified malachite green method for determination of inorganic phosphate. J Biochem (Tokyo). 99:1465-1472. (1986).

20. H.D. White. Special instrumentation and techniques for kinetic studies of contractile systems. In Methods in Enzymology, vol. 85, D.W. Frederiksen and L.W. Cunningham ed. Academic Press, New York   698-724 (1955).

21. R.F. Siemankowski, M.O. Wiseman, H.D. White. ADP dissociation from actomyosin subfragment 1 is sufficiently slow to limit the unloaded shortening velocity in vertebrate muscle. Proc.Natl.Acad.Sci.USA 82: 658-662 (1985)

22. J.J. Katz. Chemical and biological studies with deuterium. American Scientist 48: 544-580 (1960).

23. A.J. Fisher, C.A. Smith, J.B. Thoden, R. Smith, K. Sutoh, H.M. Holden, I. Rayment.   X-ray structures of the myosin motor domain of Dictyostelium discoideum complexed with MgADP.BeFx and MgADP.AlF4-. Biochemistry.   34:8960-8972 (1995).

24. T. Yamada, H. Shimizu, M. Nakanishi, M. Tsuboi. Environment of the tryptophan residues in a myosin head: a hydrogen-deuterium exchange study.   Biochemistry.   20:1162-1168 (1981).

25. J.T. Finer, R.M. Simmons, J.A. Spudich. Single myosin molecule mechanics: piconewton forces and nanometre steps. Nature.   368:113-119 (1994).

26. M. Barany.   ATPase activity of myosin correlated with speed of muscle shortening.J Gen Physiol.   50: 197-218 (1967).

27. I. Amitani, T. Sakamoto, T. Ando. Link between the enzymatic kinetics and mechanical behavior in an actomyosin motor.   Biophys J.   Jan;80(1):379-397 (2001).

28. R.K. Cook, D. Root, C. Miller, E. Reisler, P.A. Rubenstein. Enhanced stimulation of myosin subfragment 1 ATPase activity by addition of negatively charged residues to the yeast actin NH2 terminus.   J Biol Chem. 268:2410-2415. (1993).

29. K. Sutoh, M. Ando, K Sutoh, Y.Y. Toyoshima.   Site-directed mutations of Dictyostelium actin: disruption of a negative charge cluster at the N terminus. Proc Natl Acad Sci U S A.   88:7711-7714   (1991).

30. M. Johara, Y.Y. Toyoshima, A. Ishijima, H. Kojima, T. Yanagida, K.Sutoh.   Charge-reversion mutagenesis of Dictyostelium actin to map the surface recognized by myosin during ATP-driven sliding motion. Proc Natl Acad Sci U S A.   90:2127-2131 (1993).

31. R.H. Crosbie, C. Miller, P. Cheung, T. Goodnight, A. Muhlrad, E. Reisler.   Structural connectivity in actin: effect of C-terminal modifications on the properties of actin. Biophys J. 67:1957-1964 (1994).

32. A.S. Rovner, Y. Freyzon, K.M. Trybus. Chimeric substitutions of the actin-binding loop activate dephosphorylated but not phosphorylated smooth muscle heavy meromyosin. J Biol Chem.   270:30260-30263 (1995)

33. T. Hozumi, M. Miki, S. Higashi-Fujime Maleimidobenzoyl actin: its biochemical properties and in vitro motility. J Biochem (Tokyo).   119:151-156 (1996).

34. T. Barman, M. Brune, C. Lionne, N. Piroddi, C. Poggesi, R. Stehle, C. Tesi, F. Travers, M.R. Webb. ATPase and shortening rates in frog fast skeletal myofibrils by time-resolved measurements of protein-bound and free Pi. Biophys J.   74:3120-3130 (1998).

35. H.L. Sweeney, S.S. Rosenfeld, F. Brown, L. Faust, J. Smith, J. Xing, L.A. Stein, J.R.Sellers.   Kinetic tuning of myosin via a flexible loop adjacent to the nucleotide binding pocket. J Biol Chem.   273:6262-6270 (1998).

36. B. Brenner, M. Schoenberg, J.M. Chalovich, L.E. Greene, E.Eisenberg. Evidence for cross-bridge attachment in relaxed muscle at low ionic strength.   Proc Natl Acad Sci U S A.   79:7288-7291 (1982)

DISCUSSION

Ogata: What is the difference of viscosity in your solutions?

Chaen: The viscosity of D_2O is about 25% higher than that of H_2O.

Pollack: It is known that the contractile proteins contain substantial hydration layers. Is it possible that D_2O and H_2O hydrate differently, and therefore have direct effects on the contractile proteins?

Chaen: I think it might be possible. Because of stronger hydrogen bond in D_2O than H_2O, protein conformation would change more markedly in D_2O than in H_2O to compensate the entropy decrease due to structuring of water.

Cecchi: In our experiments with intact fibers (Cecchi, Colomo and Lombardi, J. Physiol. 1981, 317), we found no change in V_{max} in D_2O-Ringer. However, there was a considerable increase in the curvature of the P-V relation. This means that the shortening velocity was reduced even with a small load. Is it possible that the decrease in the sliding velocity you found means that the sliding was somewhat loaded?

Chaen: In our in vitro motility assay, actin filaments slide on myosin under unloaded condition.

Sugi (Comment after Discussion): Dr. Cecchi's experiment was performed with single intact muscle fibers, while our experiment is done with in vitro motility assay systems. So, the results between the two experiments can not be readily compared with each other.

III. MOLECULAR BASIS FOR REGULATORY MECHANISM OF MUSCLE CONTRACTION

NMR STRUCTURAL STUDY OF TROPONIN C C-TERMINAL DOMAIN COMPLEXED WITH TROPONIN I FRAGMENT FROM AKAZARA SCALLOP

Fumiaki Yumoto[1,2], Koji Nagata[3], Kyoko Adachi[4], Nobuaki Nemoto[5], Takao Ojima[6], Kiyoyoshi Nishita[6], Iwao Ohtsuki[2] and Masaru Tanokura[1]

1. INTRODUCTION

Scallop muscle has been demonstrated to possess both myosin-linked and actin-linked systems[1-3] (Fig. 1), even though molluscan muscles were known to be regulated only by the myosin-linked regulatory system mediated through Ca^{2+}-binding to myosin light chains[4-6]. Recently, the physiological significance of the coexistence of the two systems in scallop adductor muscle was investigated using CDTA-treated scallop myofibrils[1]. Actin-linked (Troponin-linked) system has been well known as the regulatory system in the muscle contraction of vertebrate striated muscles[7]. It is regulated by troponin in a Ca^{2+} dependent manner. Troponin contains three distinct components, i.e., a Ca^{2+} binding component (TnC), an inhibitory component troponin I (TnI), and a tropomyosin-binding component troponin T (TnT). TnC contains two independent Ca^{2+} binding domains, each of which consists of two EF-hand motifs[8]. Vertebrate striated muscle TnCs bind three or four Ca^{2+} ions in a molecule and act as the Ca^{2+} sensor of muscle contraction associated with the binding and release of one or two Ca^{2+} ion(s) in the N-terminal domain[9, 10, 11]. The N-terminal domain has, thus, been called the regulatory domain and contains one or two low affinity Ca^{2+}-binding sites (Sites I and II)[12]. On the other hand, the C-terminal domain has been called the structural domain and contains two high-affinity sites (Sites III and IV). They also bind Mg^{2+} and are called as Ca^{2+}/Mg^{2+} sites.

[1]Department of Applied Biological Chemistry, Graduate School of Agricultural and Life Sciences, University of Tokyo, Tokyo; Japan [2]School of Medicine, The Jikei University, Tokyo; Japan [3]Biotechnology Research Center, University of Tokyo, Tokyo; Japan [4]Marine Biotechnology Institute, Shimizu, Shizuoka; Japan [5]Varian Technologies Japan, Tokyo; Japan [6]Laboratory of Biochemistry and Biotechnology, Graduate School of Fisheries Science, Hokkaido University, Hakodate, Hokkaido; Japan

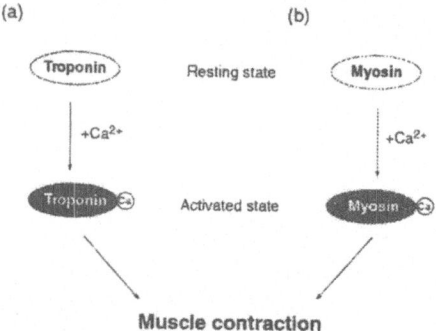

Fig.1 Two pathways of regulatory system in scallop striated muscle (a) Troponin-linked (Actin-linked) regulation (b) Myosin-linked regulation

2. Akazara scallop troponin C

Invertebrate muscles also have troponin molecules and their TnCs bind less Ca^{2+} than vertebrate ones, because they have lost the Ca^{2+}-binding ability at several sites due to the replacement of amino acids critical to coordination of Ca^{2+}. Akazara scallop (*Chlamys nipponensis akazara*) is one of the invertebrates and its striated adductor muscle contains troponin that works as a Ca^{2+} regulator of contraction[13, 14]. Akazara troponin also contains three components and its TnC binds only one Ca^{2+} ion at the C-terminal EF-hand motif (site IV) [12, 13, 14]. This character suggests that the molecular mechanism of Ca^{2+}-regulation of scallop TnC is somehow different from that of vertebrate striated muscle troponin Cs. In fact, vertebrate skeletal and cardiac muscle troponin Cs require the binding of Ca^{2+} to N-terminal sites (sites I & II) on TnC to switch contraction.

X-ray crystallography and NMR spectroscopic studies on structure of TnC have been revealed that TnC has two globes separated by a central linker[15-18]. These studies also show that the regulation of its function as Ca^{2+} sensor is achieved by the conformational change from closed-form to open-form with exposing the hydrophobic patch[19, 20]. This structural change of N-domain is thought to facilitate the relief of inhibition of acto-myosin ATPase activity by TnI. This region of TnI binds to Ca^{2+}-bound N-domain, and its amphiphilic C-end locates in the hydrophobic cavity[20]. After the releasing its Ca^{2+} from N-domain, on the other hand, it returns to the closed-form and hence it pushes the amphiphilic-helix of the C-end of TnI inhibitory region out of the hydrophobic patch[21]. In addition to the mechanism, mainly worked by N-domain, the role of the C-domain in vertebrate striated muscle TnC is also studied[20, 22-24].

While many things are known about the structure of vertebrate fast skeletal and cardiac muscle TnC[26-33], including the binding sites with other components, little structural information are available at present, except the coordination structure of single Ca^{2+}-binding site on Akazara scallop TnC[33]. And the mechanism of the regulation of vertebrate TnC as a Ca^{2+} sensor is not applicable to the scallop TnC since the N-domain of scallop TnC is incapable of binding Ca^{2+}.

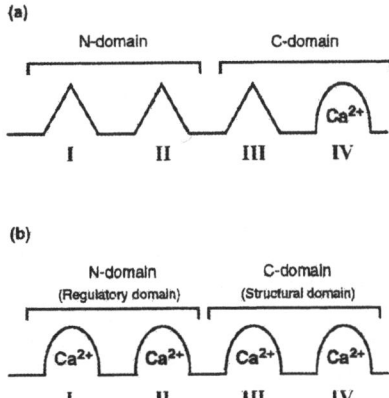

Fig.2 Ca²⁺-binding sites of TnCs. (a) Ca²⁺-binding site of Akazara scallop TnC (b) Ca²⁺-binding sites of vertebrate fast skeletal TnC.

3. NMR analysis

To analyze the Ca^{2+}-dependent structural change and functional regulation of scallop TnC, we are conducting structural analysis of scallop TnC C-domain, contains two EF-hand motifs (sites III and IV), bound to $TnI_{129-183}$ by multidimensional NMR spectroscopy.

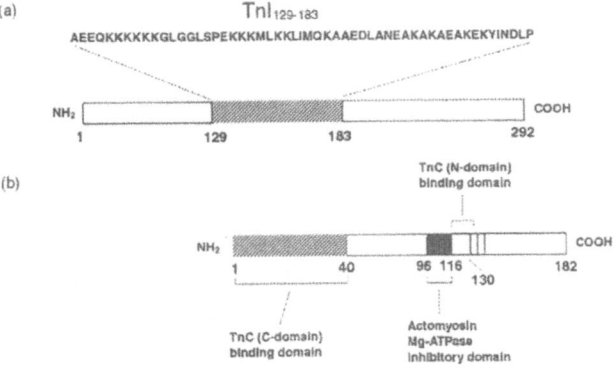

Fig. 3 Domain structure of troponin Is (a) Akazara scallop troponin I has an N-terminus that is approximately 130 residues longer than that of vertebrate. (b) Chicken fast skeletal muscle troponin I. The roles of the domains are characterized as shown.

Scallop TnI$_{129-183}$ is the peptide corresponding to the N-terminal region of the vertebrate fast skeletal muscle TnIs, which binds to the C-domain of TnC[34] (Yumoto *et al.*, unpublished result). TnC and TnC C-domain only tend to make an aggregate on the concentration in the range of mM. The aggregation is suppressed upon the complex formation with TnI$_{129-183}$.

To prepare the sample for NMR study, purified TnC C-domain and TnI$_{129-183}$ mixed and dialyzed against a buffer containing 10 mM Mops-KOH pH 6.8, 10 mM CaCl$_2$, 1 M KCl, 2 mM dithiothreitol and 6 M urea. The sample was then dialyzed consecutively against KCl solution of 500 mM, 250 mM, and 100 mM, each containing 10 mM Mops-KOH pH 6.8, 10 mM CaCl$_2$, 2 mM dithiothreitol. After dialysis, the complex solution was concentrated by Centriprep YM-3 (amicon) and applied to Superdex 75 26/60 column (amersham) equilibrated with the final buffer of the dialysis. Complex of TnC C-domain with TnI$_{129-183}$ is confirmed by the 15% SDS-polyacrylamide gel electrophoresis.

The NMR sample contained 2.9 mM ^{13}C, ^{15}N-TnC C-domain/non-labeled TnI$_{129-183}$ complex in 95% H$_2$O and 5% D$_2$O at pH 6.80, containing 100 mM KCl, 10 mM CaCl$_2$ and 10 mM deuterated DTT. The pH values were not corrected for the isotope effect of ^{2}H$_2$O and the values described were direct pH readings. All NMR experiments were performed on a Varian Unity Inova 750 spectrometer. Backbone and side chain ^{1}H, ^{13}C and ^{15}N resonances of TnC C-domain, complexed with the TnI$_{129-183}$, were assigned using ^{15}N-^{1}H HSQC, and HN(CO)CA, HNCA, CBCA(CO)NH, HNCACB, (HCA)CO(CA)NH, HNCO spectra. Data were processed with NMRPipe[35] and were analyzed with Sparky[36].

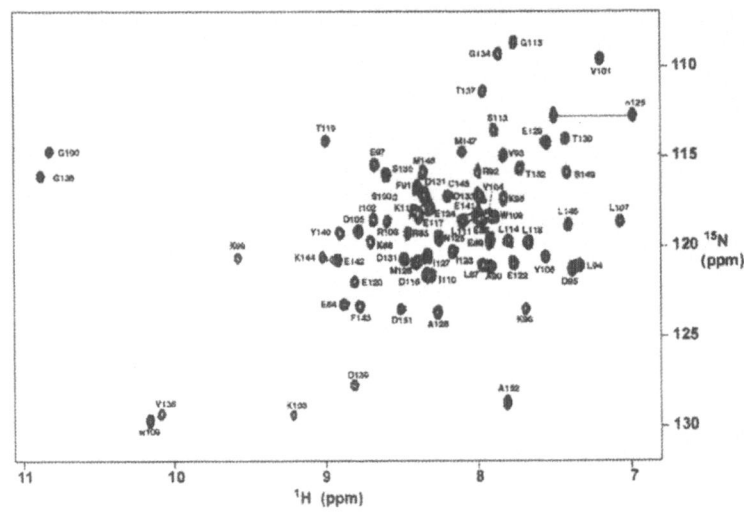

Fig. 4 The two-dimensional ^{15}H-^{1}H HSQC spectrum of ^{13}C,^{15}N Akazara scallop TnC C-domain complexed with nonlabeled TnI$_{129-183}$. The capital letters indicate the assignments for the backbone ^{15}N-^{1}H groups. Indole NH cross-peak from Trp109 residue is marked as w109, and cross-peaks connected by line correspond to side chain NH$_2$ groups of Asn125 residues.

The backbone amide resonances are completely assigned with the exception of N-terminal three residues those were not observed due to rapid exchange with solvent. Downfield shifts are observed for the amide protons of Gly_{100} and Gly_{136}, as a consequence of hydrogen bonding in Ca^{2+} binding loops III and IV.

Three-dimensional structure calculation is in progress.

REFERENCES

1. Shiraishi, F., Morimoto, S., Nishita, K., Ojima, T., and Ohtsuki, I. Effects of removal and reconstitution of myosin regulatory light chain and troponin C on the Ca^{2+}-sensitive ATPase activity of myofibrils from scallop striated muscle. *J. Biochem.* **126**, 1020-1024 (1999).

2. Ojima, T., and Nishita, K. Troponin from akazara scallop striated adductor muscles. *J. Biol. Chem.* **261**, 16749-16754 (1986).

3. Ojima, T., and Nishita, K Isolation of troponins from striated and smooth adductor muscles of akazara scallop. *J. Biochem.* **100**, 821-824 (1986).

4. Szent-Györgyi, A. G., and Szentkiralyi, E. M. The light chains of scallop myosin as regulatory subunits. *J. Mol. Biol.* **74**, 179-203 (1973).

5. Kendrick-Jones, J., Lehman, W., and Szent-Györgyi, A. G. Regulation in molluscan muscles. *J. Mol. Biol.* **54**, 313-326 (1970).

6. Kendrick-Jones, J., Szentkiralyi, E. M., and Szent-Györgyi, A. G. Regulatory light chains in myosins. *J. Mol. Biol.* **104**, 747-775 (1976).

7. Ebashi, S., Endo, M., and Ohtsuki, I. In: *CALCIUM as a CELLULAR REGULATOR* Carafoli, E., and Klee, C. B., ed., Oxford Univ. Press, New York, pp.579-595 (1999).

8. Zot, A. S., and Potter, J. D. Structural aspects of troponin-tropomyosin regulation of skeletal muscle contraction. *Ann. Rev. Biophys. Chem.* **16**, 535-539 (1987).

9. Collins, J. H., Potter, J. D., Horn, M. J., Wilshire, G., and Jackman, N. The amino acid sequence of rabbit skeletal muscle troponin C: gene replication and homology with calcium-binding proteins from carp and hake muscle. *FEBS Lett.* **36**, 268-272 (1973).

10. van Eerd, J. P., and Takahashi, K. Determination of the complete amino acid sequence of bovine cardiac troponin C. *Biochemistry* **15**, 1171-1180 (1976).

11. Wilkinson, J. M. Troponin C from rabbit slow skeletal and cardiac muscle is the product of a single gene. *Eur. J. Biochem.* **103**, 179-188 (1980).

12. Nishita, K., Tanaka, H., and Ojima, T. Amino acid sequence of troponin C from scallop striated adductor muscle. *J. Biol. Chem.* **269**, 3464-3468 (1994).

13. Ojima, T., Tanaka, H., and Nishita, K. Cloning and sequence of a cDNA encoding Akazara scallop troponin C. *Arch. Biochem. Biophys.* **311**, 272-276 (1994).

14. Ojima, T., Koizumi, N., Ueyama, K., Inoue, A., and Nishita, K. Functional Role of Ca^{2+}-Binding Site IV of Scallop Troponin C. *J. Biochem.* **128**, 803-809 (2000).

15. Herzberg, O., and James, M. N. G. Structure of the calcium regulatory muscle protein troponin-C at 2.8 Å resolution. *Nature* **313**, 653-659 (1985).

16. Satyshur, K. A., Rao, S. T., Pyzalska, D., Drendel, W., Greaser, M., and Sundarlingam, M. Refined structure of chicken skeletal muscle troponin C in the two-calcium state at 2 resolution. *J. Biol. Chem.* **263**, 16620-16628 (1988).

17. Herzberg, O., and James, M. N. G. Refined crystal structure of troponin C from turkey skeletal muscle at 2.0

A resolution. *J. Mol. Biol.* **203**, 761-779 (1988).

18. Slupsky, C. M., and Sykes, B. D. NMR solution structure of calcium-saturated skeletal muscle troponin C. *Biochemistry* **34**, 15953-15964 (1995).

19. Gagné, S. M., Tsuda, S., Li, M. X., Smillie, L. B., and Sykes, B. D. *Nat. Struct. Biol.* **2**, 784-789 (1995).

20. Tripet, B., Eyk, V. E., and Hodges, R. S. Mapping of a second actin-tropomyosin and a second troponin C binding site within the C terminus of troponin I, and their importance in the Ca^{2+}-dependent regulation of muscle contraction. *J. Mol. Biol.* **271**, 728-750 (1997).

21. Vassylyev, D. G. Takeda, S., Wakatsuki, S., Maeda, K., and Maeda, Y. Crystal structure of troponin C in complex with troponin I fragment at 2.3-A resolution. *Proc Natl Acad Sci U S A* **95**, 4847-4852 (1998).

22. Mercier, P., M. X., and Sykes, B. D. Role of the structural domain of troponin C in muscle regulation: NMR studies of Ca^{2+} binding and subsequent interactions with regions 1-40 and 96-115 of troponin I. *Biochemistry* **39**, 2902-2911 (2000).

23. Mercier, P., Spyracopoulos, L., and Sykes, B. D. Structure, dynamics, and thermodynamics of the structural domain of troponin C in complex with the regulatory peptide 1-40 of troponin I. *Biochemistry* **40**, 10063-10077 (2001).

24. Gasmi-Seabrook, G., Howarth, J. W., Finley, N., Abusamhadneh, E., Gaponenko, V., Brito, R. M., Solaro, R. J., and Rosevear, P. R. *Biochemistry* **38**, 8313-8322 (1999).

25. Spyracopoulos, L., Li, M. X., Sia, S. K., Gagne, S. M., Chandra, M., Solaro, R. J., and Sykes, B. D. Calcium-induced structural transition in the regulatory domain of human cardiac troponin C. *Biochemistry* **36**, 12138-12146 (1997).

26. Houdusse, A., Love, M. L., Dominguez, R., Grabarek, Z., and Cohen, C. *Structure* **5**, 1695-1711 (1997).

27. McKay, R. T., Pearlstone, J. R., Corson, D. C., Gagné, S. M., Smillie, L. B., and Sykes, B. D. Structure and interaction site of the regulatory domain of troponin-C when complexed with the 96-148 region of troponin-I. *Biochemistry* **37**, 12419-12430 (1998).

28. McKay, R. T., Tripet, B. P., Pearlstone, J. R., Smillie, L. B., and Sykes, B. D. Defining the region of troponin-I that binds to troponin-C. *Biochemistry* **38**, 5478-5489 (1999).

29. Li, M. X., Spyracopoulos, L., and Sykes, B. D. Binding of cardiac troponin-I147-163 induces a structural opening in human cardiac troponin-C. *Biochemistry* **38**, 8289-8298 (1999).

30. Blumenschein, T. M., Tripet, B. P., Hodges, R. S., and Sykes B. D. Mapping the interacting regions between troponins T and C. Binding of TnT and TnI peptides to TnC and NMR mapping of the TnT-binding site on TnC. *J Biol Chem.* **276**, 36606-36612 (2001).

31. Abbott M. B., Dong, W. J., Dvoretsky, A., DaGue, B., Caprioli, R. M., Cheung, H. C., and Rosevear, P. R. Modulation of cardiac troponin C-cardiac troponin I regulatory interactions by the amino-terminus of cardiac troponin I. *Biochemistry* **40**, 5992-6001 (2001).

32. Dvoretsky, A., Abusamhadneh, E. M., Howarth, J. W., and Rosevear, P. R. Solution Structure of Calcium-saturated Cardiac Troponin C Bound to Cardiac Troponin I. *J Biol Chem.* **277**, 38565-38570 (2002).

33. Yumoto, F., Nara, M., Kagi, H., Iwasaki, W., Ojima, T., Nishita, K., Nagata, K., and Tanokura, M. Coordination structures of Ca^{2+} and Mg^{2+} in Akazara scallop troponin C in solution. FTIR spectroscopy of side-chain COO^- groups. *Eur. J. Biochem.* **268**, 6284-6290 (2001).

34. Tanaka, H., Ojima, T., and Nishita, K. Amino acid sequence of troponin-I from Akazara scallop striated adductor muscle. *J. Biochem.* **124**, 304-310 (1998).

35. Delaglio, F., Grzesiek, S., Vuister, G. W., Zhu, G.., Pfeifer, J., and Bax, A. NMRPipe: a multidimensional spectral processing system based on UNIX pipes. *J. Biomol. NMR.* **6**, 277-293 (1995).

36. Goddard, T. D. and Kneller, D. G., SPARKY 3, University of California, San Francisco; http://www.cgl.ucsf.edu/home/sparky/

DISCUSSION

Gordon: Does the C terminal of scallop TnC have an open structure in the presence of Ca but in the absence of scallop TnI? I realize that it is difficult to obtain the NMR spectra from isolated C terminal of TnC, but what happens with whole scallop TnC?

Tanokura: There is a change in the structure of C-terminal of TnC to an open conformation when it binds Ca in the presence of scallop TnC. But we cannot know the conformation of C-terminal of TnC by NMR, because it aggregates under the NMR measurement conditions. We are presently trying to analyze the structure of whole scallop TnC in the complex with TnI fragment. NMR spectra of whole scallop TnC is improved as compared with C-terminal of TnC, but look difficult to be analyzed.

CALCIUM BINDING TO TROPONIN C
AS A PRIMARY STEP OF
THE REGULATION OF CONTRACTION
A microcalorimetric approach

Kazuhiro Yamada*

SUMMARY

Microcalorimetric titration studies of EF-hand Ca-binding proteins (troponin C, calmodulin and parvalbumins) resulted in the notion that Ca binding to the "active" Ca site, which is involved in the regulation of contraction, induces a characteristic anomalous enthalpy and heat-capacity changes indicating an exposure of hydrophobic residues to the solvent, which enables the proteins to interact with their targets. There is a good agreement between the results of the calorimetric and the structural studies in frog and chicken skeletal troponin C. In both species one of the N-terminal low-affinity Ca-sites is the "active" Ca site regulating muscle contraction. The results from calorimetry have shown, however, that the situation in rabbit skeletal troponin C may be more complex. Moreover, in both calorimetric and structural studies, the situation in cardiac troponin C is quite different. These results suggest the need for further studies to elucidate the mechanism of regulation by Ca. These characteristic changes do not occur in Ca-buffering proteins.

INTRODUCTION

Around 1980s we performed enthalpy titrations with calcium (Ca) by using titration microcalorimetry on a number of Ca-binding proteins in muscle and other tissues including troponin C (see Yamada, 1999). These studies were initiated to explain initial heat production of muscle. We applied microcalorimetric titration methods for these problems because heat of Ca binding to the regulatory proteins was thought to be one of the sources of the initial heat production during muscle contraction, because troponin and

---

* Department of Physiology, Oita Medical University, Oita 879-5593 Japan

other Ca-binding proteins have multiple Ca-binding sites and also because enthalpy changes associated with the Ca binding are related to structural changes of the proteins involved.

During calorimetric titrations, although the Ca-binding reactions are mostly exothermic, we noticed an anomalous heat absorption mostly in a single site in many classes of the Ca-binding proteins studied. Therefore the calorimetric results were also analyzed in terms of structural changes by observing heat-capacity changes on binding Ca. It was found that the phase of the anomalous heat absorption corresponded to an anomalous increase in heat capacity. In structural terms these anomalous changes are associated essentially with the "opening" of the molecule and the exposure of hydrophobic residues to the solvent phase. These characteristic changes are important functionally because these enable the protein to interact with other proteins such as troponin I, the inhibitory component of troponin complex, and the "target" for calmodulin.

A general rule emerged that the characteristic anomalous change in enthalpy and heat capacity accompanies Ca binding in regulatory proteins such as skeletal troponin C and calmodulin, but not in Ca buffering proteins such as parvalubumin. Later similar ideas seem to have been introduced in structural studies (Skelton et al., 1994).

METHODS

After extraction metal-free proteins were prepared either by chelating-resin methods (see Yamada and Kometani, 1982) or by TCA methods (Tanokura and Yamada, 1984). Calorimetric titrations of Ca-binding proteins with Ca^{2+} or Mg^{2+} were carried out by using an LKB batch microcalorimeter (Yamada and Kometani, 1982). The solutions contained 0.1 M KCl and 10–20 mM Pipes-NaOH at pH 7.0. Ca^{2+} or Mg^{2+} was titrated by means of a motor-driven micrometer and syringes (Yamada and Kometani, 1982).

Figure 1 shows an example of such titrations, and also shows an example of the anomalous heat absorption, with Ca titrations of calmodulin (Tanokura and Yamada, 1983), and Fig. 2 the block diagram of the titration microcalorimeter.

RESULTS OF MICROCALORIMETRIC STUDIES ON CALCIUM-BINDING PROTEINS

Troponin C (TnC)

Troponin C (TnC) is one of the proteins of EF-hand family and is the calcium-binding regulatory component of troponin complex in skeletal and cardiac muscles (Ebashi and Endo, 1968). TnC binds four Ca ions and the signal of Ca binding to TnC is transmitted

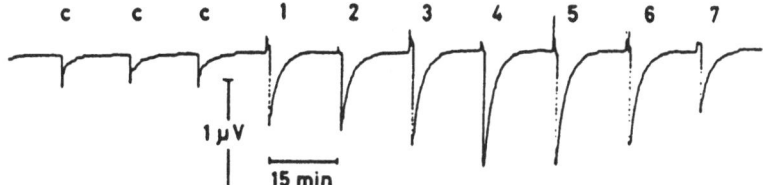

Figure 1. Calorimetric records during enthalpy titrations with Ca of bovine brain calmodulin in 0.1 M KCl and 20 mM Pipes-NaOH (pH 7) at 25°C.

to the thin filament to activate contraction (Ebashi and Endo, 1968)).   Binding of Ca to troponin and TnC produces heat (Yamada et al., 1976; Yamada, 1978; see Yamada, 1999).

Rabbit skeletal troponin C

When TnC extracted from the rabbit skeletal muscle was titrated with Ca we noticed several phases of heat produced (Yamada and Kometani, 1982).   Typically a substantial amount of heat is produced on Ca binding initially (Fig. 3).   The heat- capacity change, which can be obtained by temperature dependence of ΔH, are substantially negative (Fig. 4).   Quite conspicuously, however, there is a phase where heat produced becomes very small or even heat is absorbed; this phase is also characterized by an increase in heat capacity.   Thus the calorimetric titration of TnC extracted from rabbit skeletal muscle showed a phase of the conspicuous anomalous enthalpy and heat-capacity changes (characteristic positive enthalpy and heat capacity changes).   These seem to be affected by the presence of 1 mM Mg.   Yamada and Kometani (1982) argued that the characteristic phase of the positive heat-capacity change was associated with Ca binding to one of the two high-affinity Ca sites of the C-terminal domain.   Because the characteristic positive heat-capacity change is associated mostly with the proteins having regulatory functions as will be shown below, this characteristic phase of Ca binding should be involved in the regulatory mechanism of contraction.

Thermodynamic analysis of calorimetric results on skeletal TnC in more detail revealed that during the phase of the anomalous enthalpy and heat-capacity changes described above hydrophobic residues became more exposed to water (a gain of 33 mole of water per mole of TnC).   At the same time the molecular structure became less ordered.   In other phases of titrations a sequestering of the hydrophobic residues into the interior regions of the TnC molecule, as well as a tightening of the overall structure occured (Yamada and Kometani, 1982).   As will be described below, however, in cardiac TnC no such anomalous phase is seen (Kometani and Yamada, 1983).   Some of the characteristic features of the Ca-binding proteins studied by microcalorimetry are listed in Table 1.

Mg will be replaced by Ca at Ca-Mg sites.   The conformation of Mg-bound proteins is not necessarily the same with that of the Ca-bound ones as will be shown below.   The effects of Mg on the "active" Ca response are also listed in Table 1.

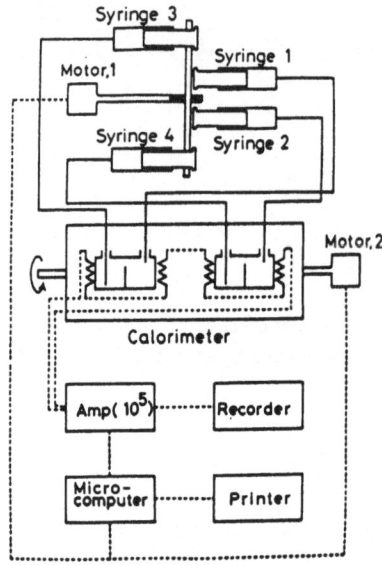

Figure 2. Block diagram showing a batch microcalorimeter with a twin gold cells and titration apparatus. Motor 1 drives syringes 1 and 2 for adding Ca, and motor 2 rotates the calorimeter block for mixing solutions in the cells.

Frog skeletal troponin C

Among vertebrate skeletal muscles fast skeletal muscles of the frog have been taken to operate basically similar to those of mammalians including the rabbit. However, the amino acid composition of TnC in frog skeletal muscle has been shown to be distinctly different from that of rabbit (van Eerd et al., 1978).

Calorimetric studies of TnC extracted from bullfrog skeletal muscle also showed the phase of the characteristic anomalous positive enthalpy and heat-capacity changes as in rabbit skeletal muscle TnC. Unlike with rabbit skeletal TnC, however, in frog skeletal TnC the characteristic phase of enthalpy and heat capacity was associated with one of the two low-affinity Ca-binding sites (Imaizumi et al., 1987; Imaizumi et al., 1990; Imaizumi and Tanokura, 1990).

The difference between the rabbit and the bullfrog TnC has posed an interesting problem. Li et al. (1995) have shown by titration structural studies using NMR that the Ca binding to the regulatory N-domain of TnC from chicken skeletal muscle, which has more than 90 % sequence identity with the frog, also occurs in a stepwise manner (see also DISCUSSION). There may be fundamental differences in TnC of skeletal muscles between the rabbit and the frog in terms of the "active" Ca-binding site involved in the regulation of contraction if the characteristic anomalous phase of the thermodynamic functions is concerned with the regulation of contraction.

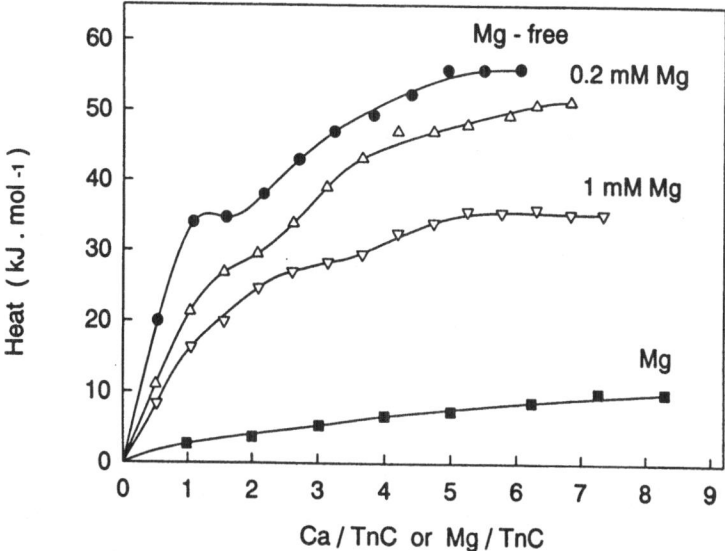

Figure 3. Heat produced per mole of rabbit skeletal troponin C against concentration ratios of Ca or Mg and troponin C at pH 7 at 15°C. Reproduced with permission from Yamada, 1999.

Cardiac troponin C

The regulation of contraction in cardiac muscle is mediated also by the troponin-tropomyosin system incorporated in the thin filament as in the skeletal muscle. Although the amino acid sequence of cardiac TnC is 70 % identical to that of the skeletal TnC, there are significant differences in the first 40 residues (van Eerd and Takahashi, 1975). As a consequence the N-terminal calcium site I is inactive.

Cardiac TnC exhibited a decrease in enthalpy and a decrease in heat capacity with Ca binding (Kometani and Yamada, 1983). Also only two high-affinity Ca-binding sites were detected. Therefore the process of Ca binding to the low-affinity Ca-site, if present, should be thermally neutral. Therefore, quite on the contrary to skeletal TnC, calcium binding to bovine cardiac TnC did not show any sign of the anomalous enthalpy and heat-capacity changes observed associated with the calcium binding to skeletal TnC as described above (Kometani and Yamada, 1983). As will be seen below cardiac TnC is the only one exception of the general rule (see DISCUSSION).

Calmodulin

Calmodulin (CaM) is one of the Ca-binding proteins and plays important roles in many intracellular functions. CaM regulates a number of fundamental cellular activities, and also mediates the activation of a number of intracellular enzymes. The tertiary structure of CaM is very similar to that of the skeletal TnC. Results of enthalpy

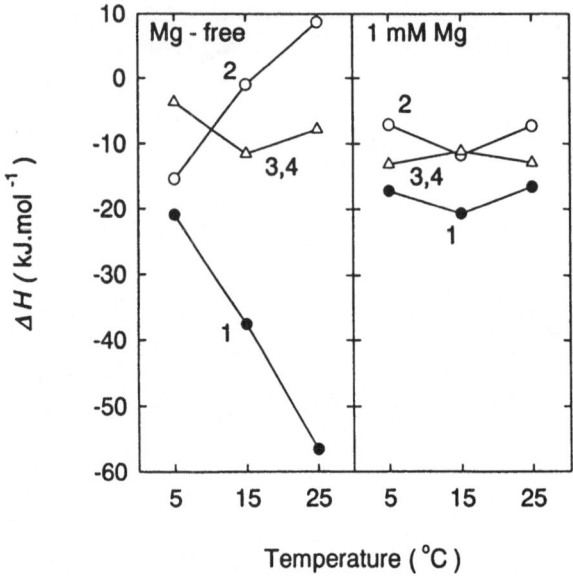

Figure 4. Temperature dependence of ΔH on Ca binding to rabbit skeletal troponin C. Numbers 1 and 2 refer to the two high-affinity Ca-Mg sites, and 3 and 4 the two low-affinity Ca-specific sites. Reproduced with permission from Yamada, 1999.

titrations of calmodulin with calcium indicated that the binding of Ca to CaM in the absence of Mg is an endothermic reaction and thus is driven by a large entropy changes at pH 7 and at $25°C$ (Tanokura and Yamada, 1983). It was also shown that Ca displaces Mg at the four Ca-binding sites. The enthalpy titration profile of CaM with Ca in the absence of Mg can be divided into two stages; the first stage is characterized by a substantial absorption of heat and the second stage by even a greater heat absorption. Although absolute values of enthalpy changes are much smaller than those for TnC, Ca binding to calmodulin shares the characteristic features of the anomalous Ca-binding sites of TnC, i.e. positive enthalpy and positive heat-capacity changes.

As discussed above Ca binding to CaM extracted from various sources (bovine brain and wheat germ, Tanokura and Yamada, 1983; 1984; 1993) is endothermic, and the heat-capacity changes are negative but characteristically small. Also binding Ca to CaM is driven by large entropy changes. These can be taken to be similar to the characteristics associated with the Ca binding to proteins with regulatory functions.

As has been described above Ca binding to every site of CaM is an endothermic process, while Ca binding to CaM-trifluoperazine(TFP) complex is exothermic (Tanokura and Yamada, 1984; 1985b; 1986). The results indicated that the endothermic process of Ca binding to CaM enables the Ca-CaM complex to activate various targets, but the Ca binding to the CaM-TFP complex cannot interact with any target.

Table 1.   Identification of the "active" Ca sites in troponin C and other EF-hand Ca-binding proteins

	Source	Active Ca site	Hydrophobic exposure	Effect of Mg
Skeletal TnC	rabbit[a]	C–terminal site III or IV	strong	strong
	bullfrog[b]	N–terminal site I or II		weak
Cardiac TnC	bovine[c]	N–terminal site II	weak	weak
Calmodulin	bovine brain[d]	site I ~ IV	moderate	weak
	wheat germ[e]			
CaM–TFP	bovine brain[f]	none	–	–
Parvalbumin	bullfrog[g]	none	–	–
	toad[h]			

a, Yamada and Kometani, 1982; b, Imaizumi and Tanokura, 1990; c, Kometani and Yamada, 1983;

d, Tanokura and Yamada, 1983, 1984; e, Tanokura and Yamada, 1993; f, Tanokura and Yamada, 1985b, 1986;

g, Tanokura and Yamada, 1985a, 1987; h, Tanokura, Imaizumi and Yamada, 1986.

Parvalbumins

Parvalbumin is a protein found in the cytosol of muscle cells, and is thought to act as a Ca buffer to ensure fast response for Ca sensitivity.   The binding of Ca to parvalubumins are exothermic and the heat-capacity changes are substantially   negative (Tanokura and Yamada, 1985a; 1987; Tanokura et al., 1986).   This can be taken to be characteristic of Ca binding to Ca-buffering proteins, because parvalbumins have never been found to interact with any proteins.

These results suggest that the anomalous positive enthalpy as well as positive heat-capacity changes are characteristic of the regulatory functions of Ca-binding proteins.

DISCUSSION

High-resolution structures of some of the Ca-binding proteins have been determined. The crystals of turkey skeletal muscle TnC studied by Herzberg and James (1988) contained Ca in C-terminal sites III and IV but not in N-terminal sites I and II.   They showed that the cation-filled C-terminal sites occupy a rather open structure with hydrophobic regions exposed to the solvent.   Based on these structures Herzberg et al. (1986) proposed that the N-terminal Ca-specific domains of TnC may take up a conformation similar to that exhibited by the two cation-filled C-terminal sites on Ca binding (HMJ model).   This idea has been substantiated later by Slupsky and Sykes (1995) and by Gagne et al. (1995), but by using chicken skeletal TnC.

The results and the argument described above had an impact because the exposure of hydrophobic regions may enable interactions to occur with the inhibitory component of troponin, troponin I, thereby activates contraction. The results of our microcalorimetric studies have already shown definitely that the exposure of the hydrophobic residues to the solvent occurs only associated with one specific site of the four Ca-binding sites of TnC.

The structure of the Ca-saturated N-terminal domain of skeletal TnC (NTnC, residues 1-90) originated from the reconstituted chicken TnC has been determined by solution structural studies using NMR (Slupsky and Sykes, 1995). Ca-saturated NTnC showed an "open" structure and the hydrophobic pocket was exposed to the solvent compared with the apo N-terminal domain (Slupsky et al., 1995). Moreover, Ca binding to the regulatory N-domain (NTnC) occured in a stepwise manner (Li et al., 1995). The results of these structural studies are rather similar to those of our calorimetric studies on bullfrog skeletal than to those of rabbit skeletal TnC. The calorimetry on rabbit skeletal TnC has shown that the stepwise binding of Ca occurs on high-affinity Ca-sites of C-terminal domain, and that the massive exposure of the hydrophobic residues also occurs on one of the Ca sites of the C-terminal domain (Yamada and Kometani, 1982).

The amino-acid sequence of TnC from frog skeletal muscle has been determined and compared with the TnC from various sources (van Eerd et al., 1978). Frog TnC has the same number of amino-acid residues as chicken TnC, which has three amino-acid residues more than rabbit TnC. There are 21 amino-acid substitutions and additions between frog and chicken TnC, and 20 between frog and rabbit. There are apparently no large differences in structure, however there can be a subtle difference in structure and function so that a shift may exist in the specific "active" calcium site between the rabbit and the frog skeletal TnC.

Based on various reasons it has been deduced that the low-affinity Ca-sites of N-terminal domain are related to the regulation of contraction (Potter and Gergely, 1975). In the thin filament of muscle, TnC is complexed with other components of troponin and also with the components of the filament. Therefore the interactions among these components may easily affect the characteristics of the components themselves. However, even after associated with other components, one of the C-terminal site that shows the anomalous characteristics of TnC should retain its fundamental nature. In rabbit skeletal TnC, therefore, although the C-terminal domain containing the "active" Ca sites is responsible for the interaction with troponin I, it is probable that, when constituted in the thin filament, the binding of Ca to one of the N-terminal Ca-specific site is involved in "sensing" the Ca signal. In frog skeletal TnC the situation is thought to be simpler.

Skelton et al. (1988) compared apo and Ca-loaded calbindin D_{9k} with the apo N-terminal domain as well as the Ca-loaded C-terminal domain of TnC. Based on the results of these studies they argued that intracellular buffer proteins, such as calbindin D_{9k}, need only bind Ca efficiently, and do not require conformational changes and corresponding solvent exposure of hydrophobic residues specific to 'sensor' proteins such as TnC and CaM. This argument has been taken to be a general rule for Ca-binding proteins by Sia et al. (1997). Results of our microcalorimetric studies on parvalbumins are in general agreement with this notion (Tanokura and Yamada, 1985; Yamada, 1999).

In structural studies (Sia et. al., 1997) whether or not Ca is bound to a specific site is self-evident, while in microcalorimetric titrations this factor could be inaccurate and should be discussed with some caution. It is unlikely, however, that such an error was introduced and affected the identification of the class of site involved, partly because the amount of Ca introduced to the reaction solutions in the microcalorimetric cells should be an order less than the amount of TnC (see Yamada and Kometani, 1982).

Cardiac troponin C and the mechanism of the thin filament regulation

TnC on the thin filament of the cardiac muscle plays a key role in the regulation of contraction as in the skeletal muscle. However, there exists a crucial difference in that site I in cardiac TnC is inactive structurally. Results on cardiac TnC, both by calorimetry and by NMR, do not agree with the general rule that Ca sites are needed that causes the molecule to expose hydrophobic residues to interact with troponin I (see Kometani and Yamada, 1983; Sia et al., 1997). Results of our microcalorimetric studies have shown that the exposure of the hydrophobic residues to the solvent may be weak if present (Kometani and Yamada, 1983). Sia et al. (1997) demonstrated that, in contrast to models described above, structural changes are quite different in cardiac TnC on Ca binding. They have shown that, in the structure of Ca-saturated cardiac TnC, the regulatory domain is compact possibly as a consequence of an inactive Ca-binding site I. There could be a fundamental difference in the mechanism of the regulation of contraction by Ca between the cardiac and the skeletal TnC.

ACKNOWLEDGEMENTS

I express my sincere gratitude to Professor Setsuro Ebashi for advice and encouragements during the course of these studies, and to late Professor Hidenobu Mashima for supporting these studies in early stages. I am particularly indebted to my former colleagues; Dr. Kaoru Kometani, Dr. Masaru Tanokura and Dr. Masamoto Imaizumi.

REFERENCES

Ebashi, S., and Endo, M., 1968, Calcium and muscle contraction. *Prog. Biophys Mol. Biol.* **18**: 123.

van Eerd, J.-P., and Takahashi, K. 1975, The amino acid sequence of bovine cardiac troponin-C. Comparison with rabbit skeletal troponin-C. *Biochem. Biophys. Res. Commun.* **64**: 122.

van Eerd, J.-P., Canopy J.-P., Ferraz, C., and Pechere, J.-F., 1978, The amino-acid sequence of troponin C from frog skeletal muscsle. *Eur. J. Biochem.* **91**: 231.

Gagne, S.M., Tsudsa, S., Li, M.X., Smillie, L.B., and Sykes, B.D., 1995, Structures of the troponin C regulatory domains in the apo and calcium-saturated states. *Nat. Struct. Biol.* **2**: 784.

Herzberg, O., Moult, J., and James, M.N.G., 1896, A model for the Ca^{2+}-induced conformational transition of troponin C. *J. Biol. Chem.* **261**: 2638.

Herzberg, O., and James, M.N.G., 1988, Refined crystal structure of troponin C from turkey skeletal muscle at 2.0 A resoltion. *J. Mol. Biol.* **203**: 761.

Imaizumi, M., and Tanokura, M., 1990, Heat capacity and entropy changes of troponin C from bullfrog skeletal muscle induced by calcium binding. *Eur. J. Biochem.* **192**: 275.

Imaizumi, M., Tanokura, M., and Yamada, K., 1987, A calorimetric study on calcium binding by troponin C from bullfrog skeletal muscle. *J. Biol. Chem.* **262**: 7963.

Imaizumi, M., Tanokura, M., and Yamada, K., 1990, Calorimetric studies on calcium and magnesium binding by troponin C from bullfrog skeletal muscle. *J. Biochem.* **107**: 127.

Kometani, K., and Yamada, K., 1983, Enthalpy, entropy and heat capacity changes induced by binding of calcium ions to cardiac troponin C. *Biochem. Biophys. Res. Commun.* **114**: 162.

Li, M.X., Gagne, S.M., Tsuda, S., Kay, C.M., Smillie, L.B., and Sykes, B.D., 1995, Calcium binding to the regulatory N-domain of skeletal muscle troponin C occurs in a stepwise manner. *Biochemistry,* **34**: 8330.

Potter, J.D., and Gergley, J., 1975, The calcium and magnesium binding sites on troponin and their role in the regulation of myofibrilar ATPase. *J. Biol. Chem.* **250**: 4628.

Sia, S.K., Li, M.X., Spyracopoulos, L., Gagne, S.M., Liu, W., Putkey, J.A., and Sykes, B.D.,1997, Structure of cardiac muscle troponin C unexpectredly reveals a closed regulatory domain.   *J. Biol. Chem.* **272**: 18216.

Skelton, M.J., Kordel, J., Akke, M., Forsen, S., and Chazin, W.J. 1994, Signal tranduction versus buffering activity in Ca^{2+}-binding proteins. *Nat. Struct. Biol.* **1**: 239.

Slupsky, C.M., Kay, C.M., Reinach, F.C., Smillie, L.B., and Sykes, B.D., 1995, Calcium-induced dimerization of troponin C: Mode of interaction and use of trifluoroethanol as a denaturant of quaternary structure. *Biochemistry,* **34**: 7365.

Tanokura, M., Imaizumi, M., and Yamada, K., 1986, A calorimetric study of Ca binding by parvalbumin of the toad (Bufo): distinguishable binding sites in the molecule. *FEBS Lett.* **209**: 77.

Tanokura, M., and Yamada, K., 1983, A calorimetric study of Ca and Mg binding by calmodulin. *J. Biochem.* **94**: 607.

Tanokura, M., and Yamada, K., 1984, Heat capacity and entropy changes of calmodulin induced by calcium binding. *J. Biochem.* **95**: 643.

Tanokura, M., and Yamada, K. 1985a, A calorimetric study of Ca binding to two major isotypes of bullfrog parvalbumin. *FEBS Lett.* **185**: 165.

Tanokura, M., and Yamada, K., 1985b, Effects of trifluoperazine on calcium binding by calmodulin. A microcalorimetric study. *J. Biol. Chem.* **260**: 8680.

Tanokura, M., and Yamada, K., 1986, Effects of trifluoperazine on calcium binding by calmodulin. Heat capacity and entropy changes.   *J. Biol. Chem.* **261**: 10749.

Tanokura, M., and Yamada, K. 1987, Heat capacity and entropy changes of the two major isotypes of bullfrog (Rana catesbeiana) parvalbumins induced by calcium binding. *Biochemistry* **26**: 7668.

Tanokura, M., and Yamada, K., 1993. A calorimetric study of Ca^{2+} binding by wheat germ calmodulin. Regulatory steps driven by entropy. *J. Biol. Chem.* **268**: 7090.

Yamada, K., 1978, The enthalpy titration of troponin C with calcium. *Biochim. Biophys. Acta* **535**: 342.

Yamada, K., 1999, Thermodynamic analyses of calcium binding to troponin C, calmodulin and parvalubumins by using microcaloimetry. *Molec. Cell. Biochem.* **190**: 39.

Yamada, K., and Kometani, K., 1983, The changes in heat capacity and entropy of troponin C induced by calcium binding. *J. Biochem.* **92**: 1505.

Yamada, K., Mashima, H., and Ebashi, S., 1976, The enthalpy changes accompanying the binding of calcium to troponin relating to the activation heat production of muscle. *Proc. Japan Acad.* **52**: 252.

DISCUSSION

Woledge: How can you separate the hydrophobic and vibrational components of the entropy changes?

K. Yamada: The hydrophobic component is calculated for the heat capacity changes we observed using the empirical relations.

Woledge: Can you give us a scale for these hydrophobic changes, –a comparison perhaps with some protein/ protein interaction?

K. Yamada: It is possible to express the results in terms of number of water molecules included or excluded from the surface area.

ISOLATION AND CHARACTERIZATION OF 190K PROTEIN FROM AORTA SMOOTH MUSCLE

Li-Juan Yan, Naoki Yoshinaga, Naoki Niida, and Yoh Okamoto[*]

ABSTRACT

Molecular assemblies of actin and myosin for the contractility of smooth muscle are quite different from those of striated muscle. Another striking difference is that vascular smooth muscle has a potential to transform to migratory synthetic cell type. At this point of view, smooth muscle cell has properties similar to those of non-muscle. In fact, myosin Ic, a single headed unconventional myosin, was identified in aorta smooth muscle. During the studies on myosin Ic, we have found another calmodulin related 190kDa protein. This protein binds to calmodulin irrespective on calcium ion and to F-actin in an ATP independent manner. Furthermore, the F-actin binding stoichiometry diminished to half upon the addition of exogenous calmodulin. Partial amino acid sequence indicated a high homology to those of GRD (GTPase Related Domain) of human brain IQGAP1. Western blot analysis using anti-human IQGAP1 antibody also indicated a strong cross-reactivity with the protein. We have tested the protein with respect to the characteristic F-actin gelation by IQGAP1. In the presence of cdc42 and GTPγS, 190kDa protein could cause a high viscosity of F-actin. These data indicate a close similarity to human brain IQGAP1. The presence of IQGAP1 in aorta smooth muscle suggests contributions for cellular processes such as actin reorganization during contraction-relaxation cycle, association of cytoskeletal structure to cell membrane, organelle movement.

---

[*]Correspondence:Yoh Okamoto, Muroran Institute of Technology, Department of Applied Chemistry, Muroran, Hokkaido 050-8585, Japan.   E-mail:yoh@mmm.muroran-it.ac.jp

INTRODUCTION

Contractility of smooth muscle has been known as a process that is regulated by cytosolic Ca^{2+} concentration following calmodulin dependent phosphorylation of myosin light chain. This phosphorylation activates actomyosin ATPase resulting in contraction. On the other hand, the contractility and the extents of light chain phosphorylation is not always to be a linear function of intra-cellular Ca^{2+} concentration.[1] Under this situation, GTP-dependent contractility at moderate Ca^{2+} concentration has been reported such as GTP bound rho protein induced enhancement of light chain phosphorylation.[2] By now roh and its effectors dependent process has been proposed for the regulation of myosin light chain phosphorylation.[3, 4] This mechanism also appears to be operated in non-muscle cells for the motility.[5]

In the case of smooth muscle cell, there is no sarcomere, intracellular contractile unit structure, in contrast to striated muscle. Filamentous assembly of smooth muscle myosin II depends on the phosphorylation of regulatory light chain. The actin filament in smooth muscle also appears to exist in less ordered structure such as dense body. Filament formation of myosin and actin and the stability might be much more dynamic nature than that of striated muscle. The regulatory mechanism for proper alignment of contractile proteins is not well known.

Here we report an isolation and identification of 190K protein as IQGAP1 for the first time from vascular smooth muscle. The aorta 190K protein could cause F-actin crosslinking in a small G-protein and GTPγS dependent manner. These data suggest that aorta 190K protein is a possible candidate contributing reorganization of actin filament within smooth muscle.

MATERIALS AND METHODS

Smooth muscle of porcine aorta media was minced and washed with 20mM $MgCl_2$, 1mM EGTA, 10mM MOPS (pH 7) and stocked in 50% (v/v) glycerol at -80℃.   Actin from rabbit skeletal muscle was prepared according to the method of Spudich-Watt.[6] Polyclonal antibody against aorta 190K protein was raised in rabbit by immunization of 190K protein extracted from SDS-PAGE.   Calmodulin has been either prepared from Scallop testis[7] or purchased bovine brain one from Sigma Chem. Co.

Q-Sepharose FF and SP-SepharoseFF ion exchange resins, Sepharose CL4B resins and glutathion-Sepharose was purchased from Amersham Bioscience. Glutathion and other chemicals were from Wako Chem. Co. ATP, GTPγS, GDP were from Roche Molecular Biochemicals. GST- cdc42 was from Calbiochem and Cytoskeleton Co. Anti IQGAP1 antibodies were purchased from Santa Cruz Biotechnology and Becton Dickinson Inc.

RESULTS

Purification of 190K Protein from Porcine Aorta Smooth Muscle

Extraction has been performed using high salt buffer containing 1mM EGTA, 1mM $MgCl_2$, 5% sucrose, 1mM ATP, 20mM Tris Cl (pH 7.5), 0.1mM PMSF, 0.1 mM

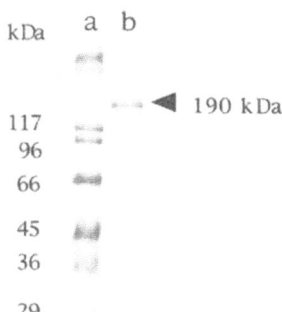

Figure 1. Aorta 190K protein purified by Q-Sepharose ion exchange column chromatography. Pooled fraction of 190K protein after Sepharose CL4B chromatography has been applied to Q-Sepharose ion exchange column as described in the text. 190K protein was eluted at ~0.25M NaCl. (a) molecular weight markers, (b) 190K protein.

DFP. Most of myosin II has been removed by SP-Sepharose chromatography at first. The fractions containing 190K was on to Sepharose CL4B size exclusion chromatography. The 190K protein enriched fractions were applied to Q-Sepharose ion exchange chromatography. The column has been developed by a linear gradient elution of NaCl from 0.2 M to 0.45 M.   The 190K protein was eluted at 0.25M NaCl as shown in Figure 1. Beginning with 200g of porcine aorta media, the final yield of 190K protein has been ~1mg.

Calmodulin Affinity Chromatography of 190K Protein

The 190K protein has been co-purified with small amounts of 17K protein. Calmodulin could be associated with 190K protein. In order to test this possibility, binding affinity of 190K protein to Calmodulin-Sepharose has been examined. Calmodulin-Sepharose bind 190K protein in the presence or absence or calcium ion even in a presence of 1% Triton X100. It is, therefore, established that 190K protein retains calcium insensitive calmodulin binding site, which is characteristic properties of the 190K protein.

Effect of Calmodulin on 190K Protein Binding to F-actin

In order to test contribution of 190K protein for the contractility of aorta, F-actin binding affinity has been examined. Without F-actin the 190K protein has been recovered in a supernatant after ultracentrifugation in 0.25 M NaCl at pH 7.5. On the contrary, a large amount of 190K protein was recovered in the pellet when F-actin was added.

With increasing concentration of calmodulin, the amount of 190K protein bound to F-actin has been diminished to   ~50% of that observed in the absence of exogenous calmodulin. This inhibition by calmodulin was calcium independent manner.

218

Y. OKAMOTO *EI AL.*

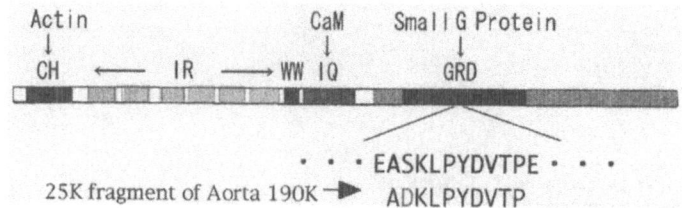

Figure 2. Amino acid sequence and the homology search. Proteolytic fragment (~25kDa) of aorta 190K protein was sequenced. Homologous sequence found in GTPase related domain (GRD) of human brain IQGAP1.

Primary Structure and the Homology Search

Amino acid sequence of the 190K protein is necessary for further characterization and comparing with those of known proteins. The internal primary sequence has been analyzed after limited proteolytic cleavage of 190K protein. A 25kDa fragment has been generated after V8 protease treatment. The amino terminal sequence has been searched in a protein sequence library. We found a well fitted sequence in small G-protein related domain (GRD) of human brain IQGAP1[8] as shown in Fig. 2. IQGAP1 has been known as an actin binding ~190K protein widely expressed in non-muscle cells.[9]

Immunochemical Cross Reactivity with Anti-Human Brain IQGAP1 and Indications of the small G- protein

For further characterization of aorta 190K protein, immunochemical analyses were done using anti- human brain IQGAP1 antibody. Both ELISA and western blot have shown that the antibody recognizes well aorta 190K protein. Furthermore, cdc42 and/or Rac1 have been detected in an early step of 190K protein isolation at size exclusion chromatography. Both types of small G-proteins were known as physiological targets of IQGAP1.[9]

Falling Ball Viscometry of F actin and the Effect of 190K Protein

Characteristic property of IQGAP1 has been known as F-actin crosslinking activity depending on a presence of cdc42 with GTPγS.[10] Aorta 190K protein has been tested about this property. As predicted from the experiment previously done using IQGAP1 from non-muscle cell, aorta 190K protein caused remarkable increase of F-actin viscosity which is observed in the presence of cdc42 and GTPγS but not with GDP.

Binding Affinity between Cdc42 and Aorta 190K Protein in the Presence of GTPγS

In order to verify direct interaction between aorta 190K protein as a putative IQGAP1 from muscle cell and cdc42, we have tried to trap 190K protein to GST-cdc42 attached to glutathion-Sepharose resin in the presence of GTPγS. The 190K protein binds to glutathion-Sepharose in the presence of GST-cdc42 and GTPγS but not in the absence.[11, 12]

DISCUSSION

It is accepted that fundamental mechanisms of contraction are common for striated and smooth muscle. In spite of common concept of mutual sliding between actin and myosin necessary for contraction, the architectures of the contractile apparatus are distinct with each type of muscle. In the case of smooth muscle, there is no sarcomere like unit structure within the cell. Moreover, phosphorylation of the myosin light chain regulates the filament formation upon the activation by intracellular Ca^{2+} increase. This calcium regulatory mechanism is further modulated by GTP dependent amplification.

On the other hand, vascular smooth muscle cell is known to undergo transformation the contractile cell type to migratory synthetic one in the case of endothelial cell damage. This might have implied an existence of unconventional myosin necessary for the cell migration as well known in non-muscle cell motility. As a result for first step to examine this possibility, we have identified myosin Ic (formerly Iβ) from porcine aorta smooth muscle.[13] The subcellular localization in A10 cell have shown a diverse distribution including some ten percent in the cytosol.[14] During the course of this study, we have found another calmodulin related 190K protein.

In this report, we have shown the purification method and the binding affinity with calmodulin, F-actin and a small G-protein, cdc42. Furthermore, the 190K protein could cause F-actin crosslinking in a GTPγS dependent manner. The partial amino acid sequence indicates a close similarity to those of human brain IQGAP1. All of these data indicate that aorta 190K protein is IQGAP1 identified in muscle for the first time. It should be noted that the yield of aorta 190K protein, from tissue is comparable with that reported for that of IQGAP1 from adrenal gland. [10]At present, the exact function of IQGAP1 in vascular muscle cell is not known. Several pioneer works about IQGAP1 in non- muscle cells suggest the functions as a cytoskeletal reorganizer concerning the migration and cell-cell adhesion.[9] It is interesting to see if corresponding functions exist in 6vascular smooth muscle cell. Particularly, contribution of crosslinking actin filaments by IQGAP1 is attractive possibility for the contraction-relaxation mechanism and its regulation. Slow tension development and its maintenance of smooth muscle could be influenced by such contribution. In any case, it is keen to clarify the regulatory mechanism of binding between IQGAP1 and F-actin modulated by cdc42, guanine nucleotides, calcium and calmodulin. It might be critical information to find out the physiological significance of IQGAP1 in vascular smooth muscle cell. Biochemical screening of binding partner and histochemical study of smooth muscle IQGAP1 are in progress.

ACKNOWLEDGEMENTS

We are greatly indebted for the technical assistance of Tomoko Honda and Sachiko Hatsukaiwa during this study.

REFERENCES

1. Bradley, A. B. & Morgan, K.G.   *J. Physiol.* **385**, 437-448 (1987)
2. Kitazawa, T., Masuo, M., & Somlyo, A.P. *Proc.Natl.Acad.Sci.USA* **88**, 9307-9310 (1991)
3. Amano, M., Ito, M., Kimura., K, Fukata, Y., Chihara, K., Nakano, T., Matsuura, Y., & Kaibuchi, K. *J.Biol.Chem.* **271**, 20246-20249 (1996)
4. Somlyo, A. *Nature* **389**, 908-911 (1997)
5. Guiliano, K. A. & Taylor, D, L., *Curr. Opn. Cell Biol.* **7**, 4-12(1995)
6. Spudich, J.A. & Watt, S. J.   *J. Biol. Chem.* **246**, 4866-4871 (1971)
7. Yazawa, M., Sakuma,M. & Yagi,K.   *J.Biochem.* **87**, 1313-1320 (1980)
8. Weissbach, L., Settleman, J., Kalady, M.F., Snijders, A.J., Murthy, A.E., Yan, Y.X., & Bernards, A. *J. Biol. Chem.* **269**, 20517-20521 (1994)
9. Kaibuchi, K.,Kuroda, S. & Amano, M.   *Ann. Rev. Biochem.* **68**, 459-486 (1999)
10. Bashour, A.M., Fullerton, A.T., Hart, M.J., Bloom, G.S.   *J. Cell Biol.* **137**, 1555-1566   (1997)
11. Kuroda, S., Fukata, M., Kobayashi, K., Nakafuku, M., Nomura, N., Iwamatsu, A., & Kaibuchi, K. *J. Biol. Chem.* **271**, 23363-23367 (1996)
12. Ho,Y-D., Joyal, J. L., Li, Z. & Sacks, D. B.   *J.Biol. Chem.* **274**, 464-470 (1999)
13. Hasegawa, Y., Kikuta, T. & Okamoto, Y.   *J. Biochem.* **120**, 901-907 (1996)
14. Hasegawa, Y. Tsuwaki,S.,Yamada,N., Araki,S.,Kimura, S.,Sugawara,J., Yamamoto,K. & Okamoto, Y. *J. Biochem.* **124**, 421-427 (1998)

DISCUSSION

Morano: You observed 190K protein in A10 cells. Have you found the same protein also in normal smooth muscle cells?

Okamoto: This 190K protein has been isolated from contractile aorta smooth muscle.

Ikebe: Does IQGAP1 have Rac/cdc 42 GAP activity?

Okamoto: We have not determined it yet.

ter Keurs: Do your viscosity data tell you whether IQGAP1 binds to one or more F-actins?

Okamoto: Surely, this aorta IQGAP1 crosslinks two or more F-actins.

SEVERAL ASPECTS OF CALCIUM REGULATOR MECHANISMS LINKED TO TROPONIN

Iwao Ohtsuki*, Sachio Morimoto, and Fumi Takahashi-Yanaga

1. INTRODUCTION

The contraction of vertebrate striated muscle is regulated by Ca^{2+} through specific regulatory proteins, troponin and tropomyosin. The contractile interaction of myosin and actin is suppressed by troponin-tropomyosin in the absence of Ca^{2+}. On increasing the Ca^{2+}-concentration, this suppression is removed to activate contraction.[1,2] Structurally, troponin and tropomyosin distribute along the entire length of thin filament with a 38 nm periodicity. Two head-to-tail filaments of fibrous tropomyosin molecules, each binding to troponin at its specific region, run almost in register along the grooves of actin double strands. The action of Ca^{2+} on troponin is mediated to actin molecules through tropomyosin.

Troponin consists of three different components, troponins C, I and T.[2] Troponin C is a Ca^{2+}-binding component; troponin I is an inhibitory component of the contractile interaction; and troponin T is a tropomyosin-binding component. The contractile interaction of myosin and actin in the presence of tropomyosin is inhibited by the inhibitory action of troponin I and this inhibition by troponin I is removed by the neutralizing action of troponin C at almost all Ca^{2+} concentrations. Troponin T makes the neutralizing action of troponin C fully sensitive to Ca^{2+} and, in addition, slightly elevates the maximum level of contraction. The Ca^{2+}-sensitivity is therefore conferred on the contractile interaction only in the concomitant presence of the three components of troponin along with tropomyosin.

This article reviews our studies on the several aspects of the Ca^{2+}-regulatory mechanisms of the striated muscle contraction: 1) the functional consequences of the mutations in cardiac troponin components found in genetic disorders; 2) the roles of troponin in the Ca^{2+}-regulation of molluscan striated muscle contraction; and 3) the structural aspect of the inhibitory interaction of troponin I with tropomyosin-actin.

Department of Pharmacology, Faculty of Medicine, Kyushu University, Fukuoka 812-8582, Japan
*Present address; Department of Physiology, The Jikei University School of Medicine, Minato-ku, Tokyo 105-8461, Japan.

2. FUNCTIONAL CONSEQUENCES OF THE MUTATIONS OF CARDIAC TROPONIN COMPONENTS

2.1. Exchange of troponin components in skinned fibers and myofibrils

We previously reported that troponin C_I_T complex was replaced by added troponin T in skinned muscle fibers and myofibrillar preparations.[3,4] Troponin T, which was added to incubating solution in an excess amount at slightly acidic conditions, was incorporated into the myofibrillar matrix in exchange of the intrinsic troponin C_I_T. In this condition, the interaction between troponin and tropomyosin in the myofibrillar structure would be relatively weak and thus the whole complex of intrinsic troponin could be exchanged, according to the principle of mass action, by the troponin T added to the incubating solution. As a result, both troponins C and I disappeared from skinned fibers or myofibrils. During the course of troponin T-treatment, the resting level of contraction at low Ca^{2+} increased gradually with small concomitant decrease in the maximum contraction. The Ca^{2+}-sensitivity also gradually increased, whereas the cooperativity of Ca^{2+}-activation was depressed to some extent. The contraction was finally desensitized to Ca^{2+} at the level of about 70 % of the maximal activity. The Ca^{2+}-insensitive contraction of the troponin T-treated preparations was completely inhibited by reconstituting troponin I, and the original Ca^{2+}-sensitive contraction was recovered by further reconstitution with troponin C. This finding has enabled us to examine the characteristic properties of isoforms or recombinant proteins of the three troponin components under physiological conditions.[5] This procedure was applied for detecting delicate alterations in Ca^{2+}-activated contraction caused by mutations in genes for human cardiac troponin components.

2.2. Analyses of mutations in genes for human cardiac troponins T and I associated with familial hypertrophic cardiomyopathy

Familial hypertrophic cardiomyopathy (HCM) is an autosomal dominant heart disease. Genetic analyses of this disorder have revealed that the HCM is caused by mutations in genes for various cardiac sarcomeric proteins, including ,-myosin heavy chain, myosin light chain, troponin T, troponin I, ·-tropomyosin, myosin-binding protein C, and connectin.[6] As to the mutations in genes for troponin components, seventeen mutations for troponin T and eight mutations for troponin I have been reported. A mutation for cardiac troponin C in patients with HCM was also reported recently.[7]

2.2.1. *Troponin T mutations*

In 1994, two missense mutations (Ile79Asn, Arg92Gln) and one splice donor site mutaion in intron15 (G_1_A) in human cardiac troponin T were reported.[8] In order to explore the functional consequences of the mutations in the Ca^{2+}-regulation of contraction, the effect of the Ile79Asn, Arg92Gln mutations on the skinned cardiac muscle fibers was first examined by employing the troponin-exchange procedure described in the preceding section.[9] We found that both mutations had a Ca^{2+}-sensitizing effect without any changes in the maximum force and cooperativity of the Ca^{2+}-activated force generation. The splice donor site mutation in intron 15 caused aberrant splice donor products encoding two truncated mutants of troponin T; one lacks the C-terminal 14 residues ($TnT_{\not 14}$) and the other lacks C-terminal 21 residues due to the replacement of the C-terminal 28 residues with 7 novel residues ($TnT_{\not 28(+7)}$). The truncated troponin T mutants both increased the Ca^{2+}-sensitivity and decreased the cooperativity but did not affect the maximum contraction.[10] The difference in functional consequences between these mutations was indicated by the finding that the phosphorylation of troponin I by cAMP-dependent protein kinase caused the decrease in maximum contraction of the fibers reconstituted with $TnT_{\not 28(+7)}$, whereas the maximum level of the fiber tension containing $TnT_{\not 14}$ was not affected by the phosphorylation.

The characteristic outcome of two missense mutations, Phe110Ile and Glu244Asp, was that these mutations significantly increased the maximum contraction without affecting cooperativity.[11] The Glu244Asp mutation also increased the Ca^{2+}-sensitivity of the contraction, whereas the Phe110Ile mutation, associated with the patients with relatively benign prognosis, did not affect the Ca^{2+}-sensitivity. The Arg278Cys mutation caused a decrease in the maximum contraction of the skinned fibers with a small increase in the Ca^{2+}-sensitivity.[12] The deletion mutation, $\not E$Glu160, had an enhancing effect on the Ca^{2+}-sensitivity without any changes in the cooperativity or maximum contraction.[13] Preliminary examinations also indicated that the mutations Arg92Leu, Arg92Trp, Arg94Leu, Glu163Arg and Glu163Lys, had the effect of increasing the Ca^{2+}-sensitivity without any changes in the cooperativity and the maximum activity, whereas the Arg278Pro mutation decreased both the cooperativity and maximum contraction with a slight increase in Ca^{2+}-sensitivity.

In summary, the mutations of troponin T associated with the HCM are classified into five groups as follows; i) marked increase in Ca^{2+}-sensitivity, no changes in cooperativity or maximum contraction: Ile79Asn, Arg92Gln/Leu/Trp, Arg94Leu, Ala104Val, Arg130Cys, $\not E$Glu160, and Glu163Arg/Lys: ii) marked increase of Ca^{2+}-sensitivity and maximum contraction, no change in cooperativity: Glu244Asp; iii) marked increase in Ca^{2+}-sensitivity, a decrease in cooperativity, and no change in maximum contraction: $TnT_{\not 14}$ and $TnT_{\not 28(+7)}$; iv) marked increase in maximum contraction, no changes in Ca^{2+}-sensitivity or cooperativity: Phe110Ile; v) slight increase in Ca^{2+}-sensitivity, decrease in cooperativity and maximum contraction: Arg278Cys/Pro. We also investigated the effects of troponin T mutations on the Ca^{2+}-activated myofibrillar ATPase activity reconstituted with the following troponin T mutants: Ile79Asn, Arg92Gln, Phe110Ile, $\not E$Glu160, Glu244Asp and Arg278Cys.[14,15] The results were essentially consistent with those obtained by the studies employing skinned fibers. These considerations strongly indicate that the Ca^{2+}-sensitization is a common effect caused by mutations in troponin T gene associated with the HCM. The Ca^{2+}-sensitizing effects of mutations of troponin T have also been reported from other laboratories on the following mutations: (skinned fiber experiments) Ile79Asn, Arg92Gln, Phe110Ile, $\not E$Glu160, Glu163Lys, and Arg278Cys; (myofibrillar ATPase) Ile79Asn, Arg92Gln, $\not E$Glu160.[16,17] The effects of five troponin T mutants were examined on the inhibitory action of troponin I and the neutralizing action of troponin C.[15,18] The inhibitory action of troponin I, as estimated by the amount of troponin I necessary for the 50% inhibition (IC_{50}), was disturbed by the 4 mutations (Ile79Asn, Arg92Gln, Phe110Ile, Gle244Asp), while it was little affected by the Arg278Cys mutation. The neutralizing action by troponin C was modulated only by Phe110Ile and Glu244Asp by increasing both the maximum activation and the amount of troponin C necessary for the 50% activaition (EC_{50}). Since the mutations Ile79Asn, Arg92Gln and Glu244Asp had a marked Ca^{2+}-sensitizing effect, it is probable that the impairment of the inhibitory activity is closely related to the Ca^{2+}-sensitizing processes due to these mutations. Small changes in both the inhibitory action of troponin I and neutralizing action of troponin C by the Arg278Cys mutation may explain the small Ca^{2+}-sensitization by this mutation. The EC_{50} for the neutralizing action of troponin C was enhanced by the Phe110Ile and Glu244Asp mutations, suggesting that the enhanced neutralizing activity is closely related to the elevation of maximum contraction.

2.2.2.   *Troponin I mutations*

Eight mutations of troponin I have been reported to be associated with HCM: Arg145Gly, Arg145Gln, Arg162Trp, ¢Lys183, Ser199Gln, Gly203Ser, Lys206Gln, and ¢exon 8.[19,20] We examined the effect of six mutations (Arg145Gly, Arg145Gln, Arg162Trp, ¢Lys183, Gly203Ser, Lys206Gln) on the Ca^{2+}-activated contraction of skinned fibers and myofibrillar ATPase activity.[21,22] Five of six mutations, except the Gly203Ser mutation, had the Ca^{2+}-sensitizing effect. The Arg145Gly, Arg145Gln and Arg162Trp mutations showed the Ca^{2+}-sensitizing effect with increased minimum contraction and no changes in the maximum level of contraction or cooperativity, whereas the ¢Lys183 and Lys206Gln mutations caused an increase in Ca^{2+}-sensitivity without detectable changes in the minimum and maximum levels or cooperativity. The Gly203Ser mutation showed a tendency to increase the Ca^{2+}-sensitivity, though statistically insignificant, with slight increase in the minimum level and no change in the maximum level of contraction or cooperativity. The consistent feature is that the four mutations elevated the minimum level of contraction, suggesting that the inhibitory activity of troponin I is impaired by these mutations.

It is interesting to note that most changes in the parameters of Ca^{2+}-activated contraction caused by mutations of human cardiac troponins T and I, except the increase in the maximum contraction level caused by Phe110Ile and Glu244Asp mutations of troponin T, are the same as those observed in the Ca^{2+}-activated contraction of skinned fibers on partially removing troponins C and I by troponin T-treatment.[3] This would reflect that most mutations essentially weaken the regulatory function of troponin through various minute alterations in the interactions among thin filament proteins.

2.3. Analysis of the Lys210-deletion mutation in human cardiac troponin T found in familial dilated cardiomyopathy

A mutation in the cardiac troponin T gene (Lys210-deletion) was reported to cause a different type of cardiomyopathy, familial dilated cardiomyopathy (DCM).[23] Examination of the skinned fiber tension and myofibrillar ATPase activity revealed that this mutation caused Ca^{2+}-desensitiziation of contraction, while cooperativity and the maximum and minimum contraction levels were not affected.[13]

2. 4. Summary

Familial hypertrophhic cardiomyopathy(HCM) is an autosomal dominant cardiac disease, caused by mutations in the genes for a variety of cardiac sarcomeric proteins including   seventeen   mutations in troponin T,   eight mutations   in troponin I and one

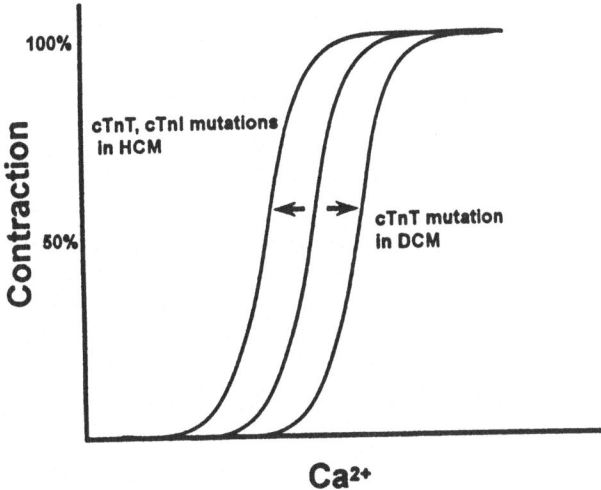

Figure 1. Scheme illustrating the modulation of the Ca^{2+}-sensitivity of contraction by the mutations in genes of human cardiac troponins T and I associated with familial hypertrophic (HCM) and dilated (DCM) cardiomyopathies.

mutation in troponin C. In order to explore the functional characteristics of these mutations in the genes for troponin components in the Ca^{2+}-regulation of contraction, we first examined the effect of two missense mutations of troponin T on the permeabilized cardiac muscle fibers by employing the troponin-exchange technique and found that both mutations had a Ca^{2+}-sensitizing effect without affecting the maximum force or cooperativity (1998).[9] Succeeding examinations of most troponin T mutations reported have clarified that Ca^{2+}-sensitization is a common effect caused by mutations in troponin T. Examination of the effect of the mutations in troponin I also showed that five of six mutations had the Ca^{2+}-sensitizing effect.   Essentially the same results were obtained on the ATPase activity of the myofibrils reconstituted with mutants of troponin T and I. Recently a novel mutation in the cardiac troponinT gene (Lys210-deletion) was reported to cause an another type of cardiomyopathy, familial dilated cardiomyopathy (DCM). Examinations of both the skinned fiber tension and myofibrillar ATPase activity revealed that this mutation caused Ca^{2+}-desensitization of contraction. These findings strongly suggest   that the modulation of the Ca^{2+}-sensitivity of contraction is the   critical functional consequence caused by mutations in genes of human cardiac troponin T and troponin I leading to HCM or DCM.(Fig. 1).[6,13]

3.   REGULATORY ROLE OF TROPONIN IN SCALLOP STRIATED MUSCLE CONTRACTION

In molluscan striated muscle, the Ca^{2+}-regulation linked to myosin has long been considered to be the Ca^{2+}-regulatory mechanism for contraction.[24] Ca^{2+} binds to the essential light chain of scallop myosin. The Ca^{2+}-activated ATPase of myofibrils is desensitized to Ca^{2+} on removal of the regulatory light chain of myosin (RLC) by treatment with divalent cation chelators. At the same time, troponin was also isolated from scallop striated muscles and its biochemical properties were investigated.[25]

Troponin from the scallop striated muscle, in the presence of tropomyosin, showed an activating effect on the ATPase activity of Perry's desensitized myofibrils without significant change in Ca^{2+}-sensitivity, while tropomyosin alone did not affect the myofibrillar ATPase activity that was regulated through the myosin-linked Ca^{2+}-regulation. The relation of troponin-linked regulation to myosin-linked regulation was examined as follows.[26] Both myosin-linked and troponin-linked regulations of myofibrillar preparations of scallop striated muscle were desensitized to Ca^{2+} by removing the RLC and troponin C through treatment with a strong divalent cation chelator, CDTA. The ATPase level of the desensitized myofibrils was approximately half the maximum activity of the intact myofibrils regardless of Ca^{2+}-concentrations. The ATPase of the desensitized myofibril was activated only at higher Ca^{2+}-concentration by reconstituting troponin C, whereas it was inhibited at low Ca^{2+}-concentrations by RLC. This activating effect of troponin C was observed in the physiological temperature range of 5 to 15 °C but was not detected at 25 °C. These findings strongly indicate that the physiological contraction of scallop striated muscle is regulated by both myosin- and troponin-linked Ca^{2+}-regulations.

The essential processes of the regulatory mechanism of the contraction of vertebrate striated muscle by troponin are the suppression of contraction by troponin at low Ca^{2+} and the de-suppression through Ca^{2+}-action on troponin. At higher Ca^{2+}-concentrations, troponin activates the contraction at slightly higher levels.[2,5] In the case of scallop, however, troponin-linked regulation only activates the contraction at high Ca^{2+}-concentrations, while the contraction at low Ca^{2+} is suppressed by the myosin-linked regulatory mechanism. In this respect, the activating action of troponin, which works only slightly in the Ca^{2+}-activated contraction of vertebrate striated muscle, is the predominant mechanism of the troponin-linked Ca^{2+}-regulation in scallop striated muscle.

There are several differences in the properties of troponin between vertebrate and scallop striated muscles. Scallop troponin C binds only one Ca^{2+} ion at its C-terminal region (site IV). This is in sharp contrast with the finding that vertebrate fast skeletal troponin C has four Ca^{2+}-binding sites: two regulatory low affinity sites in the N-domain and two high affinity sites in the C-domain. Functional properties of Ca^{2+}-binding site IV of scallop troponin C were recently investigated.[27] The coordination structure of Ca^{2+} and Mg^{2+} in the scallop troponin C was also studied.[28]

The Ca^{2+}-regulatory mechanism by troponin in scallop striated muscle is therefore considerably different from the mechanisms in vertebrate striated muscle. Further detailed study will clarify the entire feature of the Ca^{2+}-regulatory mechanisms of muscle contraction in the animal kingdom

4. PERIODIC BINDING OF TROPONIN I TO TROPOMYOSIN-ACTIN FILAMENT

The inhibitory activity of troponin I represents the suppression of the contractile interaction between myosin and actin by troponin-tropomyosin in the absence of Ca^{2+}.

The inhibitory position of troponin I in the thin filament is generally considered to be determined through the binding of troponin T to tropomyosin. However, it has been also shown that without troponin C and T, troponin I fully exerts its inhibitory action on actomyosin in the presence of tropomyosin, while troponin I shows no significant inhibitory activity in the absence of tropomyosin.[2] This suggests that troponin I itself has a certain structural relationship with tropomyosin along the actin double strands.[2]

A recent immunoelectron microscopic investigation demonstrated that troponin I or the binary troponin C_I complex binds to tropomyosin-actin filaments with a 38 nm periodicity, even in the absence of troponin T.[29] Since the length of 38 nm just corresponds to the repeating intervals of filamentous tropomyosin molecules, this finding indicates that troponin I is located at the specific region of each tropomyosin molecule (Fig. 2). This suggests that each troponin I forms a complex with both actin and tropomyosin axially at the specific region of tropomyosin. The specific region of tropomyosin must bind directly to troponin I and/or enhance the affinity of the specific actin to troponin I among seven actins in each 38 nm period. The above considerations strongly indicate that the seven actin molecules in the 38 nm period are by no means equal in terms of the affinity to troponin I; rather, specific actin molecules probably have a higher affinity to troponin I than other actins in the 38 nm period.

In the native thin filament, two head-to-tail tropomyosin filaments, which bind troponin at the specific region of each tropomyosin molecule, run axially almost in register along the two grooves of actin double strands.[1,2,30] The formation of relatively narrow transverse anti-troponin I striations along the bundles of filaments of troponin I-tropomyosin-actin strongly suggests that two troponins I and hence two tropomyosins in each period align almost in register. It is very plausible that two tropomyosin molecules by themselves bind cooperatively to the grooves of actin double strands in the 38 nm period; troponin I may be involved in this cooperative binding of two tropomyosin molecules.[2]

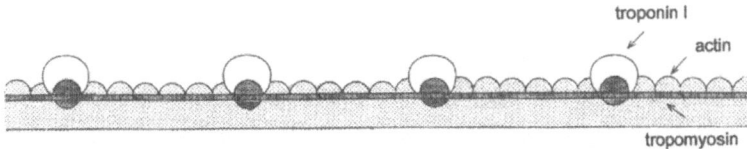

Figure 2. Periodic binding of troponin I to tropomyosin-actin filament. In this schematic illustration, one of two strands of tropomyosin-actin is shown for the sake of simplicity. Troponin I forms a complex with the actin (shaded in the figure) located at a specific region of each tropomyosin and hence distributes at regular intervals of 38 nm; that is the 7 actin length or repeating period of the head-to-tail filaments of tropomyosin molecules lying in the grooves of actin double strands.

REFERENCES

1. S. Ebashi, M. Endo, and I. Ohtsuki,   Control of muscle contraction. *Q. Rev. Biophys.* **2**, 351-384 (1969)
2. I. Ohtsuki, K. Maruyama, and S. Ebashi, Regulatory and cytoskeletal proteins of vertebrate skeletal Muscle. *Adv. Prot. Chem.* **38**, 1-68 (1986)
3. M. Hatakenaka, and I. Ohtsuki, Effect of removal and reconstitution of troponins C and I on the Ca^{2+}- activated tension development of single glycerinated rabbit skeletal muscle fibers. *Eur. J. Biochem.* **205**, 985-993 (1992)
4. F. Shiraishi, M. Kambara, and I. Ohtsuki, Replacement of troponin components in myofibrils,   *J. Biochem.* **111**, 61-65 (1992)
5. I. Ohtsuki, Troponin components and calcium ion regulation of myofibrillar contraction in skeletal muscle, in"*Calcium as Cell Signal*" edited by K. Maruyama, Y, Nonomura, and K. Kohama (Igaku-shoin Tokyo • New York, 1995) pp. 36-42
6. D. Fatkin, and R. M. Graham, Molecular mechanism of inherited cardiomyopathies. *Physiol. Rev.* **82**, 945-980 (2002)
7. B. Hoffmann, H. Schmidt-Traub, A. Perrot, K. J. Osterziel, and R. Gessner. First mutation in cardiac troponin C, L29Q,in a patient with hypertrphic cardiomyopathy. *Human mutation.* **17**. 524 (2001)
8. L. Thierfelder, H. Watkins, C. MacRae, R.Lamas, W. McKenna, H. P. Vosberg, J. G. Seidman,   and C. E. Seidman. --Tropomyosin and cardiac troponin T mutations cause familial hypertrophic cardiomyopathy. *Cell*, **77**, 701-712 (1994)
9. Morimoto, F, Yanaga, R, Minakami, and I. Ohtsuki, Ca^{2+}-sensitizing effects of the mutations at Ile-79 and Arg-92 of troponin T in hypertrophic cardiomyopathy. *Am J Physiol.* **275**,C200-C207 (1998)
10. H. Nakaura, S. Morimoto, F. Yanaga, M. Nakata, H. Nishi, T. Imaizumi, and I. Ohtsuki, Functional changes in troponin T by a splice donor site mutation that causes hypertrophic crdiomyopathy. *Am. J. Physiol.* **277**, C225-C232 (1999)
11. H. Nakaura, F. Yanaga, I. Ohtsuki, and S. Morimoto, Effects of missense mutations Phe110Ile and Glu244Asp in human cardiac troponin T on force generation in skinned cardiac muscle fibers. *J. Biochem.* **126**, 457-460 (1999)
12. S. Morimoto, H. Nakaura, F. Yanaga, and I. Ohtsuki, Functional consequences of a carboxy terminal missense mutation Arg278Cys in human cardiac troponin T, *Biochem. Biophys. Res. Commun.* **261**, 79-82 (1999)
13. S. Morimoto, Q-W. Lu, K. Harada, F. Takahashi-Yanaga, R. Minakami. M. Ohta, T. Sasaguri, and I. Ohtsuki, Ca^{2+}desensitizing effect of a deletion mutation delta-K210 in cardiac troponin T that causes familial dilated cardiomyopathy. *Proc. Natl. Acad. Sci. USA* **99**. 913-918 (2002)
14 K. Harada, F. Takahashi-Yanaga, R. Minakami, S. Morimoto, and I. Ohtsuki, Functional consequences of the deletion mutation ∤Glu160 in human cardiac troponin T, *J. Biochem.* **127**, 263-268 (2000)
15. F. Yanaga, S. Morimoto, and I. Ohtsuki, Ca^{2+} sensitization and potentiation of   the maximum level   of myofibrillar ATPase activity caused by mutations of troponin T found in familial hypertrophic cardiomyopathy. *J. Biol. Chem.* **274**, 8806-8812 (1999)
16. D. Szczesna, R. Zhang, J. Zhao, M. Jones, G. Guzman, and J. D. Potter. Altered regulation of cardiac muscle contraction by troponin T mutations that cause familial hypertrophic cardiomyopathy. *J. Biol. Chem.* **275**, 624-630 (2000)
17. L. S. Tobacman, D. Lin, C. Butters, C. Landis, N.Back, D. Pavlov, and E. Homsher. Functional consequences of troponin T mutations found in hypertrophic cardiomyopathy. *J. Biol. Chem.* **274**, 28363- 28370 (1999)
18. F. Takahashi-Yanaga, I. Ohtsuki, and S. Morimoto, Effects of troponin T mutations in hypertrophic cardiomyopathy on regulatory function of other troponin subunits. *J. Biochem.* **130**, 127-131 (2001)
19. A. Kimura, H. Harada, J. E. Park, H. Nishi, M. Satoh, M. Takahashi, S. Hiroi, T. Sasaoka, N. Ohbuchi, T. Nakamura, T. Koyanagi, T. H. Hwang, J. A. Choo,K. S. Chung, A. Hasegawa, R. Nagai, O. Okazaki, H. Nakamura, M. Matsuzaki, T. Sakamoto, H. Toshima, T. Imaizumi, and T. Sasazuki. Mutations in the cardiac troponin I gene associated with hypertrophic cardiomyopathy. *Nature Genet.***16**, 379-382 (1997)
20. Familial Hypertrophic Cardiomyopathy Mutation Database. http://w.w.w.angis.org.au./Databases/Heart/
21. F. Takahashi-Yanaga, S. Morimoto, and I. Ohtsuki. Effect of Arg145Gly mutation in human cardiac troponin I found in familial hypertrophic cardiomyopathy. *J. Biochem.* **127**, 355-357 (2000)
22. F. Takahashi-Yanaga, S. Morimoto, K. Harada, R. Minakami, F. Shiraishi, M. Ohta, Q-W. Lu, T. Sasaguri, and I. Ohtsuki, Functional consequences of the mutations in human cardiac troponin I gene found in

familial hypertrophic cardiomyopathy. *J. Mol. Cell. Cardiol.* **33**, 2095-2107 (2001)

23. M. Kamisago, S. D. Sharma, S. R. DePalma, S. Solomon, P. Sharma, B. Mcdonough, L. Scott, M. P. Mullen, P. K. Woolf, E. D. Wigle, J. G. Seidman, and C. E. Seidman. Mutations in sarcomeric protein genes as a cause of dilated cardiomyopathy. *N. Engl. J. Med.* **343**, 1688-1696 (2000)

24. A. G. Szent-Györgyi, V. N. Kalabokis and C. L. Perrealt- Micale, Regulation by molluscan myosins. *Mol. Cell. Biochem.* **190**, 55-62 (1999)

25. T. Ojima, and K. Nishita, Troponin from *Akazara* scallop striated adductor muscle. *J. Biol. Chem.* **261**, 16749-16754 (1986)

26. F. Shiraishi, S. Morimoto, K. Nishita, T. Ojima, and I. Ohtsuki, Effects of removal and reconstitution of myosin regulatory light chain and troponin C on the Ca^{2+}-sensitive ATPase activity of myofibrils from scallop striated muscle. *J. Biochem.* **126**, 1020-1024 (1999)

27. T. Ojima, N. Koizumi, K. Ueyama, A. Inoue, and K. Nishita. Functional role of Ca^{2+}-binding site IV of sallop troponin C, *J. Biochem.* **128**, 803-809 (2000)

28. F. Yumoto, M. Nara, H. Kagi, W. Iwasaki, T. Ojima, K. Nishita, K. Nagata, and M. Tanokura. Coordination structure of Ca^{2+} and Mg^{2+} in Akazara scallop troponin C in solution. FTIR spectroscopy of side-chain carboxy grouf. *Eur. J. Biochem.* **268**. 6284-6290 (2001)

29. I. Ohtsuki, and F. Shiraishi, Periodic binding of tropnin C · I and troponin I to tropomyosin-actin filaments. *J. Biochem.* **131**, 739-743 (2002)

30. I. Ohtsuki, Localization of troponin in thin filament and tropomyosin paracrystal. *J. Biochem.* **75**, 753-765 (1974)

DISCUSSION

ter Keurs: Is the prognosis of these particular changes due to mutations in the troponin T determined by arrhythmia or heart failure?

Ohtsuki: Prognosis is an overall judgement based on the various clinical symptoms, including arrhythmia or heart failure. But characteristic features caused by the HCM mutations of cardiac troponin T gene reported are mild or clinically undetectable hypertrophy and a high frequency of sudden death.

ter Keurs: Do you think that the leftward shift of the force-pCa curves with the hypertrophic cardiomyopathy rules out a naive hypothesis that dilation causes reduced force, which in turn causes hypertrophy?

Ohtsuki: Yes, I think so.

Morano: You used recombinant proteins. What kind of "tag" have you used?

Ohtsuki: We have not used any tag in preparing recombinant human troponin components.

IV. STRUCTURAL CHANGES DURING CONTRACTION

X-RAY INTERFERENCE EVIDENCE CONCERNING THE RANGE OF CROSSBRIDGE MOVEMENT, AND BACKBONE CONTRIBUTIONS TO THE MERIDIONAL PATTERN

H.E. Huxley[1], M. Reconditi[2], A. Stewart[1], T. Irving[3]

INTRODUCTION

Interference fringes on the 14.5 nm meridional reflection (M3) are generated by diffraction from the arrays of myosin crossbridges in the two halves of each thick filament. The thick filaments are all constructed in identical fashion (or nearly so), and all have H-zones of the same width, so that the interference distance between the two diffracting arrays in each filament is the same, and so is the pattern each gives. The separation between the fringes is inversely proportional to this interference distance (about 900 nm) so that a very high-resolution camera is necessary to resolve them. At the M3 reflection, one is seeing fringes of approximately the 62nd order, so that small changes in interference distance are magnified by this factor, and, for instance, a 1% change in distance will shift the position of the fringes by more than one half a fringe width.

In a contracting muscle, the application of a small, rapid, length decrease results in an approximately synchronous large decrease in the intensity of the M3 reflection, which has been interpreted in terms of some displacement or change of tilt of the attached, tension-generating cross-bridges (Huxley et al, 1981, 1983). Since the intensity drop is so large – down to 20% or less of the initial intensity – it is very likely that a high proportion of the initial intensity must come from such attached crossbridges. Indeed, it is very plausible that this should be the case, since the regular helical packing arrangement of the myosin heads around the thick filament backbone is lost during contraction. This disorder would very likely affect the axial position of unattached heads too, if only because of the free rotation, believed to be permitted, of the S_1 subunit relative to S_2. In the case of the attached, tension-bearing heads, however, their orientation will be confined between closer limits, and the S_2 connections to the thick filament backbone will all be aligned with the backbone periodicity

---

[1] Corresponding Author: Brandeis University, Waltham, Massachusetts 02454 U.S.A.
[2] University of Florence, Italy
[3] BioCAT, Illinois Institute of Technology, Chicago, IL 60616

and will be pulled tight by tension, so they will contribute more strongly to the 14.5 nm meridional reflection.   Absolute values of intensity are not known at present.

 The interference distance can be thought of, in a crude way, in terms of the position of the center of mass of all the diffracting myosin heads, in each half of the sarcomere. If the crossbridge model of force-development during contraction is correct, then in a quick-release, the attached myosin heads would move inwards, towards the center of the sarcomere, by an amount equal to the relative sliding distance of the actin filaments further into the array of myosin filaments. On the tilting lever arm model, the catalytic domain of S_1 remains fixed on its specific binding site to actin, and this movement is accomplished – indeed, produced – by a tilting of the lever arm (the light-chain binding region) of the myosin head. Since the distal end of S_2, to which the lever arm is attached, remains fixed in axial position (on this model), such a tilting movement must produce an inward movement of the catalytic domains moving the actin towards the H-zone. When this happens very quickly, before there is time for a significant number of heads to detach or attach, there must be an inward shift of the whole center of mass of the attached heads, reducing the interference distance slightly, and producing a significant shift of the fringe pattern on the M3 reflection. This will most easily be seen as a change in the relative heights of the two peaks as the fringe pattern moves across the profile of the M3 reflection. These properties of the system were first recognized by Linari *et al.* (2000).

 In practice, it is convenient and appropriate to think of the interference as happening between corresponding pairs of heads in the two halves of each filament. One member of the first pair is the outermost attached head of one half filament, the other member being the innermost attached head (nearest to the H-zone) in the other half of the same filament. The second pair will consist of the next outermost (14.5 nm closer in) head in one half on the filament and the next to innermost in the other half. Each of these pairs will have the same separation and will give the same set of fringes, and the 14.5 nm axial separation between the successive pairs will cause the fringes to sample a peak corresponding to a 14.5 nm spacing, the width of whose envelope depends on the length of filament overlap. (This formulation makes it easier, in computation, to take account of the fact that the two arrays in each thick filament are related by a two-fold rotation axis, and not merely by a translation).

EARLIER RESULTS

 The behaviour of the reflections and their fringes have been studied in great detail in both single fibers (Lombardi, et al. 2000; Piazzesi et al, 2001; Irving, et al, 2000, 2001, 2002) and in whole muscles (Huxley, et al. 2000, 2001, 2002). The experiments have been carried out using synchrotron radiation, at the Grenoble Storage Ring, and on the BioCAT beam-line at the APS Storage Ring at Argonne USA. In each case, large changes were seen, during quick releases, in both the relative intensity of the two peaks, and in their total intensity. The latter is a measure of the intensity of the reflection, which would have been produced by the heads in each half sarcomere on their own.

 In the case of the whole muscle experiments, the patterns were recorded in 1 or 2 millisecond time windows centered about 2 milliseconds after the initiation of the release (i.e., at T_2). The initial ratio of the peaks, in the isometric state, was about 0.8:1, with the outer peak (i.e., the higher angle component) having the lower intensity. The ratio decreased steadily as the extent of release was increased, reaching a minimum value of about 0.3, when the extent of relative sliding of actin past myosin filaments in each half sarcomere was estimated to be about 6-7 nm. The total intensity of the reflections decreased as observed previously, with very little change for smaller releases, and then a steady decrease at larger releases. When the behaviour was observed in the same muscle specimen over a range of different releases, it could be seen that there was in fact a definite increase in intensity at first, for shortenings up to about 1.5 nm, by approximately 10%. This would be consistent with a lever arm whose initial orientation placed it beyond the point of maximal alignment with the catalytic domain (crudely, 'beyond the perpendicular'). It would come into alignment with small amounts of shortening, and then become progressively further and further out of alignment and tilted on the other side of perpendicular as shortening progressed (as pointed out by Irving).

Thus the observations provided very strong evidence both for an inward movement of the catalytic domains (presumably attached to actin) and a tilting movement of the lever arms, and thus gave strong support to the tilting lever arm – crossbridge model. However, quantitatively, a discrepancy was observed, in that the extent of the ratio change was less than would be expected if all of the reflection was being generated by the attached heads alone. This can be most clearly recognized by the fact that the observed ratio of the intensity of the two peaks does not decrease to zero with larger releases, as it should when the outer fringe lies over the first zero of the transform of the array of heads.

The discrepancy can be accounted for very readily if there is a second component in the reflection, some structure who diffraction pattern is similarly sampled by interference fringes, but which remains fixed in position during a quick release. Quantitatively, this second component would need to have a scattering power approximately the same as that of the moving head, and be positioned about 1.5 nm on the H-zone side of moving heads when they are in their isometric position. All the changes in the reflections can then be explained by the tilting model rather accurately, and this was the case in both the whole muscle and the single fiber experiments. A possible identity for the fixed component might be the second head of those myosin molecules whose other head was attached to actin (perhaps thereby stabilizing the position of the second head, as compared to the myosin molecules which had neither head attached to actin). However, there are a number of complications to this simple picture, which will be described below.

EVIDENCE FOR OTHER COMPONENTS CONTRIBUTING TO THE MYOSIN MERIDIONAL PATTERN

The myosin pattern shows a number of higher order reflections, at spacings corresponding to higher orders of the 14.5 nm axial repeat of the crossbridges. The most prominent of these is at 7.25 nm, and is designated as M6, recognizing the helical nature of the myosin filament structure, with a helical period of 43.5 nm and the sets of crossbridges occurring at one third of this (14.5 nm), giving the strong meridional reflection referred to as M3 and discussed in the previous section. Other meridional reflections occur at the third and the fifth order of this axial repeat, and are designated M9 and M15. They are also sampled by the fringe system.

One can use the same molecular model which correctly predicted the behaviour of the M3 reflection to predict the contribution of the crossbridges to the higher order reflections. This model consists of two myosin S_1 subunits (using the Rayment (1993) structure), sharing a common origin at the start of S_2, and in the isometric case have one lever arm (for the attached head) at an angle of about 55° away from its position in the Rayment-Holmes (1993) rigor structure (for myosin S1 attached to actin), and the other lever arm at about 45° away from the rigor position, and fixed in that position when the attached head moves and its lever arm angle changes. These structures (using weighted components in the α-carbon positions) are then projected on the filament axis and used to calculate the meridional diffraction pattern.

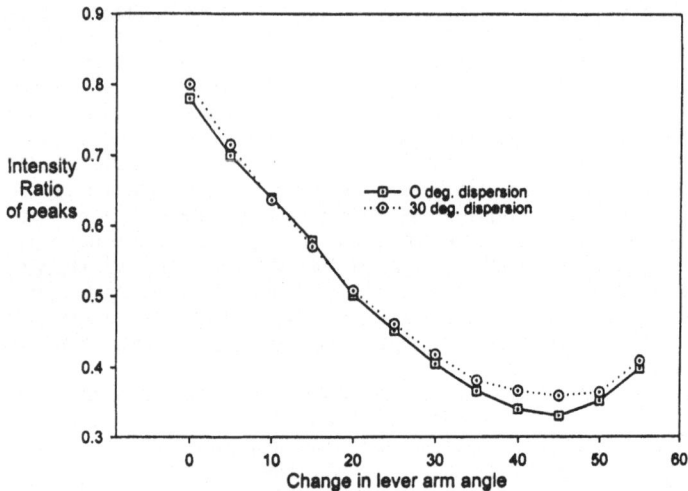

Figure 1. Variation of the intensity ratio of the two interference peaks on the M3 reflection with lever arm angle calculated for zero dispersion (all lever arms at same angle) and 30° dispersion (lever arm angles distributed evenly between –30° and + 30° of the angle shown in the plot. In the examples shown, the 0° dispersion plot assumes that the fixed contribution to the M3 reflection is 3/2 times that from the moving heads, while in the case of ± 30° dispersion, the two contributions are assumed to be equal. Slightly different assumptions could eliminate the slight discrepancy at the lower ratios, which are within the variation found between individual muscles. At present, there are no independent means of determining the fractional contribution of fixed components to this reflection.

Table 1. Variation of total intensity of M3 and M6 reflections as a function of dispersion of lever arm angles, calculated for standard model (see text)

	± 1% Dispersion	± 10° Disp.	± 20° Disp.	± 30° Disp.	± 40° Disp.	± 50° Disp.
M3	100	89.9	67.5	42.7	22.9	10.0
M6	27.5	19.0	6.93	1.92	.69	1.0
M6 (As % M3)	27.5%	21.1%	10.3%	3.8%	3.0%	10%

Effects of Dispersion

A factor which becomes important at this stage is the dispersion in lever arm angles of the attached heads. Since the actin and myosin axial periodicities are incommensurate, it is not possible for all the myosin heads to be attached with the same lever arm angle. Also, since the helical periods are also incommensurate, azimuthal alignment also becomes a factor. A proportion of heads could attach at approximately the same lever arm angles, but as the proportion of all the heads which attach in the isometric state is unknown, we are not in a position to estimate the minimum average lever arm angle dispersion that there could be. If helical considerations are ignored, then with an actin subunit spacing of 5.5 nm, a myosin head is always within $\pm$ 2.75 nm of an actin site, and a dispersion of about $\pm$ 17° in the lever arm angle would be sufficient. But all azimuthal positions on one side of an actin filament may not be accessible, and one does not know how many actin filaments a given myosin head is able to explore.

A further complication is that the length of the working stroke of a myosin crossbridge may be greater than the range of positions necessary for attachments to form. We do not know whether, in the isometric state, the myosin heads are attached at all stages of their working stroke – as in a snapshot of a shortening muscle – or whether these is a predominance of them near the beginning of the stroke, as is often assumed.

Thus the question of dispersion can only be treated at present in an empirical way, by computing what effect various degrees of dispersion should have on the observed pattern, using the standard model and applying the dispersion to both heads (as a first approximation), and comparing the results with what is actually observed.

In the case of the M3 reflection, the presence of relatively large amounts of dispersion (up to $\pm$ 30° or more in lever arm angle) has little effect on the ability to model the changes in peak ratio with average lever arm angle (Fig. 1). Naturally, the absolute intensity will be smaller at larger dispersions, but that is not a quantity we are in a position to measure. Small adjustments need to be made to the assumed H-zone width to give the same starting ratio, but again, that is a quantity for which we do not have an independent measure, at least of the required accuracy. So observations on the M3 alone are uninformative about the amount of dispersion, and even the assumption of zero gives a perfectly good match with the actual behaviour.

In the case of the M6 reflection, the situation is different, since the ratio of the M6 intensity expected from the myosin heads, to that of the M3, depends strongly on the amount of dispersion present. As Table I shows, this percentage ratio will vary from 27.5% at zero dispersion to 1.92% at $\pm$ 30° dispersion. However, measurements of the observed ratio (which is about 10% in isometric contraction, before correction for differences in sampling by the sphere of reflection, and differences in the profile of the underlying transforms) cannot be used directly to estimate the dispersion. It is apparent from the form of the interference fringes on the M6 reflection that a major part of the observed reflection must be generated by structures other than the crossbridges. As can be seen from Figure 2, the predicted profile of the M6 reflection has the smaller of the two interference peaks on the lower angle side irrespective of the dispersion, whereas in practice it always occurs on the high angle side. This means that there must be a 7.25 nm periodicity present, which does not arise from the myosin heads that we have modeled to account for the behaviour of the 14.5 nm reflection. Instead, it must come from some backbone structure in the thick filaments, since, although the phase of the fringes differs from that produced by myosin heads, their spacing shows that they are derived from diffracting systems of the same characteristics as the heads on thick filaments (i.e., two sets of repeating structures with their centers about 900 nm apart).

We cannot estimate the size and position of the repeating structure in the backbone directly from the isometric pattern alone for, *a priori*, we do not know the dispersion of the crossbridges. However, we have observed that characteristic changes occur in the M6 reflection when we apply different amounts of quick release to the muscle. The ratio of the peaks only changes by a small amount, usually from about 0.55 in isometric contraction to about 0.45 in larger releases, but the total intensity of the peaks increases substantially, by up to nearly 40% with 4-5 nm of filament sliding. This can be used to estimate the two contributions. That there is such a large increase does not necessarily imply the head contribution to intensity is a sizeable fraction of that from the backbone, for it is the amplitude of the two components that has to be summed, and then

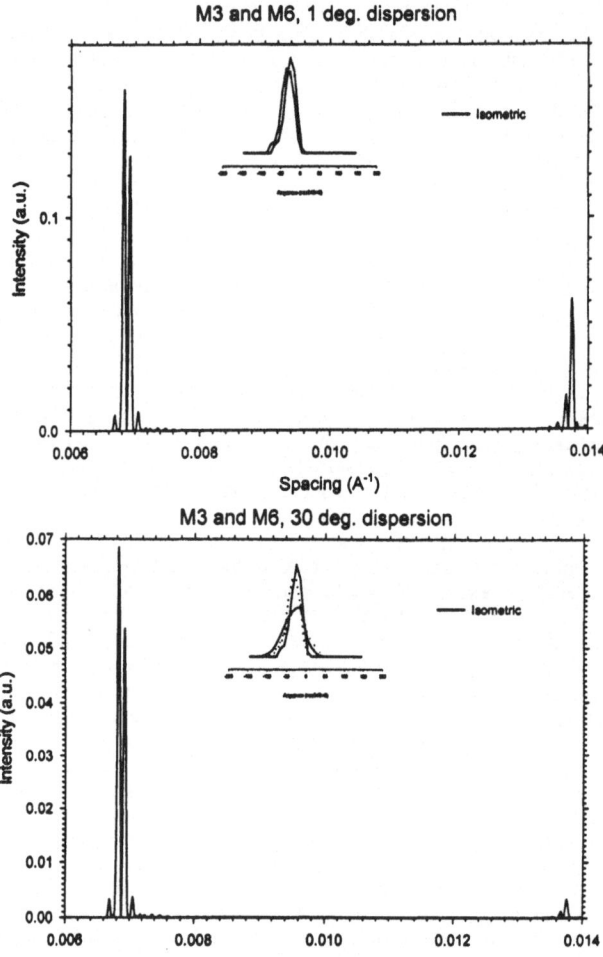

Figure 2: Calculated interference profiles for the M3 and M6 reflections with the standard model (zero dispersion), with lever arms at 55° and 48° from the rigor position (see text). The H-zone width is chosen to give the appropriate ratio (0.7 – 0.8) between the height of the two peaks in the M3 reflection. A large decrease in the M6 intensity is produced when dispersion of the lever arm angles is introduced, but the relative positions of the larger and smaller peaks is unaffected and is in all cases the reverse of that seen experimentally.

squared to give the resulting intensity. Thus if two components of a reflection would individually give intensities of 100 and 4, the total intensity would be 144, if they were summed in phase (10 + 2 = 12). In practice, the phase of the head contribution, as well as its amplitude, changes with release, and the result can only be estimated by numerical methods, i.e., by selecting a number of different amounts of dispersion and choosing values for the amplitude and phase of the fixed contribution from the backbone which will best match the observed change in the intensity and fringe pattern of the M6 reflection with different amounts of release.

This is a complicated process which will be described in detail elsewhere, but, in brief, quite good agreement can be obtained if one assumes a dispersion of ± 23°, i.e. that all lever arm angles within this range will occur with equal probability, and that the intensity of the backbone contribution is approximately 8 times that from the heads, in isometric muscle.

M9 and M15 Reflections

These reflections show similar properties to those analyzed on M6. The profiles which would be generated by the crossbridges are very different from those actually observed, and both reflections increase in intensity with release of the muscle, though by smaller amounts than M6. The intensity changes can be modeled in a similar way to those from M6, and can be accounted for using a similar lever arm dispersion, and backbone contributions whose intensity is 12.8 times and 71.4 times that of the heads' contribution at the M9 and M15 reflections respectively.

THE BACKBONE

The work summarized above shows that the myosin filament backbone contains a fixed structure which has structural periodicities of 14.5/2 nm, 14.5/3 nm, and 14.5/5 nm. This shows that its minimum underlying repeat is 14.5 nm, as indeed would be expected, since the backbone contains the LMM parts of the myosin molecules which are arranged with a 14.5 nm axial periodicity in isometrically contracting muscle. The question therefore arises as to whether this backbone structure actually contributes to the observed 14.5 nm meridional reflection. All that can be deduced from the M6, M9 and M15 reflections is that there are density components which can be represented by cosine waves with the appropriate periodicities. There is no absolute requirement that the first order term is also present, i.e. that the structure must also contain density represented by a cosine wave corresponding to a 14.5 nm periodicity. However, given that this must be the underlying physical repeat of the positions of the LMM rods, it would be surprising if they did not generate *some* reflection with this periodicity (early X-ray diagrams from oriented LMM by Cohen and Szent-Györgyi (1960) do show a weak 14.5 nm and a strong 7.25 nm reflection). In that case, at least part of the fixed component of the M3 reflection, which we have modeled as coming from another myosin head, might in fact come from the backbone.

This is not necessarily fatal to the idea that the second unattached head of tension-generating myosin molecules is involved in the reflection. Once the idea of dispersion in head positions is admitted, there is no reason why the dispersion of the two heads should be the same; indeed, it seems highly improbably that they would be so, and likely that the second head would have the larger dispersion, and that some additional contribution would be needed. However, it does complicate the picture, and one wonders if even the (presumably very disordered) completely unattached pairs of myosin heads might also make some small contribution.

What determines the value of the 14.5 nm Period?

In a resting muscle, the spacing of the M3 reflection is close to 14.3 nm, and it increases by about 1.5% when the muscle contracts. The higher order myosin meridional reflections also all increase by approximately the same amount (apart from M9, which appears to be a composite of two reflections). Since most of the intensity of these higher order reflections comes from the filament backbone, the change in periodicity must occur in the backbone too, rather than being some function of mismatch between the myosin head positions and the actin filaments to which they attach. The natural way to think of the change is that the backbone structure undergoes some kind of co-operative change when the heads, or some of them, move away from the original helical positions and interact with actin. That structural change results in the 1.5% increase in backbone periodicity, and since the force generating heads are being held with the S_2 rods under tension and connected to positions which must now have the new periodicity, the average axial position of the heads will show the same 1.5% increase in period.

During quick releases, the backbone reflections show the changes in spacing expected from the elasticity of the myosin filaments, as previously determined (Huxley et al, 1994). The individual peaks of the M3 reflection show larger spacing changes, as the fringes move across the profile of the transform of the array of crossbridges, but the underlying transform changes spacing by a lesser amount. This is still a larger spacing change than seen in the backbone reflections, since about half of the reflection comes from myosin heads attached to an actin filament, which changes its axial repeat spacing by a larger percentage amount than myosin during a tension change.

ACKNOWLEDGEMENTS

This work was support in part by NIH Grant 5 R01 AR43733-03.

REFERENCES

Huxley, H.E., Reconditi, M., Stewart, A., & Irving, T. "Interference changes in the 14.5 nm reflection during rapid length changes". *Biophys J.* 78, 134A, 2000.

Huxley, H.E., Reconditi, M., Stewart, A., Irving, T., & Fischetti, R. "Use of X-ray interferometry to study crossbridge behavious during rapid mechanical transients." *Biophys J.* 80, 266A, 2001.

Huxley, H.E., Reconditi, M., Stewart, A., & Irving, T. "Crossbridge and backbone contributions to interference effects on meridional X-ray reflections." *Biophys J.* 82, 5A, 2002.

Huxley H.E., Simmons R.M., Faruqi A.R., Kress M., Bordas J., & Koch M.H. "Millisecond time-resolved changes in x-ray reflections from contracting muscle during rapid mechanical transients, recorded using synchrotron radiation." *Proc Natl Acad Sci U S A.* 78, 2297-301, 1981.

Huxley H.E., Simmons R.M., Faruqi A.R., Kress M, Bordas J, & Koch MH. "Changes in the X-ray reflections from contracting muscle during rapid mechanical transients and their structural implications." *J Mol Biol.* 169, 469-506, 1983.

Huxley H.E., Stewart A., Sosa H., & Irving T. "X-ray diffraction measurements of the extensibility of actin and myosin filaments in contracting muscle." *Biophys J.* 67, 2411-21, 1994.

Linari M., Piazzesi G., Dobbie I., Koubassova N., Reconditi M., Narayanan T., Diat O., Irving M., & Lombardi V., "Interference fine structure and sarcomere length dependence of the axial x-ray pattern from active single muscle fibers." *Proc Natl Acad Sci U S A.* 97, 7226-31. 2000.

Lombardi, V., Piazzesi, G., Linari, M., Vanicelli-Casoni, M.E., Lucci, L., Boesecke, P., Narayanan, T., & Irving, M. "X-ray Interference Studies of the working stroke in single muscle fibers." *Biophys J.* 78, 134A, 2000.

Piazzesi, G., Reconditi, M., Linari, M., Lucii, L., Sun, Y., Koubassova, N., Boesecke, P., Narayanan, T., Irving, M., & Lombardi, V. "X-ray interference study of myosin head motions during rapid length changes of single muscle fibers." *Biophys J.* 80, 509A, 2001.

Irving, M., Reconditi, M., Linari, M., Lucii, Y., Narayanan, T., Boesecke, P., Stewart, A., Fischetti, R., Irving, T., Piazzesi, G., & Lombardi, V. "X-ray interference measurements of myosin head motions during isotonic shortening of skeletal muscle fibers." *Biophys J.* 82, 371A, 2002.

Piazzesi G., Reconditi M., Linari M., Lucii L., Sun Y.B., Narayanan T., Boesecke P., Lombardi V., & Irving M. "Mechanism of force generation by myosin heads in skeletal muscle." *Nature* 415, 659-62, 2002.

Rayment, I., Holden, H.M., Whittaker, M., Yohn, C.B., Lorenz, M., Holmes, K.C., & Milligan, R.A. "Structure of the actin-myosin complex and its implications for muscle contraction." *Science* 261, 58-65, 1993

Rayment I., Rypniewski W.R., Schmidt-Base K., Smith R., Tomchick D.R., Benning M.M., Winkelmann, D.A., Wesenberg G., & Holden H.M. "Three-dimensional structure of myosin subfragment-1: a molecular motor." *Science* 261, 50-8, 1993.

Szent-Györgyi, A.G., Cohen, C., & Philpott, D.E. "Light Meromyosin Fraction I: A Helical Molecular from Myosin." *J. Mol. Biol.* 2, 133-142, 1960.

DISCUSSION

Cecchi: From my understanding, the changes in the intensity of the fringes are caused by shifting of the total S1 mass moving away or towards the M-line, therefore from these movements you cannot say if the tail is tilting or not.

Huxley: The mass movement causes the changes in the relative intensity of the two fringes. The tail tilting causes the changes in the total intensity of the reflection (approximately, the sum of the intensities of the two fringes), by the expected amount.

Pollack: We measured the compliance of isolated actin filaments and isolated myosin filaments using the filament length change by subjecting the filament to tension ranging from zero to full isometric force P_0. The filament extension with P_0 was of the order of 1.5%. If these compliances show up during the quick release, is it possible that the compliances alone might account for the observed cross-bridge tilting?

Huxley: In an intact muscle during contraction, we have found that the actin and myosin filaments change their axial repeats by approximately 0.35% and 0.2% respectively, for a tension change equal to P_0. This would not account for the changes seen. The 1.5 % lengthening of the myosin filaments during activation is not related to tension, but to a structural change in the backbone of the filaments, probably related to the configuration of the cross-bridge.

Pollack: Is it possible that the 1.5 % lengthening of the thick filament arises because titin becomes stiffer during activation, and stretches the thick filament?

Huxley: Possibly. But in the intact muscle during contraction, the cross-bridge repeat changes much less during quick release, as mentioned above. Also, my understanding of Bagni & Cecchi's results is that titin becomes stiffer at whatever length it has when the muscle is activated, but does not generate significant tension at that length. Extra tension is only generated when it is stretched.

Yu: (On M3 spacing changing from 143Å to 145 Å) I would like to comment that in skinned rabbit psoas muscle, M3=143 Å under when ADP.Pi is bound at the active site of myosin (i.e. A·M·ADP.Pi and M·ADP.Pi). In other states in the ATP hydrolysis cycle is 144.3 Å (e.g. A·M·ATP and M·ATP etc.). So, the change in spacing appears to depend on the state of myosin not on strong binding to actin.

Huxley: Yes, and John Wray has found that similar spacing changes can be seen in relaxed mammalian muscle at different temperatures, when the equilibrium constant in the phosphate cleavage reaction of ATP on myosin changes.

Holmes: What happens between relaxed and active muscle −143Å goes to 145 Å?

Huxley: This seems to be a change in periodicity of the myosin backbone, brought about by the loss of the helical packing of the myosin heads around its surface, as they interact with actin. The axial repeat of the heads is determined by the backbone, and the M3 and the M5 reflections each change by approximately 1.5 %, the latter reflection coming almost entirely from the backbone.

MODELING ANALYSIS OF MYOSIN-BASED MERIDIONAL X-RAY REFLECTIONS FROM FROG SKELETAL MUSCLES IN RELAXED AND CONTRACTING STATES

Kanji Oshima, Yasunori Takezawa, Yasunobu Sugimoto, Maya Kiyotoshi, and Katsuzo Wakabayashi*

ABSTRACT

Analysis of the myosin-based meridional intensity data in the X-ray diffraction patterns of live frog skeletal muscles was performed to propose a more precise model for a myosin crown periodicity and an axial disposition of two-headed crossbridges along the thick filament in a sarcomere. Modeling studies revealed that the thick filament has a mixed structure of two different periodicities of the myosin crossbridge crown arrangement and that the crown periodicity and the axial disposition of crossbridges are altered when muscle goes from the relaxed state to the contracting state. Factors that primarily affect the meridional intensities were examined.

1. INTRODUCTION

X-ray diffraction patterns from relaxed frog skeletal muscles show a series of layer-line reflections indexed to a crystallographic period of 42.9 nm (14.3 nm x 3), the dominant features of which are attributed to the "perturbed" helical structure of the three-stranded thick filament. Thick filaments are ca. 1.6 μm long with myosin crossbridges located at ca. 50 crown levels in each half of the filament with an average repeat of 14.3 nm. Layer-line features remain in the X-ray diffraction pattern when muscle contracts isometrically but with an increased periodicity of 43.5 nm (14.5 nm x 3). In a model of thick filament forming the three-stranded helix with a regular crossbridge crown repeat of

---

*Kanji Oshima, Yasunori Takezawa, Yasunobu Sugimoto, Maya Kiyotoshi and Katsuzo Wakabayashi, Division of Biophysical Engineering, School of Engineering Science, Osaka University, Toyonaka, Osaka 560-8531, Japan.

14.3 nm within a 42.9 nm-period, the meridional reflections can appear only on the myosin-based layer lines indexed as multiples of three. However, a series of the meridional reflections indexed as orders of 42.9 nm (or 43.5 nm) are observed (see Huxley and Brown, 1967) and have been thought to originate from a systematic perturbation in the axial repeat of the crossbridge crowns within the crystallographic period (Yagi et al., 1981; Squire et al., 1982). All these meridional reflections are sampled by the closely-spaced diffraction peaks arising from interference between the two symmetrical halves of thick filament centered on the M-line in a sarcomere (Huxley and Brown, 1967; Haselgrove, 1975).

Analysis of the myosin-based meridional intensity data with high angular resolution has allowed us to propose a more precise model for a myosin crown periodicity and an axial disposition of two-headed myosin crossbridges along the thick filament in a sarcomere. Based on our recent modeling studies we briefly report the alteration of the crossbridge arrangement when muscle goes from the relaxed state to the contracting state, and discuss factors that primarily affect these meridional intensities.

2.   MATERIALS AND METHODS

2.1. Specimen and X-ray Diffraction

Live frog sartorius muscles were used for X-ray studies. X-ray diffraction experiments were performed using collimated synchrotron X-rays (wavelength of 0.15 nm) at the beamline 15A at the Photon Factory, Tsukuba, Japan. Two-dimensional X-ray diffraction patterns from relaxed and isometrically contracting muscles were recorded with a storage phosphor area detector (an image plate) at the specimen-to-detector distance of ca. 2.4 m as described previously (see Wakabayashi et al., 1994). The intensity distributions on the meridian in the X-ray patterns were traced. The myosin-based meridional reflections were selected and their integrated intensities in the narrow radial range of 0-0.0136 nm^{-1} were measured. The fine splitting of the second and fifth order meridional reflections with a period of 42.9 nm in the relaxed X-ray patterns was measured by assuming a gaussian model for the peaks on a background.

2.2. Modeling Calculations

Myosin filament-based meridional intensities for one-dimensional models were computed: the Fourier transform of a three-dimensional filament model projected onto the fiber axis was calculated assuming that the contributions of backbone, C-proteins, M-line structure and other accessary proteins have a minor effect on the meridional intensities. For modeling the crossbridge arrangement along the thick filament we used the intensities of the second to the eleventh order reflections with the basic period of 42.9 nm in the relaxed state and of 43.5 nm in the contracting state. According to the electron microscopic data, ca. 50 crossbridge crowns locate in each half of the thick filament with an average repeat of 14.3 nm and in the center of the filament the bare zone missing the crossbridges is ca. 160 nm long (see Squire, 1981). In the fitting calculation of intensities, the bare zone length was varied in the relaxed and contracting states. A two-headed crossbridge structure projected onto the fiber axis was modeled by assuming two gaussian profiles. Ten independent parameters were allowed to vary (see below). The most probable values of these parameters were determined by searching the best fit of the calculated intensities of the meridional reflections to those observed to minimize the R-factor defined as

$$R = \frac{1}{N} \sum_{i=2}^{N} \frac{|I_{i,\,obs} - k I_{i,\,cal}|}{I_{i,\,obs}}$$

where k is the scale factor and the summation of i is performed over the second to the eleventh order meridional reflections ($N=10$) with the 42.9-nm period where all these reflections were treated to have equal weights. In the present fitting calculations, the first order reflection was omitted because its meridional intensity could not be accurately measured due to overlap with other reflections, but its feature was taken into account for the criterion of fit.

3.   RESULTS AND DISCUSSION

We investigated the intensity distributions along the meridian of the X-ray diffraction pattern from live frog sartorius muscles at the rest length of sarcomere. As mentioned above we treated only a distribution of projected electron density arising from the crossbridges. In the X-ray patterns with high angular resolution from relaxed muscles, the fine sampling period on the meridional reflections except for the multiples of three, the so-called "forbidden reflections" (Huxley and Brown, 1967), corresponds to the distance between the centers of two regions in the whole filament where the crown levels are systematically perturbed (the perturbed regions). Detailed analysis of the fine splitting of the second and the fifth order meridional reflections showed that the average sampling period was ca. 790 nm ± 90 nm, indicating that the perturbed regions of crossbridge crown levels occupy the central zones of crossbridge arrays. The crowns with a regular repeat of 14.3 nm were assumed to locate both in the proximal and distal parts (the regular regions). The estimated distance between centers of two perturbed regions is close to the distance (ca. 730 nm) between two C-protein zones by electron microscopy (see Squire, 1981).

To analyze a perturbed structure of a crossbridge crown arrangement along the thick filament in relaxed and contracting states, we simulated the intensities of meridional reflections. In the modeling we set the unit cell comprising of three successive crossbridge crown levels and assumed that every repeat of the crowns in the unit cells of the perturbed region was different. In the regular region there is no displacement in the triplets. For examination of the axial disposition of two-headed crossbridges along the thick filament, the one-dimensional projected electron density profile of each crossbridge crown was approximated by two gaussian functions. Fixing the area of a gaussian function, its width (defined by 4σ-value) representing the axial extent of the density corresponding to a myosin head was used as a variable parameter. In our calculations the axial disposition of each head in a crossbridge was allowed to be different in the perturbed and regular regions. Thus, the meridional diffraction patterns were determined by ten independent parameters: the shifts of the distance of each crown from the 14.3 nm- or 14.5 nm-repeat in the triplets in the perturbed regions; the width of each gaussian profile and its center-to-center distance between two gaussians representing each crown; the center-to-center distance between two perturbed regions; and the number of crown triplets in the perturbed regions.

Figure 1 compares the observed intensities and the calculated ones from the model giving the lowest value of the R-factor. Our calculated intensity curves in both states revealed the appearance of the reflection orders of the 42.9 nm-period and the characteristic splitting on these reflections due to interference effects between two halves of the filament, confirming that the thick filament has a mixed structure of two different periodicities of the myosin crossbridge crown arrangement, consistent with the findings of Malinchik and Lednev (1992). Our simulation indicated that a fairly distinct reflection appears on the meridian at $1/42.9$ nm^{-1} from the crossbridge arrangement from relaxed muscles and the first four forbidden reflections are much weakened during contraction as observed (Fig.1). Recently we have observed the remnant of the first meridional reflection at $1/43.8$ nm^{-1} in the X-ray pattern from contracting muscle, although it seemed to overlap with other nearby reflections in the relaxed state (Takezawa et al., unpublished). The best fit to the experimental X-ray intensity data gave a model for the crossbridge arrangement in the relaxed and contracting states (Fig.2). In the relaxed state, in each crossbridge zone there are ca. 300 nm-long perturbed regions with 7 triplet levels of 42.9 nm. Inside and outside the regions there are ca. 130 nm-long and 290 nm-long

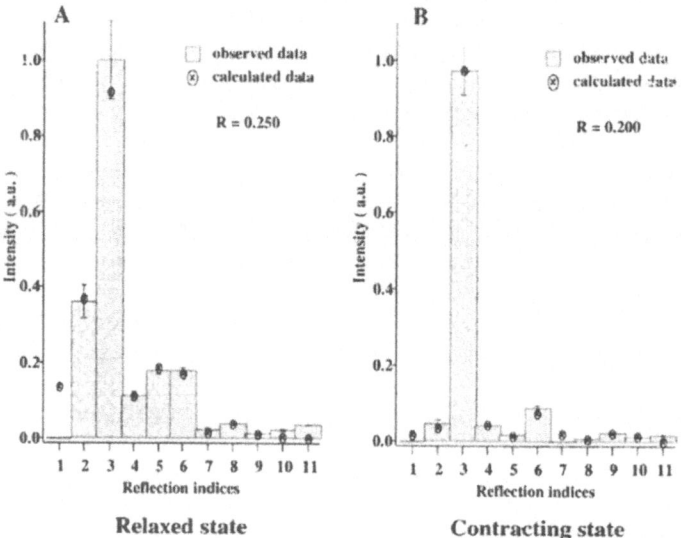

Figure 1. Comparisons of the best-fit calculated intensities (⊗) and the observed intensities (bar graphs) of the myosin-based meridional reflections with a 42.9 nm- or 43.5 nm-period in the relaxed and contracting states. A, relaxed state; B, contracting state. The intensity of the third order reflection in the relaxed state is normalized to 1.0. In A and B, R denotes the minimum value of the discrepancy factor. Vertical lines on bar graphs are associated S.D..

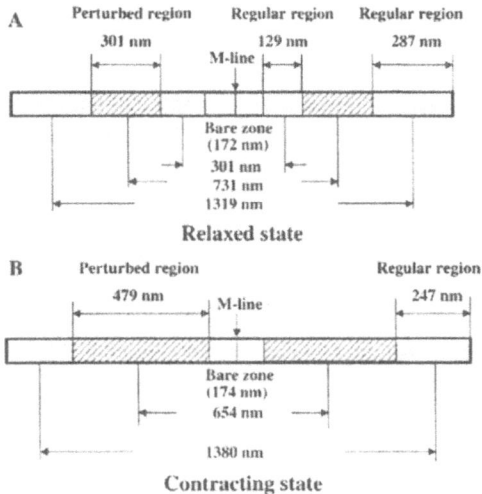

Figure 2. Distributions of the regular and perturbed regions of myosin crossbridges along the thick filament derived from the best-fit models in the relaxed and contracting states. A; relaxed state, B; contracting state.

regular regions, respectively. In the contracting state, the perturbed regions are ca. 480 nm long and occupy the position from the first (neighboring to the M-line zone) to the 33th crossbridge levels while the regular regions occupy the outside positions containing 17 crossbridge levels. Thus, when muscle contracts, in each half of the filament the region of perturbation becomes longer to ca. 1.6 times while the regular region in total becomes shorter to ca. 0.6 times respective crossbridge regions in the relaxed state. The center-to-center distance between the perturbed regions in both sides of the bare zone in the contracting state is shorter than that in the relaxed state. The distances among crossbridge crowns in the triplets are 15.3 nm, 17.3 nm and 10.3 nm in the relaxed state and 13.5 nm, 13.5 nm and 16.5 nm in the contracting state (Fig.3). The crown distance in the regular region is 14.3 nm in the relaxed state and 14.5 nm in the contracting state.

As depicted in inlets of Fig. 3, in the regular region of the relaxed state, the axial widths of the projected density corresponding to myosin heads in a crossbridge are similar to each other. The total width is ca. 15 nm and the axial distance between centers of two densities is ca. 6 nm. In the perturbed region the projected density of one head is sharper than that of the other head, forming an asymmetric profile as a whole. The distance between two density peaks is ca. 3 nm and the overall profile is much sharper than in the regular region. In the contracting state the projected density profiles of a crossbridge are the same in both regions, being slightly sharper than that of a crossbri.' in the perturbed region of the relaxed model. Thus, the two heads of a crossbridge are axially flaring in both states and the two-headed crossbridge in the contracting state orients more perpendicularly to the fiber axis than in the relaxed state. The result on the

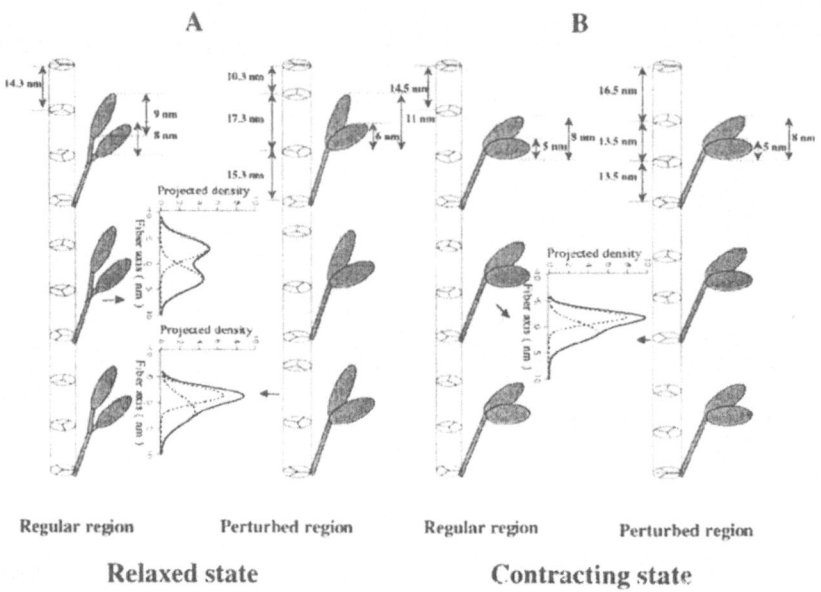

Figure 3. Summary of the axial disposition of two-headed myosin crossbridges in the regular and perturbed regions along the thick filament in the best-fit relaxed and contracting models. A; relaxed state, B; contracting state. In A and B each crossbridge crown repeat in a triplet of the perturbed regions is shown and are also shown the axial density profiles of a two-headed myosin crossbridge projected onto the fiber axis in the regular and perturbed regions (inlets).

projected density profile of a crossbridge in the contracting state may be similar to that derived by the Fourier synthesis using the first to fourth order meridional reflections with a 14.5 nm-repeat (Juanhuix et al., 2001). Our present model of a crossbridge arrangement in the relaxed state should be compared with earlier models of Malinchik and Lednev (1992) and Wakabayashi et al. (1998).

It is interesting to note that in our modeling the length of the perturbed region becomes longer in the contracting state than in the relaxed state, despite the fact that the forbidden meridional reflections become much weaker in the contracting state. Our simulation reveals that the weakened forbidden reflections in the contracting state arise from the displacements of each crown repeat in the triplets becoming relatively smaller and the projected densities of all crossbridges becoming sharper than in the relaxed state. These changes of the crossbridge arrangement along the thick filament during contraction may be assumed to be favorable for optimum interaction with actin.

Using a one-dimensional model of a crossbridge arrangement along the thick filament, we have demonstrated that various factors affect the myosin-based meridional intensities. Close examination revealed that the R-factor is primarily influenced by the crossbridge shifts from the regular repeat in the triplets together with changes in the projected density profile of a two-headed crossbridge. Any changes of both lengths of the regular and perturbed regions also affect markedly the R-factor. Forbidden meridional reflections increase or recover and three-multiple reflections decrease in intensity when muscle shortens from isometric contraction (Yagi et al, 1993; Bordas et al, 1999). The interpretation of the meridional intensities is not straightforward. The above factors may be needed to take into account for the explanation of intensity changes of the meridional reflections when rapid length changes are applied to a contracting muscle. The results of our detailed analysis will be published elsewhere.

ACKNOWLEDGEMENTS

The authors thank Drs. T. Kobayashi and H. Tanaka for kind help with physiological and X-ray experiments at the Photon Factory, and Dr. T. C. Irving for critical reading of the manuscript. This work was partly supported by the Special Coordination Funds for Promoting Science and Technology, and Grant-in-Aid for Scientific Research (B) (No. 13480220) of the Ministry of Education, Culture, Sport, Science and Technology from the Japanese Government.

4. REFERENCES

Bordas, J., Svensson, A., Rothery, M., Lowy, J., Diakun, G. P., and Boesecke, P., 1999, Extensibility and symmetry of actin filaments in contracting muscle, *Biophys. J.* **77**:3197-3207.

Haselgrove, J. C., 1975, X-ray evidence for conformational changes in the myosin filaments of vertebrate striated muscle, *J. Mol. Biol.* **92**:113-114.

Huxley, H. E., and Brown, W., 1967, The low-angle x-ray diagram of vertebrate striated muscle and its behaviour during contraction and rigor, *J. Mol. Biol.* **30**:383-434.

Juanhuix, J., Bordas, J., Campmany, J., Svensson, A., Bassford, M. L., and Narayanan, T., 2001, Axial disposition of myosin heads in isometrically contracting muscles, *Biophys. J.* **80**:1429-1441.

Malinchik, S. B., and Lednev, V. V., 1992, Interpretation of the x-ray diffraction pattern from relaxed skeletal muscle and modelling of the thick filament structure, *J. Muscle Res. Cell Motil.* **13**:406-419.

Squire, J. M., 1981, The Structural Basis of Muscle Contraction, *Plenum Press, London,* pp. 344-362.

Squire, J. M., Harford, J. J. , Edman, A. C., and Sjostrom, M., 1982, Fine structure of the A-band in cryo-sections. III. Cross-bridge distribution and the axial structure of the human C-zone, *J. Mol. Biol.* **155**:467-494.

Wakabayashi, K., Sugimoto, Y., Tanaka, H., Ueno, Y., Takezawa, Y., and Amemiya, Y., 1994, X-ray diffraction evidence for the extensibility of actin and myosin filaments during muscle contraction, *Biophys. J.* **67**:2422-2435.

Wakabayashi, K., Sugiyama, H., Yagi, N., Irving, T. C., Iwamoto, H., Horiuti, K. , Takezawa, Y., Sugimoto, Y., Iino, S., Kim, D-S., Amemiya, Y., Yamamoto, S., and Ando, M., 1998, High-resolution x-ray diffraction of muscle using undulator radiation from the TRISTAN main ring at KEK, *J. Synchrotron Rad.* **5**:280-285.

Yagi, N., O'Brien, E. J., and Matsubara, I., 1981, Changes of thick filament structure during contraction of frog striated muscle, *Biophys. J.* **33**:121-138.

Yagi, N., Takemori, S., and Watanabe, M., 1993, An x-ray diffraction study of frog skeletal muscle during shortening near the maximum velocity, *J. Mol. Biol.* **231**:558-677.

MYOSIN FILAMENT STRUCTURE AND MYOSIN CROSSBRIDGE DYNAMICS IN FISH AND INSECT MUSCLES

John M. Squire, Hind A. AL-Khayat, Jeffrey J. Harford, Liam Hudson, Tom C. Irving, Carlo Knupp, Ngai-Shing Mok & Michael K. Reedy

1. INTRODUCTION - THE PROBLEM

The muscle crossbridge power stroke on actin appears to involve a change in angle between the actin-attached motor domain and the neck region of the myosin heads (the 'tilting neck hypothesis'). However, this mechanism has not been proved beyond doubt and a reasonable question to ask is how actual proof might be achieved. This is essentially a structural question. The question can be put as 'what are the molecular shapes that the myosin head adopts during the crossbridge cycle on actin?'. A further question that also needs answering is 'how do the biochemical stages of the actin-myosin ATPase cycle map onto the structural changes that are seen?'. The purpose of the present paper is to address how the first question might be answered and to start to make suggestions about the second.

Structural questions require to be answered by structural techniques. Although great insights into the possible structural states of the myosin head come from the results of protein crystallography (e.g. Rayment *et al*, 1993; Houdusse *et al*, 1999; Dominguez *et al*, 1998), the functional question really relates to what actually happens in intact muscle, where there are mechanical constraints on crossbridge action. Ideally, therefore, the structural probes need to be applied to intact muscles or fibres. Those methods that come immediately to mind are X-ray diffraction, electron microscopy and the use of probes attached to various parts of the actin-myosin system. Each of these methods has

John Squire, Hind AL-Khayat, Carlo Knupp, Jeffrey Harford, Liam Hudson, Ngai-Shing Mok, Biological Structure & Function Section, Biomedical Sciences Division, Faculty of Medicine, Imperial College London, Exhibition Road, London SW7 2AZ, UK. Tom Irving, BioCAT, Dept. Biological, Chemical and Physical Sciences, Illinois Institute of Technology, Chicago, IL 60616, USA. Michael Reedy, Dept of Cell Biology, Duke University, Durham, NC 27710, USA.

its own advantages and disadvantages. X-ray diffraction methods can be applied to intact, contracting muscles in real time (see Reviews in Harford & Squire, 1997; Squire, 1998; 2000). The problem is that X-ray diffraction patterns need to be interpreted and modelled unambiguously to provide useful information and this is rarely easy. Electron microscopy of muscle has the great advantage that what is obtained is a direct image of the contractile machinery. The main disadvantages are that uncertainty always remains about the degree of preservation of the native structure actually achieved, that the micrographs obtained are selective, that they can only ever provide a snapshot of dynamic events (even using sophisticated rapid freezing methods) and that the resolution in the micrograph images is limited. Electron microscopy is currently made much more effective by new cryo-fixation methods, by the application of 3D reconstruction methods, especially tomography (e.g. Schmitz *et al*, 1997; Taylor *et al*, 1999) and by the use of techniques of molecular docking into electron density maps (e.g. Chen *et al*, 2002). However, the problems of preservation and selectivity remain. Probes, especially the bifunctional fluorescent probes used by Hopkins *et al* (2002), provide useful orientational information on different key sites in the contractile machinery, but at present their interpretation is still ambiguous. For example there is a two-fold ambiguity in probe direction relative to the fibre axis direction. And, since the method is essentially invasive, one needs to be sure that what is observed using chemically modified proteins is truly representative of what normally occurs in native muscle. In addition, if there is actually a mixture of several states, then identifying the separate components using polarised fluorescence measurements is difficult.

Collectively, results from these various techniques have already shown that the putative swinging of the lever arm relative to a 'fixed' motor domain on actin is *consistent* with the observations. However, we would argue that there is a great difference between what is *consistent* and what is *actually proved*. This paper discusses how the 'tilting neck hypothesis' might actually be proved or disproved in the case of muscle contraction. Here we present the case for using X-ray diffraction.

2.   EVALUATING THE LOW-ANGLE DIFFRACTION METHOD

2.1   What can be done by X-ray diffraction

Since all available structural techniques have their limitations, there can be little doubt that each method on its own needs to be tested against the others as much as possible. Similar results obtained by more than one method go a long way towards substantiating both the results obtained and the methods being used.

This paper is primarily about the application of X-ray diffraction methods to intact or skinned muscle preparations. The biggest single problem with X-ray diffraction methods is that of interpreting the observed diffraction data. Where possible our diffraction results are therefore tested against data from other methods, particularly electron microscopy. Modern X-ray methods permit the recording of well-resolved low-angle X-ray diffraction patterns from living muscle during electrically stimulated 'physiologically normal' contractions on a time-scale (milliseconds or better) at which important molecular events in the crossbridge cycle are thought to occur. However, results from different muscle types are not all as easy as each other to interpret. In any structural method which involves recording information averaged over the size of the object being studied, the simplest structures to interpret are usually those with the best

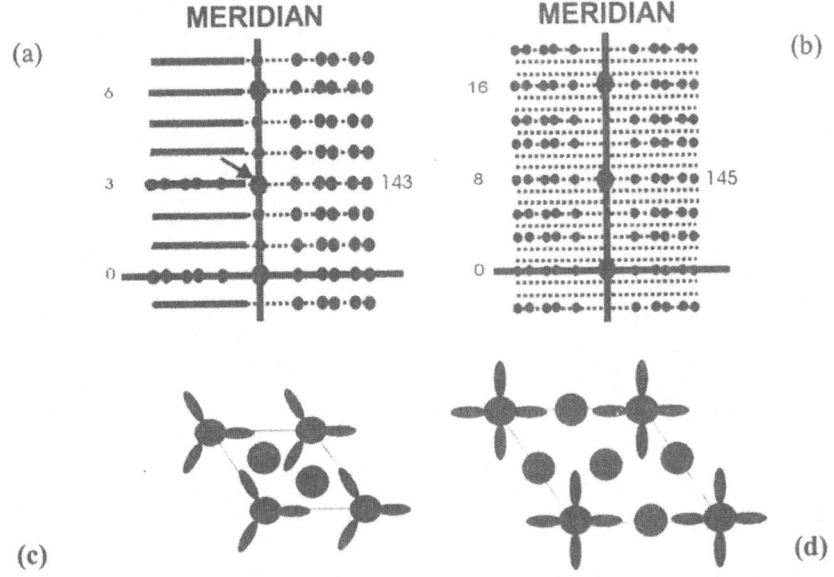

Figure 1. (a) and (b) are simulations of the low-angle X-ray diffraction patterns from vertebrate muscle (a) and insect flight muscle (b). (a) shows layer-lines which are orders of a 3 x 143 Å repeat. The meridional reflection on the third layer-line is the well-known M3 reflection (arrowed) at 143 Å. The left part of (a) shows layer-lines that are largely unsampled, as in patterns from e.g. frog and rabbit muscles. The right part of (a) shows well-sampled layer-lines as in patterns from bony fish muscles. This is because bony fish muscles have a simple lattice A-band shown in (c), where the 3-fold objects represent myosin filaments looking down the filament axis and the solid circles represent actin filaments. (b) shows the layer-lines at orders of a 1160 Å repeat in patterns from insect flight muscle, with a strong meridional peak on the 8th order at 145 Å. These are beautifully sampled due to the simple lattice A-band structure in this muscle (d). Here the 4-fold objects represent the myosin filaments and the circles represent actin.

order. For this reason, we are using X-ray diffraction methods applied to only the most highly ordered striated muscles (insect flight muscle, IFM, and bony fish muscle, BFM). In these cases alone, interpretation of the observed diffraction data, although demanding, is relatively straightforward. This is in comparison to data from the muscles of higher vertebrates (e.g. frog or rabbit) and all invertebrate striated muscles other than IFM, all of which contain considerable in-built disorder in their myofibrils. Our aim is to define the actin-myosin structures *in situ*, and if possible, to determine the kinetics of the transitions between the crossbridge structural states, that are found in the highly ordered striated muscles, IFM and bony fish muscle.

2.2 Advantages of 'Simple Lattice' Muscles.

The myofibrils of insect flight muscle and bony fish muscle appear relatively highly ordered compared with all other striated muscles because, unlike these other muscles, IFM and BFM myofibrils contain myosin filaments whose crossbridge arrays are all identically oriented across each A-band – they have so-called 'simple lattices' (Fig. 1). This compares with the superlattices present in the A-bands of higher vertebrate muscles (Luther and Squire, 1980; Luther *et al*, 1996) or the disordered lattices that occur in other invertebrate muscles (see Squire, 1981: Chapter 8). Simple lattice muscles have the unique advantage that the X-ray layer-lines coming from the myosin filaments (horizontal lines in Fig 1) are crossed by vertical row-lines showing the presence of good three-dimensional order. Just as diffraction patterns from protein crystals can be solved to give electron density maps to which atomic arrangements can be fitted, so also well-sampled low-angle X-ray diffraction patterns from muscle can be rigorously interpreted in terms of molecular conformations. The observed diffraction spots from a protein crystal are such that the electron density in the crystal unit cell can be calculated by a method known as Fourier synthesis. This sums a series of terms each of which contains information for each observed diffraction spot, namely the amplitude ($\propto\sqrt{I}$, where I is the measured diffraction spot intensity) and *phase* information. In both cases the intensities of different diffraction spots can be determined experimentally. In protein crystallography, however, the *phases* of the reflections can also be determined experimentally, although not directly. The difference is that in the case of low-angle X-ray diffraction data there is usually no experimental way of determining the *phases*. What has to be done in this case is to test and modify possible model structures against the observed intensity data until a good match is obtained.

2.3   Resolution and Sensitivity Tests

The next problem with low-angle X-ray diffraction is one of resolution. In the case of protein crystallography, diffraction data to a resolution of about 3 Å or better (ideally 1 Å) are used to provide electron density maps which locate atoms in proteins with high precision. In the case of low-angle X-ray diffraction methods applied to whole muscle, we are often dealing with diffraction information in the very different spacing range 430 to 27.5 Å. Our work described below used myosin layer-line data to a resolution of 65 Å. There is a big difference between 65 Å and the 1 to 3 Å used by protein crystallographers. Is the application of low-angle X-ray diffraction a meaningless exercise?   Is the resolution just too low to be useful?   The answer to this is clearly no, or we wouldn't be doing it, but for some it may not be immediately obvious why this is so. The reason is essentially as follows. If we know nothing about the structures of the component molecules in the actin-myosin system, particularly the myosin head and actin monomer, then the low-angle X-ray diffraction data can only provide general information about filament spacings and surface lattice geometry and only rough indications of mass distributions. This was the case in the muscle field before 1990, as was summarised and expounded in Squire (1981). *The publication of the actin monomer and filament structures in 1990 (Kabsch et al, 1990; Holmes et al, 1990) and the publication of the myosin S1 structure by Rayment and his colleagues in 1993 (Rayment et al, 1993a) totally transformed this situation.* The point is that when the basic shapes of the molecules in a system are already known, then low-angle X-ray diffraction is extremely sensitive to the three-dimensional disposition of these molecules.   The intensities of X-ray reflections of nominal 'resolution' in the 430 to 65 Å range are in fact sensitive to movements of as little as a few Å by molecules of known shape. Figures 2, 3 & 4 show examples of this. In one case (Fig. 2) a simple series of 143 Å-spaced densities, perhaps representing crowns of myosin heads along a vertebrate muscle myosin filament with a 430 Å axial repeat, have had every third crown shifted axially in the same direction by a small distance Δx.   The intensities of various orders of the 430 Å axial repeat

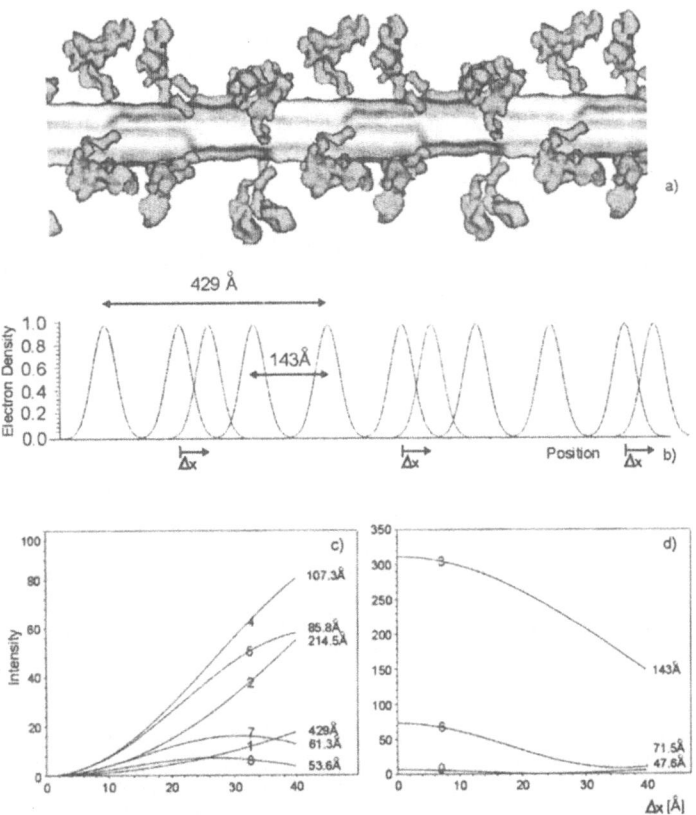

Figure 2. (a) The crossbridge array in bony fish muscle determined by Hudson *et al* (1997).   (b) Example of an axial shift by Δx of every third level of heads in (a) and the effect of this on various low-angle meridional X-ray reflections from muscle (c,d), all orders of 429 Å.   The usual strong meridional orders of M3 (labelled 3) at 143 Å are shown in (d) and the so-called 'forbidden' reflections in (c) are meridional reflections that become apparent when a true 429 Å repeat is established.

(e.g. the third order at 143 Å; the well-known M3 reflection) have then been computed as a function of Δx. This clearly shows that a movement of every third crown by as little as 10 Å has a major effinal order. ms of molecular For example the meridional orders M3 (143 Å), M6 (72 Å) and M9 (48 Å) all drop by several % (Fig. 2(d)) for a 10 Å crown shift, and the in between orders (the 'forbidden' meridional reflections from a perfect crossbridge helix; M1, M2, M4, M5 etc), which have zero intensity when Δx is zero, all increase in intensity rapidly to give quite appreciable intensities (Fig. 2(c)) even for 10 or 20 Å shifts Δx of every third peak.

Another example tests whether the low-angle region of the X-ray diffraction pattern is sensitive to crossbridge tilt. We have previously published a model for the myosin filament arrangement in relaxed fish muscle (Hudson *et al*, 1997). This represents the arrangement of myosin heads using the Rayment *et al* (1993a) head shape, and a possible hinge between the myosin motor and neck regions, that most closely fits the observed X-ray diffraction data. With such a model it is possible to calculate 'model' *phases* for the observed X-ray reflections and then to compute, as protein crystallographers do, an electron density map of the structure. This is actually what is shown in Figure 2(a). However, since the model amplitudes ($\sqrt{I_{calc}}$), and the observed amplitudes ($\sqrt{I_{obs}}$), are never exactly the same, one can compute an electron density map using the intensity differences ($\sqrt{I_{obs}} - \sqrt{I_{calc}}$) and the model phases to indicate where any discrepancies might lie. This is known as a *Fourier difference synthesis*; it is sensitive to differences between the model structure and the true structure. A map of this kind is shown as Figure 4 in Squire *et al* (1998). This same approach can also be used to determine how sensitive the low-angle X-ray diffraction pattern is to movements of the myosin heads. The test that we have done is simply to take the fish muscle myosin filament structure in Figure 2(a) and to move one of the heads in each pair on only one crown level (out of the three crowns in the 429 Å repeat) by tilting the head up from its modelled resting positions. The diffraction pattern from this new structure (the repositioned head model) was then computed to give new intensity values $I_{calc(moved)}$ which could be compared with the best fit resting model values I_{calc} discussed above. Is such a head movement detectable using low-angle diffraction data? In fact, the Fourier difference map (Fig. 3) calculated using the model *phases* and the amplitude difference $\sqrt{I_{calc}} - \sqrt{I_{calc(moved)}}$ shows positive electron density peaks where the new head positions are (in the repositioned head model, i.e. in positions where mass was not present in the original best-fit model) and it shows negative electron density peaks in the positions where the heads originally were. It is very clear from the Fourier difference electron density map that significant electron density has moved to the appropriate new positions, just as might have been hoped. Taking this a step further to a model of the whole unit cell of, for example, IFM, including the actin filaments, if such a head movement actually occurred during a transition from the already-modelled relaxed state A to some new state B (perhaps a pre-powerstroke interaction with actin), we can hope to detect and locate the new head position by calculating the Fourier difference map using the phases from state A and the amplitude differences between state A and state B. In addition the method appears to be sensitive to movements of only a small fraction of the total head number.

To summarise, low-angle X-ray diffraction patterns, despite the large apparent spacings of the layer-lines, are extremely sensitive to small (Å) movements of the myosin heads and even to relatively small tilts of just a few of the myosin heads out the total head population. The results in Figure 3 show movements of only 3 heads out of a total of 18 heads in the 430 Å axial repeat of the myosin filament, but the method easily detects this.

Another test refers to the actin filament structure. We show in Figure 4 that in the case of actin filaments, the actin layer-line intensities are acutely sensitive to the movements of even the smallest of the actin subdomains, namely subdomain 2, relative to the rest of the actin molecule (see brief discussion of this in AL-Khayat *et al*, 1995 and in Harford and Squire, 1997). An azimuthal swing of actin sub-domain 2 (Fig. 4(a)) by as little as 20° (equivalent to a displacement of the centre of mass of sub-domain 2 by as little as 10 Å) can change the intensity of the actin layer-lines, including the first at 370 Å and the second at 185 Å, by as much as 30%.

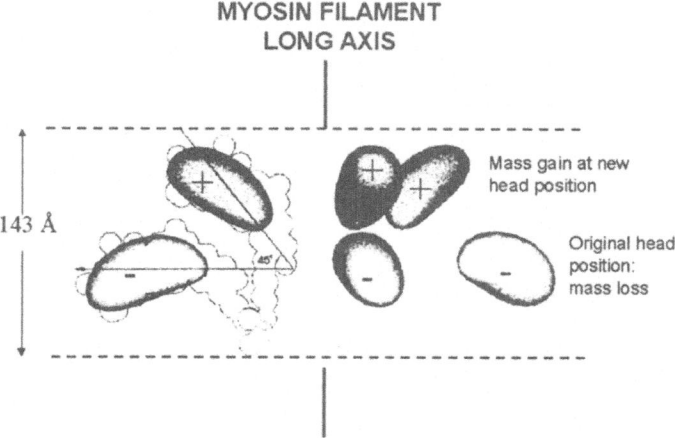

Figure 3. Illustration of the sensitivity of the low-angle X-ray diffraction method to single head movements in one of each of the three myosin head pairs in only one crown out of three myosin head crowns in a 429 Å repeat as in Fig. 1(a). A very clear shift of mass is observed from the lower densities (mass loss –ve peaks) to the upper densities (mass gain +ve peaks) using Fourier difference synthesis. The peaks coincide exactly with the original and new positions of the heads (fine outlines).

The present work also shows that the fit to the observed myosin layer-lines from both fish muscle and insect flight muscle is exquisitely sensitive even to bends between the motor domain and the neck of the myosin head. In this way, as discussed later, we have determined a new myosin head conformation in relaxed insect flight muscle.

2.4 Assessing the X-ray Diffraction Modelling Method by Electron Microscopy

The myosin filament model shown in Figure 2(a) was based entirely on rigorous modelling of the low-angle X-ray diffraction data from resting bony fish muscle (plaice fin; Hudson *et al*, 1997) using the Rayment head shape and variants of it (Rayment *et al*, 1993a). The modelling consisted of defining the possible filament structure in a 3-fold symmetric myosin filament in terms of a number of parameters (e.g. tilt, slew and rotation for each independent head, filament radius etc), changing these parameters using a simulated annealing method, and finally applying local refinement methods to define the structure with the best fit to the observations. The structure in Figure 2(a) gave a pleasingly low (3%) R-factor (the R-factor compares the observed and calculated diffraction patterns in an objective way), it showed the remarkable sensitivity of the fit to small movements of the myosin heads (Squire *et al*, 1998) and it automatically generated a model in which the myosin heads fit snugly together without physically overlapping, even though an overlapping constraint had not been applied in the modelling.

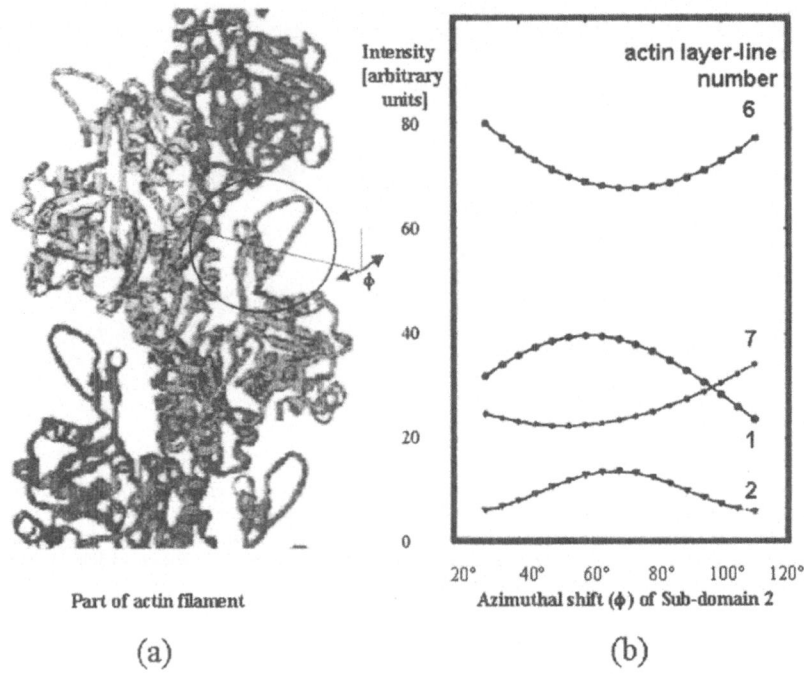

Part of actin filament

(a)

Azimuthal shift (ϕ) of Sub-domain 2

(b)

Figure 4.   (a) Illustration of part of the actin filament indicating sub-domain 2 (circled) and azimuthal angle ϕ. (b) Computed intensity changes on various actin layer-lines (orders of roughly 360 Å) as a function of sub-domain 2 azimuthal position as in (a).

Nevertheless, despite the apparent uniqueness and reliability of the fit, it is always as well to try to confirm or test the result using a totally independent method. In this case, we studied electron micrographs of negatively-stained A-segments from bony fish muscle (Cantino *et al*, 2000). Measurements on these micrographs were consistent with the X-ray model within the limitations of the two methods. We also studied electron micrographs of isolated fish muscle myosin filaments prepared by Dr. R. Kensler (Kensler and Stewart, 1989) and in the first instance carried out conventional 3D helical reconstruction of the myosin filaments from these images (Eakins *et al*, 2002). This approach is not strictly valid, since the myosin head array, as seen in Figure 2(a), has a periodic perturbation along it; crowns within the 430 Å repeat are actually of three non-identical kinds. Helical reconstruction methods, by their in-built assumption of exact helicity, make all crown levels the same. Even so, our preliminary results using this method produced reconstructions which were consistent with the X-ray diffraction model, allowing for the helical averaging. Further work on this is making use of the single particle analysis approach (van Heel *et al*, 2000) in which the differences in the three crown structures are preserved (AL-Khayat *et al*, in preparation – poster presented at the 11th London Muscle Conference, September 20th, 2002). Once again early results indicate myosin head arrays that are consistent with the X-ray diffraction model. So far the low-angle X-ray diffraction modelling approach appears to be reliable.

2.5   Conclusions about the Low-Angle Diffraction Method

In summary, the low-angle X-ray diffraction patterns from muscle are sensitive to molecular domain movements in the actin monomer, to changes in general tilt of the myosin heads and even to changes in the shape of the myosin heads around a pivot between the motor domain and the neck. These patterns can also be successfully modelled in a systematic way using the simulated annealing and refinement approach (Hudson *et al*, 1997) and the results so obtained appear to be compatible with independent evidence from electron microscopy. Since this method can also be applied to intact, living, contracting muscle with very high time-resolution, it provides a very sensitive and rigorous method for monitoring structural changes during the crossbridge cycle. The following paragraphs summarise what we are finding out using this approach and they illustrate the great potential of the time-resolved low-angle X-ray diffraction method.

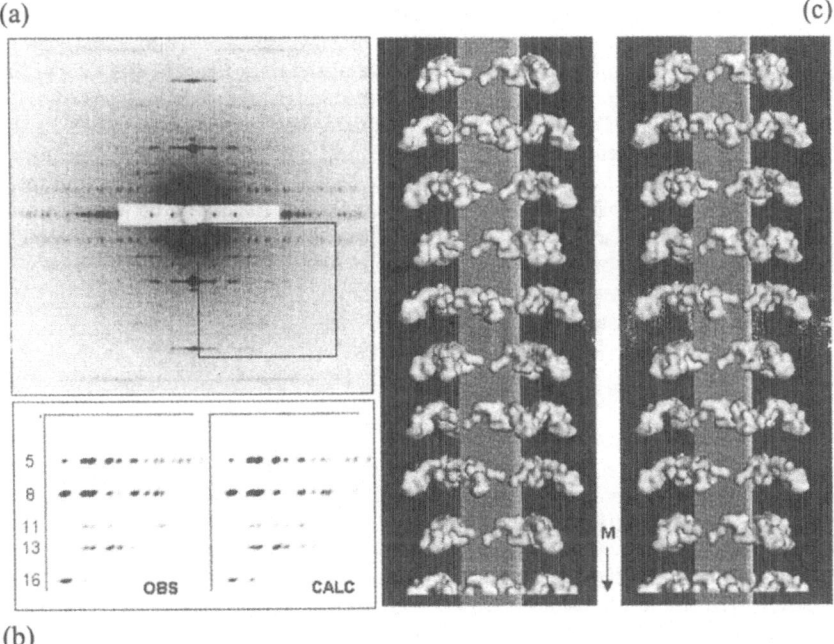

(a)

(c)

(b)

Figure 5. The low-angle X-ray diffraction pattern from relaxed insect (*Lethocerus*) flight muscle extending axially to just beyond the 59 Å layer-line and showing the lower-right quadrant (boxed) that is illustrated in (b).   (b, left) is the stripped, observed intensity data and (b, right) is the computed data from the head arrangement in (c).   Numbers at the lower left in (b) label the myosin only layer-lines as orders of the 1160 Å thick filament repeat.   (c) shows a stereo pair of our new structure for resting insect flight muscle thick filaments. The polarity of the filament is inferred by comparison with the IFM thick filament reconstructions of Morris *et al* (1991), where the M-band position is known (AL-Khayat *et al*, ms submitted).

3.   NEW RESULTS AND DISCUSSION

3.1   Crossbridge Organisation in Relaxed Insect Flight Muscle

The approach applied so successfully to the myosin layer-lines in low-angle X-ray diffraction patterns from resting bony fish muscle has now been applied rigorously to the equivalent X-ray diffraction patterns from resting insect flight muscle (Fig. 5(a); Reedy *et al*, 2000). Details of this new modelling are being published elsewhere (AL-Khayat *et al*, 2002, ms submitted). What we do here is pick out particular features of the new model that are relevant to our general discussion on diffraction methods and what can be gleaned from them about the crossbridge cycle.

Our procedure was as follows: From the low-angle X-ray pattern of relaxed insect (*Lethocerus*) flight muscle, five myosin-only layer-lines (62 reflections) were modelled to 6.5 nm resolution by computer-simulated annealing, giving the same best-fit thick-filament model (R-factor 9.7%; see Fig. 5(b)), regardless of the starting structure chosen. Each run searched ~5000 different structures, generated by varying likely atomic models, positions and shapes of myosin heads over 13 parameters, always converging to the same lowest R-factor model. The best head shape differed from all known X-ray crystallography structures of S1, but was clearly more similar to nucleotide-bound than to nucleotide-free S1 structures. The two heads in one myosin molecule (Fig. 5(c)) are in different configurations, one projecting outwards from the thick filament surface and the other wrapping around the filament surface and placed behind the projecting head from a neighbouring molecule. The motor domain of the outer head is in close proximity to the neighbouring actin filament and almost has the correct orientation for actin attachment. It is as though the inner head is holding the outer head in the appropriate position for a small radial movement and immediate actin binding. The interaction between the two heads (Fig. 6) is such that the Essential Light Chain (ELC, residues 5-71) of the outer head is close to the ATP-binding pocket (residues 271-327) of the inner head. This could possibly block the ATPase of the two heads in the Ca^{2+} -free state of the myosin. We come back to this point later.

3.2   A Possible Crossbridge Cycle in IFM

In a typical crossbridge cycle in IFM (Fig. 7) it can be envisaged that the relaxed projecting head, already oriented almost exactly as needed for actin attachment, will attach to actin, release ADP and Pi and end up in the conventional nucleotide-free rigor conformation on actin. Figure 7(b) shows what would happen to the head on actin. In fact the neck part of the head would swing through an almost purely axial trajectory to give an axial lever arm displacement of very close to 100 Å in going to the nucleotide-free state. The whole arrangement appears well poised to engage directly in the rapid oscillatory mode of active insect flight muscle.

3.3   Resting Head Arrangements in Different Muscles

The head arrangement in relaxed fish muscle is different to that in IFM. The best modelled head shape in resting bony fish muscle myosin filaments (Fig. 2(a); Fig. 8(a)) is with the neck almost in the Rayment *et al* (1993a) nucleotide-free conformation. The interaction of adjacent heads is also very different from that in insect. In fish muscle the two heads from the same myosin molecule appear to stabilise each other (Fig. 8(a)),

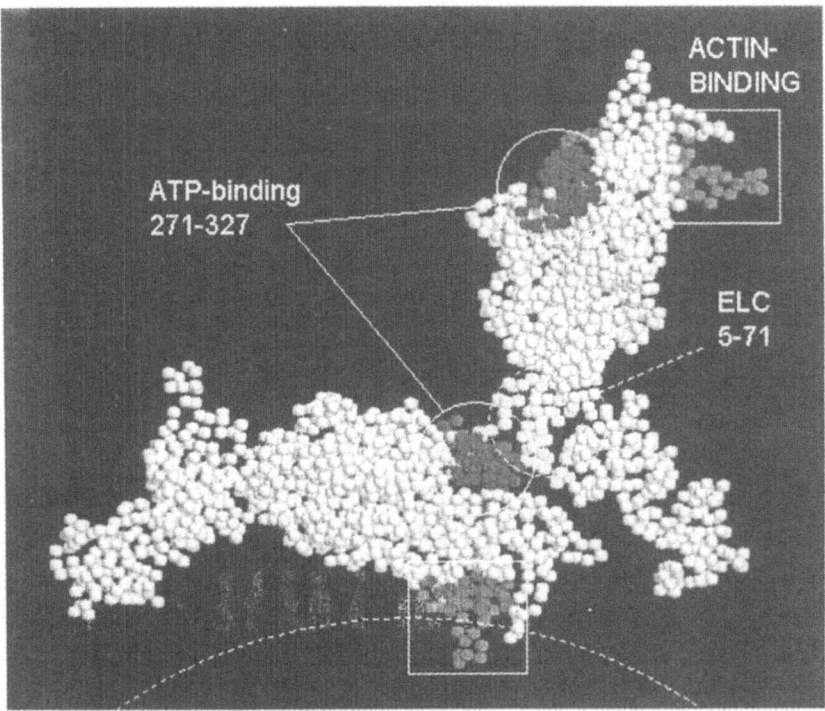

Figure 6.   Illustration of the projecting and supporting heads from two adjacent myosin molecules on the myosin filament surface (lower dashed curved line) of the insect flight muscle myosin filament model illustrated in Fig. 5(c).   The view is looking from the M-band towards the Z-line. For details see text.

rather than heads from adjacent myosin molecules interacting as in insect muscle (Fig. 8(b)).

However, in both cases the heads appear to be organised such that their motor domains are close to the orientation needed to attach to actin in the conventional motor domain docking position (Rayment *et al*, 1993b).   In insect flight muscle only the projecting heads, that is half the total number, are set to interact with actin. It is also known that even in rigor *Lethocerus* muscle only about 80% of the heads can attach to actin (Lovell *et al*, 1981).   The situation in vertebrate muscle is very different (Fig. 8(a)). Here the motor domains of almost all the heads are in suitable positions for actin attachment.   It is also known that, unlike IFM, in rigor vertebrate muscle 100% of the heads attach to actin (Lovell *et al*, 1981).   These results together suggest a number of conclusions: (i) That active IFM may have relatively few heads attached compared with active vertebrate muscle.   (ii) That different muscles, despite having myosin filament backbones built up according to related geometrical schemes (Squire, 1973; 1986), have the heads on the myosin filaments in different configurations that may be optimised for differing functional needs of the muscle. These might be, for example, participation in stretch-activation versus Ca^{2+}- or phosphorylation-dependent activation.

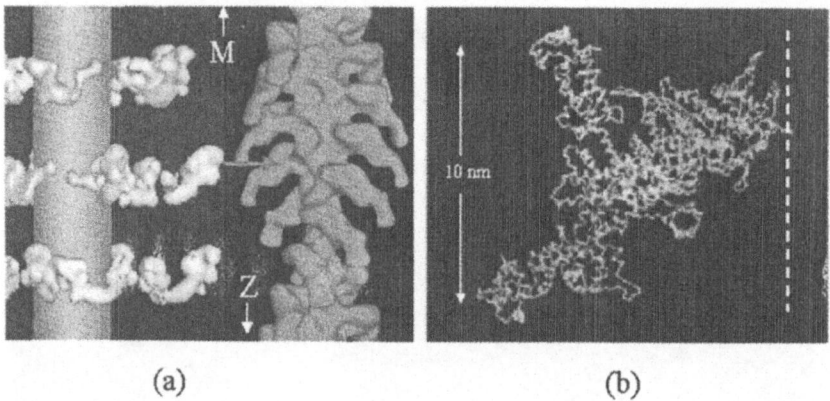

(a)   (b)

Figure 7. Part of the head arrangement in Fig. 5(c) inverted to put the M-band direction towards the top
and next to a 3D reconstruction of an actin filament labelled with S1. In the muscle these two filaments would
be much closer together. However, it shows that the motor domain only needs to move out radially, actually
by only about 20 Å, to attach to a suitably oriented adjacent actin monomer. Once attached the neck domain
of the head could swing to the nucleotide-free conformation. The difference between these two structures is
illustrated in (b). The neck domain of the head tilts axially to give a movement of its outer end of about
100 Å.

Such a conclusion reinforces the suggestions of Padron *et al* (1998) and Offer *et al*
(2000) where there appears to be yet another configuration of myosin heads, this time in
tarantula thick filaments. Finally, (iii) some resting muscles (e.g. fish) have resting
heads in the nucleotide-free shape (Rayment *et al*, 1993a), despite the presence of
nucleotide, whereas others (e.g. IFM) have their heads in a shape similar to the putative
M.ADP.Pi shape of Houdusse *et al* (2000) and Dominguez *et al* (1998). Xu *et al* (1999)
have previously suggested that the M.ADP.Pi state is necessary for good helical ordering
of the myosin heads even in relaxed vertebrate muscle.

3.4   Time-Resolved X-ray Diffraction Studies of Contracting Fish Muscle

With clear results coming from low-angle X-ray diffraction studies both about
myosin head shape and myosin head configuration in resting muscle, we are currently
seeking to interpret our 1 ms time-resolved X-ray diffraction results from contracting fish
muscle (e.g. Harford and Squire, 1992; 1997; Squire, 2000) in terms of possible myosin
head movements and shape changes. These time-resolved diffraction results have already
indicated that for proper interpretation they require the existence of at least two
structurally different states of attachment of myosin heads on actin (Harford and Squire,
1992; Squire, 1997). On present evidence we believe it is likely that these two
structurally different states are more or less equivalent to the weak- and strong-binding
states referred to by many others (see e.g. Holmes *et al*, 1996; Squire, 1997; Houdusse
and Sweeney, 2001). These may well be similar to the two shapes seen in Figure 7(b)
with the neck having two different tilts on the motor domain. The current analysis can
also provide estimates of the relative populations of these two states and the amount of
force generated by each. Details of this analysis will be presented elsewhere (Knupp,
Harford, Mok & Squire, ms in preparation).

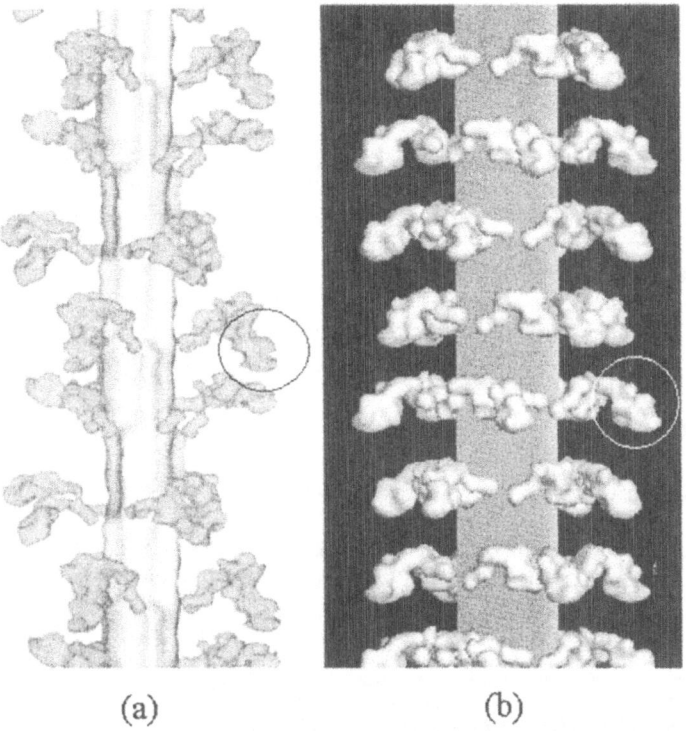

(a)   (b)

Figure 8.   Comparison of the resting head arrangements on the myosin filaments in (a) bony fish muscle (Hudson *et al*, 1997) and (b) *Lethocerus* flight muscle (AL-Khayat *et al*, ms submitted).   Although the head shapes are modelled to be different, the two structures contrive to get their head motor domains in similar configurations relative to actin (i.e. in the right position to attach to actin in the Rayment *et al* (1993b) conformation), as shown by the circled heads.   The M-band direction is downwards in this view.

4.   CONCLUSION

The purpose of this paper has been to show the real power of the low-angle X-ray diffraction method when used to study muscle contraction, especially when it is applied to the uniquely highly ordered structures in bony fish muscles and insect flight muscles. The advantage of this approach is that X-ray diffraction can be carried out on intact, normally functioning, muscle, that it has very high time-resolution in the millisecond time regime or better, that the diffraction data can be analysed in rigorous ways to provide meaningful structural information, that the sensitivity of the method is very great now that the atomic arrangements within the main components of the contractile machinery are known, and finally, that the populations of heads attached to actin in active muscle can be determined, as can the relative tensions that they produce.

How these structural states relate to the biochemistry of the actin-myosin ATPase cycle (Hibberd and Trentham, 1986) will still need to be defined.

ACKNOWLEDGEMENTS

We are indebted to Dr. E. P. Morris for many helpful discussions and for making available his electron micrograph information on isolated insect flight muscle myosin filaments. We thank Bruce Baumann for assistance in collecting fibre X-ray diffraction patterns from IFM at the Argonne/ APS/ BioCAT beamline. We acknowledge specific support to JMS for the insect work from a UK BBSRC project grant (# 28/S10891) and for the fish work from the Wellcome Trust (#061729). MKR was supported by NIH AR-14317. CCP13 software was developed as part of UK BBSRC/ EPSRC funded projects (e.g. # 28/B10368 & 28/B15281). Use of the Advanced Photon Source was supported by the U.S. Department of Energy, Basic Energy Sciences, Office of Energy Research, under Contract No. W-31-109-ENG-38. BioCAT is a U.S. National Institutes of Health-supported Research Centre RR08630.

REFERENCES:

AL-Khayat, H.A., Yagi, N. & Squire, J.M., 1995, Structural changes in actin-tropomyosin during muscle regulation:

  Computer modelling of low-angle X-ray diffraction data. *J. Mol. Biol.* **252**, 611-632.

Cantino, M. E., Brown, L. D., Chew, M., Luther, P. K. and Squire, J. M., 2000, A-band architecture in vertebrate

  skeletal muscle: polarity of the myosin head array. *J. Mus. Res. CellMotil.* **21**, 681-690.

Chen, L.F, Winkler, H., Reedy, M.K., Reedy, M.C. and Taylor, K.A., 2002, Molecular modelling of averaged rigor

  crossbridges from tomograms of insect flight muscle. *J. Struct. Biol.* **138**, 92-104.

Dominguez, R., Freyzon, Y., Trybus, K. M. and Cohen, C., 1998, Crystal structure of a vertebrate smooth muscle

  myosin motor domain and its complex with the essential light chain: visualization of the pre-power stroke state. *Cell*

  94, 559-571.

Eakins, F., AL-Khayat, H. A., Kensler, R. W., Morris, E. P. and Squire, J. M., 2002, 3D Structure of Fish Muscle

  Myosin Filaments. *J. Struct. Biol.* **137**, 154-163.

Harford, J. and Squire, J. M., 1997, Time-resolved diffraction studies of muscle using synchrotron radiation. *Rep. Prog.*

  Phys. **60**, 1723-1787.

Harford, J. J. and Squire, J. M., 1992, Evidence for structurally different attached states of myosin cross-bridges on actin

  during contraction of fish muscle. *Biophys. J.* **63**, 387-396.

Hibberd, M. G. and Trentham, D. R., 1986, Relationships between chemical and mechanical events during muscular

  contraction. *Ann. Rev. Biophys. & Biophys. Chem.* **15**, 119-161.

Holmes, K. C., 1996, Muscle proteins-their actions and interactions. *Curr. Opin Struct. Biol.* **6**, 781-789.

Holmes, K. C., Popp, D., Gebhard, W., and Kabsch, W., 1990, Atomic model of the actin filament. *Nature* **347**, 44-49.

Hopkins, S. C., Sabido-David, C., van der Heide, U. A., Ferguson, R. E., Brandmeier, B. D., Dale, R. E.,

  Kendrick-Jones, J., Corrie, J. E., Trentham, D. R., Irving, M., and Goldman, Y. E., 2002, Orientation

  changes of the myosin light chain domain during filament sliding in active and rigor muscle. *J. Molec. Biol.*

  318, 1275-1291.

Houdusse, A., Kalabokis, V.N., Himmel, D., Szent-Gyorgyi, A.G., and Cohen, C., 1999, Atomic structure of

  scallop myosin subfragment S1 complexed with MgADP: A novel conformation of the myosin head. *Cell* **97**,

  459-470

Houdusse, A., and Sweeney, H. L., 2001, Myosin motors: missing structures and hidden springs. *Curr. Opin Struct.*

  Biol. **11**, 182-194.

Houdusse, A., Szent-Gyorgyi, A. G., and Cohen, C., 2000, Three conformational states of scalop myosin S1. Proc.

Natl Acad Sci. U S A **97**, 11238-11243.

Hudson, L., Harford, J. J., Denny, R. C., and Squire, J. M., 1997, Myosin head configuration in relaxed fish muscle: resting state myosin heads must swing axially by up to 150 A or turn upside down to reach rigor. *J. Molec. Biol* **273**, 440-455.

Irving, M., Piazzesi, G., Lucii, L., Sun, Y. B., Harford, J. J., Dobbie, I. M., Ferenczi, M. A., Reconditi, M., and Lombardi, V., 2000, Conformation of the myosin motor during force generation in skeletal muscle. *Nat. Struct. Biol*

7, 482-485.

Kabsch, W., Mannherz, H.G., Suck, D., Pai, E.F., and Holmes, K.C., 1990, Atomic structure of the actin: DNase I

complex. *Nature* **347**, 37-44

Kensler, R. W., and Stewart, M., 1989, An ultrastructural study of crossbridge arrangement in the fish skeletal muscle

thick filament. *J. Cell Sci.* **94**, 391-401.

Lovell, S.J., Knight, P.J., and Harrington, W.F., 1981, Fraction of myosin heads bound to thin filaments in rigor filaments from insect flight and vertebrate muscles. *Nature*, **293**, 664-666.

Luther, P. K., and Squire, J. M., 1980, Three-dimensional structure of the vertebrate muscle A-band. II. The myosin

filament superlattice. *J. Molec. Biol* **141**, 409-439.

Luther, P. K., Squire, J. M., and Forey, P. L., 1996, Evolution of myosin filament arrangements in vertebrate skeletal

muscle. *J. Morphol* **229**, 325-335.

Morris, E. P., Squire, J. M., and Fuller, G. W., 1991, The 4-stranded helical arrangement of myosin heads on insect

(*Lethocerus*) flight muscle thick filaments. *J. Struct. Biol* **107**, 237-249.

Offer, G., Knight, P.J., Burgess, S.A., Alamo, L., and Padron, R., 2000, A new model for the arrangement of myosin

molecules in tarantula thick filaments. *J. Mol. Biol* **298**, 239-260.

Padron, R., Alamo, L., Murgich, J., and Craig, R., 1998, Towards an atomic model of the thick filaments of muscle. *J. Mol. Biol* **275**, 35-41.

Rayment, I., Rypniewsky, W. R., Schmidt-Bäse, K., Smith, R., Tomchick, D. R., Benning, M. M., Winkelmann, D. A.,

Wesenberg, G., and Holden, H. M., 1993a, Three-dimensional structure of myosin subfragment-1: a molecular motor. *Science* **261**, 50-58.

Rayment, I., Holden, H. M., Whitaker, M., Yohn, C. B., Lorenz, M., Holmes, K. C., and Milligan, R. A., 1993b, Structure of the actin-myosin complex and its implications for muscle contraction *Science* **261**, 58-65.

Reedy, M. K., Squire, J. M., Baumann, B. A. J., Stewart, A., and Irving, T. C., 2000, X-ray Fibre Diffraction of the Indirect Flight Muscle of *Lethocerus indicus*. In Advanced Photon Source User Activity: Report 2000 (Argonne, IL,

Argonne National Laboratory).

Schmitz, H., Reedy, M. C., Reedy, M. K., Tregear, R. T., and Taylor, K. A., 1997, Tomographic three-dimensional reconstruction of insect flight muscle partially relaxed by AMPPNP and ethylene glycol. *J. Cell Biol* **139**, 695-707.

Squire, J.M., 1973, General model of myosin filament structure III: molecular packing arrangements in myosin

filaments. *J. Mol. Biol* **77**, 291-323.

Squire, J.M., 1981, The Structural Basis of Muscular Contraction Plenum Press, NY.

Squire, J.M., 1986, Muscle myosin filaments: Internal structure and crossbridge organisation.

Comments Molec. Cell Biophys. **3**, 155-177.

Squire, J.M., 1997, Architecture and function in the muscle sarcomere. *Curr. Opin Struct. Biol* **7**, 247-257.

Squire, J. M., 1998, Time-resolved X-ray diffraction In *Current Methods In Muscle Physiology*, H. Sugi, ed. (Oxford,

Oxford Univ. Press), pp. 241-285.

Squire, J. M., 2000, Fibre and Muscle Diffraction In *Structure and Dynamics of Biomolecules*, E. Fanchon, E. Geissler,

L.-L Hodeau, J.-R Regnard, and P. Timmins, eds. (Oxford, UK, Oxford Univ. Press), pp. 272-301.

Squire, J. M., Cantino, M., Chew, M., Denny, R., Harford, J., Hudson, L., and Luther, P., 1998, Myosin rod-packing schemes in vertebrate muscle thick filaments. *J. Struct. Biol.* **122**, 128-138.

Taylor, K. A., Schmitz, H., Reedy, M. C., Goldman, Y. E., Franzini-Armstrong, C., Sasaki, H., Tregear, R. T., Poole, K.
   J. V., Lucaveche, C., Edwards, R. J., Chen, L.F, Winkler, H., and Reedy, M.K., 1999, Tomographic 3-D reconstruction of quick frozen, Ca^{2+}-activated contracting insect flight muscle. *Cell* **99**, 421-431.

Van Heel, M., Gowen, B, Matadeen, R., Orlova, E.V., Finn, R., Pape, T., Cohen, D., Stark, H., Schmidt, R., Schatz, M., and Patwardan, A., 2000, Single particle electron microscopy: towards atomic resolution. *Quart. Rev. Biophys.* **33**, 307-369.

Xu, S., Gu, J., Rhodes, T., Belknap, B., Rosenbaum, G., Offer, G., White, H., and Yu, L. C., 1999, The M.ADP.P(i) state is required for helical order in the thick filaments of skeletal muscle. *Biophys. J.* **77**, 2665-2676.

DISCUSSION

Cecchi: Since you are working on whole muscle, I presume there is some shortening during the tetanic tension development and this could change the 10 and 11 intensities.

Squire: As you say, we were using whole muscle and therefore could not measure the sarcomere length directly. However there was only a very small change in interfilament spacing in these experiments, and the change that occurred was complete in a few milliseconds, being consistent with little change in sarcomere.

Cecchi: In frog muscle, there is a lattice compression due to cross-bridge force. If this also occurs in your muscle it could obscure a possible expansion due to shortening.

Squire: We have no evidence for that in fish muscle.

Ranatunga: Your modeling showed possible interaction between adjacent cross-bridges, in insect muscle. Could that be a mechanism for stretch activation?

Squire: Yes, it is possible but still speculative. Heads from adjacent myosin molecules in a crown interact with one head being at lower radius and behind a protecting head from the neighboring molecule. The essential light chain of the outer head interacts with the ATP binding site of the inner head. A small increase in calcium level is enough to set up insect flight muscle to become active. Perhaps the outer heads can bind actin at calcium, but stretch moves them away from the inner head such that the cross-bridge cycle can proceed. Other mechanisms can also be envisaged based on this new structure.

USE OF SINUSOIDAL LENGTH OSCILLATIONS TO DETECT MYOSIN CONFORMATION BY TIME-RESOLVED X-RAY DIFFRACTION

Giovanni Cecchi, M. Angela Bagni, Barbara Colombini, Christopher C. Ashley, Heinz Amenitsch, Sigrid Bernstorff, and Peter J. Griffiths[*]

1. INTRODUCTION

In contrast to many other theories of muscle contraction, the proposal of H.E. Huxley (Huxley, 1969), that force generation by actomyosin systems results from an active tilting of the S1 moiety of myosin, defined a structural change as the central event of the force generation process. As a result, a large range of techniques for probing structural changes in myosin have been applied to the study of contraction, but have yielded only equivocal support for the tilting S1 theory (Yanagida, 1981; Cooke et al., 1982; Tanner et al., 1992). Uniquely, because of the highly ordered structure of striated muscles, X-ray diffraction provides a powerful probe of structural events in this tissue, and the enormous advances over the last 40 years in both detector technology and intensity of X-ray sources available have permitted time-resolved X-ray diffraction studies of intact, working muscle cells to advance from the time domain of hours to that of microseconds.

The meridional X-ray reflection at 14.5nm from active muscle is thought to sample the structure of actin-bound S1 moieties, and undergoes large changes in intensity (I_{M3}) during the elastic and synchronized power stroke components of the force response induced by a rapid length change of contracting muscle (Huxley et al., 1983; Irving et al., 1992). I_{M3} reaches maximum intensity ($I_{M3\ max}$) following a small release; for larger releases and for stretches, I_{M3} falls. These changes in I_{M3} are thought to arise partially from active changes in tilt of the lever arm domain of S1, since crystallographic studies of S1, mimicking different stages of the power stroke by binding different ATP analogues at the active site, show different orientations of the lever arm domain (Houdusse and Sweeney, 2001), and optical probes attached to this domain indicate angular

[*] G.C, M.A.B. and B.C., Dipartimento di Scienze Fisiologiche, Università degli Studi di Firenze, Viale G.B. Morgagni 63, Florence I-50134, Italy; C.C.A. and P.J.G. University Laboratory of Physiology, Parks Road, Oxford OX1 3PT, U.K.; H.A. Institute of Biophysics and X-ray Structure Research, Austrian Academy of Sciences, Schmiedlstraße 6, A-8042 Graz, Austria. S.B. Sincrotrone Elettra Trieste S.C.p.A., S.S. 14-km. 163.500, AREA Science Park, I-34012 Basovizza TS, Italy.

Molecular and Cellular Aspects of Muscle Contraction
Edited by H. Sugi, Kluwer Academic/Plenum Publishers, 2003

displacement occurs during the power stroke (Corrie et al., 1999) whilst probes on the motor domain have failed to detect tilting in this part of the molecule (Yanagida, 1985). In addition, I_{M3} changes are also thought to arise from elastic tilting or bending of the lever arm (Irving et al., 1992).

We have explored the relation between length changes and I_{M3} using sinusoidal perturbations, which allow continuous monitoring of I_{M3} during changes in fibre length and permit higher time resolution of I_{M3} signals than is possible with steps.

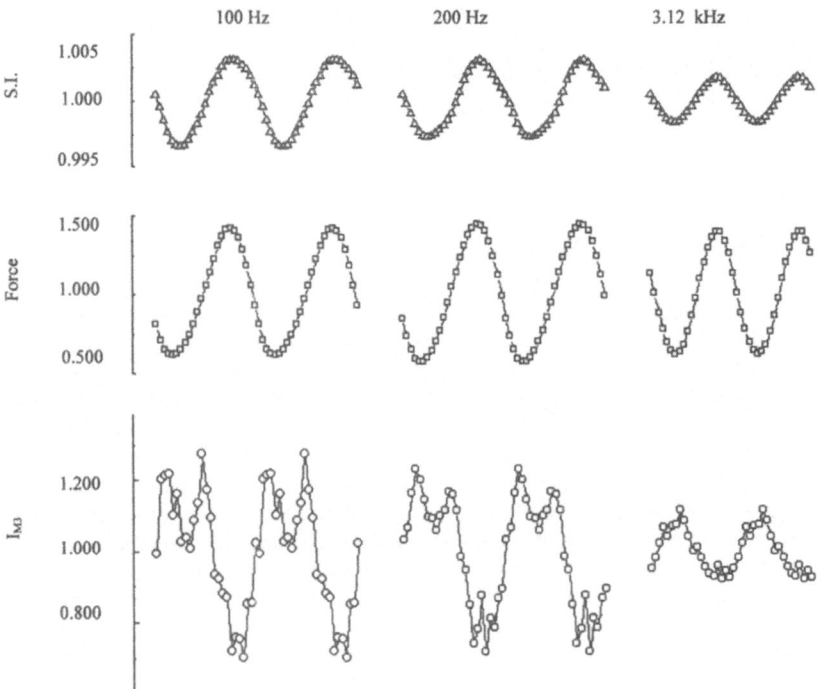

Figure 1. Sarcomere length, force and I_{M3} changes during sinusoidal length oscillations at various frequencies. Amplitude of length oscillation chosen to obtain about the same peak to peak change in force at each frequency. I_{M3} signals were averaged over one oscillation period, but plotted over two periods to clarify waveform shape. Force is normalised to isometric tension, I_{M3} to maximum values during oscillations. X-ray data was collected at beamline SAXS (Elettra, Trieste, Italy). Radiation (λ= 0.15nm, beam dimensions 0.5mm x 3mm) passed through the chamber via Kapton windows (thickness 12µm) positioned within 0.2mm of the fibre. Meridional intensity was captured on a 1D delay line X-ray detector (camera length 2.46m). X-ray spectra were collected at intervals of one twentieth of a length oscillation period (i.e. from 500 to 16µs) by a controlling computer. X-ray data was averaged over a 500ms oscillation period per tetanus and over 20-100 tetani.

2. EFFECTS OF FREQUENCY OF SINUSOIDAL LENGTH OSCILLATIONS

The recovery of tension following a length step, which is thought to represent a synchronized S1 power stroke, approximates to an exponential process with a rate constant of the order of $1000s^{-1}$. We therefore applied length oscillations to bundles of 1 to 8 intact fibres from tibialis anterior muscles of Rana temporaria over a range of frequencies from 100Hz to 3kHz, where the contribution of the power stroke to the force response should be varying considerably, while adjusting oscillation amplitude to maintain a constant peak to peak change in force (and therefore constant elastic displacement of the lever arm). Representative results showing force and I_{M3} from these experiments are shown in fig. 1. It can be seen that at 3kHz oscillation frequency (as in general at a frequencies higher than 1kHz) responses approximated closely to a sinusoid, in phase with length changes, with peak intensity reached at maximum shortening (Dobbie et al., 1998; Griffiths et al., 1998). At frequencies less than 1kHz, the I_{M3} signal became a distorted sinusoid, a second intensity minimum appearing at maximum shortening (Bagni et al., 2001). This minimum became progressively deeper as frequency was reduced and at 200Hz and 100Hz (fig. 1) forming a distinct double peak in the I_{M3} signal at the point at which I_{M3} was maximal above 1kHz.

Since distortion was frequency dependent and elastic displacement of S1 was maintained constant at all frequencies, this showed that distortion depended on the contribution of the power stroke to the force signal at a given frequency. When the amplitude of oscillations imposed on the preparation was reduced, distortion of I_{M3} was also reduced. In order to interpret these findings, we designed a simulation of force responses and I_{M3} signals with which we could investigate the effects of different parameters on the I_{M3} signal during oscillations.

3. SIMULATION OF FORCE AND I_{M3} RESPONSE

To simulate the force response to oscillations we used a system similar to that described in Ford et al. (1977). Equations describing the behavior of this (viscoelastic) system, including the dependence of power stroke kinetics on fibre length, were solved using a Runge-Kutta algorithm (Griffiths et al., 2002). For sinusoidal input to the system, ten cycles of oscillation were permitted to allow the decay of transient states. The quality of simulation was evaluated by comparison of the phase relations between force and fibre length, and of the apparent stiffness of simulated and experimental responses at different frequencies, measured as the amplitude of an instantaneous release just sufficient to discharge isometric tension (y_o, see table 1). Agreement was good between these indices of simulation performance, though at high frequencies the mean tension during simulations was slightly elevated in comparison to experimental findings, so mean simulated tension was adjusted in such cases to correspond to the experimental data. To simulate I_{M3} signals, we first used a construction of overlapping spheres (fig. 2) to simulate S1.

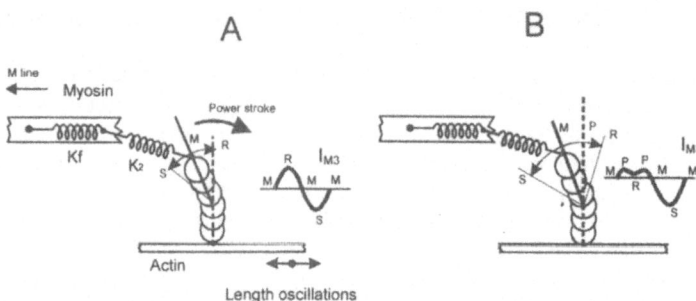

Figure 2. Simulation of I_{M3} changes during sinusoidal oscillations with a simplified S1 model, to small (A) and great (B) length oscillations. Double peak distortion is present only in B when myosin tail during oscillations passes through the point at which I_{M3} is maximal (tail perpendicular to filament axis).

While the physical similarity of this structure to S1 is limited, at 14.5nm resolution its behavior may be expected to be very similar, and it makes no assumptions about the disposition and orientation at which S1 binds to actin. Typically, we used 6 spheres of 2.5 nm radius, designating the upper 3 spheres to represent the lever domain, and allowing these spheres to tilt rigidly to produce an angular displacement of the lever relative to the axis of alignment of the lower 3 spheres (Bagni et al., 2001).

The extent of this displacement was dependent on muscle load, taking the compliance in series with S1 to be half the total compliance of the muscle. We then

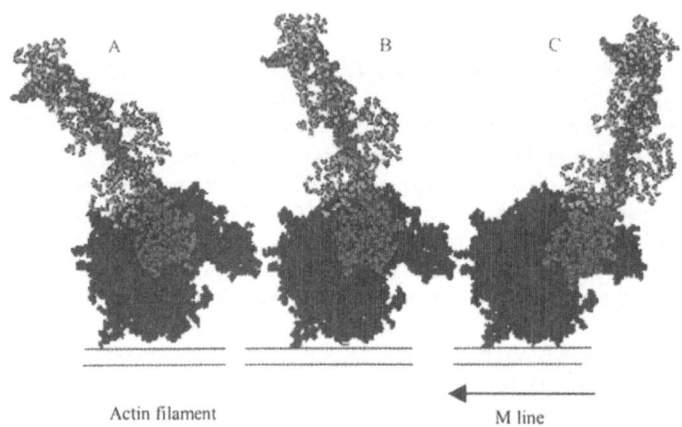

Figure 3. Myosin disposition under various conditions. A, 9nm away from Rayment structure in the stretch direction; B, mean disposition during sinusoidal oscillations at tetanus plateau; C, nucleotide free (Rayment structure).

computed the squared modulus of the Fourier transform of 49 structures of this kind on either side of the M-line, separated by 14.5nm, to obtain I_{M3}. Subsequently, we simulated I_{M3} using the Fourier transform of the α-carbon chains (see fig. 3) of chicken striated muscle S1 structure determined by Rayment et al. (1993a), orientated with respect to actin as suggested in Rayment et al. (1993b). As shown in fig. 4 the agreement between the two types of simulation was excellent.

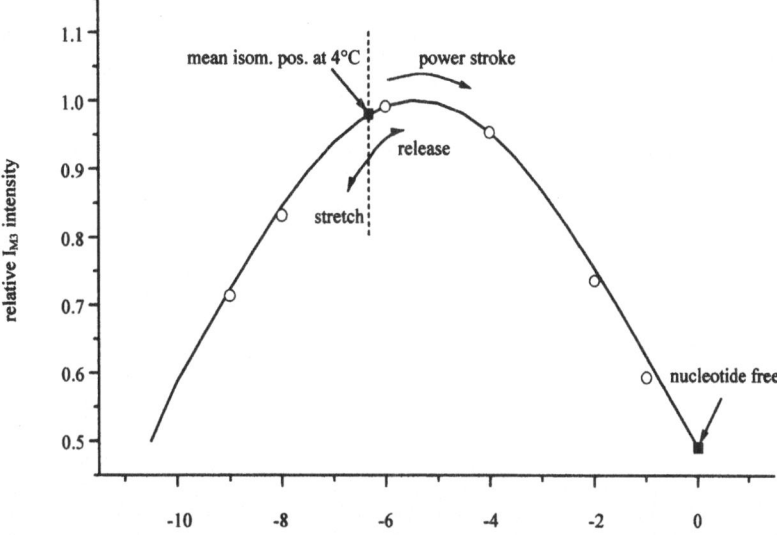

Figure 4. Dependence of I_{M3} from displacement of the lever arm tip respect to Rayment conformation, indicated on the figure as nucleotide free. Continuous line: model with carbon atoms; empty circles: model with 6 spheres. Mean position during oscillations is shifted by 0.82nm with respect to $I_{M3\,max}$ in the stretch direction.

In these simulations, the distortion of the I_{M3} signal during oscillations can be accounted for by the lever passing through its orientation at I_{M3max} during the part of the sinusoid over which the fibre shortens. This can explain the dependence of the distortion on oscillation frequency; at low frequencies, the power stroke has a greater time in which to proceed towards completion, and the total displacement of the tail domain is sufficient to carry it through its $I_{M3\,max}$ disposition. At high frequencies, the power stroke has insufficient time to proceed appreciably, its contribution to lever displacement is suppressed, and total lever movement is predominantly elastic and too small to pass through $I_{M3\,max}$. This effect is illustrated in table 1, where simulated total lever displacement for an oscillation amplitude giving a peak to peak force oscillation equal to isometric tension at each frequency is shown. For example, at 100Hz, axial displacement of the lever tip is 7.1nm during oscillations, of which 76% is accounted for by the power stroke. In contrast, at 2.8kHz, total displacement is only 2.8nm, 26% of which is now accounted for by the power stroke event.

4. LEVER ARM TILT DURING ISOMETRIC CONTRACTIONS

If $I_{M3\,max}$ is reached at particular disposition of the lever arm, by measuring the lever arm displacement required to reach $I_{M3\,max}$ we may define the relative disposition of the arm for a variety of S1 states. This was done by using the force response obtained during oscillations to drive the simulations of lever arm tilting. The simulated I_{M3} signal was then matched to that observed experimentally by varying the mean tilt of the lever arm. When the correct tilt was found, the axial displacement of the lever arm tip required to reach its tilt at $I_{M3\,max}$ was calculated (Δy). Because absolute changes in I_{M3} differed between simulations and experimental data, we could only match data in which I_{M3} distortion was present (i.e. data obtained at frequencies below 1kHz); undistorted I_{M3} signals could be matched by any lever orientation for which the $I_{M3\,max}$ tilt was not reached during oscillations, and therefore Δy was undefined in the higher frequency range. The estimates of lever tilt at different frequencies are given in table 1. As can be seen, there is little indication of any dependence of the measured tilt on the frequency of applied oscillations. The overall mean Δy for the range 100Hz to 1kHz was 0.82nm.

Table 1. Frequency dependence of muscle stiffness and lever disposition

Frequency (Hz)	(N)	Δy (nm)	Simulated y_o (nm)	Measured y_o (nm)	Total lever motion (nm)	Active fraction (%)
100	1	0.91	7.7	7.9	7.1	76
200	5	0.72 ± 0.19	6.8	6.9	5.9	68
400	1	0.65	6.2	5.8	4.7	60
1000	4	0.96 ± 0.16	5.3	4.6	3.8	48
2800	7	0.68 ± 0.21	4.6	-	2.8	26

5. LEVER TILT DEPENDS ON FORCE PER ACTIN-BOUND S1

When the temperature of a muscle fibre is increased, its force development increases without a corresponding rise in the number of actin-bound S1 moieties (Ford et al., 1977). It follows that at higher temperatures S1 generates more force. This increased force should extend series compliance further than during contraction at low temperature, and hence the tilt of S1 should be altered. As a test of the tilting lever power stroke theory, we determined Δy over a range of temperatures, to determine whether the predicted change in tilt could be detected. Because $I_{M3\,max}$ was reached following a release, where the lever arm is actively tilting to perform additional power stroke work to re-extend series compliance, it was expected that a higher force per S1 at higher temperatures would tilt the lever closer to its $I_{M3\,max}$ orientation. We therefore examined the I_{M3} response to oscillations at 1kHz, since this frequency is the highest at which distortion could be observed at low temperature, and therefore any change in lever tilt towards $I_{M3\ max}$ would be expected to produce a marked increase in I_{M3} distortion. Fig. 5 shows experimental and simulated records of I_{M3} in the same fibre bundle at 4, 12 and 18 °C. In accordance with the predicted shift towards $I_{M3\ max}$ as temperature increased, we found a marked increase in distortion as temperature rose. We then used the simulation to test whether this effect could be accounted for solely by an increase in the speed of power stroke kinetics due to the high temperature, but were unable to duplicate the observed I_{M3} responses in this way as can be seen on fig. 6.

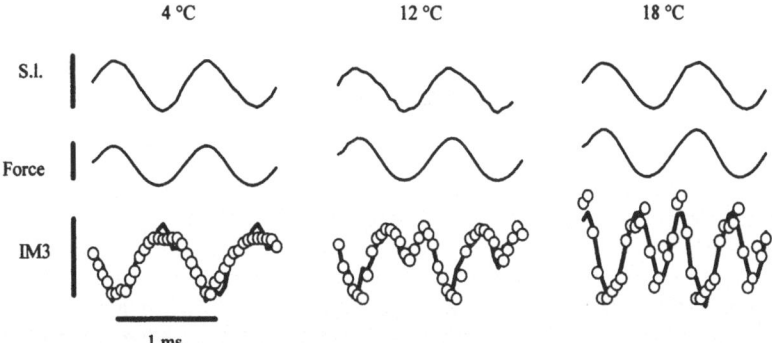

Figure 5. Comparison of experimental (empty squares) and calculated I_{M3} changes (lines) during oscillations (1 kHz frequency) at various temperatures. The double peak distortion increases with temperature and at 18 °C I_{M3} changes appear almost doubled in frequency. Bar calibrations represent: at 4°C, 6nm/hs, P_0 , 0.2 $I_{M3\ max}$, at 12 °C, 5nm/hs, 0.86P_0 ,0.16 I_{M3max}, at 18°C, 7.9nm/hs,1.05P_0, 0.22 IM_{3max}.

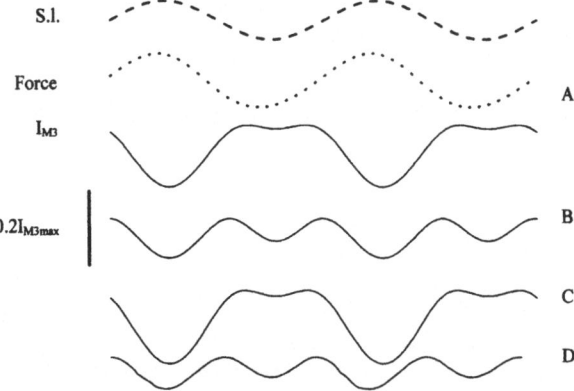

Figure 6. Simulation of the effects of temperature on I_{M3} changes during sinusoidal length oscillation (thick continuous lines) at 1kHz frequency. A and B: simulation of the response at 4°C and 18°C; C same tail position as at 4°C, but kinetics of quick recovery as at 18°C; D, position as in B, but kinetics as at 4 °C.

Instead, we found that the increased distortion was well accounted for by a reduction in Δy from 0.96nm at 4°C to 0.23nm at 22°C.

Since the corresponding force increase over this temperature range was $0.28P_0$, this enabled us to estimate the series stiffness against which S1 was acting as $0.28P_0/(0.96-0.23)$nm $= 0.38P_0.\text{nm}^{-1}$, which is rather lower than the known stiffness of the myofilaments acting in series with S1 (approximately $0.5 \ P_{0}.\text{nm}^{-1}$) and suggested the presence an additional elasticity within S1 (Griffiths et al., 2002). However, the data from table 1 point to a smaller value of Δy at 4°C (0.82nm), which gives a series stiffness of $0.47 \ P_0.\text{nm}^{-1}$, locating almost all cross-bridge related compliance within the S1 lever arm.

6. DETERMINATION OF Δy AT HIGH FREQUENCIES

Our simulations suggested that the absence of distortion of the I_{M3} signal at high frequencies resulted from the suppression of the power stroke contribution to lever movement as frequency increased. As the frequency of oscillation approaches the speed of the power stroke, the power stroke contributes less to displacement of the lever arm, which becomes increasingly accounted for by elastic distortion, and total displacement becomes smaller and insufficient to reach $I_{M3 \ max}$. If this interpretation is correct, increasing oscillation amplitude should be sufficient to restore distortion by replacing the lost power stroke contribution to lever displacement by a larger elastic contribution. However, this was not feasible with tibialis fibres because of intrinsic limitations in performance of the stretcher motor. To overcome this difficulty, we applied high frequency oscillations to fibres from the dorsal interossei muscle of *Rana temporaria*, which are 30% shorter than tibialis., and therefore experience a proportionally larger sarcomere length change for the same amplitude of applied oscillations. As shown by fig. 6, at 2.8 kHz with this preparation, when using an oscillation amplitude great enough, we were able to observe the same distortion present at lower frequencies in tibialis fibres (fig. 7a). Again, we simulated the I_{M3} response, and obtained a value for Δy of 0.68nm (table 1). Because oscillations at this frequency caused a small rise in mean force ($\leq 0.05P_o$), which would cause the lever to move closer to $I_{M3 \ max}$, we corrected Δy for this elevation of tension, obtaining a value of 0.81nm at P_o, very close to the mean value from lower oscillation frequencies. Reducing the amplitude of the length oscillations causes the distortion to disappear (fig. 7b)

7. RELATION TO OTHER STUDIES

Wakabayashi et al. first reported distortion of I_{M3} signals during length oscillations (Wakabayashi et al., 1986; Mitsui et al., 1994), but the oscillation frequencies applied did not exceed 10Hz, where distortion might be partially accountable for by cross-bridge detachment. In contrast, experiments at high frequencies (3kHz, Dobbie et al., 1998; 1kHz, Griffiths et al., 1998) failed to detect distortion in I_{M3} signals. Our findings now show that this was the result of application of insufficiently large oscillation amplitudes.

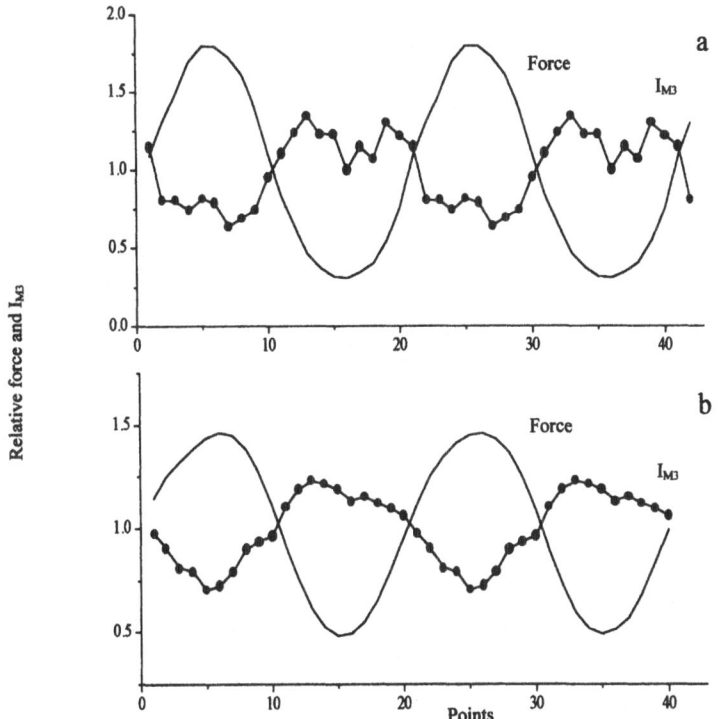

Figure 7. Experimental records at 2.8kHz oscillation frequency at two different oscillation amplitude. In **a** the oscillations (6.44nm/hs p-p amplitude) produced a peak to peak force change of 1.5P_0. In **b** the amplitude was 4.3nm/hs and the force change was about equal to P_0.

The fact that distortion decreases so markedly over the frequency range corresponding to the quasi-exponential power stroke recovery of force after a length step, and the restoration of distortion by increased oscillation amplitude at oscillation frequencies >1kHz, where power stroke contributions to I_{M3} are strongly suppressed, is cogent evidence for similar lever tilting both as a result of the power stroke and as a result of the lever's intrinsic elasticity, and for the summation of these contributions to angular displacement of the arm in response to a load change.

Our simulations locate lever elasticity in the lever-motor domain junction. Although it is unlikely that the remainder of the actomyosin complex is completely rigid, there are grounds for location of a major source of compliance in this region (Houdusse and Sweeney, 2001), and since the displacement of the lever arm accompanying the additional force development at higher temperatures seems almost entirely explicable in terms of extension of filament compliance, this seems good evidence for allocation of S1 compliance chiefly to the lever arm domain.

Our estimate of Δy would rotate the lever arm through an angle of 35° relative to the rigor state of Rayment et al. (1993a), 0.82nm from the $I_{M3\,max}$ orientation.

From this position, a muscle shortening of 6.3nm per half sarcomere would be required to reach the Rayment orientation, consistent with the size of release just sufficient to prevent any power stroke recovery of tension in contracting muscle. This orientation is ca. 1nm closer to the $I_{M3\,max}$ orientation than the disposition proposed by Irving et al. (2000) at the same temperature, evaluated by step length changes. Although oscillations can elevate mean tension, moving the lever closer to $I_{M3\,max}$, such an elevation was corrected for at very high oscillation frequencies and was virtually absent at lower frequencies. We therefore find this to be an unlikely explanation for this discrepancy. It is possible that, since step length changes sample the relation between I_{M3} and fibre length chiefly before an after the step, and not continuously as with oscillations, they may be susceptible to additional factors influencing the intensity signal such as disorder. These effects are unlikely to account for intensity changes in the oscillation signal, since the double peak at maximum shortening can only be convincingly accounted for by tilting of the lever arm through its $I_{M3\,max}$ orientation.

8. CONCLUSIONS

By use of $I_{M3\,max}$ to define a reference structure for S1, we are able to determine the lever arm disposition during isometric contraction. At 4°C, this disposition corresponds to a rotation of the lever through 35° from its rigor tilt. Our findings show that this disposition depends on the force being exerted by S1, in agreement with the proposed mechanism of contraction whereby force results from a torque generated by the tilting of part or all of the S1 moiety of myosin. We also find that almost all S1 compliance is associated with the lever arm.

9. REFERENCES

Bagni, M.A., Colombini, B., Amenitsch, H., Bernstorff, S., Ashley, C.C., Rapp, G., and Griffiths, P.J., 2001, Frequency dependent distortion of meridional Intensity changes during sinusoidal length oscillations of activated skeletal muscle, *Biophys. J.*, **80**:2809-2822.

Cooke, R., Crowder, M.S., and Thomas, D.D., 1982, Orientation of spin labels attached to cross-bridges in contracting muscle fibres, *Nature*, **300**:776-778.

Corrie, J.E.T., Brandmeier, B.D., Ferguson, R.E., Trentham, D.R., Kendrick-Jones, J., Hopkins, S.C., van der Heide, U.A., Goldman, Y.E., Sabido-David, C., Dale, R.E., Criddle S., and Irving M., 1999, Dynamic measurement of myosin light-chain-domain tilt and twist in muscle contraction, *Nature*, **400**:425-430.

Dobbie, I., Linari, M., Piazzesi, G., Reconditi, M., Koubassova, N., Ferenczi, M.A., Lombardi, V., and Irving, M., 1998, Elastic bending and active tilting of myosin heads during muscle contraction, *Nature*, **396**:383-387.

Ford, L.E., Huxley, A.F., and Simmons, R.M., 1977, Tension responses to sudden length changes in stimulated frog muscle fibres near slack length, *J. Physiol.*, **269**:441-515.

Griffiths, P.J., Amenitsch, H., Ashley, C.C., Bagni, M.A., Bernstorff, S., Cecchi, G., Colombini, B., and Rapp, G., 1998, Studies on the 14.5nm meridional X-ray reflection during length changes of intact frog muscle fibres, *Adv. Exp. Med. Biol.*, **453**:247-257.

Griffiths, P.J., Bagni, M.A., Colombini, B., Amenitsch, H., Bernstorff, S., Ashley, C.C., and Cecchi, G., 2002, Changes in myosin S1 orientation and force induced by a temperature increase, *Proc. Nat Acad. Sci. USA.*, **99**:5384-5389.

Houdusse, A., and Sweeney, H.L., 2001, Myosin motors: missing structures and hidden springs, *Curr. Opin. Struct. Biol.*, **11**:182-194.

Huxley, H.E., 1969, The mechanism of muscle contraction, *Science*, **164**:1356-1366.

Huxley, H.E., Simmons, R.M., Faruqi, A.R, Kress, M., Bordas, J., and Koch, M.H.J., 1983, Changes in the X-ray reflections from contracting muscle during rapid mechanical transients and their structural implications, *J. Mol. Biol.*, **169**:469-506.

Irving, M., Lombardi, V., Piazzesi, G., and Ferenczi, M.A., 1992, Myosin head movements are synchronous with the elementary force-generating process in muscle, *Nature*, **357**:156-158.

Irving, M., Piazzesi, G., Lucii, L., Sun, Y-B., Harford, J.J., Dobbie, I.M., Ferenczi, M.A., Reconditi, M., and Lombardi, V., 2000, Conformation of the myosin motor during force generation in skeletal muscle, *Nature Struct. Biol.*, 7:482-485.

Mitsui, T., Wakabayashi, K., Wang, En-Z., Iwamoto, H., Tanaka, H., Kobayashi, T., Hamanaka, T., Amemiya, Y., Sugi, H., Ohshima H., and Hiraoka, W., 1994, Frequency dependence of the variation of the x-ray diffraction pattern from tetanised frog skeletal muscle during sinusoidal length changes, in: *Synchrotron Radiation in the Biosciences*, ed., B. Chance, Clarendon Press, Oxford, 460-472.

Rayment, I., Rypniewski, W.R., Schmidt-Base, K., Smith, R., Tomchick, D.R., Benning, M.M., Winkelman, D.A., Wesenberg, G., and Holden, H.M., 1993a, Three-dimensional structure of myosin subfragment-1: a molecular motor, *Science*, **261**:50-58.

Rayment, I., Holden, H.M., Whittaker, C.B., Yohn, C.B., Lorenz, M., Holmes, K.C., and Milligan, R.A., 1993b, Structure of the actin-myosin complex and its implications for muscle contraction, *Science*, **261**:58-65.

Tanner, J.W., Thomas, D.D., and Goldman, Y.E., 1992, Transients in orientation of a fluorescent cross-bridge probe following photolysis of caged nucleotides in skeletal muscle fibres, *J. Mol. Biol.*, **223**:185-203.

Wakabayashi, K., Tanaka, H., Kobayashi, T., Amemiya, Y., Hamanaka, T., Nishizawa, S., Sugi, H., and Mitsui T., 1993, Time-resolved X-ray study of effect of sinusoidal length change on tetanised frog muscle, *Biophys. J.*, **49**:581-584.

Yanagida, T., 1981, Angles of nucleotides bound to cross-bridges in glycerinated muscle fiber at various concentrations of _-ATP, _-ADP and _-AMP.PNP detected by polarized fluorescence, *J. Mol. Biol.*, **146**:539-560.

Yanagida, T., 1985, Angle of active site of myosin heads in contracting muscle during sudden length changes, *J. Muscle Res. Cell Motil.*, **6**:43-52.

DISCUSSION

Sugi: Around 1985, Drs. Wakabayashi, Mitsui and I also applied sinusoidal oscillations (~10 Hz) to isometrically contracting muscle, and observed that the 14.3 reflection intensity changed sinusoidally with a frequency two times the applied oscillation (Wakabayashi et al., Biophys. J. 49: 581-584). We interpreted the results in terms of two myosin heads; one head is fixed in position while the other head moves around the former head. How do you think about our interpretation based on myosin two heads?

Cecchi: I know you performed sinusoidal oscillation experiments with a much lower frequency than those we used. In our condition, I think our interpretation is more appropriate than yours.

ORIENTATION AND MOTION OF MYOSIN LIGHT CHAIN AND TROPONIN IN RECONSTITUTED MUSCLE FIBERS AS DETECTED BY ESR WITH A NEW BIFUNCTIONAL SPIN LABEL

Toshiaki Arata, Motoyoshi Nakamura, Hidenobu Akahane, Tomoki Aihara, Shoji Ueki, Kazunori Sugata, Hiroko Kusuhara, Masashi Morimoto, and Yukio Yamamoto*

ABSTRACT

Using electron spin resonance, we have studied dynamic structures of myosin neck domain and troponin C by site-directed spin labeling. We observed two broad but distinct orientations of a spin label attached specifically to a single cysteine (cys156) on the regulatory light chain (RLC) of myosin in relaxed skeletal muscle fibers. The two probe orientations, separated by a 25° axial rotation, did not change upon muscle activation, but orientational distributions became narrower substantially, indicating that a fraction of myosin heads undergoes a disorder-to-order transition of the myosin light chain domain upon force generation and muscle contraction. These results provide insight into the mechanism how myosin heads move their domains to translocate an actin filament.

Site-directed spin-labeling was achieved by cysteine residues of human cardiac troponin C (TnC). Spin dipole-dipole interaction showed that free TnC undergoes a global structural change (extended-to-compact) by Ca^{2+} or Mg^{2+}. The spectra from the spin labels at N-terminal half domain were broad and almost identical in parallel and perpendicular orientations of fiber, suggesting that the N-terminal of TnC molecule is flexible or disoriented with respect to the filament axis. We also succeeded, for the first time, in fixing the newly-synthesized bifunctional spin label rigidly on TnC molecule in solution (either in $\pm Ca^{2+}$), giving a promise that we can determine the precise coordinate of the spin principal axis on protein surface.

* Toshiaki Arata, Motoyoshi Nakamura, Tomoki Aihara, Shoji Ueki, Kazunori Sugata, and Hiroko Kusuhara, Department of Biology, Graduate School of Science, Osaka University, Toyonaka, Osaka 560-0043, Japan. Hidenobu Akahane, Masashi Morimoto, and Yukio Yamamoto, Graduate School of Human and Environmental Studies, Kyoto University, Kyoto 605-8501, Japan.

Molecular and Cellular Aspects of Muscle Contraction
Edited by H. Sugi, Kluwer Academic/Plenum Publishers, 2003

1. INTRODUCTION

The ultimate goal in biological functions of molecular motor protein actin-myosin complex and Ca-regulatory protein troponin-tropomyosin-actin complex in muscle has been to find force-generating or switching protein structural changes.   The ESR spectroscopy provides dynamics, orientation, and spatial information of spin-labels attached to the region of protein.   To see how motor and regulatory proteins work in muscle, we have extensively studied on spin-labeled myosin light chain and spin-labeled troponin both exchanged into glycerinated muscle fibers.   These results provide insight into the mechanism how myosin heads or troponin move their domains to function in muscle.   We will extend these studies to the other molecular function systems such as microtubule (tubulin)-based motor kinesin or dynein and transporting machinery in membrane.

2. MATERIALS AND METHODS

We used myosin RLC from rabbit skeletal muscle which contains two cysteines that react differentially and achieve specific spin-labeling at C-terminal domain (cys156) as will be described elsewhere and is expected to bind the heavy chain properly in a parent rabbit muscle when exchanged with native RLC.   Spin-labeled RLC was prepared previously[1]: it was purified from rabbit skeletal myosin which had been labeled with a cysteine-specific spin probe, 4-maleimido-2,2,5,5-tetramethyl-1-pyrrolidinyloxy (MSL) by controlling an incubation time to 18 hr.   Rigor buffer contained 20 mM imidazale, 5 mM $MgCl_2$, 1 mM EGTA. Relaxation and contraction buffers contained, in addition, 5 mM ATP, 30 mM creatine phosphate, and 0.25 mg/ml creatine phosphokinase; in contraction, pCa was 4.0. Ionic strength was adjusted to 200 mM with KCl, and pH was adjusted to 7.0.

Human cardiac troponin C (TnC) was purified from *E. coli.* expressing TnC with mutational substitution to cysteine residue, and labeled with mono-(MSL) or bifunctional spin label (BSL) at excess molar ratio. BSL was synthesized as will be described elsewhere.

Skinned fiber bundles, approximately 0.5 mm in diameter, were prepared from rabbit proas muscle fibers and were extracted removing the native RLC or TnC and labeled with a spin-labeled RLC or TnC as described previously[1] except that the temperature was kept at 4°C for TnC exchange.

High-resolution detection of the LC domain or TnC probe orientation was obtained from X-band ESR spectra of the fiber bundles placed at 15 - 20°C in a capillary that is set from a side hole in a specially-made TM_{110} cavity[2] with a continuous perfusion system connected to the capillary as described previously[3].   Spectral simulations and nonlinear least-squares fits were performed on Igor Pro Software (WaveMetrics, Inc., Lake Oswego, Oregon, USA).

3. RESULTS AND DISCUSSION

3.1. Two Distinct Spectral (Orientational) Components of LC Domain Probe

ESR has the orientational resolution needed to detect multiple orientations of nitroxide spin labels, because each spin label orientation corresponds to a unique spectral

Table 1. Spin Label Orientation of Myosin Light Chain with Respect to Fiber Axis[a]

Fit	Two Gaussians			Two Gaussians + random		
parameter	center[c]	full-width[b,c]	fraction[d]	center[c]	full-width[b,c]	fraction[d]
Rigor	48°	42°	0.63	50°	41°	0.41
	75°	41°	0.37	75°	40°	0.09
				random		0.50
Relaxation	48°	41°	0.51	50°	40°	0.27
	75°	40°	0.49	75°	41°	0.23
				random		0.50
Contraction	48°	36°	0.51	50°	40°	0.28
	73°	24°	0.49	75°	25°	0.22
				random		0.50

[a] Data were taken from more than 3 fiber preparations. [b] Full-width of half maximum. [c] Data have the standard deviations of ± 3-4°. [d] Data have the standard deviations of ± 0.02-0.03 except for fixed fraction of random orientation (0.50).

lines. ESR spectra of skinned muscle fibers spin-labeled specifically at RLC were measured in parallel orientation to the magnetic field. In rigor and relaxation, the spectra showed those similar to randomly-oriented sample. However, extensive spectrum analysis showed that they are resolved into different relative intensities of two broad but distinct angular distributions. Digital subtraction of any two experimental spectra [i.e. relax − 0.81(rigor) or rigor − 0.77(relax)] unambiguously resolved the same two distinct spectral components, each of which was uniquely simulated as a single Gaussian population having central angle of 50° or 75° and full-width of ~40° with respect to the fiber axis. Therefore, the spectra in three physiological states are described by different linear combinations of the same two distinct spectral components corresponding to two distinct orientational populations (Table 1).

3.2. Two Orientations of LC Probe in Relaxation, Rigor and Active Contraction

For each physiological state, the relative contributions of spectral component 1 and 2 to the experimental spectrum determine the mole fractions, p1 and p2. In relaxation, component 1 (center 48°± 4° full-width 42°± 3°) and 2 (center 75°± 3° full-width 41°± 4°) are equally populated (p1 = 0.51 ± 0.03). In the absence of ATP (in rigor), the spectrum shows a marked increase in 1 (p1 = 0.63 ± 0.03). Baker et al.[4] showed that a disordered population (~40%) was required to completely describe the experimental spectra of all three physiological states. When about 50% of randomly-oriented component, p3=0.5, was supplemented for spectrum fitting, two relative populations were determined similarly. In relaxation, component 1 (center 50°± 4° full-width 40°± 3°) and 2 (center 75°± 3° full-width 41°± 4°) are equally populated (p1 = 0.27 ± 0.02, p2=0.23 ± 0.02). In rigor, the spectrum shows a remarkable increase in 1 (p1 = 0.41 ±

0.02, p2=0.09 ± 0.02).   However, it is difficult to tell which fit is better.

Calcium activation changed the spectral properties of component 2.   Digital subtraction of two experimental spectra [i.e. active − 0.81(rigor)] provided a distinct spectral component, which was simulated as a single Gaussian population having central angle of 75° and full-width of ~25° with respect to the fiber axis. The full-width decreased markedly without changing population.   Component 1 (center 48°± 4° full-width 36°± 4°) and 2 (center 73°± 4° full-width 24°± 4°) are equally populated (p1 = 0.50 ± 0.02).   In the second model, component 1 (center 50°± 4° full-width 40°± 4°) and 2 (center 75°± 4° full- width 25°± 3°) are equally populated (p1 = 0.28 ± 0.02, p2 = 0.22 ± 0.02) with random component (p3=0.50).

Control experiments indicate that the probe nanosecond mobility does not change with respect to changes in the physiological state.

1.2. Implication from LC Probe Orientation for Mechanism of Muscle Contraction

The present study provided insight into the mechanism of muscle contraction from (i) there exist two distinct conformations of myosin light chain domain in intact muscle fibers, and (ii) one of them shifts to rigid conformation upon muscle activation.   In the relaxed state, the catalytic domain of myosin head is assumed disordered and dissociates from actin, so the two LC domain orientations, 1 and 2, are probably ordered by interactions with the core of the myosin filament.   The LC domain structure in the orientation around 50° is stabilized and becomes dominant when the catalytic domain binds strongly to actin in rigor state.   Upon calcium activation, the two LC domain orientations are similarly ordered by interactions with the myosin filament.   However, the structure in the orientation around 75° is stabilized and shows a twice narrower angular distribution.   Myosin heads with the LC domain orientation around 75° would undergo a weak-to-strong actin-binding transition on muscle activation, as a result of filament sliding and force generation. This force-generating heads may be different from rigor cross-bridges because different LC orientations are stabilized in rigor and on activation.

3.4. Mobility, Orientation and Site-Proximity of Troponin C Studied by Using a Bifunctional Spin Label

Site-directed spin-labeling was achieved by cysteine residues of human cardiac troponin C (TnC). By using spin dipole-dipole interaction, the distances between two labeled amino-acid residues (35-84, 84-159) could be monitored successfully from amplitude decrease or broadening of ESR spectra.   When Ca^{2+} or Mg^{2+} was added and removed, the large spectral changes were observed reversibly, indicating a marked decrease in the distance between 84- and 159-residues in cation-bound state.   This is the first finding by ESR that free TnC in solution undergoes a global structural change (extended-to-compact) by Ca^{2+} or Mg^{2+} binding as consistent with the results by NMR[5] and X-ray scattering[6,7] in solution.

The spin labels attached to 35- and 84-residues were rotationally mobile (Ca-independent) in solution but became weakly and strongly immobilized, respectively, on the protein surface at moderate (nanosecond) and slow (>submicrosecond) time scale when TnC were exchanged into the thin filament in the rabbit skeletal fibers. When the reconstituted fibers were set in a perfusion capillary parallel and perpendicular to the magnetic field, the spectra were broad and almost identical (either in relaxed or rigor state), suggesting that the N-terminal structure (at least near 35- and 84-residues) of TnC molecule is flexible or disoriented with respect to the filament axis at moderate to slow time scale.   The reports from fluorescence probe[8] and from spin probe in rigor muscle[9] also showed considerable disorder of the N-domain although they are not directly compared with our study.   This flexibility may be important for TnC to function as a molecular switch on responding to changes in Ca concentration.

We are investigating Ca^{2+} induced structural change of TnC using a new bifunctional spin label (BSL) which was synthesized for the purpose of getting more precise information from probes attached rigidly at known coordinate of TnC.   ESR spectra showed a large restriction of spin label motion when the BSL replaced a monofunctional spin label at a pair of amino acid residues: 84- and 91-residues or 84- and 98-residues. Rotational correlation time increased to about 7 nanoseconds which corresponds to that of rigid-body rotation of TnC molecule[9].   Therefore, we succeeded, for the first time, in fixing the spin label rigidly on TnC molecule in free state (either in $\pm Ca^{2+}$).   This provides a promise that we can determine the precise coordinate of the spin principal axis of BSL on protein surface.   This study is now in progress to know the orientation of the vector between two amino acid residues on TnC relative to the fiber axis and to determine the coordinates of the TnC molecule on the actin filament.

ACKNOWLEDGEMENTS

We are grateful to Drs. Yuichiro Maeda and Soichi Takeda for generously giving us human TnC gene and helpful discussions. We also thank Dr. Katsuzo Wakabayashi for continuous discussion.   Finally, we are supported by Special Coordination Funds for promoting Science and Technology and by Grants-in-Aid for Scientific Research, from Ministry of Education, Science, Technology, Culture and Sports of Japanese Government.

REFERENCES

1.   T. Arata, Orientation of spin-labeled light chain 2 of myosin heads in muscle fibers, *J. Mol. Biol.* 214, 471-478 (1990).
2.   T. Arata, The use of spin probes (Chapter 9), in: *Current Methods in Muscle Physiology*, edited by H. Sugi (Oxford University Press, 1998), pp. 223-239.
3.   T. Arata, and H. Shimizu, Spin-label study of actin-myosin-nucleotide interactions in contracting glycerinated muscle fibers, *J. Mol. Biol.* 151, 411-437 (1981).
4.   J. E. Baker, I. Brust-Mascher, S. Ramachandran, L. E. LaConte, and D. D.Thomas, A large and distinct rotation of the myosin light chain domain occurs upon muscle contraction, *Proc. Natl. Acad. Sci. U S A.* 95, 2944-2949 (1998).
5.   C. M. Slupsky, and B. D. Sykes, NMR solution structure of calcium-saturated skeletal muscle troponin C, *Biochemistry* 34, 15953-15960 (1995).
6.   S. R. Hubbard, K. O. Hodgson, and S. Doniach, Small-angle x-ray scattering investigation of the solution structure of troponin C, *J. Biol. Chem.* 263, 4151-4158 (1988).
7.   T. Fujisawa, T. Ueki, and S. Iida, Structural change of the troponin C molecule upon Ca2+ binding measured in solution by the X-ray scattering technique, *J. Biochem. (Tokyo)* 105, 377-383 (1989).
8.   D. A. Martyn, M. Regnier, D. Xu, and A. M. Gordon, Ca^{2+}- and cross-bridge-dependent changes in N- and C-terminal structure of troponin C in rat cardiac muscle. *Biophys. J.* 80, 360-370 (2001).
9.   H. C. Li, and P. G. Fajer, Orientational changes of troponin C associated with thin filament activation, *Biochemistry* 33, 14324-14332 (1994).

DISCUSSION

Katayama: Doesn't intramolecular cross-linking bifunctional EPR probe affect the function of TnC?

Arata: Yes, Ca-regulation is damaged sometimes. So, we have to carefully choose a pair of cysteine substituted by mutation.

MOLECULAR ORGANIZATIONS OF MYOFIBRILS OF SKELETAL MUSCLE STUDIED BY ATOMIC FORCE MICROSCOPY

Takenori Yamada, Yuki Kunioka, Jun'ichi Wakayama, Momoko Aimi, Yu-suke Noguchi, Nao Akiyama, and Taisuke Kayamori

1. INTRODUCTION

Skeletal muscle fiber is composed of a bundle of myofibrils which shows a characteristic striation pattern coming from a periodic array of sarcomeres.[1] In each sarcomere, the hexagonal lattices of myofilaments composed of actin and myosin filaments are inter-digitated. The contractile force is produced by the interaction of myosin heads extruded from the myosin filaments with the actin filaments, which causes the two filaments slide with each other. The produced force is transferred to the both ends of the sarcomere via these filaments to Z-bands which mechanically link the adjacent sarcomeres.

For muscle fibers to stably produce force, it is essentially important that the myofilament lattice structures are kept intact in each sarcomere during the force development. It is generally considered that the Z-band and the M-line are the major components stabilizing the sarcomere structure by bundling the myofilament lattice. However only limited data are available about the mechanical characteristics of molecular architecture of sarcomere.

The atomic force microscope (AFM) is a recently developed technology to capture the images of specimen by scanning the tip of flexible cantilever over the specimen and, based on the deflection signals of the cantilever, up-and-down contours of the specimen are constructed to produce images.[2] The AFM technique can also be applied to study the mechanical characteristics of various biological specimen based on the deflection signals of the cantilever produced by its tip making contact with the specimen. In the present studies we applied the AFM technology to study the mechanical organizations of myofibrils of skeletal muscle.

Department of Physics (Biophysics Section), Faculty of Science, Science University of Tokyo, Shinjuku-ku, Tokyo 162-8601, Japan.

Molecular and Cellular Aspects of Muscle Contraction
Edited by H. Sugi, Kluwer Academic/Plenum Publishers, 2003

2. EXPERIMENTAL METHODS

2.1. Materials

Single myofibrils were prepared from glycerinated muscle fibers of rabbit psoas as described previously.[3] Fiber bundles were dissected and skinned in a relaxing solution (see the composition below) containing 1.0% Triton X-100 and stored in a 50% glycerol in a relaxing solution at -20°C for 5-12 weeks before use. A short bundle of glycerinated fibers was cut into pieces with scissors, put in a relaxing solution, and homogenized with a Polytron homogenizer. The suspension of myofibrils thus prepared contained single myofibrils and used for experiments.

The calcium activated neutral protease (CANP) treatment of myofibrils was made by incubating myofibril preparations in a rigor solution containing 0.15 mg/ml of CANP and the reaction was arrested by adding the final concentration of 10 mM EGTA.

2.2. AFM measurements

The assembly of the atomic force microscope is schematically depicted in Fig. 1A. A flexible cantilever equipped in the AFM assembly deflects upward and downward by the force applied to it. By making the tip of the flexible cantilever contact with a specimen or scanning over it, the deflection signals of the cantilever were processed by use of the computer.

The AFM apparatus used in the present studies was incorporated in an inverted optical microscope (NV2500; Olympus Optical Co., Tokyo, Japan), the cantilever of which could be located at an appropriate locus of specimen under the optical microscope. Commercially available cantilevers made of silicon nitride were used (the spring constant, 0.02 N/m; the tip radius, ca. 50 nm). In some experiments were used the modified cantilevers having a ZnO whisker (the tip diameter, ca. 20 nm). The ZnO whisker was

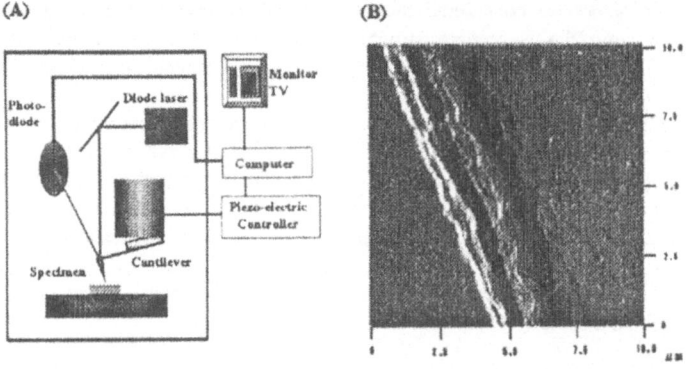

Figure 1. (A) Schematic drawing of AFM assembly with its computer-assisted control system. The up-and-down deflections of AFM cantilever were detected by the photo-diode system. (B) Typical AFM image of single myofibrils of skeletal muscle in rigor state.

further treated with the amino-silanation reaction so as to its surface get positively charged and adhere to negatively charged materials.

For AFM experiments, myofibrils were attached to the coverslip which was the bottom of a modified culture dish. A drop of myofibril suspension prepared as above was put in the culture dish and stood for ca. 10 min to make myofibrils attached to the coverslip. Unattached myofibrils were washed out before experiments. To change the physiological conditions of myofibrils, the bathing solution in the culture dish was replaced by an appropriate solution.

2.2.1. Imaging of myofibrils

AFM images of myofibrils were captured with the AFM by scanning the cantilever over the myofibril by the tapping mode imaging in which the distance between the tip of cantilever and the specimen was kept constant by adjusting the height of the cantilever via the feedback loop and the feedback signals were used to construct myofibril images.

2.2.2. Force-distance curves

When the tip of cantilever is approached to or retracted from a specimen, the cantilever deflects by making mechanical contacts with a specimen (Fig. 2A). The trace of the deflection signals vs. the cantilever height is called the force-distance curve (Fig. 2B). The transverse stiffness of myofibril can be determined based on the force-distance curves obtained by approaching the AFM cantilever to myofibril preparations attached to coverslip. When the cantilever is lowered toward a myofibril, the tip of the cantilever eventually deflects upward as it hits the surface of the myofibril. The magnitude of the deflection depends on the indentation of the myofibril. The stiffness of the myofibril is determined as the applied force divided by the indentation. The applied force is calculated from the deflection of cantilever based on its spring constant. The indentation is determined as the distance obtained by subtracting the cantilever deflection from the height of cantilever lowered after its tip hits the surface of myofibril.

The rupture processes of myofibrils can be examined also based on the force-distance curves obtained by retracting the AFM cantilever away from a myofibril. Firstly the tip of cantilever is pressed to the surface of myofibrils to make it adhere to some surface components. When the cantilever is retracted, it deflects downward if the tip picks up myofibril components. Associated with the rupture of the picked-up components from main components, sudden upward deflections would take place in the downward deflections in the force-distance curve (Fig. 2B). Based on such rupture processes, the molecular organizations of myofibrils can be analyzed.

2.3. Composition of the bathing solutions

The composition of the bathing solutions was as follows (in mM); rigor solution: K^+-propionate, 109; $MgCl_2$, 5; EGTA, 10; Imidazole, 20; relaxing solution: K^+-propionate, 93; $MgCl_2$, 5; EGTA, 10; ATP, 5; Imidazole, 20; +AMPPNP solution: K^+-propionate, 121; $MgCl_2$, 5; $CaCl_2$, 2.3; EGTA, 2; AMPPNP, 2.5; Imidazole, 20. The pH of all solutions was adjusted to 7.0. ATP, AMPPNP, and CANP were purchased from

(A)   (B)

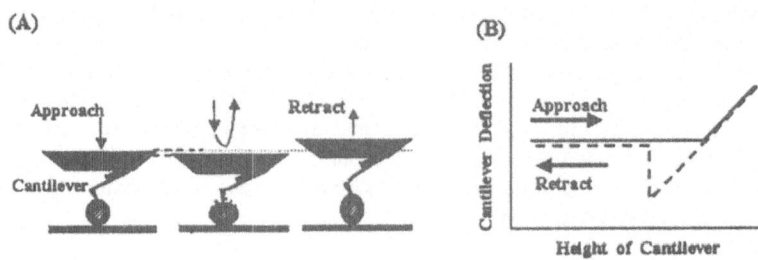

Figure 2. (A) Schematic drawings showing the approach and retraction of AFM cantilever over myofibrils. (B) Force-distance curves produced by approaching and retracting AFM cantilever as in (A).

Sigma Chemical Co. (St. Louis, MO, USA). Other chemicals were of analytical grade and purchased from Wako Pure Chemicals (Osaka, Japan).

3. RESULTS

3.1. AMF imaging of myofibrils

By scanning the cantilever over a single myofibril firmly attached to coverslip, AFM images of the myofibril were captured by the tapping mode imaging. A typical AFM image of single myofibrils in rigor solution is shown in Fig. 1B. The AFM images of the single myofibrils were ca. 1 μm in thickness and in width, having characteristic periodical structures, similar to the striation pattern of sarcomere structure observed under optical microscope. The repeating unit structure was ca. 2 μm in length containing filamentous components. The length of the repeating unit was almost equal to the sarcomere spacing of rabbit skeletal muscle at around slack length, where the tips of thick filaments almost make contact with Z-bands.[3]

3.2. Transverse stiffness measurements

The force-distance curves for myofibrils attached to coverslip were measured by positioning the cantilever tip at various loci of myofibrils under the optical microscope. The typical force-distance curves obtained at two different loci of myofibrils in rigor state are shown in Fig. 3A, in which the force-distance curve obtained for coverslip is included for reference. It can be seen that the upward slopes for myofibrils are substantially less steep compared with that for coverslip. This indicates that the myofibrils got indent by being pressed by the tip of cantilever. Further the slope of the force-distance curves was different at these two loci of myofibril, indicating the transverse stiffness is not uniform along myofibril.

To examine the distribution of transverse stiffness, we measured similar force-distance curves at an interval of 0.1 μm along myofibrils, and the stiffness determined at each locus. As the upward slopes were not linear, the stiffness of myofibrils was determined at 25-60 nm of the indentations of myofibril preparations considering the lattice spacing of myofilaments.[1]   The stiffness distribution of myofibril thus obtained is

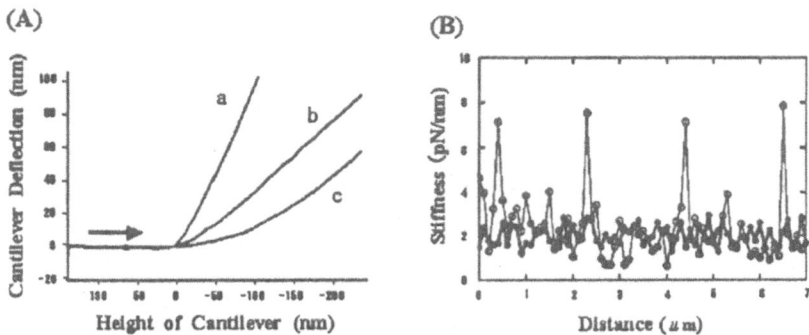

Figure 3. Transverse stiffness measurements of skeletal myofibrils in rigor state. (A) Typical force-distance curves for intact myofibrils. (a) Coverslip for reference, (b, c) intact myofibrils at two different loci. (B) Transverse stiffness distributions of myofibrils. (O) Intact myofibril, (●) CANP-treated myofibril.

shown in Fig. 3B. It can be seen that the rigid bands having the width of ca. 0.1 µm are located at the interval of ca. 2 µm along myofibril. As the striation having the same spacing was observed for the myofibrils under optical microscope, the above band structures could be either Z-bands or M-lines. To identify the rigid band structures, myofibrils attached to coverslip were treated with CANP and their force-distance curves similarly examined. CANP is known to specifically destruct Z-bands[4], which was confirmed in the present studies by observing under optical microscope that Z-bands disappeared from myofibrils by the CANP treatment. The distribution of the transverse stiffness for CANP-treated myofibrils thus obtained is included in Fig. 3B. It can clearly be seen that the rigid bands selectively disappeared, indicating that they were Z-bands.[5]

As it is well known that the stiffness of muscle fibers changes depending on the physiological conditions, we examined how the transverse stiffness of myofibrils changes depending on the physiological conditions of the bathing solution. The transverse stiffness distributions of myofibrils in rigor, +AMPPNP, and relaxing solutions are shown in Fig. 4. It can clearly be seen that the overall transverse stiffness of myofibrils significantly changes as the physiological condition was altered. The transverse stiffness of the overlap region of myofibrils was 4.6±0.7, 2.6±0.5, and 1.0±0.2 pN/nm in rigor, +AMPPNP, and relaxing solutions respectively.[6]

3.3. Measurements of rupture processes of myofibrils

A modified AFM cantilever having a tip of ZnO whisker, treated with the amino-silanation reaction, was pressed to a myofibril attached to coverslip and then retracted. Characteristic downward deflections of cantilever having sudden upward deflections were observed in the force-distance curves as can typically be seen in Fig. 5A. This clearly indicates that,   as the cantilever was retracted, the tip of cantilever picked up some surface components of myofibrils and mechanically peeled them off from the main structure. The obtained data were analyzed by the rupture force and the rupture distance as defined in Fig. 5B. The results thus examined at various loci of myofibrils are

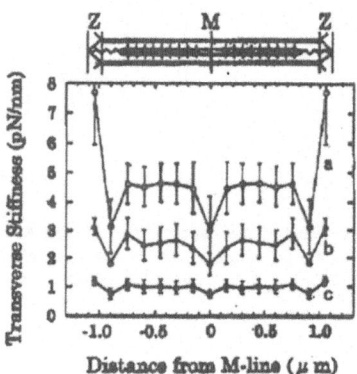

Figure 4. Transverse stiffness distributions along sarcomere of skeletal myofibrils in various bathing solutions. (a) Rigor, (b) +AMPPNP, and (c) relaxing solutions. The upper panel is a schematic drawing of the overlap between actin and myosin filaments at slack length. Reproduced from (6) by copyright permission of Elsevier Science.

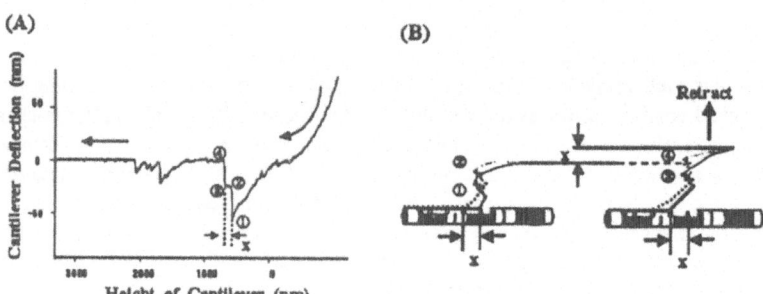

Figure 5. (A) Typical force-distance curve showing rupture processes produced when a modified AFM cantilever was pressed on a skeletal myofibril and then retracted. (B) Schematic drawings of the rupture processes of a filamentous component from a myofibril by the retraction of cantilever. The rupture distance x is determined as shown in (A).

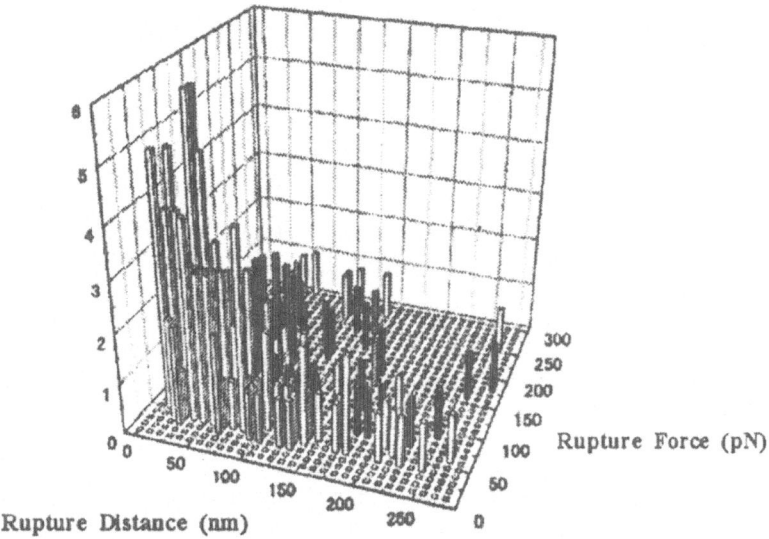

Figure 6. Histogram showing the distributions of the rupture force and the rupture distance obtained from the force-distance curves like shown in Fig. 5A.

summarized as a histogram in Fig. 6. It can be seen that the rupture distance is heavily distributed in 50-100 nm and the rupture force in 30-150 pN. The rupture signals could not be observed over the rupture distance of 1-2 μm.

4. DISCUSSION

4.1. Z-disc as the most rigid sarcomere structure of myofibril

The AFM images of myofibrils clearly showed the sarcomere-like structures, the repeated arrays having a spacing of ca. 2.0 μm. In the transverse stiffness distribution along myofibrils, rigid bands were located at the same interval of ca. 2 μm while, in CANP-treated myofibrils, the rigid bands were completely absent. As CANP selectively disrupts Z-bands, this result clearly indicates that the rigid band is Z-band. Notably the transverse stiffness of M-lines seen in the stiffness distribution of sarcomere is almost comparable in magnitude to the overlap regions between actin and myosin filaments (Fig. 4), suggesting that M-lines are not very rigid. These results are in accord with the observation that, by stretching rigor muscle fibers, M-lines completely disappeared while Z-bands remained almost intact, although skewed slightly.[7]

4.2. Anisotropic stiffness of the overlap region of myofibril

The results shown in Fig. 4 clearly indicate that the transverse stiffness of myofibrils changes depending on the physiological conditions. This can be compared with similar dependence of the longitudinal stiffness of muscle fibers on the physiological conditions.

To compare the longitudinal and the transverse components of the stiffness of myofibrils, the transverse stiffness obtained above was transformed to the corresponding Young's modulus by employing the following Hertz equation.[8]

$$\delta = \lambda \sqrt[3]{\frac{9P^2}{64R}\left(\frac{1-\sigma^2}{E}\right)^2}$$

where δ, the indentation; E, the transverse Young's modulus of myofibril; σ, Poisson's ratio (=0.5); P, the force applied to cantilever; R, the radius of cantilever tip (=50 nm); λ, Lame's constant (=2.1).

The Young's modulus of myofibrils in the transverse direction thus calculated was 84, 38, and 12 kPa in rigor, +AMPPNP, and relaxing solutions respectively.[6] These values can be compared with the results for single myofibrils of flight muscle, 94, and 40 kPa in rigor and relaxed states respectively.[9] These values are comparable in magnitude to those estimated based on the osmotic compressions of muscle fibers; 39-89 kPa for rigor fibers[10, 11] and 16 kPa for relaxed fibers.[10]

On the other hand the longitudinal Young's modulus for single myofibrils was 10, 2, 0.2 MPa in rigor, +AMPPNP, and relaxing solutions respectively.[12] The corresponding value for glycerinated rabbit muscle fibers was 29 MPa for rigor state.[13]

It should be remarked that the Young's modulus of myofibrils are far greater for the longitudinal direction than that for the transverse direction.[6, 9] In other word the sarcomere structure of myofibrils is anisotropic in which the longitudinal structures is more rigid than the transverse structures. This suggests that some major component(s) in the overlap region has anisotropic mechanical characteristics. It is established that the longitudinal stiffness dominantly comes from the myosin heads interacting with actin filaments.[14] It was further pointed out that the longitudinal and the transverse stiffness roughly change in a proportional fashion.[12] These results suggest that the actomyosin complexes formed in the overlap region dominantly contribute to the two stiffness (and also their anisotropy) of myofibrils.

4.3. Filamentous networks inter-connected in myofibril

The histogram of the rupture force and the rupture distance shown in Fig. 6 indicates that various major components are inter-connected in sarcomere having the rupture distance in the range of 50-100 nm with the rupture force of 30-150pN. These rupture processes have the rupture force comparable in magnitude to the following rupture processes; i.e. the force to break the rigor complex of actomyosin, ca. 15 pN[15], the force to break actin filament from α-actinin, 18 pN[16], the force to break actin filament by stretching, ca. 108 pN[17], and the force to extend connectin (or titin) filament, 150-300 pN.[18] Considering that the rupture distances were found to be at most 1-2 μm and that the sarcomeres are ca. 2 um in length and composed of filament lattices[1], the results of the above rupture experiments strongly suggest that the AFM cantilever picked up filamentous components in sarcomere structures, such as thin and thick filaments, connectin (titin), nebulin and so on, although not identified at the present stage. As the maximal force produced per one thin filament is ca. 110 pN[1], the present results indicate that various filamentous components are inter-connected in sarcomere structures with the mechanical strength comparable or sufficient to sustain the force produced in the actomyosin filament lattice.

5. SUMMARY

By applying AFM technology, we studied mechanical characteristics of myofibrils of skeletal muscle. The obtained results indicate that (1) the Z-band is the most rigid sarcomere component stabilizing the myofibril structures, (2) various filamentous components are inter-connected in sarcomere with sufficient mechanical strength to support the contractile force, and (3) the molecular structure of the overlap region between actin and myosin filaments is anisotropic. In any case the present studies clearly indicate that the AFM technique is a powerful tool to investigate the mechanical characteristics of sarcomere structure of muscle fiber.

6. ACKNOWLEDGEMENT

This study was supported in part by a Grant-in-Aids for Promotion of Science (C-12680660) from the Ministry of Education, Science, Sport, and Culture of Japan (to T.Y.).

REFERENCES

1. C. R. Bagshaw. *Muscle Contraction* (Chapman & Hall, Tokyo, 1993).
2. W. Heckl, and O. Marti. Atomic force microscopy. In: *Proceedings in scanning probe microscopies*, ed. by R. J. Colton et al. (John Wiley & Sons, New York, 1998), pp. 85-148.
3. K. Yuri, J. Wakayama, and T. Yamada. Isometric contractile properties of single myofibrils of rabbit skeletal muscle. *J. Biochem.* 124: 565-571 (1998).
4. M. K. Reddy, J. D. Etlinger, M. Rabinowitz, D. A. Fischman, and R. Zak. Removal of Z-lines and _-actinin from isolated myofibrils by a calcium-activated neutral protease. *J. Biol. Chem.* 250: 4278-4284 (1975).
5. J. Wakayama, Y. Yoshikawa, T. Yasuike, and T. Yamada. Atomic force microscopic evidence for Z-band as a rigid disc fixing the sarcomere structure of skeletal muscle. *Cell Struct. Funct.* 25: 361-365 (2000).
6. Y. Yoshikawa, T. Yasuike, A. Yagi, and T. Yamada. Transverse elasticity of myofibrils of rabbit skeletal muscle studied by atomic force microscopy. *Biochem. Biophys. Res. Comm.* 256: 13-19 (1999).
7. S. Suzuki, and H. Sugi. Extensibility of the myofilaments in vertebrate skeletal muscle as revealed by stretching rigor muscle fibers. *J. Gen. Physiol.* 81: 531-546 (1983).
8. R. J. Roark. *Formulas for Stress and Strain* (McGraw-Hill, New York, 1965).
9. L. R. Nyland, and D. W. Maughan. Morphology and transverse stiffness of Drosophila myofibrils measured by atomic force microscopy. *Biophys. J.* 78: 1490-1497 (2000).
10. Y. Umazume, and N. Kasuga. Radial stiffness of frog skinned muscle fibers in relaxed and rigor conditions. *Biophys. J.* 45: 783-788 (1984).
11. D. W. Maughan, and R. E. Godt. Radial forces within muscle fibers in rigor. *J. Gen. Physiol.* 77: 49-64 (1981).
12. T. Yamada, Y. Yoshikawa, and J. Wakayama. Longitudinal and transverse stiffness of single myofibrils of rabbit skeletal muscle in various physiological states. *Biophys. J.* 76: A160 (1999).
13. K. Tawada, and M. Kimura. Stiffness of glycerinated rabbit psoas fibers in the rigor state. *Biophys. J.* 45: 593-602 (1984).
14. L. E. Ford, A. F. Huxley, and R. M. Simmons. The relation between stiffness and filament overlap in stimulated frog muscle fibres. *J. Physiol. (Lond.)* 311: 219-249 (1981).
15. T. Nishizaka, R. Seo, H. Tadakuma, K. Kinosita, and S. Ishawata. Characteirzation of single actomyosin rigor bonds: load depencence of lifetime and mechanical properties. *Biophys. J.* 79: 962-974 (2000).

16. H. Miyata, R. Yasuda, and K. Kinosita. Strength and lifetime of the bond between actin and skeletal muscle α-actinin studied with an optical trapping technique. *Biochim. Biophys. Acta* **1290**: 83-88 (1996).

17. A. Kishino, and T. Yanagida. Force measurements by micromanipulation of a single actin filament by glass needles. *Nature* **334**: 74-76 (1988).

18. M. Rief, M., Gautel, F. Oesterhelt, J. M. Fernandez, and H. E. Gaub. Reversible unfolding of individual titin immunoglobulin domains by AFM. *Science* **276**: 1109-1112 (1997).

DISCUSSION

Ranatunga: Your results showed that longitudinal stiffness was 10-20 times greater than transverse stiffness (in myofibrils). Is that true? What does it mean?

Yamada: Yes, it is based on the Young's modulus. May be due to non-isotropic nature of cross-bridges and other structures in two directions.

Maughan: Did you consider the possibility that the S2 hinge flexibility accounts for the low transverse stiffness compared to the 20-fold greater longitudinal stiffness?

Yamada: Yes, it could be a possibility, but other parts of the filament lattice could also have anisotropic nature.

Pollack: Did you do any measurements at different sarcomere length, where lateral interfilament stacing changes?

Yamada: Now under investigation.

Gonzalez-Serratos: Have you done similar experiments with preparation where the sarcolemma is present?

Yamada: No, my work is only with myofibrils from skinned preparations.

THREE-DIMENSIONAL STRUCTURAL ANALYSIS OF INDIVIDUAL MYOSIN HEADS UNDER VARIOUS FUNCTIONAL STATES

Eisaku Katayama*, Norihiko Ichise, Naoki Yaeguchi, Tsuyoshi Yoshizawa, Shinsaku Maruta,and Norio Baba

1. INTRODUCTION

Half a century has passed since the dedicated studies on the contraction mechanisms of muscle began, with considerable knowledge on its molecular architecture. Two major hypotheses were raised very early, one, "sliding filament theory", [1, 2] and the other, "crossbridge theory". [3] The former was readily accepted, because the phenomenon was apparently visible under optical microscope. The latter, however, has been hindered from thorough experimental proof even now, though nothing other than crossbridges connect thick and thin filaments enabling force development. The original idea postulated the rowing movement of actin-bound myosin head coupled with ATP hydrolysis, but it was later replaced by swinging of the "lever-arm" moiety, [4] according to the discovery of intramolecular bending by X-ray crystallography. [5-8] One of the major reasons for such persistent difficulty to prove this simple hypothesis might be the lack of means to directly observe the actual structural change of working crossbridges with time and spatial resolution enough to visualize the fine details of the molecular nano-machine. Though the crystal structure of each component; actin[9, 10] and myosin subfragment-1 (S1) with or without various nucleotides, [5-8] was determined ten years ago, none of their complexed form was solved nor might be the subject matter for easy crystallization.

---

* Eisaku Katayama and Norihiko Ichise, Division of Biomolecular Imaging, Institute of Medical Science, The University of Tokyo, Minato-ku, Tokyo 108-8639 Japan. Naoki Yaeguchi and Norio Baba, Department of Electric Engineering, Kogakuin University, Hachioji, Tokyo 192-0015 Japan. and Tsuyoshi Yoshizawa and Shinsaku Maruta, Department of Bioengineering, Faculty of Engineering, Soka University, Hachioji, Tokyo 192-8577 Japan

The most important message from the new concept is that the behavior of individual molecules might be different from each other and that some new important information could be revealed only by measuring unaveraged properties of each single molecule, separately.   A number of amazing results fascinated the researchers in other fields and prompted them to apply new techniques to various fields and materials including live cells.   Thus, single molecule physiology has already become one of the most powerful and indispensable approaches in current biophysical sciences.   The final aim of our study is to establish the means to obtain dynamic structural information of single protein molecules in function under physiological conditions with spatial and time resolution high enough to elucidate the underlying mechanism of force development by the molecular-motor.

2.   STRATEGIES USED FOR THE SPECIFIC PURPOSE

Authentic means of current structural biology; i.e. X-ray crystallography or multi-dimensional NMR analysis, collects the data from a vast number of target particles to be averaged both in time and space, and apparently is not applicable to "single molecule" matter.   The third approach, electron microscopy, is unique in terms that it has a potential to visualize the structure of individual molecules, though the spatial resolution might not be as high as that of the others.   In order to obtain the structural information of functioning actomyosin motor compatible to single molecule physiology, we have been utilizing quick-freeze deep-etch replica electron microscopy with mica-flake technique[13] to capture transient three-dimensional (3-D) configuration of myosin crossbridges supporting actin movement *in vitro*.   This experimental system is suitable for such purpose, since a variety of molecular events extensively studied under fluorescence microscope could be instantaneously arrested within one millisecond and the structure of individual protein molecules under almost any physiological conditions might be clearly visualized with very high contrast metal-shadowing.   In the previous paper,[14] we showed, at first, the feasibility of our experimental system by a well-known structural change of myosin heads upon binding of nucleotides.   Two pear-shaped heads of heavy meromyosin (HMM) were strongly kinked in the presence of ATP or ADP/ inorganic vanadate (Vi), whereas they were almost straight or only slightly curved in the absence of nucleotide, in accordance with the results of S1 crystallography.   Then, we observed the structure of HMM crossbridges which support sliding actin-filaments, together with that in rigor-complex.[14]   We learned there that the configurations of myosin heads, especially those during sliding, have too much diversity and that the only way to study their 3-D structure might be to analyze each particle separately.   Since the resolution of freeze-replica images often appeared high enough to visualize subdomain constitution of "individual" protein molecules, we started our project by reconstruction of 3-D surface profile of replica specimen using computer-tomography with conventional back-projection method.[15]   Replica specimens are extremely tolerant to high-dose electron beam irradiation and it should be possible to take many micrographs of the same field, a mandatory procedure for 3-D reconstruction. As the first test material, we reconstructed the image of single molecule from tilt-series micrographs of HMM complexed with ADP/Vi[15] and compared it with various X-ray atomic models of S1 or its truncated fragments obtained by that time[5-7] (regulatory light-chain portion was artificially added in ref. 7). Though the result was not bad as a preliminary trial, we seriously realized the impaired resolution along Z-axis by the effect of unavoidable "missing data range problem", that is specific for electron microscopic tomography due to limited specimen tilt-angles. In order to overcome the difficulty, we developed a new algorithm of 3-D reconstruction, by introducing height determination of corresponding points in multiple tilt-images using precise parallax measurements, which was devised originally for the topographic measurement of the bacterial flagellar apparatus.[16]

As a complementary approach to characterize the 3-D structure of the target particles, we simulated the replica images of given protein molecules from their atomic coordinates, using light-rendering computer software. Since the image contrast of replica specimen arises by heavy metal shadowing, virtual physical model in the shape of protein particle was placed in cyber-space, and its image was rendered as illuminated by a number of surrounding light-sources from the appropriate elevation angles. We were encouraged by initial trials with actin filament and actomyosin rigor complex, which not only showed generally good matching to real replica images, but also represented some subtle differences in the configurations in between them, suggestive for the feasibility and usefulness of such unconventional approach. Hence, 3-D reconstruction is suitable for the study of gross molecular envelopes and simulation might work nicely to define detailed features of molecular surface profile at the final stage of fine adjustment. We applied these strategies to examine 3-D structural features of myosin heads under a variety of conditions; free in solution with various bound nucleotides, and/or associated with actin filament.

3. MYOSIN CROSSBRIDGE CONFIGURATION DURING ACTIN SLIDING

Figure 1 indicates a gallery of images of myosin heads supporting actin filaments just after addition of ATP, so that the filaments could be captured while sliding. Unlike rigor crossbridges presenting rather uniform configuration, working crossbridges exhibited a great variety of structure both in shape and attachment angles to actin, though all of them were associated with actin through only one of two heads. It was clear that any method to take the average does not work properly here in this occasion. We also noticed that many crossbridges were bent to hold actin filament inside its curvature. Because we knew that HMM head sharply kinks by the addition of ATP, and we actually observed free HMM particles of such shape on the background (see the particles shown by arrowheads), we naturally supposed, at first, that actin-filaments might be held inside this curvature. If we examine known X-ray structure of nucleotide-bound S1, however, putative actin binding site should come rather outside the curvature of bent S1 body.

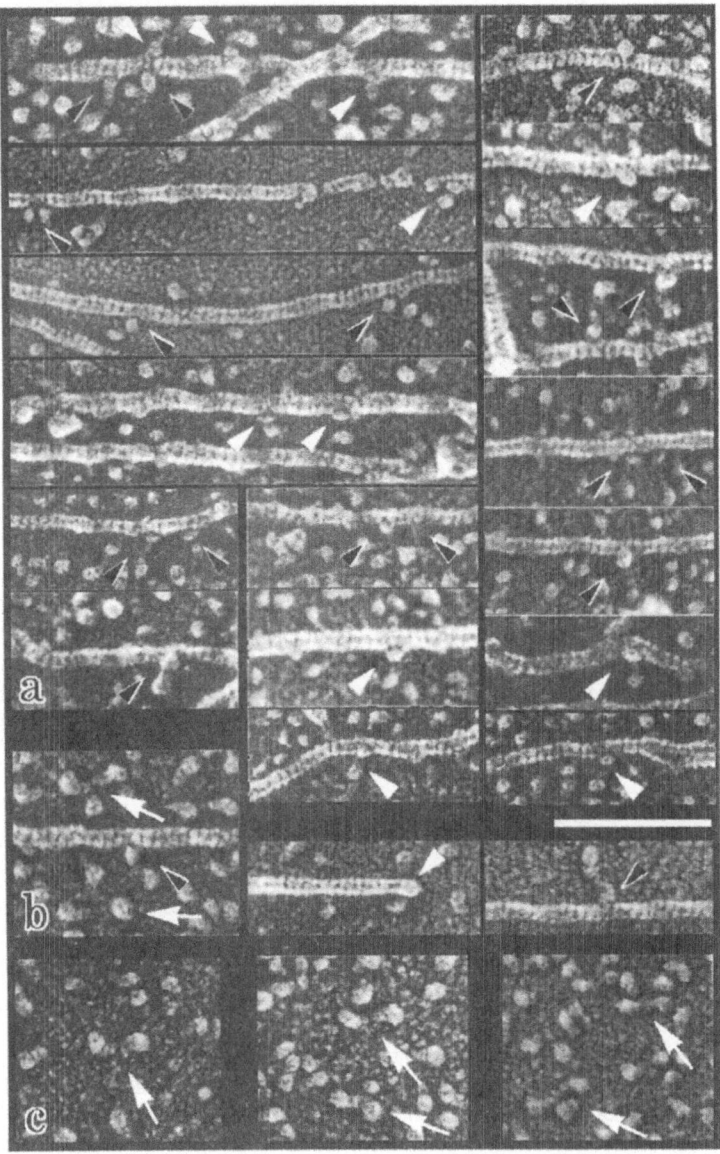

Figure 1.   Gallery of HMM crossbridge structure supporting actin filaments under *in vitro* sliding assay conditions in the presence of ATP.   Note that all the HMM particles are attached to actin through only one of two heads.   Many crossbridges (indicated by white arrowheads) are holding actin filaments in the inner side of the main body curvature.   White arrows exhibit free HMM particles on the background in a kinked configuration. Scale bar indicates 100 nm.   Modified from Fig.5(a) in J. Mol. Biol. 278, 349-367 (1998).

We reconstructed some free HMM particles with kinked configuration and confirmed that actin filament should bind outside the curvature (data not shown).   According to recent swinging lever-arm model, [4, 8] the motor-domain is presumed to stay attached to actin in the same configuration throughout its working stroke, and the lever-arm moiety moves along the direction almost parallel to actin filament axis.   Thus, actin-holding configuration we observed very often in actual replica images cannot reconcile with any of the transient states of swinging lever-arm model.

4. SEARCHES FOR THE CHEMICAL STATE CORRESPONDING TO THE CROSSBRIDGE CONFIGURATION OBSERVED DURING ACTIN SLIDING

An important feature of authentic swinging lever-arm mechanism is that each chemical state of myosin head in its ATP-hydrolysis cycle corresponds to certain structural state of the crossbridge configuration, and that the mechano-chemical coupling between them is essentially tight.[4-8]   Since the above crossbridge configuration we observed during sliding does not seem to match with any one of crystal structures so far reported, we sought for the possibility if myosin head could take such unusual configuration only transiently, during its ATP-hydrolysis.   Maruta et al. have been characterizing the properties of various myosin/ADP/metallo-fluoride complexes and correlated each of them to certain chemical substep.[17-19]   Along that line, we examined the structure of HMM complexed with various nucleotide analogues, using quick-freeze deep-etch replica electron microscopy.   What we found was that the structure of HMM heads in each molecular species was actually the mixture of two or three different configurations (elongated, globular and bent) and that the difference among chemical states might merely be the distribution of these structures (data not shown).   However, none of them represented the new configuration we found.   Thus, we extended our search to chemically modified myosin.   Vertebrate skeletal myosin has two highly reactive thiol groups called SH1 and SH2 whose chemical modification greatly affects myosin's intrinsic enzymatic activity.   X-ray structure of S1 revealed that they are actually located at both ends of a single α-helix in the heart of myosin.   From this structural feature, it is apparent that they cannot come close each other as long as that helix is kept stable.   It has been reported that these two thiols react each other to be cross-linked by certain bifunctional reagents under limited conditions in the presence of ADP. [20, 21]   p-Phenylenedimaleimide (pPDM) is one of such reagents whose span between reactive groups is approximately 12Å.   With electron microscopy, each head of pPDM cross-linked HMM/ADP was bent and appeared similar to that of usual Vi-type kinked head, at first sight.   We noticed, however, that the polarity of the head curvature was now opposite to that we have observed previously for ATP-bound or ADP/Vi-bound HMM.   This cannot be a simple mirror image of Vi-type configuration, because the segment connecting S1 to S2 appeared strangely crooked or twisted, whereas smoothly curved Vi-form S1 continued quite naturally to S2 moiety (Fig.2).

To characterize possible new configuration more in detail, we examined the structure of those particles by 3-D reconstruction and computer-simulation as above. Because our purpose here was to determine which side of S1 molecule is facing toward outside of the curvature, we laid our stress on the simulation of surface profiles of S1 and

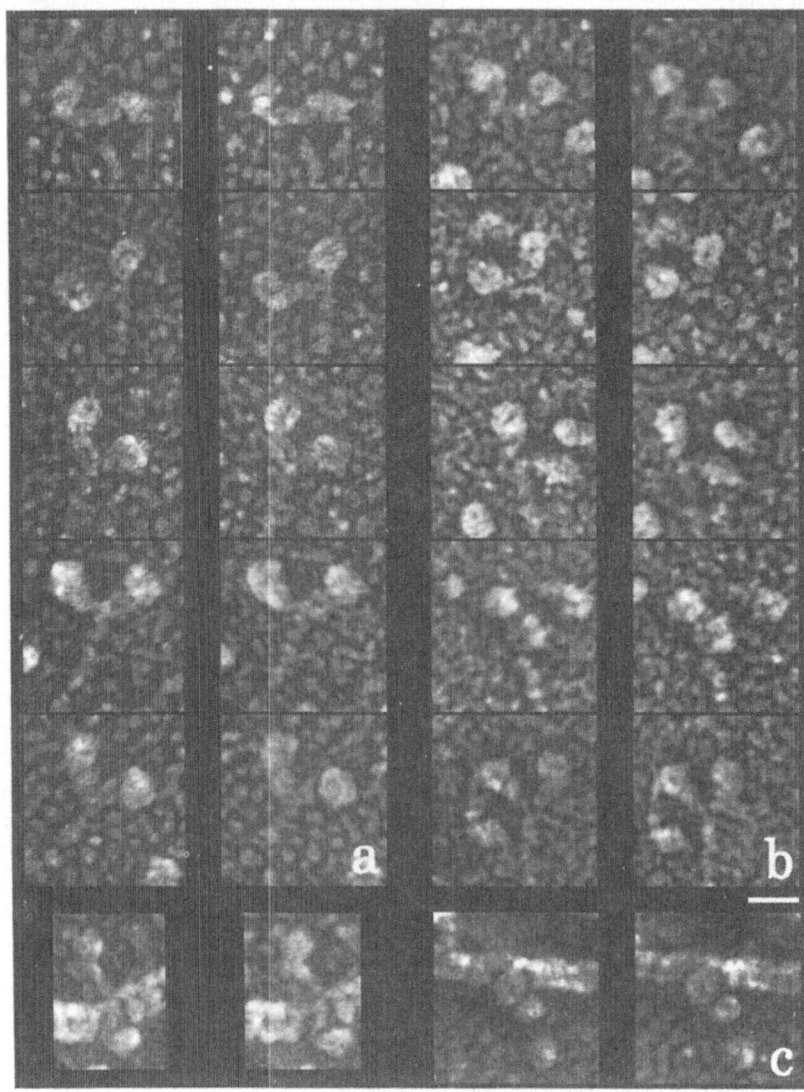

Figure 2. Gallery of HMM particles showing two kinds of bent-configurations. a) HMM/ADP/Vi; b) pPDM-HMM/ADP; c) actin attached HMM in the presence of ATP. All the particles are exhibited as stereo-pairs. Each head of HMM is curved in reversed polarity in a) and b). Note that the junctional piece of S1 and S2 are irregularly twisted for pPDM-HMM. Scale bar indicates 20 nm.

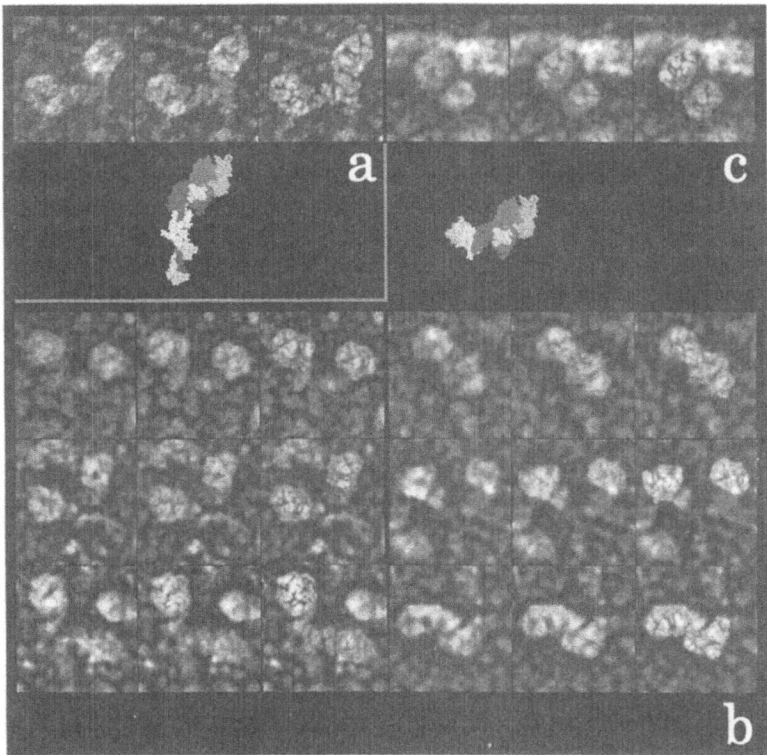

Figure 3. Matching of computer-simulated images to replica images. Artificial images were overlayed (right frames) on the background of real replica images (left frames) so that a set of corresponding landmark features of both images match to each other. Center frames exhibit half-and half-blended images of both. a) HMM/ADP/Vi: b) pPDM-HMM/ADP: c) actin attached HMM in the presence of ATP.

compared them with replica images actually observed under electron microscope. As the first step, we searched for the configuration of S1 in which the lever-arm comes closer to the opposite side of scallop Vi-form, by somehow changing the S1 backbone. Since glycine is the only residue accepted generally as a rotational hinge, the torsional angle of appropriate glycyl residues along heavy chain was selected and rotated every 10 degrees from its original value. There were not many cases that meet such requirement, because of the clash during rotation of the main-chain. In the configuration we selected as a candidate, the lever-arm orientation was about 90 degrees deviated from its original structure. Then we generated various artificial images by light-rendering the modified model placed in the orientation as postulated and compared them with actual replica images of pPDM-HMM. By examining the correlation between the positions of the subtle surface features of the real images (details of the method will be published elsewhere) with those simulated, we could finally determine almost the right orientation of the molecules. Fig. 3 exhibits some of such examples, in comparison with Vi-type and actin attached myosin head. In each set of frames, real replica images on the left panel were merged with the artificial images on the right to produce half and half-blended images in the center panel. Matching of major surface features was judged good enough, at least for the present purpose, though more complete quantitative evaluation of the matching should be the most important issue for the next step. Thus, simulated images of the particles placed in appropriate orientation matched well with various replica images, and we might be able to say safely that the actin-binding site of pPDM-HMM heads certainly faces toward the inner side of the curvature. Since actin-bound HMM head (Fig. 3; upper panel) represented essentially similar surface features to that of pPDM-HMM, we concluded that one of the most abundant actin-bound configuration observed during sliding would presumably be a short-lived transient configuration of myosin, in which SH-loop might be disorganized.

5. BIOCHEMICAL EVIDENCE FOR p-PDM CROSSLINKED MYOSIN TAKING UNDESCRIBED CONFIGURATION

In the most recent report on the X-ray structure of scallop S1,[22] authors state that the conformation of pPDM cross-linked scallop S1 is very close to that of scallop ADP-form, though their atomic coordinates are not yet open to public. So, we checked biochemically, whether the structure of our material; i.e. pPDM cross-liked vertebrate S1, is similar to any other configuration. According to the X-ray atomic model of S1, essential light chain (ELC) twines around the proximal portion of the lever-arm and has a unique cysteinyl residue; Cys 177 (for ELC1) or Cys 136 (for ELC2), that directly faces to the motor-domain. Because of such geometry, the fluorescent dye attached to that cystein might be expected to senseitively report the environmental change according to the formation of intermediates in ATPase cycle. We put 6-bromoacetyl-2-dimethyl aminonaphthalene (BD) specifically to the cysteinyl residue and monitored its fluorescent spectra of pPDM-S1, as well as S1-ADP/metallofluoride complex. Fig. 4 exhibits the fluorescence spectra taken from various molecular species. It is clear that the spectrum of pPDM-S1, or the environment of the dye, is distinct from that of any other molecular species, suggesting that pPDM-S1/ADP would take a unique configuration.

According to the interpretation of Cohen's team, pPDM-S1/ADP structure, together with S1/ADP-form, is classified as actin-detached configuration that is not directly involved in force development processes.[22] They found that pPDM crosslinked Cys-693 (SH2) with Lys-705 in scallop S1, and not Cys-703 (SH1) as in vertebrate S1. Though the latter two resudues are very close in the primary sequence, their azimuth angles along the helix are almost opposite. Such difference in reactivity might be the reflection of some delicate functional difference between vertebrate and scallop myosin. If we consider the frequent occurrence of this configuration but attached to actin during sliding, we cannot help postulating the role of this unusual configuration as more active one that would participate in the crossbridge cycle.

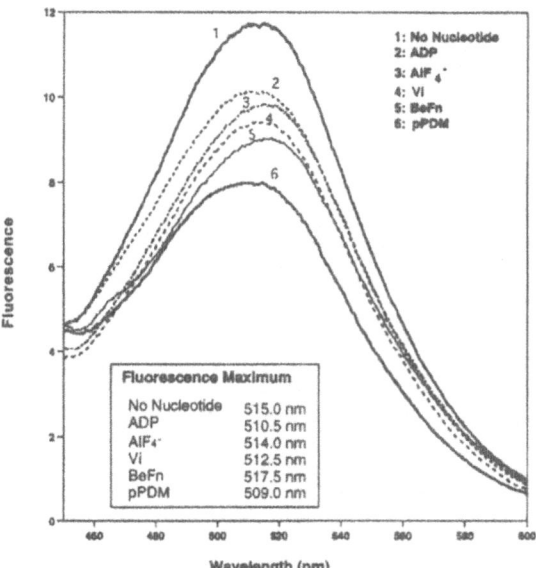

Figure 4. Fluorescence emission spectra of pPDM-crosslinked BD-S1 and BD-S1/ADP/Pi analogue ternary complexes. Cys177 (for ELC1) or Cys136 (for ELC2) was specifically labeled with a fluorescent dye (BD). The SH1 and SH2 of the labeled S1 were crosslinked by pPDM. Ternary complexes with ADP·Pi analogues were formed by incubation of BD-S1 with 1mM ADP and 1mM phosphate analogues of AlF$_4$, BeFn or Vi in 120 mM NaCl, 30 mM Tris-HCl pH 7.5, and 2 mM MgCl$_2$ at 25 °C for 3h in the dark. The emission spectra were recorded through 400-600 nm wavelength range by excitation at 390 nm.

REFERENCES

1. A.F. Huxley, and R. Niedergerke, Structural changes in muscle during contraction. Interference microscopy of living muscle fibres, Nature, 173, 971-973 (1954).

2. H.E. Huxley, and J. Hanson, Changes in the cross-striations of muscle during contraction and stretch and their structural interpretation, Nature, 173, 973-976 (1954)

3. A.F. Huxley, and R.M. Simmons, Proposed mechanism of force generation in striated muscle, Nature, 233, 533-538 (1971).

4. R.D. Vale, and R.D. Milligan, The way things move: looking under the hood of molecular motors, Science, 288, 88-95 (2000).

5. I. Rayment, W.R. Rypniewski, K. Schmidt-Bäse, R. Smith, D.R. Tomchick, M.M. Benning, D.A. Winkelman, G. Wesenberg, and H.M. Holden, Three-dimensional structure of myosin subfragment-1: A molecular motor, Science, 261, 50-58 (1993).

6. A.J. Fisher, C.A. Smith, J. Thoden, R. Smith, K. Sutoh, H.M. Holden, and I. Rayment, X-ray structures of the myosin motor domain of dictyostelium discoideum complexed with MgADP. BeFx and MgADP. AlF$_4$, Biochemistry, 34, 8960-8972 (1995).

7. R. Dominguez, Y. Freyzon, K.M. Trybus, and C. Cohen, Crystal structure of vertebrate smooth muscle myosin motor domain and its complex with the essential light chain: Visualizatioin of the pre-power stroke state, Cell, 94, 559-571 (1998).

8. A. Houdusse, A.G. Szent-Gyorgyi, and C. Cohen, Three conformational states of scallop myosin S1, Proc. Natl. Acad. Sci.USA, 97, 11238-11243 (2000).

9. W. Kabsch, H.G. Mannherz, D. Suck, E.F. Pai and K.C. Holmes, Atomic structure of the actin: DNase I complex, Nature, 347, 37-44 (1990).

10. K.C. Holmes, D. Popp W. Gebhard, and W. Kabsch, Atomic model of the actin filament, Nature, 347, 44-49 (1990).

11. J. Kron, and J.A. Spudich, Fluorescent actin filaments move on myosin fixed to a glass surface, Proc. Natl. Acad. Sci. USA, 83, 6272-6276 (1986).

12. Y. Harada, A. Noguchi, A. Kishino, and T. Yanagida, Sliding movement of single actin filaments on one-headed myosin filaments, Nature, 326, 805-808 (1987).

13. J.E. Heuser, Procedure for freeze-drying molecules adsorbed to mica flakes, J. Mol. Biol., 169, 155-195 (1983).

14. E. Katayama, Quick-freeze deep-etch electron microscopy of the actin-heavy meromyosin complex during the *in vitro* motility assay, J. Mol. Biol., 278(2), 349-367 (1998).

15. E. Katayama, G. Ohmori and N. Baba, Three-dimensional image analysis of myosin head in function as captured by quick-freeze deep-etch replica electron microscopy, Adv. Exp. Med. Biol., 453, 37-45 (1998)

16. E. Katayama, T. Shiraishi, K. Oosawa, N. Baba and S. Aizawa, Geometry of the flagellar motor in the cytoplasmic membrane of *Salmonella Typhimurium* as determined by stereo-photogrammetry of quick-freeze deep-etch replica images, J. Mol. Biol., 255, 458-475 (1996).

17. S. Maruta, G.D. Henry, B.D. Sykes and M. Ikebe, Formation of the stable myosin-ADP-aluminium fluoride and myosin-ADP-beryllium fluoride complexes and their analysis using 19F-NMR, J. Biol. Chem., 268, 7093-7100 (1993).

18. S. Maruta, Y. Uyehara, K. Homma, Y. Sugimoto and K. Wakabayashi, Formation of the myosin-ADP-gallium fluoride complex and its solution structure by small-angle synchrotron X-ray scattering, J. Biochem., 125, 177-185 (1999).

19. S. Maruta, T. Aihara, Y. Uyehara, K. Homma, Y. Sugimoto, and K. Wakabayashi, Solution structure of myosin-ADP-MgFn ternary complex by fluorescent probes and small-angle synchrotron X-ray scattering, J. Biochem., 128, 677-684 (2000).

20. E. Reisler, M. Burke, S. Himmelfarb, and W.F. Harrington, Spatial proximity of the two essential sulfhydryl groups of myosin, Biochemistry, 13, 3837-3840 (1974).

21. J.A. Wells and R.G. Yount, Chemical modification of myosin by active-site trapping of metal-nucleotides with thiol crosslinking reagents, Methods. Enzymol., 85, 93-116 (1982).

22. D.M. Himmel, S. Gourinath, L. Reshetnikova, Y. Shen, A.G. Szent-Gyorgyi, and C. Cohen, Crystallographic findings on the internally uncoupled and near-rigor states of myosin: further insights into the mechanics of the motor, Proc. Natl. Acad. Sci. USA, 99, 12645-12650 (2002).

HELICAL ORDER IN MYOSIN FILAMENTS AND EVIDENCE FOR ONE LIGAND INDUCING MULTIPLE MYOSIN CONFORMATIONS

Leepo C. Yu, Sengen Xu, Jin Gu, Howard D. White, and Gerald Offer[*]

1. INTRODUCTION

The basic processes of muscle contraction are well understood: it is a result of cyclic interactions between myosin and actin, driven by the energy of ATP hydrolysis. Since the availability of the crystal structures of the contractile proteins, and with the advent of single molecule assays, the field has made great strides in understanding the underlying processes. However, the details of the mechanism of transduction of chemical to mechanical energy still remain largely unresolved. One of the obstacles is that most of the studies at the molecular level are based on isolated, *in vitro* systems, e.g. the atomic structure of the myosin head is known but not its complex with actin and EM reconstruction is based on isolated filaments. The link between the information obtained from the *in vitro* systems and the actual processes occurring in intact muscle is still largely missing. The aim of our efforts is to provide such a link.

The biochemistry of ATP hydrolysis by actin-myosin in solution can be described by the scheme shown below (Scheme I):

$$
\begin{array}{ccccccc}
M{\bullet}ATP & \leftrightarrow & M{\bullet}ADP{\bullet}P_i & \leftrightarrow & M{\bullet}ADP & \leftrightarrow & M \\
\updownarrow & & \updownarrow & & \updownarrow & & \updownarrow \\
A{\bullet}M{\bullet}ATP & \leftrightarrow A{\bullet}M{\bullet}ADP{\bullet}P_i & \leftrightarrow & A{\bullet}M{\bullet}ADP & \leftrightarrow & A{\bullet}M
\end{array}
$$

We have been systematically determining the relation between the structures of the filaments in muscle and the biochemical states found in solution. X-ray diffraction data from each of the eight intermediate states in Scheme (I) have been analyzed quantitatively with emphasis on the weak binding states in relaxed muscle, since the rigor and the A.M.ADP states have been studied extensively by others earlier (e.g.. Huxley, 1968; Takezawa et al., 1999).

[*] Leepo C. Yu, Sengen Xu, Jin Gu, and Gerald Offer, Laboratory of Muscle Biology, NIAMS, NIH, Bethesda, MD 20892 USA. Howard D. White, Eastern Virginia Mediacal School, Norfolk, VA 23501, USA. Gerald Offer, University of Bristol, University Walk, Bristol BS8 1TD, UK.

Molecular and Cellular Aspects of Muscle Contraction
Edited by H. Sugi, Kluwer Academic/Plenum Publishers, 2003

Specifically we have shown that (1) in a relaxed muscle, the myosin head in the A.M.ATP state is attached to actin in a wide range of orientations while maintaining some degree of helical distribution on the thick filament. The results suggested a high degree of flexibility in the actomyosin complex. Furthermore, the binding site on actin differs from that found for binding in the rigor state. The data also revealed that the second (detached) head of myosin is located close to the surface of the thick filament backbone (Xu et al., 2002; Gu et al., 2002). (2) For the A.M.ADP.P$_i$ state, although the results are not yet complete, the binding characteristics are clearly different from those before hydrolysis (Xu et al., 2001). The myosin heads appear to be better ordered on the thick filament in this attached state. The implications of this finding are not yet fully explored. (3) In order to get a fuller picture of the cross-bridge cycle we also need to understand the structure(s) of the detached myosin. One way to study this is by stretching the muscle to non-overlap sarcomere length. One striking feature here is the disorder⇌order transition in the thick filament, which was first observed by Wray (Wray, 1987) and by T. Wakabayashi (Wakabayashi et al., 1988). The remainder of this presentation will focus on our recent results addressing the underlying process of this remarkable transition and the relationship between it and the conformational changes in the myosin head.

Effects of Temperature and Ligand on Helical Order in Myosin Filaments

Mammalian and avian myosin filaments show a remarkable dependence on temperature. Above 20°C the X-ray diffraction pattern of relaxed rabbit skeletal muscle in the presence of ATP is similar to that of frog muscle indicating that the myosin heads are helically arranged on the surface of the filament with an axial translation of 143 Å and a repeat of 429 Å. However, on cooling, the X-ray pattern indicates increasing disorder (Wray, 1987; Wakabayashi et al., 1988; Schrumpf and Wray, 1992; Xu et al., 1997; Xu et al., 1999). The degree of order is also markedly dependent on the ligand bound to myosin. Disorder occurs at all temperatures in the absence of nucleotide or with bound ADP, GTP or ATPγS, while CTP promotes order more than ATP (Xu et al., 1999).

Wray initially suggested that the helical order might require ADP and P$_i$ rather than ATP to be bound at the active site (Wray, 1987). The M.ATP ⇌ M.ADP.P$_i$ equilibrium is temperature dependent, so that at low temperatures M.ATP predominates and at high temperatures, M.ADP.P$_i$. In apparent support of this hypothesis we found only small differences between K_{hyd}^{app}, the equilibrium constant for the hydrolytic step for rabbit myosin subfragment-1 in solution and K_{ord}^{app}, the equilibrium constant for the disorder-order transition in relaxed rabbit psoas muscle. Similar results were found for a broad range of nucleotides including CTP, ATPγS and GTP.

An alternative hypothesis for explaining the effect of ligands and temperature on order is that they both influence myosin conformation and that helical order requires only one of the conformations. In our previous experiments (Xu et al., 1999), the effect of temperature was not tested independently since changes in temperature were accompanied by changes in bound ligand due to hydrolysis. In the present study, we used ligands that are non-hydrolyzable to discriminate between these hypotheses.

2. MATERIALS AND METHODS

Muscle Preparation and Solutions

Muscle fibers stretched to non-overlap

All experiments were performed on chemically skinned bundles of rabbit M psoas major stretched to non-overlap ($\sim$ 4.2 μm), so that only detached myosin heads were studied and the affinity for nucleotide was not modulated by actomyosin interactions ((Frisbie et al., 1997;Frisbie et al., 1998;Resetar and Chalovich, 1995;Xu et al., 1999)).

Solutions

The following solutions were used:

(1) Relaxing (MgATP) solution contained 2 mM MgATP, 2 mM MgCl$_2$, 2 mM EGTA, 5 mM DTT, 10 mM imidazole, 10 mM creatine phosphate, 133 mM potassium propionate, pH 7.0, ionic strength (μ) = 170 mM. To complete the ATP-backup system 109 u/ml of creatine phosphokinase (CPK) was present.

(2) Rigor solution contained 2.5 mM EGTA, 2.5 mM EDTA, 10 mM imidazole, 5 mM DTT, 150 mM potassium propionate, pH 7.0, μ= 170 mM.

(3) NTP-free solution contained 2 mM EGTA, 4 mM MgCl$_2$, 10 mM imidazole, 50 mM glucose, 147 mM potassium propionate, 5 mM DTT, 2 u/ml hexokinase, 0.25 mM Ap5A, pH 7.0, μ= 170 mM.

(4) AMPPNP-containing solution was made by adding 2 mM AMPPNP to the NTP-free solution with the final ionic strength adjusted to 170 mM; similarly for the ADP.Vi-containing solution by adding 2 mM ADP and 3 mM Na$_3$VO$_4$; for the ADP.AlF$_4^-$ containing solution by adding 2 mM ADP, 2 mM AlCl$_3$ and 10 mM NaF; for the ADP.BeF$_x^-$ containing solution by adding 2 mM ADP, 2 mM BeCl$_2$ and 10 mM NaF. Stock ortho-vanadate solution was prepared according to (Goodno, 1982). BeCl$_2$, AlCl$_3$ and NaF solutions were prepared according to (Werber et al., 1992).

The fiber bundles were immersed in solutions containing non-hydrolyzable ligand for 20-30 minutes before taking the X-ray patterns. The temperature of the bathing solution in the chamber was maintained at pre-set temperatures $\pm$1°C. The temperature ranged between 5 - 35°C.

X-ray Source, Camera, and Detector System

The experiments were performed at beamline X27C (Advanced Polymer PRT) at the National Synchrotron Light Source (NSLS), Brookhaven National Laboratory (BNL), Upton, NY (for details see (Xu et al., 2002)).

To correct for contributions from the thin filament and the thick filament backbone to myosin layer lines, difference patterns were obtained: the patterns obtained in the absence of nucleotide were subtracted from those in the presence of ligands.

The layer lines in the difference patterns index on the 430 Å repeat and therefore arise from the myosin filaments. Consequently, the intensities of the myosin layer lines are directly proportional to the square of the mass of the myosin heads in the helical array scattering coherently. The equilibrium constant of the disorder⇌order transition is defined as

$$K_{ord}^{app} = \frac{a}{1-a}$$

where a is the fraction of the heads in the ordered state. In our earlier study on the disorder$\rightleftharpoons$order transition of the myosin filament, the fraction of heads in the ordered state in ATP at 25°C was determined to be 0.93 (Xu et al., 1999). Hence, to derive the fraction in the ordered state in a ligand at any temperature T°C, the amplitude obtained under these conditions was divided by the amplitude obtained in the presence of ATP at 25°C and multiplied by 0.93.

The dependence on temperature of the equilibrium constants for the disorder$\rightleftharpoons$order transition or a conformational change can be expressed by the van't Hoff relation:

$$\ln K = \frac{-\Delta H^{o}}{RT} + \frac{\Delta S^{o}}{R}$$

where K is the equilibrium constant, T is the absolute temperature, ΔH^{o} is the enthalpy of the transition, ΔS^{o} is the entropy, and R is the gas constant. The dependence on temperature of these equilibrium constants were fit using the non-linear least-squares fitting routines in the software package Scientist (Micromath, Ogden, Utah).

3. RESULTS

Effect of Temperature on Helical Order

(1) In the presence of transition state analogues

ADP.V_i and ADP.AlF$_4$ are analogues of the transition state of ATP hydrolysis. According to the hypothesis that helical order requires hydrolysis products bound at the active site, in the presence of the ADP.Vi and ADP.AlF$_4$ the heads should be fully ordered at all temperatures. At temperatures ~25°C, we did indeed observe strong myosin layer lines indexing on a ~430 Å repeat extending out to the 6th layer line (Fig 1). This is similar to the result of Takemori (Takemori et al., 1995) who showed that in *frog* muscle the patterns with ATP and ADP.Vi were similar. However, on cooling to lower temperatures, e.g. 5°C, the layer lines became very weak, showing very little order (Fig. 1). The effect of cooling was reversible (data not shown).

(2) In the presence of ATP analogues

ADP.BeF$_x$ and AMPPNP were used as the ATP analogues (Fisher et al., 1995; Frisbie et al., 1998). According to Wray's original hypothesis (Wray, 1987), there should be no helical order of the myosin heads in the presence of these analogues. At 15°C, the intensity of the myosin layer lines with both of these ATP analogues was low

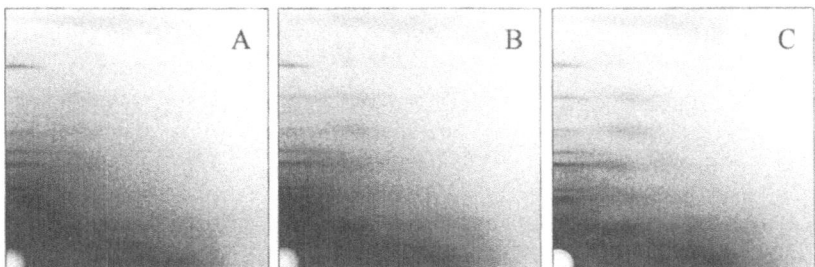

Figure.1. X-ray diffraction patterns obtained from a single bundle of rabbit psoas muscle fibers with sarcomere length 4.0 μm in the MgADP.VO$_4$ solution at (A) 5°C, (B) 15°C and (C) 25°C. Myosin layer lines increased with temperature. Note: The solution background has been subtracted.

and on further cooling to 5°C the myosin layer lines became very weak, showing little helical order (intensity too weak to be measured accurately). However, at temperatures $\geq$25°C with either of these ATP analogues we observed clear myosin layer lines indexing on a ~430 Å repeat extending out to the 6th layer line.

(3) Intensity profiles indicate the same helical structures.

The intensity profiles in the direction *parallel* to the meridian of the six myosin layer lines obtained at 25 °C are shown in Fig. 2a and compared with that in ATP. The intensity distributions *along* the first myosin layer line are shown in Fig. 2b and also compared with that in ATP. Although the intensities in the presence of the analogues are all lower than in ATP, the profiles are similar and the positions of the main peaks remain the same, indicating that the same helical structure is present but with varying degrees of order.

(4) Analysis of the van't Hoff plots

Fig. 3 compares the dependence on temperature of K_{ord}^{app} in the presence of analogues with that in ATP. The curves are non-linear and tend to plateau at higher temperatures suggesting that more than one transition is occurring. In Figure 3 we also plotted the recent solution results of Málnási-Csizmadia et al. (Málnási-Csizmadia et al., 2000) who studied the fluorescence change from a mutant of *dictyostelium* S1 (W501+) with a single tryptophan in the relay loop. In the presence of ATP, the following equilibria exist:

$$M + ATP \rightleftharpoons M.ATP \rightleftharpoons M^{\dagger}.ATP \rightleftharpoons M^{*}.ATP \rightleftharpoons M^{*}.ADP.Pi \quad ....(1)$$

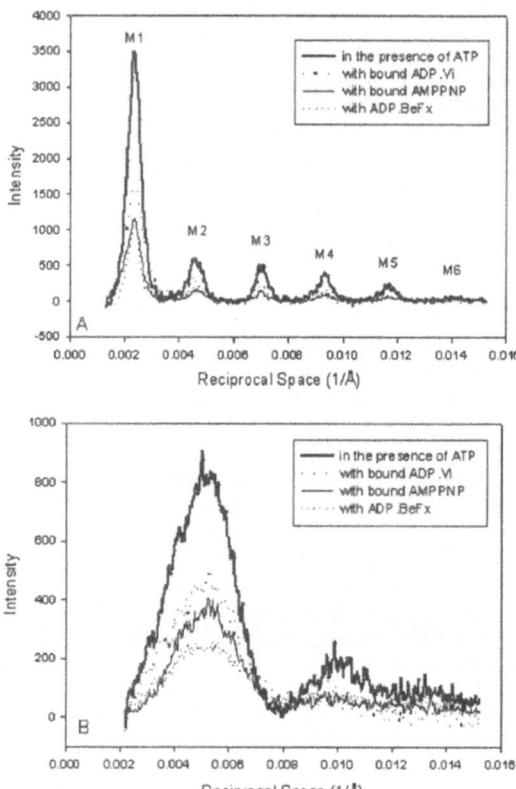

Figure 2. Comparison of the intensities of the myosin layer lines in the presence of different ligands at 25°C. (a) profiles of integrated intensity in a slice *parallel* to the meridian from R = 0.00280 Å^{-1} to R = 0.00795 Å^{-1} (b) profiles of integrated intensity *along* the first myosin layer line (i) in the presence of ATP, (ii) with bound ADP.Vi, (iii) with bound AMPPNP, (iv) with bound ADP.BeFx .

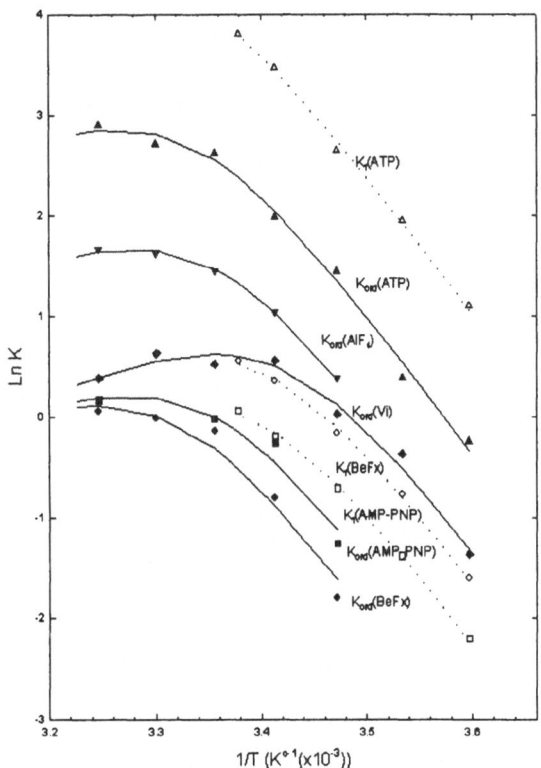

Figure 3. van't Hoff plots showing the effect of temperature on the disorder⇌order transition in rabbit myosin filaments (K_{ord}) and on the conformational change in the W501+ mutant of dictyostelium S1 measured by fluorescence (K_f) (data of Málnási-Csizmadia (2001)). The experimental points for the disorder⇌order transition are (i) in the presence of ATP (ii) with bound ADP.AlF$_4$ (iii) with bound ADP.Vi (iv) with bound AMPPNP (v) with bound ADP.BeFx. Those for the conformational change are (vi) in the presence of ATP (vii) with bound AMPPNP (viii) with bound ADP.BeFx . The lines (solid for X-ray data; dot for fluorescence data) are theoretical lines with ΔH_{AB} = -154 kJ/mol and ΔH_{BC} = +116 kJ/mol with ΔS_{AB} and ΔS_{BC} values chosen to best fit the data.

where M is the apo conformation, $M^\dagger$ is a conformation with slightly reduced fluorescence, and $M^\bullet$ is a conformation with high fluorescence. The second equilibrium is essentially irreversible, so the $M^\dagger$ and $M^\bullet$ conformations dominate.

The hydrolysis of ATP at the active site and the conformational change from $M^\dagger$ to $M^\bullet$, previously supposed to be coincident, are separate processes which are not absolutely coupled. So after ATP hydrolysis the heads are largely in the $M^\bullet$ conformation, but before hydrolysis both $M^\dagger$ and $M^\bullet$ conformations coexist. The high fluorescence $M^\bullet$ state can be assigned to be in the closed conformation determined by X-ray crystallography (Geeves and Holmes, 1999), while the $M^\dagger$ state tentatively can be assigned to the "detached" crystal conformation (Houdusse et al., 1999; Houdusse et al., 2000). Málnási-Csizmadia et al. showed that there was a temperature dependence of $K_{\dagger\bullet}^{app}$, the equilibrium constant for transitions between the low fluorescence conformation ($M^\dagger$) and the high fluorescence conformation ($M^\bullet$), depending on the ligand bound at the active site. The van't Hoff plot of $K_{\dagger\bullet}^{app}$ for each ligand is non-linear and remarkably similar to that of K_{ord}^{app} obtained in fibers.

The simplest model that can accommodate this non-linearity is one in which there are two disordered states, A and B, assumed to be of equally low fluorescence, in equilibrium with an ordered, high-fluorescence state C. Because the conformational data was obtained by observing the fluorescence of a single tryptophan residue in the relay loop, the implication is that the high fluorescence state C, predominant at high temperatures, originates from the $M^\bullet$ (closed) conformation and the A and B states from conformations different from C.

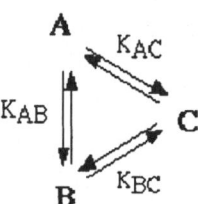

We distinguish between A and B by defining A as having the higher enthalpy so that ΔH_{AB} is negative.

Remarkably, good fits were obtained to the data with identical enthalpy changes (ΔH_{AB} and ΔH_{BC}) for all ligands and for both the conformational transition and the disorder⇌order transition. ΔH_{AB} is ~-154 kJ/mol and ΔH_{BC} is ~+116 kJ/mol (implying ΔH_{AC} is ~-38 kJ/mol) for all ligands. The large enthalpy difference between A and B states suggests that they have significantly different conformations. The differences in free energy (and thus equilibrium constant) characteristic of each of the ligands and myosins were assigned to differences in the entropy values (ΔS_{AB} and ΔS_{BC}). At low temperature (~5°C) the predominant state is the low enthalpy state B. As the temperature is raised, the proportion of the higher enthalpy states A and C increases at the expense of B, so the helical order and fluorescence intensity increase. But as ΔH_{AC} is small, the ratio of [A]/[C] is relatively insensitive to temperature so even at high temperatures, C coexists with A explaining why the van't Hoff plots of K_{ord}^{app} and $K_{\dagger\bullet}^{app}$ plateau. In the presence of ATP the fraction in the A state is small even at high temperatures (for details of the model, see (Xu et al., 2003)).

4. DISCUSSION

The results that helical order in the thick filament was markedly temperature dependent in the presence of the non-hydrolysable ATP analogues and the transition state analogues argue strongly that the disorder⇌order transition is not *directly* associated with nucleoside triphosphate hydrolysis, but rather with change of myosin conformation.

Furthermore, the data indicate that it is the closed myosin conformation that is required for helical order. We note that the series of ligands with increasing effectiveness in promoting helical order we have found in this and previous work is: none ~ ADP ~ ATPγS ~ GTP < ATP (unhydrolyzed) ~ AMPPNP ~ ADP.BeF$_x$ < ADP.Vi < ADP.AlF$_4$ < ADP.P$_i$ < CDP.P$_i$. This closely parallels the relative order of ligands at promoting the formation of the closed conformation. With ADP and ATPγS the M† conformation of the myosin predominates even at higher temperature (Malnasi-Csizmadia et al., 2000). With these nucleotides we have observed only disorder. The equilibrium constant $K_{\dagger*}^{app}$ for the transition between M† and M* for ADP.BeF$_x$ and AMPPNP at 5°C was 0.20 and 0.11 and at 25°C it was 1.74 and 1.06 respectively (Málnási-Csizmadia et al., 2000). In the presence of ADP.V$_i$ and ADP.AlF$_4$, based on fluorimetric measurements, the closed conformation M* is favored more than ADP.BeF$_x$, consistent with our findings. In the presence of ATP at 25 °C the closed conformation is strongly favored; under these conditions a high level of helical order was found. Evidence is thus compelling that the M* (closed) conformation is required for helical order.

It might be argued that helical ordering of the thick filament involves an additional step beyond the formation of the closed conformation, for example the interaction of heads in the closed conformation with the heads, head-tail junctions or tails of neighboring myosin molecules in the filament. However, the enthalpy changes for the transitions between myosin conformations (ΔH_{AB}, ΔH_{BC} and ΔH_{AC}) are independent of the bound ligand, and they are the same for the disorder⇌order transition in filaments and for the conformational changes in solution. This implies that the same underlying processes are involved. Since no filament structure is present in the solution studies, it is unlikely that disorder⇌order transition in the thick filament involves processes other than conformational changes in myosin. Quantitative modeling also supports the idea that conformational change in myosin is directly coupled to disorder⇌order. (For detailed analysis and modeling, see Xu et al., (2003)). Therefore, it is reasonable to conclude that no additional step is required for helical ordering.

The greater ordering when the myosin heads are in the closed conformation may be because the heads are inherently stiffer in that conformation, reducing the thermal fluctuation of the heads and thereby the helical arrangement of the myosin molecules in the thick filament becomes clearly observable. The idea of stiffness change is consistent with the crystal structures of S1, where the four subdomains appear to be only loosely linked in the "detached" conformation, and possibly also in the open conformation, but closely interact in the closed conformation (Houdusse et al., 2000). It is important in this context to note that even in the ordered state the heads are by no means completely rigid but show a degree of thermal disorder (Lowy et al., 1991; Malinchik et al., 1997); in the model of Malinchik et al. (Malinchik et al., 1997) for the "ordered" state the root mean square isotropic displacement was ~20 Å.

One outcome of this study is that with each of the analogues bound at the active site, three states, A, B and C coexist in equilibrium, the proportions varying with temperature. The crystal structures of myosin have largely been interpreted to mean that the conformation of myosin was absolutely determined by the nature of the ligand bound (see for example (Fisher et al., 1995;Houdusse et al., 2000)). However, several recent X-ray crystallographic studies have given results that do not fit this simple picture. The *dictyostelium* motor domain with ADP.BeFx bound has been crystallized in the closed (Geeves and Holmes, 1999) as well as the open conformation (Fisher et al., 1995). Similarly, scallop S1 with ADP bound has been crystallized both in the "detached" and open conformations (Houdusse et al., 2000). Thirdly, the chicken smooth motor domain attached to the essential light chain domain crystallized in the closed conformation whether ADP.AlF$_4$ or ADP.BeF$_x$ were bound (Dominguez et al., 1998). These results, however, can be explained by the concept that one ligand can induce multiple conformations, i.e. the idea "one ligand, one conformation" is inadequate. Rather, myosin appears to conform to the idea of "one ligand, multiple conformations".

Another consequence of this study and that of Urbake and Wray (Urbanke and Wray, 2001) and Málnási-Csizmadia et al. (Málnási-Csizmadia et al., 2000) is that temperature primarily affects conformation rather than the equilibrium of the ATP hydrolysis step. Hence, the coupling between myosin conformation and ATP hydrolysis is not absolute, as had been formerly thought.

In summary, by studying the helical order of myosin heads in the thick filament we have demonstrated that the global conformation of unmodified myosin in a physiological system is sensitive to temperature and the nature of the ligand bound to the active site. Our results strongly support the view that myosin exists in several conformations, the proportions of which depend on temperature as well as the ligand bound. Helical order in the thick filament is a signature for only one of these, the closed conformation (Geeves and Holmes, 1999). Together with the solution studies (Jahn et al., 1999; Málnási-Csizmadia et al., 2000; Málnási-Csizmadia et al., 2001; Urbanke and Wray, 2001), our findings provide a consistent explanation of why certain complexes of myosin with nucleotide crystallize in more than one conformation.

ACKNOWLEDGMENT

The authors wish to thank Drs. John Wray, Clive Bagshaw, Andras Málnási-Csizmadia and Ralph Yount for helpful discussions. We also thank the staff at beamline X27C at the National Synchrotron Radiation Laboratory for their expert assistance.

REFERENCES

Dominguez,R., Y.Freyzon, K.M.Trybus, and C.Cohen. 1998. Crystal structure of a vertebrate smooth muscle myosin motor domain and its complex with the essential light chain: visualization of the pre- power stroke state. *Cell* 94:559-571.

Fisher,A.J., C.A.Smith, J.B.Thoden, R.Smith, K.Sutoh, H.M.Holden, and I.Rayment. 1995. X-ray structures of the myosin motor domain of Dictyostelium discoideum complexed with MgADP.BeFx and MgADP.AlF4-. *Biochemistry* 34:8960-8972.

Frisbie,S.M., J.M.Chalovich, B.Brenner, and L.C.Yu. 1997. Modulation of cross-bridge affinity for MgGTP by Ca2+ in skinned fibers of rabbit psoas muscle. *Biophys. J.* 72:2255-2261.

Frisbie,S.M., S.Xu, J.M.Chalovich, and L.C.Yu. 1998. Characterizations of cross-bridges in the presence of saturating concentrations of MgAMP-PNP in rabbit permeabilized psoas muscle. *Biophys. J.* 74:3072-3082.

Geeves,M.A. and K.C.Holmes. 1999. Structural mechanism of muscle contraction. *Annu. Rev. Biochem.* 68:687-728.

Goodno,C.C. 1982. Myosin active-site trapping with vanadate ion. *Methods Enzymol.* 85 Pt B:116-23.:116-123.

Gu,J., S.Xu, and L.C.Yu. 2002. A model of cross-bridge attachment to actin in the A*M*ATP state based on x-ray diffraction from permeabilized rabbit psoas muscle. *Biophys. J* 82:2123-2133.

Houdusse,A., V.N.Kalabokis, D.Himmel, A.G.Szent-Györgyi, and C.Cohen. 1999. Atomic structure of scallop myosin subfragment S1 complexed with MgADP: a novel conformation of the myosin head [In Process Citation]. *Cell* 97:459-470.

Houdusse,A., A.G.Szent-Györgyi, and C.Cohen. 2000. Three conformational states of scallop myosin S1. *Proc. Natl. Acad. Sci. USA* 97:11238-11243.

Huxley,H.E. 1968. Structural difference between resting and rigor muscle: evidence from intensity changes in the low-angle equitorial X-ray diagram. *J. Mol. Biol.* 37:507-520.

Jahn, W., Urbanke, C., and Wray, J. Fluorescence temperature jump studies of myosin S1structure. Biophys.J 76, A146. 1999.

Lowy,J., D.Popp, and A.A.Stewart. 1991. X-ray studies of order-disorder transitions in the myosin heads of skinned rabbit psoas muscles. *Biophys. J.* 60:812-824.

Malinchik,S., S.Xu, and L.C.Yu. 1997. Temperature-Induced Structural changes in the Myosin Thick filament of skinned Rabbit Psoas Muscle. *Biophys. J.* 73:2304-2312.

Málnási-Csizmadia,A., D.S.Pearson, M.Kovacs, R.J.Woolley, M.A.Geeves, and C.R.Bagshaw. 2001. Kinetic resolution of a conformational transition and the ATP hydrolysis step using relaxation methods with a Dictyostelium myosin II mutant containing a single tryptophan residue. *Biochemistry* 40:12727-12737.

Málnási-Csizmadia,A., R.J.Woolley, and C.R.Bagshaw. 2000. Resolution of conformational states of Dictyostelium myosin II motor domain using tryptophan (W501) mutants: implications for the open-closed transition identified by crystallography. *Biochemistry* 39:16135-16146.

Resetar,A.M. and J.M.Chalovich. 1995. Adenosine 5'-(gamma-thiotriphosphate): an ATP analog that should be used with caution in muscle contraction studies. *Biochemistry* 34:16039-16045.

Schrumpf,M. and J.Wray. 1992. Structural effects of Al-F and Be-F as analogues of Pi in skeletal muscle myosin. *J. Mus. Res. & Cell. Mot.* 13:254a.

Takemori,S., M.Yamaguchi, and N.Yagi. 1995. An X-ray diffraction study on a single frog skinned muscle fiber in the presence of vanadate. *J. Biochem. (Tokyo.)* 117:603-608.

Takezawa,Y., D.S.Kim, M.Ogino, Y.Sugimoto, T.Kobayashi, T.Arata, and K.Wakabayashi. 1999. Backward movements of cross-bridges by application of stretch and by binding of MgADP to skeletal muscle fibers in the rigor state as studied by x-ray diffraction. *Biophys. J* 76:1770-1783.

Urbanke,C. and J.Wray. 2001. A fluorescence temperature-jump study of conformational transitions in myosin subfragment 1. *Biochem. J* 358:165-173.

Wakabayashi,T., T.Akiba, K.Hirose, A.Tomioka, M.Tokunaga, C.Suzuki, C.Toyoshima, K.Sutoh, K.Yamamoto, T.Matsumoto, K.Sacki, and Y.Amemiya. 1988. Temperature induced changes of thick filament and location of the functional site of myosin. *In* Molecular Mechanism of Muscle Contraction. H.Sugi and G.H.Pollack, editors. Plenum Publishing Co., New York. 39-48.

Werber,M.M., Y.M.Peyser, and A.Muhlrad. 1992. Characterization of stable beryllium fluoride, aluminum fluoride, and vanadate containing myosin subfragment 1-nucleotide complexes. *Biochemistry* 31:7190-7197.

Wray,J. 1987. Structure of relaxed myosin filaments in relation to nucleotide state in vertebrate skeletal muscle. *J. Mus. Res. & Cell. Mot.* 8:62a.

Xu, S, Gu, J, Melvin, G., and Yu, L. C. 2001. Evidence that the conformation of the actomyosin complex with bound ADP.Pi (the A.M.ADP.Pi state) differs from that in the A.M.ATP state. Biophys.J 80, 267a.

Xu,S., S.Malinchik, D.Gilroy, Th.Kraft, B.Brenner, and L.C.Yu. 1997. X-ray diffraction studies of cross-bridges weakly bound to actin in relaxed skinned fibers of rabbit psoas muscle. *Biophys. J.* 73:2292-2303.

Xu,S., G.Offer, J.Gu, H.D.White, and L.C.Yu. 2003. Temperature and ligand dependence of conformation and helical order in myosin filaments. *Biochemistry,* in press.

Xu,S., J.Gu, G.Melvin, and L.C.Yu. 2002. Structural characterization of weakly attached cross-bridges in the A*M*ATP state in permeabilized rabbit psoas muscle. *Biophys. J* 82:2111-2122.

Xu,S., J.Gu, T.Rhodes, B.Belknap, G.Rosenbaum, G.Offer, H.White, and L.C.Yu. 1999. The M.ADP.P(i) state is required for helical order in the thick filaments of skeletal muscle. *Biophys. J* 77:2665-2676.

EVIDENCE FOR THE INVOLVEMENT OF MYOSIN SUBFRAGMENT 2 IN MUSCLE CONTRACTION

Haruo Sugi[*], Tsuyoshi Akimoto,and Takakazu Kobayashi

1. INTRODUCTION

Muscle contraction results from alternate formation and breading of cross-links between the myosin head (subfragment 1; S-1), extending from the thick filament and a neighboring thin filament (Huxley A. F., 1957; Huxley H. E., 1969). The energy for contraction is supplied by ATP hydrolysis. Since the ATPase activity and actin binding site are localized in the S-1 region of myosin, S-1 is commonly believed to play a major role in muscle contraction. In fact, *in vitro* motility assay experiments have shown that S-1 alone is sufficient to produce force and move actin filaments. However, the ATP-dependent actin-myosin sliding observed in the assay systems is not the same as that actually taking place in muscle.

On the other hand, it has been proposed that melting and shortening in the proteolytically sensitive hinge region, lying between the short subfragment 2 (S-2) and light meromyosin segments of the myosin tail, contributes to force generation in muscle (Tsong et al., 1979; Applegate and Reisler, 1983; Ueno and Harrington, 1986, 1987). In support of this hypothesis, polyclonal anti-S-2 antibody has been shown to reduce Ca^{2+}-activated isometric force in glycerinated skeletal muscle fibers, while ATPase activity of the fibers and the initial unloaded shortening velocity of isolated myofibrils undergo little change (Lovell et al., 1988; Harrington et al., 1990). It has also been shown that, in the presence of antibody directed against a 20-amino acid peptide segment within the hinge region of cardiac myosin, movement of actin filaments in an *in vitro* motility assay is suppressed, while ATPase activity of myofibrils and purified S-1 remained unchanged (Margossian et al., 1991).

---

[*] Department of Physiology, School of Medicine, Teikyo University, Itabashi-ku, Tokyo 173-0003, Japan

In spite of the above reports, the possible involvement of S-2 region of myosin has not been widely accepted, while a number of *in vitro* experiments are being done using S-1 or heavy meromyosin (HMM) fragment of myosin, in which the S-2 region is not included. In this article, we will describe additional evidence for the involvement of myosin S-2 shortening in muscle contraction, which have been obtained in our laboratory using anti-S-2 polyclonal antibody, prepared in the laboratory of Harrington (Ueno and Harrington, 1984; Lovell et al., 1988).

2. EFFECT OF ANTI-S-2 ANTIBODY ON THE CONTRACTION CHARACTERISTICS IN DEMEMBRANATED RABBIT PSOAS MUSCLE FIBERS

In 1991, I was asked by Dr. Harrington to use polyclonal anti-S-2 antibodies prepared in his laboratory in our muscle mechanics experiments, and started to investigate the effect of anti-S-2 antibody on the contraction characteristics of demembranated rabbit psoas muscle fibers with interesting results (Sugi et al., 1992).

Single demembranated rabbit psoas muscle fibers (diameter, 40 - 80 μm) were mounted horizontally in an experimental chamber (0.1 ml) between a force transducer (compliance 0.1 mm / N, resonant frequency 5 kHz; Akers, Horten, Norway; AE801) and a servomotor (controlled by JCCX101 control unit; General Scanning, Watertown, MA; G100PD) by glueing both ends to the extension of the transducer and the servomotor with collodion. The servomotor contained a displacement transducer (differential capacitor) sensing the position of the motor arm. The compliance of the motor arm (length, 10 mm) at the point of attachment of the fiber was ≈ 0.2 mm / N when the servomotor system was operating in the length clamp mode. The sarcomere length of the fiber was measured by use of optical diffraction with HeNe laser light. The fiber was initially kept in relaxing solution at its slack length (≈ 3 mm) with sarcomere lengths of 2.2 - 2.3 μm, and maximally activated with contracting solution. All experiments were performed at room temperature (18 °C -20 °C).

2.1. Stiffness Measurement

Muscle fiber stiffness was continuously determined by applying small sinusoidal length changes (1 kHz) of fixed peak-to-peak amplitude (≈ 0.1 % of fiber length) with the servomotor and measuring the amplitude of resulting force changes. Typical muscle fiber stiffness and isometric force records in response to contracting solution are shown in Fig.1 (Insets). After each application of contracting solution, the fiber could be made to relax completely with relaxing solution, so that both stiffness and force records always started to change from "zero" baselines. As shown in Fig. 1, muscle fiber stiffness and isometric force always changed in parallel with each other, indicating that the stiffness-isometric force relation remains unchanged in the presence of anti-S-2 antibody. Since steady isometric force attained is progressively reduced with time after administration of the antibody, the antibody may progressively reduces the number of cross-bridges generating isometric force.

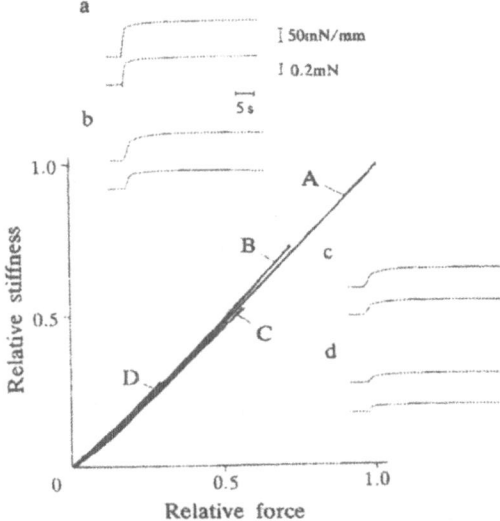

Figure 1. Relation between muscle fiber stiffness and isometric force during Ca^{2+}-activated force development of a single muscle fiber before (curve A) and 30 (curve B), 60 (curve C), and 90 (curve D) min after administration of anti-S-2 antibody (1.5 mg / ml). Both stiffness and force are expressed relative to control values in the absence of antibody. (Insets) Stiffness versus force curves A, B, C, and D were obtained from stiffness (upper traces) and force (lower traces) records a, b, c, and d, respectively (Sugi et al., 1992).

2.2 Force-Velocity Relation

The servomotor system operated either in the length control mode or in the force control mode. First, the system was in the length control (length clamp) mode so that the fiber contracted isometrically when it was maximally activated in contracting solution. After the fiber developed steady isometric force, the servomotor system was switched to the force control mode, and a ramp decrease in force (= load) from the steady force to zero was applied by feeding a ramp force decrease signal from the waveform generator to the servomotor system. The resulting shortening of the fiber was recorded in the digital wave memory together with the ramp decrease in force, and force-velocity (P-V) relation was obtained and displayed on the X-Y plotter after data processing with the microcomputer. The rate of force decrease was ≈ 5.5 mN / s.

The P-V relations in maximally activated fibers were determined either in the absence or the presence of antibody by applying ramp decreases in force from steady isometric forces to zero. Since the force (= load) on the fiber was continuously changing with time, shortening velocity was determined by averaging the first-time derivative of fiber length record for each consecutive time segment (duration, 1 ms) during the course of fiber shortening. As shown in Fig. 2A, the P-V curves, obtained at various levels of steady isometric force, were double hyperbolic in shape as with the P-V curve of single frog muscle fibers (Edman, 1988; Iwamoto et al., 1990), while the maximum shortening velocity remained unchanged in spite of the marked antibody-induced reduction of steady isometric force to < 30 % of control value. The P-V curves

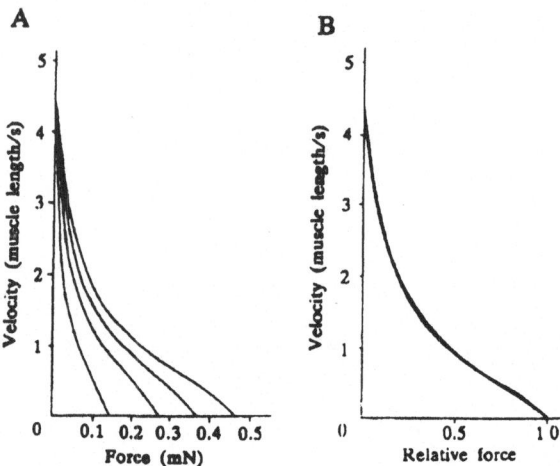

Figure 2. Effect of anti-S-2 antibody on *P-V* relation in a Ca^{2+}-activated single fiber. (A) *P-V* curves obtained (control) and 30, 60, 90 min after administration of anti S-2 antibody. Both velocities and forces are expected in absolute values. Note that maximum shortening velocity remains unchanged in spite of marked reduction of steady isometric force. (B) The same *P-V* curves in which forces are expressed relative to their respective steady initial isometric forces. Note that all curves are identical in shape(Sugi et al., 1992).

were, however, found to be identical in shape when velocities were replotted against forces expressed relative to their respective steady isometric forces (Fig. 2B), indicating that the *P-V* curves were scaled in proportion to steady isometric forces at which ramp force decreases were applied. As the maximum fiber shortening velocity remains unchanged by the reduction in number of cross-bridges interacting with actin (Huxley, 1957; Edman, 1979), these results again indicate that the antibody reduces the number of active cross-bridges by binding to the S-2 region of myosin molecules; the remaining 'native' cross-bridges continue to interact with actin without changing their kinetic properties.

3. DISSOCIATION OF ISOMETRIC FORCE DEVELOPMENT AND ATPASE ACTIVITY INDUCED BY ANTI-S-2 ANTIBODY

Mg-ATPase activity of the fibers during Ca^{2+}-activated isometric force development was recorded by the decrease of NADH during cleavage of ATP (Tanaka et al., 1977). A small fiber bundle consisting of two or three muscle fibers was mounted horizontally between the force transducer and a stainless-steel rod in the sample compartment (≈ 0.36 ml) of a dual-wavelength spectrophotometer (model 156; Hitachi) with a sample monochrometer at 340 nm and a reference monochrometer at 400 nm, so that the decrease of NADH was measured from the difference in absorbance between 340 and 400 nm. To both relaxing and contacting solutions, 0.25 mM NADH / 1.25 mM phosphoenolpyruvate / pyruvate kinase (50 units / ml) lactic dehydrogenase (50 units / ml) / 10 mM NaN_3 / 50 μM quercetin / oligomycin (1μg / ml) was added. The light path length through the sample compartment was 10 nm, and solutions in the

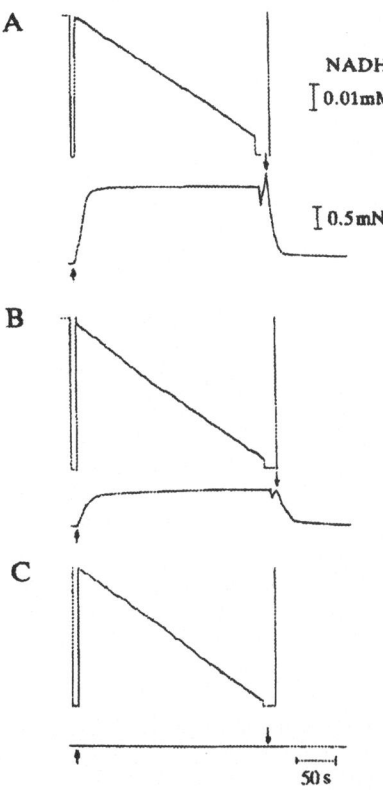

Figure 3. Simultaneous recordings of Mg-ATPase activity (upper traces) and Ca²⁺-activated isometric force development (lower trace) of a small fiber bundle consisting of three fibers before (A), and 100 (B) and 150(C) min after administration of anti-S-2 antibody (1.5 mg /ml). Note that the slope of ATPase records does not change appreciably, even when Ca²⁺-activated force development is reduced to zero (C). Decrease of ATPase trace shows decrease of NADH absorbance. Times of application of contracting solution and relaxing solution are indicated by upward and downward arrows, respectively (Sugi et al., 1992).

compartment were constantly stirred with a magnetic stirrer. The outputs of the spectrophotometer and the force transducer were fed to the digital oscilloscope and displayed on the X-Y plotter (Sugi et al., 1992).

Typical example of simultaneous recordings of Mg-ATPase activity and isometric force development of Ca²⁺-activated fibers are presented in Fig. 3. In both the absence (Fig. 3A) and presence (Fig. 3C, D) of anti-S-2 antibody, Mg-ATPase activity of relaxed fibers was very small and was not significantly different from the rate of spontaneous of NADH concentration in the sample compartment without the fibers (≈ 0.02 μM / s). Therefore, the result that the slope of ATPase records showed no appreciable changes when Ca²⁺-activated force development was reduced even to zero (Fig. 3C) indicates no appreciable effect of anti-S-2 antibody on Mg-ATPase of Ca²⁺-activated fibers. Similar results were obtained with five different fiber bundles.

4. BIDIRECTIONAL FUNCTIONAL COMMUNICATION BETWEEN S-1 AND S-2 IN DEMEMBRANATED RABBIT PSOAS MUSCLE FIBERS

The results described in the preceding sections strongly suggest the involvement of S-2 region in contraction of demembranated rabbit psoas muscle fibers. A question, that arises at this stage, is whether the catalytic and / or actin binding site, located in the S-1 head of a myosin molecule, can actually communicate with the S-2 region. To prove this functional communication between the S-1 and the S-2 in each myosin molecule in the fiber, we further performed the experiments on the demembranated muscle fibers with the following results (Kobayashi et al., 1998).

4.1. Rigor Force Development is Inhibited by Preceding Antibody Application

In the absence of anti-S-2 antibody, the fibers could be put into rigor state by the removal of external ATP. The magnitude of steady isometric rigor force was 50 - 60 % of the control steady Ca^{2+}-activated force obtained from one and the same fiber. During the development of rigor force, muscle fiber stiffness also increased in parallel with isometric rigor force as with Ca^{2+}-activated isometric force. Figure 4A shows typical muscle stiffness versus Ca^{2+}-activated force and muscle stiffness versus rigor force relations. Figure 4B shows muscle fiber stiffness versus isometric rigor force relation obtained during development of rigor force at various times after antibody application. The fiber was first put into rigor state in rigor solution, made to relax in relaxing solution containing the antibody (1.5 mg / ml), and then subjected to alternate application of rigor and relaxing solutions, all containing the antibody (1.5 mg / ml) to produce steady rigor forces at various times after application of antibody. Though the magnitude of steady isometric rigor force relation during the development of rigor force remained unchanged.

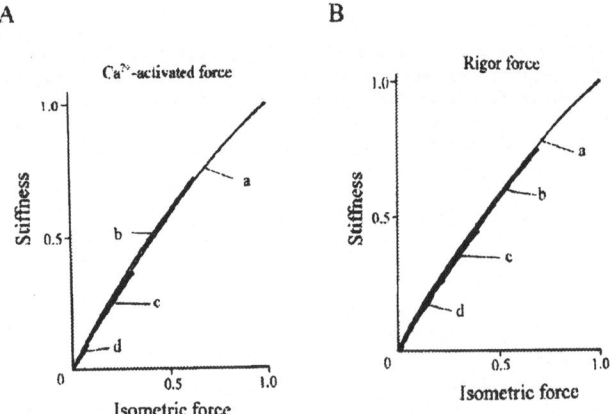

Figure 4. Muscle fiber stiffness versus isometric force relation during the development of Ca^{2+}-activated (A) and rigor (B) forces before (curves a) and at 30 (curves b), 60 (curves c) and 90 min (curves d) after application of the antibody (1.5 mg / ml). Both stiffness and force values are expressed relative to the control values. Data in A and B were obtained from different fibers (Kobayashi et al., 1998).

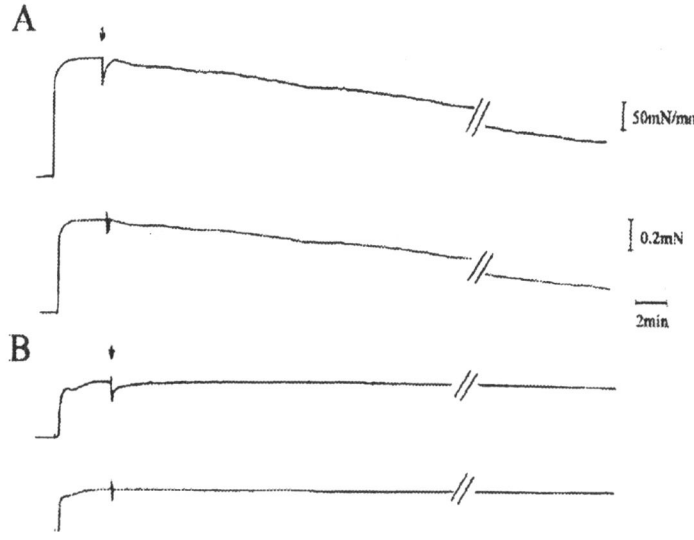

Figure 5. Typical stiffness (upper traces) and force (lower trace) records when the antibody was applied to the fiber after development of steady Ca²⁺-activated force (A) and after development of rigor force (B). Note that both stiffness and force decreases gradually on application of antibody in a Ca2+-activated fiber, while both stiffness and force in a rigor fiber remain unchanged at least for 90 min in the presence of antibody. Arrows indicate time of application of antibody (3mg /ml). Each record was interrupted for 40 min as indicated by a pair of oblique bars (Kobayashi et al., 1998).

4.2. Rigor Force Already Developed in the Absence of Antibody is not Inhibited by Subsequent Antibody Application

When the antibody (1.5 - 3 mg / ml) was applied to the fiber after the development of steady Ca²⁺-activated force, it produced parallel reduction of isometric force and muscle fiber stiffness (Fig. 5A). If, on the other hand, the antibody was applied to the fiber after the development of steady rigor force, both the rigor force and rigor stiffness remained unchanged at least for more than 120 min in the presence of antibody (1.5 - 3 mg / ml) (Fig. 5B). The force and stiffness also remained unchanged in fibers put into rigor state in the absence of antibody for more than 120 min, indicating that the antibody had no effect on both the force and stiffness in rigor fibers. The fibers that had been in rigor state for more than 120 min in the presence of antibody could be made to relax in relaxing solution without antibody, they developed full rigor force on application of rigor solution. Since the antibody binds to the S-2 region in an irreversible manner, this result may be taken to indicate that the antibody does not bind to the S-2 region in rigor fibers.

4.3. Bidirectional Communication Between S-1 and S-2 Regions of A Myosin Molecule

The results described in this section, therefore, can be summarized that (1) the antibody gradually binds to the S-2 region of relaxed fibers to progressively reduce rigor force development by reducing the number of cross-bridges forming rigor linkages with actin, and (2) the antibody does not bind to the S-2 region of rigor fibers, in which all cross-bridges have already formed rigor linkages with actin.   If this interpretation is correct, the S-1 and the S-2 regions in each myosin molecule can actually communicate each other.   The inhibitory effect of antibody binding to the S-2 region on the development of Ca^{2+}-activated and rigor forces (Figs. 4 and 5A) results from communication from the S-2 antibody binding site to the S-1 actin binding site, while the ineffectiveness of antibody binding in the fibers with rigor actin-myosin linkages (Fig. 5B) results from communication from the S-1 actin binding site to the S-2 antibody binding site.   Their functional communication can therefore be regarded as bidirectional.

5. DYNAMIC ELECTRON MICROSCOPY OF ATP-INDUCED CROSS-BRIDGE MOVEMENT IN LIVING MYOSIN FILAMENTS

A most straightforward way is for studying the mechanism of muscle contraction is to record the movement of individual cross-bridges on the thick filament with an electron microscopy at sufficiently high magnifications.   Though cellular functions, such as development, growth, and differentiation, are very readily impaired by electron beam irradiation (critical electron dose, 10^{-9} - 10^{-5} C / cm^2), crystalline structures of various biomolecules are known to be resistant to much higher electron doses (Butler and Hale, 1981).   This indicates the possibility of studying dynamic structural changes of living biomolecules in an electron microscope, using a gas environmental (hydration) chamber (EC), a device to keep biological specimens in wet state in an electron microscope (Fukami et al., 1991).   With the above technique, Suda et al. (1992) determined the critical electron dose for the reduction of ATP-induced myofibrillar shortening to be 5×10^{-4} C / cm^2.

Based on the above studies, we attempted to use the EC for studying the ATP-induced cross-bridge movement in muscle thick filaments.   After a number of trials over 5 years, we succeeded in recording the ATP-induced cross-bridge movement in living synthetic thick filaments as described below (Sugi et al., 1997).

5.1. Gas Environmental Chamber

As shown in Fig. 6, the EC used in the present study is a small cylindrical compartment (diameter, 2.0 mm; height, 0.8 mm) with upper and lower windows to pass electron beam.   Each window is covered with a thin carbon sealing film (thickness, 15 - 20 nm) held on a copper grid with nine apertures (diameter, 0.1 mm) (Fukami et al., 1991).   The specimen placed on the lower carbon film is kept wet by constantly circulating the air saturated with water vapor (pressure, 60 - 80 torr; temperature, 26 - 28 °C) through the chamber.   The vapor flow rate (0.1 - 0.2 liter / min) can be adjusted in

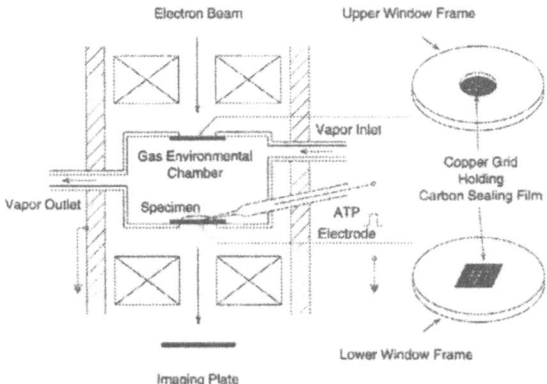

Figure 6. Schematic diagram of the EC. The upper and lower windows are converted with carbon sealing films held on copper grids with nine apertures. The interior of the EC is constantly circulated with water vapor to keep the specimen placed on the lower carbon film in wet state. The EC contains an ATP-containing microelectrode to apply ATP to the specimen iontophoretically. The image of the specimen is recorded on the imaging plate (Sugi et al., 1997).

such a way that the thin layer of ATP-free experimental solution around the specimen is in equilibrium with vapor pressure in the EC. The EC contains an ATP-containing microelectrode with its tip immersed in the experimental solution. The EC is attached to a 200 kV transmission electron microscope (JEM 2000EX; JEOL).

5.2. Thick Filament Specimen

The specimen used was synthetic thick filaments (myosin-paramyosin core complex), in which rabbit skeletal muscle myosin was bound to the surface of synthetic paramyosin filaments (diameter, 50 - 200 nm; length 10 - 30 μm) prepared from molluscan smooth muscle. The synthetic thick filaments were prepared from rabbit skeletal muscle myosin and paramyosin extracted from the anterior byssal retractor muscle of *Mitillus edulis*. The advantages of using the synthetic thick filaments are; (i) they are stiff and tend to form nearly straight rods when placed on the carbon film; (ii) the filament profile is clearly distinguished from the back-ground due to their large diameter; and (iii) because myosin molecules are bound around the paramyosin core in many rows, most cross-bridges on the upper side of the filaments are free from possible constraints due to their attachment to the carbon film.

Colloidal gold particles (diameter, 15 nm; coated with protein A; E Y Laboratories) were attached to the cross-bridge as position markers. using a site-directed antibody (IgG) to the junctional peptide between 50- and 20-kDa segments of myosin heavy chain (Sutoh et al., 1989) The synthetic thick filaments showed an Mg-ATPase activity ($\approx$ 0.13 s-1, 28 °C), which was not appreciably affected after mixing with the antibody. A small drop of the experimental solution ($\approx$ 5 μl) containing the filaments was put onto the carbon film in the EC and blotted with filter paper. The final quantity of the solution remaining on the carbon film might be as small as 11^{-6} ml, though the value is admittedly crude. The filaments with the gold position markers attached on the cross-bridges were also observed after negative staining with uranyl acetate and rotary shadowing with platinum at an angle of 32° (BAF400D, Balzers).

5.3. Recording of Filament Image

To avoid electron beam damage to the specimen, observation and recording were made with a total incident electron dose below 10^{-4} C / cm^2, being well below the critical dose for the reduction of ATP-induced myofibrillar shortening. For this purpose, the filaments were observed with extremely weak beam intensities below 5×10^{-5} A / cm^2 (measured with a Faraday cup on the microscope screen; AFC20, JEOL), so that observation and focusing of the filaments on the microscope screen required enormous skill. The actual beam intensity through the filaments with a magnification of 10,000× was $5 \times 10^{-13} \times (10,000)^2 = 5 \times 10^{-5}$ A / cm^2. As soon as the fold particles located on the upper surface of the filaments were brought in focus, electron beam was stopped except for the time of recording. The filament images were recorded with an imaging plate system (PIX system, JEOL) with a magnification of 10,000×. The imaging plate was 10.2 × 7.7 cm in size (2,045 × 1,536 pixels), and had a sensitivity ≈ 60 times that of x-ray film. The exposure time was 0.1 s with a beam intensity of 1 - 2 × 10^{-12} A / cm^2. Due to the limitation of total incident electron dose, the recording was made only twice.

5.3. Analysis of Imaging Plate Records

The filament images recorded with a magnification of 10,000× were analyzed with an image processor (Nexus Qube system, Nexus, Tokyo). In this condition, the pixel size on the imaging plate records was 5 × 5 nm, while the average number of electrons reaching a single pixel during the exposure time was ≈ 30. Reflecting this electron statistics, each fold particle image on the record consisted of 4 - 15 pixels. Each imaging plate record was divided into subframes containing 512 × 480 pixels, and each subframe was observed on the monitor screen (26.5 × 20 cm). Particles suitable for analysis were selected after an appropriate binarization procedure, i.e., the procedure to determine each particle configuration consisting of pixels, with electron counts above a chosen level. They consisted of 5 - 10 pixels with shapes not markedly influenced by the level of binarization used.

The center of mass position for each selected particle was determined as the coordinates (two significant figures) within a single pixel where the center of mass position was located, and these coordinates representing the position of the particles (and therefore the position of the cross-bridges) were compared between the two different imaging plate records. The absolute coordinates common to the two records were obtained based on the position of natural markers (bright spots on the carbon film, see Fig 2A). The distance (D) between the two center of mass positions (with the coordinates x_1, y_1, and x_2, y_2, respectively) was calculated as,

$$D = \sqrt{\left(x_1 - x_2\right)^2 + \left(y_1 - y_2\right)^2}.$$

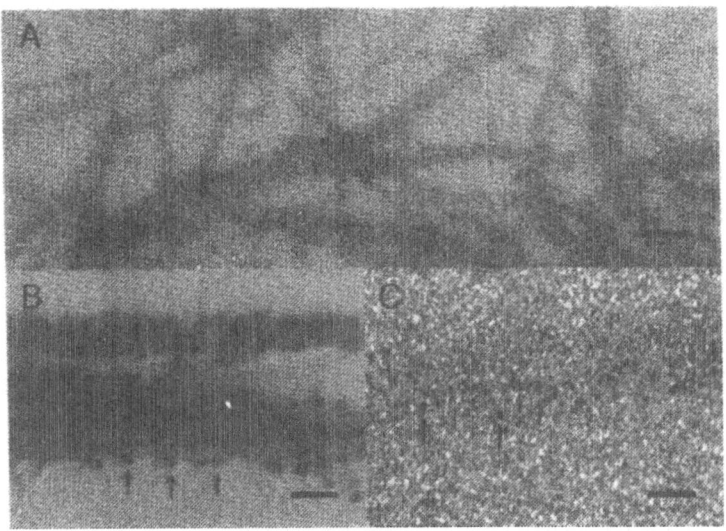

Figure 7. Typical images of the synthetic thick filament (myosin-paramyosin core complex). (A) Imaging plate record of the thick filaments with gold particles attached to the cross-bridges, taken with a magnification of × 10,000. (Bar = 500 nm). (B) Conventional electron micrograph of the filaments after negative staining with uranyl acetate and rotary shadowing with platinum (thickness, 2 nm). (Bar =100 nm) (C) Enlarged imaging plate record showing part of a thick filament with gold particles on it. (Bar = 100 nm). Arrows in B and C indicate gold particles (Sugi et al., 1997).

5.4. Stability of the Cross-bridge Position in the Absence of ATP

A typical imaging plate record of the synthetic thick filaments is shown in Fig. 7A. The gold particles attached to the cross-bridges are clearly seen as discrete dark spots on the filaments, which are also readily distinguished from the background. As shown in Fig. 7B, the filaments are actually covered by the cross-bridges with the gold particles attached to them. An enlarged image of a thick filament with the particles on it is shown in Fig. 7C. To examine whether the particle (and the cross-bridge) positions are stable or fluctuate with time, we compared the particle positions between two records of the same filaments taken at an interval of 3-5 min.

Among 141 particles examined on three different pairs of records, 87 particles showed no significant position changes ($D < 5$ nm), and 50 particles showed only small position changes (5 nm $< D < 10$ nm). These results indicate that, though the position of each cross bridge is expected to fluctuate due to thermal motion, its mean position time averaged for 0.1 s does not change appreciably with time in the ATP-free experimental solution.

5.5. ATP-induced Cross-bridge Movement

We examined possible cross-bridge movement in response to ATP application to the

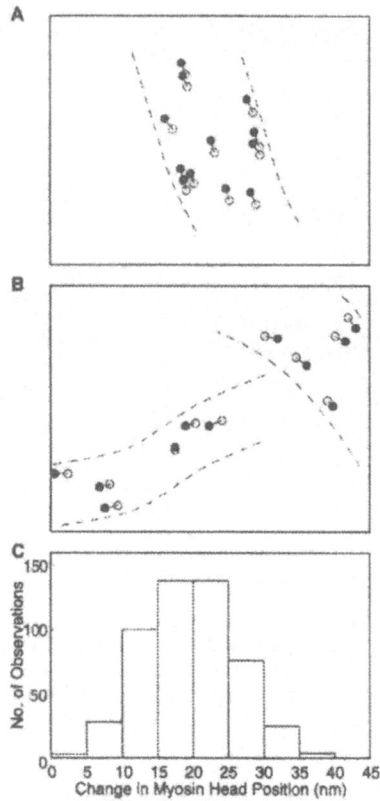

Figure 8. Movement of the individual cross-bridge on the filaments in response to ATP application. (A and B) Examples showing changes in the center of mass position of the same particles after ATP application. Filled and open circles (diameter, 15 nm) are also drawn around the center of mass positions of the same particles in the records before and those after ATP application, respectively. (C) Histogram showing the distribution of distance between the two center of mass positions of the same particles in the two records, representing the amplitude of the ATP-induced cross-bridge movement (Sugi et al., 1997).

filaments, by taking two records of the same filaments, one before and the other after ATP application. Based on the time of ATP diffusion from the ATP-containing microelectrode to the filaments (< 30 s), the second records were taken at 60 s after the on set of current pulse to the electrode, while the first records were taken 2-3 min before ATP application.

As schematically illustrated in Fig. 8A, B, the center of mass position of each particle was found to move by up to 30 nm in the direction parallel to the filament long axis after ATP application, indicating the marked ATP-induced movement of individual cross-bridges.   Fig. 8C is a histogram showing the distribution of the amplitude of ATP-induced cross-bridge movement, constructed from 512 measurements on three different pairs of records.   The histogram exhibits a peak at ≈ 20 nm.   The average amplitude of ATP-induced cross-bridge movement was 19.6 ± 0.3 nm (mean $\pm$ S.D., n = 512).

To further exclude the possibility that the cross-bridge movement is associated with any drift of the filaments, the contour of the filaments (especially the end of the filaments, on which the cross-bridge movement was recorded), was determined from the contrast-enhanced filament image after appropriate binarization and smoothing procedures, and the position for each outermost pixel along the filament contour was compared before and after ATP application.   Irrespective of the level of binarization, there was a significant fraction (> 25%) of pixels whose position remained unchanged by ATP application, and such pixels were distributed at fairly regular intervals (every 3-5 consecutive pixels).   The position of other outermost pixels changed randomly after ATP application.   At the filament ends, the difference in position of the outermost pixels were 5-15 nm along the filament axis, and the pixel position changes were frequently opposite to the direction of cross-bridge movement.   These results indicate that the filament drift, if any, may not seriously affect the results obtained.

The ATP-induced cross-bridge movement was not observed in the filaments whose ATPase activity had been completely eliminated by N-ethylmaleimide.   We also applied ADP to the filaments with a microelectrode containing ADP, but observed no appreciable cross-bridge movement in response to ADP application.   These results are consistent with the view that the ATP-induced cross-bridge movement is associated with reaction between the cross-bridge and ATP.

5.6. Discussion

Our findings described above are summarized: (i) in the absence of ATP, the position of the individual cross-bridges (time averaged for 0.1 s) does not change appreciably with time (Fig. 7); (ii) on application of ATP, the individual cross-bridges move by ≈ 20 nm along the filament long axis (Fig. 9); and (iii) application of ADP dose not produce the cross-bridge movement.   These results indicate that the ATP-induced cross-bridge movement found in the present study is a step in the cyclic interaction between actin, myosin, and ATP that is responsible for muscle contraction.   When ATP released from the microelectrode reaches the cross-bridge, it reacts rapidly with the cross-bridge (M) to form the complex M·ADP·Pi, having the average lifetime of > 10 s due to its slow Pi release (Lymn and Taylor, 1971).   Though some cross-bridges would repeat the ATPase reaction cycle more than once, the majority of the cross-bridges in the filament record may be in the state of M·ADP·Pi, suggesting that the cross-bridges movement is coupled with reaction, M + ATP → M·ADP·Pi.   The present finding that the movement coupled with ATP hydrolysis can take place in the absence of the thin filament is consistent with the suggestion that the cross-bridge position in relaxed muscle changes depending on the state of bound nucleotide (Wray, 1987).

Since our experimental system does not include actin filaments, the ATP-induced cross-bridge movement observed in the present study is likely to correspond to the cross-bridge preparatory stroke, which is the same in amplitude as, but different in direction from, the cross-bridge powerstroke.   To further clarify the cross-bridge movement associated with ATP hydrolysis, we are currently studying the ATP-induced cross-bridge movement using bipolar synthetic filaments consisting only of rabbit skeletal muscle myosin.

The large amplitude of the cross-bridge movement (≈ 20 nm) may not readily fit into the contraction model only based on the cross-bridge rotation, and would include shortening of myosin S-2 region.

6. CONCLUSION

In this article, we have presented evidence that, in addition to the cross-bridges, i. e. the S-1 head of myosin molecule, the myosin S-2 region plays an essential role in producing force and motion in vertebrate skeletal muscle fibers.   Concerning the experiments on the effect of anti-S-2 antibody including the dissociation of ATPase activity and isometric force development (Figs 1 - 4), (Sugi et al., 1992), however, the results may be accounted for not only by the inhibitory effect of antibody binding on the shortening of S-2 region, but also by the possible effect of antibody binding to reduce flexibility of cross-bridge - myosin filament backbone junctional region, which may be necessary to produce muscle contraction(Lauzon et al., 2001).

On the other hand, the bidirectional communication between the S-1 and the S-2 regions as revealed by use of anti-S-2 antibody (Figs. 5 and 6) (Kobayashi et al., 1998) unambiguously indicates that "functional" signals are actually transmitted between the S-1 and the S-2 regions many nanometers distant from each other.   It remains to be investigated whether such signals travel along some unknown molecules running along the cross-bridge - myosin filament backbone junction or along α-helix of myosin heavy chain.

The strongest evidence for the involvement of myosin-S-2 shortening in muscle contraction is our results of dynamic electron microscopy of living thick filaments (Sugi et al., 1997).   Although the ATP-induced cross-bridge movement of ≈ 20 nm (Figs. 7 - 9) is likely to be the preparatory stroke, which may precede each powerstroke in each cross-bridges, the magnitude of the former should be the same as that of the latter, if the cross-bridges repeat their powerstroke during contraction.   Finally, we emphasize that much more attention should be paid to the function of myosin-S-2 region, which seems at present to be almost totally ignored (A. F. Huxley, 2000).

7. REFERENCES

Applegate, D., Reisler, E., 1983, Crossbridge release and alpha-helix-coil transition in myosin and rod minifilaments, *J. Mol. Biol.* 169(2): 455 - 468.

Butler, E. P., and Hale, K. F., 1981, *Dynamic Experiments in the Electron Microscope*, North Holland, Amsterdam, 457 pp.

Edman, K. A., 1988, Double-hyperbolic force-velocity relation in frog muscle fibres, *J. Physiol.* 404: 301 -321.

Fukami, A., Fukushima, K., and Kohyama, N., 1991, Observation technique for wet clay minerals using film-sealed environmental cell equipment attached to high-resolution electron microscope, in: *Microstructure of Fine-Grained Sediments from Mud to Shale*, R. H. Bennett, W. R. Bryant, and M. H. Hulbert, eds., Springer, Heidelberg, pp. 321 - 331.

Harrington, W. F., Karr, T., Busa, W. B., and Lovell, S. J., 1990, Contraction of myofibrils in the presence of antibodies to myosin subfragment 2, *Proc. Natl. Acad .Sci. U S A.* 87(19): 7453 - 7456.

Huxley, A. F., 1957, Muscle structure and theories of contraction, *Prog. Biophys. Chem.* 7: 255 - 318.

Huxley, A. F., 2000, Mechanics and models of the myosin motor, *Phil.Trans. R. Soc. Lond. B* 355(1396): 433 - 440.

Huxley, H. E., 1969, The mechanism of muscular contraction, *Science.* 164(886): 1356 - 1365.

Kobayashi ,T., Kosuge, S., Narushima, K., Sugi, and H., 1998, Evidence for two distinct cross-bridge populations in tetanized frog muscle fibers stretched with moderate velocities, *Biochem. Biophys. Res.*

Commun. 249(1): 161 - 165..

Lauzon, A. M., Fagnant, P.M., Warshaw, D.M., and Trybus, K. M., 2001, Coiled-coil unwinding at the smooth muscle myosin head-rod junction is required for optimal mechanical performance, *Biophys. J.* 80(4):1900 - 1904.

Lovell, S., Karr, T., and Harrington, W. F., 1988, Suppression of contractile force in muscle fibers by antibody to myosin subfragment 2, *Proc. Natl. Acad. Sci. U S A.* 85(6): 1849 - 1853.

Lymn, R. W., and Taylor, E. W., 1971, Mechanism of adenosine triphosphate hydrolysis by actomyosin, *Biochemistry.* 10(25): 4617-4624.

Margossian, S.S., Krueger, J.W., Sellers, J.R., Cuda, G., Caulfield, J.B., Norton, P., and Slayter, H.S., Influence of the cardiac myosin hinge region on contractile activity, *Proc. Natl. Acad. Sci. USA.* 88(11): 4941 - 4945.

Suda, H., Ishikawa, A., and Fukami, A., 1992, Evaluation of the critical electron dose on the contractile activity of hydrated muscle fibers in the film-sealed environmental cell, *J. Electron Microsc.* 41(4): 223-229.

Sugi, H., Kobayashi, T., Gross, T., Noguchi, K., Karr, T., and Harrington, W. F., 1992, Contraction characteristics and ATPase activity of skeletal muscle fibers in the presence of antibody to myosin subfragment 2, *Proc. Natl. Acad. Sci. U S A.*89(13): 6134 -6137.

Sugi, H., Akimoto, T., Sutoh, K., Chaen, S., Oishi, N., and Suzuki ,S., 1991, Dynamic electron microscopy of ATP-induced myosin head movement in living muscle thick filaments, *Proc. Natl. Acad. Sci .USA.*94(9):4378-82.

Tsong, T.Y., Karr, T., and Harrington, W. F.,1979, Rapid helix--coil transitions in the S-2 region of myosin, Proc. Natl. Acad. Sci. USA. 76(3): 1109 - 1113.

Ueno, H., and Harrington, W. F., 1984, An enzyme-probe study of motile domains in the subfragment-2 region of myosin, *J. Mol. Biol.* 180(3): 667 - 701.

Ueno, H., and Harrington, W. F., 1986, Local melting in the subfragment-2 region of myosin in activated muscle and its correlation with contractile force, *J. Mol. Biol.* 190(1): 69 - 82.

Ueno, H., and Harrington, W. F., 1987, Cross-linking within the thick filaments of muscle and its effect on contractile force, *Biochemistry.* 26(12):3589 - 3596.

Wray, J. S., 1987,Structure of relaxed myosin filaments in relation to nucleotide state in vertebrate skeletal muscle, *J. muscle. Res. Cell Motil.* 8: 62 (abstr.).

DISCUSSION

Holmes: How can you get imaging plate pictures with sharp contrast without staining the specimen?

Sugi: The imaging plate system works as a signal processor, and we can very much enhance the contrast of the pictures taken.

Alpert: You showed that binding of anti-S2 antibody to the S2 portion of myosin molecules results in a complete dissociation of mechanical response and the ATPase activity in skinned muscle fibers. How do you think about the mechanism?

Sugi: I once discussed this with Dr. Sutoh, who prepared antibody to myosin head in our work, and we were forced to the conclusion that there should be a transmission of information along the amino acid helical structure. But I have no idea about such mechanism.

Alpert: I had some additional thoughts about your antibody experiments. In our laboratory, it has been shown that, if the flexibility of myosin head-rod junction is reduced by leucine zipper, the unitary displacement of actin-myosin sliding was reduced to one-tenth compared to control myosin (Lauzon et al., Biophys. J. 80: 1900-1904,

2001). It seems possible that your anti-S2 antibody acts like leucine zipper, thus eliminating myosin powerstroke with the ATPase activity remains unchanged.

Sugi: Thank you for your comments. I will consider this possibility in our future work.

V. NONMUSCLE MOTILE SYSTEMS

A STUDY OF LAMELLIPODIAL MEMBRANE DYNAMICS BY OPTICAL TRAPPING TECHNIQUE: IMPLICATION OF MOTOR ACTIVITY IN MOVEMENTS

Hidetake Miyata[*]

1. INTRODUCTION

1.1. Lamellipodial Movements and Actin Polymerization

Cell locomotion is an important activity in many cellular phenomena such as wound healing, morphogenesis and development (Stossel, 1993). The cell locomotion can be divided into three distinct, but consecutive steps: protrusion of the cell front, adhesion of that portion to substrate and contraction of the tail portion of the cell. The membrane protrusion is indispensable in the locomotion, and the mechanism that drives this movement has been a subject of a number of studies (for reviews, see Condeelis, 1993; Mitchson and Cramer, 1996; Lauffenburger and Horwitz, 1996; Borisy and Svitkina, 2000). Since lamellipodium is bordered by an elastic cell membrane and is highly anisotropic in shape, it should be supported by some intracellular architecture to maintain its morphology. Electron microscopic studies have established that in lamellipodium crisscrossing meshwork of actin filaments resides immediately beneath the cell membrane with their plus end (the end where polymerization occurs in vivo) oriented toward the membrane (Small et al., 1978; Small, 1988). Hence, actin filament plays a major role in supporting lamellipodial shape.

Actin filaments not only support the shape of lamellipodia, but also actively mold their shape. Previous studies have shown that actin polymerization occurs between the actin network and the cell membrane, concomitant with the protrusion of lamellipodia.

*Hidetake Miyata, Physics Department, Graduate School of Science,
  Tohoku University, Aramaki, Aoba-ku, Sendai, Miyagi 980-8578, Japan.

As a consequence of the polarity of actin filaments in the network, they elongate toward the cell membrane to make the membrane protrude. This leads to a production of mechanical work (Cooper, 1991), because the cell membrane is elastic and usually a lateral tension exists in the membrane (Sheetz and Dai, 1996), which should be overcome when the protrusion occurs. A series of theoretical studies by Hill has shown that growing polymer can perform a mechanical work against the load (Hill, 1981; Hill and Kirschner, 1982). The ultimate driving force in this work production is the difference of the chemical potential between the actin monomer in the solution and that in the filament. The argument was based on thermodynamics and hence, the relation between the external load and rate of polymer elongation (= velocity of the membrane protrusion) was not investigated.

In the cell actin polymerization is a spatially and temporally regulated process (Svitkina and Borisy, 2000). For example, a series of intracellular biochemical events, beginning with a growth factor binding to its receptor, results in the lamellipodial development around the activated receptor. In lamellipodia actin filaments start growing from the tip of the preexisting filaments, or from the side of filaments; in the latter case a protein complex called Arp2/3 nucleates the actin polymerization and form a branched structure which is similar to what is observed in the cell (Pantaloni et al., 2001). Eventually, the tip of the growing filament abuts the cell membrane, and further polymerization becomes impossible unless a gap is created between the filament tip and the cell membrane (Stossel, 1993).

Two mechanisms of gap creation have been postulated: one is fluctuation based mechanism (Peskin et al., 1993; Mogliner and Oster, 1996). These mechanisms depend on the passive and stochastic formation of the gap and will be discussed later (see Fig.4). Another is myosin-based mechanism: as depicted in Fig.1, the sliding of cell membrane-bound myosin (possibly myosin I) over the actin filament network fixed to the substrate results in the forward movements of the membrane. The gap thus created is assumed to be immediately followed by the actin polymerization (Sheetz et al., 1992; Welch et al., 1997). Note that in other organisms, other mechanisms may function. For example, contraction of the tail of the cell driven by actin-myosin interaction in the cortical layer which is coupled to osmotic swelling of actin gel in lamellipodia. In certain type of large amoebae expansion of actin cortex coupled with the tail contraction may be the driving force for another type of membrane protrusion, pseudopod (Condeelis, 1988). Thus, the mechanism of membrane protrusion may not be uniform.

Effort to understand the polymerization dynamics of actin in the cell has also made a great progress. These studies are carried out to understand the above-described spatial and temporal regulation of actin polymerization, and are belived to help understand the membrane dynamics. The role of several regulatory proteins in lamellipodial dynamics has now been fairly well understood. As described above, the crisscrossing meshwork of actin filaments is formed by the branching polymerization mediated by Arp2/3 complex (Pantaloni et al., 2001). Nucleation of actin polymerization under the physiological high ionic strength, which induces rapid polymerization in vitro, is usually suppressed by sequestration of actin monomer by thymosin β4; polymerization of the sequestered monomer is promoted by profilin (Pollard et al., 2000). The monomer

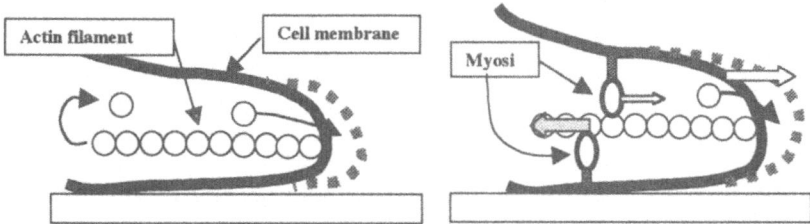

Fig.1 Proposed mechanisms of membrane protrusion
Left, actin polymerization-based mechanism. For simplicity, actin filament is represented with a single stranded linear polymer. For this mechanism to operate, a gap should be generated for monomer insertion (*curved arrows*). Fluctuation-based gap generation mechanisms are described in more detail in Fig.4. *Right*, myosin-based mechanism. The lower myosin fixed to the substrate drives the rearward movement of the actin meshwork (*hatched arrow*), while the movement of the membrane-bound myosin (*white arrow*) drives the membrane forward thereby generates the gap between the filament tip and the cell membrane.

required for polymerization seems to be supplied by depolymerization at the minus end of actin filament, which is located at 竹 the cell interior (Cramer, 1999).
Thus in the cytoplasm, actin monomer circulates from the minus end to plus end. This flux of actin is to be balanced by the treadmilling occurring from plus to minus end in individual actin filaments (Pantaloni et al., 2001).

The early studies of lamellipodial dynamics were carried out by optical microscopy (Abercrombie et al., 1970) and the analysis of the movement was limited by the spatial resolution of optical microscope (ie., ~300 nm). The measurements have revealed that lamellipodium protrudes at a velocity of ~100 nm/s and withdraws at a similar velocity. Recently, an atomic force microscope has been utilized to probe the cell periphery (Rotch et al., 1999), and the minute movements (~1 μm) of the active protruding edge of the cell have been detected, but the time resolution (~1 sec) may not be sufficient to fully resolve the detail of membrane dynamics.

1.2. Bacterial Movements

There is another system that has been postulated to utilize the thermal fluctuation for producing the mechanical work. Certain kind of bacteria (*Listeria*, for example) propel in cytoplasm at ~300 nm/s (Tilney and Portnoy, 1985). Extensive studies during a past decade have established that actin polymerization is necessary for the bacterial propulsion. Since cytoplasm is a viscous fluid, ~0.2 pN force is exerted on a bacterium, the size of which is 1.5 μm × 0.5 μm, when it propels at the above speed. Hence, actin polymerization produces the mechanical work. Quite interestingly, the regulatory proteins involved in the propulsive process seem to come from the host cell and are largely in common with those utilized in the membrane dynamics (Cossart and Lecuit, 1998). Therefore, it has been proposed that the mechanism of membrane protrusion and that of bacterial propulsion are quite similar.

1.3. Reconstituted System

Several reconstituted systems have been developed to investigate the possibility of the work production by actin polymerization: one system is giant liposomes encapsulating monomeric actin. When the encapsulated actin monomers were allowed to polymerize, the liposome shape changed (Miyata and Hotani, 1992; Cortese et al., 1989) and under certain condition protrusive growth occurred (Miyata et al., 1999). The rate of protrusive growth, 200-500 nm/s, was comparable to the rate of elongation of actin in vitro. Thus, the growth of actin filaments in the liposome seemed to drive the deformation of the membrane. Since the pure lipid membrane is an elastic entity and an entropy-driven tension exists (Helfrich and Servuss, 1984), this result indicates that the elongation of the actin filaments produced a mechanical work (the elastic one and the one against the membrane tension), although its amount could not be estimated due to the difficulty in measuring the lateral membrane tension, an origin of the load.

Another system is plastic beads coexisting with actin and several motility-related proteins (eg., Arp2/3, actin capping protein, actin depolymerizing factor etc.). Thus, in aqueous solution propulsion of plastic beads by actin polymerization in the presence of the motility-related proteins has been demonstrated and requirement of individual regulatory components and their relative importance have been determined (Loisel et al., 1999). Because of the simplicity of the system, the rate of work production can be estimated from the speed and the size of the bead. This clearly establishes that the polymerization-driven propulsion of bacteria and beads is purely physical phenomenon. An important point is that the knowledge of previous in vitro studies on actin polymerization dynamics (on and off rate constants, affinity between actin and actin binding proteins) has been found to be applicable to explain the speed of bead movements. Thus, the study of the mechanism of actin polymerization-related motility is more advanced with the bacterial system.

1.4. In Vivo Studies

A recent study of in vivo function of Ena/VASp, a regulatory protein of actin polymerization (Bear et al., 2002), has provided evidence that is consistent with the prediction of the mechanism based on the thermal fluctuation of actin filaments (the elastic Brownian ratchet theory, see Fig.4). In this study increasing cellular concentration of Ena/VASp enhanced the rate of lamellipodial protrusion but reduced the stroke of the protrusion. Electron microscopic observation has indicated that in the overexpressing cells the free ends of actin filaments were significantly longer than the wild type cell and the filaments were aligned parallel to the cell periphery. Hence, Ena/VASp promoted actin polymerization thereby increased the length of the free ends, but longer free ends became susceptible to the external load, the membrane tension in this case, thereby resulting in the filaments aligned parallel to the cell periphery (Bear et al., 2002).

On the other hand, a recent measurement of the motion of *Listeria* in the cell by optical trapping-based particle tracking method with high spatial and temporal resolution has demonstrated a result that is inconsistent with the fluctuation-driven mechanism (Kuo and McGrath, 2000). The analysis has shown that the bacteria intermittently moved with 5.4 nm steps with remarkably small position fluctuation, which has led the researchers to conclude that the movement occurred along growing actin filaments and was driven by a mechanism that can generates a force that is stronger than what is expected from the polymerization-driven mechanisms.

2. EXPERIMENTAL STUDY OF LAMELLIPODIAL DYNAMICS

As described above, while it is established in reconstituted systems the polymerization-driven movements can produce a mechanical work, it is yet to be proved in vivo if this mechanism plays a primary role in the generation of the work. Therefore, it is important to perform detailed investigations of the movements that might be driven by actin polymerization. In this context we have been studying the movements of lamellipodia. As described above, quantitative measurements of lamellipodial movements have been performed by optical microscopy or atomic force microscopy, but either spatial or temporal resolution may not be sufficient to approach the mechanism of lamellipodial movement. We anticipated that the optical trapping technique might be suitable for this purpose, because it has been applied to the measurements of motor proteins (Finer et al., 1994).

2.1. Experimental Details and Results

The measuring system is schematically shown in Fig.2. To create an optical trap, a 1064 nm laser beam was focused with NA1.3, 100× objective lens. The trap stiffness, ranging over 0.024 to 0.090 pN/nm, was determined from thedistribution of bead position in the trap potential as previously described (Miyata et al., 1994; Svoboda and Block, 1994). For the cell measurement, a 1 μm polystyrene bead was captured and brought into contact with the periphery of the cell which had spread on a coverslip coated with poly-L-lysine. In order to enhance the lamellipodial motility, phorbol ester (PMA) was added at 0.5 μM. The phase-contrast image of the trap-held bead was recorded

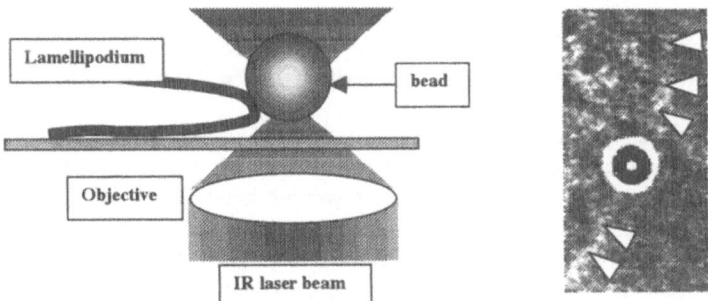

Fig.2 _ *Left panel*, principle of the measurement. *Right panel*, a phase contrast micrograph of the cell and the bead: *white arrowheads* indicate the edge of lamellipodium, and the circle in the middle is the 1 μm bead.

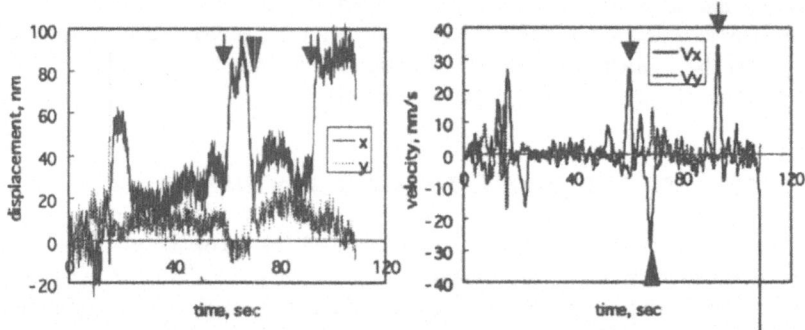

Fig.3 _Example of time course of the bead displacement and velocity. *Left panel*, bead displacement, where x (black trace) is taken in the direction of membrane protrusion. *Right panel*, the bead velocities in x and y direction derived from the smoothed displacement data (not shown). The protrusive and withdrawal movements occurred at the time indicated respectively with downward arrow and arrowhead in the *left panel*. The downward arrowhead and the upward arrowhead in the *right panel* indicate the forward and withdrawal velocities.

for 1-2 min,and the position of the bead was determined every 33 ms from the calculation of the centroid of the phase contrast image of the bead using an NIH image-based program written by Dr. Akira Goto (Physics Department, Graduate School of Science, Tohoku University). This calculation provided x-t and y-t traces. These data were smoothed over 1 sec, and the bead velocity at each time point (v_x, v_y), was determined from the smoothed data by calculating the value $\Delta x/2\Delta t$ and $\Delta y/2\Delta t$, respectively, where Δx and Δy are the differences of the bead coordinates corresponding to an appropriate time difference, $2\Delta t$. The force from the trap, f_x and f_y, were calculated as $\kappa \times x$ and $\kappa \times y$, where κ is the trap stiffness.

A difficulty in the measurement of lamellipodial dynamics is the adhesion of the bead to the cell surface. The adhesion often caused centripetal transport of the bead (Choquet et al., 1997), which may be driven by a different mechanism than that of membrane dynamics. Coating the bead with bovine serum albumin reduced the adhesion, but still considerable fraction (~50%) of the bead was bound to the surface. After each experiment the trap was turned off to check if the bead floated away and only the data of non-adherent bead were analyzed. Another difficulty was that the bead sometimes moved in the direction parallel to the cell periphery for unknown reason: we only analyzed the movements which occurred in the direction within ±45° from the normal to the cell periphery.

Fig.3 shows example of the displacement and the velocity of the bead. As reported elsewhere (Takahashi et al., to be published), the bead movements away from the cell (forward movements, as indicated with downward arrows in the *left panel*) ranged over 50 to 100 nm, and occurred with non-uniform velocity (acceleration in the beginning and deceleration in the end, downward arrows in the *right panel*). Following the forward movement the bead sometimes exhibited rearward movement (indicated with an arrowhead in the *left panel*). This movement also occurred with acceleration and deceleration (upward arrowhead in the *right panel*). The movements were significantly

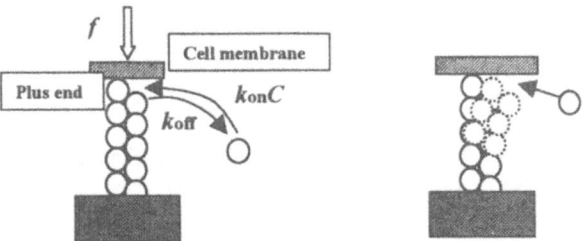

Fig.4 An illustration of the two types of fluctuation-based membrane protrusion models.
Left, Brownian ratchet model assuming the fluctuation of the cell membrane (*top gray box*); *right*, the elastic Brownian ratchet model assuming the fluctuation-driven bending of an actin filament: the bent filament goes back straight, but since the membrane is abutted by the newly polymerized monomer, it is pushed forward by the filament whereby the elastic energy stored in the bent filament is converted into the mechanical work. Since an actin filament is double-stranded, a gap of the size half that of the monomer (= δ; ~2.7 nm) must be created for the polymerization of actin monomer in the left case, whereas a smaller gap is sufficient in the right case. The minus end of the filament is assumed to be blocked. The stall force that is necessary to stop the filament elongation has been calculated to be ~5 pN for the Brownian ratchet model (left) and ~7 pN for the elastic Brownina ratchet model (right).

suppressed by cytochalasin D, which inhibits actin polymerization in vitro (Cooper, 1987) and peripheral membrane motion of fibroblasts (Schafer et al., 1998). Hence, actin polymerization must play an important role in the mobility of lamellipodia, as has been assumed (Cooper, 1991; Condeelis, 1993; Mitchson and Cramer, 1996). We have also observed that the maximum protrusive velocity tended to decrease with increasing trapping force. Quite interestingly, the v_{max} of the forward and that of rearward movements exhibited a correlation (Takahashi et al., to be published). _

2.2. Interpretation and Models

As has been mentioned above, the gap has to be created between the cell membrane and the plus end of actin in order for the polymerization to occur. The mechanism that utilizes the thermal fluctuation has been proposed (Peshkin et al., 1993; Mogliner and Oster, 1996; Fig.4). The theory predicts that the velocity of the membrane depends on the concentration of actin monomer C and the external force (load) f in the following manner:

$$V = \delta \times \{k_{on}C \times E(f, k_B T) - k_{off}\},$$

where δ is the extension of actin filament per each addition of actin monomer, k_{on} and k_{off} are the on and off rate of actin polymerization and $k_B T$ is the Boltzmann factor. E is the monotonously decreasing function of f.

Accroding to the above equation, the change in the forward velocity of the bead (=

protrusive velocity of the membrane) may be explained by assuming that during the acceleration the actin monomer concentration, C increased, while during the deceleration it decreased.   The deceleration could also result from the increase in the external force, f (= the trapping force which increases with the advancement of the bead).   The decrease in the maximum protrusive velocity with increasing trapping force is also consistent with the theory's prediction.   On the other hand, since the mechanism of the withdrawal movement has not been treated in the polymerization-based theory, the experimentally observed withdrawal movement and the existence of correlation between the velocities of successive protrusion and withdrawal of the membrane seem to require some other explanation.

In order to fully understand the observed membrane dynamics, we must assume the existence of some mechanistic relation between the protrusion and withdrawal movements.   As described above, the similarity in the magnitudes of the protrusive and withdrawal velocities has been observed (Abercrombie, 1970; Sheetz et al., 1992; Bear et al., 2002), and involvement of myosin has been postulated to explain the observation (Sheetz et al., 1992).   Bear et al. have discussed a possibility that the withdrawal movements can be a result of the buckling of the filament immediately beneath the cell membrane, while the protrusive movement is driven by the polymerization.   However, we consider that the observed rearward movement is active and is difficult to explain by the passive mechanism (Takahashi et al., to be published).   With all these considerations we speculate that the mechanism including the myosin-driven membrane movement (Fig.1, *right panel*) most consistently explains our observations.

Is there any place in the membrane protrusion where the fluctuation plays a role? The decrease in the maximum protrusive velocity with increasing trapping force may reflect the involvement of fluctuation: the external force may suppress the myosin mobility, thereby reducing the size of the gap.   However, even if the movement by myosin was not inhibited by the external load, the gap size can be reduced by the trap-held bead, because the cell membrane itself is a flexible entity: when the membrane is moved forward and a gap is created, it will execute thermal fluctuation, because the bending modulus of the cell membrane, $\sim10^{-19}$ J (Evans, 1983), is small enough to allow ~10 nm out-of-plane fluctuation for a tension free, (0.25 μm)^2 membrane section (Helfrich and Servuss, 1984).   It is possible that the bead placed in front of the membrane acted as an obstacle to reduce the degree of fluctuation.   This notion is particularly important when one considers the regulatory aspects of the membrane dynamics.   It is conceivable that the membrane tension, which will be easily controlled by myosin II activity in the cell cortex, may alter the membrane fluctuation thereby regulate the protrusive velocity (Sheetz and Dai, 1996).   Further investigation should be performed to evaluate this notion.

3. FUTURE DIRECTION

The optical trapping method has provided data that invoke reconsideration of the mechanism of membrane protrusion.   The high spatial and temporal resolution was essential to probe the membrane movements.   Although most of our observation can be explained by the previously postulated myosin-based membrane dynamics, the fluctuation may play a role in the regulation of the protrusive velocity. Future experiments should aim at the clarification of the role of the fluctuation in the protrusive process. It will be also interesting to reconsider the role of fluctuation in the bacterial propulsion in the cell.

4. ACKNOWLEDGEMENT

This work was supported by grants from Takeda Science Foundation and Sumitomo Foundation, and Grant-in-Aid from Ministry of Education, Culture, Science and Sports, Japan.

5. REFERENCES

Abercrombie, M., Heaysman, J.E.M., and Pegrum, S.M., 1970, The locomotion of fibroblasts in culture, *Exp. Cell Res.* **59**:393-398.

Bear, J.E., Svitkina, T.M., Krause, M., Schafer, D.A., Luoreiro, J.J., Strasser, G.A., Maly, I.V., Chaga, O.Y., Cooper, J.A., Borisy, G.G., and Gertler, F.B., 2002, Antagonism between Ena/VASP proteins and actin filament capping regulates fibroblast motility, *Cell* **109**:509-521.

Borisy, G.G., and Svitkina, T.M., 2000, Actin machinery pushing the envelope, *Curr. Opin. Cell Biol.* **12**:104-112.

Choquet, D., Felsenfeld, D.P., and Sheetz, M.P., 1997, Extracellular matrix rigidity causes strengthening of integrin-cytoskeleton linkages, *Cell* **88**:39-48.

Condeelis, J., 1993, Life at the leading edge: the formation of cell protrusions, *Annu. Rev. Cell Biol.* **9**:411-444

Condeelis, J., Hall, A., Bresnick, A., Warren, V., Hock, R., Bennet, H., and Ogihara, S., 1988, Actin polymerization and pseudopod extension during amoeboid chemotaxis, *Cell Motil. Cytoskeleton* **10**: 77-90.

Cooper, J.A., 1987, Effects of cytochalasin and phalloidin on actin, *J. Cell Biol.* **105**:1473-1478.

Cooper, J.A., 1991, The role of actin polymerization in cell motility, *Annu. Rev. Physiol.* **53**:585-605.

Cortese, J.D., Schwab III, B., Frieden, C., and Elson, E.L., 1989, Actin polymerization induces a shape change in actin-containing vesicles, *Proc. Natl. Acad. Sci. USA* **86**:5773-5777.

Cossart, P., and Lecuit, M., 1998, Interactions of *Listeria monocytogenes* with mammalian cells during entry and actin-based movement: bacterial factors, cellular ligands and signaling, *EMBO J.* **17**:3797-3806.

Cramer, L.P., 1999, Role of actin-filament disassembly in lamellipodium protrusion in motile cells revealed using the drug jasplakinolide, *Curr. Opin. Cell Biol.* **9**:1095-1105.

Evans, E.A., 1983, Bending elastic modulus of red blood cell membrane derived from buckling instability in micropipette aspiration tests, *Biophys. J.* **43**:27-30.

Finer, J.T., Simmons, R.M., and Spudich, J.A., 1994, Single myosin molecule mechanics: piconewton forces and nanometer steps, *Nature* **368**:113-119.

Helfrich, W., and Seruvuss, R.-M., 1984, Undulations, steric interaction and cohesion of fluid membranes, *Il Nuovo Cimento* **3D**:137-151.

Hill, T.L., 1981, Microfilament or microtubule assembly or disassembly against a force, *Proc. Natl. Acad. Sci. USA*, **78**:5613-5617.

Hill, T.L., and Kirshcner, M.W., 1982, Subunit treadmilling of microtubules or actin in the presence of cellular barriers: possible conversion of chemical free energy into mechanical work, *ibid*, **79**:490-494.

Kuo, S.C., and McGrath, J.L., 2000, Steps and fluctuations of *Listeria monocytogenes* during actin-based motility, *Nature* **409**:1026-1029.

Lauffenburger, D.A., and Horwjtz, A.F., 1996, Cell migration: a physically integrated molecular process, *Cell* **84**:359-369.

Loisel, T.P., Boujeman, R., Pantaloni, D., and Carlier, M.-F., 1999, Reconstitution of actin-based motility of *Listeria* and *Shigella* using pure proteins, *Nature* **401**:613-616.

Mitchson, T.J., and Cramer, L.P., 1996, Actin-based cell motility and cell locomotion, *Cell* **84**:371-379.

Miyata, H. and Hotani, H., 1992, Morphological change in liposomes caused by polymerization of encapsulated actin and spontaneous formation of actin bundles, *Proc. Natl. Acad. Sci. USA*, **89**: 11547-11551.

Miyata, H., Hakozaki, H., Yoshikawa, H., Suzuki, N., Kinosita, K. Jr., Nishizaka, T., and Ishiwata, S.-I., 1994, Stepwise motion of an actin filament over a small number of heavy meromyosin molecules is revealed in an in vitro motility assay, *J. Biochem.* (Tokyo) **115**:644-647.

Miyata, H., Nishiyama, S., Akashi, K-i., and Kinosita, K. Jr., 1999, Protrusive growth from giant liposomes driven by actin polymerization, *Proc. Natl. Acad. Sci. USA*, **96**:2048-2053.

Mogliner, A., and Oster, G., 1996, Cell motility driven by actin polymerization, *Biophys. J.*, **71**:3030-3045.

Pantaloni, D., LeClainche, C., and Carlier, M.-F., 2001, Mechanism of actin-based motility, *Science* **292**: 1502-1506.

Peskin, C.S., Odell, G.M., and Oster, G.F., 1993, Cellular motions and thermal fluctuations: the Brownian ratchet, *Biophys. J.* **65**:316-324.

Pollard, T.D., Blanchion, L., and Mullins, R.D., 2000, Molecular mechanisms controlling actin filament dynamics in nonmuscle cells, *Annu. Rev. Biophys. Bimol. Struct.* **29**:545-576.

Rotch, C., Jacobson, K., and Radmacher, M., 1999, Dimensional and mechanical dynamics of active and stable edges in motile fibroblasts investigated by using atomic force microscopy, *Proc. Natl. Acad. Sci. USA.* **96**: 921-926.

Schafer, D.A., Welch, M.D., Machesky, L.M., Bridgman, P.C., Meyer, S.M., and Cooper, J.A., 1998, Visualization and molecular analysis of actin assembly in living cells, *J. Cell Biol.* **143**:1919-1930.

Sheetz, M.P., Wayne, D.B., and Pearlman, A., 1992, Extension of filopodia by motor-dependent actin assembly. *Cell Motil. Cytoskeleton*, **22**:160-169.

Sheetz, M.P., and Dai, J., 1996, Modulation of membrane dynamics and cell motility by membrane tension. *Trends Cell Biol.* **6**:85-89.

Small, J.V., 1988, The actin cytoskeleton, *Electron Microsc. Rev.* **1**:155-174.

Small, J.V., Isenberg, G., and Celis, J.E., 1978, Polarity of actin at the leading edge of cultured cells. *Nature*,

272:638-639.

Stossel, T.P., 1993, On the crawling of animal cells, *Science.* **260**:1086-1094.

Svoboda, K., and Block, S.M., 1994, Biological applications of optical forces, *Annu. Rev. Biophys. Biomol. Struct.* **23**:247-285.

Takahashi, F., Higashino, Y., and Miyata, H., *Biophysical Journal*, to be published.

Tilney, L.G., and Portmoy, D.A., 1985, Actin filaments and the growth, movement, and spread of the intracellular bacterial parasite, *Listeria monocytogenes, J. Cell Biol.* **109**:1597-1608.

Welch, M.D., Mallavarapu, A., Rosenblatt, J., and Mitchson, T.J., 1997, Actin dynamics in vivo, *Curr. Opin. Cell Biol.* **9**:54-61.

DISCUSSION

Pollack: It is known that there is a stream of cytoplasm toward the leading edge. Do you think this could be a mechanism?.

Miyata: It is a possible mechanism and we cannot rule out that possibility.

ACTIN DYNAMICS IN NEURONAL GROWTH CONE REVEALED WITH A POLARIZED LIGHT MICROSCOPY

Kaoru Katoh, Fumiko Yoshida, Ryoki Ishikawa[*]

1. INTRODUCTION

More than 100 years ago, S.R. y Cajal[1] discovered an enlargement of cytoplasm at the tip of neurite through a careful observation of the fixed nervous tissues. He named this structure as growth cone. He suggested a possibility that the growth cones might guide growing dendrites and axons to their targets. Since then, this hypothesis has been supported by neuroscientists. It is established that the growth cone crawls around and finds suitable path for neuronal elongation[2-4]. Present research interests in this field are path finding by the growth cone and its motile mechanism[5-8].

The growth cone consists of thin and flat lamellipodium with radially aligned finger-like filopodia that protrude at the leading edge (Fig. 1)[9-12]. Lamellipodia move like moving veils between filopodia and induce advance of leading edge[13]. Filopodia move sideways and back-and-forth[14,15]. Filopodia are supposed to have some sensory system for the path finding, because contact of a single filopodium is enough to recognize suitable path for neuronal elongation[16-18].

The fine structures of growth cones were revealed with electron microscopy[9-12]. The growth cone periphery is rich in filamentous actin (f-actin). The f-actin makes two types of structures: actin meshwork that supports lamellipodia and actin bundles that

[*] Kaoru Katoh, Neuroscience Research Institute, National institute of Advanced Industrial Science andTechnology (AIST), 1-1-1 Umezono, Tsukuba, 305-8568 JAPAN.
PRESTO, JST, 6F 9-6 Konya-Imamachi, Kumamoto, 860-0012 JAPAN.
Fumiko Yoshida, Neuroscience Research Institute, AIST, 1-1-1 Umezono, Tsukuba, 305-8568 JAPAN.
Ryoki Ishikawa, Department of Pharmacology, Gunma University School of Medicine, Maebashi, Gunma, 371-8511 JAPAN.

Molecular and Cellular Aspects of Muscle Contraction
Edited by H. Sugi, Kluwer Academic/Plenum Publishers, 2003

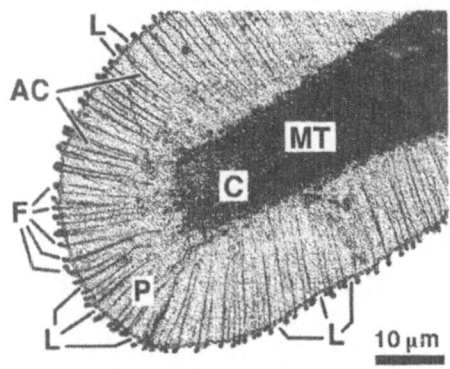

Figure 1 Schematic view of a growth cone. C: central domain. P: peripheral domain.
L: lamellipodia. F: filopodia. MT: microtubules. AC: actin bundle.

associate with filopodia. Microtubules are rare in the peripheral region but abundant on the central region of the growth cone. The localization of f-actin in the periphery implies a possibility that actin-based structures participate in the growth cone motility. This possibility is also supported by pharmacological and physiological experiments.

The actin dynamics[19-21] was, therefore, examined with light microscopes by using fluorescence labeling methods. Most remarkable movement in the growth cone was retrograde flow of filamentous actin from leading edge toward the central domain. Similar movements were demonstrated either by photo bleaching or photo activating fluorescence label on the actin filaments in many cells (fibroblasts[22,23]; keratocytes[24]; growth cones[25-27]). Forscher and coworkers reported the relationship between the retrograde flow and growth cone motility[28-31].

Although the fluorescence imaging revealed a certain aspect of actin dynamics, there are limitations specific to this method. For example, the fluorescent imaging requires to load fluorescent dyes or to express fluorescent protein molecules in the target cells. These fluorescence labels can be toxic or may interfere with normal cell functions. Furthermore, fluorescence visualizes only the structures labeled specifically. All the other unlabeled structures are invisible. If we can visualize dynamics of filamentous actin with a new method, it should contribute on the analysis of growth cone movements.

Primary objective of our research is to know the motile mechanism of the growth cones. We, therefore, visualized actin dynamics directly with a new type of polarized light microscope (LC-polscope)[32] and analyzed the movement with computer. Moreover, to examine the role of actin associate proteins on the actin dynamics, we visualized the actin associate proteins by labeling with GFP. Here, we summarize our results obtained recent years.

2. MATERIALS AND METHODS

2.1. Optical set-up of the microscope

To directly observe actin bundles in living unstained growth cones non-invasively, we measured and displayed optical anisotropy (birefringence) of cellular structures. Birefringence is an optical property caused by ordered molecules. Several careful authors have observed birefringence of filamentous actin (f-actin) with highly sensitive polarized-light microscope[33,34], although birefringence of f-actin is so small that f-actin rich region in muscle is almost invisible and called I-band (isotropic band).

We, therefore, visualized actin dynamics with a LC-polscope (Fig.2)[32], which is a new type of highly sensitive polarized-light microscope controlled by the computer and electro-optical module. Two improvements have been made in the optical path of traditional polarized light microscope: 1) replacement of the traditional compensator by an electrically controlled universal compensator made from a liquid crystal, and 2) replacement of the linear analyzer by a left circular analyzer. By sending electrical commands from computer to the universal compensator, we can make any kinds of polarized light waves without moving any parts of the microscope.

Regularly, the LC-polscope works as follows. A CCD camera attached on the microscope records a set of images that include an image illuminated by the circular polarized light and three images by the elliptically polarized light. These captured images are transferred to a computer in order to compute two image maps, birefringence retardation map and azimuth map. In the birefringence retardation map (here after, called the 'retardation image'), the gray value of each pixel represents birefringence retardation, while, in the azimuth map, a gray value represents an azimuth angle. In a bundle of protein filaments, birefringence retardation is proportional to the number of filaments in the bundle[35] and azimuth angle is parallel to the orientation of the filaments

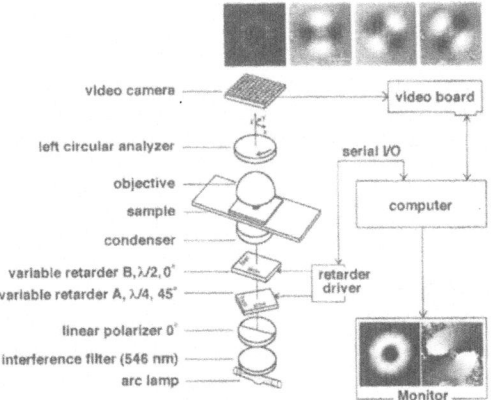

Figure 2 Optical set-up of LC-polscope. Top row: Original images captured at four settings of universal compensator. Right bottom: Two computed images (left: retardation map, right: azimuth map). (Reprinted from *Nature* **381** p811 (1996) with permission).

in the bundle.   Thus, we can estimate number of filaments and its orientation with this method.

A list of instrument parts used in the current study: An Olympus AX-70 microscope equipped with plan-apochromat oil-immersion objective lens that is selected as strain free optics (from KS Olympus, Tsukuba JAPAN); a mercury arc lamp followed by an Ellis light scrambler (Technical Video, Woods Hole, MA); a narrow band path filter (546nm, 10nm FWHM Omega optical Brattleboro, VT); LC-polscope unit with computer (CRi, Cambridge MA); CCD camera for Polscope (DC-330, MTI-Dage, Michigan City, IN); And a CCD camera for recording images of fluorescence (Micro MAX 512 BFT, Roper Scientific, NJ).

2.2. Data analysis

To make a kymograph on the computer, we used NIH image.   Thin trips are cut from a stack of time-lapse images and made into kymograph by placing them next to each other in chronological order.   The kymograph is automatically processed with a NIH image software with home-made macro.

2.3. Cell Culture and Preparation of the Samples

Aplysia BAG cell neuron: Primary cell culture of *Aplysia* BAG cell neurons were prepared according to the methods of previous authors[36,37] and cultured on the cover slips bathed in artificial sea water.   Typically, growth cones appear 6-8 hours after plating the cells and disappear within 36 hours.

NG-108 cells (gifts of Prof. Hideaki TANAKA, Kumamoto University):
Cells were cultured in DMEM cell culture media (GIBCO) supplemented with 10 % FBS, penicillin/streptomycin, and 1 X H.A.T. at 37 °C and 5% CO_2.   3-4 days before observation, these cells were treated with 1 mM dibutyryl cyclic AMP, which induces generation of axons and growth cones.   1 day before the observation, the cells were briefly trypsinized and replated on a coverslips coated with poly-lysine.   To visualize actin associate proteins, cells were transfected with a plasmid that encodes GFP-fascin, GFP-tropomyosin or GFP-alpha-actinin, by using lipofection.

For observation with high-resolution microscope, a coverslip with cultured cells was mounted on the glass slide with thin spacers (150 um) and sealed to avoid evaporation. After sealing, the cells could be observed for several hours without visible damages.

3. RESULTS AND DISCUSSION

3.1. Actin Dynamics in the Growth Cone

Fig.3 shows cellular structures observed in the growth cone[38].   Filopodia are associated with actin bundles.   Parts of actin meshwork that is stretched by surface tension of cell membrane are observed as birefringence patches in the lamellipodia. Intrapodia (actin comets) appear in the lamellipodial region near the central domain. Stress fibers are not observed in growth cone region of cultured neurons.   NG108 cells

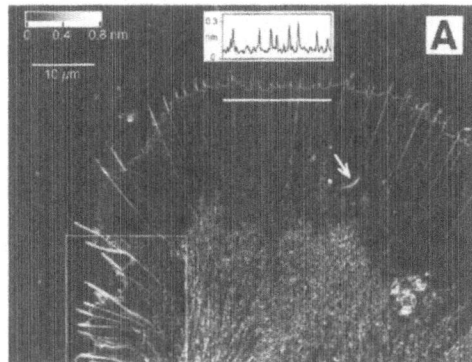
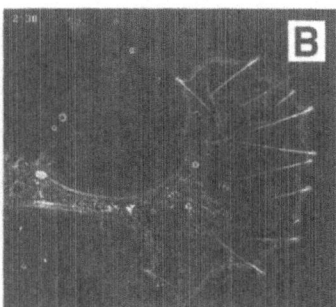

Figure 3   Growth cones revealed with LC-polscope.   **A:** *Aplysia* BAG cell neuron..   Arrow: intrapodia   Plot profile of retardation at the white line region is shown above.   (Reprinted from *Mol Biol Cell* **10** p197-210 (1999) with permission.)   **B:** NG-108 cell

(hybridoma between neuroblastoma and glioma) also show cytoskeletal structures similar to the cultured neurons, especially in the growth cone regions.

3.1.1. Movement of Actin Bundles Associated with Filopodia

Dynamics of actin bundles is correlated with filopodial behaviors that include generation, degeneration, sideway motion, back-and-forth movement and fusion with adjacent filopodia.

3.1.1.a. Generation of filopodia and actin bundles

Generation of filopodia and of actin bundles was observed near the leading edge (Fig.4).   The creation of new filopodia is preceded by the appearance of a birefringence spot at the leading edge (5sec-image in Fig.4).   Azimuth angle of the birefringence spot, which indicates orientation of the short filament, is almost perpendicular to the leading edge.   The spot then grows into an actin bundle, of which angle is similar to the azimuth angle of the birefringence spot.   The new bundle looks like pushing the surface membrane and form new filopodium.   The new filopodia and actin bundles are often tilted against the leading.

3.1.1.b. Sideway motion of filopodia and actin bundles

The tilted filopodia and actin bundles move sideways, sometimes crossing over several adjacent actin bundles[15].   Some of the filopodia change their angle parallel to the orientation of the retrograde flow, stop their lateral travel, and become stable filopodia.   On the other hand, stable filopodia can be unstable, traveling filopodia.   Switching from stable to traveling filopodia is started with the tilt of filopodia against the leading edge.

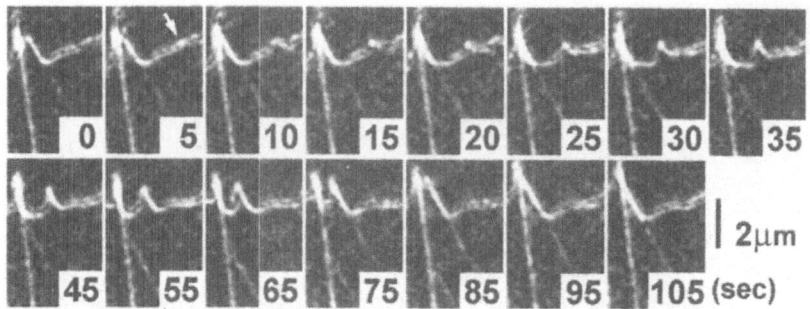

Figure 4 Generation of filopodia and actin bundle
Generation of filopodia is proceeded by a birefringent spot (arrow in the
5-sec image) at the leading edge.

3.1.1.c. Fusion of filopodia with adjacent filopodia

The tilted traveling filopodia frequently fuse with one of the other filopodia (Fig.4).
The fusion starts near the tip of the tilted traveling filopodia. A branching of the
filopodia is, then, formed. As the fusion progresses, the branching point moves down at
a constant rate similar to the retrograde flow and enters into the lamellipodial region as a
fused zone of actin bundles. The number of actin filaments in the fused zone, which is
estimated from the birefringence retardation, is almost equal to the sum of the number of
filaments in each fiber.

3.1.2 Movement of Actin-based structures in lamellipodia

Retrograde flow of actin-based structure is reported in the neuronal growth cones by a
combination of fluorescently labeled-actin molecules and photobleaching experiments.
We directly visualized the retrograde flow of the native actin-based structures.

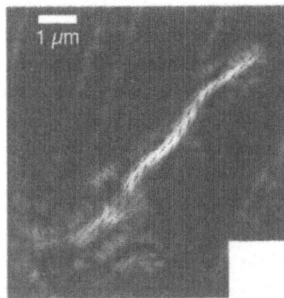

Figure 5 Intrapodia (reprinted from *Mol Biol Cell* **10** p197-210 (1999)
with permission).

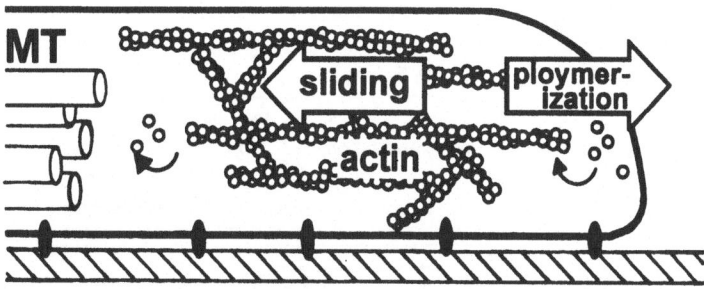

Figure 6 Hypothetical actin dynamics in the growth cone. F-actin is polymerized at the leading edge and moved to the central domain. If the polymerization rate is larger than sliding rate (rate of retrograde flow), the growth cone advances.

The retrograde flow is recognized as the movement of a specific form of the actin based structures in the polscope movies. All the structures related to the radial actin bundles, such as fused zone or branching point of the radial bundles, show retrograde flow. Other structural elements that are recognized as birefringence patches between the radial actin bundles also showed retrograde flow.

3.1.2.b. Intrapodia (actin comets)

Spike-like, fast-moving aggregates spontaneously appear in the lamellipodial region[38] (See the arrow in Fig.3A). Rochlin et al.[39,40] named these structures intrapodia. The intrapodia were seemingly propelled by a highly birefringent tail. They look like the actin-rich tail induced by intracellular pathogens (Listeria[41]), by extracellular polycation-coated beads (inductopodia[42]) or by spontaneously in many cells (actin comets[43,44]). The intrapodia also participate in the retrograde flow (Fig.5).

3.2. Motile Mechanism of Filopodia and their Actin Bundles

A variety of actin dynamics were observed in the growth cone with the polscope. The complicated movements of actin bundles are analyzed in this section.

3.2.1 Kymographic Analysis

A simple model of the actin turnover and growth cone motility is shown in Fig.6. Filamentous actins are polymerized at their ends near the leading edge, moved towards the central domain by the retrograde flow and depolymerized near the central domain. This model gives an overview of actin dynamics in growth cones. The continuous polymerization and rearward movement of actin bundles makes the portions within actin bundles in the temporal order (Fig.7A)[45]. Portion of actin bundles near the leading edge is polymerized at the latest time point, while the portion far from the leading edge is polymerized earlier.

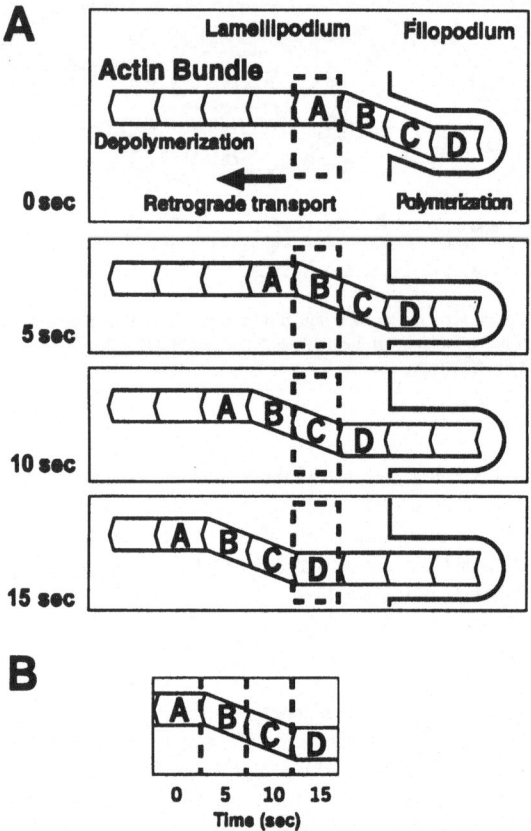

Figure. 7   Scheme of retrograde flow of actin bundle (A) and kymographic analysis (B).
A: A time series of image flames showing a radial actin bundle inside a filopodium and lamellipodium.
Retrograde flow moves the actin bundle from the filopodium.   Actin units are continuously added to the tip
of filopodia.   B: A kymograph assembled from a portion of image shown by dashed line in A.   An actin
bundle in the kymograph is similar in its shape to the bundle in the last frame (15sec).   (Reprinted from
PNAS **96** p7928-7931 (1999) with permission.).

According to the model above, we constructed kymographs from thin image strips
that recorded the position and make-up of actin bundles near the leading edge.   The
strips were automatically placed next to each other in temporal order with a computer
(Fig.7B).   The bundle portion in the kymograph is expected to reflect the state and
location of the bundle portion in the lamellipodium.   We hypothesized that the shape of
actin bundles in the kymograph should be the same as the actin bundles in the whole
lamellipodia recorded at the latest time point of the kymograph, if shape of the actin
bundles is determined near the leading edge and moved to the central domain keeping
their shape (compare 15 sec-image in Fig 7A with Fig 7B).   We, therefore, examined the
relationship between behavior of individual actin bundles immediately proximal to the

filopodia and the shape of the actin bundles in the whole lamellipodium, by comparing a kymograph with an image of lamellipodium.

Fig.8A shows lamellipodia from which thin strips were cut off to make kymograph (Fig8B). The 15-actin bundles in the lamellipodium (Fig.8C) are similar in their shape and make-up to that of assembled kymograph (Fig.8B). This finding suggests that the shape of the bundles were determined near the leading edge and moved toward the central domain with keeping their shapes and make-up. That is, arrangement and shape of actin bundles in lamellipodia is a kind of live 'kymograph' that records filopodial behavior in the past several minutes.

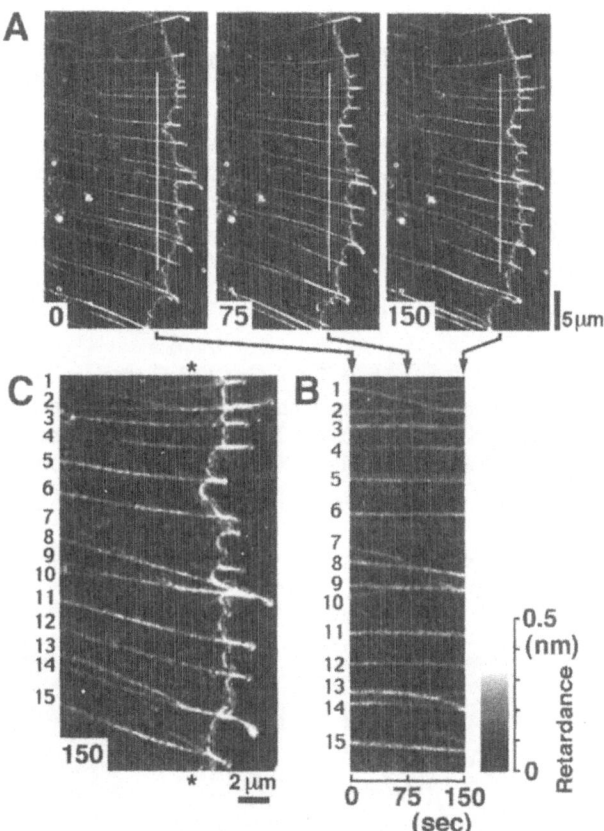

Figure 8 Dynamics of actin bundle arrangement in the lamellipodium of the growth cone and assembly of a kymograph. A: Frames of a time-lapse movie of a growth cone. Numbers bottom left shows time in sec. White line shows a position of the image strips that were used to assemble kymograph in B.
B: A kymograph assembled from the image strips of 31 consecutive frames. C: Magnified image of the lamellipodium at 150 sec frame in A. (Reprinted from *PNAS* **96** p7928-7931 (1999) with permission.)

3.3. Live Imaging of Actin Associate Proteins in the Growth Cones

We visualized dynamics of actin associate proteins in the growth cones by using fluorescence of GFP. GFP-fascin (Fig. 9), GFP-tropomyosin, GFP-alpha-actinin, GFP-caldesmon and GFP-drebrin are transiently overexpressed in the growth cones of neuroblastoma. Distribution of this GFP-actin associate proteins are similar to the distribution revealed by the immunochemical methods, suggesting that over expression of these proteins does not induce any pathological effects on the growth cones of neuroblastoma. We are analyzing dynamic behavior of these actin associate proteins.

4. CONCLUSION

We have visualized dynamics of native unstained actin bundle in living cells with the LC-polscope and showed that a history of filopodial behavior is recorded as a shape of actin bundles. That is, shapes of the actin bundles were determined near the leading edge, when actin bundles were formed. To know the regulatory proteins needed for the formation of filopodia, we combined LC-polscope with fluorescent microscope. By using GFP-imaging, we will analyze a relationship between distribution of actin associate proteins and actin dynamics.

5. ACKNOWLEGEMENTS

We are grateful to Dr. Shinya Inoué, Dr. Rudolf Oldenbourg and Peter J.S. Smith in Marine Biological Laboratory for collaboration in the early part of these studies. Latter half of this work was supported by Industrial Technology Research Grant Program in '00 from New Energy and Industrial Technology Development (NEDO) of JAPAN to KK.

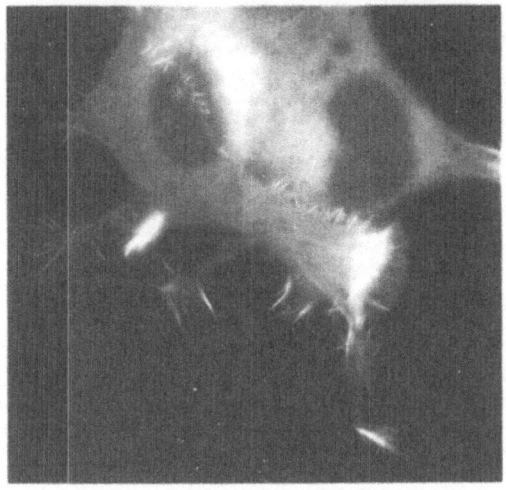

Figure 9   Distribution of GFP-fascin in NG-108.

6. REFERECES

1. Cajal, S.R.y A quelle époque apparaissent les expansions des cellules nerveuses de la moelle épinière du poulet? *Anat. Anz.* **5** 609-613 (1890).

2. Letourneau, P.C., Kater, S.B., and Macagno, E.R. (eds.) *The Nerve Growth Cone*, (Raven Press, New York, 1991):..

3. McCaig, C.D. *Nerve Growth and Guidance*, (Portland Press. London: 1996)

4. Heidemann, S.R. Cytoplasmic mechanism of axonal and dendritic growth in neuron. *Int. Rev. Cytol.* **65** 235-296. (1996)

5. Ming GL, Wong ST, Henley J, Yuan XB, Song HJ, Spitzer NC, Poo MM Adaptation in the chemotactic guidance of nerve growth cones. *Nature* **417** 411-418 (2002)

6. Xiang Y, Li. Y, Zhang Z, Cui K, Wang S, Yuan XB, Wu CP, Poo MM, Duan SM Nerve growth cone guidance mediated by G protein-coupled receptors. *Nature Neurosci.* **5** 843-848 (2002)

7. Gomez T.M., Spitzer N.C. Regulation of growth cone behavior by calcium: new dynamics to earlier perspectives *J. Neurobiol* **44** 174-183 (2000)

8. Gomez T.M., Spitzer N.C. In vivo regulation of axon extension and pathfinding by growth-cone calcium transients. *Nature* **397** 350-355 (1999)

9. Bridgman, P.C. and Dailey M.E. The organization of myosin and actin in rapid frozen nerve growth cones. *J. Cell Biol.* **108** 95-109. (1989)

10. Yamada K.M., Spooner, B.S. and Wessells, N.K. Axon growth: role of microfilaments and microtubules. *Proc. Natul Acad. Sci. USA* **66** 1206-1212. (1970)

11. Yamada K.M., Spooner, B.S. and Wessells, N.K. Ultrastructure and function of growth cones and axons of cultured nerve cells. *J. Cell Biol.* **49** 614-634. (1971)

12. Tosney, K.W. and Wessells, N.K. Neuronal motility: the ultrastructure of veil and microspikes correlates with their motile activities. *J. Cell Biol.* **61** 389-411. (1983)

13. Goldberg D.J. and Burmeister, D.W. Stages in axon formation: observation of growth of Aplysia axons in culture using video-enhanced contrast-differential interference contrast microscopy. *J. Cell Biol.* **103** 1921-1931. (1986)

14. Bray, D. and Chapman. K. Analysis of microspike movements on the neuronal growth cone. *J. Neurosci.* **5** 3204-3213 (1985).

15. Oldenbourg R, Katoh K. Danuser G. Mechanism of lateral movement of filopodia and radial actin bundles across neuronal growth cones. *Biophys. J.* **78** 1176-1182. (2000).

16. O'Connor TP, Duerr JS, Bentley D. Pioneer growth cone steering decisions mediated by single filopodial contacts in situ. *J. Neurosci.* **10** 3935-3946 (1990)

17. Davenport RW, Dou P, Rehder V, Kater SB, A sensory role for neuronal growth cone filopodia. *Nature* **361** 721-724 (1993)

18. Gomez TM, Robles E, Poo MM, et al. Filopodial calcium transients promote substrate-dependent growth cone turning. *Science* **291** 1983-1987 (2001)

19. Okabe S. and Hirokawa N. Actin dynamics in growth cones. *J. Neurosci.* **11** 1918-1929. (1991)

20. Small VJ Lamellipodia architecture: actin filament turnover and the lateral flow of actin filaments during motility. *Semin. Cell Biol.* **5** 157-163 (1994)

21. Welch M.D., Mallavarapu, A., Rosenblatt, J., and Mitchison, T.J. Actin dynamics in vivo. *Curr. Opin. Cell Biol.* **9** 54-61. (1997)

22. Wang Y.L. Exchange of actin subunits at the leading edge of living fibroblasts: possible role of treadmilling. *J. Cell Biol.* **101** 597-602. (1985)

23. Watanabe N. and Mitchison TJ Single-molecule speckle analysis of actin filament turnover in lamellipodia. *Science* **295** 1083-1086 (2002)

24. Theriot J.A. and Mitchison T.J. Actin filament dynamics in locomoting cells. *Nature* **352** 126-131. (1991)

25. Smith SJ Neuronal cytomechanics: the actin-based motility of growth cone *Science* **242**, 708-715. (1988)

26. Welnhofer, E.A. Zhao, L. and Cohan I. Actin dynamics and organization during growth cone morphogenesis in Helisoma neurons. *Cell Motil. Cytoskeleton*, **37** 54-71. (1997)

27. Mallavarapu, A. and Mitchison, T.J. Regulated actin cytoskeleton assembly at filopodium tips controls their extension and retraction. *J. Cell Biol.* **146** 1097-1106 (1999)

28. Forscher P. and Smith, S.J Actions of cytochalasins on the organization of actin filaments and microtubules in a neuronal growth cone. *J. Cell. Biol.* **107** 1505-1516 (1988)

29. Lin C-H. and Forscher P., Cytoskeletal remodeling during growth cone-target interactions. *J. Cell Biol.* **121** 1369-1383 (1993).

30. Lin C-H. and Forscher P.,   Growth cone advance is inversely proportional to retrograde F-actin flow. *Neuron* **14** 763-771. .(1995)
31 Lin C-H, Espreafico, E.M., Mooseker, M.S. and Forscher, P   Myosin drives retrograde f-actin flow in neural growth cone. *Neuron* **16** 769-782 (1996).
32. Oldenbourg R.   A new view on polarization microscopy.   *Nature* **381** 811-812
33. Maeda Y.   Birefringence of oriented thin filaments in the I-bands of crab striated muscle and comparison with the flow birefringence of reconstituted thin filaments.   *Eur. J. Biochem* **90** 113-121 (1978)
34. Soranno, T. and E. Bell.   Cytostructural dynamics of spreading and translocating cells. *J. Cell Biol.* **95** 127-136 (1982.).
35. Tran P. Salmon E.D., Oldenbourg R.   Quantifying single and bundled microtubules with the polarized light microscope.   *Biol Bull.* **189** 206. (1995)
36. Kaczmarek, L.K., Finbow, M., Revel, J.P., and Strumwasser, F.   The morphology and coupling of Aplysia bag cell within the abdominal ganglion and in cell culture.   *J. Neurobiol.* **10**, 535-550. (1979).
37. Knox, R. J., E. A. Quattrocki, J. A. Connor and L. K. Kaczmarek.   Recruitment of Ca2+ channels by protein kinase C during rapid formation of putative neuropeptide release sites in isolated Aplysia neurons. *Neuron* **8** 883-889. (1992)
38. Katoh, K., Hammar K., Smith P.J.S., and Oldenbourg, R.   Birefringence imaging directly reveals architectural dynamics of filamentous actin in living growth cones.   *Mol. Biol. Cell.* **10** 197-210. (1999)
39. Rochlin, M.W., Dailey, M.E and Bridgman P.C.   Intrapodia: novel actin-rich structures that advance through nerve growth cone lamellipodia.   *Mol Biol. Cell* **8** 256a. (1997)
40. Rochlin, M.W., Dailey, M.E and Bridgman P.C.   Polymerizing microtubules activate site-directed f-actin assembly in nerve growth cones   *Mol. Biol. Cell* **10** 2309-2327. (1999)
41. Dabiri G.A., Sanger J.M., Portnoy D.A. and Southwick F.S.   Listeria monocytogenes moves rapidly through the host cell cytoplasm by inducing directional actin assembly.   *Proc. Natl. Acad. Sci. USA* **87** 6068-6072. (1990).
42. Forscher P., Lin, C-H., and Thompson, C.   Novel form of growth cone motility involving site directed actin filament assembly.   *Nature* **357** 515-518. (1992)
43. Ma L, Cantley LC, Janmey PA, Kirschner MW: Corequirement of specific phosphoinositides and small GTP-binding protein Cdc42 in inducing actin assembly in Xenopus egg extracts. *J Cell Biol* **140** 1125-1136. (1998).
44. Rohatgi R, Ma L, Miki H, Lopez M, Kirchhausen T, Takenawa T, et al.   The interaction between N-WASP and the Arp2/3 complex links Cdc42-dependent signals to actin assembly. *Cell* **97** 221-231 (1999).
45. Katoh, K., Hammar K., Smith P.J.S., and Oldenbourg, R. Arrangement of radial actin bundles in the growth cone of Aplysia bag cell neurons shows the immediate past history of filopodial behavior.   *Proc. Natl Acad. Sci. USA* **96** 7928-7931. (1999)

DISCUSSION

Lännergren: What kind of proteins accounts for the adhesion to the glass surface?

Katoh: Some adhesion molecules are supposed to be involved in the attachment.

Ishiwata: The bundle of actin filaments appeared to move laterally in added "to the polymyosin" in the filopodia. What is the motive force for this lateral movement of actin bundles?

Katoh: Side way motion of the actin bundles can be explained by three parameters: the rate of retrograde flow, the rate of polymerization, and the orientation of the polymeryzation. We proved that with a computer simulation.

Pollack: Why is there retrograde flow of actin, in situations in which actin should be moving forward?

Katoh: Neuronal growth cones repeats trials and errors, to find most suitable path for neuronal elongation. If they happen to elongate in the wrong direction, they have to shrink in order to correct elongation path. Retrograde flow may take part in such a process.

THEMODYNAMIC MECHANO-CHEMISTRY OF SINGLE MOTOR ENZYMES

Katsuhisa Tawada, Yasuhiro Imafuku and Neil Thomas[*]

1. INTRODUCTION

Muscle contraction is powered by the chemical free energy of ATP hydrolysis. In other words, the chemical free energy of ATP hydrolysis is used to perform mechanical work by muscle. Myosin is the ATPase, an enzyme which catalyzes the ATP hydrolysis for contraction in muscle. It is now well established by single-molecule measurements that a single molecular motor such as myosin and kinesin performs mechanical work upon hydrolyzing an ATP molecule. What is the immediate source of the energy in order for a single molecular motor to perform mechanical work? Is the immediate energy source the chemical free energy of ATP hydrolysis catalyzed by a single molecular motor?

The mechanical work performed by a single molecular motor is a quantity that can be defined for an individual molecular motor. In contrast, the chemical free energy of ATP hydrolysis is not such a quantity but a thermodynamic quantity: this energy is defined by a function containing the concentrations of ATP, ADP and Pi in solution. How to relate the mechanical process of single molecular motors to the thermodynamics of ATP hydrolysis catalyzed by them is one of the most fundamental problems on the molecular motor research. This is the purpose of this paper.

2. ATP HYDROLYSIS REACTION

Consider the following ATP hydrolysis reaction:

[*] Katsuhisa Tawada and Yasuhiro Imafuku, Department of Biology, Kyushu University Graduate School of Sciences, Fukuoka 812-8581, Japan. Neil Thomas, Department of Physics, Birmingham University, Birmingham B15 2TT, England.

$$ATP \underset{k_-}{\overset{k_+}{\Leftrightarrow}} ADP + P_i \tag{1}$$

For simplicity, we do not include water in the above reaction scheme. The rate of the forward reaction is $k_+ = k'_+[ATP]$ and the rate of the reverse reaction is $k_- = k'_-[ADP][P_i]$. Their ratio is given by

$$R_t = k_+/k_- = k'_+[ATP]/k'_-[ADP][P_i] \tag{2}$$

The free energy change of the chemical reaction (1) is given by

$$\Delta G = \Delta G^{\ominus} + RT\ln([ADP][P_i]/[ATP]) \tag{3}$$

where $\Delta G^{\ominus}$ is the standard free energy change of the ATP hydrolysis reaction and is related to its equilibrium constant as follows:

$$K_{eq} = k'_+/k'_- = \exp(-\Delta G^{\ominus}/RT) \tag{4}$$

From the above two equations, we have

$$R_t = (k'_+/k'_-)\exp(\Delta G^{\ominus}/RT)\exp(-\Delta G/RT) = \exp(-\Delta G/RT) \tag{5}$$

where equation (4) is used.

At equilibrium, $\Delta G = 0$, and hence $R_t = 1$. This means that the forward reaction rate is balanced by the reverse reaction rate at equilibrium (The principle of the detailed balance). If $\Delta G \leq 0, R_t \geq 1$, and hence the reaction (1) proceeds from left to right. If $\Delta G \geq 0, R_t \leq 1$, and hence the reaction (1) proceeds from right to left.

3.   ENZYME-CATALIZED REACTION

Here we consider ATP hydrolysis reaction catalyzed by an enzyme as follows.

$$E + ATP \underset{k_{-1}}{\overset{k_{+1}}{\Leftrightarrow}} E \cdot ATP \tag{6a}$$

$$E \cdot ATP \overset{k_{+2}}{\underset{k_{-2}}{\Leftrightarrow}} E + ADP + P_i \qquad (6b)$$

where E is an enzyme, $k_{+1} = k'_{+1}[ATP]$ and $k_{-2} = k'_{-2}[ADP][P_i]$.

At equilibrium, the principle of the detailed balance requires that $k_{+1}[E]_e = k_{-1}[E \cdot ATP]_e$ and $k_{+2}[E \cdot ATP]_e = k_{-2}[E]_e$, or that

$$k'_{+1}[ATP]_e[E]_e = k_{-1}[E \cdot ATP]_e \qquad (7a)$$

$$k_{+2}[E \cdot ATP]_e = k'_{-2}[ADP]_e[P_i]_e[E]_e \qquad (7b)$$

where $[\bullet]_e$ is the concentration at equilibrium. From equations (7), we have the equilibrium equation:

$$k'_{+1}k_{+2}/k_{-1}k'_{-2} = [ADP]_e[P_i]_e/[ATP]_e = k'_+/k'_- = K_{eq} \qquad (8)$$

Note that the effective rate constants of the forward and reverse reactions in the presence of enzyme are proportional to $k'_{+1}k_{+2}$ and $k_{-1}k'_{-2}$, respectively, with a same proportionality constant. Their ratio is equal to the ratio of the rate constants of the forward and reverse reactions in the absence of enzyme; the enzyme does not shift the equilibrium between reactants and products.

The ratio of the forward reaction rate to the reverse reaction rate at non-equilibrium is given by

$$R_t = k_{+1}k_{+2}/k_{-1}k_{-2} = \left(k'_{+1}k_{+2}/k_{-1}k'_{-2} \right)\left([ATP]/[ADP][P_i] \right) \qquad (9)$$

From Eqs.(3), (4), (8) and (9), we obtain

$$R_t = \exp(-\Delta G/RT) \qquad (10)$$

Note that Eq.(10) is the same as Eq.(5). At equilibrium, $\Delta G = 0$ and hence $R_t = 1$. If $\Delta G \leq 0, R_t \geq 1$, and hence the reaction (6) proceeds from left to right. If $\Delta G \geq 0, R_t \leq 1$, and hence the reaction (6) proceeds from right to left.

The enzyme-catalyzed reaction (6) contains one reaction intermediate. If an enzyme-catalyzed reaction contains more intermediates, we again obtain the same equation as Eq.(10) for the ratio of the forward reaction rate to the reverse reaction rate.

4. PROCESSIVE AND TIGHTLY-COUPLED MOLECULAR MOTORS

Here we consider ATP-driven movement of a single molecule of myosin-V, to which a load (f) is applied (Figure 1). The stepping motion of myosin-V motor is processive and tightly-coupled to ATP hydrolysis (= advancing one step of a fixed step size (u_0) upon hydrolyzing one ATP molecule). The ratio of the forward reaction rate to the reverse reaction rate of ATP hydrolysis catalyzed by myosin V as shown in figure 1 is given by

$$R_t = k(f)_{+1}k(f)_{+2}k(f)_{+3}k(f)_{+4}/k(f)_{-1}k(f)_{-2}k(f)_{-3}k(f)_{-4}$$
$$= \left(k(f)'_{+1}k(f)_{+2}k(f)_{+3}k(f)_{+4}/k(f)_{-1}k(f)_{-2}k(f)'_{-3}k(f)'_{-4}\right)\left([ATP]/[ADP][P_i]\right)$$

$$(11)$$

where some of the rate constants may depend on the load f, which is model-dependent. The chemical free energy change owing to the hydrolysis of one ATP molecule is

$$\Delta g = \Delta g^{\ominus} + kT\ln\left([ADP][P_i]/[ATP]\right)\qquad(12)$$

where k is Boltzmann constant. From Eqs.(11) and (12), we have

$$R_t = Z(f)\exp\left(-\Delta g/kT\right)\qquad(13)$$

The explicit form of $Z(f)$ is given below.

When a motor moves one step (u_0) forward against a load (f), it performs work of $u_0 f$. When the amount of this work energy is equal to the chemical free energy (-Δg) of ATP hydrolysis, the motor will be in a stationary state: on average, it neither moves nor splits ATP ($R_t = 1$). The motor stays at the same position while keeping an isometric tension. Therefore we find $Z(f) = \exp\left(-u_0 f/kT\right)$, and

$$R_t = \exp\left(-\left[\Delta g + u_0 f\right]/kT\right)\qquad(14)$$

If $u_0 f \le -\Delta g, R_t \ge 1$, then the molecular motor splits ATP (catalyzes spontaneously occurring ATP hydrolysis) and steps forward, performing work on a load. However, if $u_0 \ge -\Delta g, R_t \le 1$, then the molecular motor steps backwards, and work is done on the motor, resulting in the synthesis of ATP from ADP and Pi.

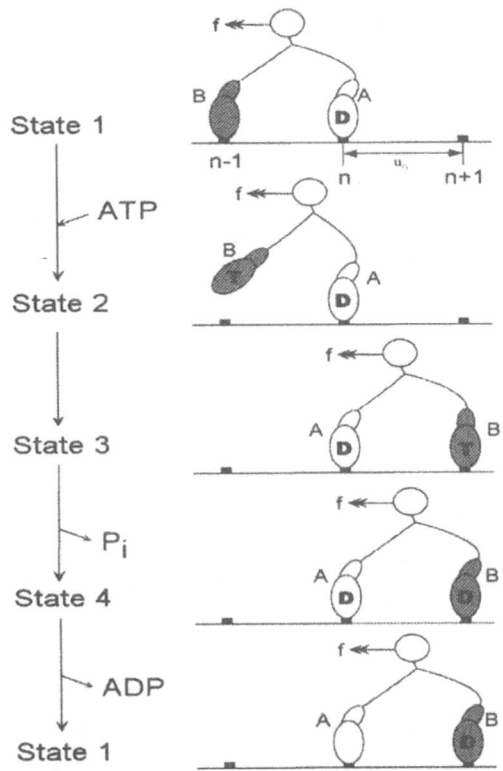

State 1

↓ ← ATP

State 2

↓

State 3

↓ ← P$_i$

State 4

↓ ← ADP

State 1

Figure 1. Four state cycles for two-headed myosin-V motor (Rief et al., 2000).

5. ROLE OF ATP HYDROLYSIS

Here we consider the cycle of a molecular motor as shown in figure 2. As the motor works in a cycle, there is no change in its internal energy for the complete cycle:

$$\Delta U = \Delta work + \Delta heat$$
$$= 0$$

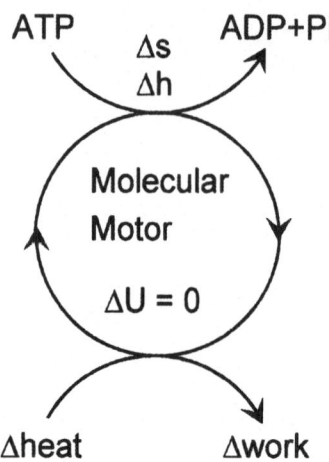

Figure 2. Cycle of a molecular motor and the first law of thermodynamics: $\Delta U = \Delta work + \Delta heat$, where U is the internal energy of the motor. Δs and Δh are the entropy and enthalpy changes owing to the ATP hydrolysis.

Hence, the first law of thermodynamics (which is applied, strictly speaking, to an ensemble of motors) requires that the work done per cycle, $-\Delta work$ ($=u_0 f$), equals to the heat absorbed by the motor from the surroundings ($\Delta heat$). (The motor absorbs heat in order to conduct thermally activated transitions that perform mechanical work on the load.) Heat absorption by the motor decreases the entropy of the surroundings by $u_0 f/T$ per step, while there can be no change in the entropy of the motor itself for a complete cycle. Hence this process cannot occur spontaneously on its own, as required by the second law of thermodynamics. However, the forward stepping of the motor is linked to a spontaneous process of ATP hydrolysis (the entropy and enthalpy changes owing to which are referred to as Δs and Δh). The entropy of the system hence increases by $\Delta s - \Delta h/T - u_0 f/T$. In order for a motor to perform work on a load by catalyzing the ATP hydrolysis, the second law of thermodynamics requires that $\Delta s - \Delta h/T - u_0 f/T \geq 0$. Since $\Delta g = \Delta h - T\Delta s$, we obtain $u_0 f \leq -\Delta g$. This is the same condition, which we have obtained with Eq.(14).

As shown by the above argument, the immediate source of energy in order for a molecular motor to perform mechanical work is not the chemical free energy from ATP hydrolysis, but the heat absorbed by the motor from the surroundings (Thomas et al., 2001).

6. ·MECHANICAL CHARACTERISTICS OF MOLECULAR MOTORS

If the load-dependence of the rate constants in a molecular motor model such as that shown in figure 1 are known, we can derive motor's characteristic equations such as a force-velocity equation and its dependence on the ATP concentration (Thomas et al., 2001 and 2002). Since motor's stepping motion is tightly-coupled to the spontaneously occurring ATP hydrolysis reaction, the motor's stepping motion is a biased random walk as theoretically shown (Thomas et al, 2001). The randomness parameter, which is defined as the ratio of the variance of the stepping distance to the average stepping distance, is expressed by a general equation with thermodynamic parameters (reaction rate constants) (Thomas et al, 2001).

7. CONCLUDING REMARKS

We have shown that a single processive and tightly-coupled molecular motor performs mechanical work by using the heat absorbed by the motor from the surroundings upon catalyzing spontaneously occurring ATP hydrolysis.

8. REFERENCES

Rief, M., Rock, R. S., Mehta, A. D., Mooseker, M. S., Cheney, R. E., and Spudich, J. A., 2000, Myosin-V stepping kinetics: a model of processivity, *Proc. Natl. Acad. Sci. USA* **97**: 590-593.

Thomas, N., Imafuku, Y., and Tawada, K., 2001, Molecular motors: thermodynamics and the random walk, *Proc. Roy. Soc. Lond.* B **268**: 2113-2133.

Thomas, N., Imafuku, Y., Kamiya, T., and Tawada, K., 2002, Kinesin: a molecular motor with a spring in its step, *Proc. Roy. Soc. Lond.* B (in press)

VI.   EXCITATION-CONTRACTION COUPLING

COOPERATIVE ACTIVATION OF SKELETAL AND CARDIAC MUSCLE

A. M. Gordon, A. J. Rivera, C-K. Wang, and M. Regnier[*]

1. INTRODUCTION

Both skeletal and cardiac muscles show a steep force-pCa relationship indicative of cooperative activation, but there are differences in some of the underlying mechanisms of this cooperativity. As we have discussed previously (Gordon *et al.*, 2000), these give rise to significant differences in the properties of skeletal and cardiac muscle that are important for their various physiological roles and methods of control. Cardiac contractions occur spontaneously and rhythmically, driven by the cardiac pacemaker cells in the SA node, with spread of electrical activity from cell to cell. This activates cardiac cells in sequence to eject blood allowing the heart to function as a periodic pump. Since each cell contracts on each beat, variations in cardiac output occur with variations in heart rate and the strength of contraction on each beat. Through intrinsic and extrinsic regulation via the autonomic nervous system, the rate and strength of each contraction can be regulated to meet the circulatory needs. In contrast, skeletal muscle contraction is controlled through the central nervous system as motor units, defined as a motoneuron and the muscle fibers it innervates. Although force varies with frequency of stimulation of each motor unit, the major means of regulation is by recruitment of motor units, a mechanism unavailable to the heart cells.

The cellular contractile properties that contribute to these differences in physiological control include the high apparent cooperativity in skeletal muscle cell activation brought about by (1) extensive interactions along the thin filament so that activation of one structural thin filament unit (7 actins, 1 troponin, 1 tropomyosin) (A_7TnTm) by Ca^{2+} binding or strong cross-bridge binding spreads easily to help activate neighboring units and by (2) strong cross-bridge binding which aids in activating each unit. This leads to a steep relationship between $[Ca^{2+}]$ and force and maximal activation with saturating Ca^{2+} levels. In contrast, cardiac muscles show less activation by Ca^{2+} alone within the regulatory unit, less interaction between neighboring units along the thin

[*] A. M. Gordon, Department of Physiology and Biophysics, University of Washington, Seattle, WA 98195. A. J. Rivera, C-K. Wang, and M. Regnier, Department of Bioengineering, University of Washington, Seattle, WA 98195.

filament so that activation spreads less easily to neighboring units, but still show cooperative activation by strongly attached cross-bridges within the regulatory unit and enhanced Ca^{2+} binding brought on by strongly attached cross-bridges. The result in cardiac muscle is that Ca^{2+} does not fully activate thin filaments even at maximal Ca^{2+} making the contraction more dependent on other factors that affect cross-bridge attachment and interactions between regulatory proteins, properties that can be modulated by protein phosphorylation or by changes in sarcomere length. Furthermore, the Ca^{2+} released on each heart beat can also be controlled. In this paper we will describe experiments which demonstrate these different properties and discuss possible mechanisms underlying them.

Activation of the thin filament takes place when Ca^{2+} binds to troponin-C (TnC), followed by interactions between the troponin subunits, tropomyosin and actin, and resulting in tropomyosin movement along the surface of the thin filament. This exposes sites of myosin binding to actin, greatly increasing the probability of strong, force-generating binding of myosin to actin (Gordon *et al.* 2000). The overlap of tropomyosin molecules along the thin filament makes possible the cooperative interaction between troponin-tropomyosin-actin regulatory units along the thin filament.

2. METHODS AND RESULTS

To investigate the interaction of regulatory proteins along the thin filament, we relied on a technique of substituting TnC's that are deficient in Ca^{2+} binding to the N-terminal triggering sites for the native TnC in demembranated muscle cells. This substitution leaves the troponin, tropomyosin, actin interactions intact, but prevents Ca^{2+} binding to that unit and thus prevents Ca^{2+} activation of that unit. Activation within that unit must therefore come from neighboring units spreading activation through interaction of overlapping tropomyosins.

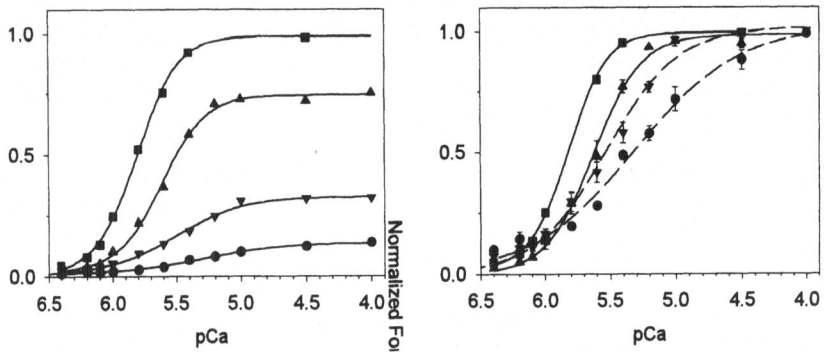

Figure 1. Force-pCa relations in rabbit psoas muscle fibers after extraction of endogenous TnC and reconstitution with mixtures of sTnC and xxsTnC. In the figure on the left, the data are shown for reconstitution with mixtures containing 100% sTnC (), 60% sTnC (), 20% sTnC (), and 15 % sTnC (). Force has been normalized to F_{max} for these fibers. In the figure on the right, this data have been normalized to the maximum force for that fiber after reconstitution to better visualize changes in pCa_{50} and n_H.

Substitution of the skeletal form of this TnC, deficient in N-terminal Ca^{2+} binding (xxsTnC), into fast skeletal muscle fibers (rabbit psoas) produces differing force-pCa curves depending on the fraction of TnC's substituted (sTnC/(sTnC + xxsTnC)). (A more complete discussion of this technique and the controls can be found in Regnier et al. (2002).) Figure 1 (left) shows that as the fraction of xxsTnC is increased, the force achieved at maximal Ca^{2+} decreases, the Ca^{2+} sensitivity (pCa_{50}) shifts to the right, and the slope of the curve decreases. These latter two points can be more easily seen in figure 1 (right) where the data in figure 1 (left) is normalized to the maximum for each condition and replotted. In figure 2, the maximum Ca^{2+} activated force (F_{max}) is plotted as a fraction of the possible active units (the fraction of sTnC/(sTnC + xxsTnC)). The slope of this curve is greater than one and consistent with Ca^{2+} binding to each unit activating more than the 7 actins of that unit, possibly up to 10-12 actins (see Regnier et al., 2002, for a more complete discussion of this point). In figure 3, we replot, for this data, the changes in the Hill coefficient obtained from a fit of the Hill equation to the data (a measure of cooperativity) and the changes in the Ca^{2+} sensitivity (pCa_{50}) as a function of the fraction of the added sTnC, the fraction of activatable units. As can be seen, the Hill coefficient, n_H, does not change significantly until the fraction of activatable units falls below 50%, that is when the probability of each unit having an activatable unit on at least one side falls. Extrapolation of each of the curves to near 0% activatable units gives the apparent properties of the isolated regulatory unit, uncoupled from neighboring units by having TnC's that cannot bind Ca^{2+} to activate. The properties of the isolated unit obtained by this extrapolation are a Ca^{2+} sensitivity (pCa_{50}) of 5.38 (4.2μM) and n_H of near 1.0, implying little cooperativity in the isolated unit. This implies that for fast skeletal muscle, a major source of the cooperativity in activation comes from the interaction between units along the thin filament through overlapping tropomyosins

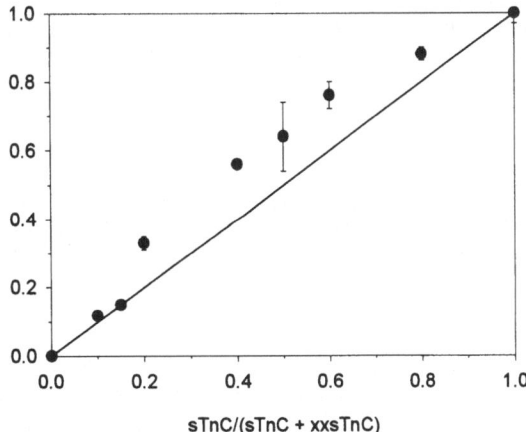

sTnC/(sTnC + xxsTnC)

Figure 2. Dependence of the maximum force in the fibers shown in Figure 1 after reconstitution as a function of the fraction of sTnC in the reconstitution mixture. Since sTnC and xxsTnC appear to bind with equal probability, this axis represents the fraction of sites reconstituted with sTnC thus able to bind Ca^{2+}. The straight line indicates what would be expected in the maximum force were proportional to the sTnC content.

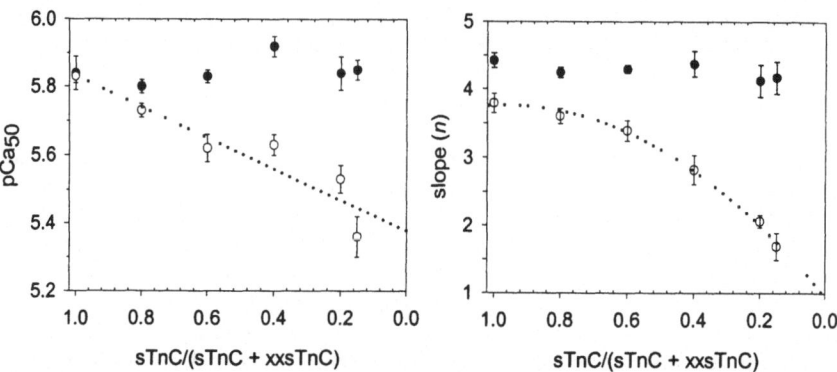

Figure 3. Dependence of pCa_{50} (left) and n_H (right) from the Hill equation fits to the force-pCa data shown in Figure 1on the fraction of sTnC in the reconstitution mixtures (o) compared to the control values in the fiber before extraction (●). The pCa_{50} data on the left for the reconstituted fibers is fit with a linear equation, $y = y_0 + ax$, with $a = 0.45 \pm 0.09$ and $y_0 = 5.38 \pm 0.09$. The n_H data on the right for the reconstituted fibers is fit with the quadratic formula, $y = y_0 + ax + bx^2$ where $y_0 = 0.95 \pm 0.11$.

brought on by Ca^{2+} activation and strong cross-bridge binding to neighboring units. It however does not argue that the whole skeletal muscle thin filament is switched on as a unit.

The comparable experiments in cardiac muscle are more difficult. Others (Putkey *et al.*, 1989; Huynh *et al.*, 1996) and we have produced the cTnC which is deficient in Ca^{2+} binding to the single N-terminal Ca^{2+} triggering site (xcTnC or CBMII). However, because it is difficult to extract all of the native cTnC, soaking the extracted preparation in a relaxing solution with a given % xTnC does not produce that fraction of cTnC sites without Ca^{2+} triggering sites and does not guarantee random placement of the mutant xcTnC. Experiments in which cTnC is extracted so that the maximal Ca^{2+} activated force is decreased to 20-30% of the initial maximum and reconstituted with 100% xcTnC, the maximum force remains only 20-30% of the initial force and there is little change in the pCa_{50} or n_H (Figure 4). The implication is that there is less spread of activation between units along the thin filament in cardiac muscle. In support of this interpretation that there is a distinct difference between regulation in skeletal and cardiac muscle and not just a difference in experimental protocol is the data on partial extraction of skeletal muscle by Moss *et al.*, 1985, and our own data. Unlike cardiac muscle, partial extraction of sTnC so that the maximum force is reduced to 20-30% with no reconstitution (Moss *et al.*, 1985) or with reconstitution with 100% xxsTnC (own data) in skeletal fibers yields large changes in pCa_{50} and n_H, similar to complete extraction followed by reconstitution mixtures of sTnC:xxsTnC that produce ~20% of pre-extracted F_{max}. In any case, new techniques are now available for total exchange of whole cTn (Brenner *et al.*, 1999; Köhler *et al.*, in press) reconstituted from specified Tn subunits, so that the possibility is now available for a more precise determination of the properties of the isolated regulatory unit in cardiac muscle and the spread of activation along the thin filament.

Without this spread in activation along the thin filament, cardiac muscle depends more on activation by strongly attached cross-bridges in the individual regulatory unit to

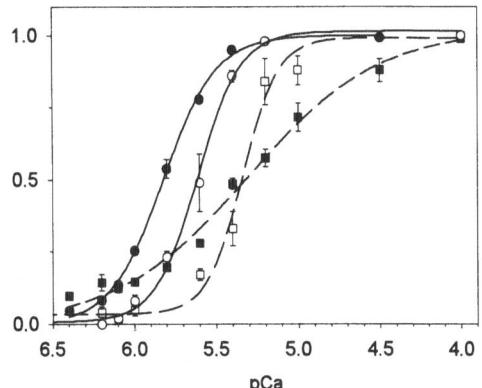

Figure 4. Comparison of TnC extraction and reconstitution protocol on force-pCa curves in skeletal (filled symbols) and cardiac (open symbols) muscle. The pre-extraction, control curves (circles) and extraction-reconstitution curves (squares) are compared for rabbit psoas fibers and a trabecula from PTU treated rats. The rabbit psoas fibers were completely TnC extracted and reconstituted with a 15% sTnC:85% xxsTnC mixture giving 14% of the maximum force. The trabeculae from PTU treated rats were TnC extracted down to a 20% maximum force and reconstituted with a 100% xcTnC solution. The comparative pCa_{50} and n_H for the skeletal fibers are 5.83 ± 0.02 and 4.0 ± 0.1 respectively for the control fibers and 5.32 ± 0.04 and 1.7 ± 0.2 for the extracted/reconstituted fibers. For the cardiac trabeculae, the pCa_{50} and n_H are 5.61 ± 0.03 and 3.2 ± 0.2 respectively for the control trabeculae and 5.27 ± 0.04 and 2.9 ± 0.3 for the extracted/reconstituted trabeculae. In each case, the data was normalized to the force for that condition so that the shapes of the curves can be more easily compared.

activate. Evidence that cardiac muscle is more dependence on cross-bridge attachment to activate comes from experiments in which cross-bridge attachment is modified. Cross-bridge attachment can be enhanced by low ATP slowing dissociation or by adding using dATP as a substraite for ATP. Figure 5 shows that as cross-bridge attachment is enhanced by increasing the cycling rate (dATP for ATP) (Regnier *et al.*, 1998), the maximum Ca^{2+} activated force is enhanced much more in cardiac muscle than skeletal muscle (see Regnier *et al.*, 2000, for a more complete discussion of these results). That there are more cross-bridges attached in cardiac muscle is shown by the observation that the preparation stiffness changes along with the force (Regnier *et al.*, 2000. Furthermore, when cross-bridge attachment is decreased by the addition of BDM, in cardiac muscle maximum force is decreased by the same fractional extent in control and partial xcTnC exchanged cardiac muscle. In contrast, in skeletal muscle, BDM decreases force proportionately much more in skeletal muscle in which 85% of the sTnC has been replaced by xxsTnC to isolate regulatory units than in control muscle. This implies that the without the near neighbor cooperativity, skeletal muscle relies on cross-bridges more for activation as does cardiac muscle.

3. DISCUSSION AND CONCLUSIONS

These results suggest that in normal skeletal muscle strong cross-bridge attachment can increase the spread of activation along the thin filament and enhance Ca^{2+} activation

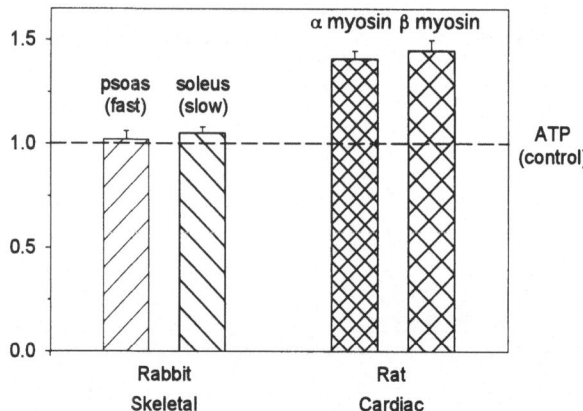

Figure 5. Increase in the maximum force in pCa 4.0 with substitution of 2 deoxy-ATP for ATP in rabbit skeletal muscle (on left, psoas and soleus muscle fibers) and rat cardiac trabeculae from normal animals expressing mostly α myosin or PTU treated animals expressing mostly β myosin. The maximum forces are normalized to the value with ATP.

within the individual regulatory unit giving a steep force-pCa relationship and activation over a narrow range of $[Ca^{2+}]$. In cardiac muscle, these strongly attached cross-bridges not only can increase Ca^{2+} binding to cTnC but can enhance force in regulatory unit making the maximum force sensitive to the modulation of strong cross-bridge attachment, making cardiac muscle more sensitive to other control factors which affect cross-bridge attachment. It is this functional difference which gives each cardiac cell and the heart as a synchronous contracting unit the large dynamic range required by the functioning, regulated heart.

One of the possible reasons for the difference between fast skeletal and cardiac muscle activation lies in the differences in TnC. As discussed (Gordon *et al.*, 2000), cardiac TnC has just one N-terminal Ca^{2+} binding site and may not undergo the complete conformational change at maximal Ca^{2+}, requiring binding to cTnI for the full transition to the E-F hand structure (Sia *et al.*, 1997). This might be the reason why maximal Ca^{2+} activation of cardiac muscle does not completely activate and why strongly attached cross-bridges can enhance activation. To test this hypothesis, when either cTnC or skeletal TnC mutated to decrease Ca^{2+} binding to site I are substituted into skeletal muscle for the native sTnC, the maximal Ca^{2+} activated tension is now less than that with the native sTnC. Furthermore dATP now enhances this maximal Ca^{2+} activated force as it does in cardiac muscle. However, differences in TnC cannot fully account for the differences since slow skeletal muscle shows no enhance force with dATP (see Figure 5) even though it has the cardiac TnC isoform.

The TnC is not the only isoform difference between fast skeletal and cardiac muscle as TnI, TnT, Tm and myosin are all different. Myosin is not a likely suspect as both cardiac isoforms show the same cooperative activation and enhancement of maximal Ca^{2+} activated force by dATP (Figure 5). Tropomyosin is particularly suspect since in the absence of Tn, its equilibrium position is more toward the inhibited state of the thin filament than is that for skeletal Tm (Lehman *et al.*, 2000). Studies of transgenic animals with switched tropomyosins suggest that there can be a shift in Ca^{2+} sensitivity between the αand the β homodimers (Wolska et al, 1999), but there was less indication of a change in n_H. Experiments with different Tn subunits can not be done as it is now possible to exchange whole Tn into skeletal (Brenner *et al.*, 1999) and cardiac muscle (Köhler et al, in press) to further define the contributions of each subunit.

These observations suggest that in skeletal muscle cooperative interaction between neighboring RUs are extensive. This cooperative spread of activation along the thin filament helps ensure complete activation of individual muscle fibers during contraction. This is consistent with control of skeletal muscle by some summation of force in individual fibers, but with recruitment of motor units playing a major role in control. In cardiac muscle cooperative interactions between neighboring RUs are less. Because of this, thin filament activation is more dependent on strong cross-bridge binding to shift Ca^{2+} sensitivity and, even at high levels of Ca^{2+}, the thin filament can be more completely activated by increasing cross-bridge binding. Since cardiac muscle is activated phasically during each heartbeat, activation depends on the release of Ca^{2+} to activate, but the response depends on the extent of strong cross-bridge binding. This allows for a greater capacity to modulate force at the cellular level in cardiac muscle as is required by having to modulate force in each cell since all cells are active during each beat. This makes cardiac cellular function very dependent on factors that can modulate cross-bridge attachment such as protein phosphorylation and sarcomere length.

4. ACKNOWLEDGEMENTS

We thank Dr. P. Bryant Chase for assistance with the initial experiments and to Mandy Bates for doing important control experiments. We also thank Drs. Larry Tobacman and Vicci Korman for solution measurements of Tn binding to Tm-actin filaments, Drs. E. Homsher and D. A. Martyn for critical comments and Ying Chen, Robin Mondares, Carol Freitag, Martha Mathiason, and Claire Zhang for excellent technical assistance. This work was supported by USA NIH HL65497.

5. REFERENCES

Brenner, B., Kraft, T., Yu, L.C., and Chalovich, J. M., 1999, Thin filament activation probed by fluorescence of N-((2-(Iodoacetoxy)ethyl)-N-methyl)amino-7-nitrobenz-2-oxa-1,3-diazole-labeled troponin I incorporated into skinned fibers of rabbit psoas muscle, *Biophys. J.* 77:2677.

Gordon, A. M., Homsher, E., and Regnier, M., 2000, Regulation of contraction in striated muscle, *Physiol. Rev.* **80**:853.

Huynh, Q., Butters, C. A., Leiden, J. M. and Tobacman, L. S., 1996, Effects of cardiacthin filament Ca^{2+}: statistical mechanical analysis of a troponin C site II mutant, *Biophys. J.* 70:1447.

Köhler, J. Chen, Y., Brenner, B., Gordon, A. M., Kraft, T., Martyn, D. A., Regnier, M. Rivera, A. J., Wang, C-K., and Chase, P. B., in press, Famililal hypertrophic cardiomyopathy mutations K183Δ, G203S and K206Q in cardiac troponin I enhance filament sliding, *Physiol. Genom.*

Lehman, W., Hatch, V., Korman, V. Rosol, M., Thomas, L., Maytum, R., Geeves, M.A., Van Eyk, J. E., Tobacman, L. S., and Craig, R., 2000, Tropomyosin and actin isoforms modulate the localilzation of tropomyosin strands on actin filaments, *J. Mol. Biol.* **302**:593.

Moss, R. L., Allen, J. D., and Greaser, M. L., 1985, The effects of partial extraction of TnC upon the tension-pCa relationship in rabbit skinned skeletal muscle fibers, *J. Gen. Physiol.* **86**:585.

Putkey, J. A., Sweeney, H. L., and Campbell, S. T., 1989, Site-directed mutation of the trigger calcium-binding sites in cardiac troponin C, *J. Biol. Chem.* **264**:12370.

Regnier, M., Rivera, A. J., Wang, C-K., Bates, M. A., Chase, P.B., and Gordon, A. M., 2002, Thin filament near-neighbour regulatory unit interactions affect rabbit skeletal muscle steady-state force-Ca^{2+}relations, *J. Physiol.* **543**:485.

Regnier, M., Martyn, D. A., and Chase, P. B., 1998, Calcium regulation of tension redevelopment kinetics with 2-deoxy-ATP or low [ATP] in rabbit skeletal muscle, *Biophys. J.* **74**:2005.

Regnier, M., Rivera, A. J., Chen, Y., and Chase, P. B., 2000, 2-deoxy ATP enhances contractility of rat cardiac muscle. *Circ. Res.* **86**:1211.

Sia, S.K., Li, M. X., Spyracopoulos, L., Gagn'e, S. M., Putkey, J. A., and Sykes, B. D., 1997, Structure of cardiac muscle troponin C unexpectedly reveals a closed regulatory domain, *J. Biol. Chem.* **272**:18216.

Wolska, B. M. Keller, R. S., Evans, C. C., Palmiter, K. A., Phillips, R. M., Muthuchamy, M., Oehlenschlager, J., Wieczorek, D. F., deTombs, P. P., and Solaro, R. J., 1999, Correlation between myofilament response to Ca^{2+} and altered dynamics of contraction and relaxation in transgenic cardiac cells that express β-tropomyosin, *Circ. Res.* **84**:745.

DISCUSSION

Ishiwata: Are there any conditions at which the regulatory units (RU) become more active? In other words, do you think that pure actin filaments are in the most activated state or can they be more activated by the RU under some conditions.?

Gordon: No. Pure actin filaments are not the most active. We and others have shown in the in vitro motility assay that regulated filaments slide faster than pure actin.

Gonzalez-Serratos: Since cooperativity, according to your proposal, depends on the attachment of the cross bridge, wouldn't you expect a relationship between cooperativity and the degree of overlap between thin and thick filaments?

Gordon: No. For cardiac muscle with less cooperativity along the thin filament, filament overlap per se should not affect cooperativity. Of more importance is the decreased filament separation at longer sarcomere lengths and enhanced cross-bridge binding.

Rall: How can the number of strong cross bridges be increased in the heart in vivo?

Gordon: By increasing sarcomere length and decreasing myofilament spacing. By increasing C-protein phosphorylation and phosphorylation of the thin filament proteins.

Stehle: Can the different effect of desoxy-ATP refers to the different reactivity? Could changes of F_{max} in the two fiber types which occur on substitution of ATP by desoxy ATP in cardiac and skeletal fiber also arise because of changes in cross-bridge turnover?

Gordon: We think that the changes in cross-bridge kinetics favor an increase in strong cross-bridge attachment. In the case of cardiac muscle, this appears to cause greater thin filament activation.

Winegrad: With your scheme in cardiac muscle, cooperativity occurs within the unit of 7 actins and does not extend to adjacent functional units. This would mean that cooperativity does not occur until at least 2 cross-bridges are attached per unit. This is close to the number commonly believed to be near the maximum number of attached cross bridges during maximum isometric force. How does your scheme of cooperativity operate in view of this consideration?

Gordon: Our data suggest that there is much less cooperativity between units in cardiac thin filament than in skeletal muscle. However, there are reasons to think that there is some cooperativity, since we see changes in calcium sensitivity with TnC extraction in cardiac muscle even though the Hill coefficient does not change to a large extent. Thus, the regulatory unit maybe somewhat larger and the number of possible attached cross-bridges can be greater.

Sweeney: Since the soleus does show the large range cooperativity while the heart does not, why do you feel that the TnC isoform plays a deterministic role?

Gordon: We feel that the TnC isoform plays some role, although other Tn and Tm isoforms must be involved.

Sweeney: What about nebulin? It is the most obvious difference between the thin filaments of skeletal vs. cardiac muscle.

Gordon: We have not looked at it, but nebulin could be disrupted in skinned fibers as a way to look at that question.

MEASUREMENT OF FORCE DEVELOPED BY A SINGLE CARDIAC MYOCYTE USING NOVEL CARBON FIBERS

Seiryo Sugiura, So-ichiro Yasuda, Hiroshi Yamashita, Kaoru Kato, Yasutake Saeki, Hiroko Kaneko, Yoshihisa Suda, Ryozo Nagai, and Haruo Sugi

1. ABSTRACT

In order to study the mechanical activity of a single cardiac myocyte under a wide range of load, we have developed a novel force measurement system using carbon fibers. Newly fabricated Graphite Reinforced by Carbon (GRC) fibers greatly facilitate the firm attachment of cell membrane to the fibers. A pair of fibers was attached to both ends of the cell; the rigid fiber as a mechanical ground and the compliant fiber for the strain gauge. By connecting the compliant fiber to the piezoelectric translator and applying the position signal to the driver, we could make the myocyte contract under isometric condition. Feedback control of the system also enabled us to study the relation between work output and the load. This system can be a useful tool in studying the mechanical activity of the cardiac myocyte under genetic as well as pharmacological interventions.

2. INTRODUCTION

Significance of the study on isolated cardiac myocyte is twofold. First, it provides us with an opportunity to study the mechanical property of heart without complications inherent in the multicellular preparation such as the heterogeneity in electrical activation and contraction [1]. Second, the function of isolated myocytes can be easily modified by gene trasfer thus forming an experimental model for the gene therapy of heart disease.

Seiryo Sugiura. Author Institute of Environmental Studies, Graduate School of Frontier Sciences, University of Tokyo, Tokyo 113-0033, Japan, So-ichiro Yasuda, Hiroshi Yamashita, Ryozo Nagai, Department of Cardiovascular Medicine, Graduate School of Medicine, University of Tokyo, Tokyo 113-8655, Japan, Kaoru Kato, Supermolecular Division, Electrotechnical Laboratory, Tsukuba 305-8568, Japan, Yasutake Saeki, Department of Physiology, Dental School, Tsurumi University, Yokohama 230-0063, Japan, Hiroko Kaneko, Tsukuba Materials Information Laboratory, Co. Ltd., Tsukuba 305-0051, Japan, Yoshihisa Suda, Mitsubishi Pencil Co. Ltd., Fujioka 375-8501, Japan, Haruo Sugi, Department of Physiology, School of Medicine, Teikyo University, Tokyo 173-8605, Japan

Molecular and Cellular Aspects of Muscle Contraction
Edited by H. Sugi, Kluwer Academic/Plenum Publishers, 2003

However, to characterize the performance of the cardiac myocyte over a wide range of loading conditions which intact heart may actually encounter in diseased state, not only untethered but also loaded cells should be examined.

To this end, several techniques for the cell attachment have been devised. Fabiato et al. [2] immobilized the cell with glass micro-needles. By selecting the preparation consisting of the three cells connected in series and impaling the outer two cells, they could leave the middle cell intact and successfully measured the twitch tension but the success rate was reported to be very low. Suction micropipette technique was developed by Brady et al. [1] and further improved by his colleagues. With this system using the intercalated disk as the main point of attachment, one can measure the isometric force of electrically stimulated cardiac myocyte without introducing significant deformation to the cell. However, this method also requires technical expertise. Adhesives such as fibrin glue or poly-L-lysine can also be used to attach the cell end but large portion of the cell surface is involved in the attachment and the significant preparation time is required. A relatively easy alternative is the carbon fiber technique developed by Gannier and Bernengo [3]. This technique has disadvantage of its low compliance which allows the full activation of the cells only at short sarcomere length (auxotonic contraction), but they could make the cell attachment easily without damaging the cell membrane.

In this study we adopted the carbon fiber technique but reduced its compliance by feedback system to successfully measure the twitch force of an isolated cardiac myocyte over a wide range of load applied to it. We also modified the surface of the carbon fiber to facilitate the cell attachment thus improved the success rate of the experiment. This system can be used for the detailed evaluation of the cardiac myocyte performance of which is modulated not only pharmacologically but also genetically.

3. Method

3.1 Fabrication of carbon fibers

As was suggested by Garnier et al. [3] adhesion of carbon fiber to the cell surface is mediated by static charge. Based on this hypothesis, novel carbon rod composed of both amorphous carbon and carbaon graphite (Graphite Reinforced by Carbon: GRC) was fabricated [4]. In contrast to the ordinary carbon fibers, GRC fiber with carbon graphite has a highly organized structure with its charged residues sticking out in the same direction thus maximizing the net surface charge. Compliance of the fiber was determined according to the method by Yoneda [5]. Briefly, we pushed the tip of a carbon fiber against the tip of the glass micro-needle the compliance of which was calibrated by suspending a thin copper wire weight at the tip. By comparing the displacement of the tip, we calculated the compliance of the fibers.

3.2 System configuration

A pair of carbon fibers were used for cell attachment. The stiff one with larger diameter (30 μm, compliance 0.015-0.02μN/μm) served as the rigid support of the cell and the thin compliant one (diameter 7 μm, compliance 5.5 – 7.5 μN/μm) worked as a strain gauge. Each fiber was placed and glued inside a holder made of a glass micropipette. The holder for thick fiber was mounted directly on a micro-manipulator, but the other one was connected to a Piezo-electric translator (Physik Instrumente, Germany). The image of the preparation viewed by a CCD camera coupled with an inverted microscope (IMT-2, Olympus Tokyo, Japan, objective 40 X) for sarcomere length measurement. The same image was split by a half mirror and the thin fiber was projected a pair of photodiode (S4201, Hamamatsu Photonics). The output of one component of the diode was subtracted from the other to give the displacement signal of the fiber with which we can calculate the force (Fig 1) [6]. The signal was digitized and stored at 1 KHz by a data acquisition system (MacLab, ADInstruments, Australia). To reduce the compliance of the system, the position signal was applied to the feed back circuit which drives the Peizo-electric translator.

3.3 Cell preparation and attachment

Adult rat ventricular myocyte was isolated from 4-week-old male Wistar rat. Animals were anesthetized with sodium pentobarbital and hearts were removed quickly. Aorta was cannulated and the retrograde perfusion was performed with HEPES-Tyrode solution containing Collagenase (pH 7.4, 37 °C). After the perfusion, the heart was cut

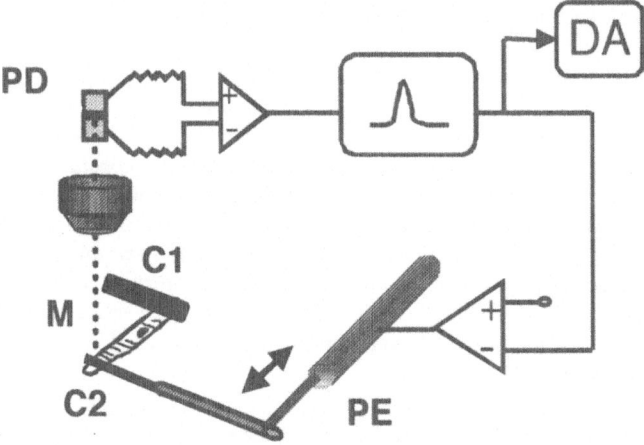

Figure 1. Schematic diagram of the force measurement system. A sing le myocyte (M) is suspended between the two carbon fibers (C1, C2). The image of the compliant fiber is projected onto a photodiode sensor to give the motion signal. This signal is also applied to feedback circuit to controla the piezoelectric (PE) device.

into small pieces. The pieces were incubated and the trypsin inhibitor was added. After the undigested cells were removed, the cells were resuspended in HEPES-Tyrode solution (Ca 1.8mM, 37 °C) and used for the assay.

3.4 Force measurement while varying the compliance

Experimental chamber was made of acrylic plate with a coverglass bottom and a pair of platinum plate was attached on each end for field stimulation. In each experiment, a single rod shaped and striated cell arranged perpendicularly to the fibers was selected. By lowering the fibers and making contact with the cell surface gently, myocyte was firmly attached to the fibers. After the attachment was established, cells were stimulated at 0.2 Hz. To study the myocyte contraction under a wide range of load, we changed the feedback gain to alter the compliance of the system.

4. RESULTS AND DISCUSSION

4.1 Carbon fiber as a handle for cell manipulation

To study the single myocyte contraction under load, various techniques have been devised for the cell attachment. In these method researchers pierce, suck, or glue to attach the cell surface for the manipulation of the cell. Compared to these methods, the currently used carbon fiber technique does not require expertise facilitating the high success rate. Because only the gentle contact is required to hold the cell, we can manipulate the intact cell without causing irritability to the cell membrane. Furthermore, the attachment force at the cell interface can surely support the force generated by a single rat ($\sim$0.7 µN in this study) cardiac myocyte not only in twitch contraction but also under maximum activation for rat ($\sim$1.1 µN in this study).

4.2 Feedback reinforcement of the compliance

Another drawback of the carbon fiber system is its high system compliance which allows us to study the myocyte contraction only under auxotonic condition. To avoid this problem, we introduced the feedback control of the fiber to enhance the stiffness and successfully measured the isometric force developed by a single cardiac myocyte (Fig.2). Recently, Tasche et al. [7] reported a novel force transducer based on the atomic force microscope sensor. Their system has very high resonance frequency (> 10 KHz) and high stiffness (10 - 1000 N/m) thus ideal for the basic mechanical study e.g. quick release-stretch. However, the use of glue for cell attachment requires experimental time and technical expertise.

4.3 Force measurement while varying the compliance

Adjusting its work output in response to the load is one of the fundamental property of muscle cell. Taking advantage the compliance altering ability of the system,   we could

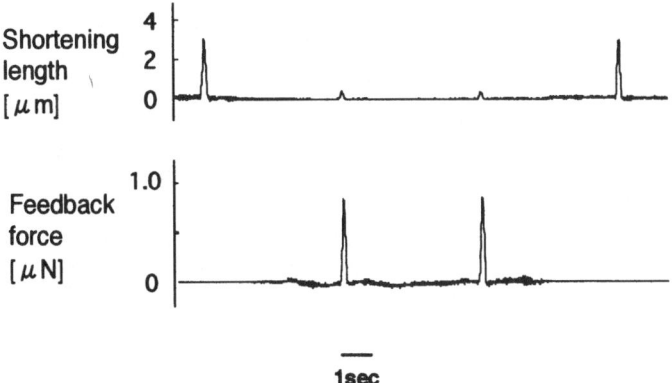

Figure 2 Two mode of contraction. In the first and the last contractions, myocyte is allowed to shorten. In the second and the third contractions, feedback loop is closed and the myocyte contracts under isometric conditions.

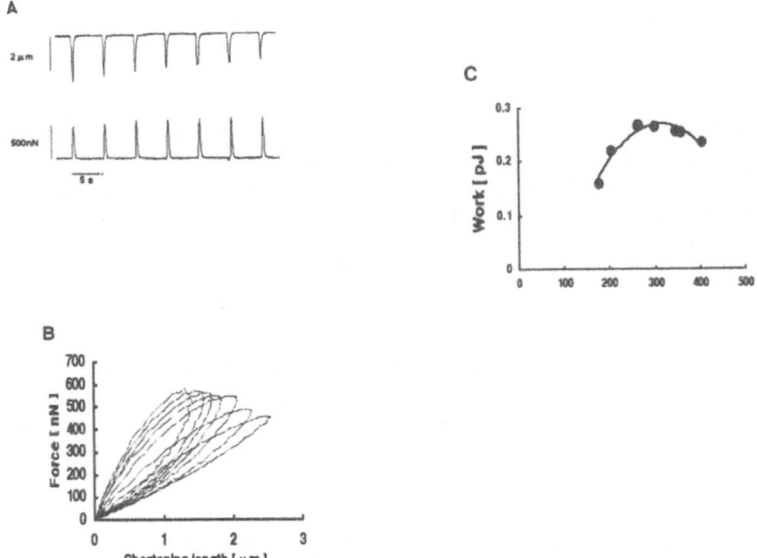

Figure 3 Myocyte contraction while changing the load. A. as the load was increased, the shortening distance decreases and the developed force increases. B. force-length loops while varying the load. C. the relation between the load and the work-output.

obtain the relation between the work output and the compliance (load) of the conttacting cardiac muscle cell. Under controlled condition (without any inotropic interventions), a single rat cardiac myocyte developed 0.73 ± 0.17 μN against a light load. As the load was increased the shortening distance became smaller to reach the isometric condition (Fig. 3A) and the trajectory formed by these two parameters was oval shaped(Fig. 3B). From these data, we plotted the relation between the load (compliance of the fiber) and the external work (the area circumscribed by the force and myocyte length trajectory) (Fig. 3C). The relation was similar to that obatained with muscle preparation showing the physiological property of the system. In congestive heart failure, the heart is known to work against abnormally high load [8]. With our system, one can study whether failing myocardium is functionally adapted to high load or not. This line of information will surely contribute to the pathophysiology of heart failure.

4.4 Summary

In summary, we have developed a carbon fiber based force measuring system for isolated cardiac myocyte. This system can vary the load applied to the myocyte in twitch contraction. We could evalute the inotrophic effect of the drug as well as the effect of load on work output of the cardiac muscle.

REFERENCES

1. A.J. Brady. Mechanical properties of isolated cardiac myocytes. Phyiol. Rev. 71:413-428 (1991).

2. A. Fabiato. Myoplasmic free calcium concentration reached during the twitch of an intact isolated cardiac cell and during calcium-induced release of calcium from the sarcoplasmic reticulum of a skinned cardiac cell from the adult rat or rabbit ventricle. J Gen Physiol. 78(5):457-97 (1981).

3. D. Garnier. Attachment procedures for mechanical manipulation of isolated cardiac myocytes: a challenge. Cardiovasc Res. 28:1958-1964 (1994).

4. S.I. Yasuda, S. Sugiura, N. Kobayakawa, H. Fujita, H. Yamashita, K. Katoh, et al. A novel method to study contraction characteristics of a single cardiac myocyte using carbon fibers. Am J Physiol Heart Circ Physiol. 281(3):H1442-6 (2001).

5. M. Yoneda. Force exerted by a single cilium of Mytilus Edulus. I. Exo Biol. 37:461-468 (1960).

6. S. Kamimura. Direct measurement of nanometric displacement under an optical microscope. Applied Optics. 26:3425-3427 (1987).

7. C. Tasche, E. Meyhofer, B. Brenner. A force transducer for measuring mechanical properties of single cardiac myocytes. Am J Physiol. 277(6 Pt 2):H2400-8 (1999).

8. H. Asanoi, S. Sasayama, T. Kameyama. Ventriculoarterial coupling in normal and failing heart in humans [published erratum appears in Circ Res 1990 Apr;66(4):1170]. Circ Res. 65(2):483-93 (1989).

DISCUSSION

Gonzalez-Serratos: Do you think that the increase in fluorescence signal under low load is caused by a movement artifact in the light detector?

Sugiura: No, I don't think so. We used ratio-metric indicator to eliminate motion artifact. In one of my slide, we checked that motion-artifact is absent.

Cecchi: You said that you could not get complete isometric conditions. Is this because you could not increase the gain of the feedback system or because full isometric tension detaches the myocyte from the carbon fiber?

Sugiura: In the experiment you are talking about, we intentionally lower the gain to make the feedback incomplete. In isometric experiment, the shortening length was less than 1 μm long.

ter Keurs: What is the force per cross sectional area in these cells?

Sugiura: About 5mN/mm^2, which is low compared to the values reported in skinned myocyte.

ter Keurs: Is it possible that force is underestimated because a large number of myofibrils are not attached to cell?

Sugiura: That is a possibility.

ter Keurs: What is the break force for the bond between the cell and the carbon fiber?

Sugiura: More than twice as large as the isometric force under intotropic intervention with isoproterenal.

THE SPECIAL STRUCTURE AND FUNCTION OF TROPONIN I IN REGULATION OF CARDIAC CONTRACTION AND RELAXATION

R. John Solaro*

1. SUMMARY

In this chapter I review evidence for a pivotal role of the sarcomeric thin filament protein, troponin I, in cardiac muscle activation and its modulation by covalent modifications, sarcomere length, and intracellular pH. This evidence demonstrates that the cardiac variant of troponin I (cTnI), which is the only isoform expressed in the adult myocardium, has unique structure and function that are specialized for extrinsic and intrinsic control of cardiac contraction and relaxation.

2. THE ROLE OF TROPONIN I IN THIN FILAMENT ACTIVATION

Figure 1 illustrates a current concept of the interactions among thin filament protein in a regulatory unit of cardiac sarcomeres during diastole and how these interactions change in the transition to systole. In diastole the reaction of myosin heads (crossbridges) with the thin filament is impeded by the position of tropomyosin (Tm) and possibly the state of actin (See Solaro, 2001; Solaro et al., 2002 for reviews). This disposition of actin-Tm is imposed by the diastolic troponin complex, consisting of cTnC, cTnI, and troponin T (cTnT). cTnC has two lobes connected by an alpha-helical stalk. The N-lobe contains a "regulatory" relatively low affinity Ca-binding domain, and the C-lobe contains two relatively high affinity Ca/Mg domains (Holroyde et al. 1980; Pan and Solaro, 1987). In diastole the cellular Ca concentration is below the threshold for binding to the N-lobe regulatory site of TnC, however the slowly exchanging metal binding

*Department of Physiology and Biophysics (M/C 901), University of Illinois at Chicago, College of Medicine, Chicago, IL 60612

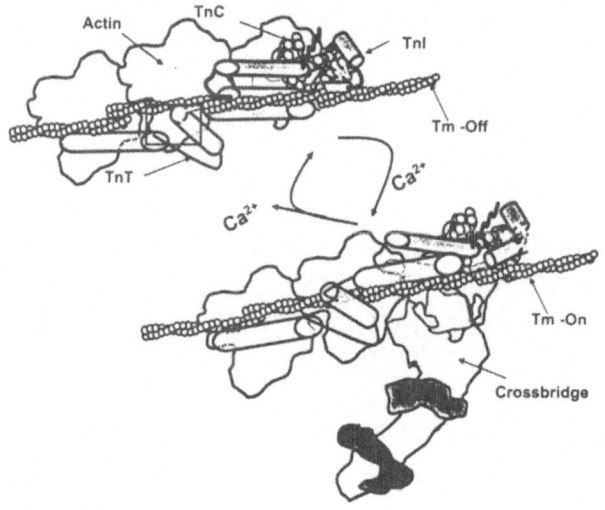

Figure 1. Schematic illustration of the role of troponin and tropomyosin in the transition from relaxed to active myofilaments. The top panel shows the diastolic state and the bottom panel shows the systolic state with a force generating crossbridge reacting with actin. Tropomyosin (Tm) is shown blocking sites for crossbridge interaction with actin (hatched areas) when Ca^{2+} is not bound to a regulatory site at the N-terminus of troponin C (TnC). Ca^{2+}-binding to TnC sets into a motion a series of protein-protein interactions leading to movement of Tm away from the sites where crossbridges may interact. See text for discussion.

sites at the C-lobe of the TnC are likely to be occupied by either Mg or Ca. In diastole, a near N-terminal region of cTnI binds to the cTnC C-lobe. This region is likely to be comprised of cTnI amino acids 39-58, which exhibit nanomolar affinity for TnC (Ferrieres et al., 2000). Spatial relationships within the cTnI-cTnC complex have also been determined using NMR studies (Krudy et al., 1994; Finley et al., 1999). In one set of studies, we (Krudy et al., 1994) demonstrated that residues 33-80 were sufficient to account for chemical shifts in the methionines of cTnC induced by cTnI (30-211). The methionines were restricted to the C-lobe of cTnC, and these data provided strong evidence for an anti-parallel arrangement of cTnI and cTnC in the Tn complex. The C-terminal "head" of cTnT also binds to N-terminal regions of cTnI, while also interacting with the N-lobe of cTnC (Potter et al., 1995; Blumenstein et al., 2001). Apart from its interaction with the C-lobe of cTnC in diastole, cTnI binds to actin through a negatively charged region termed the inhibitory peptide (Ip). We (Rarick et al., 1997) also reported that residues 152-199 of cTnI, which are C-terminal to the Ip are essential for full inhibition of the actin-myosin interaction. These reactions of cTnI and cTnC act together with cTnT to hold actin-tropomyosin in a blocked state in which force generating reactions of crossbridges with actin are not permitted. As illustrated in Figure 1, this inhibition is released when Ca binds to the N-lobe of C-TnC. The binding of Ca opens a hydrophobic pocket in the N-lobe of cTnC that reacts with a C-terminal peptide of cTnI and triggers sarcomeric activation by inducing movements of cTnI and Tm that permit force generating reactions of crossbridges with actin. A unique aspect of a functional effect of cTnI in contrast to fsTnI is its role in the Ca-dependant induction of the open state of the N-lobe of CTnC. Our studies comparing the structure of the N-lobe of cTnC and fsTnC free in solution showed that whereas Ca-binding to the N-lobe of fsTnC resulted in a significant exposure of a hydrophobic patch of amino acids, Ca-binding to cTnC did not. Subsequent studies using NMR structural analysis (Sia et al., 1997) or fluorescence resonance energy transfer techniques (Dong et al., 1999) demonstrated that binding of cTnI is required for induction of the open state of cTnC by Ca-binding to the N-lobe. This requirement for cTnI binding indicates a mechanism of triggering of sarcomeric activation that is unique for the heart myofilaments.

The binding of crossbridges may also activate the thin filament by moving Tm to a position nearer the groove of the thin filament formed by the actin helices. This crossbridge dependant activation is important as a determinant of the relatively steep, cooperative, relation between steady-state Ca concentration and steady-state tension (reviewed in Moss and Buck, 2001). The cooperativity is rooted in local effects of cross-bridges on a particular regulatory unit or in spread of activation to near neighbor regulatory units. The mechanism for these cooperative effects of crossbridges dependent activation include enhanced Ca-binding to cTnC and long range effects on the position of Tn and the state of activation (Tobacman, 1996; Gordon et al., 2000; Solaro, 2001).

3. TROPONIN I AND NEUROHUMORAL SIGNALING IN THE HEART

3.1 The protein kinase A pathway

There are unique aspects of the role of cTnI in modulation of thin filament activation by signaling cascades activating protein kinase A (PKA). A prominent unique feature of cTnI is an N-terminal extension of some 30 amino acids that is not present in fast skeletal (fsTnI) or slow skeletal (ssTnI) variants. This is depicted in Figure 1 as the wavy line at the N-terminus of cTnI. This N-terminal extension of cTnI contains serial Ser residues that are substrates for protein kinase A (PKA). Extensive evidence (reviewed in Solaro, 2001) indicates that these sites are important in the enhanced relaxation of heart muscles during the high heart rates associated with beta-adrenergic stimulation. Transgenic (TG) mouse hearts in which ssTnI fully replaced cTnI demonstrate a blunted relaxant effect of adrenergic stimulation (Fentzke et al., 2000). Moreover recent studies (Wolska et al., 2002) employing a mouse model in which ssTnI is expressed in a phospholamban null background demonstrates a blunted relaxant effect of beta-adrenergic stimulation, when compared to the phospholamban deficient hearts. The mechanism for these effects of PKA dependent phosphorylation includes a enhanced off rate for Ca-binding with the regulatory site of cTnC (Robertson et al., 1982), and enhanced crossbridge cycling (Saeki et al., 1990; Kentish et al., 2001; Strang et al., 1994), but this latter effect remains controversial (Janssen and deTombe, 1997; Hoffman and Lange, 1995).

Structural studies provide evidence on the mechanism by which phosphorylation of the N-terminal extension of cTnI influences cTnC. These studies, which employed nuclear magnetic resonance spectroscopy (NMR), fluorescent probes, and fluorescence resonance energy transfer (FRET), indicated that the N-terminal extension of cTnI is highly flexible and able to interact with the N-lobe of cTnC. The conformational change induced by phosphorylation of Ser 23, 24 in the N-terminal extension of cTnI itself has been probed by insertion of IAANS at a Cys residue engineered into position 5 (Dong et al., 1997). The results of studies of emission spectrum, quantum yield, and intensity decay together with acrylamide quenching indicated that phosphorylation induces a folded conformation, and decreases the axial ratio of the peptide. To further probe the structural significance of the N-terminal extension, Dong et al., (1997a) constructed the following two cTnI mutants each containing a single cysteine: (1) a full-length cTnI mutant (S5C/C81I/C98S) and (2) a truncated cTnI mutant (S9C/C50I/C67S) in which the N-terminal 32 amino acid residues were deleted. The truncated mutant bound cTnC with an affinity only 1.4 lower than that of the full-length cTnI mutant, indicating that the N-terminus of cTnI contributes but a small effect to the binding of cTnI to cTnC. Using FRET analysis, which determined the distribution of the separation of Cys-5 labeled with IAANS from Trp 192, we estimated that phosphorylation of cTnI induces a decrease of 9-12 angstroms in the mean distance between these sites located at the N- and C-terminal portion of cTnI. This FRET analysis was extended in studies by Dong et al., (2000) defining structural relations between cTnI and cTnC. In these studies, distance distributions were determined between Trp192 in the C-terminus of cTnI and fluorescent probes attached at Cys engineered into sites at position 5, 40, 81, 98, 115, 133, 150, 167. The results indicated that cTnI in the binary complex with cTnC displays an open and extended conformation. With phosphorylation of Ser23, Ser24 distances between residues 40-167 with respect to Trp 192 did not change. The data also indicated that the N-terminal extension of cTnI is highly flexible in the cTnI-cTnC complex and either not bound or loosely bound by the C-terminus of cTnC. Although our initial hypothesis (Wattanapermpool et al., 1995) was that phosphorylation induced a new state of the cTnI-cTnC complex by promoting an interaction with the N-terminus of cTnC, the bulk of evidence from our subsequent studies (Gaponenko et al., 1999; Finley et al., 1999) as well as from other laboratories (Ward et al., 2002) indicates the opposite. Gaponenko et al., (1999) and Finley et al., (1999) investigated the effects of cTnI on the dynamics of cTnC conformational changes in the regulatory domain of cTnC by using 15-N transverse relaxation rates and chemical shift mapping. When a peptide comprised of amino acids 1-80 of cTnI was phosphorylated at Ser23 and Ser24 or mutated to S23D, S24D, the conformational equilibrium of cTnC shifted to that of cTnC in the free, Ca-saturated state. The binding affinity of free cTnC for Ca is relatively low compared to cTnC in the binary complex with cTnI (Holroyde et al., 1980). Thus, this transition in cTnC induced by phosphorylation of cTnI provides a structural basis for decreases in myofilament cTnC binding when cTnI is phosphorylated. These data also fit with functional measurements, which demonstrated that force generating myofilaments (Chandra et al., 1999) and reconstituted myofilaments splitting ATP (Noland et al., 1995) have a reduced sensitivity to Ca when the native cTnI is exchanged with a mutant missing the N-terminal extension. These studies strongly indicate that the *dephosphorylated* N-terminal extension of cTnI interacts with the N-terminus of cTnC producing a state of relatively high cTnC Ca-affinity and phosphorylation releases the peptide from cTnC, thereby lowering the Ca-affinity. Further support for this conclusion comes from a study by Ward et al., (2002) that has advanced our understanding of the mechanism of the transmission of the phosphorylation signal from the N-terminus of cTnI to the regulatory domain of cTnC. From studies using deletion and site directed mutagenesis of cTnI, Ward et al., concluded that the residues 1-15 have no major impact on transmission of the phosphorylation signal. However, deletion of Ala-Pro-Ala making up residues 16-18 abolished the functional effects of phosphorylation of Ser 23 and Ser 24. Moreover, further deletion of residues 16-29 mimicked the effect of phosphorylation. On the basis of these data, Ward et al. (2002) proposed that the open state of the C-lobe of cTnC is promoted by the cTnI residues 16-29, and that phosphorylation promotes the closed state thereby desensitizing the myofilaments to Ca and enhancing the off-rate for Ca binding to cTnC. Although it is not clear exactly which residues on cTnC are affected by amino acids 16-29, it is not necessary that the interaction be directly with site II. Conformational changes in the far N-terminal region of cTnC, containing the N-helix, although not a Ca binding site may alter the energetics of the structural changes induced by binding of Ca to site II, the N-lobe regulatory site. Evidence for an important role of the N-helix of cTnC comes from studies on an E41A mutant of fsTnC (Li et al., 1997). As is the case with cTnC, fsTnC(E41A) remains closed following saturation of the metal binding sites with Ca. Moreover, site II of fsTnC containing the E41A mutation had about a 10 fold reduction in Ca-affinity, indicating that the conformation of the N-helix affects site II structure and function. Chemical shift changes induced by Ca occurred over the entire sequence of the N-lobe of cTnC further supporting the idea that Ca^{2+}-binding to site II perturbs structure of the N-helix (Li et al.,1997; Spyracoupoulos et al., 1997).

3.2 The Protein Kinase C Pathway

Signaling through the protein kinase C (PKC) cascade also involves special aspects of the structure function relations of cTnI. The near-terminal region of cTnI contains Ser 43 and Ser 45 that are the substrates for protein kinase C (PKC). As discussed below, functional effects of these sites also appear to have a special role in heart muscle. The inhibitory peptide of cTnI also is unique in that it contains a Thr at position 144 that is a Pro in fsTnI and ssTnI. As discussed below, phosphorylation of this site by PKC also has been demonstrated to affect myofilament activity. Investigations initiated in the laboratory of J F Kuo (Katoh et al., 1983; Liu et al., 1989) provided the first evidence that the activity of striated muscle myofilaments might be affected by PKC dependent phosphorylation of cTnT and cTnI. Using reconstituted preparations of thin filaments and heavy meromyosin (HMM), Kuo and colleagues (Noland et al. 1989; 1993) further reported that PKC dependent phosphorylation of both cTnI and cTnT was able to decrease maximum acto-myosin HMM ATPase rate and to inhibit the affinity of crossbridges for the thin filament. Subsequent studies identified the sites of PKC dependent phosphorylation on TnI as Ser 43, Ser 45, and Thr 144. Joint studies with Kuo's laboratory and our laboratory (Noland et al.,1995; Noland et al.,1996) used a mutational analysis to demonstrate that the depressant effect on actomyosin ATPase rate by phosphorylation of cTnI was due largely to phosphorylation of Ser 43 and Ser 45 rather than Thr 144. On the basis of this evidence, we generated a transgenic mouse model in which a mutant cTnI (cTnI-S43A, S45A) missing the PKC sites was expressed in the heart (MacGowan et al., 2001). We estimated that about 50% of the native cTnI was replaced by cTnI- cTnI-S43A, S45A. Total TnI content was the same in wild-type and transgenic mice. Hearts of the transgenic mice expressing cTnI- cTnI-S43A, S45A showed no gross anatomical or histological abnormalities. Studies on isolated hearts and myofilaments from these mice have provided further insights into potential significance of PKC dependent phosphorylation of Ser 43 and Ser 45 of cTnI. Isolated transgenic hearts developed isovolumic pressure that was not statistically different between controls and transgenic hearts. Moreover rates of pressure raise (+dp/dt max) and fall (-dp/dt max) were the same. One major difference found between control and transgenic hearts was an increase in the ratio of developed pressure to oxygen consumption. However, at elevated levels of perfusate Ca that we used to increase PKC activity, compared to controls, the TG-cTnI-S43A, S45A hearts demonstrated a significant decrease in peak systolic intracellular Ca of 35% , a 32% decrease in diastolic Ca and prolongation of the transient. We interpreted these results as evidence that the alteration in cTnI has a feed back effect on cellular Ca fluxes. When stressed by global ischemia (which we presume activated PKC), the TG hearts were more susceptible to ischemic contracture. These findings suggest that the PKC sites on cTnI are phosphorylated in the basal state. The findings also fit generally with the *in vitro* effects of PKC dependent phosphorylation on the Ca-force relation of cardiac myofilaments. The prediction from the *in vitro* studies is that removal of the PKC sites at Ser 43 and Ser 45 should blunt the ability of PKC dependent phosphorylation to depress tension at a given level of intra-cellular Ca. This would result in a higher susceptibility to ischemic contracture as well.

To further explore the importance of Ser 43 and Ser 45 in the signaling cascades linked to other agonists, we compared the effects of stimulation of myocardial preparations from wild type (WT) and TG- cTnI-S43A, S45A hearts (Pyle et al., 2002). We measured phosphorylation, tension, stiffness, and ATPase rate of detergent extracted fiber bundles. As expected, theses studies revealed that basal phosphorylation levels of cTnI were lower in TG vs. wild-type (WT) myofilaments. Compared to controls the transgenic fiber bundles demonstrated an increase in tension cost (ATPase rate/tension developed). These data indicate that phosphorylation of Ser 43 and Ser 45 may reduce tension cost. However, this interpretation must be viewed with caution inasmuch as there were increased levels of phosphorylation in cTnT in the TG-cTnI-S43A, S45A myofilaments. This result fits with earlier work (Montgomery et al., 2001) in which we demonstrated an interaction among the PKC-dependent sites on cTnI and cTnI. Translation of these findings to the oxygen cost of pressure development is difficult in that there are changes in the Ca fluxes associated with the TG- cTnI-S43A, S45A hearts that will affect overall energy balance (MacGowan et al., 2001) For example, our determination of the Ca-pressure relation, indicated a 35% reduction in peak systolic Ca, which predicts a 17.5% reduction in ATP hydrolysis. This compared well with measured 15% reduction in the ratio of developed pressure to myocardial oxygen consumption.

Several lines of evidence indicate that the near N-terminal region of cTnI, which contains Ser 43 and Ser 45, interacts with the C-lobe of cTnC. This region of cTnC is generally referred to as structural rather than regulatory, but this designation does not seem appropriate in view of evidence that alterations at this interface, such phosphorylation, significantly affect tension. Data of Calvert et al. (2000) also suggested a role for the C-lobe of cTnC in modulating tension. They reported that the interactions at the C-terminal domain of cTnC are responsible for most of the free energy changes and is 8 fold stronger in the presence of Mg than in the presence of Ca. Solution structures of the Ca-saturated C-lobe of cTnC free and in the presence of the near N-terminal domain of cTnI comprised of residues 33-80 clearly demonstrate binding determined using hetero-nuclear multidimensional NMR (Gasmi-Seabrook et al., 1999). The binding interface involves hydrophobic interactions with an open pocket of the C-lobe. This binding induces the largest chemical shift changes in the loop region connecting helix F and helix G that contains highly conserved residues. Increases in the response to Ca-induced by the sarcomeric activator, EMD 57033, appear due to alterations at the interface between the C-lobe of cTnC and the near N-terminus of cTnI. Wang et al., (2001) determined from the NMR structure of a 1:1 complex of EMD 57033 and the C-lobe of cTnC that EMD docks in the hydrophobic pocket and makes several key contacts with the protein. Wang et al. (2001) determined the amino acid residues interacting with the protons of EMD. These amino acids were, in fact, the same or close to those perturbed by binding of the peptide cTnI-33-80 as measured by Gasmic-Seabrook et al. (1999). It is therefore not surprising that EMD was completely displaced by a peptide of cTnI comprised of amino acids 34-71, but not by a peptide comprised of amino acids 128-147. We think this provides strong evidence that the near N-terminal peptide of cTnI shares a common binding site with EMD and may therefore influence tension differentially depending on the state of phosphorylation of Ser 43 and Ser 45. By extrapolation from studies on skeletal muscle Tn components, it is apparent that a region of cTnT may also bind to the C-lobe of cTnC (Blumenschein et al., 2001). In the case of fsTn components, a fsTnT peptide comprised of amino acids 160-193 (close to the corresponding the adult isoform of cTnT that also contain sites of PKC dependent phosphorylation), binds to the C-lobe of cTnC in a region separate from that of a fsTnI peptide, amino acids 1-40. Although these regions of TnI and TnT do not compete for the same binding site they are close to each other on the surface of TnC and may affect each other. This may provide a structural basis for our observation that alterations in Ser 43, Ser 45 of cTnI affect the state of phosphorylation of cTnT.

4. TROPONIN I AND LENGTH DEPENDENT MYOFILAMENT ACTIVATION

There is emerging evidence that cTnI plays a central role in the molecular basis of the major intrinsic regulator of cardiac function, Starling's Law, in which the pressure developed by the ventricle increases as the end diastolic volume increases. It has been generally agreed for some time that this dependence of pressure on end diastolic volume is related to the length-tension properties of the myofilaments, i.e. the effect of length on the number of force generating crossbridges reacting with the thin filament (Allen and Kentish, 1985). Increases in sarcomere length not only increase maximum tension, but also myofilament Ca-sensitivity. This length dependence of activation is especially important in the heart, which activates at sub-maximal levels. Theories for why Ca-activation depends on sarcomere length have centered on alterations in inter-filament spacing, which decreases with increases in sarcomere length, thus increasing the local concentrations of crossbridges. Increases in local concentration of crossbridges in the vicinity of a region of the thin filament will increase the probability of a force generating interaction as well as the probability of cooperative activation of the thin filaments by strongly bound crossbridges. The cooperative activation could involve feedback effects of bound crossbridges on the affinity of cTnC for Ca within a functional unit and/or the cooperative spread of activation along the thin filament i.e. interaction energy among functional units (Fitzsimons and Moss, 1998; Gordon et al., 1999). Cooperative activation may vary independently of inter-filament spacing, and this may explain why alterations in inter-filament spacing do not always correlate well with alterations of length dependence of activation (Konhilas et al., 2002). These variations in cooperative activation of the thin filament appear to involve troponin. Exchange of cTnC in heart myofilaments with fsTnC does not appear to modify length dependent activation (Mcdonald et al., 1995). However, switching of cTnT with developmentally related isoforms (Akella et al., 1995) or with mutant forms linked to familial hypertrophic cardiomyopathy (Chandra et al., 2001) does alter length dependent activation. Komukai and Kurihara, (1997) also proposed a role for protein phosphorylation involving cTnI in length dependent activation. We (Arteaga et al., 2000) extended these findings by the demonstration that a specific exchange of cTnI with ssTnI in mouse cardiac thin filaments induced a reduction in length dependent activation. This result provided the first direct evidence that cTnI is a key element in the molecular signaling associated with length dependent activation. Using chimeras of cTnI and ssTnI, Smith et al., (2002) have reported preliminary data indicating that difference in the C-terminal region of TnI may be largely responsible differential effects of the cardiac and slow skeletal variant on length dependent activation.

A role for cTnI in the feedback effects of strong crossbridges on cTnC Ca-affinity was also suggested from data demonstrating that length dependent activation of slow skeletal myofilaments, which contain cTnC, is not associated with an increase in Ca-affinity (Wang and Fuchs, 1994). Lehrer (1994) suggested that the number of crossbridges in the weak binding state may be modulated by the state of TnI. Moreover, TnI and TnT, both of which are asymmetric molecules extending across 2-3 actins (Fig. 1), could in fact contribute significantly to steric blocking of the actin-crossbridge reaction monomers (Squire and Morris, 1998; Solaro and Rarick, 1998). Interestingly, Kajiwara et al., (2000) have reported that PKA-dependent phosphorylation of cTnI depresses length dependent activation in cardiac myofilaments. However, Konhilas et al. (2003) came to the opposite conclusion in their report that PKA dependent phosphorylation of cardiac myofilaments *enhanced* length dependent activation. Moreover, compared to myofilaments in a basal state of phosphorylation, myofilaments phosphorylated with PKA showed in an increase in inter-filament lattice spacing. In these experiments, however, both myosin binding protein C and cTnI were phosphorylated. In the study by Kajiwara et al. (2000) length dependence of crossbridge dependent activation was also studied by varying MgATP concentration at pCa 9.0. Increases in sarcomere length shifted this relation to the right. Either phosphorylation of cTnI or extraction of cTnC eliminated this right shift of the pMgATP-relative tension relation. These data point to possible PKA-dependent alterations of the interaction of the N-lobe of cTnI with the CTnC C-lobe (which would predominate at pCa 9.0) by phosphorylation of the N-terminus of cTnI.

5. SPECIAL ROLE OF TROPONIN I IN EFFECTS OF ACIDOSIS ON CARDIAC CONTRACTION

Effects of intracellular pH on contractility are dominated by altered response to Ca and there is evidence of a central role of cTnI in the mechanism for the altered response. Following initial studies comparing effects of pH on pCa-force relations of fast, slow and cardiac muscle by Donaldson et al. (1978), we (Solaro et al. 1986) demonstrated a relative insensitivity of adult dog heart myofilaments to acidic pH when compared to myofilaments from neonatal dog hearts. This study of Solaro et al. pointed to thin filament alterations downstream of cTnC inasmuch as the lack of effect of acidic pH on the neonatal myofilaments could not be attributed to thick filament proteins or cTnC. Subsequent studies by Saggin et al. (1989) identified a neonatal/embryonic isoform of TnI as identical to ssTnI, and Martin et al. (1991) reported a close correlation between pH sensitivity and expression of ssTnI in developing rat heart myofilaments. Moreover, these in vitro findings with isolated myofilaments appeared to translate to the intact heart in which we (Solaro et al., 1988) reported that the fall in tension associated with declining pH in neonatal muscle was less than the fall in adult muscle. Using the using the photo protein aequorin as a measure of intracellular Ca, Solaro et al. (1988) reported that hypercapnic acidosis induced either no change or a rise in the peak amplitude of Ca-transients in both adult and neonatal heart muscles.   With this information in hand, we were motivated to generate the ssTnI-TG transgenic mouse, in which cTnI was completely replaced with ssTnI (Fentzke et al., 1999).   We (Wolska et al., 2001) compared twitch tension of the TG-ssTnI papillary muscles to WT preparations during a pulse of elevated carbon dioxide.   The hypercapnic episode induced a sustained and significant depression in tension in the WT papillary muscles, whereas the tension developed by ssTnI-TG papillary muscles fully recovered following a transient fall. Studies by Wolska et al., (2001) showed that a drop from pH 7.0 to pH 6.5 induced less of a depression of Ca -activated tension generated by myofilaments of TG mouse hearts compared to WT myofilaments.   These results using a transgenic approach agreed the work of Morimoto et al. (1999), who exchange ssTnI with cTnI *in vitro* in rabbit skinned fiber bundles, and by Westfall and Metzger (2001), who used adenoviral transfer to express ssTnI in rat cardiac myocytes.

Studies employing chimeras with regions from cTnI, ssTnI, and fast skeletal TnI have been carried out in order to determine the localization of the domains accounting for the differences in effects of acidic pH on myofilament function.   The general conclusion of these studies (Li et al., 2001;Westfall and Metzger, 2001) was that isoform specific differences in C-terminal regions of cTnI and perhaps the Ip are responsible for altered effects of pH on the various muscle types.   Our studies (Li et al., 2001) supported the hypothesis the hypothesis that differences in charged amino acids in C-terminal regions of cTnI outside the Ip are the main determinants of the differential effects of acidic pH on cardiac and skeletal myofilaments. This conclusion fit with our previous data (El-Saleh and Solaro, 1988) demonstrating   conformational changes that occurred in a   region surrounding Cys 133 of fsTnI (corresponding to Ser 135 in ssTnI and Ser 166 in cTnI). Taking a lead from these findings, Dargis et al., 2002 compared the effects of pH on the pCa-ATPase activity relation of preparations of thin filaments (reconstituted with either WT cTnI or a mutant, cTnI-A162H) reacting with rabbit skeletal myosin S-1. Substitution of cTnI-A162H induced a complete reversal of the desensitization of Ca-activation at pH 6.5 to the Ca-sensitivity at pH 7.0.   Moreover, introduction of His residue at the homologous Ala at 130 of fsTnI induced a deactivation similar to that occurring with cTnI. With a decrease from pH 7.0 to pH 6.5, the charge on His (pKa ~ 6.5) would become more positive and, as pointed out by Dargis et al. (2002), would likely interact more strongly than the analogous Ala residue of WT cTnI with negative residues in the N-lobe of cTnC.   Although these changes in TnI affected alterations in Ca-sensitivity induced by acidic pH, they did not affect minimum or maximum ATPase rate.   Thus it is unlikely that the single point mutation would reverse the well-known effects of acidic pH on maximum tension.   In the case of the ssTnI-TG myofilaments, maximum tension is depressed by acidic pH to the same extent as WT myofilaments (Wolska et al. 2001).   Even so, a hypercapnic acidosis that substantially reduced tension of WT papillary muscles had no effect on ssTnI-TG muscles.

Interestingly, Fukuda et al., (2001) reported that acidosis enhances length dependence of Ca-activation of tension in rat skinned cardiac muscle preparations. These results indicate that at acidic pH may impair the ability of strong crossbridges to activate the thin filament.   Morimoto et al., (2001) reported that crossbridge dependent activation (varied by reducing the MgATP concentration at pCa 9.0) was depressed in acidosis, and that this effect of acidic pH was attenuated when cTnI-cTnT-cTnC was exchanged with ssTnI-cTnT-cTnC in skinned fiber bundles from rat heart. Morimoto et al. (2001) found that whereas strong crossbridges could activate myofilaments containing either cTnI-cTnT or ssTnI-cTnT, but lacking cTnC, the activation was the same at pH 7.0 and pH 6.2. These results indicate: 1) that strong crossbridges can reverse the inhibition of the thin filament by TnI, presumably by releasing the cTnI Ip and C-terminal binding sites from actin in a mechanism that is unaffected by a change from pH 7.0 to pH 6.2, and 2) that activation of the thin filament by strong crossbridges depends on a pH dependent interaction of the C-terminal lobe of cTnC with the N-terminal domain of TnI, which is different for ssTnI and cTnI. It remains unknown whether the N-terminal differences in charged residues at Arg 92, Lys 95 and Lys 98 of ssTnI (Thr 123, Ala 126, and Thr 129 of cTnI) are important in this difference. Differences in human TnI at Ala 65 of cTnI, which is His 34 at the homologous position of ssTnI may be of special significance. As discussed above, cTnI has a special role in modulation of Ca-dependent activation myofilament activity by changes in pH, and here the differences between cTnI and ssTnI in deactivating effects of acidic pH appear to be due entirely to a similar Ala to His difference in the C-terminus of TnI.

6. ACKNOWLEDGEMENTS

I am grateful to my many colleagues for collaborations permitting the study of cardiac troponin I at many levels of organization. The work described in this review from my laboratory was supported in part by grants from the NIH NHLBI and AHA.

7. REFERENCES

Akella, A.B., Ding, X.L., Chen, R., and Gulati, J.,1995, Diminished Ca sensitivity of skinned cardiac muscle contractility coincident with troponin T-band shifts in the diabetic rat. *Circ. Res.* **76**:600-606.

Allen, D.G. and Kentish, J.C., 1985, The cellular basis of the length-tension relation in cardiac muscle. *J. Mol. Cell Cardiol.* **17**:821-840.

Arteaga, G.M., Palmiter, K.A., Leiden, J.M., and Solaro, R.J. ,2000, Attenuation of length-dependent activation in myofilaments of transgenic mouse hearts expressing slow skeletal troponin I. *J. Physiol.* (London) **526**:541-549.

Blumenschein, T.M.A., Tripet, B.P., Hodges, R.S., Sykes, B.D. ,2001, Mapping the interacting regions between troponins T and C. *J. Biol. Chem.* **276**:36606-366612.

Calvert, M.J., Ward, D.G., Trayer, H.R., Trayer, I.P. ,2000, The importance of the carboxyl-terminal domain of cardiac troponin C in Ca-sensitive muscle regulation. *J. Biol. Chem.*, **275**:32508-15.

Chandra, M., Kim, J.J., Solaro, R.J., 1999, An improved method for exchanging troponin subunits in detergent skinned rat cardiac fiber bundles. *Biochem. Biophys. Res. Commun.* **263**:219-23

Chandra, M, Rundell, V.L., Tardiff, J.C., Leinwand, L.A., de Tombe, P.P., and Solaro, R.J. ,2001, Ca- activation of myofilaments from transgenic mouse hearts expressing R92Q mutant cardiac troponin T. *Am. J. Physiol. Heart Circ. Physiol.* **280**:H705-13.

Dargis, R., Pearlstone, J.R., Barrette-Ng, I., Edwards, H., Smillie, L.B. ,2002, Single mutation (A162H) in human cardiac troponin I corrects acid pH sensitivity of Ca2+-regulated actomyosin S1 ATPase. *J. Biol .Chem.* **277**:34662-5

Donaldson, S.K., Hermansen, L., and Bolles, L. ,1978, Differential, direct effects of H+ on Ca2+ -activated force of skinned fibers from the soleus, cardiac and adductor magnus muscles of rabbits. *Pflugers Arch.* **376**:55-65.

Dong, W-J., Xing, J., Villain, M., Hellinger, M., Robinson, J.M., Chandra, M., Solaro, R.J., Umeda, P.K., and Cheung, H.C. (1999) Conformation of the regulatory domain of cardiac muscle troponin C in its complex with cardiac troponin I. *J. Biol. Chem.* **274**:31382-90.

Dong, W-J., Chandra, M., Xing, J., Solaro, R.J. and Cheung, H.C. ,1997, Conformation of the N-terminal segment of a monocysteine mutant of Troponin I from cardiac muscle. *Biochemistry*. **36**: 6745-6753.

Dong, W-J., Chandra, M., Xing, J., She, M., Solaro, R.J. and Cheung, H.C. ,1997a, Phosphorylation-induced distance change in a cardiac muscle Troponin I mutant. *Biochemistry*. **36**: 6754-6761.

Dong, W.J., Xing, J., Chandra, M., Solaro, R.J., Cheung, H.C. ,2000, Structural mapping of single cysteine mutants of cardiac troponin I. *Proteins* **41**:438-447.

El-Saleh, S. and Solaro, R.J. (1988) Troponin I enhances acidic pH induced depression of Ca-binding to the regulatory sites in skeletal troponin C. *J. Biol. Chem.* **263**:3274-3278.

Fentzke, R.C., Buck, S.H., Patel, J.R., Lin, H., Wolska, B.M., Stojanovic, M.O., Martin, A.F., Solaro, R.J., Moss, R.L., and Leiden, J.M. (1999) Impaired cardiomyocyte relaxation and diastolic function in transgenic mice expressing slow skeletal troponin I in the heart. *J. Physiol.* **517**:143-157.

Ferrieres, G., Pugniere, M., Mani, J.C., Villard, S., Laprade, M., Doutre, P., Pau, B., Granier, C.,2000, Systematic mapping of regions of human cardiac troponin I involved in binding to cardiac troponin C: N- and C-terminal low affinity contributing regions. *FEBS Lett* **479**:99-105.

Finley, N., Abbott, M.B., Abusamhadneh, E., Gaponenko, V., Gasmi-Seabrook. G., Howarth. J.W., Rance, M., Solaro, R.J., Cheung, H.C., Rosevear, P.R. ,1999, NMR anaylsis of cardaic troponin C-troponin I complexes: Effects of phosphorylation. *FEBS Lett.* **453**:107-112.

Fitzsimons, D.P. and Moss, R.L. (1998) Strong binding of myosin modulates length-dependent Ca2+ activation of rat ventricular myocytes. *Circ. Res.* **83**:602-607.

Gasmi-Seabrook, G., Howarth J.W., Finley, N., Abusamhadneh, E., Gaponenko, V., Brito, R.M.M., Solaro, R.J., Rosevear, P.J. ,1999, Solution structure of the C-terminal domain of cardiac troponin C complexed with the N-terminal domain of cardiac troponin I. *Biochemistry* **38**:8313-8322.

Gaponenko, V., Abusamhadneh, E., Abbott B., Finley, N., Gasmi-Seabrook, G., Solaro, R.J., Rance, M., Rosevear, P.R. ,1999, Effects of TnI phosphorylation on conformational exchange in the regulatory domain of cardiac troponin C. *J. Biol. Chem.* **274**:16681-16684.

Gordon, A.M., Homsher, E., and Regnier, M. ,2000, Regulation of contraction in striated muscle. *Physio. Rev.* **80**:853-924.

Hoffman, P.A .and Lange, J.H., III. ,1995, Effect of phosphorylation of troponin I and C protein on isometric tension and velocity of unloaded shortening in skinned single cardiac myocytes from rats. *Circ. Res.* **74**:718-726.

Holroyde, M.J., Roberston, S.P., Johnson, J.D., Solaro, R.J. and J.D. Potter. ,1980, The Ca^{2+} and Mg^{2+} binding sites on cardiac troponin and their role in the regulation of adenosine triphosphatase. *J. Biol. Chem.* **255**:11688-11693.

Janssen, P.M.L. and deTombe, P.P. ,1997, Protein kinase A does not alter unloaded velocity of sarcomere shortening in skinned rat cardiac trabeculae. *Am. J. Physiol.* H2415-H2422.

Katoh, N., Wise, B.C., and Kuo, J.F. ,1983, Phosphorylation of cardiac troponin inhibitory subunit (troponin I) and tropomyosin-binding subunit (troponin T) by cardiac phospholipid-sensitive Ca^{2+}-dependent protein *Biochem. J.* **209**:189-195.

Kajiwara, H., Morimoto, S., Fukuda, N., Ohtsuki, I., Kurihara, S. ,2000, Effect of troponin I phosphorylation by protein kinase A on length-dependence of tension activation in skinned cardiac muscle fibers. *Biochem. Biophys. Res. Commun.* **272**:104-10

Kentish, J.C., McCloskey, D.T., Layland, J., Palmer, S., Leiden, J.M., Martin, A.F., and Solaro, R.J. ,2001, Phosphorylation of troponin I by protein kinase A accelerates relaxation and cross-bridge cycle kinetics in mouse ventricular muscle. *Circ. Res.* **88**:1059-1065.

Komukai, K. and Kurihara, S. ,1997, Length dependence of Ca(2+)-tension relationship in aequorin-injected ferret papillary muscles. *Am. J. Physiol.* **273**:H1068-H1074.

Konhilas, J.P., Irving, T.C., de Tombe, P.P. ,2002, Length-dependent activation in three striated muscle types of the rat. *J. Physiol.* **544**:225-36.

Konhilas, J.P., Irving, T.C. Wolska, B.M., Jweied, E.E., Martin, A. F., Solaro, R.J., deTombe, P.P. ,2003, Troponin I in the heart: influence on length dependent activation and interfilament spacing. *J. Physiol.* (In Press).

Krudy, G.A., Kleerekoper, Q., Guo, X., Howarth, J.W., Solaro, R.J., Rosevear, P.R. ,1994, NMR studies delineating spatial relationships within the cardiac troponin I-troponin C complex. *J. Biol. Chem.* **269**:23731-5.

Lehrer, S. S. ,1994, The regulatory switch of the muscle thin filament: Ca2+ or myosin heads? *J. Muscle Res. Cell Motil.* **15**:232-6.

Li, M.X., Gagne, S.M., Spyracoupoulos, L., Kloks, C.P.A.M., Audette, G., Chandra, M., Solaro, R.J., Smillie, L.B., Sykes, B.D.,1997, NMR Studies of Ca^{2+}-binding to the regulatory domains of cardiac and an E41A skeletal muscle troponin C reveal emportance of site 1 to energetics of the enduced structural changes. *Biochemistry* **36**:12519-12525.

Li, G., Martin, A.F., and Solaro, R.J. ,2001, Localization of regions of troponin I important in deactivation of cardiac myofilaments by acidic pH. *J. Mol. Cell Cardiol.* **33**:1309-1320.

Li, M.X., Spyracopoulos, L., and Sykes, B.D. ,1999, Binding of cardiac troponin-I147-163 induces a structural opening in human cardiac troponin-C. *Biochemistry* **38**:8289-98.

Liu, J.D., Wood, J.G., Raynor, R.L., Wang, Y.C., Noland, T.A. Jr., Ansari, A.A, Kuo, J.F. ,1989, Subcellular distribution and immuno-cytochemical localization of protein kinase C in myocardium, and phosphorylation of troponin in isolated myocytes stimulated by isoproterenol or phorbol ester. *Biochem. Biophys. Res. Commun.* **162**:1105-10

MacGowan, G.A., Du, C., Cowan, D.B., Stamm, C., McGowan, F.X., Solaro, R.J., Koretsky, A.P., and Del Nido, P.J. ,2001, Ischemic dysfunction in transgenic mice expressing troponin I lacking protein kinase C phosphorylation sites. *Am. J. Physiol. Heart Circ. Physiol.* **280**:H835-H843

Martin, A.M., Ball, K., Gao, L., Kumar, P.K., and Solaro, R.J. ,1991, Identification and functional significance of troponin I isoforms in neonatal rat heart myofibrils. *Circ. Res.* **69**:1244-1252.

Martyn, D.A. and Gordon, A.M. ,2001, Influence of length on force and activation-dependent changes in troponin c structure in skinned cardiac and fast skeletal muscle. *Biophys. J.* **80**:2798-808.

Mcdonald, K.S, Field, L.J., Parmacek, M.S., Soonpaa, M., Leiden, J.M., and Moss, R.L., 1995, Length dependence of Ca2+ sensitivity of tension in mouse cardiac myocytes expressing skeletal troponin C. *J. Physiol.* **483**:131-139.

Montgomery, D.E., Chandra, M., Huang, Q-Q., Jin, J.-P., and Solaro, R.J. ,2001, Transgenic incorporation of fast skeletal troponin T into cardiac myofilaments blunts PKC-mediated depression of force. *Am. J. Physiol. (Heart)* **280**:H1011-H1018.

Morimoto, S., Harada, K., and Ohtsuki, I. ,1999, Roles of troponin isoforms in pH dependence of contraction in rabbit fast and slow skeletal and cardiac muscles. *J. Biochem. (Tokyo)* **126**:121-9.

Morimoto, S., Ohta, M., Goto, T., and Ohtsuki, I. ,2001, A pH-sensitive interaction of troponin I with troponin C coupled with strongly binding cross-bridges in cardiac myofilament activation. *Biochem. Biophys. Res. Commun.* **282**:811-5.

Moss, R.L. and Buck, S.H., 2001, Regulation of Cardiac Contraction by Calcium, in: *Handbook of Physiology: Section 2. The Cardiovascular System. Vol 1. The Heart,* E. Page, H. Fozzard, R. J. Solaro, eds.) Oxford University Press, New York, pp. 420-454.

Noland, T.A., Jr., Raynor, R.L., and Kuo, J.F. ,1989, Identification of sites phosphorylated in bovine cardiac troponin I and troponin T by protein kinase C and comparative substrate activity of synthetic peptides containing the phosphorylation sites. *J. Biol. Chem.* **264**:20778-20785.

Noland, T.A., Jr. and Kuo, J.F. ,1993, Protein kinase C phosphorylation of cardiac troponin I and troponin T inhibits Ca^{2+}-stimulated MgATPase activity in reconstituted actomyosin and isolated myofibrils, and decreases actin-myosin interactions. *J. Mol. Cell. Cardiol.* **25**:53-65.

Noland, T.A. Jr., Raynor, R.L., Jideama, N.M., Guo, X., Kazanietz, M.G., Blumberg, P.M., Solaro. R.J., and Kuo, J.F. ,1996, Differential regulation of cardiac actomyosin S-1 MgATPase by protein kinase C isozyme-specific phosphorylation of specific sites in cardiac troponin I and its phosphorylation site mutants. *Biochemistry* **35**:14923-14931.

Noland, T.A., Guo, X., Raynor, R.L., Averyhart-Fullard, V., Jideama, N.M., Solaro, R.J., and Kuo, J.F. ,1995, Cardiac troponin I mutants: phosphorylation by protein kinases C and A and regulation of Ca^{2+}-stimulated MgATPase of reconstituted actomyosin S-1. *J. Biol. Chem.* **43**:25445-25454.

Pan B-S. and Solaro, R.J. ,1987, Calcium binding properties of troponin C in detergent skinned heart muscle fibers. *J. Biol. Chem.* **262**:7339-7349.

Potter, J.D., Sheng, Z., Pan, B.S., Zhao, J. A., 1995, A direct regulatory role for troponin T and a dual role for troponin C in the Ca2+ regulation of muscle contraction. *J Biol Chem* **270**:2557-62.

Pyle, W.G., Sumandea, M.P., Solaro, R.J., deTombe, P.P., 2002, Troponin I serines 43/45 and regulation of cardiac myofilament function. *Am. J. Physiol. (Heart Circ. Physiol.)* **283**:H1215-1224.

Rarick, H.M., Tu, X., Solaro, R.J., and Martin, A.M. (1997) The C-terminus of cardiac troponin I is essential for full inhibitory activity and Ca^{2+}-sensitivity of rat myofibrils. *J. Biol. Chem.* **272**:26887-26892.

Robertson, S.P., Johnson, J.D., Holroyde, M.J., Kranias, E., Potter, J.D., and Solaro, R.J. ,1982, The effect of TnI phosphorylation on static and kinetic Ca binding by cardiac TnC. *J. Biol. Chem.* **257**:260-263.

Saeki, Y., Shiozawa, K., Yanagisawa, K., and Shibata, T., 1990, Adrenaline increases the rate of cross-bridge cycling in rat cardiac muscle. *J. Mol. Cell Cardiol.* **22**:453-460.

Saggin, L., Gorza, L., Ausoni, S., and Schiaffino, S., 1989, Troponin I switching in the developing heart. *J. Biol. Chem.* **264**:16299-302.

Sia, S.K., Li, M.X., Spyracopoulos, L., Gagne, S.M., Liu, W., Putkey, J.A., Sykes, B.D. ,1997, Structure of cardiac muscle troponin C unexpectedly reveals a closed regulatory domain. *J. Biol. Chem.* **272**:18216-21.

Smith, S.H., Versluis, J.P., Martin, A.F., Solaro, R.J., de Tombe, P.P. ,2002, Role of Troponin I in the Sarcomere Length Dependence of Calcium Sensitivity in Skinned Rat Trabeculae *Circulation* **106(II)**:101.

Solaro, R.J., Lee, J., Kentish, J., and Allen, D.A. ,1988, Differences in the response of adult and neonatal heart muscle to acidosis. *Circ. Res.* **63**:779-787.

Solaro, R.J.,2001, Modulation of cardiac myofilament activity by protein phosphorylation. *Handbook of Physiology: Section 2. The Cardiovascular System. Vol 1. The Heart* (E. Page, H. Fozzard, R. J. Solaro, Eds.) Oxford University Press, New York, pp 264-300.

Solaro, R.J. and Rarick, H.M. ,1998, Troponin and tropomyosin: proteins that switch on and tune in the activity of cardiac myofilaments. *Circ. Res.* **83**: 471-480.

Solaro, R.J., Kumar, P., Blanchard, E.M., and Martin, A.M. ,1986, Differential effects of pH on Ca^{2+} activation of myofilaments of adult and perinatal dog hearts: Evidence for developmental differences in thin filament regulation. *Circ. Res.* **58**:721-729.

Solaro RJ, Wolska, BM, Arteaga G, Martin AF, Buttrick P, deTombe, P. ,2002, Modulation of Thin Filament Activity in Long and Short Term Regulation of Cardiac Function. in: *Molecular Control Mechanisms in Striated Muscle Contraction,* R.J. Solaro and R.L. Moss, eds., Kluwer Academic Publishers, Dordrecht, Netherlands pp. 291-327.

Spyracoupoulos, L., Li, M.X., Sia, S.K., Gagne, S.M., Chandra, M., Solaro, R.J., Sykes, B.D. ,1997, Calcium-induced structural transition in the regulatory domain of human cardiac troponin C. *Biochemistry* **36**:12138-46.185.

Squire, J.M. and Morris, E.P. ,1998, A new look at thin filament regulation in vertebrate skeletal muscle. *FASEB J.* **12**:761-771.

Strang, K.T., Sweitzer, N.K., Greaser, M.L., and Moss, R.L., 1994, Beta-adrenergic receptor stimulation increases unloaded shortening velocity of skinned single ventricular myocytes from rats. *Circ. Res.* **74**:542-549.

Tobacman, L.S. ,1996, Thin filament-mediated regulation of cardiac contraction. *Ann. Rev. Physiol.* **58**:447-481.

Wang, X., Li, M., Spyracopoulos, L., Beier, N., Chandra, M., Solaro, R.J., and Sykes, B.D. 2001, Structure of the C-domain of human cardiac troponin C in complex with the Ca^{2+}-sensitizing drug EMD 57003. *J. Biol. Chem.* **276**:25456-25466.

Wang. Y.P, and Fuchs, F. (1994) Length, force, and Ca(2+)-troponin C affinity in cardiac and slow skeletal muscle. *Am. J. Physiol.* **266**:C1077-C1082.

Ward, D.G., Cornes, M.P., Trayer, I.P. ,2002, Structural consequences of cardiac troponin I phosphorylation. *J. Biol. Chem.* **277**:41795-41801.

Wattanapermpool, J., Guo, X. and Solaro, R.J. ,1995, The unique amino-terminal peptide of cardiac troponin I regulates myofibrillar ATPase activity only when it is phosphorylated. *J. Mol. Cell. Cardiol.* **27**:1383-1391.

Westfall, M.V. and Metzger, J.M. ,2001, Troponin I isoforms and chimeras: tuning the molecular switch of cardiac contraction. *News Physiol. Sci.* **16**:278-81.

Wolska, B.M., Arteaga, G.M,, Pena, J.M., Phillips, R.M., Sahai, S., de Tombe, P.P., Martin, A.F., Leiden, J.M., Kranias, E.G., Solaro, R.J.,2002, Expression of Slow Skeletal Troponin I in Phospholamban Knockout Mice Alters the Relaxant Effect of β-Adrenergic Stimulation. Circ Res **90**:882-888.

Wolska, B.M., Vijayan, K., Arteaga, G.M., Konhilas, J.P., Phillips, R.M., Kim, R., Naya, T., Leiden, J.M., Martin, A.F., de Tombe, P.P., and Solaro, R.J., 2001, Expression of slow skeletal troponin I in adult transgenic mouse heart muscle reduces the force decline observed during acidic conditions. *J. Physiol.* **536**:863-70.

DISCUSSION

Rall: What is the mechanism of the TnI phosphorylation decrease in the maximum force development in the isolated heart cell?

Solaro: Possibly a decrease in cross-bridge affinity for actin. The mechanism is not completely known, however.

Rall: Is the same effect of TnI phosphorylation observed in intact heart?

Solaro: Yes, the result has been published.

Pfitzer: Does the phosphorylation of the PKC sites depend on the isoforms of PKC?

Solaro: This is not completely known, but TnI can be directly phosphorylated by PKC epsilon and PKC-beta.

Winegrad: Have you measured actomyosin ATPase activity and force in the same preparation, in which you have produced the change in amino acids? If so, do the changes always go in the same direction?

Solaro: In the case of TnT charge substitutions, the changes in ATPase rate go in the same direction. We don't have enough data to make conclusions regarding TnI charge changes

Winegrad: Of interest is the question of whether efficiency of contraction changed by phosphorylation of TnI or TnT?
Solaro: It is predicted that phosphorylation of TnI at Thr 144 would change tension cost.

ter Keurs: Can you speculate why the velocity of sliding is modified so dramatically with modifications of TnI?

Solaro: The answer is not clear, but data comparing transgenic myofilaments missing Ser 43, 45 by substitution with sea indicate that ADP release step may be involved.

ter Keurs: What are the physiological conditions under which PKC is active *in vivo*?

Solaro: This could occur with alpha-adrenergic receptor and endothelin receptor activation.

PHARMACOLOGICAL CALCIUM SENSITIVITY MODULATION OF CARDIAC MYOFILAMENTS

J.C. Rüegg[*]

1. INTRODUCTION

As is well known, force generation and contractility of the heart may be controlled by calcium by way of regulating crossbridge attachment and the transition of attaching crossbridges into a force-generating state. These calcium activation processes do not occur in an all-or-none manner, but may be graded e.g. by altering the degree of calcium occupancy of troponin C (e.g. Pan and Solaro, 1987). Calcium occupancy depends on the free calcium concentration in the myoplasm in the range of 0.1 to 10 μM, but also, of course, on the calcium affinity of troponin C. Thus, at an intermediate cytosolic Ca^{2+} concentration allowing force to reach half maximum activation, any increase in calcium affinity would increase calcium occupancy and hence force and contractility. Thus, a lower Ca^{2+} concentration (or for that matter a higher pCa_{50} value) would then be required to reach half maximum activation under these conditions. This we refer to as Ca^{2+} sensitization of the myofilaments. The term calcium sensitivity modulation then is taken to mean a change in the pCa_{50} value, i.e. in the level of ionised calcium required for 50% activation of the myofilament activity normalized to the maximum contractile activity at saturating levels of calcium.

In recent years, it has become clear that the calcium sensitivity of the heart is altered, and often decreased, under pathological conditions such as, for instance, the stunned myocardium, caused by reperfusion injury (e.g. Kusuoka et al., 1990). Therefore, the development of positive inotropic cardiac drugs that increase calcium sensitivity rather than the intracellular concentration of free calcium has been an interesting novel concept. This principle had been discovered by Herzig and colleagues (1981a, c.f. also Solaro and Rüegg, 1982) and since then a large number of novel calcium-sensitizing drugs had been developed by pharmaceutical industry, and their mechanism of action had been investigated in vivo as well as on a molecular pharmacological level (c.f. Teramura and Yamakado, 1998, for review). As pointed out by Rüegg (1998) in the future, novel

[*] Department of Physiology and Pathophysiology, University of Heidelberg, D-69120 Heidelberg, Germany.

calcium-sensitizing drugs may play an emerging role in increasing the force of the failing heart, and already some of these calcium sensitizers (e.g. levosimendan, c.f. Haikala and Pollesello, 2000) may replace classical positive inotropic drugs such as dobutamin or PDE-inhibitors in the treatment of acute failure. The latter drugs increase the power of the heart by increasing the free calcium concentration in the myoplasm while at the same time decreasing the calcium sensitivity of the myofilaments. In the following some molecular, physiological, and clinical aspects of pharmacological up- and down-regulation of calcium sensitivity will be briefly reviewed.

2. DOWN-REGULATION OF CA^{2+} SENSITIVITY BY PHOSPHODIESTERASE INHIBITORS AND CATECHOLAMINES

β-adrenergic receptor stimulation by catecholamines elicits its positive inotropic effect by enhancing the Ca^{2+} influx into the cytosol due to an increased level of cytosolic cAMP, while inhibitors of the cAMP-inactivating enzyme phosphodiesterase (PDE) such as amrinone and milrinone may exert their positive inotropic action by imitating the effect of β-adrenergic receptor stimulation as they prevent cAMP breakdown. Significantly, treatment of cardiac myofibrils with cAMP and cAMP-dependent protein-kinase (PKA) alters the response to the calcium activator due to a phosphorylation of troponin I causing a decrease in the calcium sensitivity of the myofibrillar actomyosin-ATPase (Ray and England, 1976). Similarly, cAMP-dependent protein kinase causes a marked reduction of isometric contractile force of skinned cardiac fibres while maximum force was not affected (Herzig et al., 1981b). As also recently reviewed by Kögler and Rüegg (1997), two phosphorylatable adjacent serine residues (Ser^{23} and Ser^{24}) are present in the cardiac-specific amino terminal extension of human cardiac troponin (Mittmann et al., 1990) which are sequentially phosphorylated when treated with the catalytic subunit of proteinkinase A (Mittmann et al., 1992). Obviously the phosphorylation of both adjacent serine residues is required to induce the observed calcium-desensitizing effects of proteinkinase A. Thus, Zhang and colleagues (1995) reconstituted mutant troponin I in which one of the two serine residues was replaced by a (non-phosphorylatable) alanine into skinned cardiac fibres and showed that the phosphorylation of one serine residue alone is not sufficient to cause calcium desensitization. Likewise, unpublished experiments carried out in our laboratory (c.f. Kögler and Rüegg, 1997) showed that reconstitution of skinned cardiac fibres with mutant cardiac troponin I carrying a $Ser^{23}Ala/Ser^{24}Asp$ or a $Ser^{23}Asp/Ser^{24}Ala$ replacement failed to cause the calcium-desensitizing effect previously observed with the $Ser^{23}Asp/Ser^{24}Asp$ mutant. In these latter studies Dohet and colleagues (1995) were able to show that one could mimick the effect of double phosphorylation on calcium sensitivity by reconstituting skinned cardiac fibres with mutant troponin I carrying two residues of aspartic acid instead of the two serine residues. Taken at their face value these results might be taken to mean that it is the negative charge introduced by phosphorylation that is causing the observed effect on calcium sensitivity. Finally it should be noted that after reconstitution with (non-phosphorylatable) N-terminally truncated cardiac troponin treatment of skinned cardiac fibres with the catalytic subunit of PKA did not decrease calcium sensitivity (Wattanapermpool et al., 1995). As C-protein did become phosphorylated under these conditions its phosphorylation is probably not causally involved in the cAMP-induced calcium sensitivity modulation of cardiac myofibrils.

The decrease in calcium sensitivity caused by cAMP-dependent phosphorylation of cardiac troponin I may be important for accelerating cardiac relaxation which involves three consecutive steps: (1) the free calcium is lowered to basal levels (about $0.1\mu M$) by the calcium pump of the sarcoplasmic reticulum. (2) Ca^{2+} dissociates from the single calcium-specific site of cardiac troponin C. (3) crossbridges detach from actin. As shown by Saeki and colleagues (1997), _-adrenergic receptor stimulation increases the rate of crossbridge detachment by activating cAMP-dependent protein kinase that also causes (by phosphorylating troponin I, TnI) a decrease of calcium affinity of TnC and hence an increase of the calcium off-rate thereby accelerating dissociation of Ca^{2+} from its calcium-specific binding sites (Saeki et al., 2001). Because of this the rate of relaxation of skinned fibres by rapid removal of calcium (using a UV-flash activated photolabile caged calcium chelator) is markedly enhanced. Furthermore the calcium pump will not have to lower the level of intracellular free calcium to the same extent, as the threshold of mechanical calcium activation will also be increased after cAMP-dependent phosphorylation of troponin I. Hence, cAMP-dependent stimulation of the calcium pump and troponin I phosphorylation act synergistically to increase the rate of relaxation under conditions of β-adrenergic receptor stimulation or following the therapeutic application of cardiotonic drugs that inhibit phosphodiesterase III. The more rapid relaxation of the heart that may, at least partly, be due to cAMP-dependent reduction in myofibrillar calcium sensitivity is of critical importance for the heart, since it improves the so called diastolic function: If relaxation of the heart is more rapid and more complete, the filling of the ventricles is improved. Furthermore, since coronary blood flow peaks early in diastole, everything which prolongs the relative duration of the diastole will be beneficial for coronary blood supply to the myocardium. All of this is particularly important when the heart beat is accelerated, but also under conditions of pathological impairment of cardiac filling, which may be an early indication of left ventricular dysfunction and chronic heart failure (c.f. Sys and Brutsaert, 1995).

3. UP-REGULATION OF CA²⁺ SENSITIVITY BY CARDIOTONIC CALCIUM SENSITIZERS

The first calcium sensitizer studied was sulmazole (Herzig et al., 1981). It decreased the Ca^{2+} concentration required for half maximum isometric tension development in skinned cardiac fibres, but also stimulated Ca^{2+} binding and contractile ATPase of cardiac myofibrils (Solaro and Rüegg, 1982). Thus, it enhanced the calcium affinity of the myofibrillar calcium-binding protein (troponin C) and therefore, presumably, the Ca^{2+} occupancy of the calcium regulatory protein at intermediate levels of cytosolic free calcium (c.f. Rüegg, 1986). This is one (of several) possible mechanisms of calcium sensitization. Until the end of the decade many more calcium sensitizers were developed by pharmaceutical industry, but all of them were not very specific as they also inhibited phosphodiesterase III (PDE III).This was not really surprising, since in fact calcium sensitizers were originally designed as PDE III inhibitors and their calcium-sensitizing action was discovered only later, sometimes accidentally (see Teramura and Yamakado,

1998, for review). PDE-inhibitors have the advantage of being strong inotropes and vasodilators, and they also improve the diastolic function of the heart for the reasons discussed above. However, while improving the quality of life for a short while, PDE-inhibitors such as milrinone will not be beneficial in the long term: As shown by Packer and colleagues (1991) they shorten life expectancy. Thus, not only typical PDE III inhibitors but many of the calcium-sensitizing drugs became also suspect, as their positive inotropic action could (in vivo) to a large extent be ascribed to their PDE-inhibiting action rather than to calcium sensitization and it was not possible so far to dissociate PDE inhibition from calcium sensitization.

The situation changed when E. Merck (Darmstadt) developed a novel calcium sensitizer, EMD 57033, an enantiomer of the thiadiazonone derivative EMD 53998 (5-[1-(3,4-dimethoxybenzoyl)-1,2,3,4-tetrahydro-6-quinolyl]-6-methyl-3,6-dihydro-2H-1,3,4-thiadiazin-2-one) that had no PDE-inhibitory activity, but nevertheless activated the contractile structures (Solaro et al., 1993). In intact (surviving) cardiac preparations such as papillary muscles, EMD 53998 may enhance cardiac force with little if any increase in the intracellular Ca^{2+} transient resulting from the rapid release of Ca^{2+} from and its subsequent reuptake by the sarcoplasmic reticulum (Lee and Allen, 1991). In this way it was possible to show, for the first time, that the contractile force of the heart may be enhanced largely by increasing the Ca^{2+} sensitivity of the myofilaments. However, these experiments also revealed a new problem: The rate of relaxation of twitching intact preparations was lower in the presence than in the absence of the drug. This was perhaps not unexpected, as any increase in Ca^{2+} affinity of troponin would necessarily decrease the Ca^{2+} off-rate of this regulatory protein (see Teramura and Yamakado, 1998, for review). If occurring in the beating heart, such an effect would surely harm the diastolic cardiac function as discussed above. As many calcium sensitizers, e.g. pimobendan (Boehringer Ingelheim) and MCI-154 (Mitsubishi-Tokyo Pharmaceuticals Inc.) inhibited PDE activity and increased the Ca^{2+} affinity of troponin C, the question arose whether it was at all possible to design a positive inotropic calcium sensitizer devoid of PDE-inhibitory effects that, nevertheless, did not impair diastolic function and was not arrhythmogenic? It turned out that a novel calcium sensitizer, levosimendan (Orion Corp. Espoo, Finnland), apparently fulfils these criteria of an "ideal" calcium sensitizer (c.f. Haikala and Pollesello, 2000). Thus levosimedan increased the force of human heart preparations, even if the intracellular Ca^{2+} transient was not detectably increased (Hasenfuss et al., 1998). Notably, relaxation time of these strip preparations was not prolonged, and even the diastolic function of the whole heart was not impaired (Janssen et al., 2000). Furthermore, (Ca^{2+} activated) arrhythmias were not occurring, presumably because even in systole the intracellular Ca^{2+} concentration and cAMP levels remained comparatively low, as levosimedan barely inhibited phosphodiesterase III when applied in a low dose (Singh et al., 1999). At higher concentrations however, levosi-mendan is also a PDE inhibitor, but importantly, unlike many other cardiotonic agents, levosimedan does not increase the frequency of the heart beat (Slawsky et al., 2000).

The calcium-sensitizing and troponin C-binding properties of levosimendan were discovered by Haikala and colleagues (1995) when the drug was found to be retained in a Ca^{2+}-dependent manner by cardiac-troponin C HPLAC columns (high performance liquid affinity chromatography) suggesting that levosimendan is bound in a calcium-dependent manner. It is known that cardiac troponin C (cTnC) exists in two forms, a "closed" and an "open" one which are thought to be in rapid equilibrium. In the absence of Ca^{2+} the "closed" one predominates, while at high Ca^{2+} levels it is the "open" (active) configuration that predominates (Pääkönen et al., 2000). Heteronuclear NMR studies and small angle x-ray scattering experiments by Sorsa et al. (2001) confirmed that levosimendan bound to a pocket in the N-terminal domain of calcium-saturated cTnC, and it is hypothesized that the bound drug prolongs the average life span of the "open" active form of cTnC, presumably, however, without preventing calcium to detach from its binding site (Piero Pollesello, private communication). In other words: levosimendan would not be expected to decrease the Ca^{2+} off-rate, which is in agreement with experimental finding by Edes et al. (1995) showing that levosimendan is not increasing the calcium affinity of cTnC. Note that at a given intermediate free calcium concentration, i.e. at intermediate calcium occupancy of TnC, two calcium-bound conformations of TnC (open and closed ones) coexist and these may (for instance) be equally populated, but in rapid equilibrium. Levosimendan binds and stabilizes the active (open) rather than the inactive calcium-TnC complex, thereby prolonging the lifetime of the former. Therefore, the active TnC species would become more populated than the inactive calcium-bound one and the degree of contractile activation would increase, even if the free calcium concentration and the calcium affinity (i.e. the extent of TnC calcium occupancy) remains constant. Though levosimendan does not alter Ca^{2+} affinity, it is nevertheless found to augment the *apparent* Ca^{2+} sensitivity, as it increases contractile activation at intermediate, but not at saturating levels of free calcium. As mentioned above, levosimendan does apparently not lower the Ca^{2+} off-rate and it will therefore not impair the diastolic heart function, also, of course, because it is bound to the Ca^{2+}-occupied species of cTnC only.

Unlike in cardiac muscle, in smooth muscle levosimendan causes a decrease rather than an increase of the sensitivity of myofilaments to calcium thereby dilating coronary blood vessels (Bowman et al., 1999). But it also dilates coronary arteries (and even reduces myocardial infarct size) via activation of K(ATP) channels (Kersten et al., 2000). In this way, the blood supply to the myocardium is increased even though the myocardial metabolism is not enhanced (Lilleberg et al., 1998). Because of its vasodilating action the mean blood pressure and hence the afterload of the heart (and presumably wall tension) may be even (slightly) decreased despite the fact that levosimendan increases the stroke volume (due to both the positive inotropic action and the reduction of the afterload). Furthermore it is interesting to note that levosimendan increases the metabolic rate of the heart very little (c.f. Lilleberg et al., 1998). Thus, from the energetic point of view, levosimedan compares favourably with other cardiotonic agents, particularly those that produce an inotropic effect by rising the levels of cAMP and intracellular free calcium. The reason is twofold: (1) As pointed out by Saeki and colleagues (1997) compared with untreated skinned cardiac fibres in fibres treated with the catalytic subunit of cAMP-dependent protein kinase, a much higher rate of ATP splitting is required to maintain a given contractile tension, i.e. the energetic tension cost is increased. Extrapolating these findings to the situation in vivo one might expect that the tension cost for supporting a given wall stress in the ventricles is also enhanced in hearts that are stimulated by positive inotropic agents causing an increase in the level of cAMP. As the metabolic rate of cardiac ventricles is largely dependent on wall stress, the increase in tension cost caused by cAMP-dependent protein kinase may at least partly account for the comparatively high metabolic rate of hearts stimulated with e.g. dobutamine or noradrenaline. (2) A calcium sensitizer such as levosimendan does not increase the strength of the heart by increasing the intracellular level of Ca^{2+} or, for that matter, the calcium occupancy of troponin C (as the calcium affinity of TnC is not changed). In each cardiac cycle therefore the amount of calcium released from and pumped back to the sarcoplasmic reticulum need not be increased. On the other hand, "classical" inotropes such as dobutamine will greatly activate the ATP-splitting rate of the sarcoplasmic reticulum as the amount of calcium released from and returned to the SR is increased. This is particularly true, of course, if the frequency of the heart beat is also enhanced, as is the case during β-adrenergic receptor stimulation by catecholamines. Of course β-adrenergic receptor blockers would be expected to reverse this effect and to increase the Ca^{2+} sensitivity of myofilaments (that had been down-regulated by cAMP). In this respect it is interesting to note then that the therapy with adrenoreceptor blockers is also known to improve the life expectancy of patients suffering from chronic heart failure (Packer et al., 1996) that is associated with an overdriven sympatho-adrenal system (Nolan et al., 1998).

4. CONCLUSIONS AND PERSPECTIVES

As we have seen, the contractile force of the heart may be pharmacologically regulated not only by changes in the level of free calcium, as achieved by β-adrenergic stimulation or conventional positive inotropes, but also by altering the calcium sensitivity of the myofilaments in response to calcium-sensitizing drugs such as levosimendan (Orion) and pimobendan (Boehringer-Ingelheim). The latter, however, increases the calcium affinity of cardiac troponin C (cTnC), an effect which is associated with a decrease in the dissociation rate of free calcium from the calcium-specific binding sites (calcium off-rate) and hence with slowing of cardiac relaxation and impairment of the diastolic function of the heart. This negative effect is not seen in the case of levosimendan. Rather than increasing the calcium affinity of the regulatory protein, this drug raises – at intermediate levels of free calcium – the effectiveness of the calcium-cTnC-complex and hence contractile force by binding calcium-dependently to a pocket of the N-terminal domain of cTnC (see Haikala and Pollesello, 2000, for review).

Acknowledgement. I am grateful to A. Ebling for the careful preparation of the manuscript and to the Fonds für Chemie for supporting the work in my laboratory.

5. REFERENCES

Bowman, P., Haikala, H., and Paul, R. J., 1999, Levosimendan, a calcium sensitizer in cardiac muscle, induces relaxation in coronary smooth muscle through calcium desensitisation, *J. Pharmacol. Exp. Ther.* **288**:316-325.

Dohet, C., Al-Hillawi, E., Trayer, I. P., and Rüegg, J. C., 1995, Reconstitution of skinned cardiac fibres with human recombinant cardiac troponin I mutants and troponin C, *FEBS Lett.* **377**:131-134.

Edes, I., Kiss, E., Kitada, Y., Powers, F. M., Papp, J. G., Kranias, E. G., and Solaro, R. J., Effects of levosimendan, a cardiotonic agent targeted to troponin C, on cardiac func-tion and on phosphorylation and

Ca²⁺ sensitivity of cardiac myofibrils and sarcoplas-mic reticulum in guinea-pig heart, *Circ. Res.* 77:107-113.

Haikala, H., Nissinen, E., Etemadzadeh, E., Levijoki, J., and Linden, I. B., 1995, Tropo-nin C-mediated calcium sensitisation does not impair relaxation, *J. Cardiovasc. Pharmacol.* 25:794-801.

Haikala, H.,and Pollesello, P., 2000, Calcium sensitivity enhancers, *Idrugs* 3:1199-1205.

Hasenfuss, G., Pieske, B., Castell, M., Kretschmann, B., Maier, L. S., and Just, H., 1998, Influence of the novel inotropic agent levosimendan on isometric tension and calcium cycling in failing human myocardium, *Circulation* 98:2141-2147.

Herzig, J. W., Feile, K., and Rüegg, J. C., 1981a, Activating effects of AR-L 115 BS on the Ca²⁺ sensitive force, stiffness and unloaded shortening velocity (V_{max}) in isolated contractile structures from mammalian heart muscle, *Arzneim. Forsch./Drug Res.* 31:188-191.

Herzig, J. W., Köhler, H., Pfitzer, G., Rüegg, J. C., and Wölffle, G., 1981b, Cyclic AMP inhibits contractility of detergent treated glycerol extracted cardiac muscle, *Pflügers Arch. Europ. J. Physiol.* 391:208-212.

Janssen, P. M. L., Datz, N., Zeitz, O., and Hasenfuss, G., 2000, Levosimendan improves diastolic and systolic function in failing human myocardium, *Eur. J. Pharmacol.* 404:191-199.

Kersten, J. R., Montgomery, M. W., Pagel, P. S., and Warltier, D. C., 2000, Levosimen-dan, a new positive inotropic drug, decreases myocardial infarct size via activation of K (ATP) channels, *Anaesth. Analog.* 90:5-11.

Kögler, H., and Rüegg, J. C., 1997, Cardiac contractility: Modulation of myofibrillar Ca-sensitivity by _-adrenergic stimulation, *Isr. J. Med. Sci.* 33:1-7.

Kusuoka, H., Koretsune, Y., Chacko, V. P., Weisfeldt, M. L., Marban, E., 1990, Excitation-contraction coupling in postischemic myocardium. Does failure of activator Ca²⁺ transients underlie stunning? *Circ. Res.* 66:1268-1276.

Lee, J. A., and Allen, D. G., 1991, EMD 53998 sensitizes the contractile proteins to calcium in intact ferret ventricular muscle, *Circ. Res.* 69:927-36.

Lilleberg, J., Nieminen, M. S., Akkila, J., Heikkilär, L., Kuitunen, A., and Lehtonen, L., 1998, Effects of a new calcium sensitizer, levosimendan, on hemodynamics, coronary blood flow and myocardial substrate utilization early after coronary bypass grafting, *Eur. Heart J.* 19:660–668.

Mittmann, K., Jaquet, K., and Heilmeyer, Jr. L. M. G., 1990, A common motif of two adjacent phosphoserines in bovine, rabbit and human cardiac TnI, *FEBS Lett.* 273:41-45.

Mittmann, K., Jaquet, K., and Heilmeyer, Jr. L. M. G. 1992, Ordered phosphorylation of a duplicated minimal recognition motif for c-AMP-dependent proteinkinase present in cardiac troponin I, *FEBS Lett.* 302:133-137.

Nolan, J., Batin, P. D., Andrews, R., Lindsay, S. J., Brooksby, P., Mullen, M., Baig, W., Flapan, A. D., Cowley, A., Prescott, R. J., Neilson, J. M., and Fox, K. A., 1998, Prospective study of heart rate variability and mortality in chronic heart failure: results of the United Kingdom heart failure evaluation and assessment of risk trial (UK-heart), *Circulation* 98:1510-1516.

Paakkonen, K., Sorsa, T., Drakenberg, T., Pollesello, P., Tilgmann, C., Permi, P., Heikkinen, S., Kilpelainen, I., and Annila, A., 2000, Conformations of the regulatory domain of cardiac troponin C examined by residual dipolar couplings, *Eur. J. Biochem.* 267:6665-72.

Packer, M., Bristow, M. A., Cohn, J. D., Colucci, W. S., Fowler, M. B., Gilbert, E. M., and Shusterman, N. H., 1996, The effct of cravedilol on morbidity and mortality in patients with chronic heart failure, *N. Engl. J. Med.* 334:1349-1355.

Packer, M., Carver, J. R., Rodeheffer, R. J., Ivanhoe, R. J., DiBianco, R., Zeldis, S. M., Hendrix, G. H., Bommer, W. J., Elkayam, U., Kukin, M. L., 1991, Effect of oral milrinone on mortality in severe chronic heart failure. The PROMISE study research group, *New Engl. Med. J.* 325:1468-1475.

Pan, B. S., and Solaro, R. J., 1987, Calcium-binding properties of troponin C in detergent-skinned heart muscle fibers, *J. Biol. Chem.* 262:7839-7849.

Ray, K. P., and England, P. J., 1976, Phosphorylation of the inhibitory subunit of troponin and its effects on the calcium dependence of cardiac myofibril adenosine triphosphatase, *FEBS Lett* 70:11-16.

Rüegg, J. C., 1986, Effects of new inotropic agents on Ca⁺⁺ sensitivity of contractile proteins, *Circulation* 73(Suppl III):78-85.

Rüegg, J. C., 1998, Cardiac contractility: How calcium activates the myofilaments, *Naturwissenschaften* 85:575-582.

Saeki, Y., Kobayashi, T., Minamisawa, S., and Sugi, H., 1997, Proteinkinase A increases the tension cost and unloaded shortening velocity in skinned rat cardiac muscle, *J. Mol. Cell. Cardiol.* **29**:1655-63.

Saeki, Y., Takigiku, K., Iwamoto, H., Yasuda, S., Yamashita, H., Sugiura, S., and Sugi, H., 2001, Protein kinase A increases the rate of relaxation but not the rate of tension development in skinned rat cardiac muscle, *Jpn. J. Physiol.* **51**:427-33.

Singh, B. N., Lilleberg, J., Sandell, E. P., Ylönen, V., Lehtonen, L., and Toivonen, L., 1999, Effects of Levosimendan on cardiac arrhythmia: electrophysiological and ambulatory electrocardiographic findings in phase II and phase III clinical studies in cardiac failure, *Am. J. Cardiol.* **83** (12B):16(I)-20(I).

Slawsky, M. T., Wilson, S. C., Gottlieb, S. S., Greenberg, B. H., Häusslein, E., Hare, J., Hutchins, S., Leier, C. V., LeJemtel, T. H., Loh, E., Nicklas, J., Ogilby, D., Singh, B. M., and Smith, W., 2000, Acute hemodynamics and clinical effect of Levosimendan in patients with severe heart failure, *Circulation* **102**:2222-2227.

Solaro, R. J., Gambassi, G., Warshaw, D. M., Keller, M. R., Spurgeon, H. A., Beier, N., and Lakatta, E. G., 1993, Stereoselective actions of thiadiazinones on cardiac myocytes and myofilaments, *Circ. Res.* **73**:981-990.

Solaro, R. J., Rüegg, J. C., 1982, Stimulation of Ca^{2+} binding and ATPase activity of dog cardiac myofibrils by AR-L 115 BS, a novel cardiotonic agent, *Circ. Res.* **61**:290-294.

Sorsa, T., Heikkinen, S., Abbott, M. B., Abusamhadneh, E., Laakso, T., Tilgmann, C., Serimaa, R., Annila, A., Rosevear, P. R., Drakenberg, T., Pollesello, P., and Kilpelai-nen, I., 2001, Binding of levosimendan, a calcium sensitizer, to cardiac troponin C, *J. Biol. Chem.* **276**:9337-43.

Sys, S. U., and Brutsart, D. L., 1995, Diastolic significance of impaired LV systolic re-laxation in heart failure, *Circulation* **92**:3377-3380.

Teramura, S., and Yamakado, T., 1998, Calcium sensitizers in chronic heart failure: inotropic interventions – reservation to preservation, *Cardiologia* **43**:375-383.

Wattanapermpool, J., Guo, X., and Solaro, R. J., 1995, The unique amino-terminal peptide of cardiac troponin I regulates myofibrillar activity only when it is phosphorylated, *J. Mol. Cell. Cardiol.* **27**:1383-1391.

Zhang, R., Zhao, J. J., and Potter, J. D., 1995, Phosphorylation of both serine residues in cardiac troponin I is required to decrease the Ca^{2+} affinity of cardiac troponin C, *J. Biol. Chem.* **270**:30773-30780.

DISCUSSION

Morano: Is levosimendan already used as a drug in the clinic to treat patients with heart disease?

Rüegg: Yes, it is used effectively in clinical stretches: it improves the performance of the heart (systolic and diastolic cardiac function) in patients with congestive heart failure. In Sweden and other European countries, levosimendan is already available.

ter Keurs: Patients with heart failure may be treated with this class of drugs in order to improve cardiac function. However, it has recently been shown that instability of the Ryanodine receptor causes Ca^{2+} leak from the SR. Work in our lab has shown that this leak occurs in the form of repetitive diastolic Ca^{2+} release and elevates $[Ca^{2+}]i$. If this is true in humans, one should warn against the use of these drugs because they would raise left ventricular endo-diastolic pressure and worsen the condition of the patient.

Rüegg: The fact is that levosimendan does not impair diastolic function in patients and improves the conditions of the patients.

Ranatunga: Could the effects be different in different species?

Rüegg: Not yet investigated in detail.

20-HYDROXYEICOSATETRAENOIC ACID POTENTIATES CONTRACTILE ACTIVATION OF CANINE BASILAR ARTERY IN RESPONSE TO STRETCH VIA PROTEIN KINASE Cα-MEDIATED INHIBITION OF CALCIUM-ACTIVATED POTASSIUM CHANNEL

Koichi Nakayama, Kazuo Obara, Yoshiyuki Tanabe, and Tomohisa Ishikawa*

1. INTRODUCTION

Blood vessels are persistently exposed to hemodynamic forces in the form of pressure and flow. The vascular smooth muscle contracts in response to stretch/increased intraluminal pressure, and dilates in response to release/decreased intraluminal pressure (Bayliss, 1902). This autoregulatory response is myogenic in nature, and is called "Bayliss effect." To the contrary, flow/shear stress-dependent dilatation, so called "Schrezenmayar effect" (1933), caused by released vasodilators including nitric oxide from vascular endothelium or other components, can physiologically counteract the myogenic contraction. The cerebral artery is particularly sensitive to pressure and stretch, and shows myogenic contraction (Nakayama et al., 2002). Furthermore, we previously reported that large conductance Ca^{2+}-activated K^+ channel (KCa channel) blockers, including iberiotoxin, charybdotoxin, and tetraethylammonium, sensitized the canine basilar artery to mechanical stretch (Obara et al., 2001).

It is well-documented that cytochrome P450 monooxygenase metabolizes arachidonic acid, and generates vasoactive substances, such as epoxyeicosatrienoic acids and 20-hydroxyeicosatetraenoic acid (20-HETE). Epoxyeicosatrienoic acids have been considered as a candidate of endothelium-derived hyperpolarizing factor (EDHF) (Campbell et al., 1996), though it is still controversial as to the chemical structure, whereas 20-HETE is a potent and Ca^{2+}-dependent vasoconstrictor, and considered to be an endogenous mediator in the cerebral circulation (Harder et al., 1994). It has been reported that the vasoconstrictor action of 20-HETE is attributable to the protein kinase C (PKC)-mediated inhibition of KCa channel activity (Lange et al., 1997). Of at least 12 isoforms of PKC subspecies, we identified 4 isoforms (PKCα, δ, ζ, and ε) in the

*Koichi Nakayama, Kazuo Obara, Yoshiyuki Tanabe, and Tomohisa Ishikawa. Department of Pharmacology, School of Pharmaceutical Sciences, University of Shizuoka, Shizuoka 422-8526, Japan
Phone: +81-54-264-5694,   Fax: +81-54-264-5696,   E-mail: nakyamk@ys7.u-shizuoka-ken.ac.jp

Molecular and Cellular Aspects of Muscle Contraction
Edited by H. Sugi, Kluwer Academic/Plenum Publishers, 2003

canine basilar arteries (Nishizawa et al., 2000). Moreover, we found that PKCα was involved in the maintenance of delayed vasospasm after subarachinodal hemorrhage in the canine model (Nishizawa et al., 2000).

The aim of the present study is to elucidate whether PKCα plays a role in the potentiating mechanism of stretch-induced contraction by 20-HETE. Our results strongly suggest that 20-HETE inhibits KCa channel activity through activation of PKCα, whereas iberiotoxin directly blocks KCa channel (Obara et al., 2002).

2. POTENTIATING ACTION OF 20-HETE ON THE STRETCH-INDUCED CONTRACTION

The effects of 20-HETE on the canine basilar artery were compared with those of iberiotoxin, a blocker of KCa channel. The endothelial cells were mechanically removed. Rate, magnitude, and interval of stretch are the main parameters to be studied to elucidate the nature of myogenic contraction (Nakayama, 1982). A slow stretch (from initial muscle length, Li, to 1.5 Li, at a rate between 1 and 3 mm/sec, and a stimulus period of 15 min) by itself, appeared to mimic a change in mean arterial pressure, did not produce any contraction, but only a passive increase in tension. 20-HETE (1-300 nM), however, produced the contraction of the artery in response to slow stretch in a concentration-dependent manner with EC_{50} value of about 29 nM (n=5) (Figure 1). The stretch-induced contraction was inhibited by nicardipine, a 1,4-dihydropyridine Ca^{2+} channel blocker, and Gd^{3+}, a blocker of stretch-activated cation channels, indicating that the contraction was mainly attributable to activation of stretch-activated cation channels and subsequent transmembrane influx of Ca^{2+} through L-type Ca^{2+} channels (Obara et al., 2001). In the present study, 20-HETE caused a parallel shift to the left the relationship between the rate of stretch and contraction of the canine basilar artery, whereas it showed no appreciable effect on the maximum contraction. Fifty percent effective rates (ER_{50}) for stretch-induced contraction in the presence and absence of 20-HETE (100 nM) were about 0.5 mm/sec and 13 mm/sec, respectively. Thus, the artery pretreated with 20-HETE was about 30 times more sensitive to the rate of stretch than the untreated artery.

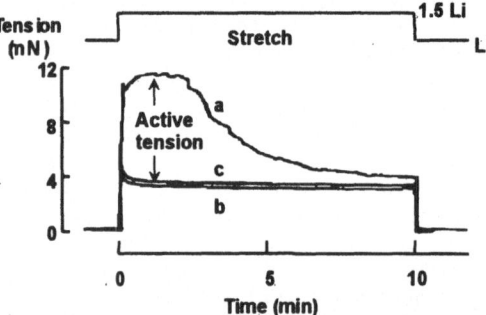

Figure 1. Effects of 20-HETE on the mechanical response of canine basilar artery to slow stretch. The artery was treated with 100 nM 20-HETE (trace a) or 100 μM papaverine (trace b) for 10 min. Isometric tension of the artery was measured in the absence (trace c) of 100 nM 20-HETE. Trace "a" was superimposed on the passive increase in tension (b) or on that without 20-HETE (c). The active tension (arrow) was produced by slow stretch at a rate of 1 mm/sec, amount of stretch from Li to 1.5 Li, and a stimulus period of 15 min.

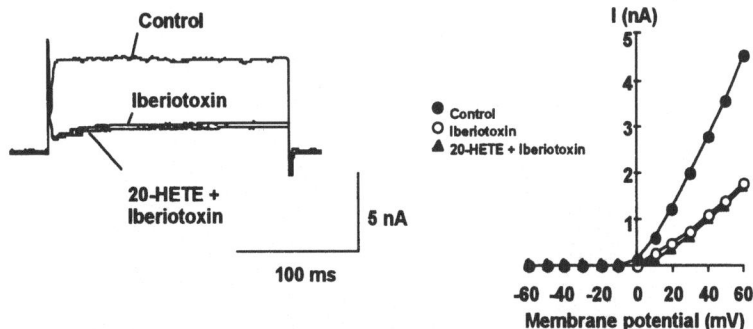

Figure 2. Effects of 20-HETE and iberiotoxin on whole-cell K^+ current. Left panel, representative tracings of outward whole-cell K^+ current elicited by depolarizing pulses to +60 mV from a holding potential of −70 mV in the absence or presence of 100 nM iberiotoxin or 100 nM 20-HETE plus 100 nM iberiotoxin. Right panel, current-voltage relationships of whole-cell K^+ current for the same cell shown in the left panel in the absence or the presence of 100 nM iberiotoxin or 100 nM 20-HETE plus 100 nM iberiotoxin. Each point represents the mean ± s.e.mean of 5 experiments.

3. INHIBITORY ACTION OF 20-HETE ON THE WHOLE-CELL K^+ CURRENT

Whole-cell K^+ current and resting membrane potential of canine basilar artery myocytes were measured with an AXOPATCH-10 amplifier by using the patch-clamp technique. 20-HETE (30-100 nM) added to the bathing medium significantly inhibited the peak whole-cell K^+ current in a concentration-dependent manner (by 37.5 ± 7.0% at 30 nM, P<0.01 versus vehicle-treated myocytes, n=4, and by about 60% at 100 nM, P<0.01, n=5). The removal of Ca^{2+} from bathing solution inhibited the peak whole-cell current by about 60%, and 20-HETE (100 nM) had no apparent effect on the Ca^{2+}-insensitive components of the whole-cell current. Iberiotoxin (100 nM) also significantly inhibited the peak whole cell current, by about 60%. 20-HETE and iberiotoxin (each 100 nM) similarly depolarized the membrane about 10 mV. However, 20-HETE (1-100 nM) combined with iberiotoxin (100 nM) produced neither additional depolarization nor enhanced inhibition of the whole-cell K^+ current (Figure 2). Also, the whole-cell K^+ current insensitive to Ca^{2+} was not affected by both 20-HETE and iberiotoxin. It has been reported that 20-HETE attenuates the KCa channel activity (Harder et al., 1994; Ma et al., 1993). Taken together with the present results, 20-HETE acts mainly as a blocker of large conductance KCa channels.

4. EFFECTS OF PKC INHIBITORS ON THE STRETCH-INDUCED CONTRACTION AND KCa CHANNEL ACTIVITY

As indicated previously, 20-HETE inhibited the activity of KCa channels *via* a mechanism involving PKC (Lange et al., 1997), we assessed the effect of calphostin C, conventional and novel PKC inhibitor on the mechanical and electrical activities.

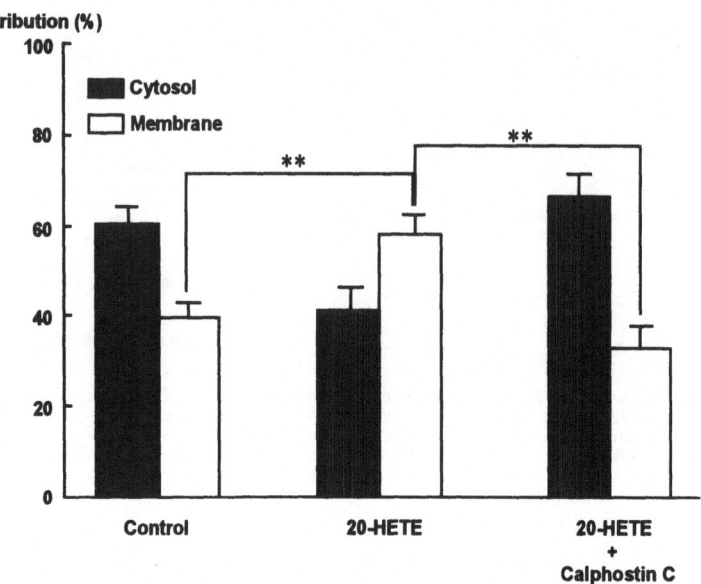

Figure 3. Effect of calphostin C on 20-HETE-induced translocation of PKCα. The artery was treated with 100 nM 20-HETE, 100 nM 20-HETE plus 1 μM calphostin C, or vehicle for 10 min. The results are expressed as a percentage of the total amount of PKCα isoform. Data are the mean ± s.e.mean of 5 individual experiments. **P<0.01 compared with the corresponding control value.

Calphostin C (1 μM) abolished the slow stretch-induced contraction augmented by 20-HETE (100 nM), whereas the inhibitor had no apparent effect on the contraction in response to stretch potentiated by iberiotoxin (100 nM). Calphostin C (1 μM) alone did not change the amplitude of the whole-cell K^+ current or the resting membrane potential. However, after treatment with calphostin C, 20-HETE failed to inhibit whole-cell K^+ current and depolarize the resting membrane potential, whereas iberiotoxin (100 nM) still significantly inhibited the whole-cell K^+ current by about 70%, and depolarized the membrane about 10 mV. These results suggest that the blockade of PKC counteracted the mechanical and electrical activities augmented by 20-HETE, whereas the blockade of PKC had no apparent effect on the augmented activities by iberiotoxin.

5. 20-HETE AND TRANSLOCATION OF PKCα ISOFORM

We identified 4 isoforms in the canine basilar artery, i.e., PKCα, δ, ζ and ε (Nishizawa et al., 2000). Furthermore, only PKCα was translocated from the cytosol to the membrane fraction by 20-HETE, indicating an activation of the kinase, and this translocation was inhibited by calphostin C (Figure 3). To the contrary, iberiotoxin had no apparent effect on the translocation of PKCα and other PKC isoforms.

6. DISCUSSION AND CONCLUSION

Several *cis*-unsaturated fatty acids, such as arachidonic acid and their metabolites, activate PKC (Hansson et al., 1986; Murakami et al., 1986; Sekiguchi et al., 1987). The activated PKC inhibited KC_a channels (Minami et al., 1993; Ribalet and Eddlestone, 1995; Zhang et al, 1995; Shipston and Armstrong, 1996), and augmented contraction of vascular tissues (Walsh et al., 1994; Lange et al., 1997). In the presence of calphostin C, 20-HETE did not inhibit the whole-cell K^+ current, indicating that the blockade of PKC counteracted the inhibitory action of 20-HETE on K^+ current. Thus calphostin C inhibited the slow stretch-induced contraction in the presence of 20-HETE. 20-HETE inhibited whole-cell K^+ current and depolarized the membrane by approximately 10 mV. These effects of 20-HETE were similar to those of iberiotoxin. However, calphostin C inhibited the action of 20-HETE, but not that of iberiotoxin. Of 12 isoforms of PKC subspecies, only $PKC\alpha$ was translocated from the cytosol to the membrane fraction by 20-HETE, whereas no translocation of PKC isoforms was produced by iberiotoxin. Therefore, it is concluded that 20-HETE inhibits KC_a channel activity through activation of $PKC\alpha$, whereas iberiotoxin directly blocks KC_a channel.

7. ACKNOWLEDMENT

The present study was supported in part by Grants-in-Aid for Scientific Research from the Ministry of Education, Culture, Sports, Science and Technology of Japan, and by grants from the Shizuoka Research and Development Foundation.

8. REFERENCES

Bayliss, W.M., 1902, On the local reaction of the arterial wall to change of internal pressure. *J. Physiol.*, **28**: 220-231.

Campbell, W.B., Gebremedhin, D., Pratt, P.F., and Harder, D.R., 1996, Identification of epoxyeicosatrienoic acids as endothelium-derived hyperpolarizing factors. *Circ. Res.*, **78**: 415-423.

Hansson, A., Serhan, C.N., Haeggstrom, J., Ingekma-Sundberg, M., and Samuelsson, B., 1986, Activation of protein kinase C by lipoxin A and other eicosanoids. Intracellular action of oxygenation products of arachidonic acid. *Biochem. Biophys. Res. Commun.*, **134**: 1215-1222.

Harder, D.R., Gebremedhin, D., Narayanan, J., Jefcoat, C., Falck, J.R., Campbell, W.B., and Roman, R., 1994, Formation and action of a P-450 4A metabolite of arachidonic acid in cat cerebral microvessels. *Am. J. Physiol.*, **266**: H2098-H2107.

Lange, A., Gebremedhin, D., Narayanan, J., and Harder, D., 1997, 20-Hydroxyeicosatetraenoic acid-induced vasoconstriction and inhibition of potassium current in cerebral vascular smooth muscle is dependent on activation of protein kinase C. *J. Biol. Chem.*, **272**: 27345-27352.

Ma, Y.H., Gebremedhin, D., Schwartzman, M.L., Falck, J.R., Clark, J.E., Masters, B.S., Harder, D.R., and Roman, R.J., 1993, 20-Hydroxyeicosatetraenoic acid is an endogenous vasoconstrictor of canine renal arcuate arteries. *Circ. Res.*, **72**: 126-136.

Minami, K., Fukuzawa, K., and Nakaya, Y., 1993, Protein kinase C inhibits the Ca^{2+}-activated K^+ channel of cultured porcine coronary artery smooth muscle cells. *Biochem. Biophys. Res. Commun.*, **190**: 263-269.

Murakami, K., Chan, S.Y., and Routtenberg, A., 1986, Protein kinase C activation by *cis*-fatty acid in the absence of Ca^{2+} and phospholipids. *J. Biol. Chem.*, **261**: 15424-15429.

Nakayama, K., 1982, Calcium-dependent contractile activation of cerebral artery produced by quick stretch. *Am. J. Physiol.*, **242**: H760-H768.

Nakayama, K., Obara, K.,Tanabe, Y., Saito, M., Ishikawa, T., and Nishizawa, S., 2002, Interactive role of tyrosine kinase, protein kinase C, and Rho/Rho kinase systems in the mechanotransduction of vascular smooth muscles. *Biorheology*, in press.

Nishizawa, S., Obara, K., Nakayama, K., Koide, M., Yokoyama, T., Yokota, N., and Ohta, S., 2000, Protein kinase Cd and a are involved in the development of vasospasm after subarachnoid hemorrhage. *Eur. J. Pharmacol.*, **398**: 113-119.

Obara, K., Koide, M., and Nakayama, K., 2002, 20-Hydroxyeicosatetraenoic acid potentiates stretch-inudced contraction of canine basilar artery *via* PKCα-mediated inhibition of K_c channel. *Br. J. Pharmacol.*, inpress.

Obara, K., Saito, M., Yamanaka, A., Uchino, M., and Nakayama, K., 2001, Involvement of different activator Ca^{2+} in the rate-dependent stretch-induced contractions of canine basilar artery. *Jpn. J. Physiol.*, **51**: 327-335.

Ribalet, B., and Eddlestone, G.T., 1995, Characterization of the G protein coupling of SRIF and beta-adrenergic receptors to the maxi K_{Ca} channel in insulin-secreting cells. *J. Membr. Biol.*, **148**: 111-125.

Schrezenmayar, A., 1933, Über regulatorisch Vorgange an Muskelarterien. *Pflügers. Arch.* **232**: 743-748.

Sekiguchi, K., Tsukuda, M., Ogita, K., Kikkawa, U., and Nishizuka, Y., 1987, Three distinct forms of rat brain protein kinase C: differential response to unsaturated fatty acids. *Biochem. Biophys. Res. Commun.*, **145**: 797-802.

Shipston, M.J., and Armstrong, D.L., 1996, Activation of protein kinase C inhibits calcium-activated potassium channels in rat pituitary tumour cells. *J. Physiol. (Lond.)*, **493**: 665-672.

Walsh, M.P., Andrea, J.E., Allen, B.G., Clement–Chomiennep, O., Cpllins, E.M., and Morgan, K.G., 1994, Smooth muscle protein kinase C. *Can. J. Physiol. Pharmacol.*, **72**: 1392-1399.

Zhang, H., Weir, B., and Daniel, E.E., 1995, Activation of protein kinase C inhibits potassium currents in cultured endothelial cells. *Pharmacology*, **50**: 247-256.

DISCUSSION

Fransen: What is the physiological role of $I_{k (Ca)}$ in smooth muscle cells?

Nakayama: Potassium channel activity of vascular smooth muscle cells, in general, plays a pivotal role of vasodilatation. Cerebral artery possesses myogenic activity despite a relatively deep membrane potential of about –60 mV, which is attributable to abundant density and high open probability of large conductance Ca-activated potassium channels (K_{Ca}). Accordingly, intracellular Ca concentration in the cerebral artery smooth muscle cells at the basal level is inherently high. Thus, the blockade of K_{Ca} channels by inhibitors, including iberiotoxin, depolarizes the membrane, and sensitizes the artery to mechanical stretch.

Takuwa: What is the action of 20-hydroxyeicosatetraenoic acid receptor-mediated on protein kinase C?

Nakayama: There is evidence that arachidonic acid metabolites, such as epoxyeicosatrienoic acids (EETs) and hydroxyeicosatetraenoic acids (HETEs), are produced in both endothelium and the vascular smooth muscles via cytochrome P450 monooxygenase. Furthermore, arachidonic acid is liberated from not only plasma membrane lipid but also intracellular structures through the actions of cytosolic phospholipase. It is a well-documented fact that the most vasoactive prostanoids, including prostaglandin E2 and F2α, thromboxane, prostacyclin, HETEs and leucotrienes, bind to GTP binding protein-coupled receptors, and produce physiological and pharmacological actions. Nevertheless, there is another possibility that the molecules of arachidonic acid released intracellularly are converted to 20-HETE by microsomal cytochrome P450 monooxygenase, and act directly on the protein kinase C.

Distinct Contractile Systems for Electromechanical and Pharmacomechanical Coupling in Smooth Muscle

Valéria Lamounier-Zepter[1], Leonidas G. Baltas[1], and Ingo Morano[1,2]

1. INTRODUCTION

The smooth muscle cells express diverse isoforms of the molecular motor Type II myosin. Three different genes coding for myosin heavy chains (MyHC) are expressed, namely one smooth-muscle specific (SM-MyHC), and two non-muscle- specific myosin heavy chain (NM-MyHC), NM-MyHCA, and NM-MyHCB, located on chromosomes 16, 22, and 17, respectively[1]. Different splice variants of the SM-MyHC are generated due to the alternatively spliced mutually exclusive exons 5b and 39[2,3]. Elimination of exon 5b in a knock-out mouse model demonstrated that a high contractile state depends on the presence of myosin with 5′-inserted heavy chains[4].

SM- and NM-MyHC isoenzymes are more closely related to each other than to striated muscle myosin[5]. They are regulated by phosphorylation/dephosphorylation on conserved sites of the 20 kDa regulatory light chain (MLC_{20}), mainly serine 19[6]. MLC_{20} could be phosphorylated by Ca2+-calmodulin dependent myosin light chain kinase (MLCK), CaM kinase II, Rho kinase, p21-activated kinase[7], and integrin-linked kinase[8].

Electromechanical coupling of bladder smooth muscle upon KCl depolarization elicits an initial phasic contraction with high shortening velocity and a subsequent tonic contraction with low shortening velocity[9]. Using a SM-MyHC knock-out mouse model, we have recently shown, that phasic contraction and subsequent tonic contraction could be generated by the recruitment of SM-MyHC and NM-MyHC, respectively[9]. Intracellular Ca^{2+} transients in smooth muscle cells of urinary bladder of SM-MyHC knock-out mice were normal[10].

In the present study, we investigated whether electromechanical and pharmacomechanical coupling could be conferred by the different SM- and NM-MyHC contractile systems present in smooth muscle cells. Therefore, we used urinary bladder preparations from our recently described SM-MyHC knock-out mouse model[9] treated either with the MLCK-inhibitor ML-7, or the protein kinase C (PKC) activator PDBu.

[1]Max-Delbrück-Center for Molecular Medicine, Berlin, and [2]Johannes-Müller Institute of Physiology, Humboldt-University (Charité), Berlin, Germany

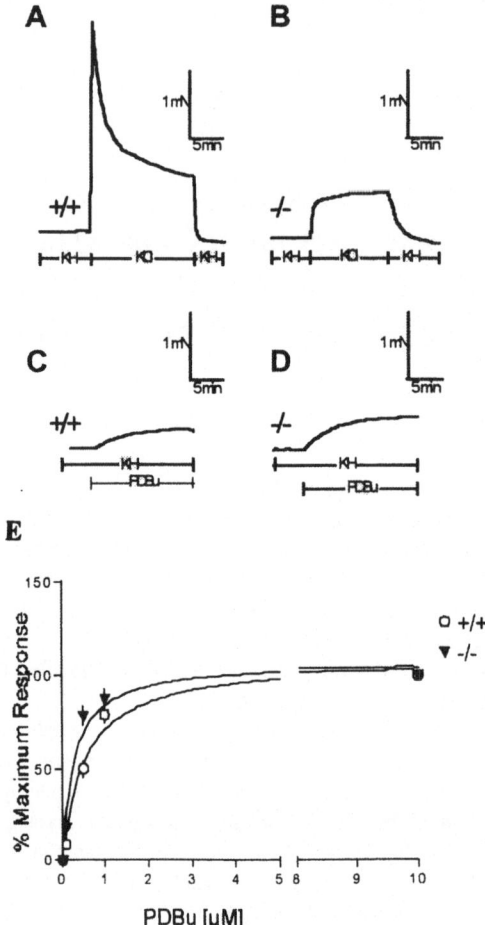

Figure 1. KCl and PDBu-induced contraction of wild-type (+/+) and homozygous SM-MyHC deficient (-/-) neonatal murine bladder preparations.
Original records of isometric force development induced by KCl in intact bladder preparations of A, wild-type (+/+) mouse; and B, homozygous SM-MyHC deficient (-/-) mice. The muscle preparations were equilibrated in Krebs-Henseleit solution (KH) for 20 min at 37°C and then stimulated in KCl depolarization solution. Stimulation with PDBu (10 µM) induced tonic contractions in both +/+ (C) and -/- (D) smooth muscle preparations.
Dose-response curve for PDBu (E) in urinary bladder smooth muscle from neonatal wild-type (+/+) (n=4) and homozygous SM-MyHC deficient (-/-) (n=5) mice. The smooth muscle preparations were incubated in Krebs-Henseleit solution for 20 min, and then stimulated with increasing doses of PDBu (10 nM, 100 nM, 0.5 µM, 1 µM, 10 µM). PDBu induced a similar response in both +/+ and -/- smooth muscle preparations.

We found that electromechanical coupling by KCl depolarization is mediated by both SM-MyHC and NM-MyHC systems. Interestingly, inhibition of MLCK by ML-7 selectively eliminated SM-MyHC-dependent phasic contraction, while tonic contraction was hardly affected. Pharmacomechanical coupling by PKC activation selectively recruits part of the NM-MyHC dependent tonic contraction.

2. MATERIALS AND METHODS

2.1. Tissue preparation

Intact urinary bladder strips were prepared from neonatal wild-type (+/+) C57BL6 mice as well as from neonatal homozygous SM-MyHC deficient (-/-) mice. The SM-MyHC deficient mice were generated by gene-targeting technology, leading to total depletion of SM-MyHC isoforms[9]. The animals were killed by decapitation and the bladders were quickly removed and placed in Krebs-Henseleit solution (119 mM NaCl, 12.2 mM glucose, 4.6 mM KCl, 25 mM NaHCO$_3$, 1.2 mM KH$_2$PO4, 1.2 mM MgSO$_4$, 2.0 mM CaCl$_2$, 2.0 mM Na-pyruvate), supplemented with 5 mM 2,3-butanedione monoxime. The bladders were excised and longitudinal strips of the posterior wall of the bladder were prepared by using micro scissors under a dissecting microscope. The strips were always approximately 1 mm long and 0.4-0.5 mm wide. All solutions were continuously oxygenated.

2.2. Force measurements

The urinary bladder preparations were mounted isometrically in the tissue bath containing 6 ml of Krebs-Henseleit solution (see 2.1.) and equilibrated for 20 min prior the experiments. The tissue bath was continuously gassed with 96% O$_2$ and 4% CO$_2$ and the temperature was kept at 37°C. Resting tension was adjusted to 0.3 mN.

Bladder preparations were activated in KCl depolarization solution (equimolar replacement of NaCl by 50 mM KCl in Krebs-Henseleit solution) or stimulated with the PKC activator phorbol 12,13-Dibutyrate (PDBu). Isometric force was recorded using a YT-recorder. For studies with ML-7 (MLCK inhibitor), the bladder preparations were initially stimulated with KCl depolarization solution, then incubated with the inhibitor for 20 minutes, and again stimulated with KCl depolarization solution without ML-7.

2.3. Genotyping

Wild-type and SM-MyHC deficient mouse tail DNAs were genotyped by PCR using primer sets specific for the wild-type and mutant allele (c.f. 9). Reactions contained 1.5 mM MgCl$_2$, 0.24 mM dNTP-Mix, 2.5 µl PCR Buffer Gibco, 0.75 UI Taq DNA-Polymerase, 0.2 µM of forward and reverse primers. PCR amplification was performed for 36 cycles of 30 sec at 94°C for denaturing, 45 sec at 62°C for annealing, and 1min at 72°C for extension. The PCR products (15 µl) were subjected to electrophoresis on a 1% agarose gel and DNA was visualized by ethidium bromide staining. Location of the products was determined by using a Low DNA Mass ladder as standard size marker.

2.4. Drugs and chemicals

The following substances were used: (5-Iodonaphthalene-1-sulfonyl)homopiperazine, HCl (ML-7 Hydrochloride), Phorbol-12,13 Dibutyrate (Calbiochem); NaCl, MgSO$_4$, NaHCO$_3$, KH$_2$PO4 (Merck); Glucose, KCl, CaCl$_2$ (Serva); Na-pyruvate, 2,3-butanedione monoxime (Sigma); MgCl$_2$, dNTP-Mix, 2.5 µl PCR Buffer Gibco (minus Mg), Taq DNA-Polymerase, Low DNA Mass ladder (Gibco BRL); agarose, ethidium bromide (Roth).

2.5. Statistical analyses

Values are expressed as means ± SEM (n = number of bladder preparations). Significance analysis was performed using Student′s t-test.

3. Results

3.1. Effects of KCl and PDBu-induced PKC activation

Activation of wild-type (+/+) neonatal bladder preparations by KCl depolarization under isometric conditions elicited an initial phasic contraction (3.35±0.3mN, n=13) and a subsequent tonic contraction (0.85±0.13 mN; n=13) (Fig. 1A). Activation of neonatal homozygous knock-out (-/-) mice by KCl depolarization, however, elicited a tonic contraction without the initial phasic force generation (0.63±0.05mN; n=13) (Fig. 1B) i.e. not statistically significant from the tonic contraction of +/+.

Incubation of neonatal bladder preparations with the phorbol ester PDBu elicited tonic contractions in both +/+ and -/- (Fig. 1C, D) neonatal bladder preparations. Cumulative doses of PDBu (10 nM, 100 nM, 500 nM, 1 µM, 10 µM) elicited contractions with EC$_{50}$ values of 0.5±0.02µM (n=4) and 0.3±0.02µM (n=5) in +/+ and -/- bladder preparations, respectively (Fig. 1E). The contractile response to maximal PDBu concentration (10 µM) was 0.36±0.04mN (n=11) in +/+ (Fig. 1C) and 0.41±0.06mN (n=11) in -/- (Fig. 1D) preparations, i.e. not significantly different. However, the PDBu induced tonic contractions of both +/+ as well as -/- bladder preparations were significantly (p<0.01) smaller than the corresponding tonic contractions elicited by KCl.

3.2. Effect of ML-7 (MLCK inhibition) on KCl-induced contraction

To examine the role of MLCK during KCl-induced smooth muscle contraction, we exposed intact bladder preparations from +/+ (Fig. 2A) and -/- mice (Fig. 2B) to the specific MLCK inhibitor ML-7[11]. Incubation with ML-7 inhibited in a concentration-dependent manner, the initial phasic contraction in bladder preparation from wild-type mice (EC$_{50}$ 5.03µM; n=7) (Fig. 2C). Inhibition of phasic contraction by ML-7 was statistically significant (c.f. Figure 2C). In contrast, the tonic contraction of bladder preparations obtained from neonatal +/+ as well as -/- mice were not affected up to a dosage of 10µM ML-7.

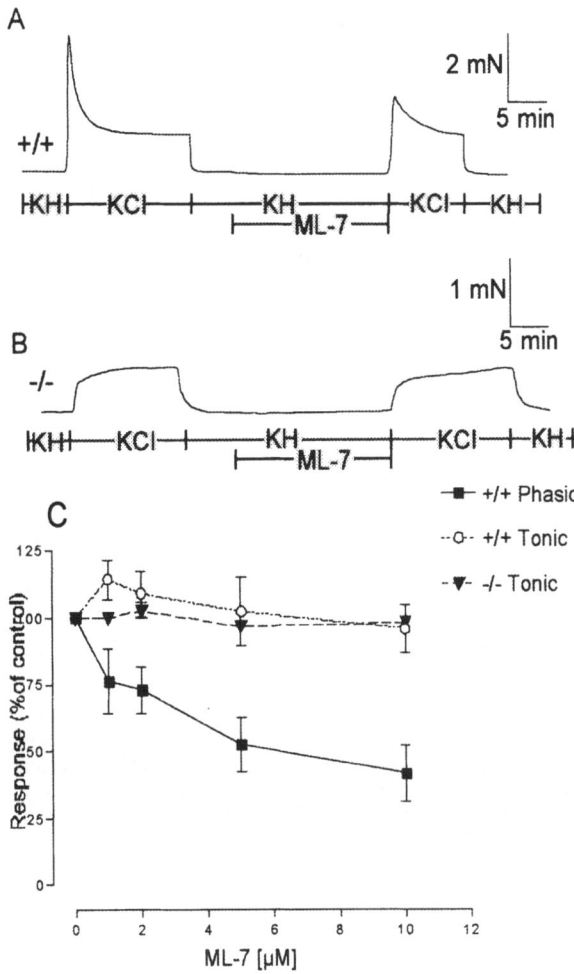

Figure 2. Effect of MLCK inhibition on KCl-induced contraction of wild-type (+/+) and homozygous SM-MyHC deficient (-/-) neonatal murine bladder preparations.

Original records of isometric force induced by KCl depolarization of intact bladder preparations from A, wild-type (+/+) mouse; and B, SM-MyHC deficient (-/-) mice, in the presence of ML-7. The smooth muscle preparations were stimulated with KCl depolarization solution for 15 min, following Krebs-Henseleit solution incubation first without, and then with ML-7 ($10\mu M$, 20 min) and again activated in KCl depolarization solution without ML-7. Incubation with ML-7 inhibited the phasic contraction of bladder preparation from wild-type mouse, but had no effect on the tonic contraction of +/+ and -/- bladder preparations.

C, Intact bladder preparations from +/+ were incubated with increasing concentrations doses of the cell permeable MLCK inhibitor ML-7 for 20 min. The effect on the initial phasic contraction (+/+ phasic) and the subsequent tonic contraction (+/+ tonic) was evaluated. Similar experiments were performed with -/- bladder preparations (-/- tonic). EC_{50} of the inhibitory effect of ML-7 on phasic contraction of +/+ was about $5.03\mu M$. Incubation with ML-7 up to $10\mu M$ had no effect on contraction of -/- preparations and on the tonic phase of +/+. Response is expressed as a percentage of the control KCl-induced isometric force prior ML-7. Each data point represents the mean±SEM of 6 (-/-) and 7 (+/+) bladder preparations. * $p<0.05$; ** $p<0.01$ if phasic contractions of +/+ animals without ML-7 is compared with contractions in the presence of ML-7.

4. Discussion

Electromechanical coupling by KCl depolarization of bladder preparations of wild-type (+/+) mice, i.e. with normal SM-MyHC and NM-MyHC expression, elicited an initial phasic contraction, followed by a tonic contraction. Using a SM-MyHC knock-out mouse model, we observed recently, that KCl depolarization of bladder preparations from homozygous SM-MyHC deficient mice (-/-), i.e. with disrupted SM-MyHC but normal NM-MyHC expression, produced tonic contraction only[9]. Thus, we proposed that phasic contraction and subsequent tonic contraction elicited upon electromechanical coupling is generated by the recruitment of SM-MyHC and NM-MyHC, respectively[9]. A primary mediator of electromechanical coupling is the Ca^{2+}-calmodulin activated myosin light chain kinase (MLCK)[6]. In this study we have shown that inhibition of MLCK by ML-7 abolished phasic contraction but did not affect tonic contraction of wild-type smooth muscle preparations up to 10μM ML-7. These data are in agreement with Murthy et al.[12] who observed inhibition of phasic but not sustained contraction in intestinal smooth muscle upon MLCK inhibition. In addition, tonic contraction of bladder preparations of -/- murine were not affected up to 10μM ML-7. Thus, electromechanical coupling of neonatal bladder smooth muscle comprises a ML-7 sensitive (Ca^{2+}-calmodulin-MLCK/SM-MyHC system) and a ML-7 insensitive (NM-MyHC) pathway.

Both wild-type and knock-out bladder preparations revealed similar intracellular Ca^{2+} transients (Fura measurements) upon KCl depolarization, i.e. an initial Ca^{2+} transient with high amplitude which rapidly declines to almost basal levels[10]. This would explain the rapid inactivation of the Ca^{2+}-calmodulin-MLCK/SM-MyHC system, which requires high Ca^{2+} for recruitment. In contrast, the tonic force generated by the NM-MyHC system, which amounts up to 25% of the initial phasic force in neonatal +/+, could be maintained with resting Ca^{2+} levels. This Ca^{2+}-independent NM-MyHC mechanism of force generation, which becomes recruited upon prolonged electromechanical stimulation and which is ML-7 insensitive, remains to be elucidated.

Pharmacomechanical coupling by the tumor-promoting phorbol esters have been shown to directly activate PKC and to generate tonic contractions in various smooth muscle preparations[12-18]. In fact, the phorbol ester PDBu induced tonic contractions in neonatal bladder preparations of both +/+ and -/- murine. Pre-incubation with the specific protein kinase C inhibitor Ro-32-0432 (5μM) completely abolished the tonic force generated upon stimulation with PDBu in both +/+ and -/- bladder preparations (unpublished observations).

Activation of protein kinase C has been shown to stimulate phosphorylation of the MLC_{20} both *in vitro* and *in vivo*[19-21]. However, the functional importance of PKC-induced MLC_{20} phosphorylation remains unclear. PKC phosphorylation of MLC_{20} was not associated with increase of ATPase activity of both smooth muscle and non-muscle myosin in the *in vitro* studies[20, 22], or with myosin movement in the *Nitella*-based motility assay[23]. In contrast, PKC-induced NM-MyHCB phosphorylation induced an increase in myosin ATPase activity[24]. Another possible pathway for PKC-induced activation of NM-MyHC is through CPI-17-dependent inhibition of myosin light chain phosphatase (MLCP)[25]. *In vitro* and *in situ* studies have shown that, PKC phosphorylates CPI-17, which subsequently inhibits MLCP[26-28], thus sensitizing the contractile apparatus of the smooth muscle cell for Ca^{2+}.

PDBu induced tonic contractions of neonatal bladder preparations from both wild-type and knock-out were significantly weaker if compared with the corresponding contractions elicited by KCl. PKC activation, therefore, may recruit only a part of the NM-MyHC contractile system present in smooth muscle cells.

In conclusion, in this study we could demonstrate, that both SM- and NM-MyHC contractile systems are activated upon electromechanical coupling. However, the SM-MyHC-dependent phasic, but not the NM-MyHC-dependent tonic contraction is ML-7 sensitive. Pharmacomechanical coupling upon protein kinase C activation through stimulation with phorbol ester recruits a fraction of the NM-MyHC but not the SM-MyHC contractile system.

5. Abstract

Electromechanical coupling by KCl depolarization of bladder preparations elicits an initial phasic and subsequent tonic contraction. Using a smooth-muscle myosin heavy chain (SM-MyHC) knock-out mouse model we could previously demonstrate, that phasic and tonic contraction of intact neonatal bladder preparations could be elicited through the recruitment of SM-MyHC and non-muscle myosin heavy chains (NM-MyHC), respectively. Inhibition of myosin light chain kinase (MLCK) by ML-7 eliminated the phasic contraction of wild-type (+/+), rather than tonic contraction of neonatal bladder strips prepared from both +/+ and homozygous SM-MyHC knock-out (-/-) mice. Pharmacomechanical coupling upon PDBu-induced activation of protein kinase C of neonatal bladder preparations elicited tonic contraction of both +/+ and -/- murine. We suggest that: i) electromechanical coupling activates both SM-MyHC and NM-MyHC systems via a ML-7 sensitive and insensitive pathway, respectively. ii) Pharmacomechanical coupling recruits part of the NM-MyHC system rather than SM-MyHC.

6. Acknowledgments

We thank D. Balzereit for excellent technical assistance. This work was supported by CAPES (Ministerio da Educacao, Brazil) BEX0559/99-7 (for VL-Z), European Commission Grant QLGI-CT-1999-51210 (for IM and LGB), and DFG Mo362/17-2 (for IM).

7. References

1.   A. Weiss and L. A. Leinwand, The mammalian myosin heavy chain gene family, *Annu. Rev. Cell Dev. Biol.* 12, 417-439 (1996).
2.   R. Nagai, M. Kuro-o, P. Babij and M. Periasamy, Identification of two types of smooth muscle myosin heavy chain isoforms by cDNA cloning and immunoblot analysis, *J. Biol. Chem.* 264, 9734-9737, (1989)

3.   P. Babij, C. Kelly and M. Periasamy, Characterization of a mammalian smooth
     muscle myosin heavy chain gene: complete nucleotide and protein coding sequence and analysis of
     the 5'end of the gene, *Proc. Natl. Acad. Sci. USA* 88, 10676-10680, (1991)

4.   G. J. Babu, E. Loukianov, T. Loukianova, G.J. Pyne, S. Huke, G. Osol, R.B. Low,
     R. J. Paul and M. Periasamy, Loss off SM-B myosin affects muscle shortening velocity and maximal
     force development, *Nat. Cell Biol.* 3, 1025-1029, (2001)

5.   J. M. Miano, P. Cserjesi, K.L. Ligon, M. Periasamy and E.N. Olson, Smooth
     muscle myosin heavy chain exclusively marks the smooth muscle lineage during mouse
     embryogenesis, *Circ. Res.* 75, 803-812, (1994)

6.   A. P. Somlyo and A.V. Somlyo, Signal transduction and regulation in smooth muscle, *Nature* 372,
     231- 236, (1994)

7.   A. R. Bresnick, Molecular mechanisms of nonmuscle myosin-II regulation, *Curr.*
     Opin. Cell Biol. 11, 26 - 33, (1999)

8.   J. T. Deng, J. E. Van Lierop, C. Sutherland and M. P. Walsh, Ca^{2+}-independent
     smooth muscle contraction. A novel funktion für integrin-linked kinase, *J. Biol. Chem.* 276 (19),
     16365-16373, (2001)

9.   I. Morano, G. X. Chai, L. G. Baltas, V. Lamounier-Zepter, G. Lutsch, M. Kott, H.
     Haase and M. Bader, Smooth-muscle myosin contraction without smooth-
     muscle myosin, *Nat. Cell Biol.* 2, 371-375, (2000)

10. M. Löfgren, E. Ekblad, I. Morano and A. Arner, Non-muscle myosin motor of
     smooth muscle *J. Gen. Physiol.*, accepted for publication (2003)

11. M. Saitoh, T. Ishikawa, S. Matsushima, M. Naka and H. Hidaka, Selective
     inhibition of catalytic activity of smooth muscle myosin light chain kinase, J. Biol. Chem.
     262 (16), 7796-7801, (1987)

12. K. S. Murthy, J. R. Grider, J. F. Kuemmerle and G. M. Makhlouf, Sustained
     muscle contraction induced by agonists, growth factors and Ca^{2+} mediated by distinct PKC isozymes,
     Am. J. Physiol. 279, G201-G210. (2000)

13. M. Yoshida, K. Nishi, J. Machida, H. Sakiyama, K. Ikeda and S. Ueda, Effects of
     phorbol ester on lower urinary tract smooth muscles in rabbits, *Eur. J. Pharmacol.*, 222, 205 - 211,
     (1992)

14. N. R. Danthuluri and R. C. Deth, Phorbol ester-induced contraction of arterial
     smooth muscle and inhibition of α-adrenergic response, *Biochem. Biophys. Res. Commun.* 125 (3),
     1103-1109, (1984)

15. P. H. Howe and A. A. Abdel-Latif, Phorbol ester-induced protein phosphorylation
     and contraction in sphincter smooth muscle of rabbit iris, *FEBS Lett.* 215 (2), 279-284, (1987)

16. H. Rasmussen, J. Forder, I. Kojima and A. Scriabine, TPA-induced contraction of
     isolated rabbit vascular smooth muscle, *Biochem. Biophys. Res. Commun.* 122 (2), 776-784, (1984)

17. Y. H. Lee, I. Kim, R. Laporte, M. P. Walsh and K. G. Morgan, Isozyme-specific
     inhibitors of protein kinase C translocation: effects on contractility of single permeabilized vascular
     smooth muscle cells of the ferret, *J. Physiol.* 517 (3), 709-720, (1999)

18. M. Castagna, Y. Takai, K. Kaibuchi, K. Sano, U. Kikkawa and Y. Nishizuka,
     Direct activation of calcium-activated, phospholipid-dependent protein kinase by tumor-promoting
     phorbol esters, *J. Biol. Chem.* 257 (13), 7847-7851, (1982)

19. M. Naka, M. Nishikawa, R. S. Adelstein and H. Hidaka, Phorbol ester-induced
     activation of human platelets is associated with protein kinase C phosphorylation of myosin light
     chains, *Nature* 306, 490-492, (1983)

20. M. Ikebe and S. Reardon, Phosphorylation of bovine platelet myosin by protein
     kinase C, *Biochem.* 29, 2713-2720, (1990)

21. S. Kawamoto, A. R. Bengur, J. R. Sellers and R.S. Adelstein, In situ
     phosphorylation of human platelet myosin heavy and light chains by protein kinase C, *J. Biol. Chem.*
     264 (4), 2258-2265, (1989)

22. M. Nishikawa, J. R. Sellers, R. S. Adelstein and H. Hidaka, Protein kinase C
     modulates in vitro phosphorylation off the smooth muscle heavy meromyosin by myosin light chain
     kinase, *J. Biol. Chem.* 259 (14), 8808-8814, (1984)

23. S. Umemoto, A. R. Bengur and J. R. Sellers, Effect of multiple phosphorylations
    of smooth muscle and cytoplasmic myosins on movement in an in vitro motility assay, *J. Biol. Chem.*
    264 (3), 1431-1436, (1989)
24. P. de Lanerolle and M. Nishikawa, Regulation of embryonic smooth muscle
    myosin by protein kinase C, *J. Biol. Chem.* 263 (19) 9071-9074. (1988)
25. A. P. Somlyo and A. V. Somlyo, Signal transduction by G-proteins, Rho-kinase
    and protein phosphatase to smooth muscle and non-muscle myosin II.
    J. Physiol. 522 (2), 177-185, (2000)
26. S. Senba, M. Eto and M. Yazawa, Identification of trimeric myosin phosphatase
    (PP1M) as a target for a novel PKC-potentiated protein phosphatase-1 inhibitory protein (CPI17) in
    porcine aorta smooth muscle, *J. Biochem.* 125, 354-362, (1999)
27. M. Eto, T. Ohmori, M. Suzuki, K. Furuya and F. Morita, A novel protein
    phosphatase-1 inhibitory protein potentiated by protein kinase C. Isolation from porcine aorta media
    and characterization, *J. Biochem.* 118 (6) 1104-1107, (1995)
28. L. Li, M. Eto, M. R. Lee, F. Morita, M. Yazawa and T. Kitazawa, Possible
    involvement of the novel CPI-17 protein in protein kinase C signal transduction of rabbit arterial
    smooth muscle, *J. Physiol.* 508 (3), 871-881, (1998)

DISCUSSION

Huxley: What are relative amounts of smooth muscle myosin (SMM) and non-muscle myosin (NMM) in the smooth muscle?

Morano: 60-70% SMM, ~30% NMM.

Pfitzer: During relaxation, there does not seem to be a decline in Ca^{2+}. How does it relax?

Morano: Intracellular calcium concentrations are lower in the relaxed than in the stimulated state.

Ikebe: Is the expression level of MLCK (myosin light chain kinase) in the knockout mice normal?

Morano: Expression levels of MLCK were found to be normal in wild-type and homozygous knock-out animals, as detected by Western-blot analysis.

Sugi: Very interesting results. In one of your early slides, you showed, in the knockout animal specimen, the initial prominent Ca^{2+} peak does not produce any mechanical response. Does this mean that the non-muscle myosin system is (relatively) insensitive to elevated $[Ca^{2+}]i$?

Morano: Yes, the non-muscle myosin system is less Ca^{2+} than the smooth muscle myosin system.

Nishimura: Are the homozygous mice look normal?

Morano: Homozygous knock-out mice were born with a normal body weight and size. However, they look pale and can be distinguished from wild-type or heterozygous mice.

Nishimura: How about heterozygous mice? Are they normal?

Morano: They are normal.

VII. MUSCLE MECHANICS

FORCE RESPONSE TO STRETCHES IN ACTIVATED FROG MUSCLE FIBRES AT LOW TENSION

M. Angela Bagni, Barbara Colombini, Francesco Colomo, Paige Geiger, Rolando Berlinguer Palmini, and Giovanni Cecchi[*]

1. INTRODUCTION

It is well known that tension development in skeletal muscle fibres upon electrical activation is preceded by an increase of muscle stiffness that begins during the latent period and continues throughout the whole tension rise in both twitch or tetanic contractions (Bressler and Clinch, 1974; Cecchi et al., 1982; Ford et al., 1986). At moderate and high tensions, fibre stiffness increase is essentially due to crossbridge attachment. However, as shown previously (Bagni et al., 1994; Bagni et al., 2002), at zero tension during the latent period and at very low tension during force development, a substantial contribution to fibre stiffness cames from some (unknown) sarcomere structure(s), outside the crossbridges, whose stiffness increases upon stimulation. The presence of this stiffness was inferred from the force response to stretches (and hold) applied to a single muscle fibre during force generation. It was found that the fast force transient produced by the stretch was followed by a period during which the tension settled to a consistent level greater than the isometric tension at the time of the stretch, until relaxation or until the fibre returned to the original length at end of the stretch. Because of this characteristic the excess of tension respect to isometric was referred to as static tension while the ratio between static tension and stretch amplitude, measured at sarcomere level, was termed static stiffness. Experiments made on tetanic contractions in Ringer containing 1-6 mM of 2,3-butanedione monoxime (BDM), an agent which strongly inhibits crossbridge formation (Horiuti et al., 1988; Bagni et al., 1992) without altering static stiffness (Bagni et al., 1994), showed that the structure responsible for static stiffness behaves with Hookean elasticity located in parallel with the crossbridges (Bagni et al., 2002). Interestingly, in both twitch or tetanic contractions, static stiffness development followed a characteristic time course distinct from that of tension and roughly similar to that of internal calcium concentration.

[*] M. Angela Bagni, Barbara Colombini, Francesco Colomo, Paige Geiger, Rolando Berlinguer Palmini, Giovanni Cecchi, Dipartimento di Scienze Fisiologiche, Università degli Studi di Firenze, Viale G.B. Morgagni, 63, I-50134, Firenze, Italy.

Molecular and Cellular Aspects of Muscle Contraction
Edited by H. Sugi, Kluwer Academic/Plenum Publishers, 2003

For this reason we speculated that static stiffness increase could be due to a sarcomere structure(s) whose stiffness would increase after the stimulation in a calcium dependent way. The experiments reported in this paper were made to investigate this possibility. We measured the static stiffness in single intact frog muscle fibres under various conditions in which isometric tension was inhibited either by reducing the calcium release by the sarcoplasmatic reticulum or directly through the inhibition on the actomyosin interaction. In addition to the effects of BDM, which confirmed previous measurements, we analyzed the static stiffness in presence of Dantrolene (DAN), Deuterium Oxide (D_2O), Methoxyerapamil (D600) and hypertonic solutions. All these agents inhibit twitch tension generation but they have a different action mechanism: Dantrolene (Takauji et al., 1975; Desmedt and Hainaut, 1977; Morgan and Bryant, 1977; Helland et al., 1988), D_2O (Allen et al., 1984; Sato and Fujino, 1987) and D600 (Eisenberg et al., 1983; Morgan et al., 1997) all depress force development mainly by reducing the calcium release, while BDM (Horiuti et al., 1988; Bagni et al., 1992) and hypertonic solution (Parker and Zhu, 1987) inhibit tension generation mainly by affecting actomyosin interaction so as to reduce crossbridge formation with little or no effect on calcium release. It was therefore possible to compare static stiffness in fibres in which similar degrees of force inhibition were obtained with and without inhibition of calcium release. This comparison allowed isolation of the effects of intracellular calcium on static stiffness.

The results confirm that static stiffness is almost unaffected by BDM even at concentrations (2.5 mM) that strongly reduced twitch tension. The same effect was obtained by bathing the fibre with hypertonic solutions. On the contrary, tension depression was accompanied by a strong static stiffness reduction when D_2O, Dantrolene or D600 were used to inhibit tension generation. This findings show that static stiffness is modulated by intracellular calcium concentration.

2. METHODS

Frogs (*Rana esculenta*) were killed by decapitation followed by destruction of the spinal cord. Single intact fibres, dissected from the tibialis anterior muscle, were mounted by means of aluminum foil clips (Ford et al., 1977) between the lever arms of a force transducer and an electromagnetic motor in a thermostatically controlled chamber provided with a glass floor for ordinary and laser light illumination. The temperature was maintained constant at 14°C (± 0.2°C). Single stimuli of alternate polarity, 0.5 ms duration and 1.5 times threshold strength, were applied transversely to the muscle fibre by means of platinum-plate electrodes. Tension was measured by means of a capacitance force transducer (natural frequency between 40 and 60 kHz) similar to that previously described (Huxley and Lombardi, 1980). Sarcomere length changes were measured using a striation follower device (Huxley et al., 1981) in a fibre segment (1.2-2.5 mm long) selected for striation uniformity in a region as close as possible to the force transducer. This eliminated the effects of tendon compliance on stiffness measurements allowing to attribute the results directly to the sarcomere structure. Resting sarcomere length was set at about 2.1 μm.

After a test of fibre viability and a measure of the isometric tetanic tension (P_0), the experiments were made on twitch responses evoked in Ringer solution and in a series of test solutions containing one of the following agents: 1) 2,3-butanedione monoxime

(BDM) at concentration of 2.5 mM; 2) Dantrolene Sodium at concentration of 6.25 μM and 3) D600 at 20 μM concentration. Experiments were also made in Deuterium Oxide Ringer (98% of water substituted with D_2O) and hypertonic solution at 1.4 normal tonicity (T) obtained by adding 50 mM of NaCl to the normal Ringer. Experiments in D_2O were made after waiting for the equilibration time (about 20 min) to allow a complete exchange of D_2O for water in the fibre. The responses in D600 were obtained in normal Ringer during the force recovery from the paralysis induced by exposing the fibre, loaded with D600, to high potassium solution, as described previously (Eisenberg et al., 1983; Morgan et al., 1997). Experiments were made on twitch contraction mainly because the relatively great and fast stretches necessary to measure the static stiffness, would quickly damage fibres developing the full tetanic tension.

As judged by light microscopy observation and by the sarcomere length signals from the striation follower, activated fibres, in all test solutions did not develop any particular sarcomere non-homogeneity upon stretching. The fibres survived after hours of experiments with stretches and fully recovered the isometric twitch tension when returned to normal Ringer, with the exception of the recovery after the exposure to D600 which was not always complete (Morgan et al., 1997). Resting fibre length, fibre cross-sectional area and resting sarcomere length (l_0) were measured under ordinary light illumination using a 10x or 40x dry objective and 25x eyepieces. The normal Ringer solution had the following composition (mM): 115 NaCl; 2.5 KCl; 1.8 $CaCl_2$; 3 phosphate buffer at pH 7.1. BDM-Ringer, Dantrolene-Ringer and D600-Ringer were obtained by adding the appropriate amount to the normal Ringer solution. Force, fibre length and sarcomere length signals were measured with a digital oscilloscope (4094 Nicolet, USA), stored on floppy disks and transferred to a personal computer for further analysis.

3. STATIC STIFFNESS MEASUREMENTS

Static stiffness was measured by applying ramp shaped stretches (amplitude 20-40 $nm.hs^{-1}$ and 0.6-0.7 ms duration) to one fibre end and measuring the force response at the other end. The short stretch duration, which resulted in a very high stretching velocity (up to 70×10^3 $nm.hs^{-1}s^{-1}$), was chosen to reduce as much as possible crossbridge cycling during the stretch itself. As shown in Fig. 1 three records were taken for each measurements: (1) isometric, (2) isometric with stretch and (3) passive response to the stretch. The isometric and the passive responses were subtracted from the isometric with stretch response to obtain the subtracted trace on which measurements were made. By subtracting the isometric record we were able to measure the static tension always on a flat baseline even when the stretch was applied on tension rise or relaxation. By subtracting the passive response we corrected for the resting tension and stiffness. This correction was, however, very small in all the experiments reported here.

As shown in Fig. 1, the static tension represents the steady force which follows to the fast force transient elicited by the stretch. The ratio between the static tension and the sarcomere elongation represents the static stiffness of the sarcomere. To describe the time course of the static stiffness development following the activation, stretches were applied in fibres at rest and at different times after the stimulus.

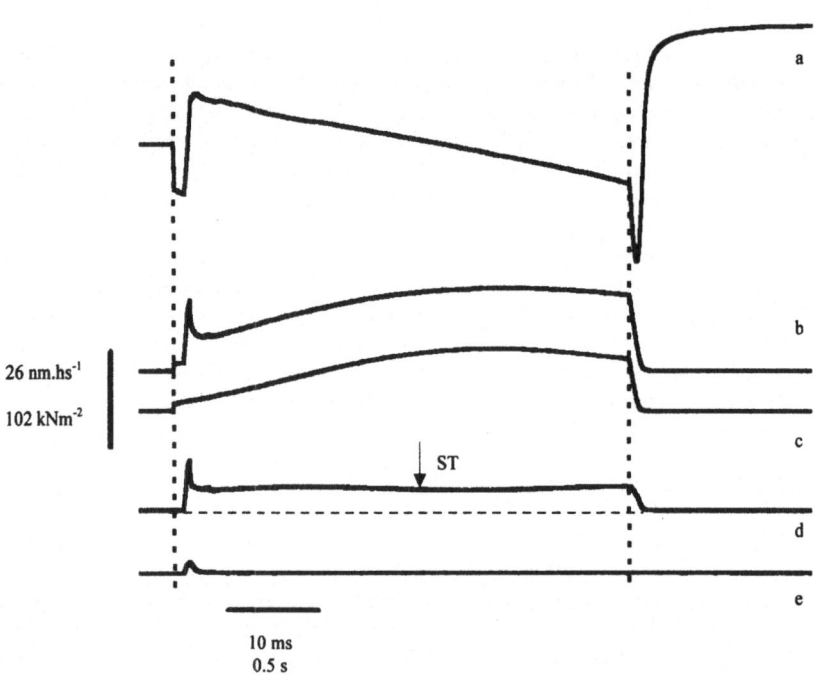

26 nm.hs^{-1}

102 kNm^{-2}

ST

10 ms
0.5 s

Figure 1. Force response to a stretch applied on the rise of a twitch. Traces: a) sarcomere length with stretch; b) isometric tension with stretch; c) isometric tension; d) subtracted trace and e) passive response to the stretch. The static tension (ST) is measured on the subtracted trace at the end of the fast tension transient as indicated by the arrow. Vertical dashed line indicates the switch from slow (1 ms/point) to fast (20 µs/point) time base. The stretch was applied 14 ms after the stimulus when force was on the rising phase of the twitch at value of 0.14 times the peak twitch force. Stretch amplitude, 30 nm.hs^{-1}; stretch duration, 720 µs.

4. RESULTS

Fig. 1 shows the force response to a fast stretch applied on the rise of twitch tension and illustrates the static tension and stiffness measurements. The subtracted trace shows that, after the fast transient, the force settles to a steady level, representing the static tension, which corresponds to 0.13 of P_0. The static stiffness obtained by dividing the static tension by the sarcomere elongation (30 nm.hs^{-1}) resulted in 0.0042 P_0/nm.hs^{-1}. Knowing that at 14 °C the total stiffness of a fully activated fibre at tetanus plateau (S_0) is about 0.2 P_0/nm.hs^{-1} (Bagni et al., 1999), it is clear that the static stiffness makes a very small contribution (2%) to the total stiffness of the fully activated fibre.

However, the static stiffness represents the whole of the stiffness increase during the latent period, and contributes substantially to the total stiffness at low tensions at the beginning of the contraction when only a few crosssbridges are formed. Previous experiments (Bagni et al., 2002) showed that static stiffness is independent of the stretch amplitude and velocity being similar to the response expected from an elastic Hookean element.

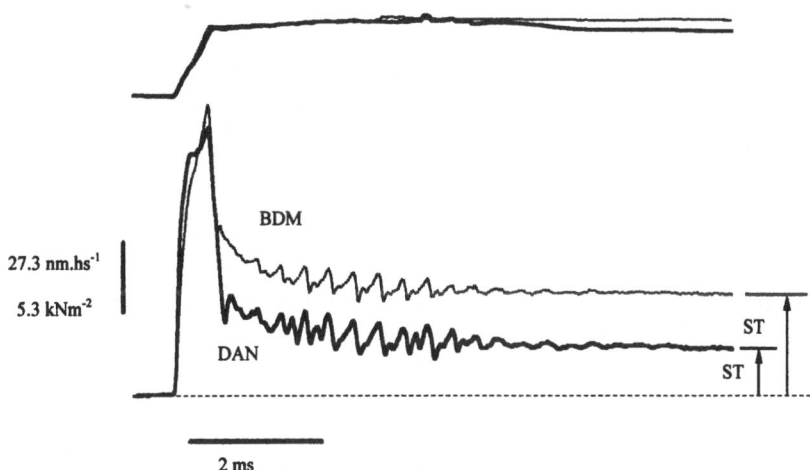

27.3 nm.hs^{-1}

5.3 kNm^{-2}

BDM

DAN

ST

ST

2 ms

Figure 2. Comparison of force response in BDM and Dantrolene. Upper traces: sarcomere length, bottom traces: force response. Subtracted traces of force response to the same stretch of about 27 nm.hs^{-1} amplitude and 680 μs duration applied on the rising phase of twitch contraction, 18 ms after the stimulus in a fibre bathed in BDM-Ringer (2.5 mM) and Dantrolene-Ringer (6.25 μM). The isometric tension in Dantrolene (about 0.25 the peak twitch force in normal Ringer) was 1.30 times greater than that in BDM-Ringer (0.19 twitch tension in normal Ringer), but the static tension in BDM was about twice that in Dantrolene-Ringer. The two force traces have been superimposed to show clearly the differences in static tension.

As reported previously (Bagni et al., 1994) with respect to normal Ringer, BDM strongly reduces the twitch tension but has almost no effect on static tension. Fig. 2 shows the force response to a stretch in a fibre in which a twitch tension inhibition of about 75% was obtained using Dantrolene, compared with the force response in BDM in which the tension inhibition was 80%. It can be seen that the static tension in Dantrolene is about 50% smaller than in BDM-Ringer, in spite of the tension being 30% higher at the time of

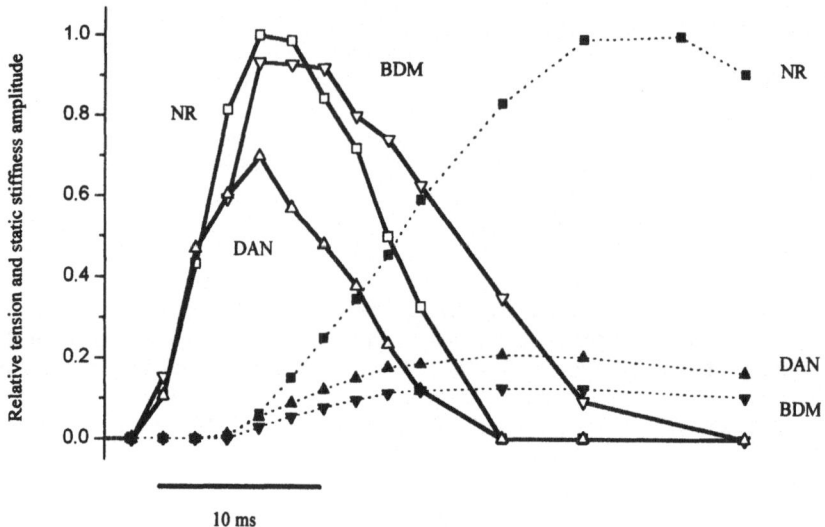

Figure 3. Effects of BDM and Dantrolene on twitch tension and static stiffness time course. Comparison of twitch tension and static stiffness time course in normal Ringer, BDM (2.5 mM) and Dantrolene (6.25 μM). Continuous lines and open symbols represent static stiffness; dashed lines and filled symbols represent tension. Squares: normal Ringer, down triangles: BDM and up triangles: Dantrolene. Tension was greatly reduced by both BDM and Dantrolene, but static stiffness was inhibited only by Dantrolene. Note the complete independence of static stiffness time course from twitch tension: static stiffness peaks at 10-12 ms after the stimulus, when tension is very low.

the stretch. Thus, in contrast to BDM, Dantrolene reduces both active and static tension. The different action of BDM and Dantrolene is also shown in Fig. 3 which reports the whole time course of twitch tension and static stiffness occurring after the stimulation.

Both Dantrolene and BDM reduce tension development but to a different extent. In agreement with our previous data in normal Ringer solution, static stiffness reached the peak 10-12 ms after the stimulus when the twitch tension was still very low (about 0.05 the twitch peak value) and fell to zero before the twitch peak. In BDM and Dantrolene-Ringer solution the static stiffness also reaches the peak value well before the tension, however while in BDM the stiffness values are almost the same as that in normal Ringer solution, with Dantrolene the static stiffness values are very reduced.

In the experiments performed in D_2O-Ringer there was a static stiffness reduction similar to that in Dantrolene-Ringer and the twitch tension was reduced to less than 10% respect to normal Ringer.

The time course of twitch tension and static stiffness in D600-Ringer is compared to that in normal and in BDM-Ringer in Fig. 4. D600 does not alter tension and static stiffness time course, but the values are strongly reduced, respect to normal and BDM-Ringer static stiffness peak value is less than 50% the control.

Fig. 5 shows the complete time course of twitch tension and static stiffness in normal Ringer and in hypertonic solution (1.4 T). Similar to BDM-Ringer, hypertonicity reduces tension without altering the static stiffness.

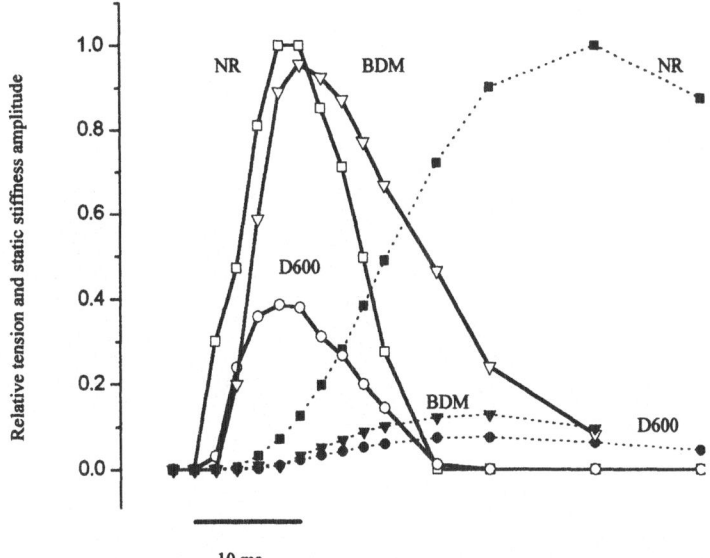

Figure 4. Comparison of twitch tension and static stiffness time course in BDM and D600-Ringer. Lines and symbols as in Fig. 3. Circles: D600 (20 µM). Similar to Dantrolene and unlike BDM, D600 strongly reduces the static stiffness.

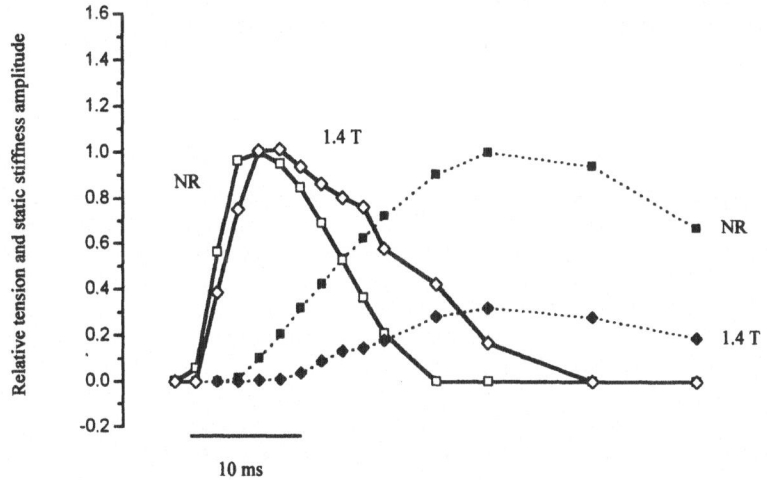

Figure 5. Effects of hypertonicity on twitch tension and static stiffness time course. Lines and symbols as in Fig. 3, diamonds represents hypertonic Ringer (1.4 T). Likewise to BDM, hypertonicity greatly affects twitch tension, but has only slight effects on static stiffness.

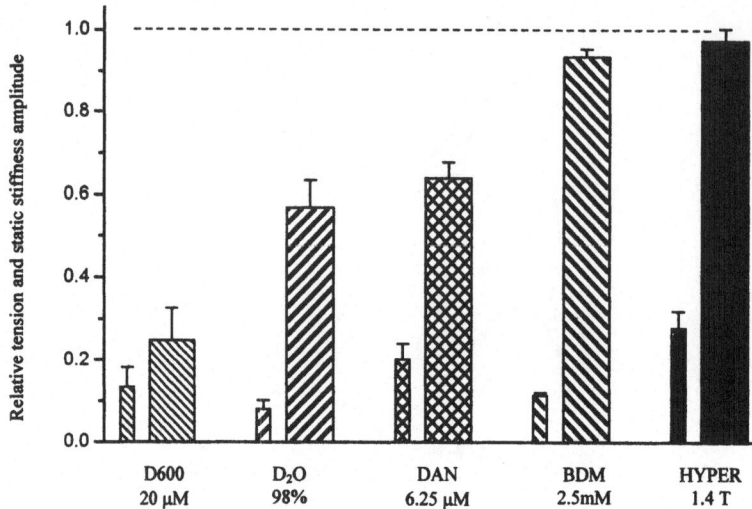

Figure 6. Mean (± S.E.M.) tension and stiffness values in presence of various tension-inhibiting agents expressed relative to normal Ringer values. Narrow columns represent tension, wide columns static stiffness. D600, D_2O and Dantrolene inhibit both tension and static stiffness, while BDM and hypertonicity inhibit tension but not static stiffness. n=4 for D600, Dantrolene and BDM; n=5 for D_2O and hypertonicity experiments.

Fig. 6 summarizes the effects of all the agents tested in all the experiments performed. It is clear that the inhibitory effect of D600, Dantrolene and D_2O on twitch tension is accompanied by the inhibition of static stiffness while BDM and hypertonic solutions inhibit tension but have almost no effect on static stiffness.

5. DISCUSSION

Previous data have shown that a small fraction of the sarcomere stiffness increase that follows activation is not attributable to crossbridge formation but to an unknown sarcomeric structure whose stiffness (static stiffness) increases upon activation following a time course well distinct from tension development (Bagni et al., 1994; Bagni et al., 2002). In order to test the possibility that static stiffness could be modulated by calcium, we have measured the effects of several tension inhibitors acting through a calcium release inhibition or through a direct inhibition on crossbridges. Our data show that the agents tested, which all inhibit twitch tension, can be separated into two groups as regarding the effect on static stiffness. The first group, including BDM and hypertonic solutions, has very little or no effect on static stiffness while the second group, including Dantrolene, D_2O and D600, inhibits static stiffness in parallel with tension.

As reported in the literature, the main effect of BDM on frog skeletal muscle at the concentrations used here, is a direct inhibition of actomyosin interaction which reduces the number of attached crossbridges (Horiuti et al., 1988; Bagni et al., 1992). BDM however does not significantly affect calcium release (Horiuti et al., 1988).

Hypertonic solutions which, for the tonicity used here (1.4 T) have an effect similar to BDM, altering crossbridge formation but not calcium release (Parker and Zhu, 1987).

The other group of agents has a different action mechanism: Dantrolene (Takauji et al., 1975; Desmedt and Hainaut, 1977; Morgan and Bryant, 1977; Helland et al., 1988), D_2O (Allen et al., 1984; Sato and Fujino, 1987) and D600 (Eisenberg et al., 1983; Morgan et al., 1997) all inhibit tension generation mainly by reducing calcium release. Moreover D_2O has an additional effect on crossbridge formation or kinetics (Cecchi et al., 1981).

Both series of agents decreases substantially twitch tension, however only those exerting their action through an inhibition of calcium release are able to reduce at the same time the static stiffness.

This effect is consistent with the idea that static stiffness changes after the stimulation are caused by the internal calcium concentration changes. The results reported here do not give further information about the structure responsible for the static stiffness, however they show the existence of a non crossbridge mechanism, by which calcium increases the stiffness of the sarcomere. It is possible, that static stiffness increase is due to a calcium dependent titin-actin interaction as it seems suggested by motility assays experiments (Kellermeyer and Granzier, 1996), or to a calcium dependent change in titin elasticity as suggested by other experiments (Tatsumi et al., 2001), but other possible sources cannot be excluded. However our findings, especially if the increase of static stiffness could be attributable to titin, should be important to understanding how the stability of the sarcomere structure is maintained at very low tension levels at the beginning of contraction.

6. REFERENCES

Allen, D.G., Blinks, J.R., and Godt, R.E., 1984, Influence of deuterium oxide on calcium transients and myofibrillar responses of frog skeletal muscle, *J. Physiol.* **354**:225-51.

Bagni, M.A., Cecchi, G., Colombini, B. and Colomo, F., 1999, Sarcomere tension-stiffness relation during the tetanus rise in single frog muscle fibres. *J. Muscle Res. Cell Motil.* **20**:469-476.

Bagni, M.A., Cecchi, G., Colombini, B., and Colomo, F., 2002, A non-cross-bridge stiffness in activated frog muscle fibres, *Biophys. J.* **82**(6):3118-27.

Bagni, M.A., Cecchi, G., Colomo, F., Garzella, P., 1992, Effects of 2,3-butanedione monoxime on the crossbridge kinetics in frog single muscle fibres, *J. Muscle Res. Cell Motil.* **13**(5):516-22.

Bagni, M.A., Cecchi, G., Colomo, F., and Garzella, P., 1994, Development of stiffness precedes cross-bridge attachment during the early tension rise in single frog muscle fibres, *J. Physiol.* **481**.2:273-278.

Bressler, B.H., and Clinch, N.F., 1974, The compliance of contracting skeletal muscle, *J. Physiol.* **237**:477-493.

Cecchi, G., Colomo, F., and Lombardi, V., 1981, Force-velocity relation in deuterium oxide-treated frog single muscle fibres during the rise of tension in an isometric tetanus, *J. Physiol.* **317**:207-21.

Cecchi, G., Griffiths, P.J., and Taylor, S., 1982, Muscular contraction: kinetics of crossbridge attachment studied by high-frequency stiffness measurements, *Science* **217**:70-72.

Desmedt, J.E., and Hainaut, K., 1977, Inhibition of the intracellular release of calcium by Dantrolene in barnacle giant muscle fibres, *J. Physiol.* **265**(2):565-85

Eisenberg, R.S., McCarthy, R.T., and Milton, R.L., 1983, Paralysis of frog skeletal muscle fibres by the calcium antagonist D-600, *J. Physiol.* **341**:495-505.

Ford, L.E., Huxley, A.F., and Simmons, R.M., 1977, Tension responses to sudden length changes in stimulated frog muscle fibres near slack length, *J. Physiol.* **269**:441-515.

Ford, L.E., Huxley, A.F., and Simmons, R.M., 1986, Tension transients during the rise of tetanic tension frog muscle fibres, *J. Physiol.* **372**:595-609.

Helland, L.A., Lopez, J.R., Taylor, S.R., Trube, G., and Wanek, L.A., 1988, Effects of calcium "antagonists" on vertebrate skeletal muscle cells, *Ann. N. Y. Acad. Sci.* **522**:259-6&.

Horiuti, K., Higuchi, H., Umazume, Y., Konishi, M., Okazaki, O., and Kitohara, S., 1988, Mechanism of action of 2,3-butanedione 2-monoxime on contraction of frog skeletal muscle, *J. Muscle Res. Cell Motil.* **9**:156-164.

Huxley, A.F., and Lombardi, V., 1980, A sensitive force transducer with resonance frequency 50 kHz, *J. Physiol.* **305**:15P-16P.

Huxley, A.F., Lombardi, V. and Peachey, L.D., 1981, A system for fast recording of longitudinal displacement of a striated muscle fibre, *J. Physiol.* **317**:12P-13P.

Kellermayer, M.S.Z., and Granzier, H.L., 1996, Calcium-dependent inhibition of in vitro thin-filament motility by native titin, *FEBS Lett.* **380**:281-286.

Morgan, K.G., and Bryant, S.H., 1977, The mechanism of action of dantrolene sodium, *J. Pharmacol. Exp. Ther.* **201**(1):138-47.

Morgan, D.L., Claflin, D.R., and Julian, F.J., 1997, The relationship between tension and slowly varying intracellular calcium concentration in intact frog skeletal muscle, *J. Physiol.* **500**.1:177-92.

Parker, I., and Zhu, P.H., 1987, Effects of hypertonic solutions on calcium transients in frog twitch muscle fibres, *J. Physiol.* **383**:615-27.

Sato, Y., and Fujino, M., 1987, Inhibition of arsenazo III Ca transient with deuterium oxide in frog twitch fibres at a resting sarcomere length, *Jpn. J. Physiol.* **37**(1):149-53.

Takauji, M., Takahashi, N., Nagai, T., 1975, Effect of dantrolene sodium on excitation-contraction coupling in frog skeletal muscle, *Jpn. J. Physiol.* **25**(6):747-58.

Tatsumi, R., Maeda, K., Hattori, A., and Takahashi, K., 2001, Calcium binding to an elastic portion of connectin/titin filaments, *J. Muscle Res. Cell Motil.* **22**(2):149-62.

DISCUSSION

Pollack: Divalent cations such as calcium cause condensation of negatively charged polymers. Conceivably, any of the sarcomeric filaments could be responsible for the static stiffness. It would be interesting to study the effect of calcium on stiffness of each of the sarcomeric filaments.

Bagni: Yes, I agree with you. It could be interesting.

Huxley: What is the relationship between the static stiffness and sarcomere length?

Bagni: We studied static stiffness in relation to sarcomere length. Lengthening of sarcomere from 2.1 μm to 2.8 μm greatly increases the static stiffness, but it for greater sarcomere length.

Gonzalez-Serratos: Have you consider the role of the costamere which are considered to transmit lateral force through the sarcolemma to the tendons?

Bagni: We have done some experiments on skinned fibers and the static stiffness is present.

Sugi: Your static stiffness that is independent of stretch amplitude remind me of the filamentary resting tension (FRT) first described by D. K. Hill. According to him, FRT exhibits the nature of viscous-like (or frictional) resistance. What has just occurred to me is the possibility that your stiffness may originate from the nebulin-titin network. As far as I remember, the position of the N-line adjacent to the Z-line shifts depending on stretch, and the N-line is now shown to be nebulin network.

Bagni: Yes, I know, you are speaking about the short-range elasticity of D.K.Hill, but our experiments were made in different conditions because the stretches we used were extremely fast. However we cannot exclude that our response would originate from the nebulin-titin network.

MOLECULAR STEP(S) OF FORCE GENERATION:

Temperature-Perturbation Experiments on Muscle Fibres

K.W. Ranatunga and M.E. Coupland[*]

1. ABSTRACT

The steady active muscle force is reduced, but the force generation induced by a standard temperature jump becomes 2-3 fold faster with increased inorganic phosphate level, $[P_i]$. The increase in the rate of force generation also exhibits saturation at higher $[P_i]$ levels and the relation is hyperbolic. These observations are consistent with a kinetic scheme where rapid P_i release by actomyosin crossbridges in muscle is preceded by the force generation step. Such a scheme accounts for the sigmoidal temperature dependence of steady active force and its sensitivity to $[P_i]$. The $[P_i]$ dependence of force recovery after stretch (positive strain) is also hyperbolic, suggesting that the "pre P_i–release force generation step" is strain-sensitive – as expected. However, length-release (negative strain) force transients are not $[P_i]$ sensitive indicating an asymmetry, but its significance and also the kinetic step underlying force recovery from negative strain remain unclear.

2. INTRODUCTION

2.1. General Background

The basic event of muscle contraction is an ATP-driven, cyclic interaction of crossbridges (myosin heads) between thick (M=myosin) and thin (A=actin) filaments in a sarcomere (Huxley, 1957; Huxley & Simmons, 1971); muscle force is generated in crossbridges during their transient attachment(s) to a thin filament. Interpretation of the steady state force (= tension) and contraction measurements from muscle is difficult because of the asynchrony of the cycling crossbridges during active muscle contraction and rapid perturbation methods have been used to induce crossbridge synchronisation. As

---

[*] Muscle Contraction Group, Department of Physiology, School of Medical Sciences, University of Bristol, Bristol BS8 1TD, UK

Molecular and Cellular Aspects of Muscle Contraction
Edited by H. Sugi, Kluwer Academic/Plenum Publishers, 2003

originally proposed by Huxley & Simmons (1971), muscle force during isometric contraction may be maintained by an equilibrium between two crossbridge states (low and high force states), so that sudden shortening (negative strain) that reduces the force will increase the forward rate constant generating force (quick-recovery) (see Huxley & Tideswell, 1996). Thus, the crossbridge force-generating step can bé directly studied by analysing the quick force recovery in the transients induced by a rapid length-release. Following the definitive description and analyses made by Ford et al, (1977) in intact frog muscle fibre, there has been significant advancement in our knowledge regarding the structural changes that accompany this quick force recovery (see Irving et al, 1992; 1995; Lombardi et al, 1992; Piazzesi et al, 2002). However, the identity of the molecular (= kinetic) step of crossbridge force generation in muscle remains unresolved.

A number of other perturbation techniques also have been used in attempts to elucidate the underlying mechanism(s) of crossbridge force generation in muscle fibres, particularly with respect to identifying the molecular step. These include release of hydrostatic pressure (P-jump, Fortune et al, 1991), sinusoidal length perturbation (Kawai & Halvorson, 1991) and rapid changes in the concentration of inorganic phosphate (P_i-jump, Miller & Homsher 1991; Dantzig et al, 1992). These particular studies led to the thesis that force generation and P_i - release are closely coupled and that force generation occurs in a kinetic step prior to the release of P_i during ATP hydrolysis by cycling crossbridges. Several studies by Kawai and colleagues (see Kawai et al, 1993; Wang & Kawai, 1997, 2001; Zhao & Kawai, 1994) on different fibre preparations and conditions and the myofibril experiments of Tesi et al (2000, 2002) have established the general validity of this kinetic scheme. However, the exact correlation between the findings from these studies and from length-perturbation experiments referred to above is not clear.

2.2. Background to Temperature-jump Experiments

Temperature perturbation (T-jump) is an obvious technique to examine the mechanism of muscle force generation, and / or its maintenance, since the active force and shortening velocity in (mammalian) muscle are very sensitive to temperature (Ranatunga, 1982). Indeed, it has been known for over half-century that maximal active force increases ~2-fold in warming from 10 °C to more physiological temperatures (>30 °C) in a range of preparations (see refs in Ranatunga & Wylie, 1983; also Ranatunga, 1994). In studies by a small number of groups, a rapid T-jump (~0.2 ms) was used to examine crossbridge kinetics in maximally Ca-activated (skinned) muscle fibres.

A T-jump induces a force rise with a characteristic time course in active muscle fibres and a component of it represents endothermic force generation (i.e. force rise occurs when heat is absorbed) in attached crossbridges (Davis & Harrington 1987b; Goldman et al, 1987; Bershitsky & Tsaturyan, 1992). However, the identity of the T-jump induced force generation and its correlation to findings from other studies remained unclear. Firstly, Davis & Rogers (1995) provided some evidence that T-jump force generation is a *de novo* process that is only indirectly coupled to P_i-release and is an isomerization of Acto-Myosin complexes after Pi-release (i.e. AM.ADP states). Secondly, the force-recovery after length release was much faster than the T-jump (Bershitsky & Tsaturyan, 2002; see also Ranatunga, 1996) or P-jump (Fortune et al, 1991; see also Vawda et al, 1999) induced force generation. The general suggestion was therefore made that the quick force recovery after length-release and force rise after T-jump (and P-jump) represent different kinetic steps. The findings from X-ray diffraction studies

(Bershitsky et al, 1997) and also the different temperature sensitivities (see Ranatunga, 1999b) were in keeping with this suggestion. Thirdly, the time course of T-jump force generation was characteristically different in fast and slow skeletal muscle fibres (and in cardiac muscle fibres, Ranatunga, 1999b) indicating some coupling with acto-myosin ATPase pathway. In a subsequent study we found that the rate of T-jump force generation is sensitive to [P_i] and that the findings could be explained if (T-jump) crossbridge force generation is closely coupled to P_i-release (Ranatunga, 1999a).

The purpose of this review is to summarise the findings from the above studies with emphasis on our temperature perturbation experiments. Reference will also be made to length perturbation experiments in mammalian muscle fibres which show that, whereas the force recovery after stretch is P_i-sensitive, that after length-release is not sensitive to added P_i (see Ranatunga et al, 2002). It will be concluded that crossbridge force generation occurs in a kinetic step prior to the release of P_i in the cycle (i.e. an isomerization of an AM.ADP.Pi complex) and it is perturbed by T-jump (and P-jump). The temperature dependence of steady active force is largely due to the temperature-effect on the force generation step in muscle (see Zhao & Kawai, 1994; Davis, 1998). As expected, this step is strain-sensitive but the strain-sensitivity is seen only with stretches (positive strain). The quick force recovery after length release, evidently, does not perturb the same step and remains to be further investigated.

3. MATERIALS AND METHODS

3.1. Mechanical Recording System

The trough system, mounted on an optical microscope stage, consisted of three ~50 _l troughs milled in a titanium block and a front experimental trough with glass windows in the front and bottom (Ranatunga, 1996). The temperature of the trough system was kept <10 °C by passing a cold antifreeze-water mixture through a jacket in the assembly whereas the front trough temperature could be independently clamped by the Peltier modules at a suitable temperature. The solution temperature in the front trough was monitored by the thermistor (for feedback) as well as by a separate small thermocouple (20-100 μm diameter, in different experiments) placed very close to the muscle fibre.

The force transducer consisted of two AE 801 elements (Akers, Norway) housed within a small brass box: one element was connected to the muscle fibre for force recording and the other acted as a dummy, forming a full bridge in order to reduce the temperature sensitivity (Ranatunga, 1999a). The force-recording silicon beam was cut to half its length so as to improve its dynamic characteristics: the natural resonant frequency was 14 kHz. The motor was built using a small permanent magnet (25 mm outer diameter) taken from a loudspeaker; the moving coil was wound round an aluminium-foil former; it was held by plastic hinges and its axial movement was monitored photo-electrically. The motor was capable of producing ramp stretches of up to 60 μm in 200 μs. The transducer hooks were made of 50-100 μm diameter Invar wires (an alloy of steel and Nickel; gift from Goodfellow, Cambridge, U.K.), used because of their low thermal coefficient of expansion (see Ranatunga, 1996).

3.2. Temperature-jump Technique

We used a near infra-red (_ = 1.32 _m) pulse of 200 _s duration (maximum power = 2 J per pulse) from a Nd-YAG laser (Schwartz Electro-Optics Inc., Florida, U.S.A.) to induce a T-jump in the front trough of the assembly described above. The energy absorption by water at this wavelength (~50 %) was such that the laser pulse that entered the trough through the front window was reflected back by the aluminium foil and raised the temperature of the 50 _l aqueous medium in the trough, and the fibre immersed in it, by typically not more than 5-6 degrees in 200 _s. As measured by a small thermocouple the raised temperature in the trough solution remained constant for ~500 ms, but the duration could be increased by using the Peltier T-clamping (see Ranatunga, 1996 for other specific details). '

3.3. Experimental Preparations, Procedures and Protocols

Bundles of fibres were obtained from the psoas muscles of adult male rabbits that were killed by an intravenous injection of an overdose of sodium pentobarbitone. The bundles were chemically skinned using 0.5 % Brij 58. A segment of a single fibre (2-4 mm in length) was mounted (using nitro-cellulose glue) between two hooks, one attached to the force transducer and the other to the motor. In length perturbation experiments, the glued fibre ends were then fixed with 8% glutaraldehyde (containing 5% Toluidine Blue), by adopting a procedure essentially similar to that described by Hilber & Galler (1998) (see Ranatunga et al, 2002). The sarcomere length change in a 0.5 mm region of the fibre near the tension transducer was monitored using He-Ne laser diffraction; the position of the first order diffraction was monitored by a diffractometer (see Ranatunga, 2001). The sarcomere length in steady activation was ~2.5 μm.

The buffer solutions contained 6-7 mM Mg-acetate, 5.5 mM ATP, 12.5 mM creatine phosphate (and 1-2 mg/ml of creatine kinase), 15-mM EGTA (relaxing solution) or Ca-EGTA (activating solution) or HDTA (pre-activating solution), 10 mM glutathione, 10 mM glycerol-2-phosphate (as a temperature-insensitive pH buffer, see Goldman et al., 1987) and ~50 mM K-acetate. K-acetate was replaced with K_2HPO_4 in P_i added solutions (i = 200 mM; pH =7.1). The solution compositions were calculated by using a computer program for solving multi-equilibria, maintaining ionic strength and - in activating solutions - the desired free Ca^{2+} concentration (~0.032 mM). Solutions also contained 4 % Dextran (mol. wt. ~ 500 kDa) to compress the filament lattice spacing in the fibres to normal dimensions (Maughan & Godt, 1979; Matsubara et al, 1985; Xu et al 1993). To determine $[P_i]$-dependence, the force transients to a standard perturbation (T-jump or stretch or length release) were examined after a fibre was activated to steady state in the presence of different concentrations of added P_i. A fibre was transferred from relaxing solution, through pre-activating to activating solution, all solutions containing 3.12, 6.25, 12.5 or 25 mM added P_i or no added P_i (control).

3.4. Intact Fibre Experiments

Intact fibre experiments referred to here were performed on small bundles of fibres isolated from the flexor hallucis brevis (FHB) muscle of adult male rats (240 - 325 g body mass). Animals were killed with an intra-peritoneal injection of an overdose (~150 mg kg^{-1} body weight) of Sodium Pentobarbitone (Euthatal, Rhône Mérieux). Whole muscle was removed from the rat foot and small bundles of ~ 5-10 intact excitable fibres were dissected from the mid-belly of the muscle under dark-field illumination. Small aluminium foil T-clips were fixed on the tendons within 0.2 mm of the fibre-ends; resting fibre length was ~2 mm and the major bundle width was ~200 _m.

A preparation was set up in a 2 ml, flow through, stainless steel chamber for isometric force recording. It was super-fused with physiological saline solution containing (mM) NaCl, 109; KCl, 5; $MgCl_2$, 1; $CaCl_2$, 4; $NaHCO_3$, 24; $NaH_2 PO_4$, 1; sodium pyruvate, 10 and 200 mg l^{-1} of bovine foetal serum. The solution was bubbled continuously with a mixture of 95 % O_2 and 5 % CO_2. The solution temperature (range 2-35 °C) was changed / controlled by means of a Peltier device and monitored with a thermocouple in the muscle trough. A bundle was directly stimulated with supra-maximal voltage pulses (<0.5 ms duration) applied to two platinum plate electrodes placed on either side and appropriate frequencies and train durations were used for recording fused tetanic contractions at different temperatures (see Ranatunga, 1982).

3.5. Some General Considerations

i). The T-jump technique we use is basically similar to that first described by Davis & Harrington (1987a,b) and by Goldman et al (1987). A T-jump is induced within the aqueous medium containing the muscle fibre and typically the amplitude of a T-jump is <5 ˙C. Bershitsky & Tsaturyan (1992; 2002) use a different technique in which a T-jump is induced by passing a brief high frequency alternating electrical current longitudinally along a muscle fibre (Joule-heating). The fibre has to be held in air and the amplitude of T-jump is calculated on the basis of muscle fibre heat capacity; the rapid temperature decay in the fibre held in air is minimised by passing a control current after the T-jump and, additionally, the method can produce large T-jumps (20-30 degrees). Bershitsky & Tsaturyan (2002) find that a Joule-T-jump force transient cannot be defined by exponential curve fitting (as used here) and probably contains only one component. Our data using small amplitude laser-T-jumps, on the other hand, indicates the occurrence of at least two exponential components in a force transient (see below). Whether the two T-jump techniques are directly comparable remains unclear at present (see Davis, 1998).

ii). It is relevant to note that, in different studies, the [Pi] in control solution with no added P_i was taken as 0 mM to ~1 mM. On the basis of our calculations (Coupland et al, 2001), the P_i concentration within active psoas fibres at 10 °C would be ~0.735 mM in control solutions (i.e. with no added P_i - due to contamination and P_i - generation after activation) and 0.735 + x mM in other activating solutions, where x = added P_i. These differences in the analyses among different studies, however, do not affect the basic conclusions made in this comparative examination.

iii). When dealing with force transients induced by different perturbations, some difficulty is encountered in identifying the homologous components. This is largely due to different preparations and different conditions used in the studies and the problem has been addressed in previous studies (see Davis & Harrington, 1993; Davis, 1998). In keeping with the nomenclature used in our previous papers (see Vawda et al, 1999; Ranatunga, 1999b; Coupland et al, 2001), the following components, or phases, will be recognised in various force transients.

Phase 1: the force change that occurs concomitant with a perturbation, where the extreme force reached will be referred to as T_1, as used in the original length perturbation experiments (Huxley & Simmons, 1971). An instantaneous drop in force (phase 1) is also seen in T-jump (and P-jump) experiments, probably induced by expansion in series elasticity. A T-jump or a P-jump induces a drop in force in rigor muscle fibres and this provides support for the expansion in some elasticity (see Goldman et al, 1987; Ranatunga et al, 1990; Ranatunga, 1994).

Phase 2: following phase 1, the force recovers quickly to T_2 force level after a length perturbation. This partial force recovery consists of two exponential components - referred to here as phase 2a (fast) and phase 2b (slow). The force generation, i.e. force rise above the pre-perturbation level, induced by a T-jump (and a P-jump) corresponds to phase 2b. In T-jump and P-jump experiments where a prominent phase 1 was seen (see Ranatunga, 1999b; Vawda et al, 1999), a quicker force recovery corresponding to phase 2a was also seen; this phase tends to partially recover the force to pre-perturbation level as in length perturbation. Both phase 1 and phase 2a are not obtained in the small amplitude T-jump experiments reported here.

Phase 3: the slower exponential component of the force transient induced by a T-jump (and a P-jump). Phases 3 and 4 of length perturbation experiments are not dealt with here.

In summary, phases 1, 2a and 2b will be recognised in the length-induced transients, where as phases 2b and 3 will be seen in the T-jump transients.

4. OBSERVATIONS AND INTERPRETATIONS

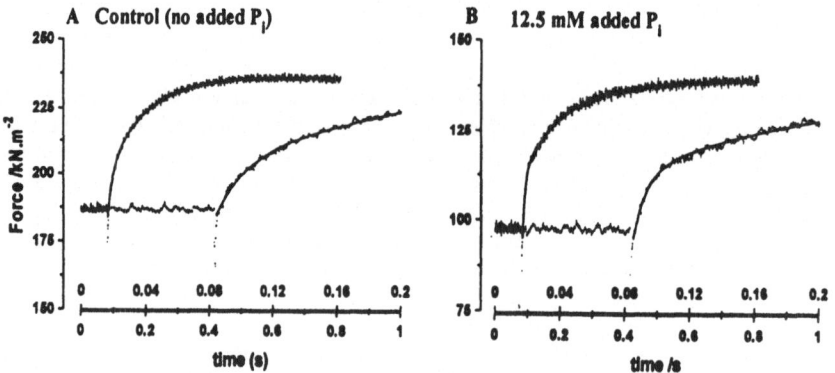

Figure 1: Sample Records of Force (= tension) transients induced by T-Jumps.
The fibre (length 2.6 mm) was maximally Ca-activated to steady state and a T-jump of ~3 °C was induced by a laser pulse at 80 ms after the beginning of each time -trace (indicated by the downward spike, an artefact, on the force trace). Each frame contains an individual force transient, illustrated at two different time scales (upper and lower labelling in the X-axis). The tansient in **Fig. 1A** was recorded in control activation (no added P_i) and that in **Fig. 1B** was recorded after the fibre relaxed and subsequently reactivated in a medium containing 12.5 mM added P_i; the final temperature was 12.1-12.5 °C. Note that the steady force before perturbation (T_0) is depressed with 12.5 mM P_i. A bi-exponential curve is fitted to each transient to separate the two phases (phases 2b and 3): the exponential rates for phases 2b and 3 were 66 s^{-1} and 9.0 s^{-1} for (A) and 123 s^{-1} and 7.5 s^{-1} for (B).

4.1. T-jump induced Force Transient and its P_i Dependence

Figure 1A shows sample records of a force transient induced by laser T-jump; the same transient is shown at two different time scales and a bi-exponential curve is fitted to them. The T-jump induced a force rise from ~190 kN/m^2 to a new steady level (~240 kN/m^2) with a characteristic time course. Figure 1B shows the force transient from the same fibre when it was subsequently reactivated in the presence of 12.5 mM added P_i. It is seen that, compared to the control, the steady force before and after the T-jump is smaller in the presence of P_i but the initial component of the force rise (phase 2b or endothermic force generation) is clearly faster in the presence of P_i.

Figure 2 shows the P_i dependence of steady active force from five fibres in such experiments. As is well known, the active force is depressed with increase of $[P_i]$ (Cooke & Pate, 1985). The force transients induced by a standard T-jump in the same experiments were analysed by curve fitting as shown in Figure 1 and the mean (± s.e.m.) apparent rate constants (reciprocal time constant, _) for the two phases are plotted against $[P_i]$ in Figure 2B. It is seen that Phase 3 rate (open symbols) shows minimal sensitivity to the level of P_i added; this may represent contribution to force rise of crossbridges going through a slower step in the cycle after the P_i-release.

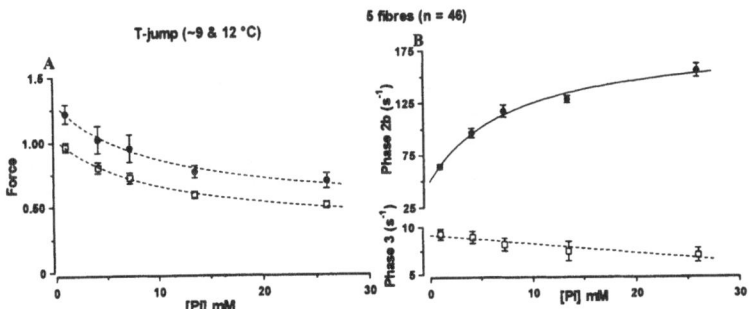

Figure 2A: [Pi] Dependence of Steady Active Force.
A fibre was maximally Ca-activated at ~9 °C (open squares) and T-jumped to ~12 °C (filled circles). Pooled steady force (mean ± s.e.m.) data collected for the two temperatures (i.e. before and after a T-jump) from 46 contractions in 5 muscle fibres are shown. Forces are normalised to that recorded in the first control active contraction in each fibre (i.e. with no added P_i at ~9 °C). A hyperbolic curve is fitted to each data set. (Note that with no added [Pi], a control solution was assumed to contain ~1 mM [Pi] due to contamination and Pi generation within the active fibre, Dantzig et al, 1992; Kawai & Halvorson, 1991). The active force depression is 40-50 % at 25 mM added [Pi] in these experiments at 9-12 °C.

Figure 2B. [Pi] Dependence of the Force Transient.
The T-jump force transients in the above experiments were analysed by curve fitting as shown in Fig. 1 and the rate (reciprocal time constant) and the amplitude of each of the two exponential components (phases 2b and 3) determined. The mean (± s.e.m., n=7-15 per symbol) rates for the two phases are plotted against [Pi]; filled symbols - data for phase 2b (= endothermic force generation, upper ordinate) and open symbols - data for phase 3 (lower ordinate). The curve fitted to the phase 2b data is a hyperbolic equation of the form, rate = k_a + $\{k_b.[Pi]/(K_D + [Pi])\}$. The curve gives 50 s^{-1} for k_a (forward rate constant of step I in scheme 1), 135 s^{-1} for k_b (backward rate of step I) and 8.0 mM for K_D (dissociation constant for Pi release, step II in scheme 1). Phase 3 rate is 5-10 s^{-1} and it shows a slight decrease with [Pi].

According to an interesting scheme proposed by Davis (1998), phase 3 may represent participation of the second myosin head in a force-holding role. On the other hand, phase 2b (or force generation) rate increases ~3-fold when added P_i is raised to 25 mM. Additionally, the data in figure 2B shows that the rate exhibits saturation at high P_i levels and the relation is hyperbolic (the curve fitted to the data).

The data in Fig. 1 and Fig. 2 illustrate four observations of interest. Firstly, a T-jump induces a force rise without any appreciable delay, indicating that crossbridge force generation itself is endothermic. Secondly, the rate of force generation (phase 2b) is sensitive to added P_i indicating that it is closely coupled to the release P_i by actomyosin crossbridges. Thirdly, the dependence of the phase 2b rate on $[P_i]$ is not linear but hyperbolic. This implies that P_i release step itself is not generating force (in which case P_i – dependence would be linear) but it involves at least two steps, a force generation step followed by rapid P_i release (for details, see Dantzig et al, 1992; Kawai & Halvorson, 1991). A force generating step after P_i release would not be very sensitive to $[P_i]$. Fourthly, the time course of T-jump force transient is clearly slower than that induced by a length-release in these fibres (see below).

As mentioned in the General Introduction, several studies using a number of different types of perturbation techniques have led to the thesis given above. The minimal scheme for actomyosin ATPase and the crossbridge cycle that explained the basic observations in these studies (see Dantzig et al, 1992; Fortune et al, 1991; Kawai & Halvorson, 1991) is given in the scheme 1 below.

Scheme 1

States	1 (low force)	2 (high force)	3 (high force)
	$\rightarrow$ M/AM.ADP.P$_i$ $\leftrightarrow$	AM*.ADP.P$_i$ $\leftrightarrow$	AM*.ADP + P$_i$ $\rightarrow$
	(k_a/k_b)	(k_c/k_d)	(k_e)
Steps	(I)	(II)	(III)

The force behaviour after a perturbation and its P_i-dependence are explained by changes in two sequential equilibria (steps I and II). The forward rate constant (k_a) of Step I is the temperature sensitive force generating step whereas its reverse rate constant k_b is not temperature sensitive. Changing the forward rate constant k_a with a Q_{10} of 3-4 (Ranatunga, 1996; corresponding to an activation enthalpy of ~80 kJ K^{-1}) simulates temperature effects. Step II is a rapid release of phosphate, where k_c is set as 10^3 s^{-1} and changing k_d simulates effect of changing $[P_i]$; the range of k_d values used corresponds to a second order rate constant of 0.6 X 10^5 M^{-1} s^{-1} for P_i binding. Step III, i.e. AM*.ADP $\rightarrow$ M/AM.ADP.P$_i$, is taken as irreversible and rate limiting (k_e, 5-10 s^{-1}) representing all the steps necessary to reprime the crossbridges for the next cycle. According to the scheme, state I (M/AM.ADP.P$_i$) is a low force state whereas states 2 and 3 (AM*.ADP.P$_i$ and AM*.ADP) are high (and equal) force states. The myosin conformational change accompanying force generation is indicated by asterisk. Thus, the sum of the fractional occupancy of states 2 and 3 is taken to be proportional to muscle force (see Dantzig et al, 1992; Kawai & Halvorson, 1991). With this scheme, the $[P_i]$ dependence of the observed rate of force recovery after a standard perturbation would be hyperbolic and given by,

$$Observed\ rate = k_a + k_b[P_i]/(Kd+[P_i]).$$

The rate will increase with $[P_i]$ and show saturation at higher $[P_i]$, as experimentally obtained for phase 2b in Fig. 2B.

4.2. Temperature Dependence of Tetanic Force in Intact Muscle

According to scheme 1, a single step (= equilibrium) in the crossbridge cycle is temperature sensitive. Whether such a simple scheme can account for temperature dependence of steady active force in muscle is examined below. Indeed, this type of analysis was first made by Davis (1998) on some of our published skinned fibre data.

Figure 3A shows fused tetanic contractions recorded from one intact fibre bundle preparation at four different temperatures and they illustrate the marked temperature sensitivity of the tetanic force in muscle. The steady-state temperature-dependence of tetanic force was determined over the temperature range ~5-35 °C in eight experiments. It was found that the force versus reciprocal temperature relation was approximately sigmoidal with half-maximal force at 9.5 °C. The results compare well with those published from whole muscle at limited temperature range (Ranatunga & Wylie, 1983) and also from diaphragm muscle strips (Hadju, 1951). If the increase of force is due to a change in single equilibrium, then the force data should show the behaviour expected from vant Hoff's relation, i.e. Log (Equilibrium Constant) _ (–_H/R)(1/T). Thus a plot of log (Eq.Const) versus reciprocal absolute temperature (1/T) should be linear with a slope given by (–_H/R) where _H is enthalpy change and R is 8.32 JK^{-1}.

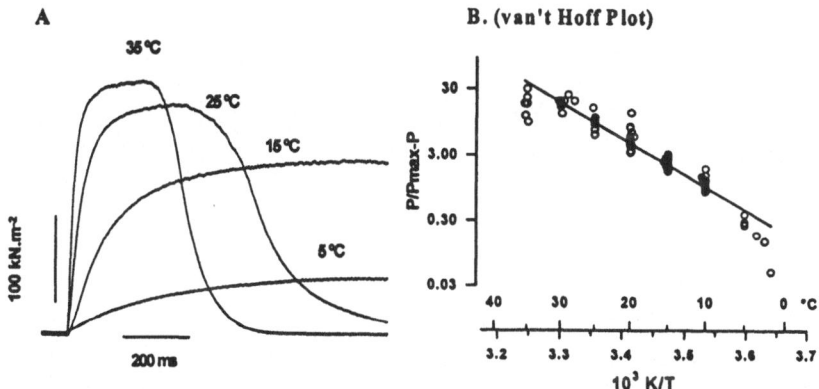

Figure 3A: Sample Isometric Tetanic Contraction Records
An intact bundle of ~5 fibres, isolated from Flexor Hallucis Brevis (FHB) muscle (fibre length ~2 mm) of rat foot, was tetanically stimulated during slow cooling and warming between 2-35 °C, to determine the temperature-dependence of steady tetanic force. Note the steady force increases markedly on heating, particularly, from 5 to 15 °C.
Figure 3B: Temperature Dependence of Tetanic Force (vant Hoff's plot)
Pooled data from 8 fibre bundles are illustrated as a *vant Hoff's plot*. The ratio "force / (maximum force minus force)" is plotted (as an approximation of the equilibrium constant) on a logarithmic ordinate against reciprocal absolute temperature (10^3 K/T) on the abscissa which is also labelled in °C (see Brown, 1957). The relation is approximately linear, as expected if the tetanic force increase on heating is due to increase of a single equilibrium (step 1 in scheme 1); _H (enthalpy) from the slope is ~130 kJ. mol^{-1}. The deviations from linearity seen at high and low temperatures may be due to the use of intact fibres (a complex cellular system).

The ratios "force / maximum force minus force" were calculated for each fibre (to approximate equilibrium constant; see Brown, 1957) and plotted on a logarithmic ordinate against 1/T on the abscissa (Fig. 3B). To a first approximation the data distribution is linear indicating that increase of tetanic force on heating is largely due to enhancement of the force generating step.

4.3. Effect of [Pi] on the Temperature Dependence of Muscle Force

According to the scheme 1 the temperature sensitive force generating step precedes the rapid release of P_i (step II) and hence, the temperature dependence of active force is expected to be P_i sensitive in a predictable manner. Figure 4A shows force data collected from 5 skinned fibres in which maximal force values during a single activation (without added P_i and with 25 mM added P_i) were recorded at a number of different temperatures over a temperature range (~5-30 °C; see Coupland et al, 2001). It is seen that the data distribution is approximately linear both for control (filled symbols) and for P_i added contractions (open symbols); however, the data with added P_i are shifted to the left (higher temperatures). As shown in Figure 4B, a similar shift in the data distribution is predicted when Scheme 1 is used to calculate forces at different temperatures with

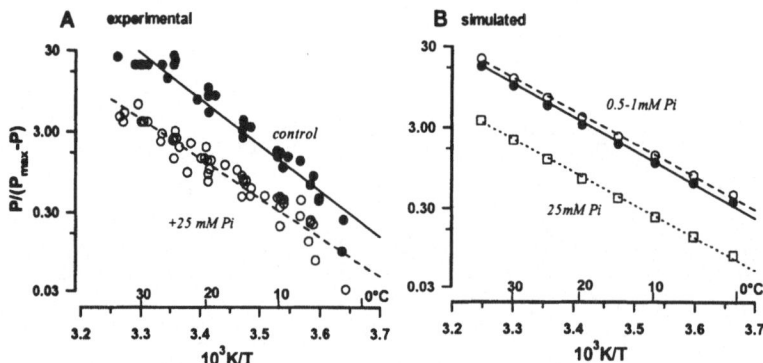

Figure 4A: Measured Active Force: Effect of P_i.
A chemically skinned fibre (rabbit psoas,) was maximally Ca-activated at <5 °C, and temperature raised in a stepwise manner by laser T-jumps and / or Peltier heating, before relaxing (see Coupland et al, 2001). The duration of a contraction was ~60s. Pooled data from five fibres in each of which force data were collected at different temperatures (range ~5-30 °C) during one control (no added P_i) and one or two test (with 25 mM P_i added) contractions. The steady forces (P) recorded at different temperatures were used to calculate P / (Pmax-P) ratios and are illustrated as vant Hoff's plots. Filled circles and solid line are from control activation. Open circles and dashed line are from activation in 25 mM P_i solutions and done prior to and / or after the control. Note that the data distribution is linear, and P_i seems to shift the relation to the left (higher temperatures).
Figure 4B: Behaviour of Simulated Force from a 3-state model (Scheme 1).
Vant Hoff's plots for calculated force (symbols) on the basis of a 3-step model with a single temperature-sensitive, force generation, step (full temperature range – 0 to 40 °C, at 5 ° intervals) for 3 different [P_i] levels. Temperature increase was simulated by increasing k_x (Q_{10} ~3) in the scheme and [P_i] was constant at 0.5 mM (open triangles), 1 mM (filled circles) and 25 (open squares) by pre-setting k_4. With added Pi, the relation is shifted to the left. The slopes correspond to _H of ~ 80-120 kJ mol^{-1} in the curves.

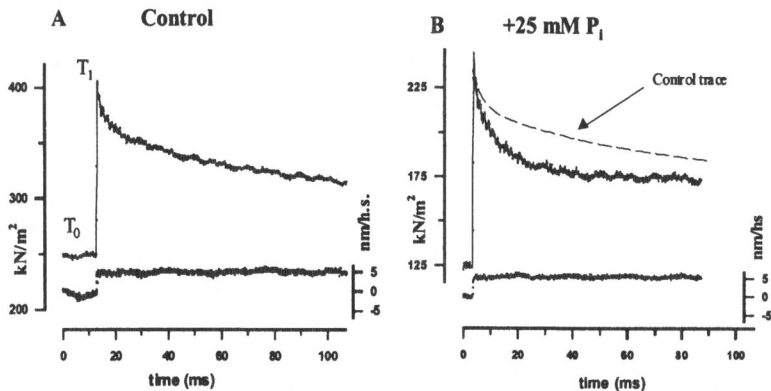

Figure 5. Effect of P$_i$ on Force Recovery after Stretch
Sample records of force transients (upper trace) from a single muscle fibre, initiated by a standard stretch (~0.5% L$_o$); lower trace is the change in half-sarcomere length (right ordinate). **Frame A** - the fibre was activated in control activating solution. **Frame B** - the fibre was activated in the presence of 25 mM added P$_i$. Records in A and B show that the initial force decay after stretch is faster in the presence of P$_i$.

different levels of added P$_i$. The discrepancies between the experimental and the simulated data may not be entirely unexpected since the actual P$_i$ levels within fibres and at different temperatures are unknown.

4.4. Strain Sensitivity of the Force Generation Step

As mentioned in the Introduction, muscle force during isometric contraction is maintained by equilibrium between low and high force crossbridge states, so that sudden shortening (negative strain) that reduces the force will perturb the equilibrium and lead to a quick rise (recovery) of tension (Huxley & Simmons, 1971). Conversely, a stretch (positive strain) that increases force would also perturb the equilibrium but lead to a quick decline of force. Applying this to Step I in Scheme 1 above (i.e. to AM.ADP.P$_i$ ↔ AM*.ADP.P$_i$), the rate of quick force recovery after a length-perturbation is expected to increase with [P$_i$] and show saturation at high [P$_i$].

Figure 5A shows a force transient (upper trace) induced by a standard stretch (5% L$_0$) when the fibre was in steady activation in the control solution; Figure 5B gives the corresponding transient induced when the fibre was activated in the presence of 25 mM added P$_i$. Figure 6A and B show from the same fibre a corresponding pair of force transients induced by a length-release. The force change induced by a length step reaches a peak during the step (T$_1$) and is followed by a partial quick recovery (T$_2$ level), as described in the original experiments on intact frog muscle fibres (Huxley & Simmons, 1971; Ford et al, 1977). The figure shows that the initial force recovery after a stretch (force decay) is faster in the presence of 25 mM added P$_i$ (compare 5A and B), whereas the quick force rise after length-release is little altered by P$_i$ (compare 6A and B).

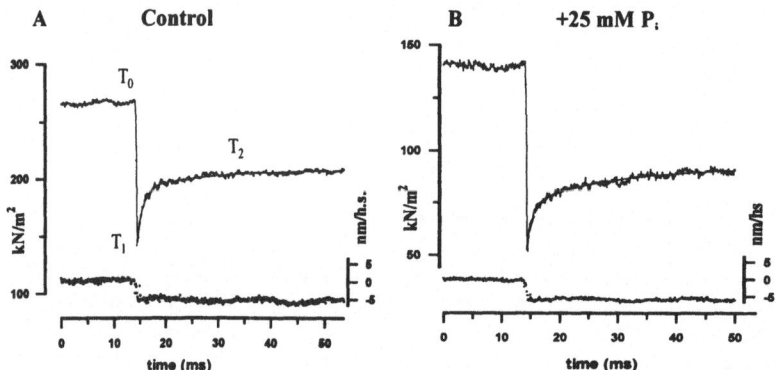

Figure 6. Effect of P_i on Force Recovery after Length release
Sample records of force transients (upper trace) initiated by a standard length release step (~0.5% L_o). **Frame A** – control and **Frame B** - when the fibre was activated in the presence of 25 mM added P_i. In contrast to the effects on stretch-induced transient, the quick recovery after length-release, is not much altered (or slower) in the presence of added P_i. (same fibre as in Fig.5).

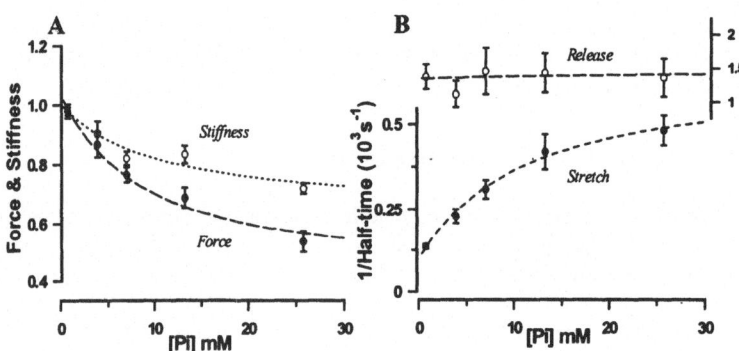

Figure 7. [P_i] - Dependence of Steady Force, Stiffness and Time Course of Force Transients
A: The mean (± s.e.m.) steady force (●) and the mean (± s.e.m.) stiffness (○) recorded from five fibres in each of which data were collected at each level of [P_i]. Stiffness was determined as the force change (i.e. as $\Delta T = T_1 - T_0$) to a standard stretch (0.4-0.5% L_0)) and both the steady force and the stiffness are plotted as ratios of that in the control solution (no added P_i). The abscissa is the [P_i] within active fibres (see Coupland et al, 2001). A hyperbolic curve is fitted to each set of data. Note that the force depression is greater than the stiffness change with increased P_i.
B: - The half-time of initial recovery of force (the decay between T_1 and T_2) was measured and the mean (± s.e.m.) reciprocal half-time for different [Pi] are shown by filled symbols, where the curve is a hyperbolic relation; it corresponded to a maximum rate of $760.s^{-1}$, a minimum rate of $99.s^{-1}$ and a half-maximal [P_i] of 11 mM. The reciprocal half-times of force recovery after length release (~$1500.s^{-1}$, Open symbols - right ordinate), show no correlation with [P_i].

Figure 7 shows pooled data from similar experiments on five fibres; Figure 7A shows the $[P_i]$ dependence of the steady force (●) and the stiffness (O, see figure legend). The steady force is decreased by about ~50 % at ~25 mM $[P_i]$, whereas the stiffness is decreased to a lesser extent, ~30% from the control. The data are consistent with the thesis that increased P_i leads to a greater reduction in the number of force generating crossbridges (force) than in the total number of attached crossbridge states (stiffness). The initial $(T_1 - T_2)$ force recovery after a length-step is normally analysed by half-time measurement. As shown in Figure 7B, the initial rate of recovery (as reciprocal half-time) after length-release (O) is high (~1500.s^{-1}) and shows no correlation with $[P_i]$. The apparent rate of force decay after a stretch (●), on the other hand, is low (~100 s^{-1}) in the control and it increases non-linearly with $[P_i]$. A hyperbolic curve could be fitted to describe the data; this' is basically as obtained in T-jump (and P-jump) experiments (see Fig. 2B).

For closer comparison with T-jump and other data, the $T_1 - T_2$ recovery in the force transients was defined by bi-exponential curve fitting, and Figure 8A shows the $[P_i]$ dependence of the rate of phases 2a and 2b and Figure 8B their amplitudes. The data shows several interesting points. Firstly, phase 2a rates from stretches and releases are similar (~1000 s^{-1}) and show little dependence on $[P_i]$. It may be argued that the phase 2a represents relaxation from non-specific strain effects on all the attached crossbridge states and / or visco-elastic relaxation in sarcomeric filaments (see Davis & Rogers, 1995). Secondly, only the rate of the phase 2b and of the stretch-induced transient (■) shows a clear hyperbolic dependence on $[P_i]$. The phase 2b rate from stretches increases from ~15 s^{-1} with no added $[P_i]$ to ~130 s^{-1} with excess $[P_i]$; $[P_i]$ for half-maximum is 11 mM.. The data are basically comparable to those obtained from P-jump, T-jump, P_i-jump and other experiments that defined a "pre-Pi-release force generation step" in psoas fibres at this temperature (see Table 1 in Ranatunga et al, 2002). The normalised amplitude of this component showed no dependence on $[P_i]$, as found in Pi-jump and P-jump experiments. Thus, the slow phase 2b component induced by stretches (positive strain) represents perturbation of the "pre-P_i-release force generation step". Thirdly, force recovery from length release is insensitive to $[P_i]$. It may be argued that such a differential P_i-sensitivity (asymmetry) indicates that stretch and length-release perturb two different molecular steps in the crossbridge (ATPase) cycle. In general, this would be consistent with the differences in their temperature sensitivities and also with the asymmetry in the time courses of stretch-induced and release-induced transients. The apparent asymmetry in the time course, however, is reduced at high $[P_i]$.

Tesi et al, (2000) reported, from myofibril experiments, that the force change (decline) obtained when $[P_i]$ was increased to a certain level was faster than the force change (rise) obtained when $[P_i]$ was decreased to the same final $[P_i]$. Our finding that an apparent asymmetry exists in the P_i dependence of the force transients induced by positive and negative strains may also be related to their finding. Kawai and colleagues (Kawai et al, 1993; Zhao & Kawai, 1994; Wang & Kawai, 1997; 2001) correlated the force generation step to phase 3 in length-release force transients, and Davis and Harrington (1993) correlated force generation to phase 2b of length-release transient. The causes of these discrepancies between studies are unclear; probably this is not a major contradiction if the rate constants of these phases are not vastly different.

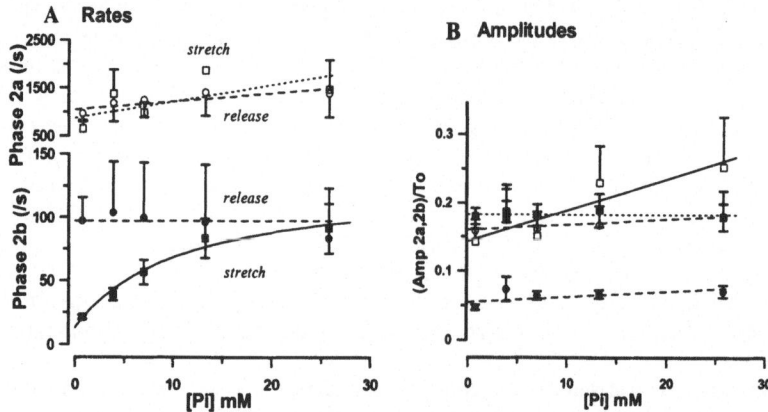

Figure 8. Phase 2a and Phase 2b Exponential Components of Force Recovery: [P_i]-Dependence

The initial recovery of force (the decay in stretches and rise in releases, between T_1 and T_2) was analysed by fitting a bi-exponential curve, extracting phases 2a and 2b.

A. The two rate constants (for Phases 2a, 2b) for different [Pi] are shown as means (± s.e.m.). Note that plus or minus error bars are not shown for some data points. The rates for Phase 2a (□, O- upper ordinate) and Phase 2b (■, ● - lower ordinate) are plotted separately for stretches (□, ■) and releases (O, ●). The slower Phase 2b from stretches (only) is correlated with [Pi] (solid curve through the points, r = 0.77) and the hyperbolic curve gives a maximum rate of $129.s^{-1}$, minimum of $13.s^{-1}$ and half-maximum [Pi] of 11 mM). Others show no correlation with [Pi] (dotted and interrupted curves).

B. The amplitudes of the two exponential components for stretches (□, ■) and releases (O, ●) are plotted separately against [P_i]; each amplitude is plotted as a ratio of steady force (T_0). Both amplitudes of releases and of phase 2b from stretches show no correlation (dotted and interrupted lines, $P > 0.05$), but the amplitude of Phase 2a from stretches is positively correlated (□, solid line, $P < 0.05$) (for details, see Ranatunga et al, 2002).

5. CONCLUSION

Many different studies have identified a "pre-P_i-release step" ($AM.ADP.P_i \leftrightarrow AM^*.ADP.P_i$ transition) as generating force in muscle fibres and, as given above, such a kinetic scheme (scheme 1) can account for a number of other observations. However, the scheme is too simplistic to accommodate all the findings. Two questions remain. Firstly, why is this step not sensitive to negative strain (length-release / rapid shortening)? One possible explanation is that during rapid shortening, operation of a more rapid molecular step elsewhere releases the negative strain in the filaments (+crossbridges) and hence shields the $AM.ADP.P_i \leftrightarrow AM^*.ADP.P_i$ transition from being exposed to negative strain. Secondly, what is the kinetic step that underlies force recovery from length release? A molecular step that is potentially negative strain-sensitive is a step coupled to ADP release (see Dantzig et al, 1991; Regnier et al, 1995; Geeves & Holmes, 1999) which, occurring after P_i-release in the cycle, would be insensitive to P_i. A marked increase of crossbridge detachment rate during shortening (negative strain) was indeed a characteristic feature of A.F. Huxley's (1957) original formulations, and used profitably in the analyses of the force-velocity relation, work-output and efficiency determinations in muscle fibres (see He et al, 1999, 2000). With small length releases, however, crossbridge detachment may not be the cause of quick force recovery (Piazzesi et al, 2002).

On the basis of our current understanding, the structural mechanism of crossbridge force generation involves a movement of the lever arm of the attached myosin head driven by a closed to open conformational change (see Geeves & Holmes, 1999). The observations above suggest that the closed to open transition in attached myosin head (= crossbridge force generation) is readily reversible and occurs before P_i - release. Ignoring differences in detail and taking all the findings together, it appears that this transition - or step(s) closely coupled to it - is endothermic (temperature sensitive), is accompanied by an increase of volume (pressure-sensitive), is reversed by P_i (P_i-sensitive) and is strain-sensitive but the strain-sensitivity is seen only with respect to positive strain (stretches).

6. ACKNOWLEDGEMENTS

We thank the Wellcome Trust Foundation for financial support of our research, Dr. Gerald Offer (Bristol) for valuable discussions and The Physiological Society and The Royal Society for permission to include data published in the Journal of Physiology and the Proceedings of the Royal Society (B).

7. REFERENCES

Bershitsky, S.Y. & Tsaturyan, A.K., (1992). Tension responses to joule temperature jump in skinned rabbit muscle fibres. *J. Physiol.* **447**: 425-448.

Bershitsky, S.Y. & Tsaturyan, A.K. (2002). The elementary force generation process probed by temperature and length perturbations in muscle fibres from the rabbit. *J. Physiol.* **540**: 971-988.

Bershitsky, S.Y., Tsaturyan, A.K., Bershitskaya, O.N., Mashanov, G.I., Brown, P., Burns, R. & Ferenczi, M.A. (1997). Muscle force is generated by myosin heads stereospecifically attached to actin. *Nature* **388**, 188-190.

Brown, D.E.S. (1957). Temperature – pressure relation in muscular contraction. In *"Influence of Tempearture on Biological Systems"* ; Edited by F.H. Johnson, pp 83–100 (American Physiological Society Inc.).

Cooke, R. & Pate, E. (1985). The effects of ADP and phosphate on the contraction of muscle fibers. *Biophysical Journal* **48**, 789-798.

Coupland, M.E., Puchert, E. & Ranatunga, K.W. (2001). Temperature dependence of active tension in mammalian (rabbit psoas) muscle fibres: effect of inorganic phosphate. *Journal of Physiology* **536**, 879-891.

Dantzig, J.A., Goldman, Y.E., Millar, N.C., Lacktis, J., & Homsher, E. (1992). Reversal of the cross-bridge force-generating transition by photogeneration of phosphate in rabbit psoas muscle fibres. *Journal of Physiology* **451**, 247-278.

Dantzig, J.A., Hibberd, M.E., Trentham, D.R. & Goldman, Y.E. (1991). Crossbridge kinetics in the presence of MgADP investigated by photolysis of caged ATP in rabbit psoas muscle fibres. *Journal of Physiology* **432**, 639-680.

Davis, J.S. & Harrington, W. (1987a). Laser temperature-jump apparatus for the study of force changes in fibers. *Analytical Biochemistry*, **161**, 543-549.

Davis, J.S. & Harrington, W. (1987b). Force generation by muscle fibers in rigor: a laser temperature-jump study. *Proceedings of National Academy of Science*, **84**, 975-979.

Davis, J.S. & Harrington, W. (1993). A single order-disorder transition generates tension during Huxley-Simmons pahe 2 in muscle. *Biophysical Journal*, **65**, 1886-1898.

Davis, J.S. & Rodgers, M.E. (1995). Indirect coupling of phosphate release to *de-novo* tension generation during muscle contraction. *Proceedings of National Academy of Science*, **92**, 10482-10486.

Davis, J.S. (1998). Force generation simplified. Insights from laser temperature-jump experiments on contracting muscle fibres. In *"Mechanisms of Work Production and Work absorption in Muscle"*; Edited by Sugi and Pollack, pp343-352 (Plenum Press, New York).

Ford, L.E., Huxley, A.F. & Simmons, R.M. (1977). Tension responses to sudden length change in stimulated frog muscle fibres near slack length. *Journal of Physiology* **269**, 441-515.

Fortune, N.S., Geeves, M.A. & Ranatunga, K.W. (1991). Tension responses to rapid pressure release in glycerinated rabbit muscle fibers. *Proceedings of National Academy of Science* **88**, 7323-7327.

Geeves, M.A. & Holmes, K.C. (1999). Structural mechanism of muscle contrcation. *Annual Reviews of Biochemistry* **68**, 687-728.

Goldman, Y.E., McCray, J. A. & Ranatunga, K.W. (1987). Transient tension changes initiated by laser temperature jumps in rabbit psoas muscle fibres. *Journal of Physiology* **392**, 71-95.

Hajdu, S. (1951). Behaviour of frog and rat muscle at higher temperatures. *Enzymologia*, **14**, 187-190.

He, Z-H., Bottinelli, R., Pellegrino, M.A., Ferenczi, M. A. & Reggiani, C. (2000). ATP consumption and efficiency of human single muscle fibres with different myosin isoform composition. *Biophysical Journal* **79**, 945-961.

He, Z-H., Chillingworth, R.K., Brune, M., Corrie, J.E.T., Webb, M.R. & Ferenczi, M. A. (1999). The efficiency of contraction in rabbit skeletal muscle fibres, determined from the rate of release of inorganic phosphate. *Journal of Physiology,* **517**, 839-854.

Hilber, K. & Galler S. (1998). Improvement of the measurements on skinned muscle fibres by fixation of the fibre ends with glutaraldehyde. *Acta Physiologica Scandinavica*, **19**, 365-372.

Huxley, A.F. (1957). Muscle structure and theories of contraction. *Progress in Biophysics* **7**, 285-318.

Huxley, A.F. & Simmons, R.M. (1971). Proposed mechanism of force generation in striated muscle. *Nature* **233**, 533-538.

Huxley, A.F. & Tideswell, S. (1996). Filament compliance and tension transients in muscle. *Journal of Muscle Research and Cell Motility* **17**, 507-511.

Irving, M., Allen, T. St. C., Sabido-David, C., Craik, J.S., Brandmeir, B., Kendrick-Jones, J., Corrie, J.E.T., Trentham, D.R. & Goldman, Y.E. (1995). Tilting of the light-chain region of myosin during step length changes and active force generation in skeletal muscle. *Nature*, **375**, 688-691.

Irving, M., Lombardi, V., Piazzesi, G. & Ferenczi, M.A. (1992). Myosin head movements are synchronous with elementary force-generating process in muscle. *Nature* **357**, 156-158.

Kawai, M. & Halvorson, H.R. (1991). Two step mechanism of phosphate release and the mechanism of force generation in chemically skinned fibers of rabbit psoas muscle. *Biophysical Journal* **59**, 329-342.

Kawai, M., Saeki, Y. & Zhao, Y. (1993). Crossbridge scheme and the kinetic constants of elementary steps deduced from chemically skinned papillary and trabecular muscles from the ferret. *Circulation Research,* **73**, 35-50.

Lombardi, V., Piazzesi, G. & Linari, M. (1992). Rapid regeneration of the actin-myosin power stroke in contracting muscle. *Nature*, **355**, 638-641.

Matsubara, I., Y. Umazume & Yagi. N. (1985). Lateral filamentary spacing in chemically skinned murine muscles during contraction. *Journal of Physiology* **360**, 135-148.

Millar, N.C. & Homsher, E. (1991).The effect of phosphate and calcium on force generation in glycerinated rabbit skeletal muscle fibers. *The Journal of Biological Chemistry*, **265**, 20234-20240.

Maughan, D.W. & Godt, R.E. (1979). Stretch and radial compression studies on relaxed skinned muscle fibers of the frog. *Biophysical Journal* **28**, 391-402.

Piazzesi, G., Reconditi, M., Linari, M., Lucil, L., Sun, Y-B, Narayan, T., Boesecke, P, Lombardi, V. & Irving, M. (2002) Mechanism of force generation by myosin heads in skeletal muscle. *Nature* **415**, 659-662.

Ranatunga, K.W. (1982).Temperature-dependence of shortening velocity and rate of isometric tension development in rat skeletal muscle. *Journal of Physiology*, **329**, 465-483.

Ranatunga, K.W. (1994). Thermal stress and Ca-independent contractile activation in mammalian skeletal muscle fibers at high temperatures. *Biophysical Journal* **66**, 1531-1541.

Ranatunga, K.W. (1996). Endothermic force generation in fast and slow mammalian (rabbit) muscle fibers. *Biophysical Journal* **71**, 1905-1913.

Ranatunga, K.W. (1999a). Effects of inorganic phosphate on endothermic force generation in muscle. *Proceedings of the Royal Society B* **266**, 1381-1385.

Ranatunga, K.W. (1999b). Endothermic force generation in skinned cardiac muscle from rat. *Journal of Muscle Research and Cell Motility* **20**, 489-490.

Ranatunga, K.W. (2001). Sarcomeric visco-elasticity of chemically skinned skeletal muscle fibres of the rabbit at rest. *Journal of Muscle Research and Cell Motility*, **22**, 399-414.

Ranatunga, K.W., Coupland, M.E. & Mutungi, G. (2002). An asymmetry in the phosphate dependence of tension transients induced by length perturbation in mammalian (rabbit psoas) muscle fibres. *Journal of Physiology*, **542**, 899-910.

Ranatunga, K.W., Fortune, N.S. & Geeves, M.A. (1990). Hydrostatic compression in glycerinated rabbit muscle fibres. *Biophysical Journal*, **58**, 1401-1410.

Ranatunga, K.W. & Wylie, S.R. (1983). Temperature-dependent transitions in isometric contractions of rat muscle. *Journal of Physiology* 339, 87-95.

Regnier, M., Morris, C. & Homsher, E. (1995). Regulation of the crossbridge transition from a weakly to strong bound state in skinned rabbit muscle fibers. *American Journal of Physiology* 269, C1532-C1539.

Tesi, C., Colomo, F., Nencini, S., Piroddi, N. & Poggesi, C. (2000). The effect of inorganic phosphate on force generation in single myofibrils from rabbit skeletal muscle. *Biophysical Journal* 78, 3081-3092.

Tesi, C., Colomo, F., Piroddi, N. & Poggesi, C. (2002). Characterization of the cross-bridge force-generating step using inorganic phosphate and BDM in myofibrils from rabbit skeletal muscles. *Journal of Physiology* 541: 187-199.

Vawda, F., Geeves, M.A. & Ranatunga, K.W. (1999). Force generation upon hydrostatic pressure release in tetanized intact frog muscle fibres. *Journal of Muscle Research and Cell Motility* 20, 477-488.

Wang, G. & Kawai, M. (1997). Force generation and phosphate release steps in skinned rabbit soleus slow-twitch muscle fibers. *Biophysical Journal* 73, 878-894.

Wang, G. & Kawai, M. (2001). Effect of temperature on elementary steps of the cross-bridge cycle in rabbit soleus slow twitch muscle fibres. *Journal of Physiology* 531, 219-234.

Xu, S., Brenner, B. & L.C. Yu. (1993). State-dependent radial elasticity of attached cross-bridges in single skinned fibres of rabbit psoas muscle. *Journal of Physiology* 461: 283-299.

Zhao, Y. & Kawai, M. (1994). Kinetic and thermodynamic studies of the cross-bridge cycle in rabbit psoas muscle fibers. *Biophysical Journal* 67, 1655-1668.

DISCUSSION

Huxley: What are relative force levels in the two high force states?

Ranatunga: Assumed to be equal in the modeling, but may not be necessarily so.

Pollack: I was impressed by large increase of tension seen on T-jump. Is this consistent with the static tension-temperature relation?

Ranatunga: Yes. Because of the sigmoidal relation between steady force and temperature, the Q_{10} of steady force values depend on the temperature value. In the range in which we did our experiments the Q_{10} is large.

RATES OF FORCE GENERATION IN *DROSOPHILA* FAST AND SLOW MUSCLE TYPES HAVE OPPOSITE RESPONSES TO PHOSPHATE

Douglas M. Swank and David W. Maughan[*]

1. INTRODUCTION

Much of the functional diversity of striated muscle that serves to meet different locomotory demands is due to different isoforms of myosin. Barany et al. (1965), who were among the first to investigate the physiological relevance of myosin isoforms, demonstrated that different speeds of contraction in various muscle types correlate with actin-activated myosin ATPase. Reiser et al. (1985), Eddinger and Moss (1987), and Sweeney et al. (1988), among others, extended these studies by showing muscle speeds correlate with myosin isoform composition. Recently, Swank et al. (2002) directly proved that myosin isoforms are the prime determinants of fiber kinetic differences by substituting the slow embryonic myosin (denoted EMB) for the native, fast adult IFM myosin (denoted IFI) in the indirect flight muscle (IFM) (**Fig 1**). Substituting the EMB isoform transformed the IFM from a muscle that generates maximum oscillatory work at high frequencies, to one that generates more work, but at much lower frequencies (**Fig. 2**, after Swank et al., 2002). The isoform substitution enhances calcium-activated isometric tension (T_o) nearly 3-fold, but reduces maximum oscillatory power (P_{max}, equal to oscillatory work times frequency) to only ~25% that of IFI fibers (Swank et al., 2002). In EMB fibers P_{max} is achieved at a frequency (~20 Hz) that is considerably lower than that at which P_{max} is achieved in IFI fibers (~150 Hz, i.e., the wing beat frequency at 15°C). At the resonant frequency of the flight system (i.e., at ~150 Hz), no power is produced, thereby explaining the loss of flight ability in the EMB lines. Thus substitution of EMB for IFI is akin to converting the muscle from a fast fiber type to a slow fiber type.

[*] Douglas M. Swank and David W. Maughan, Department of Molecular Physiology and Biophysics, University of Vermont, Burlington VT 05405 U.S.A.

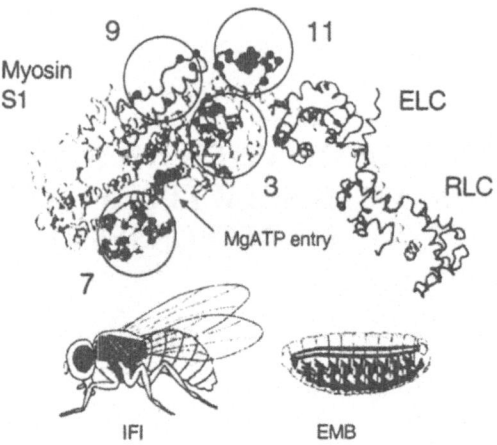

Figure 1. Atomic resolution structure of chicken myosin ST with the four *Drosophila* variable domains circled. Spheres indicate positions of non-conserved amino acid residues within these domains. The variable domains are generated by splicing of alternative exons in the mRNA from the single myosin heavy chain gene. Numbers refer to specific alternative exons. The IFI and EMB myosin isoforms differ at all four of the variable regions. IFI is expressed in the indirect flight muscle of the adult fly (bottom left) and EMB in various embryonic body wall muscles (bottom right). Modified from Swank et al. 2002.

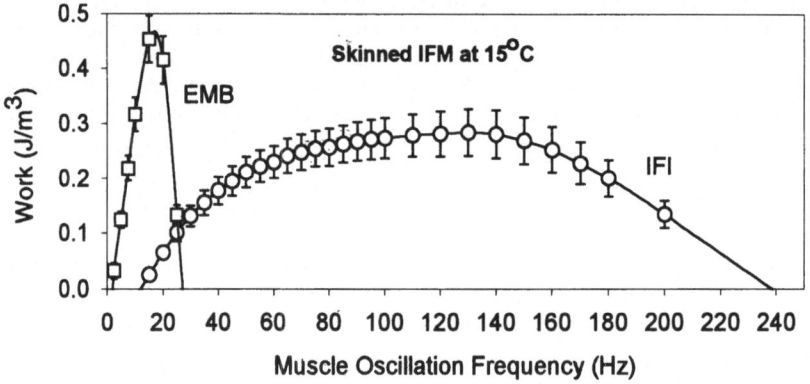

Figure 2. Work per cycle measured by small amplitude sinusoidal analysis for IFM expressing IFI and EMB myosin.

The kinetic differences observed in skinned IFM are consistent with those observed in *in vitro* motility studies (Swank et al., 2001). The sliding velocity of actin on a bed of isolated IFI is among the fastest reported for a type II myosin (6.4 μm s^{-1} at 22°C).

However, the sliding velocity of actin on EMB was 9-fold less than IFI. Laser trap studies showed no significant difference in unitary step size (Swank et al., 2001), indicating that the difference in sliding velocity must be due to cross-bridge kinetics rather than step size. The present experiments focus on the position of the rate-limiting step of the cross-bridge cycle in relation to phosphate release as a determinant of kinetic differences between IFI and EMB fibers.

2. MATERIALS AND METHODS

IFM were isolated from transgenic flies expressing IFI and EMB (Wells et al., 1996). The isolated fibers were chemically skinned with 0.5% Triton X100 in relaxing solution (pCa 8; 5 mM MgATP; 8 mM Pi), activated by calcium (pCa 4.5), and subjected to sinusoidal length perturbations (Swank et al, 2002). Isometric tension (maximal force divided by cross-sectional area) was measured prior to the length perturbations. Phosphate (Pi) concentration was varied by exchanging aliquots of activating solution (pCa 4.5) containing 32 mM or 0 mM [Pi] to obtain intermediate concentrations. [Pi] was returned to the starting concentration (8 mM) to ensure reversibility. All solutions contained an antioxidant (5 mM DTT), pH buffer (20 mM BES, pH 7), calcium buffer (15 mM EGTA), and MgATP regenerating system (15 mM creatine phosphate, 200 units/ml creatine kinase). Ionic strength (0.2 M) was adjusted with K methane sulfonate. We compared results from young adult flies from both groups (< 2 hr post-eclosion), thereby avoiding complications due to myofibril deterioration observed in older EMB flies (Swank et al., 2002).

The sinusoidal length perturbation approach is illustrated in Fig. 3 (details are reported elsewhere: Kawai & Brandt, 1980; Mulieri et al., 2002). At top left, small amplitude sinusoidal length perturbations are applied, and the phase and amplitude relation between the applied length change (_L) and resulting force change (_F) measured at different frequencies (f = 0.125-1000 Hz). The complex modulus y(f) is calculated as the ratio of tension change (_F divided by fiber cross-sectional area) and the fractional change in muscle length (ΔL/Lo, equal to 0.00125 under our standard conditions). In the exemplar Nyquist plot shown at bottom left, the viscous modulus (the 90° out-of-phase component of y(f)) is plotted versus elastic modulus (the in-phase component of y(f)). Oscillatory work at each frequency is the product of the viscous modulus and $(\Delta L/L_0)^2$. The frequency at which the viscous modulus assumes its largest negative value, f_{max}, is also the frequency at which the fiber achieves its maximum work output.

For further analysis, we deconstructed the Nyquist plot to yield components *A, B* and *C* (Fig. 3, right panel; see Mulieri et al., 2002, for details), where y(f) = A $(2\pi$ if $/\alpha)^k$ - B if $/(b+if)$ + C if $/(c+if)$. i = $\sqrt{-1}$, α = 1 Hz, and k is a unitless exponent. Coefficients A, B and C are the magnitudes of *A, B,* and *C*, expressed as mN per mm^2 fiber cross-sectional area. The characteristic frequencies of *B* and *C* are *b* and *c*, expressed in Hz. We attribute processes *B*

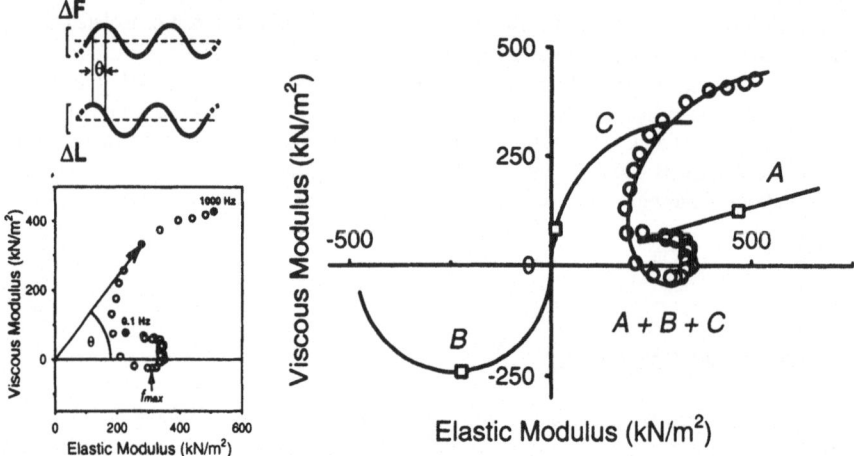

Figure 3. Sinusoidal analysis and convolution of the 3 viscoelastic components to produce the composite Nyquist plot. The complex modulus y(f) is the vector sum of the viscous and elastic moduli. Component B is work-producing; A and C, work-absorbing. Squares indicate frequency at which cross-bridges perform maximum work (i.e., at b, the characteristic frequency of process B). See text for further explanation.

and C to changes in dynamic stiffness arising from transitions between cross-bridge states (Zhao and Kawai, 1993), and process A to the viscoelastic behavior of passive elements located both inside and outside the myofibril (see Mulieri et al., 2002). Note that the viscous modulus of B is negative, which denotes a work-producing cross-bridge process, whereas the viscous modulus of C is positive, which denotes a work-absorbing cross-bridge process.

3. RESULTS AND DISCUSSION

As noted above, compared to the native IFI fibers, EMB fibers exhibit elevated calcium-activated isometric tension and oscillatory work output, but reduced power output and frequency of optimum work output. To investigate the molecular basis of these marked functional differences, we varied phosphate concentration (Zhao and Kawai, 1993). Remarkably, in IFI fibers, increasing [Pi] *elevates* T_o (**Fig. 4**). The frequency at which maximum work is produced (f_{max}) *decreases* over the range 0-16 mM (**Fig. 5**). The direction of these phosphate dependencies are *opposite* those observed in EMB fibers, which, with increased [Pi], exhibit the usual *reduction* of T_o and *elevation* of f_{max} reported in vertebrate striated muscle fiber studies (e.g., Zhao and Kawai, 1993).

To examine the kinetics of the phosphate release step in the context of a cross-bridge scheme, we deconstructed the Nyquist plots to yield the characteristic frequencies b and c of

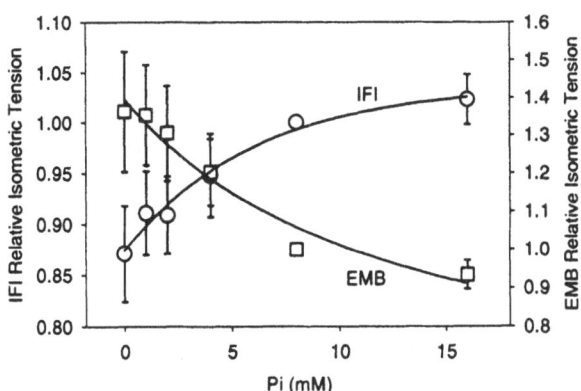

Figure 4. The effect of phosphate on isometric tension of fibers expressing IFI and EMB. Tension normalized to the level measured at 8 mM phosphate. Note different y-axis scales. At pCa 4.5 and [Pi] 8 mM, isometric tension was 1.2±0.2 mN mm^{-2} for IFI and 3.4±0.7 mN mm^{-2} for EMB (Swank et al., 2002).

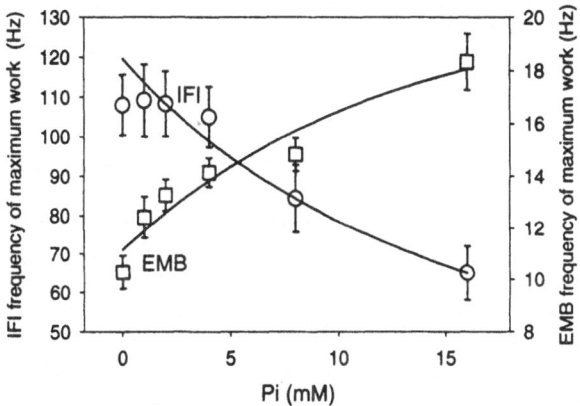

Figure 5. The effect of phosphate on the frequency at which maximum work is generated per cycle (f_{max}) in fibers expressing IFI and EMB.

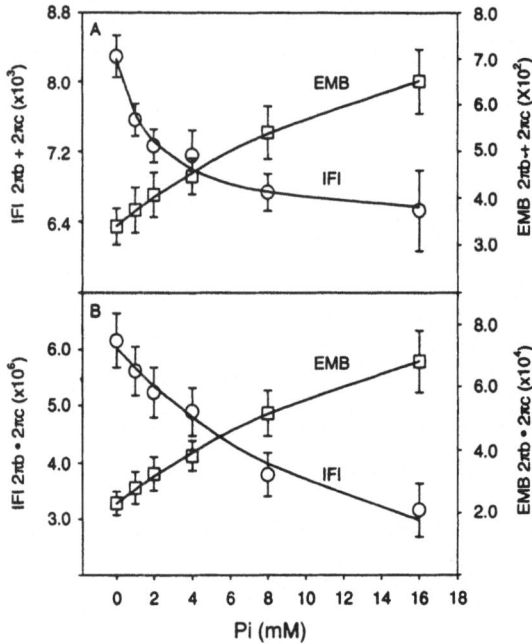

Figure 6. The effect of phosphate on the sum of rate constants $2\pi b$ and $2\pi c$ (panel A) and the product of rate constants $2\pi b$ and $2\pi c$ (panel B). Data fit with equations derived from schemes 1 and 2. For EMB, both sum and product equaled $\alpha + \beta K_P P / (1 + K_P P)$, where $\alpha = 340$ s^{-1}, $\beta = 732$ s^{-1}, and $K_P = 0.046$ mM^{-1} ($r^2 = 0.999$) for sum and $\alpha = 2.3 \times 10^3$ s^{-2}, $\beta = 1.1 \times 10^5$ s^{-2}, and $K_P = 0.043$ mM^{-1} ($r^2 = 0.999$) for product. For IFI, both sum and product equaled $\alpha + \beta / (\chi + K_P P)$, where $\alpha = 6350$ s^{-1}, $\beta = 7768$ s^{-1}, and $K_P = 0.015$ mM^{-1} ($r^2 = 0.982$) for sum, and $\alpha = 1$ s^{-2}, $\beta = 6.03 \times 10^6$ s^{-2}, and $K_P = 0.064$ mM^{-1} ($r^2 = 0.982$) for product.

the work-producing (*B*) and work-absorbing (*C*) steps of the cross-bridge cycle (see Section 2 above), from which the corresponding apparent rate constants $2\pi b$ and $2\pi c$ can be calculated (Zhao and Kawai, 1993). Because these rates are expressed as a sum ($2\pi b + 2\pi c$) and product ($2\pi b \times 2\pi c$) in the transient solutions of the cross-bridge schemes presented below, we plotted both sum and product as functions of [Pi], following the method of Kawai et al. (1993).

Fig. 6 shows that both sum and product *decrease* with [Pi] in IFI, whereas in EMB both sum and product *increase* with [Pi]. The curves appear to be rectangular hyperbolas. The direction of the phosphate dependency is consistent with that of f_{max} noted above, where, again, in EMB fibers the [Pi] dependencies of the sum and product of $2\pi b$ and $2\pi c$ are similar to those reported in vertebrate striated muscle studies, whereas in IFI fibers the dependency is reversed. This remarkable qualitative difference in phosphate dependency of the apparent rate constants of the two major force-altering processes in EMB and IFI suggests marked differences in cross-bridge cycle kinetics.

We provisionally interpret the results in terms of the following cross-bridge schemes. Scheme 1 is consistent with the hypothesis that the rate-limiting step of the cross-bridge cycle occurs *after* phosphate release in EMB, the usual assumption applied to vertebrate muscle schemes. Scheme 2 is consistent with the hypothesis that the rate-limiting step occurs *before* phosphate release in IFI.

Scheme 1: EMB

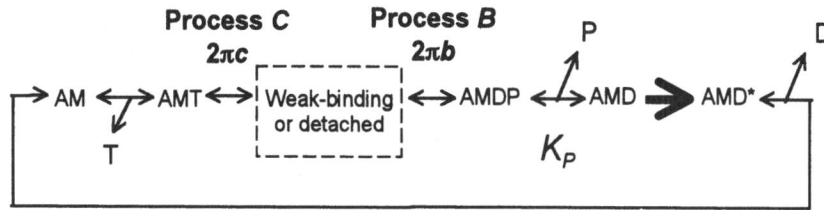

Scheme 2: IFI

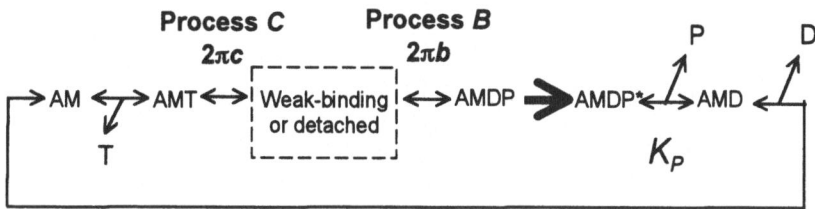

A is actin, *M* is myosin, *T* is MgATP, *D* is MgADP, and *P* is phosphate. The asterisk indicates a different conformational state. K_p is the association constant for phosphate binding (other kinetic constants omitted for simplicity). The hatched box denotes weak-binding or detached states of myosin; all other states are strong-binding. The **bold** arrow denotes the rate-limiting step of the cross-bridge cycle.

Following the method of Zhao & Kawai (1993), we obtained a transient solution of Scheme 1 (eq. 2): $2\pi b + 2\pi c = \alpha + \beta K_P P / (1 + K_P P)$, where phosphate concentration (P) is the independent variable, K_P is the association constant of P with myosin, and α and β are constants that are composites of the association constants for MgATP and MgADP ([MgATP] and [MgADP] are assumed constant). Eq. 2 is consistent with the phosphate dependency of EMB, but not IFI (Fig. 6). A non-linear least-squares fit of eq. 2 to the EMB data (solid curves, Fig. 6) yields $K_P = 0.05$ mM^{-1}. For the product of $2\pi b$ and $2\pi c$, a similar expression is obtained, yielding a similar value of K_P (0.04 mM^{-1}). Using the same approach, we obtained a transient solution of Scheme 2 (eq. 3): $2\pi b + 2\pi c = \alpha + \beta / (\chi + K_P P)$, where α, β and χ are constants. Eq. 3 is consistent with the phosphate dependency of IFI, but not EMB (Fig. 6). A non-linear least-squares fit of this expression to the IFI data (solid curve, Fig. 6) yields $K_P = 0.02$ mM^{-1}. The corresponding expression for the product of $2\pi b$ and $2\pi c$

yields $K_P = 0.06$ mM^{-1}. Averaging the two values for IFI yields a K_P of ~ 0.04 mM^{-1}, compared to ~0.05 for EMB. This result suggests that differences in kinetics between IFI and EMB may be only mildly affected by the relatively small difference in phosphate affinity, if any, a conclusion supported by the surprisingly small difference in actin-activated ATPase reported by Littlefield et al. (submitted). The corresponding value of K_P in ferret myocardium is 0.06 mM^{-1} (Kawai et al., 1993); rabbit psoas muscle, 0.07 mM^{-1} (Kawai and Halvorson, 1991); porcine myocardium, 0.10 mM^{-1} (Zhao et al, 1996), bovine myocardium, 0.14 mM^{-1} (Fujita et al., 2002); and rabbit soleus muscle, 0.18 mM^{-1} (Wang and Kawai, 1996). Note the rank order from fast to slow muscle fiber types. Compared with vertebrate striated muscle, phosphate appears to have a relatively low affinity for myosin in *Drosophila* indirect flight and embryonic muscles, with IFI having the least affinity of all.

The effect of phosphate on the amplitude of isometric force and oscillatory work can also be accounted for by schemes 1 and 2. In scheme 1, by dint of the phosphate release step occurring *before* the rate-limiting step, scheme 1 predicts that increasing phosphate concentration will push the population of force-generating cross-bridges from the AMDP state toward the weak-binding or detached state. The amplitude of isometric force- and oscillatory work output will therefore be inhibited according to scheme 1, consistent with the EMB result. In contrast, by dint of the phosphate release step occurring *after* the rate-limiting step, scheme 2 predicts that the phosphate effect will be opposite to that of scheme 1, since elevated [Pi] will push the population of cross-bridges toward the force-maintaining AMDP* state at the expense of depopulating the weak-binding or detached state, consistent with the IFI result.

In summary, the remarkable difference in phosphate response between IFI and EMB is consistent with the rate-limiting step of the cross-bridge cycle being associated with ADP release in EMB, in contrast to the rate-limiting step being associated with phosphate release in IFI. Since the affinity of phosphate for myosin is marginally less in the faster IFI fibers compared to EMB fibers, that difference is insufficient to account for the 6-fold difference in IFI and EMB fiber kinetics (f_{max}). Future studies designed to examine the response of fiber kinetics to changes in [MgATP] and [MgADP] will allow us to assess the validity of the cross-bridge schemes we have proposed. Since structural differences between the EMB and IFI myosin heads exist in only 4 specific variable regions of the S1 subunit (Fig. 1), at least one of these regions must be responsible for the kinetic differences observed in the two isoforms. Analysis of IFM/EMB myosin chimeras will allow us to map the kinetic differences to specific variable regions of myosin.

4. ACKNOWLEDGEMENTS

We acknowledge the help of William Barnes. This work was supported by grants from the National Institutes of Health (DWM: HL 68034) and the American Heart Association, Western Affiliate (DMS: 0120022Y).

5. REFERENCES

Barany, M., Barany, K., Reckard, T., and Volpe, A., 1965, Myosin of fast and slow muscles of the rabbit. *Arch Biochem. Biophys* **109**: 185-91.

Eddinger, T.J. and Moss, R.L., 1987, Mechanical properties of skinned single fibers of identified types from rat diagragm. *Am. J. Physiol* **252**: C210-C218

Fujita, H., Sasaki, D., Ishiwata, S., and Kawai, M., 2002, Elementary steps of the cross-bridge cycle in bovine myocardium with and without regulatory proteins. *Biophys. J.* **82**: 915-928.

Kawai, M., and Brandt, P.W., 1980, Sinusoidal analysis: a high resolution method for correlating biochemical reactions with physiological processes in activated skeletal muscles of rabbit, frog and crayfish. *J. Muscle Res. Cell Motil.* **1**: 279-303.

Kawai, M., and Halvorson, H.R., 1989, Role of MgATP and MgADP in the cross-bridge kinetics in chemically skinned rabbit psoas fibers. *Biophys. J.* **55**: 595-603.

Littlefield, K.P., Swank, D.M., Sanchez, B.M., Knowles, A.F., Warshaw, D.M., and Bernstein, S.I, 2002, The converter domain modulates the kinetic properties of Drosophila myosin, submitted, *Am. J. Physiol.*

Mulieri, L.A., Barnes, W., Leavitt, B.J., Ittleman, F.P., LeWinter, M.M., Alpert, N.R., and Maughan, D.W., 2002, Alterations of myocardial dynamic stiffness implicating abnormal crossbridge function in human mitral regurgitation heart failure. *Circ. Res.* **90**: 66-72.

Reiser, P.J., Moss, R.L., Giulian, G.G., and Greaser, M.L., 1985, Shortening velocity in single fibers from adult rabbit soleus muscles is correlated with myosin heavy chain composition. *J. Biol. Chem.* **260**: 9077-80.

Swank, D.M., Knowles, A.F., Sarsoza, F., Suggs, J.A., Maughan, D.W., and Bernstein, S.I. 2002. The myosin converter domain modulates muscle performance. *Nature Cell Bio.* **4(4)**, 312-316. DOI: 10.1038/ncb/776.

Swank, D.M., Bartoo, M.L., Knowles, A.F., Iliffe, C., Bernstein, S.I., Molloy, J.E., and Sparrow, J.C. 2001. Alternative exon-coded regions of Drosophila myosin heavy chain modulate ATPase rates and actin sliding velocity. *J. Biol. Chem.* **276**: 15117-24.

Sweeney, H.L., Kushmerick, M.J., Mabuchi, K., Gergely, J., and Streter, F.A. 1986. Velocity of shortening and myosin isozymes in two types of rabbit fast-twitch muscle fibers. *Am. J. Physiol* **251**: C431-C434

Sweeney, H.L., Kushmerick, M.J., Mabuchi, K., Sreter, F.A., and Gergely, J. 1988. Myosin alkali light chain and heavy chain variations correlate with altered shortening velocity of isolated skeletal muscle fibers. *J. Biol. Chem.* **263**: 9034-9.

Wang, G.Y. and Kawai. M. 1996. Effects of MgATP and MgADP on the cross-bridge kinetics of rabbit soleus muscle fibers. *Biophys. J.* **71**: 1450-1461.

Wells, L., Edwards, K. A., and Bernstein, S. I. 1996. Myosin heavy chain isoforms regulate muscle function but not myofibril assembly. *EMBO J.* **15**(17), 4454-4459.

Zhao, Y. and Kawai, M. 1996. Inotropic agent EMD-53998 weakens nucleotide and phosphate binding to cross bridges in procine myocardium. *Am. J. Physiol.* **271**: H1394-H1406.

Zhao, Y. and Kawai, M. 1993. The effect of lattice spacing change on cross-bridge kinetics in chemically skinned rabbit psoas muscle fibers. II. Elementary steps affected by the spacing change. *Biophys. J.* **64**: 197-210.

DISCUSSION

Metzger: Nice talk, David, do you know the physiological range of inorganic phosphate in fly muscle – does it get in the 10 mM range?

Maughan: I am not aware of any measures of inorganic phosphate in flight muscle of the fruit fly, but the phosphate concentration during a period of flight activity in blow flies has been reported to be 7-8 umole per gram wet weight (Sacktor et al., J. Biol. Chem. 241(3), 1966, 632). Assuming a similar range in fruit flies, it is possible that the phosphate concentration extends to 10 mM or above. We set our standard conditions at 8 mM inorganic phosphate.

Cecchi: Are there any differences in the elements that contribute to the mechanical resonance of the wing between the embryonic and the normal animal?

Maughan: Any differences in unitary stiffness between embryonic and indirect flight muscle myosin could alter the mechanical resonance of the flight system, but these differences are probably slight since the major alternations seem to be in kinetics.

DOES CROSS-BRIDGE ACTIVATION DETERMINE THE TIME COURSE OF MYOFIBRILLAR RELAXATION ?

Robert Stehle[*], Martina Krüger[*], Gabriele Pfitzer[*]

ABSTRACT

The ability of force-generating cross-bridges to activate the thin filament in cardiac muscle was tested by studying the effects of initial force and [MgADP] on force relaxation kinetics in subcellular myofibrillar bundles prepared from left ventricles of the guinea pig. Relaxation was initiated by rapidly reducing the [Ca^{2+}] from pCa 4.5 to 7.5. Initiating relaxation from lower force levels during pre-steady-state force development did not significantly accelerate the kinetics of the force decay compared to relaxations initiated from steady-state force development. This suggests that the force-generating cross-bridges which become formed during maximally Ca^{2+}-activated steady-state contractions do not maintain thin filament activation for significant enough times after Ca^{2+}-removal to exert a rate-limiting influence on force relaxation kinetics.

Adding 2 mM MgADP to solutions slowed down relaxation kinetics $\approx$ 4-fold. To differentiate whether these slower kinetics result from either (1) MgADP favoring accumulation of cross-bridges during the preceding contraction in a state of activating capability or (2) slow-down of cross-bridge turnover by the presence of the product MgADP during relaxation, the [MgADP] was either increased or removed at the time of Ca^{2+}-removal. The addition of 2 mM MgADP to activating solutions (subsequent relaxation in the absence of MgADP) slowed-down the kinetics of the initial, slow, linear force decay following Ca^{2+}-removal $\approx$1.5-fold, suggesting that the high [MgADP] during contraction favors formation of cross-bridges which contribute in rate-limiting early relaxation kinetics by transiently sustaining thin filament activation. On the other hand, the addition of 2 mM MgADP to the relaxing solution (preceding Ca^{2+}-activation in absence of MgADP) slowed-down the kinetics of the initial force decay $\approx$ 3-fold, more similar to the kinetics observed in the

---

[*] Institute of Physiology; University of Cologne, Robert-Koch-Str. 39, D-50931 Köln, Germany

continuous presence of 2 mM MgADP both before and after the Ca^{2+}-removal. This suggest that, despite some influence of cross-bridge activation, the main effect of MgADP on relaxation kinetics results from product inhibition of cross-bridge turnover. In summary, whereas under certain conditions (high [MgADP]) cross-bridge activation of the thin filament can weakly take part in rate-limiting relaxation kinetics induced by complete Ca^{2+}-removal, cross-bridge activation does not influence relaxation kinetics under more physiologically normal conditions.

1. INTRODUCTION

Relaxation of striated muscle is initiated by Ca^{2+}-dissociation from troponin C. Dissociation induces complex conformational changes involving all regulatory proteins (troponin I, troponin T, Tropomyosin), which altogether turn the thin filament to an ´off´-state, which effectively results in net-detachment of force-generating cross-bridges and thereby the force to decay. However, there is evidence that, in addition to Ca^{2+}, also strongly-bound cross-bridges can activate the thin filament in a cooperative manner ([1-3] and references cited therein). This raises the question of whether the cross-bridges which became formed during contraction can delay the switch-off of the thin filament when the Ca^{2+} is removed from the myofilaments by facilitating Ca^{2+}-binding to Troponin C and/or by allosterically inhibiting Troponin-Tropomyosin from turning off. Strongly-bound cross-bridges, if they can sustain thin filament activation after the Ca^{2+}-removal for significant times, will apparently slow-down the kinetics of the force decay because of the continued attachment of cross-bridges to the turned-on thin filament. If so, relaxation kinetics should depend on the number and the state of cross-bridges which presently generate force at the time relaxation is initiated. The more the force-generating cross-bridges are attached to the thin filament and the stronger their ability to delay the turn-off of the thin filaments, then the more the kinetics of the force decay will be slowed down. Therefore, in order to examine the activating ability of cross-bridges, our strategy was to test whether the fraction of force-generating cross-bridges and/or their distribution among strongly-bound cross-bridge states influence(s) force decay kinetics.

Previously, Poggesi & colleagues [4] and our research group [5] showed that the kinetics of force decays in skeletal and cardiac myofibrils following rapid, complete Ca^{2+}-removal are markedly biphasic: an initial slow, linear decay (rate constant k_{LIN}) lasting for a time t_{LIN} is followed by a fast exponential decay (rate constant: k_{REL}). We further showed that during the initial, slow force decay, all sarcomeres remain isometric, whereas during the subsequent fast force decay, sarcomeres relax individually in succession along the myofibril accompanied by large changes in lengths of the individual sarcomere [5]. Hence, the initial, slow force decay contains information about early changes of cross-bridge kinetics under truely isometric conditions, with k_{LIN} depending on the cross-bridge turnover rates present after Ca^{2+}-removal, and t_{LIN} reflecting the time during which strain among sarcomeres remains in a balance [4;5]. Both these parameters are reporters of early events during relaxation which should be sensitively affected by cross-bridge attachment, if transient maintenance of cross-bridge activation is of any relevance in rate-limiting mechanical relaxation kinetics.

2. METHODS

Thin, subcellular bundles of few cardiac myofibrils (2–2.5 μm in diameter, 25–70 μm in length) were prepared from left ventricles of the guinae pig and installed in an apparatus based on the principle of atomic force microscopy as described previously [5,6]. Slack sarcomere lengths were 1.98 ± 0.04 μm (mean ± SD). Before activation, bundles were pre-stretched by 15 % of their slack length. Bundles were activated/relaxed based on the technique developed by Colomo et al. [7] by rapidly translating the interface between two continuously flowing, laminar streams of solutions within deadtimes of 10 - 30 ms (illustrated in [6]).

Experiments were performed at 10°C. In order to investigate the influence of cross-bridge activation on relaxation kinetics independent from the effects of Ca^{2+}-activation, the same $[Ca^{2+}]$ was used in activating solutions (pCa 4.5) and relaxing solutions (pCa 7.5) for all experiments. Activating or relaxing solutions for experiments shown in Fig. 1 and Fig. 2. contained 10 mM imidazole, 3 mM $CaCl_2K_4EGTA$ (activating solution) or 3 mM K_4Cl_2EGTA (relaxing solution), 1 mM Na_2MgATP, 4 mM $MgCl_2$, 47.7 mM Na_2CrP, 2 mM DTT, adjusted to pH 7.0 at the experimental temperature. To prevent rephosphorylation of MgADP to MgATP, the Na_2CrP was replaced by 143 mM K-propionate in the solutions for experiments at various [MgADP] (cf. Fig. 3 and Fig. 4).

Kinetic parameters of myofibrillar relaxation (k_{LIN}, t_{LIN}, and k_{REL}; see introduction for meaning of parameters) were obtained by fitting force relaxation transients starting at the time of Ca^{2+}-removal to a function consisting of a linear and an exponential term, as described previously [6].

3. RESULTS

3.1. Effect of the fraction of force-generating cross-bridges on relaxation kinetics

If the cross-bridges which become formed during a maximally Ca^{2+}-activated steady-state contraction can rate-limit relaxation by sustaining thin filament activation after Ca^{2+}-removal, then one would expect the force to decay faster if relaxation is initiated at lower force levels. To test this hypothesis, we initiated relaxation at different pre-steady force levels during the rising phase of maximally Ca^{2+}-induced force development, i.e., by varying the time of Ca^{2+}-activation (Figure 1A). The part of the transients showing their force decays induced after Ca^{2+}-removal are shown in Figure 1B. After normalization of force transients to their initial force level at the time of the Ca^{2+}-removal, they superimpose without obvious deviations (Figure 1C). This suggests that the early kinetics of the myofibrilar force decay

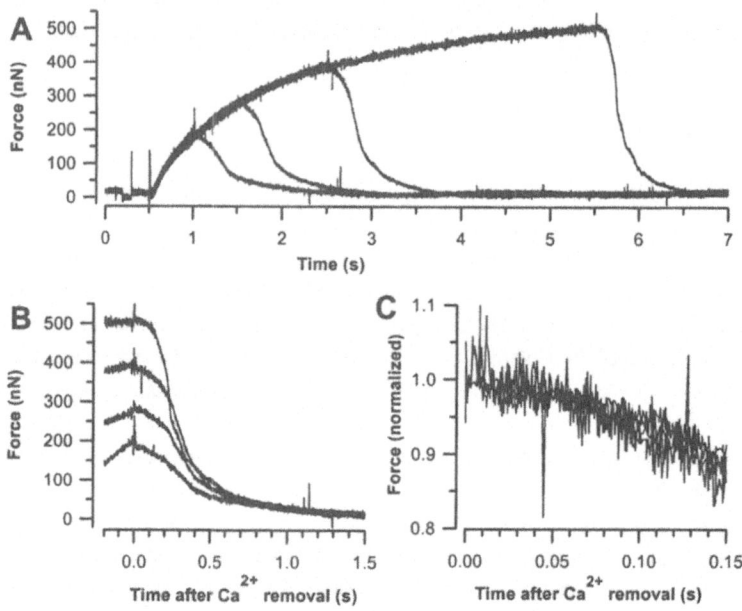

Figure 1. Kinetics of cardiac myofibrillar relaxation induced by rapidly changing the pCa from 4.5 to pCa 7.5 after different times of Ca^{2+}-activation. A. Illustration of the experimental protocol. Four transients are superimposed. In each expriement the zero force level was determined by slackening the myofibril in relaxing solution (time 0.2-0.3 s) and then the myofibril was activated by switching from pCa 7.5 to 4.5 at time 0.5 s. Relaxation was initiated at the times 1 s, 1.5 s, 2.5 s, and 5.5 s, respectively. B. Same transients as in A on shifted time scales (time of Ca^{2+}-removal in each case set to 0) to facilitate comparison of the full force decays following Ca^{2+}-removal. C. Same as in B, but each transient normalized to its force present at Ca^{2+}-removal. The present time frame in C was selected to enlarge the slow, initial, linear phases in the force decays of the four transients.

induced by rapidly reducing the $[Ca^{2+}]$ from fully activating to fully relaxing concentrations are independent of the initial fraction of force-generating cross-bridges. It is noteworthy that rapid Ca^{2+}-removal induced the force to decay early (within the first 50 ms after the removal) in all transients (Figure 1C), regardless of the preceding increase of force transients before Ca^{2+}-removal (see Figure 1B) in the relaxation experiments initiated during pre-steady-state force development. This indicates very fast switch-off kinetics of the thin filament which appear to be mostly completed very early (50 ms or even earlier) after Ca^{2+}-removal.

Force transients were fitted to obtain the kinetic parameters of the initial-slow, linear relaxation phase shown in Figure 2, where they are plotted against the relative force at the moment of Ca^{2+}-removal. The time of the slow, initial, linear force decay (t_{LIN}) showed some slight, non-significant ($P > 0.05$) tendency to be prolonged (by ≈ 20 %) in the relaxations

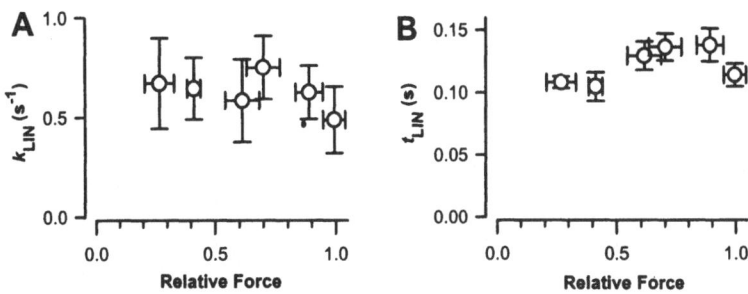

Figure 2. Dependence of the time (A) and the rate constant (B) of the initial, slow, linear force decay on the initial force present at Ca^{2+}-removal. Initial force is normalized to the steady-state force level. Data was pooled from 7 myofibrillar bundles. Each data point gives the mean ± SE of 4 – 6 experiments.

initiated at the intermediate to the high pre-steady-state force levels, compared to the relaxations initiated during the force plateau. However, any prolongation of t_{LIN} with lower initial force would contrast even to the hypothesis that the slow, linear phase of relaxation reflects a lag phase due to transient maintenance of cross-bridge activation. The results therefore suggest that under 'normal' conditions, i.e., low [MgADP] and saturating [MgATP], cross-bridge activation by force-generating cross-bridges does not delay myofibrillar relaxation induced by a sudden, complete Ca^{2+}-removal.

3.2. Effect of cross-bridges formed at high MgADP on relaxation kinetics

Figure 3 shows that compared to controls in the absence of MgADP (force relaxation transient marked by ∇), adding 2 mM MgADP to both activating and relaxing solutions strongly slowed down relaxation kinetics (force relaxation transient marked by ⌐). On average, adding 2 mM MgADP to both solutions caused similar slow downs in all force decay parameters (Figure 4): k_{LIN} decreased ≈ 4-fold, t_{LIN} prolonged 4-5-fold, and k_{REL} decreased ≈ 4-fold. These strong effects could result from: (1) MgADP accumulating cross-bridges in an AM.ADP-state of high activating capacity and/or (2) MgADP through product inhibition apparently slowing down the forward turnover rate by which cross-bridges leave force-generating states. Whereas the former activating effect (1) is determined by the present occupancy of individual cross-bridge states, the latter reduction of turnover rate (2) is defined by the present [MgADP].

Therefore, if MgADP is present during concentration and then removed together with the Ca^{2+}, it can no longer inhibit turnover kinetics. However, the presence of MgADP during only the preceding Ca^{2+}-activation will still influence relaxation kinetics by cross-bridge activation, as long as the distribution among strongly-bound, force-generating cross-bridges remains stable after MgADP-removal. The latter requirement is probably fulfilled for significant times, since we observe slow, exponential force decay kinetics (amplitudes 10 -

15 % of total force with rate constants of ≈ 1.0 s^{-1}) when MgADP is removed during Ca^{2+}-activation at pCa 4.5 (transients not shown). Therefore, to estimate the contribution of cross-bridge activation to the slowed down relaxation kinetics at high MgADP, the [MgADP] was removed simultaneously with the Ca^{2+}. Practically, this was easy to perform by adding 2 mM MgADP only to the activating solution. Figure 3 shows that the kinetics of the force relaxation transient following simultaneous removal of MgADP and Ca^{2+} (transient marked by Δ) are more similar to the kinetics of the transient obtained in continuous absence of MgADP (∇) than to the transient obtained in the continuous presence of 2 mM MgADP ($\square$). This suggests that cross-bridge activation is not the main cause for slowed-down relaxation at high [MgADP]. Nevertheless, early force relaxation kinetics were slowed down by the sole presence of MgADP before Ca^{2+}-removal (see the later onset of rapid tension fall in transient Δ compared to transient ∇ in Figure 3). As argued above, due to the absence of MgADP during relaxation, these minor effects cannot be due to product inhibition and therefore provide evidence that minor, but nevertheless highly significant ($P < 0.01$ in the case of t_{LIN}, see error bars in Figure 4) rate-limiting effects of cross-bridge activation on relaxation kinetics can be found at high [MgADP].

In order to test the contribution of product inhibition to the slow-down of force relaxation kinetics in the presence of MgADP, 2 mM MgADP were added only to the relaxing solution. Figure 3 shows that the force relaxation transient following a jump to 2 mM MgADP at Ca^{2+}-removal (marked by o) is similar to the transient obtained in the continuous presence of 2 mM MgADP ($\square$) but very different from the transient in the continuous absence of MgADP (∇). Although the increased [MgADP] after Ca^{2+}-removal will gradually form more strongly activating cross-bridges, it is unlikely that this would explain the slow-down of the relaxation, because an identical increase of [MgADP] applied during steady-state Ca^{2+}-contractions caused slow force developments with exponential rate constants of ≈ 0.7 s^{-1} (transients not shown). From this, one can estimate that only about 10 % of the cross-bridges will have changed towards the new steady-state distribution at 2 mM ADP after a time of 0.2 s, i.e., after about the time of t_{LIN} when force breaks down rapidly during relaxation in the absence of MgADP.

Figure 4 summarizes the relative effects of 2 mM MgADP during contraction and/or relaxation on the kinetic parameters of myofibrillar relaxation. It is noteworthy that, for each kinetic parameter, its relative changes (caused by the presence of MgADP either in activating solution or in relaxing solution) multiplied with each other will give a value similar to the relative change caused by the continuous presence of MgADP in both activating and relaxing solution. This is further evidence that the present experiments allow one to estimate the relative contributions of the individual effects of product inhibition and cross-bridge activation on relaxation kinetics at high [MgADP]. Figure 4 also shows that the prehistory of contraction (presence or absence of MgADP in activating solution) affects only the parameters of the initial, slow, linear force decay but not the kinetics of the rapid exponential force decay.

Relative to the active tension in the absence of MgADP, the addition of 2 mM MgADP to activating solutions increased the active tension of myofibrils to 113 ± 3 %, and the

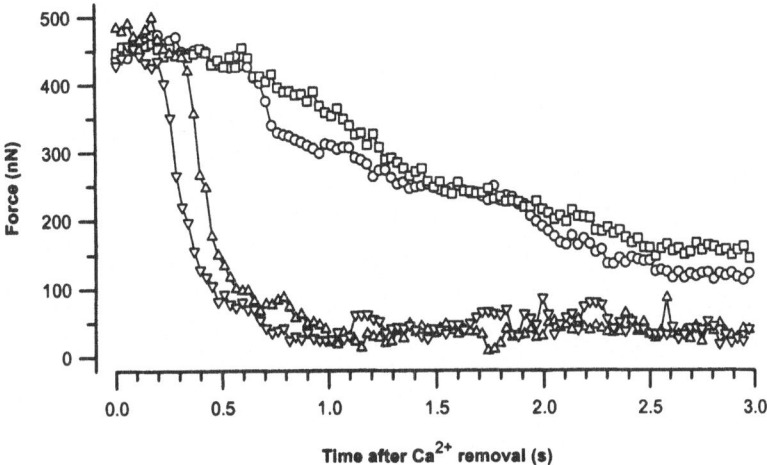

Figure 3. Effect of initial [MgADP] and final [MgADP] on cardiac myofibrillar force decay following rapid Ca^{2+}-removal (switching from pCa 4.5 to 7.5 at time = 0). Force transients were recorded: in the absence of MgADP both before and after Ca^{2+}-removal (∇); in the presence of 2 mM MgADP both before and after Ca^{2+}-removal ($\square$); in the presence of 2 mM MgADP before Ca^{2+}-removal and the absence of MgADP after Ca^{2+}-removal (Δ); and in the absence of MgADP before Ca^{2+}-removal and the presence of 2 mM MgADP after Ca^{2+}-removal (o).

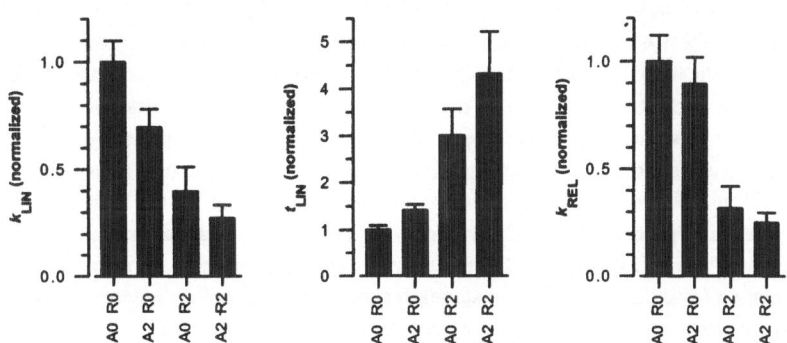

Figure 4. Relative effects of the presence of 2 mM MgADP during the preceding Ca^{2+}-activation and/or during relaxation on kinetic parameters of the cardiac myofibrillar force decay following rapid Ca^{2+}-removal. Numbers in the figure indicate the [MgADP] (in mM) added to activation solution (abbreviation: A) and relaxing solution (abbreviation: R). Kinetic parameters are normalized to their values determined in the absence of MgADP in both activating and relaxing solutions. Data was pooled from 4 myofibrilar bundles. Bars show means ± SE (n = 4-8). Note that the kinetics of the rapid exponential force decay (rate constant k_{REL}) does not significantly depend on the presence or absence of MgADP during the preceding Ca^{2+}-activation, i.e., it is determined only by the [MgADP] present after the Ca^{2+}-removal.

addition of 2 mM MgADP to relaxing solutions increased the passive tension from 4 ± 2 % to 7 ± 2 % (values are means $\pm$ SE, $n = 6$). The fact that much stronger effects of MgADP on relaxation kinetics are obtained compared to its effects on steady-state force parameters supports the conclusion that MgADP affects relaxation kinetics more so by changing the distribution within strongly-bound cross-bridge states and by effects on turnover rates than by changing the number of cross-bridges. Altogether the results suggest that, in contrast to the situation at low [MgADP], sustained activation by cross-bridges can make a minor contribution to rate-limiting the initial, slow, linear phase of isometric relaxation at high [MgADP]. Nevertheless, the fact that the relaxation kinetic parameters are more highly sensitive to the [MgADP] present after Ca^{2+}-removal than to the [MgADP] present before Ca^{2+}-removal suggests that the main mechanism causing slow relaxation at high [MgADP] is product inhibition of cross-bridge turnover kinetics defined by the free [MgADP].

4. RELATION TO PREVIOUS STUDIES

The present results suggest that cross-bridges which are formed during maximum Ca^{2+}-activation under 'normal' conditions (low [MgADP]) are not capable of sustaining thin filament activation after the removal of Ca^{2+} in a manner to exert rate-liming effects on force relaxation kinetics; whereas a small additional amount of cross-bridges becoming formed at high [MgADP] are capable of this.

Previously, Tesi et. al. [4] and our research group [5] showed that the rate constant k_{LIN} of the initial, slow linear force decay following rapid Ca^{2+}-removal from myofibrils is similar to the rate constant of force redevelopment (k_{TR}) obtained at low partial Ca^{2+}-activations. Based on the concensus that k_{TR} reflects turnover rate constants of the cross-bridge cycle [8], this was consistent with the interpretation that the kinetics of the initial, slow force decay which occurs under isometric conditions at the single sarcomere level [5] is determined by the apparent rate by which cross-bridges leave force-generating states [4,5], thus, implying a rapid, complete switch-off of the thin filament after Ca^{2+}-removal so that no new force-generating cross-bridges are formed. This interpretation was further supported by the finding that k_{LIN} was largely independent of the [Ca^{2+}] during the preceding relaxation, implying neglible influence of the initial level of force-generating cross-bridges and of preceding Ca^{2+}-activation on relaxation [4]. The present approach aims to more selectively test the influence of force-generating cross-bridges by keeping constant both the initial and final Ca^{2+}-concentrations and by varying the initial percentage of force-generating cross-bridges through varying the time of Ca^{2+}-activation. The present result, that relaxation kinetics are not accelerated when Ca^{2+} is removed at lower force levels during pre-steady-state force development, complementarily corroborates the previous conclusions [4,5] that activation by cycling force-generating cross-bridges under 'normal' conditions (low MgADP and saturating MgATP) does not rate-limit the mechanical relaxation process.

On the other hand, high [MgADP] appears to favor the accumulation of cross-bridges in a state which can transiently maintain activation after Ca^{2+}-removal and which thereby causes slower relaxation rates. Activating effects of MgADP had been described in previous studies. High [MgADP] had been shown to increase force redevelopment kinetics at submaximal levels of Ca^{2+}-activation, i.e., in the opposite direction than expected from the inhibitory effects of MgADP on cycling rates, and had therefore indicated that MgADP accumulates cross-bridges in a strongly-bound state able to enhance thin filament activation at partial 'Ca^{2+}-activations [9]. In the presence of 40 % (v/v) of the relaxing agent ethylene glycol, which reduces myofibrillar ATPase in the presence of Ca^{2+} to basal relaxing levels, MgADP activates the ATPase of myofibrils in a Ca^{2+}-dependent manner [10]. At slightly higher (> 3/1) ratios of MgADP/MgATP than used here (2/1), myofibrils and fibers start to develop active force in the absence of Ca^{2+}, with force-MgADP relations exhibiting pronounced cooperativity [11]. Altogether, these findings are clear evidence that at high [MgADP] strongly-bound cross-bridges become formed which can cooperatively turn-on the thin filament.

The question of whether activation by strongly-bound cross-bridges can influence relaxation rates had been previously investigated by flash photolysis of Ca^{2+}-chelators in skinned skeletal and cardiac fibers incubated with NEM-S1, a strongly-binding cross-bridge derivative [12;13]. Both fiber types treated with NEM-S1 showed reduced relaxation rates consistent with the interpretation that the derivative maintains the thin filament in an activated state for an extended duration after Ca^{2+}-removal [12;13]. Isolated S1 is also very effective for turning on thin filaments if it has bound MgADP, as shown by the strong cooperative binding of S1 to myofibrils in the presence of MgADP and the absence of Ca^{2+} [14]. These findings with S1-fragments support our conclusion that high [MgADP] favors the formation of a strongly-bound cross-bridge state which is able to slow down relaxation kinetics by transiently maintaining thin filament activation.

However, the exact reason why force-generating cross-bridges under more physiological buffer conditions are unable to sustain thin filament activation after Ca^{2+}-removal, whereas a certain fraction of cross-bridges formed at high [MgADP] are able to do so, remains undiscovered. Our results give reason to speculate that this qualitative difference results from comparatively higher activating capacities of certain strongly-bound, force-generating cross-bridge states. Studies in solution, e.g., revealed that more S1.ADP than S1 heads have to bind to reconstituted thin filaments to switch them on [15]. However, little is known about the extent to which such state-effects can be transformed to the strongly-bound states of cycling cross-bridges in fibers.

Despite the interesting effects of cross-bridge activation at high [MgADP] on relaxation, the present study reveals that MgADP mainly affects relaxation kinetics in a more direct manner by product inhibition and may therefore be used as a sensitive tool to probe kinetics of the cross-bridge cycle. Previous studies had shown that MgADP reduces relaxation rates in skinned skeletal fibers [16], in skinned trabeculae [17], and in skeletal myofibrils [4]. 2 mM MgADP reduced relaxation rates in skinned trabeculae of the guinea pig ≈ 5-fold [17], which would be similar to the ≈ 4-fold reduction found here in cardiac myofibrils of the same species. In rabbit psoa myofibrils, the addition of 3 mM MgADP reduced relaxation rates only ≈ 1.5 – 2-fold [4]. It could be that these differences reflect different intrinsic kinetic properties of the different MHC isoforms, i.e., β-MHC in the guinea pig ventricle and α-

MHC in the rabbit psoas. However, as also a lower background of [MgATP] had been used in our study relative to Tesi et al.s´ study [4] on psoa myofibrils, it is also possible that the different relative changes in relaxation kinetics arise from the competitive binding mechanism of MgADP and MgATP to the active site [18;19]. This means that not only the [MgADP] but also the ratio of [MgADP]/[MgATP] determine the apparent rate of cross-bridge detachment. Therefore, further experiments at different [MgADP] and [MgATP] will be needed to understand the kinetic mechanism in more detail and to manifest possible kinetic differences between MHC-isoforms.

AKNOWLEDGEMENTS

This work was supported in part by a DFG grant (SFB612-A2) and by Köln Fortune (#28/2002).

5. REFERENCES

1. Gordon, A.M., E. Homsher, and M. Regnier. Regulation of contraction in striated muscle. *Physiol Rev.* **80**, 853-924 (2000).
2. Tobacman, L.S. Thin filament-mediated regulation of cardiac contraction. *Annu.Rev.Physiol* **58**, 447-481 (1996).
3. Solaro,RJ, and H.M. Rarick. Troponin and tropomyosin: proteins that switch on and tune in the activity of cardiac myofilaments. *Circ. Res.* **83**, 471-480 (1998).
4. Tesi, C., N. Piroddi, F. Colomo, and C. Poggesi. Relaxation kinetics following sudden Ca2+ reduction in single myofibrils from skeletal muscle. *Biophys J.* **83**, 2142-2151 (2002).
5. Stehle, R., M. Krüger, and G. Pfitzer. Force kinetics and individual sarcomere dynamics in cardiac myofibrils after rapid Ca2+ changes. *Biophys J.* **83**, 2152-2161 (2002).
6. R. Stehle, M. Krüger, and G. Pfitzer. Isometric force kinetics upon rapid activation and relaxation of mouse, guinea pig, and human heart muscle studied on the subcellular myofibrillar level. *Bas.Res.Cardiol.* **226** (Suppl. 1), I/127-I/135 (2002).
7. Colomo, F., S. Nencini, N. Piroddi, C. Poggesi, and C. Tesi. Calcium dependence of the apparent rate of force generation in single striated muscle myofibrils activated by rapid solution changes. *Adv.Exp.Med.Biol.* **453**, 373-381 (1998).
8. Brenner, B. Effect of Ca2+ on cross-bridge turnover kinetics in skinned single rabbit psoas fibers: implications for regulation of muscle contraction. *Proc.Natl.Acad.Sci.* U.S.A **85**, 3265-3269 (1988).
9. Lu, Z., D.R. Swartz, J.M. Metzger, R.L. Moss, and J.W. Walker. Regulation of force development studied by photolysis of caged ADP in rabbit skinned psoas fibers. *Biophys J.* **81**, 334-344 (2001).
10. Stehle, R., C. Lionne, F. Travers, and T. Barman. Probing the coupling of Ca2+ and rigor activation of rabbit psoas myofibrillar ATPase with ethylene glycol. *J.Muscle Res.Cell Motil.* **19**, 381-392 (1998).
11. Shimizu, H., T. Fujita, and S. Ishiwata. Regulation of tension development by MgADP and Pi without Ca2+. Role in spontaneous tension oscillation of skeletal muscle. *Biophys J.* **61**, 1087-1098 (1992).
12. Patel, J.R., G.M. Diffee, X.P. Huang, and R.L. Moss. Phosphorylation of myosin regulatory light chain eliminates force- dependent changes in relaxation rates in skeletal muscle. *Biophys J.* **74**, 360-368 (1998).
13. Fitzsimons, D.P., J.R. Patel, and R.L. Moss. Cross-bridge interaction kinetics in rat myocardium are accelerated by strong binding of myosin to the thin filament. *J.Physiol* **530**, 263-272 (2001).
14. Zhang, D., K.W. Yancey, and D.R. Swartz. Influence of ADP on cross-bridge-dependent activation of myofibrillar thin filaments. *Biophys J.* **78**, 3103-3111 (2000).
15. Schaertl, S., S.S. Lehrer, and M.A. Geeves. The influence of ADP on the S1-induced switching-on of thin filaments. *J.Muscle Res.Cell Motil.* **16**, 151 (1995).

16. Lipscomb, S., R.E. Palmer, Q. Li, L.D. Allhouse, T. Miller, J.D. Potter, and C.C. Ashley. A diazo-2 study of relaxation mechanisms in frog and barnacle muscle fibres: effects of pH, MgADP, and inorganic phosphate. *Pflugers Arch.* **437**, 204-212 (1999).

17. Simnett, S.J., E.C. Johns, S. Lipscomb, I.P. Mulligan, and C.C. Ashley. Effect of pH, phosphate, and ADP on relaxation of myocardium after photolysis of diazo 2. *Am.J.Physiol* **275**, H951-H960 (1998).

18. Pate, E., and R. Cooke. A model of crossbridge action: the effects of ATP, ADP and Pi. *J.Muscle Res.Cell Motil.* **10**, 181-196 (1989).

19. Lu, Z., R.L. Moss, and J.W. Walker. Tension transients initiated by photogeneration of MgADP in skinned skeletal muscle fibers. *J.Gen.Physiol* **101**, 867-888 (1993).

DISCUSSION

Pollack: You showed that relaxation propagates from sarcomere to sarcomere. Could you speculate on the mechanism?

Stehle: I would speculate that propagation of sarcomere relaxation is a strain-transmitted mechanism. The problem is then to explain why it occurs in a sequential manner. A minimum requirement to explain this is that thin filaments behave viscoelastic.

Ranatunga: Is the duration of the isometric phase of relaxation dependent on initial sarcomere length?

Stehle: Yes, it is slightly prolonged at longer sarcomere length. It is consistent with the thesis that an imbalance in strain among individual sarcomeres terminates this phase, since sarcomere homogeneity becomes improved in activations at longer sarcomere lengths. _

ter Keurs: Allen and Kurihara have shown that rapid shortening of cardiac muscle leads to Ca^{2+} dissociation from the myofilaments. Is it possible that the lengthening, if the central sarcomere causes shortening of the adjacent two sarcomeres; leads to accelerated relaxation of these two sarcomeres; hence, they lengthen rapidly and the same happens to the next sarcomeres etc.?

Stehle: The force kinetics during the initial isometric phase of relaxation suggest that there is no reattachment of force-generating cross-bridges anymore, when the lengthening of sarcomeres start. This implies that in our myofibrils calcium dissociation had already occurred.

Gonzalez-Serratos: Is the sarcoplasmic reticulum functional in this preparation?

Stehle: No, it is not functional.

INTER-SARCOMERE DYNAMICS IN MUSCLE FIBRES
A neglected subject ?

I. A. Telley[†], J. Denoth[†], and K.W. Ranatunga[‡]

1. ABSTRACT

The sarcomere is the functional unit of muscle, and all sarcomeres are connected in series in myofibrils within a muscle fibre. From this point of view of the structure a single model consisting of a contractile, a series and a parallel element can not account for the description of a real muscle fibre. Additionally, the titin protein filament needs to be considered as a passive visco-elastic element in parallel with the contractile apparatus. Therefore, the structure of a single muscle fibre is complex due mechanical elements ("motors") operating in series and in parallel. Moreover, variability does exist in the mechanical properties along a fibre and hence a multi-segmental model is more realistic and would give rise to many new insights. By attributing a segment model to each half-sarcomere, a fibre can be constructed through rigorous coupling of these units in series and parallel. The dynamics of such a multi-segmental model is much more complex, but it can explain a variety of effects reported in standard classical mechanics experiments.

With a relatively simple mechanistic description we can show that the dynamics of such multi-sarcomere systems exhibit a variety of effects (relaxation phenomena, permanent extra-tension, biphasic force-velocity relation) and should therefore not be neglected in muscle fibre modelling. We have observed in single skinned fibre experiments that non-uniformities in sarcomere length changes are prominent during activation and relaxation.

[†] Muscle Mechanics Group, Laboratory for Biomechanics, ETH Zurich, Schlieren CH-8952, Switzerland.
[‡] Muscle Contraction Group, Department of Physiology, School of Medical Sciences, University of Bristol, Bristol BS8 1TD, UK.
For correspondance contact: telley@biomech.mat.ethz.ch

Molecular and Cellular Aspects of Muscle Contraction
Edited by H. Sugi, Kluwer Academic/Plenum Publishers, 2003

2. INTRODUCTION

Despite much advancement in our understanding at the ultra-structural and molecular levels, the working of skeletal muscle as a composite mechanical system is not fully understood and certain fundamental questions remain unresolved. One of these enigmas is how thousands of sarcomeres that line up as a series of linear motors within a single muscle cell operate during muscle contraction. Moreover, the role of passive elements such as the titin filament in a sarcomere is little understood. The interpretation of data from muscle contraction experiments is generally made in terms of cross bridge cycling but neglecting the multi-segmental mechanics of a muscle fibre.

The sarcomeres in a muscle fibre have definable force development / holding properties. They depend on the actual length and length change as well as the activation state. In single myofibrils the sarcomeres are arranged in series and therefore, as a basic feature of series-connected multi-segmental systems, the force must be the same in each sarcomere. There is no doubt that variability does exist in the mechanical properties of sarcomeres along a fibre. Several causes can be considered to produce variability, such as differences in the number of operating cross-bridges (active force), differences in the titin filaments (passive force) or calcium gradients (activation). Hence, one would expect that the dynamics of such a system is complex and non-uniform. Furthermore, one should be able to observe this non-uniformity during experiments by monitoring sarcomere lengths.

Yet in the early fifties Hill (1953), a pioneer in muscle mechanics, suggested that irregularities in the striation spacing could occur in slightly stretched fibres. He pointed out that for slow length changes on the descending limb of the length-force relation, the stiffness is negative and the homogeneity of sarcomere lengths is therefore unstable. This was cited and confirmed ten years later by Gordon et al. (1966) in their study of length dependency of force generation in striated muscle. Furthermore, sarcomere non-uniformity was considered in connection with the 'creep' phenomenon of muscle after length changes, in which tension recordings showed a slow phase in tension rise. Julian and Morgan (1979a, b) investigated the inter-sarcomere dynamics during tetanic contractions and the effects of length changes on the degree of uniformity. They found that, when the starting sarcomere length corresponded to the descending limb of the length-tension relation, most of the fibre was lengthening while small regions were shortening as the contraction proceeded: the tension 'creep' corresponded to this non-uniform behaviour. During active lengthening beyond the plateau region, the non-uniformity was seen and tension was greater after stretch than that characteristic to the longer length without stretch, a phenomenon known as 'permanent extra-tension'.

In a later study Morgan (1990) showed with a schematic static fibre model that lengthening of active muscle fibres beyond the plateau of the length-tension curve would occur very non-uniformly; it consists essentially of rapid, uncontrolled and randomly distributed elongation of individual sarcomeres until the tension is borne by passive components. This uncontrolled lengthening of some sarcomeres was called 'sarcomere popping'. In a controversial discussion Allinger et al. (1996) and Zahalak (1997) established the conditions under which sarcomeres become unstable. All these analyses were confined to a static rather than a dynamic situation. Recently, Denoth et al. (2002) reconsidered the issue of sarcomere instability and length inhomogeneity based on a more complete mechanical and mathematical analysis and in dynamic situation. This study has shown that the issue of inter-sarcomere dynamics can not be excluded for the analysis of experimental data.

There is a debate going on about the role of the titin filament and its contribution to force behaviour in a muscle fibre. Its passive force production during stretch in resting muscle fibre is accepted. There is good evidence that titin is viscoelastic rather than only elastic. Ranatunga (2001) showed that during stretch experiments with skinned and intact fibres the resting muscle tension development consists of elastic, viscoelastic and viscous components. From a mechanistic point of view the parallel arrangement of a velocity dependent contractile part and a viscoelastic passive structure can lead to a transition in the force-velocity relation of the whole system.

Research in muscle physiology and mechanics is currently concentrating on the understanding of molecular events during active and passive force development. Different experiments are being performed using single, intact or skinned muscle fibres, and interpretations are done on the level of cross-bridges. This procedure involves certain difficulties since the examined preparations are highly complex. Indeed Sugi and Tsuchiya (1998) pointed out very clearly that non-uniform sarcomere lengths (and length changes) more or less prevent interpretation of experimental data from going into elementary molecular events.

The present study addresses the problem of non-uniform sarcomere behaviour and the role of inter-sarcomere dynamics. Our theoretical considerations give some suggestions for the effect of inter-sarcomere dynamics on creep, force relaxation, extra-tension and the biphasic force-velocity relation. It appears that the relation between the characteristics of the titin filament and the actin-myosin machinery is crucial for the dynamics of a muscle fibre as a composite mechanical system.

3. THEORETICAL CONSIDERATIONS

3.1. Multi-Segmental Modelling

Emphasis of the following theoretical essay is put on the systematic treatment of sarcomeres and not on the molecular modelling of the active and passive elements in them. The half-sarcomere (h.s.) represents the mechanical unit. We address a segment model to each half-sarcomere with two parallel strands; one for the contractile apparatus, one for all the passive force-bearing structures. To account for the elasticity of the cross-bridges, the Z-line lattice and the actin and myosin filaments we introduce (in each segment) an elastic element in series to the contractile element (see Fig. 1). The force-length (F-L) and force velocity (F-V) relations of the contractile element are defined according to the sliding filament hypothesis of Huxley (1957). The amount of filament which is basically a measure for the number of force-generating cross-bridges per length overlap determines the F-L relation. It is generally divided into ascending limb, plateau region and descending limb. The dependence of cross-bridge cycling on length changes yields in the F-V relation, which can be approximated by a combined Hill-Katz curve. A sigmoidal force-calcium concentration (F-pCa) relation determines the active force development. The second strand consists of a viscoelastic element representing force generation in the passive elements spanning the half-sarcomere. We can assume that passive tension derives primarily from the extension of the titin filament. Its F-L and F-V

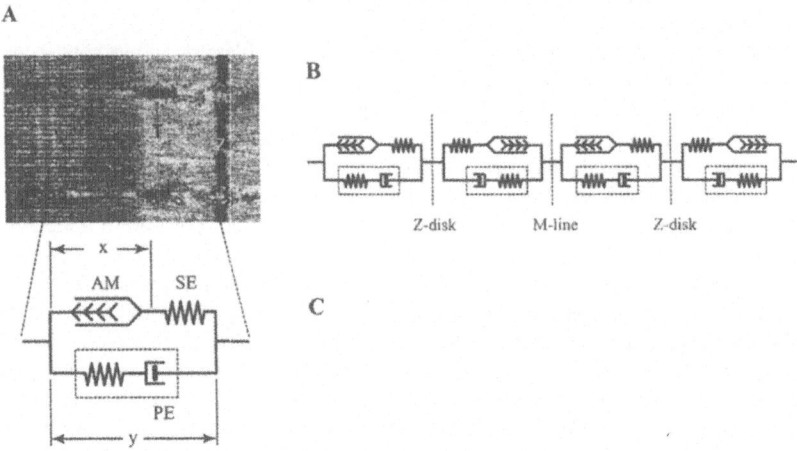

Figure 1: (A) EM micrograph and representative mechanical model of a half-sarcomere. The contractile element AM is coupled with a series element SE which represents the filament and cross-bridge elasticity. In parallel a viscoelastic element PE represents the titin filament. Although drawn in series, the viscosity and elasticity of the titin element is defined as one viscoelastic element. This is emphasised with the dashed square. **(B)** A myofibril is governed by a series of half-sarcomere models. Two segments form a sarcomere which are separated by the Z-disks. In each node between two segments the force must be the same as the external force. **(C)** Constitutive equations of the model; the segment force is the sum of the forces of the strands. The second equation describes force equality between the segments i and j.

relation are derived from data found in Kellermayer et al. (1997). The shape of the F-V relations of titin is crucial and will be discussed in a later section (see also Fig. 2). According to Denoth et al. (2002) we assume that sarcomeres have zero mass, and therefore no inertial forces occur. This leads to a simple mechanical condition for each segment (half-sarcomere) along a single myofibril: *Force generation of all segments must be equal all the time.*

In muscle mechanics two different kinds of experiments are generally performed; force-controlled experiments to determine e.g. shortening velocity, and length-controlled experiments, e.g. stretch-release. In the first case the dynamics, i.e. the length and length changes of each segment are determined independently by its individual mechanical characteristics and the external control parameter (= force). Hence, since the force of each segment is equal to the external force, the dynamics can be individually calculated. The system response is the overall length (change). In fact, this case corresponds to a system with only one half-sarcomere. In length-controlled experiments the dynamics of a segment is not independent anymore, but determined by the dynamics of all other segments and the external control parameter (= length). Mathematically speaking, the system is governed by *coupled* differential equations. A system with e.g. ten half-sarcomeres (= five sarcomeres) in series can be described by twenty coordinates (one for

A

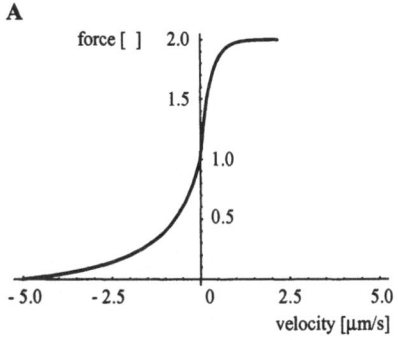

B

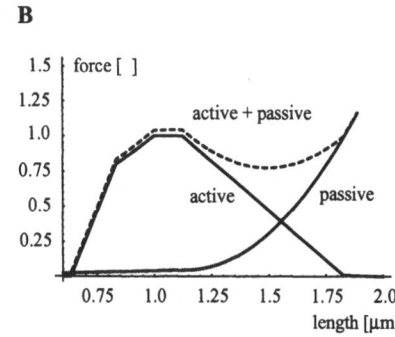

Figure 2: (A) Force-velocity relation of the contractile element in each segment. Negative velocity values denote shortening and positive values lengthening. Force is normalised to the isometric value. During fast lengthening force is assumed to double. **(B)** Force-length relation of the contractile (active force) and the titin (passive force) element. Force is normalised to the maximum plateau force. The dashed line is the sum of active and passive force. Note that 1) the passive force is not zero at any lengths and 2) the point of intersection of passive and active descending curve is not the same as the local minimum of the combined curve.

the contractile element, one for the parallel element, each) and by the external length. Hence, this system has nineteen degrees of freedom. In contrast, a system with one segment has only one degree of freedom (two coordinates less the control parameter). During experiments even though system length is held constant internal movement is possible (and observable, see section 4). This would imply that force, which is the system response, is a complex composition of the internal dynamics.

In mathematical terms of the model the dynamics of a half-sarcomere is determined by the partial derivatives of the mechanical properties (F-L, F-V and F-pCa relations) of the active *and* passive elements and eventual perturbations. Horowits and Podolsky (1987) first suggested that the positional stability of the M-band in a sarcomere is governed by the sum of the *slopes* of passive and active force-length relations. They predicted that if the positive slope of the resting (passive) force-length relation is greater in magnitude than the negative slope of the active force-length relation, no movement of thick filaments should occur during isometric activation. We can show that this is indeed a premise for stability, but not sufficient. Movement in that particular case is still possible and basically depends on the preceding perturbation (i.e. history dependence mentioned by Herzog and Leonard (2000) or Granzier and Pollack (1989)). In our understanding stability does not mean that the system is static (zero internal movement) but that length changes (= velocities) are not increasing with time. Again, this reflects our dynamic approach for understanding muscle mechanics.

3.2. The muscle Fibre: A Three-Dimensional Network

In a single muscle fibre hundreds of myofibrils are parallely aligned. The myofibrils are three-dimensionally arranged and connected through the Z-disks and M-lines by intermediate filaments (e.g. desmin between Z-disks, see also Pollack (1990)) and form a veritable network. There is good evidence that these filaments play a mechanical role. Shah et al. (2002) showed that in desmin-null mouse skeletal muscle the myofibrillar motility is increased. Z-disk showed a significantly larger displacement. They suggested that desmin plays a role in organising intact myofibrils laterally during mechanical loading tethering adjacent Z-disks. This is only possible when a certain force transmission is taking place.

Our approach for whole fibre modelling is based on the idea of networks. Linear myofibril models are connected with a spring between the segment nodes (see Fig. 3). The mechanical description of such a network is far more complex and nontrivial. In length-controlled experiments the dynamics of each segment depends not only from the segments in the same myofibril, but from all segments in the whole system. The coupling

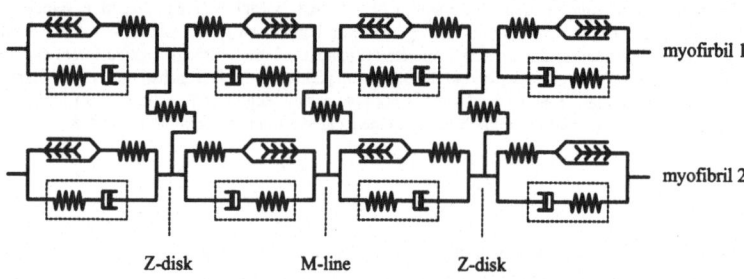

Figure 3: Schematic representation of a fibre network with two coupled linear models. The horizontal arrangement of the coupling springs points out that only force components along the myofibrils are considered. The equation of force in each node now involves coupling terms which depend on their relative displacement. Therefore, in length-controlled simulations the dynamics of a segment depends on all other segments in the network.

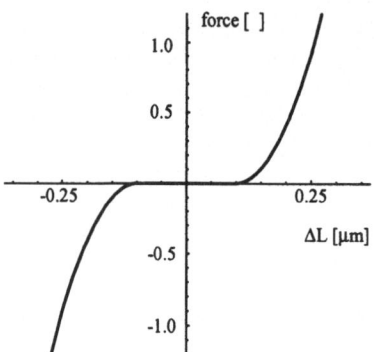

Figure 4: Proposed force-displacement relation for the intermediate filaments forming Z and M-line. For the time being we use the same properties for both kinds of filaments. The displacement ΔL is calculated by subtracting the positions of two adjacent segments in an absolute coordinate system. Note that there is a region around zero displacement ($\pm 0.1 \ \mu m$) with no force transmission.

of adjacent myofibrils involves more constraints. For the coupling we assume, for the time being, a symmetric force transmission depending on the relative displacement (see Fig. 4). Thereby, we take into account a force along the myofibril models by neglecting all transversal components. This relation is not based on measurements found in the past literature. Nevertheless, one can assume that the filamentous behaviour is similar to the unfolding process in some segments of titin (see Trombitas et al. (1998)). We adapt this idea and establish a symmetric, non-linear force-displacement relation.

3.3. Experimental Simulations

The first simulations are carried out with a linear model (one myofibril) in which the variability is defined as follows: In the active F-L relation the upper range of e.g. 5% of (normalised) plateau force is normally distributed over all half-sarcomere. Beside this systematic variability a random variability (ten fold smaller) is added. Therefore, the weakest half-sarcomere has a maximum active force capability of ~0.95 and the strongest ~1.0. In the passive F-L relation (of titin) the critical length of steep force rise is normally distributed in a defined range, for example between ½ (2.25 – 2.45) µm. The term ½ appears because we are working with half-sarcomeres. Additionally, the same variability in normalised force is added, e.g. with 5% the weakest titin element is normalised to 0.95 and the strongest to 1.0. Due to these inherent differences in force generation and the premise that the forces of half-sarcomeres in series have to be equal the starting lengths are different as well. In resting state when active force generation is zero, starting lengths are set through the titin element only. Figure 5 illustrates the definition of variability.

For the reason of time consumption the simulations are carried out with a relatively small number of segments, i.e. between ten and twenty segments. The starting conditions have to be defined by setting velocities to zero and the length of one half-sarcomere to a certain length. This will determine the lengths of all other segments. Then the devolution of the dynamics in time is calculated with variable step integration.

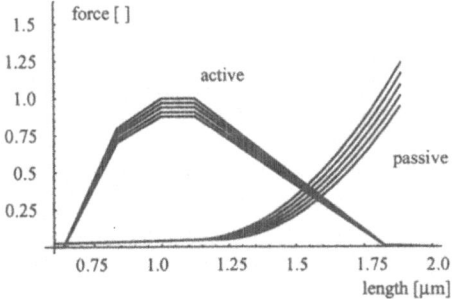

Figure 5: Force-length graph of five different segments (half-sarcomeres) with a variability of 15% in active plateau force and the passive force of the titin element. The critical length of steep force rise in the passive element is in the range of ½ (2.25 – 2.45) µm.

3.3.1. Fixed-End Contractions

Figure 6 shows length and force traces of a twenty-segment linear system during fixed-end contraction. The system is activated and deactivated with an increase and decrease in the calcium concentration, respectively. The change in pCa can be interpreted as a perturbation in force. Since the length of the system is held constant the sum of all length changes must be equal to zero. The variability causes some half-sarcomeres to shorten and others to lengthen, and both lengthening and shortening in a non-uniform manner. Some segments are even borne into the titin scaffold where force is generated passively. This lengthening has to be compensated by shortening segments which are therefore operating on the ascending limb. The splitting up in long and short segments results in a slow decrease in force ('creep').

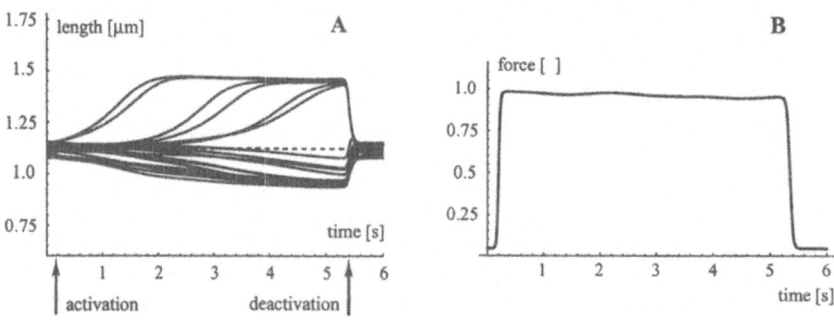

Figure 6: A Length traces of a fixed-end contraction. Activation and deactivation are indicated with arrows. The dashed line represents the mean segment length. During relaxation some short segments are stretched beyond mean length before they shorten again to a steady-state. **B** Corresponding unsteady force trace.

3.3.2. Stretch-Release Experiments

The length and force traces of a twenty-segment system during a stretch-release simulation are depicted in Figure 7. As in the previous subsection the activation of the system is controlled by a time-pCa function. Additionally, a relatively slow but large length-step of 0.25 μm h.s.$^{-1}$ (22% L_0) in one second (which gives a stretch velocity of 0.22 L_0/s) is performed. L_0 is the mean segment length before activation. Again the system reveals a large dynamics and non-uniform segment length changes. After release a few segments are stretched beyond mean segment length by a large population which is shortening slowly. The stretched segments bear the force about 50% passively. This devolution is accompanied by an unsteady force response (slow increase).

Interestingly, if one compares the dynamics of a multi-segmental model with the dynamics of only one segment with exactly the same conditions (same activation and relative stretch) the force response is completely different and contradictory at first view. In Figure 8A the dynamics of a stretch is shown. The stretch amplitude is 0.15 μm per half-sarcomere at a starting length L_0 = 1.11 μm h.s.$^{-1}$ (~14% L_0). Stretch velocity is

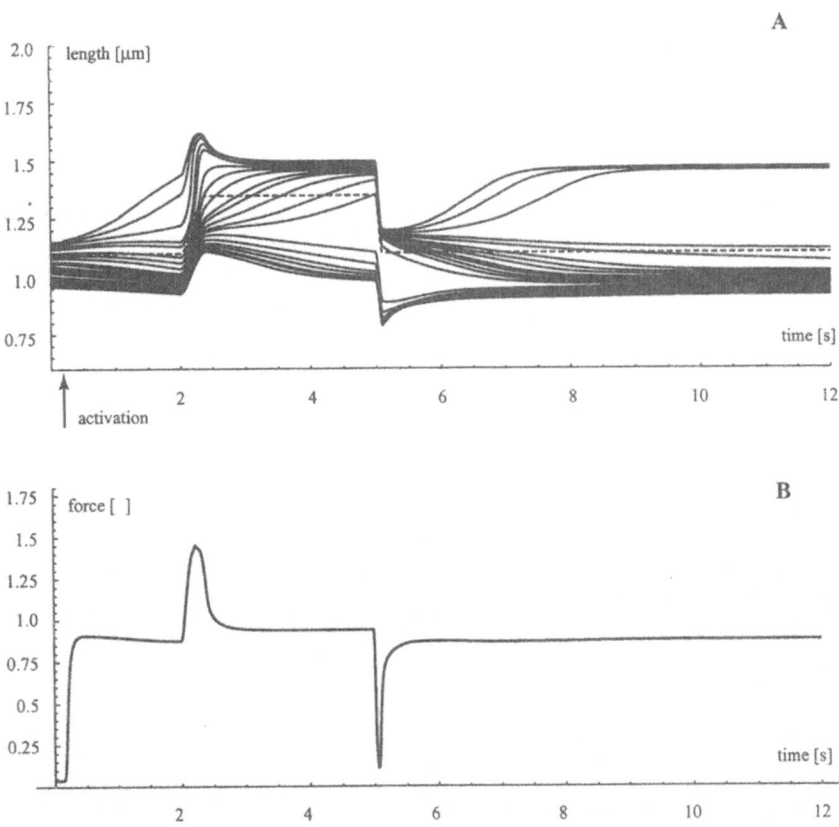

Figure 7: Force and length curves of a stretch-release simulation. **A** Length traces of a 20 segment system (ten sarcomeres). The activation (induced by a pCa step) is indicated with the arrow. The dashed line denotes the mean segment length and represents the external length control. Note that during stretch all segments are lengthening, but thereafter some are shortening and stretching others. **B** Force response of the system illustrated in A. After activation the force is decreasing slowly due to the internal dynamics ('drift'). After applying a stretch (22% L_0) a typical force relaxation occurs with a *higher* final force level than shortly after activation ('permanent extra tension'). The length release is accompanied by a force recovery.

faster than in the first simulation (~0.7 L_0/s). To illustrate the dynamics of a multi-segmental and a single-segmental system the same stretch is performed with both systems and the force responses are overlaid (Fig. 8B). Two single segments representing the weakest and the strongest half-sarcomere of the multi-segmental system are chosen. The length of a single-segmental system follows the external length and is therefore representative for the mean half-sarcomere length. The overlay graph clearly shows that in the multi-segmental system 1) the force rise during stretch is lower and has different phases, i.e. a steep rise to a peak followed by a slight decrease and a much slower rise, and 2) there is a force relaxation after stretch which is not shown by one half-sarcomere alone. Note also that after stretch the force level is even higher than the force of the strongest half-sarcomere.

A   B

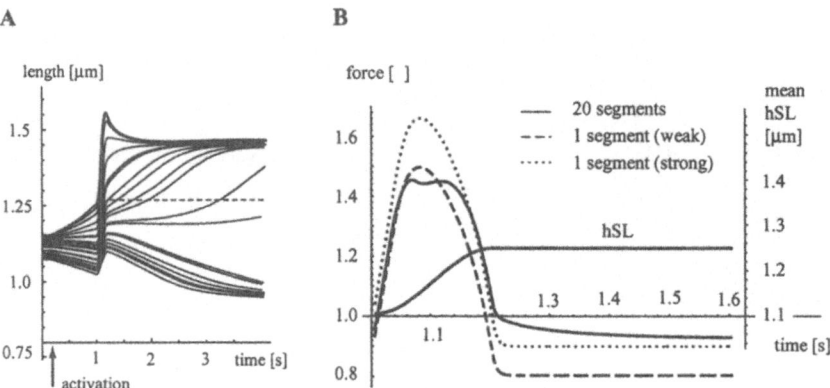

Figure 8: Comparison of the force responses of a multi-segmental (with 10% variability) and two single-segmental systems. **A** Devolution of half-sarcomere lengths during activation and stretch (0.15 μm h.s.⁻¹). The dashed line denotes the mean segment length and represents the externally controlled length. **B** Overlay of the force responses of a twenty segment (outlined), a single weak (dashed) and strong (dotted) segment during stretch. The mean half-sarcomere length (hSL) is plotted with the scale on the right hand side. Note the different phases during and the force relaxation after stretch in the multi-segmental system.

3.3.3. Simulation with Coupled Myofibrils

As introduced in section 3.2 our muscle fibre model is further advanced by coupling linear myofibril models to a veritable network. Preliminary simulations are carried out in a planar system with two linear models and uniform coupling properties. Therefore, we take the sarcomere as the segmental unit. The inter-myofibril coupling is attributed to a filamentous link between the Z-disks (i.e. the desmin filament). Figure 9 shows a sample stretch-release simulation of 1) an uncoupled system and 2) a coupled system of two myofibrils by comparison. The relaxed segment length at the beginning (2.2 μm mean), the activation, the stretch amplitude and velocity are exactly the same. Each linear model consists of only five segments (sarcomeres). A variability of 10% is implemented in the force plateau level and the elasticity of the titin element. A relatively long contraction of ten seconds is carried out to monitor the stability. A slow stretch of 0.4μm per sarcomere in 0.5 seconds (i.e. v = 0.36 L₀/s) is induced. As expected, the coupled system reveals less non-uniform behaviour. After activation the drift in lengths is smaller than in the uncoupled system. Additionally, the force response shows a slightly different shape during stretch. Due to a higher stability the segmental velocity and, consequently, the contribution of viscous forces is smaller. This is one reason why the relaxation in force is less prominent. Due to the large stretch amplitude the force-length relation needs to be considered to explain the lower force level in the coupled system after stretch. However, these rudimentary simulations show that the inter-myofibril coupling leads to certain effects in the system dynamics which need to be considered for muscle fibre modelling.

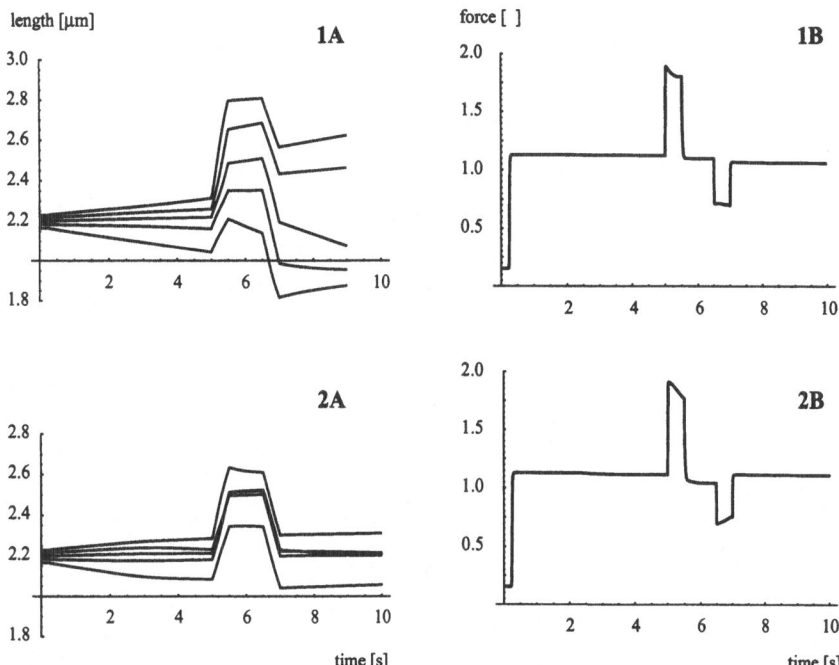

Figure 9: Force and length traces of an uncoupled (1) and a coupled (2) two-myofibril system. **A:** Length traces of the segments of one myofibril. In this case each segment is attributed to a whole sarcomere and the coupling is attributed to a filament between the Z-disks (i.e. desmin). The uncoupled system (1) clearly shows highly non-uniform segment length changes, whereas the coupled system (2) reaches a steady state before stretch. **B:** Corresponding force traces of A. Due to the small number of segments the internal movement is limited and relaxation phenomena are less prominent. However, there are differences visible in the force response during stretch.

3.4. The Mechanics of a Two-Stranded Segment Model

It is necessary to analyse the mechanics of a single segment model (half-sarcomere) in one myofibril to understand the contribution of each element to the dynamics during force- or length-controlled simulations. For the time being, we concentrate on force-controlled rather than length-controlled simulations as the latter involves coupled equation systems and demands a rigorous mathematical essay.

In a two-stranded segmental model the parallel arrangement of the force-bearing and actively force-generating elements demands that the force is the sum of the forces in the strands. It implies that the mechanical properties of stiffness and viscosity are additive as well. On the other hand we know from basic mechanics that inertia, elasticity and viscosity fully determine the time-devolution of length x and velocity $\dot{x}$. For a rough understanding of the dynamics of a segment after a fast force step we derive the devolution of the velocity of one segment from our model description. According to the formula in Fig. 1C we define the partial derivatives

$$f_{\dot{x}} = \frac{\partial}{\partial \dot{x}}\left(f_{AM}(x,\dot{x}) + f_{PE}(x,\dot{x})\right) \quad f_x = \frac{\partial}{\partial x}\left(f_{AM}(x,\dot{x}) + f_{PE}(x,\dot{x})\right) \quad \tau = \frac{f_{\dot{x}}}{f_x}$$

where the index of f denotes the differential variable. We assume for a short time interval that in a neighbourhood $\{U_{(x,\dot{x})}\}$ these partial derivatives and thus τ are constant. With t $> t_0$ we follow

$$\dot{x}(t) = \dot{x}(0) + \int_0^t exp\left(-\frac{f_x}{f_{\dot{x}}}(t-t')\right)\Delta f\,\delta(t'-t_0)\,dt' = \dot{x}(0) + \frac{\Delta f}{\tau}exp\left(-\frac{t-t_0}{\tau}\right)$$

Thereby, Δf is the step amplitude and δ is the delta-function for infinitely fast steps. The step is performed at the time t_0 with $0 < t_0 < t$. The series elastic element was neglected for sake of simplicity. This analytical term is *only* correct for constant τ. In our simulations we use numerical methods to compute the real devolution. However, we now have an idea of how the segment velocity functionally depend on both the parallel element (PE), represented by titin, and the contractile element (AM). More precisely, due to some change in time of the two partial derivatives over a large interval, τ is rather an *instantaneous* parameter than a time constant of the dynamics. Its sign tells something about stable (positive) or unstable (negative) devolution.

3.5. The Biphasic Force-Velocity Relation: A New Interpretation

It is now well established that the force-shortening velocity (F-V) relation of muscle is more complex than first assumed by pioneers like Hill (1938) (see Woledge et al. (1985)). With modern techniques it has been possible to measure the speed of shortening in single intact muscle fibres more precisely for a well-defined step in load. Edman (1988) showed that the F-V relation in intact frog fibres has a biphasic shape and a transition point at 75-80% isometric force P_0. He fitted his measured data with an extended form of Hill's (1938) hyperbolic equation. In both studies of Edman (1988) and Edman et al. (1997) it was suggested that the nature of two distinct curvatures represents the contractile behaviour of a sarcomere and lies within the cross-bridge dynamics. A four-state cross-bridge model was employed to elucidate the nature of the biphasic F-V relation. Interestingly, an increase in sarcomere length from 1.85 to 2.60 μm caused the biphasic shape almost to disappear. This effect was attributed to the decrease in width of the myofilament lattice that occurs as the sarcomere length decreases. Similar effects were observed in the same study by applying osmotic compression.

As mentioned in the previous section the parallel arrangement of the titin filament to the actomyosin motor leads to difficulties in associating the velocity dependency only to one strand. Our simulations, performed in accord to shortening velocity measurements described by Edman (1988) give several new insights.

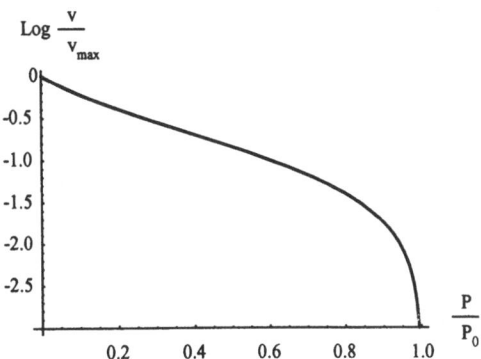

Figure 10: Logarithmic plot of the force-velocity relation after Hill (1938). This relation should not account for two phases. Nevertheless a steeper slope in the high-force range appears. It is difficult to extract changes in curvature from this kind of graph.

In Figure 10 a logarithmic plot of Hill's force-velocity equation is depicted which, of course, does not account for more than one phase. This example shows that, as a consequence, it is difficult to read out any changes in curvature and find differences to the original Hill relation. We introduce an inverse plot of force-velocity data by rewriting Hill's equation (as in enzym kinetics)

$$P = P_0 - (P_0 + a)\frac{v}{v+b} \Rightarrow \frac{P_0}{P_0 - P} \propto 1 + \frac{b}{v}$$

On the left hand side of the second equation stands the inverse of the force step relative to the isometric force. Now we have a linear relation between the inverse terms of velocity and force step. The advantage of this notation is that it accounts especially for very small velocities where P is close to the isometric force, P_0. A transition in curvature at the high-force range results in a change of the slope in the inverse plot.

For shortening velocity measurements a linear myofibril system with five segments is activated in length-controlled mode with a pCa step. After a steady force is reached the simulation is switched to force-controlled ('load-clamped') mode (see Fig. 11). During simulations all segments are operating on the plateau of the active F-L relation. Force steps in the range from 0.5% to 99% of isometric force, P_0, are performed and the shortening velocity *of the whole system* is acquired. The variability of the system was either low (1% in plateau force) or high (20% in plateau force). The titin viscosity term was defined either close to zero or much larger than the contractile velocity-term in the high-force range. The results of our simulations show that there is a systematic dependence of the high force-range (i.e. small force-steps and low shortening velocities) on variability and titin properties (see Fig. 12). A change in slope is clearly visible and, interestingly, the effect of velocity dependence of titin is opposed to that of variability. In this particular case the contribution of titin viscosity is dominant. A combination of 20% variability and a high viscosity term results in a transition to a steeper slope but less pronounced than only with high viscosity. We will show later that the effect of variability is only present if an asymmetry in the combined force-velocity relation is present. However, the transition points in the slopes occur between 0.75 P_0 and 0.82 P_0 which corresponds to the mentioned break points in the studies of Edman.

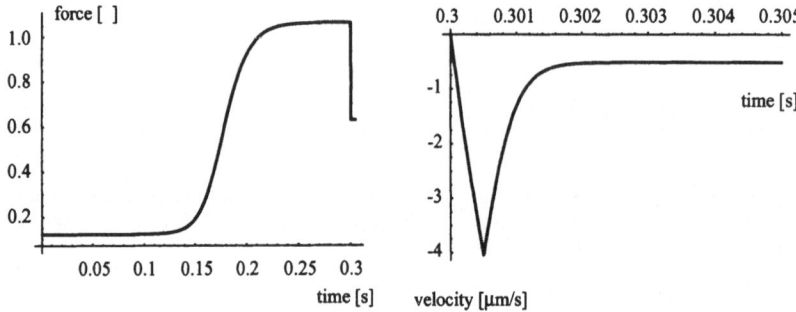

Figure 11: Left: Force trace of a five-segment system during activation and application of a force step. The relatively short time of steady force guarantees that the segments are not drifting towards the descending and ascending limb of the F-L relation and the velocities of the contractile elements are small. **Right:** Velocity of the whole system due to the force step. Negative velocities denote shortening. The peak is from the shortening of the series elastic elements. After two milliseconds a steady shortening velocity is reached.

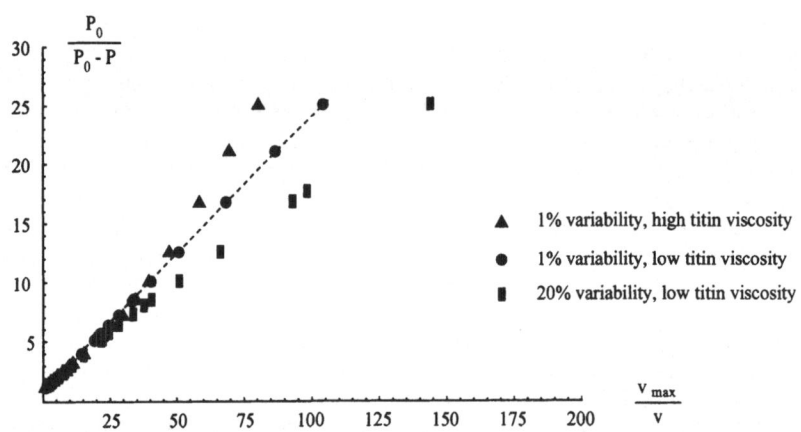

Figure 12: Inverse plots of the force-velocity relation from 'load-clamped' simulations with a five-segment system. The steady shortening velocity is measured for different force steps after length-controlled activation. The reference simulation is made with a low variability (1%) and almost no titin viscosity. It is depicted with filled circles (●) and a dashed, straight line. Clearly, there is only one phase visible. When altering the variability to 20% (filled squares ■) there is a transition in slope at ~0.75 P_0 whereby the slope is decreasing. Therefore, for a well-defined force step the shortening velocity is smaller. On the other hand, by altering titin viscosity (filled triangles ▲), we obtain a transition at ~0.82 P_0 and the slope is increasing which means that the shortening velocity is higher for the same step.

One has to be careful not to interpret the data in absolute terms of P_0 because it does not say anything about the value of P_0. With these results we want to show that by assuming titin viscosity and variability we find some effects in shortening velocity measurements which can explain certain changes in curvature of the F-V relation.

3.5.1. The Contribution of Titin

As was brought up in a recent study of Minajeva et al. (2002) the contractile element is in parallel with (visco-) elastic elements like titin which may contribute to sarcomere shortening. In their study Minajeva et al. assumed that the contribution to shortening is based on passive elasticity, and that viscous forces are opposing the shortening, which is very reasonable. In our model the viscosity of titin is implemented with the same mathematical method as the force-length-velocity relation of the contractile element, i.e. as a product of length-dependent and velocity-dependent force. Shortening induces a decrease in the passive force, not only initially (from the elastic property), but velocity-dependent by assuming a braking force on the refolding titin filament is imposed. On the other hand, the force is increased when the filament is quickly stretched.

In a first approximation the titin filament (PE visco-elasticity) is represented as a mechanical composite of a dashpot (damper) and spring in parallel but with extended, more complex properties. The force f (also denoted as P) is defined as

$$f_{PE}(x, \dot{x}) = k(x)\, g(\dot{x}) = k(x) \cdot \left(1 + \widetilde{g}(\dot{x})\right) \cong k(x) + b(x) \cdot \dot{x}$$

in which the last approximation is for a neighbourhood of zero velocity. The function $k(x)$ represents elasticity and $b(x)$ is the length dependent viscosity term of the dashpot. Figure 13A shows the force-velocity relations of the passive and active elements. Figure 13B depicts the partial derivative of $\widetilde{g}(\dot{x})$ which is, as an approximation, $b(x)$ for a fixed x around zero velocity.

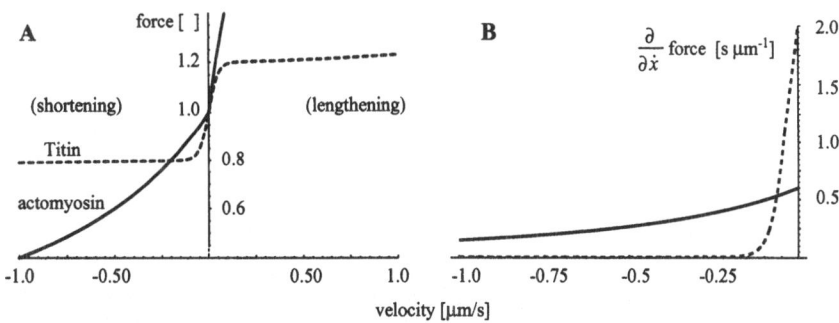

Figure 13: A Force-velocity relation of titin (dashed) and the contractile (outlined) element in comparison. Force is a product of length-dependent term and velocity-dependent term. The velocity-dependent term is equal one for v = 0, decrease for negative (shortening) velocities and increase for positive (lengthening) velocities. A sinusoidal lengthening and shortening of the titin filament would induce a hysteresis found e.g. in Kellermayer et al. (1997). **B** Derivatives of the plots in A for the shortening side (negative velocities). For the titin element (dashed) it represents the viscosity property and reflects its large contribution to force at slow speeds.

The high viscosity of titin is indicated by the steep slope close to isometric conditions ($\dot{x} = 0$). The data of Figure 12 suggests that for a given force step *the increase in shortening velocity* is higher in the system with high viscosity than in the system with low viscosity. Thus, the titin element contributes to shortening during force-controlled experiments. In their study Minajeva et al. (2002) showed that at 2.2 µm SL the passive shortening velocity for a small step release is ~ 4 µm/s which is considerably high.

3.5.2. The Effect of Non-Uniformity

To understand the role of variability in force generation during shortening velocity measurements, we concentrate on the combined force-velocity relation for shortening and lengthening (Fig. 14). It is well accepted that there is a discontinuity in the derivatives at v = 0 (first described by Katz (1939), see also Lombardi and Piazzesi (1990)). This has certain consequences for the inter-sarcomere dynamics. We consider a linear system with segments (sarcomeres or half-sarcomeres) operating *on the plateau* of the active force-length relation. During activation the system is length-controlled and the force equality demands that some segments are shortening and stretch others. The sum of the segmental velocities is zero and the measured steady force is P_0. After switching to force-controlled simulation a force step induces a change in velocities, but not all segments are shortening. For small steps (ΔP_1) some are still lengthening and the measured shortening velocity is smaller than expected. For larger steps (ΔP_2) the change in velocity state is large enough to induce shortening in every segment. Most importantly, the discontinuity in the slopes

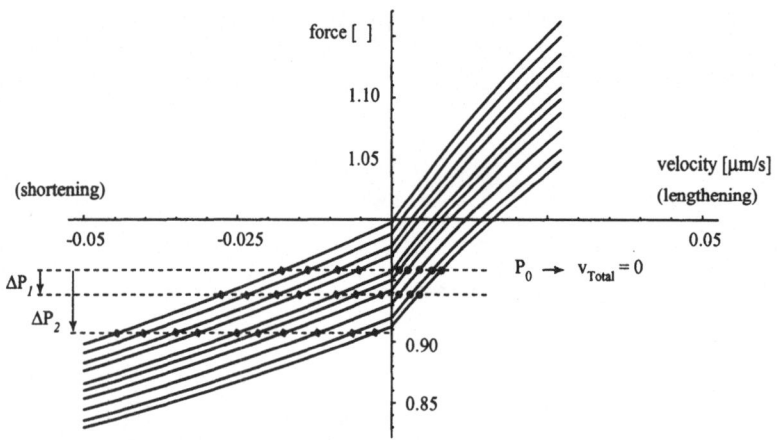

Figure 14: Force-velocity (F-V) relation of ten segments operating *on the plateau* of the force-length relation. Due to the variability and the premise of force equality some segments shorten while others lengthen during fixed-end contraction with isometric force, P_0. The measured velocity is the sum of all segmental velocities (in case of fixed-end contraction zero). When switching to force-controlled simulation and applying a force step ΔP the segmental velocities decrease (a negative velocity denotes shortening). Nevertheless, for small steps some segments still lengthen further (ΔP_1) while for larger steps all segments begin to shorten (ΔP_2). The asymmetry in the slopes of the shortening and lengthening side and, of course, the variability in force generation are the main reasons a break point of the measured F-V relation.

of the F-V relation is the reason for a point of transition and a change in curvature in the *measured* F-V relation. It is not present if the slope is continuous. However, Katz (1939) reported that the slope on the lengthening side is up to six fold larger than on the shortening side.

This idea is not new and was described e.g. by Hill (1970) in a rough calculation for only two different populations of sarcomeres. He predicted a discontinuity in the slope of the F-V relation ('break point') which depends on the distribution of weak and strong sarcomeres.

4. EXPERIMENTAL EVIDENCE

Although the laser diffraction technique is convenient to monitor fibre sarcomere length, it is an integral technique which gives an average estimate for the length of a large number of sarcomeres. In dynamic situations, e.g. relaxation, new measurement methods for a smaller population or even for individual sarcomeres have to be explored. Therefore, highly accurate and direct sarcomere length measurements have to be performed. Morgan (1994) pointed out that his predicted pattern of fast lengthening would certainly not be accurately measured with laser diffraction. Although he predicted that randomly scattered 'popped' sarcomeres were not visible with optical microscopy, but only by electron microscopy, other authors reported sarcomere length inhomogeneity and disorder in myofibrils monitored with conventional bright-field and phase-contrast techniques (see Colomo et al. (1997), Friedman and Goldman (1996), Linke et al. (1994)). Even though these studies did not particularly examine the non-uniform behaviour during activation or stretches they presented substantial evidence with these very simple techniques.

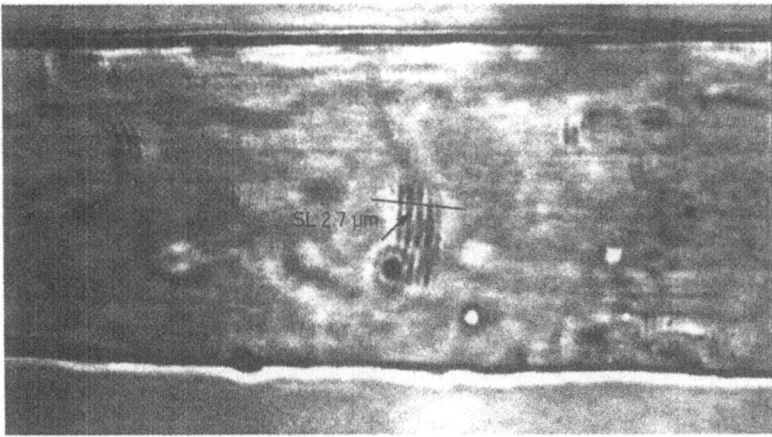

Figure 15: Image of a single skinned muscle fibre from the rabbit psoas during relaxation. Tetanic force was 185 N/mm², the fibre length ~400 μm and the fibre width 55 μm. Several large 'paches' of stretched sarcomeres with a duration of 100 – 200 milliseconds could be monitored. The sarcomere length of these patches was 2.7 ± 0.2 μm which is larger than the resting length. This indicates that in this region passive elastic structures bear active force.

We have performed preliminary experiments for direct visualisation of skinned muscle fibres. Video recording of single muscle fibres from the rabbit psoas muscle have shown that non-uniformity is well established and can be monitored with bright-field microscopy. The most noticeable non-uniform movements occur during activation and deactivation. Local irregular striation patterns that last only 100-200 milliseconds reveal the complex dynamics in a series of independent motors. The striation interval length is slightly higher than in relaxed state (Fig. 15). This indicates that passive structures bear active generated tension of sarcomeres in series. Sequential activation and deactivation can cause overstretching of whole regions in a fibre.

Single stretches with a small amplitude have less effect on the regularity of the striation and do not cause observable non-uniformity. Nevertheless, a (rather non-physiological) stretch-release experiment (up to 10% initial length) with multiple stretch cycles causes a complete disturbance of the cross striation. This is often accompanied by a complete distortion of the fibre. Uneven fibre distortion in the middle of a fibre segment and over-stretching at the ends are common. The middle part shortens and forms a 'belly' while the end parts are overstretched and reveal large 'gaps' in the striation. It seems that the structure is irreversibly changed and functionally damaged.

However, there is still a possibility for optical artefacts in the micrographs, and the interpretation of the images should be done cautiously. Since the fibre structure consists of several layers of optical grids, it is likely that a relative movement of these grids generate optical superposition effects (e.g. frequency doubling or similar). Therefore, a non-uniform change in the cross striation should not be directly interpreted as a change in sarcomere lengths. For the time being, it is referred to as regional non-uniform kinematics of the structures that cause these optical effects. However, it is enough evidence for the existence local non-uniform behaviour in muscle fibres.

5. CONCLUSIONS

The occurrence of sarcomere non-uniformity and damage in skinned fibre preparations has been a well recognised "problem" in experiments, particularly at physiological temperatures. However, non-uniformity is rarely considered to be the reason for "non-obvious" phenomena, but rather a new mechanism in the cross-bridges (e.g more attached cross-bridge states) is introduced. It is still not clear whether there is an intrinsic, built-in, sarcomere pattern arrangement; for example, whether sarcomeres at fibre ends are generally weaker, shorter etc., and how they behave during single activation, continued activation and fatigue.

Our modelling gives some quantitative accounts of the interaction of different mechanical elements both within and between sarcomeres during force transmission and / or shortening and lengthening. We present possible reasons for some of the unresolved and unexplained issues in muscle experiments. We underline that our findings do not depend on the details of the mechanical description of the cross-bridges. Non-uniformity is inherent in a system in which a population of motors operate in series and parallel and the conditions are not exactly the same for each motor. Moreover, the dynamics of such a system is highly sensitive to gradients in the mechanical properties. The presence of such gradients is undoubted and has several causes. Our preliminary results in single fibre visualisation supports our theoretical considerations and the indications in past literature that non-uniformity is present and difficult to handle.

We suggest that in muscle physiology and mechanics the understanding of inter-sarcomere dynamics needs to be promoted beside the important research at the molecular level. The combination of both areas would give a more complete understanding of how muscle operates.

6. ACKNOWLEDGMENT

We thank the Barth Fond at ETH Zurich and the Wellcome Trust Foundation (UK) for financial support during this study.

7. REFERENCES

Allinger, T. L., Epstein, M., and Herzog, W., 1996, Stability of muscle fibers on the descending limb of the force-length relation. A theoretical consideration, *J Biomech*, **29**(5):627-33.

Colomo, F., Piroddi, N., Poggesi, C., te Kronnie, G., and Tesi, C., 1997, Active and passive forces of isolated myofibrils from cardiac and fast skeletal muscle of the frog, *J Physiol*, **500** (Pt 2):535-48.

Denoth, J., Stussi, E., Csucs, G., and Danuser, G., 2002, Single muscle fiber contraction is dictated by inter-sarcomere dynamics, *J Theor Biol*, **216**(1):101-122.

Edman, K. A., 1988, Double-hyperbolic force-velocity relation in frog muscle fibres, *J Physiol*, **404**:301-21.

Edman, K. A., Mansson, A., and Caputo, C., 1997, The biphasic force-velocity relationship in frog muscle fibres and its evaluation in terms of cross-bridge function, *J Physiol*, **503** (Pt 1):141-56.

Friedman, A. L., and Goldman, Y. E., 1996, Mechanical characterization of skeletal muscle myofibrils, *Biophys J*, **71**(5):2774-85.

Gordon, A. M., Huxley, A. F., and Julian, F. J., 1966, The variation in isometric tension with sarcomere length in vertebrate muscle fibres, *J Physiol*, **184**(1):170-92.

Granzier, H. L., and Pollack, G. H., 1989, Effect of active pre-shortening on isometric and isotonic performance of single frog muscle fibres, *J Physiol*, **415**:299-327.

Herzog, W., and Leonard, T. R., 2000, The history dependence of force production in mammalian skeletal muscle following stretch-shortening and shortening-stretch cycles, *J Biomech*, **33**(5):531-42.

Hill, A. V., 1938, The heat of shortening and the dynamic constants of muscle, *Proc R Soc B*, **126**(843):136-195.

Hill, A. V., 1953, The mechanics of active muscle, *Proc R Soc B*, **141**(902):104-117.

Hill, A. V., 1970, *First and Last Experiments in Muscle Physiology*. Cambridge University Press, London, New York.

Horowits, R., and Podolsky, R. J., 1987, The positional stability of thick filaments in activated skeletal muscle depends on sarcomere length: evidence for the role of titin filaments, *J Cell Biol*, **105**(5):2217-23.

Huxley, A. F., 1957, Muscle structure and theories of contraction, *Prog Biophys Mol Biol*, **7**:255-318.

Julian, F. J., and Morgan, D. L., 1979a, The effect on tension of non-uniform distribution of length changes applied to frog muscle fibres, *J Physiol*, **293**:379-92.

Julian, F. J., and Morgan, D. L., 1979b, Intersarcomere dynamics during fixed-end tetanic contractions of frog muscle fibres, *J Physiol*, **293**:365-78.

Katz, B., 1939, The relationship between force and speed in muscular contraction, *J Physiol (Lond)*, **96**:45-64.

Kellermayer, M. S., Smith, S. B., Granzier, H. L., and Bustamante, C., 1997, Folding-unfolding transitions in single titin molecules characterized with laser tweezers, *Science*, **276**(5315):1112-6.

Linke, W. A., Popov, V. I., and Pollack, G. H., 1994, Passive and active tension in single cardiac myofibrils, *Biophys J*, **67**(2):782-92.

Lombardi, V., and Piazzesi, G., 1990, The contractile response during steady lengthening of stimulated frog muscle fibres, *J Physiol*, **431**:141-71.

Minajeva, A., Neagoe, C., Kulke, M., and Linke, W. A., 2002, Titin-based contribution to shortening velocity of rabbit skeletal myofibrils, *J Physiol*, **540**(Pt 1):177-88.

Morgan, D. L., 1990, New insights into the behavior of muscle during active lengthening, *Biophys J*, **57**(2):209-21.

Morgan, D. L., 1994, An explanation for residual increased tension in striated muscle after stretch during contraction, *Exp Physiol*, **79**(5):831-8.

Pollack, G. H., 1990, *Muscle & Molecules*. Ebner & Sons Publishers.

Ranatunga, K. W., 2001, Sarcomeric visco-elasticity of chemically skinned skeletal muscle fibres of the rabbit at rest, *J Muscle Res Cell Motil*, 22(5):399-414.

Shah, S. B., Su, F. C., Jordan, K., Milner, D. J., Friden, J., Capetanaki, Y., and Lieber, R. L., 2002, Evidence for increased myofibrillar mobility in desmin-null mouse skeletal muscle, *J Exp Biol*, 205(Pt 3):321-5.

Sugi, H., and Tsuchiya, T., 1998, Muscle mechanics I: Intact single muscle fibres, in: *Current Methods in Muscle Physiology: Advantages, Problems, and Limitations* (H. Sugi, ed.), Oxford University Press, Oxford; New York; Tokyo, pp. 3-31.

Trombitas, K., Greaser, M., Labeit, S., Jin, J. P., Kellermayer, M., Helmes, M., and Granzier, H., 1998, Titin extensibility in situ: entropic elasticity of permanently folded and permanently unfolded molecular segments, *J Cell Biol*, 140(4):853-9.

Woledge, R. C., Curtin, N. A., and Homsher, E., 1985, Energetic aspects of muscle contraction, *Monogr Physiol Soc*, 41:1-357.

Zahalak, G. I., 1997, Can muscle fibers be stable on the descending limbs of their sarcomere length-tension relations?, *J Biomech*, 30(11-12):1179-82.

DISCUSSION

Cecchi: What would you expect in the laser diffraction from the variability you assume in your model? Laser diffraction of a small segment of a fiber shows a clear and sharp peak of the first order. What would be the width of the peak expected from the variability you are assuming in your model?

Telley: We did not simulate the laser diffraction of the system. What needs to be mentioned is that the diffraction pattern does only account for the sarcomere length of a small region in a fiber. However, a fiber has several thousands of sarcomeres in series and the variability is assumed to be randomly (or systematically) distributed in the fiber. I guess that up to now it is not possible to monitor several regions of a fiber instantaneously by laser diffraction. Therefore, some of the dynamic information is missing. A better method is certainly direct sarcomere length monitoring with visualization techniques. In our simulation we assumed only twenty populations, but one can imagine these populations normally distributed in a system of thousand sarcomeres. Then the non-uniformity is not so prominent, at least by eye. To answer the question we should simulate the diffraction pattern to show whether the intensity and the width of the first order peak is considerably altered.

STRETCH-INDUCED FORCE ENHANCEMENT AND STABILITY OF SKELETAL MUSCLE MYOFIBRILS

Dilson E. Rassier[1,2], Walter Herzog[2], and Gerald H. Pollack[1]

ABSTRACT

The main purpose of the experiments presented in this chapter was to test the hypothesis that the stretch-induced force enhancement commonly observed in skeletal muscle is associated with sarcomere length instability. Single myofibrils isolated from the rabbit psoas muscle were attached to a nanolever pair for force measurement at the one end, and to a glass needle for controlled displacements at the other end. The image of the striation pattern was projected onto a linear 1024-element photodiode array, which was scanned (20 Hz) to produce a dark-light pattern corresponding to the A- and I-bands, respectively. Starting from a mean SL of ~2.55 µm, stretches of a nominal amplitude of 4 to 10% of SL, at a nominal speed of 100 nm·sec[-1] were applied to activated myofibrils ($pCa^{2+} = 4.75$). Following stretch, the isometric, steady-state force was greater by 10.9% to 45.9% than the force produced before stretch, and was greater than the force predicted at the corresponding final length. Passive force could not account for the force enhancement. Sarcomere lengths along the activated myofibrils were non-uniform, but remained constant before stretch or during the extended isometric period after stretch. Further, sarcomeres never stretched to a length beyond thick and thin filament overlap. It is concluded that sarcomeres are stable, and therefore the increased force observed after stretch must be a sarcomeric property, not associated with continuous length changes of unstable sarcomeres, as had been assumed in the past.

[1] Department of Bioengineering, University of Washington, Seattle, WA, 98195, USA

[2] Faculty of Kinesiology, University of Calgary, 2500 University Drive N.W. Calgary, Canada, T2N 1N4

Molecular and Cellular Aspects of Muscle Contraction
Edited by H. Sugi, Kluwer Academic/Plenum Publishers, 2003

1. INTRODUCTION

When an activated skeletal muscle (fiber) is stretched along the descending limb of the force-length relationship, the steady-state isometric force following stretch is greater than the isometric force at the corresponding final length [1-7]. This phenomenon – that will be called force enhancement hereupon - does not follow the predictions of the isometric force-length relationship [8], and does not allow for force predictions based on the degree of filament overlap.

Although the underlying mechanism of force enhancement is unknown, the most accepted hypothesis is that the increase in force is associated with the development of instability and non-uniformity of sarcomere lengths on the descending limb of the force-length relationship [9,10]. However, some studies have shown that sarcomere length non-uniformity cannot account for the force enhancement solely [3,11,12]. Further, the question whether skeletal muscle force production is unstable on the descending limb of the force-length relationship is strongly debated [13-16].

In the present chapter, we will briefly discuss results of mechanical experiments performed with single muscle cells suggesting that skeletal muscle force production is stable. Further, we will show novel results from experiments performed with activated myofibrils, a preparation in which all sarcomere lengths can be tracked during activation and length changes of the myofibrils. Therefore, the hypothesis that instability is responsible for force enhancement can be directly tested.

2. MECHANICAL EXPERIMENTS WITH SINGLE MUSCLE CELLS

We performed a series of mechanical experiments with single cells dissected from the frog flexor digitorum brevis muscle. In these experiments, we specifically tested one prediction that should be fulfilled if sarcomere lengths are unstable, and sarcomere length non-uniformity produces force enhancement: force following stretch cannot exceed the isometric force at the plateau of the force-length relationship.

Muscle fibers were actively stretched along the descending limb of the force-length relationship, and the isometric force produced after steady-state had been attained was compared to the forces produced at the plateau of the force-length relationship, at the initial length from which the muscle was stretched, and at the corresponding final length that was reached after the stretch.

Typical tracings of these experiments are shown in Figure 1. After a sudden increase of force during stretch, force decreased during the isometric phase following the stretch and attained a steady-state force, parallel to the tracings obtained during the purely isometric reference contractions. In Figure 1, the steady-state force produced after stretch was higher than the force produced in all other conditions. Figure 2 shows results of experiments performed on 22 fibers. For some conditions, force was ~10% higher than the force produced at the plateau of the force-length relationship.

The observation that the isometric steady-state force after active stretch is greater than the isometric force at the corresponding length has been made repeatedly [1-7]. However, if the force enhancement exceeds the isometric force at the length from which the stretch was initiated, then the force-extension curve has a positive slope on the

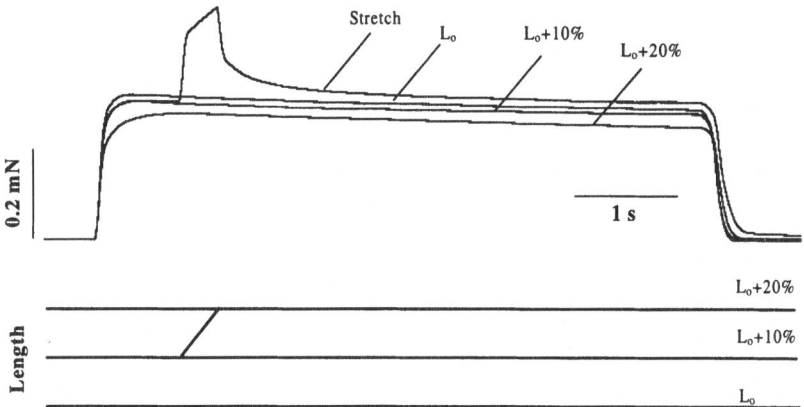

Figure 1. Force enhancement in a single muscle fiber. This fiber was stretched by 10% of optimal fiber length at 0.8 mm/s. After the stretch, force was higher than the isometric force produced at the optimal (L_o), initial (L_o + 10%) and final (L_o + 20%) lengths.

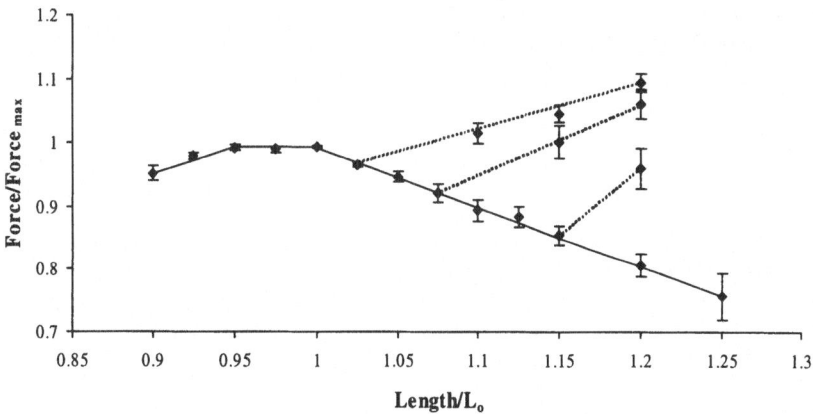

Figure 2. Mean force-length relationship during isometric contractions (solid lines) and after stretch (dotted lines). Forces were normalized relative to the maximal isometric force. Lengths were also normalized relative to the lengths of maximal isometric forces. Only active forces are shown, the passive force was subtracted from the total force records.

descending limb of the force-length relationship. A positive slope corresponds to a "hardening" behavior and a stable system. Furthermore, if force enhancement exceeds the isometric forces at optimal muscle length, the system exhibits unconditional stability, and the force enhancement cannot be explained by the development of sarcomere length non-uniformities. These could only explain force enhancement up to the isometric forces at optimal length in a steady-state situation.

Although these results indicate that single fiber force production is stable, experiments performed with single fibers only allow for an indirect interpretation of the results, as it is impossible to determine the length changes of all sarcomeres in a given preparation.

3. EXPERIMENTS WITH MYOFIBRILS

3.1. Material and Methods

3.1.1. Specimens

Small strips of rabbit *psoas* muscle were dissected and tied to wooden sticks. These samples were stored in rigor solution (see "Solutions" below) for 12 hours, after which they were transferred to a rigor/glycerol (50:50) solution for an additional 12 hours. Both procedures were carried out on ice throughout. The specimens were transferred to a fresh rigor/glycerol (50:50) solution and stored in a freezer at -20°C for 7 to 15 days. On the day of the experiment, the muscle strips were placed in a rigor solution for at least 1 hour, and then small pieces of muscle tissue (~ 2 mm length) were cut using a fine razor blade. These samples were blended (Sorvall Omni Mixer) in rigor solution using the following sequence: twice for five seconds at 1100 rpm, twice for one second at 1800 rpm, once for one second at 2500 rpm, and once for one second at 3100 rpm.

3.1.2. Solutions

The rigor solution (pH 7.4) was composed of (in mM): 50 Tris, 100 NaCl, 2 KCl, 2 $MgCl_2$, and 10 EGTA. Protease inhibitors were added to the final solution, in the following concentrations (in μM): 10 leupeptin, 5 pepstatin A, 0.2 PMSF, 0.5 NaN_3, and 0.5 DTT. The relaxing solution (pH = 7.0; pCa^{2+} = 8) was composed of (in mM): 10 MOPS, 64.4 K^+ proprionate, 5.23 Mg^{2+} proprionate, 9.45 Na_2SO_4, 10 EGTA, 7 ATP, 10 creatine phosphate. The activating solution (pH = 7.0; pCa^{2+} = 4.75) was composed of (in mM): 10 MOPS, 45.1 K^+ proprionate, 5.21 Mg^{2+} proprionate, 9.27 Na_2SO_4, 10 EGTA, 7.18 ATP, 10 creatine phosphate.

3.2. Force and sarcomere length measurements

A small amount of blended muscle mixture was placed in a test chamber whose bottom was made of a glass cover slip. The chamber was positioned on top of a moveable stage mounted on an inverted microscope (Zeiss, Axiovert 35, Germany) (Figure 3). After 5 min allowed for stabilization, some myofibrils settled onto the bottom of the

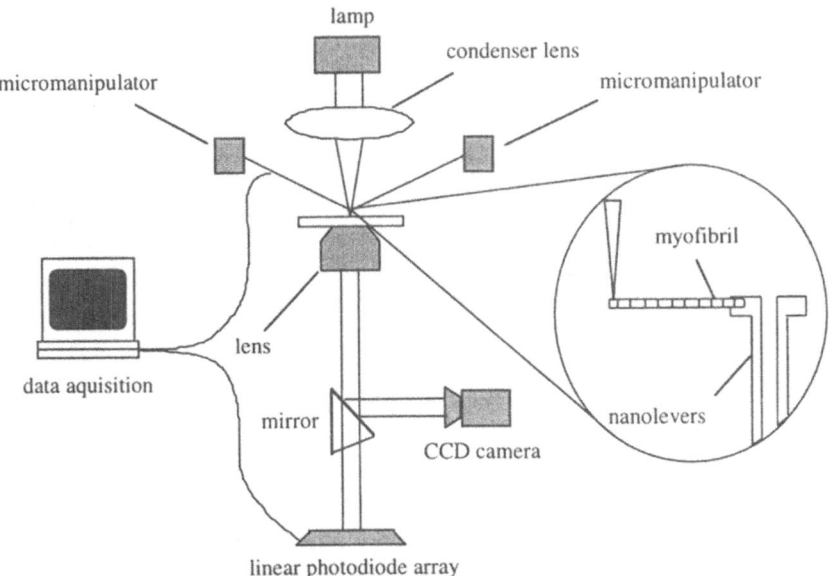

Figure 3. Apparatus used for myofibril experimentation, consisting of an inverted microscope, micromanipulators and a linear photodiode array that captures the myofibrillar striation pattern. Inset: attachment of the myofibril to a nanolever for force measurement at one end, and a glass needle for displacement control at the other end.

chamber, while most myofibrils remained in suspension. The rigor solution was slowly replaced by the relaxing solution, and the myofibrils in suspension were washed away, allowing for a clear visualization of the undisturbed myofibrils positioned at the bottom of the chamber.

Two sets of experiments were performed. In the first set (n = 5), myofibrils were attached to a glass needle at one end, and to a nanolever pair at the other. The glass needle and nanolever could be moved independently by two micromanipulators. The nanolever was used for force measurement. It consists of two parallel beams of known stiffness connected at their base. The displacement of one beam tip relative to the other is proportional to the tension exerted by the myofibril. The glass needle was connected to a motor arm that allowed for fine displacement of the needle, and therefore precise changes in myofibril length.

In order to have all sarcomeres and the two nanolevers projected onto the photodiode array, the myofibril had to be short, with a maximum of six to seven sarcomeres arranged in series. Force measurements were made in these myofibrils, but the end effects prevented systematic sarcomere length detection. When long myofibrils (more than seven sarcomeres) were used, sarcomere lengths could be measured in all sarcomeres of the myofibril. Consequently, we bypassed the measurement of force in a second set of myofibrils, so that all sarcomeres could be tracked. These myofibrils (n = 8)

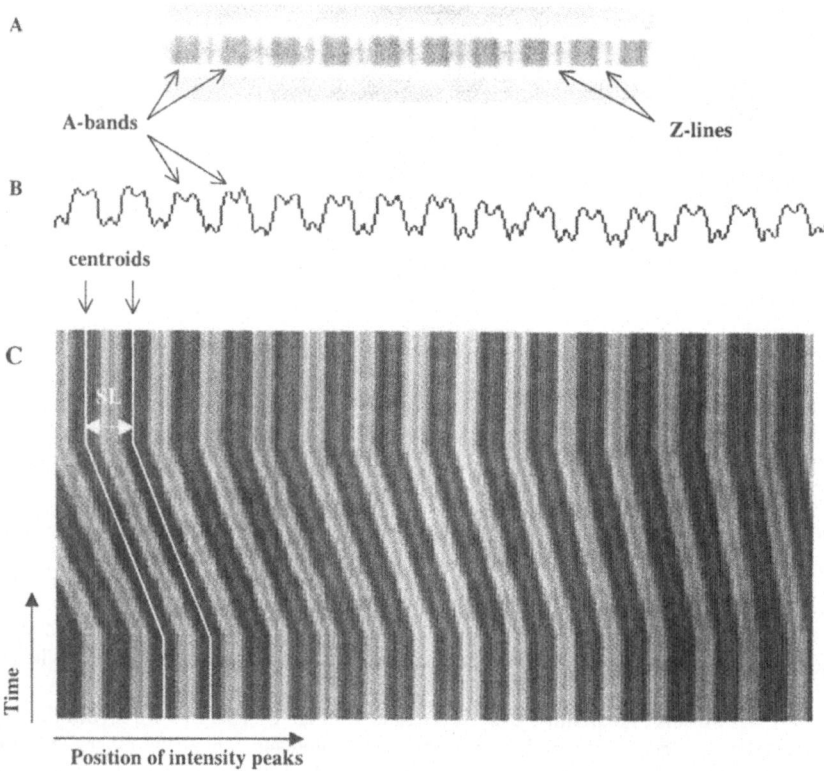

Figure 4. Measurement of myofibril striation pattern. (A) Image of a myofibril during an experiment. (B) Scan of the myofibril showing peaks associated with the dark bands. (C) Consecutive scans of the myofibril before, during, and after a stretch. Sarcomere length (SL) is measured based on the distance between the centroids of A-bands.

were attached to two glass needles, and sarcomere length changes were measured without the corresponding force measurement.

After centering the glass needles and the pair of nanolevers in the optical field, a myofibril was chosen for experimentation, based on its appearance and striation pattern. The image of the myofibril, provided by an oil-immersion phase contrast lens (Zeiss, 100X, N.A. 1.3), was projected onto a linear, 1024-element photodiode array (Reticon, Santa Clara, CA), which was scanned (20 Hz) to produce tracings of intensity vs. position along the myofibril. Because of the contrast between dark (A-band) and light (I-band) regions, the photodiode array output generates a signal representing the sarcomere-banding pattern. Sarcomere length (SL) was calculated by a minimum average risk algorithm method, with accuracy on the order of nanometers [17], based on the span between A-band centroids (Figure 4).

3.3. Experimental protocol

After obtaining a clear striation pattern, a ten-minute rest period was given before the start of the mechanical tests. The relaxing solution was replaced by activating solution, whereupon tension developed. The myofibril was kept in the activating solution for 30 to 60 sec prior to the imposition of stretch. All experiments were performed at room temperature (20°C – 22°C). Starting from a mean SL of ~2.55 μm, stretches of a nominal amplitude of 4 to 10 % (actual, 4.8 to 17.0 %) of SL were applied, with a nominal velocity of 100 nm·sec^{-1} (actual, 118.9 ±15.9 nm·sec^{-1}).

The lengths of thick and thin filaments of rabbit skeletal muscle are 1.63 μm and 1.12 μm respectively [18], and the width of the bare zone is 0.15 μm [19]. The descending limb of the force-sarcomere length relationship thus begins at 2.39 μm and extends to 3.87 μm [8], therefore all stretches were performed on the descending limb of the force-length relationship.

4. RESULTS

4.1. Force production

When activated myofibrils were stretched, force was increased (Figure 5). Once the stretch was completed and the myofibril was held isometrically, force decreased initially, and then attained a relatively constant value. In the myofibril shown in Figure 5, force measured 8 s after the end of stretch was 18.3% greater than the pre-stretch isometric force, and was 39.2% greater than the expected force at the final average sarcomere length, taking into account the increase in passive force.

In all myofibrils investigated in this study, forces following stretch were greater than the isometric forces at the pre-stretch length. Forces were always greater than the expected isometric forces at the final length, when accounting for the predicted loss in active force associated with a decrease in actin-myosin overlap and the increase in force associated with the greater passive force [20]. Table 1 shows the percent force enhancement above the initial length from the 5 myofibrils investigated in this study.

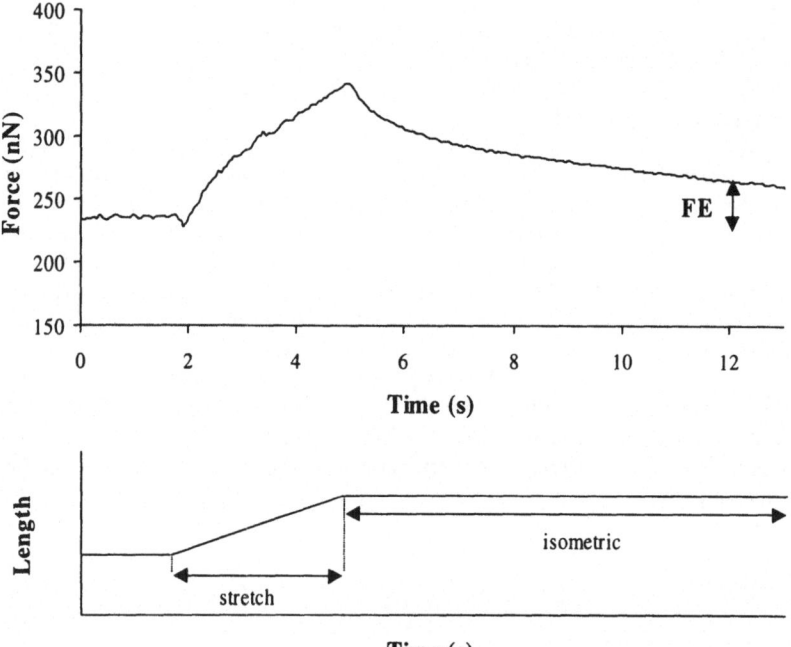

Figure 5. Force enhancement (FE) in an activated myofibril. FE was based on the isometric force produced just before stretch.

Table 1. Force enhancement (FE) in activated myofibrils following stretch (amplitude: 8% of SL; speed: 100 at nm·sec^{-1}). FE was calculated based on the isometric force produced at the initial length, from which the myofibril was stretched (F_i)

Myofibril	Initial SL (µm)	FE ($\%F_i$)
1	2.50	10.9
2	2.65	18.6
3	2.65	18.3
4	2.65	21.8
5	2.80	45.9

4.2. Sarcomere length dynamics

Upon activation, but before stretching, sarcomeres exhibited some non-uniformity, but remained at approximately constant length (Figure 6a). During stretching of the active myofibril, all sarcomeres (n = 14) elongated, albeit to a different degree. During the isometric phase following myofibril stretch, sarcomeres remained at a constant length (i.e., changed length by less than 0.1 µm), and therefore were considered stable. The mean maximal dispersion (± S.E.M.) of all sarcomeres in this myofibril was 0.026 ± 0.005, and they occurred in the shortening and lengthening directions. When the myofibril was shortened back to its initial length, all sarcomeres returned closely to their pre-stretch length.

In two other myofibrils, two and three sarcomeres, respectively, showed directional length changes. In Figure 7A, one sarcomere shortened and the other stretched during the isometric phase following myofibril stretching. These two sarcomeres were considered unstable, with the greatest length changes observed during the post-stretch phase of 0.153 µm and 0.157 µm, respectively, during the eight second stretch period.

These unstable sarcomeres were adjacent to one another and were located near the attachment point at one end of the myofibril. The combined length of these two sarcomeres was constant (Figure 7B), with a peak-to-peak range of length change of 0.0285 µm. Since these changes in sarcomere lengths occurred at one end of the myofibril exclusively, we assumed that they were caused by structural damage associated with fixation of the specimen to the glass needle. Furthermore, since the combined sarcomere length remained constant, but the individual sarcomere lengths (as measured by the distance between the centroids of A-bands) changed, it was assumed that these observed "sarcomere length instabilities" might have been associated with a shift in a single A-band.

In one myofibril, one sarcomere shortened at the onset of the imposed stretch, and then stretched for the remaining period (Figure 8). After the end of the stretch, all sarcomeres (including the one that first shortened) remained at a constant length (maximal sarcomere length changes, including all sarcomeres, during eight seconds of isometric myofibril contraction: 0.017 ± 0.001 µm). The sarcomeres then returned to the pre-stretch length upon myofibril shortening as a mirror image pathway of stretch.

From 110 sarcomeres recorded in this study, none stretched beyond myofilament overlap (i.e., SL > 3.87 µm). Three sarcomeres stretched to lengths of 3.34

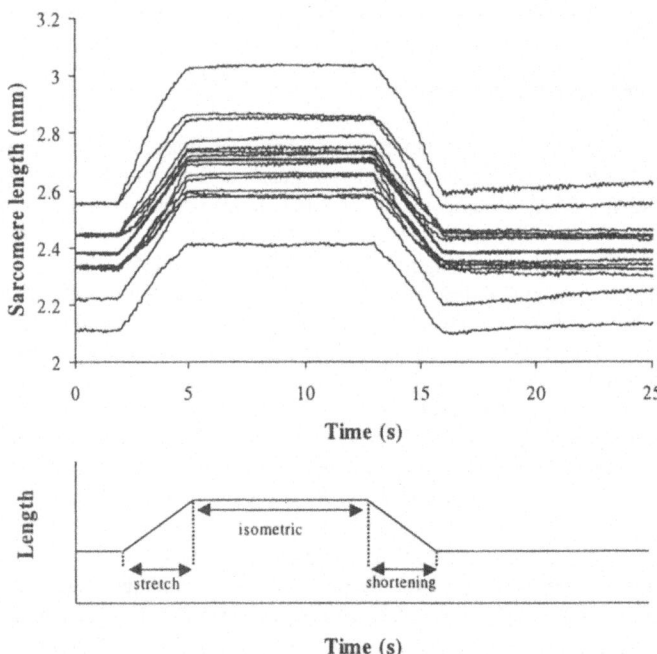

Figure 6. Behavior of individual sarcomeres during stretch in an activated myofibril. The thick line represents the average SL. The initial average SL was 2.38 μm, and the average SL following stretch was 2.70 μm. Velocity of stretch: 107.5 nm·sec[-1]. All sarcomeres remained at a constant length during the isometric period following stretch.

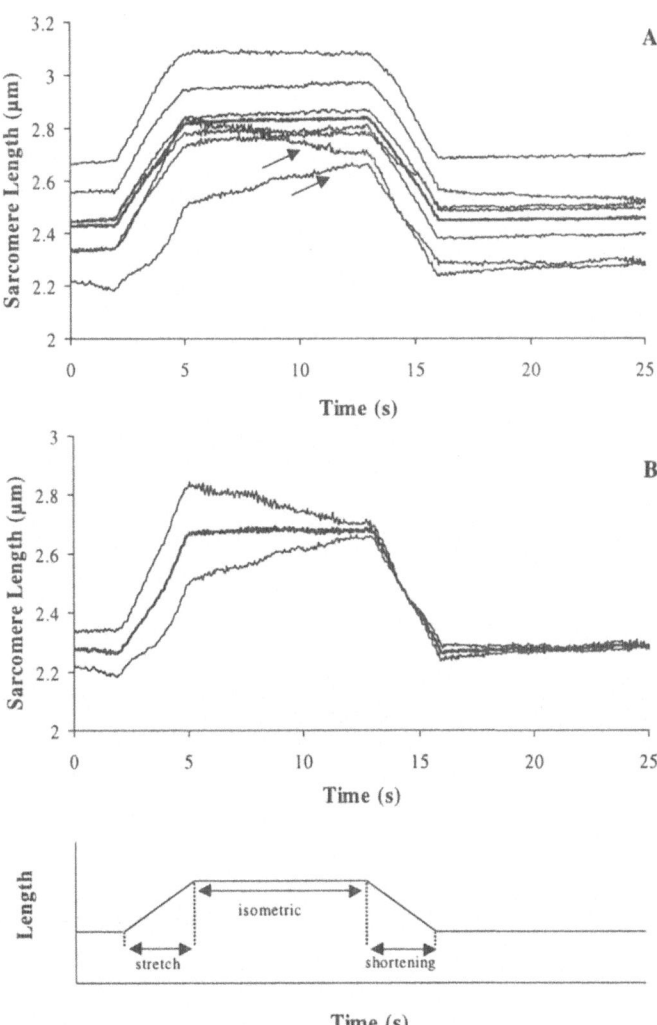

Figure 7. (A) Behavior of individual sarcomeres during stretch in another activated myofibril. The thick line represents the average SL. The initial average SL was 2.43 μm, and the average SL following stretch was 2.81 μm. Velocity of stretch: 128.1 nm·sec^{-1}. (B) Same myofibril as shown in A, but in this case only the two sarcomeres that had a directional length change (arrows) are shown. These two sarcomeres are adjacent to each other and are at the end of the myofibril, right near the attachment to the glass needle. The average length of these two sarcomeres is constant.

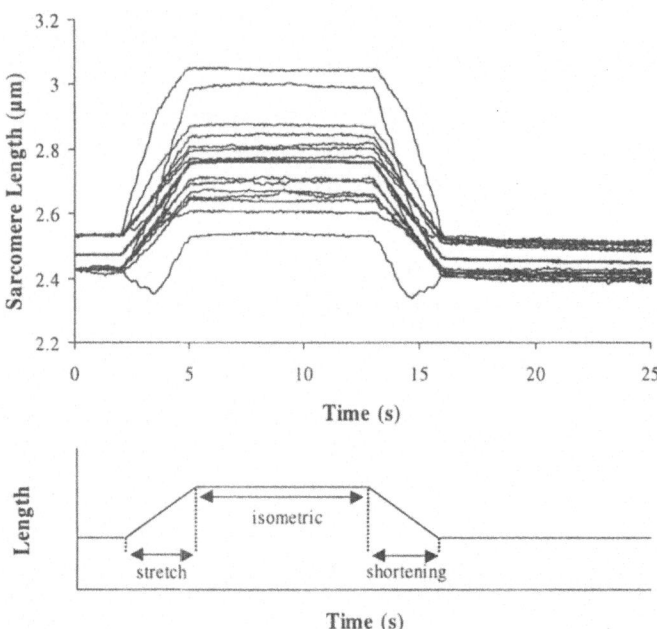

Figure 8. Behavior of individual sarcomeres during stretch in a third activated myofibril. The thick line represents the average SL. The initial average SL was 2.47 μm, and the average SL following stretch was 2.75 μm. Velocity of stretch: 93.3 nm·sec[-1].

µm, 3.33 µm, and 3.27 µm, respectively, each from a different myofibril. The remaining 107 sarcomeres never stretched beyond 3.22 µm. For sarcomere lengths smaller than 3.4 µm, the passive force contribution to the total force is small: approximately 5% or less of the active tension at full overlap of thick and thin filaments [20].

5. DISCUSSION

In experiments performed with myofibrils, we showed that sarcomeres activated on the descending limb of the force-length relationship are stable. In 105 out of 110 sarcomeres from eight myofibrils, lengths remained perfectly constant following a myofibrillar stretch on the descending limb of the force-length relationship. This finding indicates that sarcomeres arranged in series appear to be perfectly stable, even in a situation where actin-myosin overlap is decreased by stretch. This result contradicts many earlier suggestions, in which the static negative slope of the force-length relationship was interpreted as being negative for dynamic contractions as well. The sarcomeres that showed directional length changes came in clusters of 2 or 3, they were always adjacent to each other, and were always located at one end of the myofibril. Therefore we interpreted these length changes as being associated with structural damage of the myofibril caused by attachment of the myofibril to the glass needles. This interpretation is supported by the fact that these localized length changes never affected the remaining sarcomeres in the myofibril.

The steady-state isometric force following active myofibril stretch was always greater than the isometric force produced at the initial length. Since the myofibrils were stretched along the descending limb of the force-length relationship, and therefore actin-myosin overlap was decreased, the force produced after stretch must have been higher than the expected isometric force produced at the corresponding final length. This steady-state force enhancement is qualitatively the same as that observed in whole muscles [1,7,21] and single muscle fibers [2,3,9].

The mechanism underlying such force enhancement is unknown, but sarcomere instability on the descending limb of the force-length relationship has been the main explanation for the extra-force following active stretch [4,7-10]. It has been shown that, during stretch of activated muscle fibers, sarcomeres do not stretch by the same amount. Sarcomeres toward the ends of the fiber stretch less than average, while sarcomeres near the center stretch more than average [9]. Morgan [10] extended this idea by proposing a model in which "weak" sarcomeres would be stretched beyond actin-myosin overlap, a region in which they would "pop", being supported entirely by passive forces. At the same time, "strong" sarcomeres would elongate slightly during stretch. Since all sarcomeres arranged in series must bear the same steady-state force, the total force (active plus passive) in the short sarcomeres must equal the passive force in the long sarcomeres.

In the experiments described here, we provide direct evidence that force enhancement in rabbit psoas myofibril is not associated with sarcomere instability. During elongation of an activated myofibril, all sarcomeres were stretched, albeit not necessarily by the same amount, as has been shown previously in single muscle fibers [9]. However, sarcomere lengths following stretch remained constant, despite considerable differences in individual sarcomere lengths, and thus, actin and myosin overlap. Instabilities are supposed to develop very rapidly, 1.5 s after the end of the stretch [9,22]. We waited 8–10 s after stretch and did not observe instabilities, except in those exceptional cases mentioned before. Among the 110 sarcomeres tracked in eight myofibrils, we found none that was shortened, and none hat stretched beyond actin-myosin overlap to be entirely supported by passive forces.

Summarizing, each sarcomere was stretched when the myofibril was stretched along the descending limb of the force-length relationship, and the steady-state force following stretch was greater than the isometric force at the initial length. Therefore, force in each sarcomere increased as sarcomeres lost actin-myosin overlap. The small increase in passive force associated with the myofibril stretch cannot account for the increased forces following stretch [20]. These observations suggest that the observed non-uniformities in sarcomere lengths are not caused by instability, and furthermore, that force enhancement occurs in each individual sarcomere. Therefore, it seems quite possible that force enhancement is a property associated with the basic molecular mechanisms of contraction, rather than structural non-uniformities and instabilities on the fiber level.

6. REFERENCES

1. B. C. Abbott, and X. M. Aubert, The force exerted by active striated muscle during and after change of length. *J. Physiol.* **117**, 77-86 (1952).
2. K. A. P. Edman, G. Elzinga, and M. I. M. Noble, Enhancement of mechanical performance by stretch during tetanic contractions of vertebrate skeletal muscle fibres. *J. Physiol.* **281**, 139-155 (1978).
3. K. A. P.Edman, G. Elzinga, and M. I. M. Noble, Residual force enhancement after stretch of contracting frog single muscle fibers. *J. Gen. Physiol.* **80,** 769-784 (1982).
4. K. A. P. Edman, and T. Tsuchiya, Strain of passive elements during force enhancement by stretch in frog muscle fibres. *J. Physiol.* **490.1**, 191-205 (1996).
5. W. Herzog, and T. R. Leonard, The history dependence of force production in mammalian skeletal muscle following stretch-shortening and shortening-stretch cycles. *J. Biomech.* **33**, 531-542 (2000).
6. M. Linari, L. Lucii, M. Reconditi, M. E. Vannicelli. Casoni, H. Amenitsch, S. Bernstorff, and G. Piazzesi, A combined mechanical and x-ray diffraction study of stretch potentiation in single frog muscle fibres. *J. Physiol.* **526.3**, 589-596 (2000).
7. D. L. Morgan, N. P. Whitehead, A. K. Wise, J. E. Gregory, and U. Proske, Tension changes in the cat soleus muscle following slow stretch or shortening of the contracting muscle. *J. Physiol.* **522.3**, 503-513 (2000).
8. A. M. Gordon, A. F. Huxley, and F. J. Julian, The variation in isometric tension with sarcomere length in vertebrate muscle fibres. *J. Physiol.* **184**, 170-192 (1966).
9. F. J. Julian, and D. L. Morgan, The effect on tension of non-uniform distribution of length changes applied to frog muscle fibres. *J. Physiol.* 293, 379-392 (1979).
10. D. L. Morgan, An explanation for residual increased tension in striated muscle after stretch during contraction. *Exp. Physiol.* 79, 831-838 (1994).
11. H. Sugi, and T. Tsuchiya, Stiffness changes during enhancement and deficit of isometric force by slow length changes in frog skeletal muscle fibres. *J. Physiol.* **407**, 215-229 (1988).
12. L. Hill, A-band length, striation spacing and tension change on stretch of active muscle. *J. Physiol.* **266**, 677-685 (1977).
13. H. E. D. J. ter Keurs, T. Iwazumi, and G. H. Pollack, The sarcomere length-tension relation in skeletal muscle. *J. Gen. Physiol.* **72**, 565-592 (1978).
14. K. A. P. Edman, and C. Reggiani, Redistribution of sarcomere length during isometric contraction of frog muscle fibres and its relation to tension creep. *J. Physiol.* **351**, 169-198 (1984).
15. T. L. Allinger, M. Epstein, and W. Herzog, Stability of muscle fibers on the descending limb of the force-length relation. A theoretical consideration. *J. Biomech.* **29**, 627-633 (1996).

16. G. I. Zahalak, Can muscle fibers be stable on the descending limbs of their sarcomere length-tension relations? *J. Biomech.* **30**, 1179-1182 (1997).
17. F. Blyakhman, A. Tourovskaya, and G. H. Pollack, Quantal sarcomere-length changes in relaxed single myofibrils. *Biophys. J.* **81**, 1093-1100 (2001).
18. H. Sosa, D. Popp, G. Ouyang, and H. E. Huxley, Ultrastructure of skeletal muscle fibers studied by a plunge quick freezing method: myofilament lengths. *Biophys. J.* **67**, 283-292 (1994).
19. J. M. Squire, *The structural basis of muscular contraction* (Plenum Press, New York, 1981).
20. M. L. Bartoo, V. I. Popov, L. A. Fearn, and G. H. Pollack, Active tension generation in isolated skeletal myofibrils. *J. Muscle Res. Cell. Motil.* **14**, 498-510 (1993).
21. W. Herzog, and T. R. Leonard, Force enhancement following stretching of skeletal muscle: a new mechanism. *J. Exp. Biol.* **205**, 1275-1283 (2002).
22. L. M. Brown, and L. Hill, Some observations on variations in filament overlap in tetanized muscle fibres and fibres stretched during a tetanus, detected in the electron microscope after rapid fixation. *J. Muscle. Res. Cell. Motil.* **12**, 171-182 (1991).

DISCUSSION

Cecchi: You may add to the possible explanation for the long lasting force enhancement, the mechanism of static stiffness which was illustrated yesterday by Angela Bagni: a passive stiffness modulated by calcium.

Rassier: Definitely, it is possible that the static stiffness reported by Dr. Angela Bagni is partially responsible for the long lasting force enhancement. In our case, we think titin may be responsible for this static stiffness, and consequently force enhancement. In several occasions, we observe a long-lasting enhancement in passive force after stretch (and deactivation) in single skeletal muscle cells, and this passive force enhancement has characteristics that are similar to some of the characteristics of titin.

Huxley: Are your numbers close to what was shown by Dr. Angela Bagni on the static stiffness?

Cecchi: The stiffness underlying the force enhancement is about the same as the static stiffness at least at sarcomere length around 2.2 μm.

ADAPTATIONS IN TITIN'S SPRING ELEMENTS IN NORMAL AND CARDIOMYOPATHIC HEARTS

Henk Granzier, Dietmar Labeit, Yiming Wu, Christian Witt, Kaori Watanabe, Sunshine Lahmers, Michael Gotthardt, Siegfried Labeit*

1. INTRODUCTION

Titin (also known as connectin) is a giant elastic protein located in the striated-muscle sarcomere where it spans from Z-line to M-line. A large part of the I-band region of the titin molecule is extensible and functions as a molecular spring that underlies passive muscle stiffness when sarcomeres are stretched. This spring has a complex composition. In cardiac titin it consists of three extensible elements: tandem Ig segments, the PEVK segment and the N2B unique sequence. Here we discuss our recent work focused on understanding the molecular basis of titin's extensibility and in which force-extension curves were measured by using an atomic force microscope specialized for stretching single molecules. We will discuss results from recombinant proteins that represent the various elements of titin's extensible region. The obtained single molecule mechanical characteristics of titin's various spring elements explain well their measured extension in the cardiac sarcomere when stretched within their physiological length range. We also examined how titin's contribution to passive muscle stiffness may be adjusted. We discuss evidence that suggests that calcium/S100 may adjust titin-based stiffness and that phosphorylation of cardiac titin's N2B spring elements reduces titin-based passive stiffness in cardiac muscle. Finally, we show that the cardiac sarcomere of large mammals co-expresses titin isoforms and that differential splicing of titin's spring elements is a long-term mechanism of adjustment, which plays a role in passive stiffness modulation during heart disease.

* Henk L. Granzier, Yiming Wu, Kaori Watanabe, Michael Gotthardt, Sunshine Lahmers: VCAPP, Washington State University, Pullman, WA 99164-6520, USA. Dietmar Labeit, Christian Witt, and Siegfried Labeit: Anesthesiology and Intensive Operative Medicine, University Hospital Mannheim, 68135, Germany.

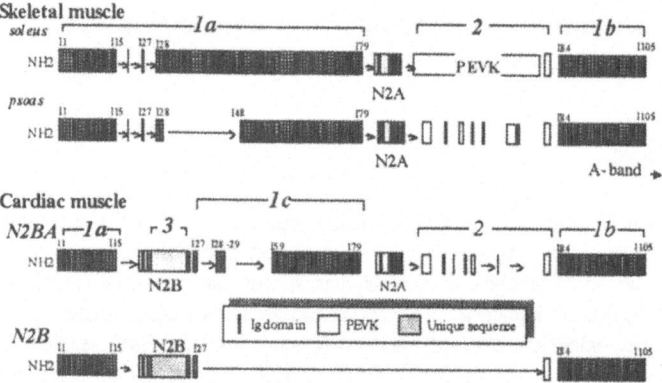

Figure 1. I-band region of titins found in various muscle types. Skeletal muscles express N2A-based titins that vary in size in different muscle types (two of many possible splice pathways are shown). Cardiac muscles express large N2BA titins and small N2B titins. All isoforms contain tandem Ig (*1*a and 1b)) and PEVK (*2*) spring elements. Furthermore, cardiac titins contain the N2B-unique sequence (N2B-Us) as a third spring element (*3*). (Based on[10].)

2. SERIALLY-LINKED SPRINGS

Titin forms a striated muscle-specific myofilament that develops passive force in response to sarcomere stretch. (For reviews see[1-7].) Titin's force is generated by an extensible segment located in the I-band region of the sarcomere. The extensible region comprises tandemly arranged immunoglobulin(Ig)-like domains (~90 amino acid domains with a seven-strand β-barrel structure [8]) making up a proximal tandem Ig segment near the Z-line and a distal tandem Ig segment near the A-band (Fig.1, *1a* and *1b*). Both tandem Ig segments are separated by another extensible region of titin, the PEVK domain (Fig. 1, *2*), so named because it is rich in proline (P), glutamate (E), valine (V) and lysine (K) residues [9]. The PEVK and the N- and C-terminally located tandem Ig segments are spring elements found in both cardiac and skeletal muscle titins, whereas the N2B-unique sequence (N2B-Us) is cardiac-specific and forms a third spring element (Fig.1, *3*).

Immunoelectron microscopy on human soleus fibers has shown that in slack sarcomeres (no external force) the tandem Ig segments and the PEVK are in a 'contracted' state' and that upon sarcomere stretch, extension of tandem Ig segments initially dominates, followed by dominating PEVK extension[11-14]. It is likely that tandem Ig extension is primarily due to unbending of linkers between folded Ig domains [11-13]. The tandem Ig segments exhibit wormlike-chain (WLC) behavior with a persistence length (measure of chain's bending rigidity) that is relatively long (~15 nm based on a recent single-molecule study [15]), explaining why tandem Ig extension dominates in moderately stretched sarcomeres where passive force is low.

When a continuous stretch is imposed on slack sarcomeres, tandem Ig segments of soleus muscle titin extend rapidly to a near constant length[11]. Further extension is halted despite the generation of high levels of passive force (in this regime PEVK extension dominates). Although unfolding of a few domains can not be excluded [16], the constant length of the tandem Ig segments in highly stretched sarcomeres of soleus muscle that can be accommodated by only folded domains indicates that large-scale Ig unfolding is unlikely to take place under physiological conditions [17]. The extensibility provided by the N2B unique sequence of cardiac titins reduces the necessity for Ig domain unfolding within the physiological sarcomere length range of cardiac muscle (see also below). The PEVK extends at relatively high forces [11,14,18] and this process is likely to result from straightening of random coil and polyproline II helices that comprise this element [19,20].

A model has emerged for extensibility of skeletal muscle titin in which the PEVK segment (acting largely as an unfolded polypeptide) and the tandem-Ig segments (containing folded Ig domains) behave as serially-linked entropic springs with low and high persistence length, respectively [12]. In short sarcomeres these springs are in a high entropy state (PEVK and tandem Ig segment are 'contracted') and upon sarcomere extension the springs straighten, lowering their conformational entropy and resulting in a force, known as entropic force. This force increases in a non-linear fashion with the segment's fractional extension (end-to-end length divided by the contour length; for details see[21,22]) and is inversely proportional to the chain's persistence length. Using persistence lengths of 15 and 2 nm for the tandem Ig and PEVK segments, respectively, allowed us to model the extensible region of soleus titin and predict the measured extension of these segments in the sarcomere [11]. Although minor deviations from measured extension are present (they may result from the arrangement and environment of titin within the sarcomere that limits the conformational space[2]) overall, predictions and measurements in soleus fibers are close. Whether this model can be applied to cardiac titin, however, requires a better understanding of the mechanical properties of the N2B unique sequence, titin's third molecular spring [23,24]. To explore the mechanical properties of the N2B unique sequence (N2B-Us) we performed atomic force microscopy on this element and compared its properties with those of the fragment I91-98 (comprising eight Ig domains from the distal tandem Ig segment), and the 188-residue PEVK domain [25]. Force-extension curves were measured by using an atomic force microscope (AFM) specialized for stretching individual molecules.

Consistent with earlier findings [26-29] the stretch curve of I91-98 displayed sawtooth-like force peaks[25,30], indicating that repetitive structural transitions occurred during stretch with each sudden force drop representing the unfolding of a β-barrel Ig domain. The unfolding force varies with stretch rate: at a rate of 50 nm/sec the mean unfolding force is 180 pN and at a rate of 1000 nm/sec the value is 250 pN[25,30]. The stretch-rate dependence of unfolding force allowed us to deduce the rate constant of unfolding at zero external force (K_u^0) and the location of unfolding barrier, or the width of the unfolding potential along the unfolding reaction coordinate (ΔX_u). Results indicate $K_u^0 = 6.1 \times 10^{-5}$ s^{-1} and $\Delta X_u = 0.25$ nm[30]. These values were used to ascertain whether unfolding of Ig domains has to be taken into account when modeling the extensible region of cardiac titin. Results indicate that in repeated stretch-release cycles the probability of domain unfolding is low and that at the upper range of the physiological force regime (~15 pN) of all titin molecules in the sarcomere only ~10% may have a single domain unfolded [25,30]. This suggests that it is justified to simulate the extensible region of cardiac titin with tandem Ig segments containing folded domains.

AFM-based stretch-release force-extension curves of both the PEVK and N2B-Us displayed little hysteresis: the stretch and release data nearly overlapped[25]. The force-extension curves closely followed wormlike chain (WLC) behavior. Thus, both the PEVK and N2B unique sequence behave as entropic springs with random coil structure. Considering the limited force resolution of the AFM (~ 5 pN), these findings do not exclude deviations from WLC behavior at low force. For example, the secondary structure predicted for the N2B-Us [25] indicates that ~1/3 of the residues form α-helical structures (the remaining residues are unstructured). Stretching of these α-helical structures may cause sequential breakage of intra-helix hydrogen bonds at forces below the resolution of the AFM [31].

By fitting the release curves of stretch-release cycles with the WLC equation we determined the persistence length (measure of chain's bending rigidity) of the PEVK and N2B-Us. Persistence length histograms indicate that the single-molecule persistence length is ~1.4 nm and ~0.65 nm for the PEVK and the N2B-Us, respectively [25].

Using the above-described molecular characteristics, we modeled the extensible region of cardiac titin (N2B isoform) as three mechanically distinct springs that are serially linked: tandem Ig segments, the N2B-Us and the PEVK. We calculated the extension of the various spring elements as function of sarcomere length. In the physiological sarcomere length range of the heart (~1.8-2.4 μm), predicted values were within experimental error of those measured, suggesting that the model parameters used in the simulations approximate those *in situ* well[25]. Because the model calculations assume that unfolding of Ig domains is absent, these findings also support the view that unfolding is not a prominent physiological process. Only at sarcomere lengths greater than ~2.4 μm does the measured tandem Ig segment extension exceed that predicted, and, thus, beyond the physiological sarcomere length range Ig unfolding is likely to take place. By having Ig domains in their folded state, the tandem Ig segments attain a long persistence length allowing them to extend under low force and to set the sarcomere length at which extension of the PEVK and N2B-Us starts to dominate.

In summary, AFM studies on N2B cardiac titin reveal that the PEVK and N2B-Us both behave as entropic springs, but with different persistence lengths. Using the obtained single molecule persistence lengths, a model of the extensible region of cardiac titin can be constructed containing three serially-linked springs (tandem Igs, PEVK and N2B-Us) that well simulates the complex extension of titin in the sarcomere.

3. TITIN-ACTIN INTERACTION

In the sarcomere titin's extensible region is located in close proximity to the actin-based thin filaments and several earlier studies have suggested that interactions between titin and actin occur (reviewed in [32]). Recently we expressed recombinant fragments representing the sub-domains comprising the extensible region of cardiac N2B titin (tandem Ig segments, the N2B splice element, and the PEVK domain), and assayed them for binding to F-actin [32]. The PEVK region bound F-actin, while no binding was detected for the other constructs. The results of Kulke et al.[33] are consistent with these findings.

F-actin contains a large patch of negatively charged residues on its exposed surface [34], and several actin-binding proteins are known to bind to this region via basic charge clusters (e.g., [35-37]). The results of our binding studies in solutions of various ionic strengths suggest that PEVK-actin interaction also includes an electrostatic component [32]. The significance of interaction between actin and the cardiac PEVK (N2B isoform) was investigated using an *in vitro* motility assay technique. The findings indicate that, as the thin filament slides relative to titin in the motility assay, a dynamic interaction between the PEVK domain and F-actin retards filament sliding[32]. We also investigated the effect of calcium on PEVK-actin interaction. Although calcium alone had no effect, S100A1, a soluble calcium-binding protein found at high concentrations in the myocardium, inhibited PEVK-actin interaction in a calcium-sensitive manner[32]. *In vitro* motility results indicate that S100A1-PEVK interaction alleviates the force that arises as F-actin slides relative to the PEVK domain . We speculate that S100A1 may provide a mechanism to free the thin filament from titin and reduce titin's passive tension during active contraction. Thus, a dynamic interaction between titin and actin contributes to passive stiffness of the sarcomere and the interaction varies with the physiological state of the myocardium.

4. POST-TRANSLATIONAL MODIFICATION

We recently showed[38] that cardiac titin can be phosphorylated by protein kinase A (PKA) and that the phosphorylation site is the N2B unique sequence. Thus, as with the well-characterized myofibrillar PKA substrates MyBP-C and TnI, titin contains a PKA-responsive domain expressed only in cardiac muscle. Interestingly, PKA-based phosphorylation of titin results in a reduction of passive tension of cardiac myocytes [38,39] that varies inversely with sarcomere length[38]. This reduction may be explained by assuming that phosphorylation destabilizes native structures within the N2B element (for example, α-helical structures, see above) causing it to extend and lower its fractional extension. This would give the tandem Ig segments and PEVK domain lower fractional extensions at a given SL, leading to a decrease in passive tension. At longer SLs, these structures will be denatured due to stretch, and phosphorylation will therefore have less impact on the extension of the N2B element and, consequently, on passive tension.

It is well established that activation of the β–adrenergic pathway increases phosphorylation of TnI and MyBP-C[40-42]. Whether phosphorylation of titin can also be enhanced by β–adrenergic agonists was investigated by using back-phosphorylation experiments. Cell suspensions were subjected to either the β-receptor antagonist propranolol or the β-receptor agonist isoproterenol, followed by rapid skinning and incubation with ^{32}P-ATP and PKA. Cell suspensions were then solubilized, electrophoresed, stained with Coomassie blue, and exposed to radiographic film. As observed for MyBP-C and Tn-I, isoproterenol treatment resulted in reduced ^{32}P incorporation[38]. Relative to propranolol, isoproterenol treatment decreased ^{32}P incorporation by 52 ± 12% (P<0.01). These findings indicate that titin is phosphorylated by PKA in response to β-adrenergic stimulation in intact myocytes. Considering that the activation of PKA via β-adrenergic stimulation constitutes a major regulatory pathway in the heart, the PKA responsive element of cardiac titins may allow modulation of diastolic function *in vivo*. We speculate that when β-adrenergic stimulation enhances the heart beat frequency and rate of contraction and relaxation, the reduction in titin's force resulting from N2B phosphorylation allows for more rapid and complete ventricular filling.

5. DIFFERENTIAL SPLICING

Titin is encoded by a single gene located in human and mouse[43] on chromosome 2. The recent genomic analysis of human titin revealed 363 exons that code for a total of 38,138 amino acid residues [43]. This includes 108 exons which code for conserved ~28-residue PEVK/PPAK-repeats, possibly corresponding to structural spring units[10,44]. Interestingly, two titin exons, M10 and novex-III, function as alternative C-termini [43], giving rise to truncated (Z1-Z2 to novex-III; Mr ~620 kDa), and full-length titins (Z1 to M10; Mr >3000 kDa). The truncated novex-3 titin isoform can integrate into the Z-line lattice but is too short to reach the A-band, and is expressed in both skeletal and cardiac muscles [43]. We have speculated that co-expression of truncated and full-length titins may adjust the titin filament system to both three- and two-fold symmetries of thick and thin filaments, respectively [43].

The I-band region of full-length titins comprising I50 to I84 is extensively differentially spliced[10,43]. Thereby, highly variable isoforms of different spring composition are generated (Fig.1). The soleus muscle expresses the largest titin isoform observed so far (M_r 3.7 MDa), while psoas muscle expresses a smaller titin (M_r 3.35 MDa) with shorter proximal tandem Ig and PEVK segments. Passive tension increases steeper with sarcomere length in psoas than in soleus fibers[10]. The differences in tandem Ig and PEVK segment length of the different muscle types provide a molecular framework to understand these mechanical differences.

The effect of tandem Ig and PEVK segment lengths on passive force can be evaluated by considering the model of passive force generation in which the tandem-Ig segments (containing folded Ig domains) and the PEVK segment (acting largely as an unfolded polypeptide) behave as serially-linked entropic springs (see above). In short sarcomeres these springs are in a high entropy state and upon sarcomere extension the springs straighten. This results in a force that is inversely proportional to the chain's persistence length and that increases in a non-linear fashion with the segment's fractional extension. The serially-linked entropic springs model of passive force development may be applied to titin isoforms by adapting the entropic forces to the fractional extensions multiplied by the contour lengths of the isoform's tandem Ig and PEVK segments. As a result of sequence differences contour lengths of tandem Ig and PEVK segments in psoas are ~100 and ~400 nm shorter, respectively than in soleus titin[10]. Thus, for a given sarcomere length, the sequence differences predict that the fractional extension is higher in psoas than in soleus muscle. It follows that entropic force is predicted to be much higher in psoas than in soleus fibers, consistent with the measured passive tension differences.

Exon 49 (containing the N2B sequence) is excluded in skeletal muscle titins, but is present in all cardiac titin isoforms[10]. The major cardiac N2B isoform results from splicing of exons 49/50 to 225, thereby excluding exons 51-224. This cardiac specific N2B isoform is co-expressed with larger cardiac-specific isoforms, the N2BA titins. N2BA cardiac titins contain in addition to exon 49 (N2B) also the exons 101-111 (coding for the so-called N2A segment). N2BA titins have also a longer PEVK segment and more Ig domains than N2B titins. The expression level of N2B and N2BA cardiac titins varies considerably. Small rodents express predominantly N2B titins whereas large mammals (including humans) co-express both isoforms at intermediate levels in their ventricles, and predominantly N2BA titin in their atria[45]. Passive stiffness of cardiac myocytes is much higher in N2B expressing myocytes than in N2BA myocytes (Figure 2). The higher passive stiffness is likely to result from the shorter I-band segment of N2B titin which, at a given sarcomere length, results in an extensible segment with a fractional extension (end-to-end length divided by the maximal length) that is higher than in N2BA titin[45].

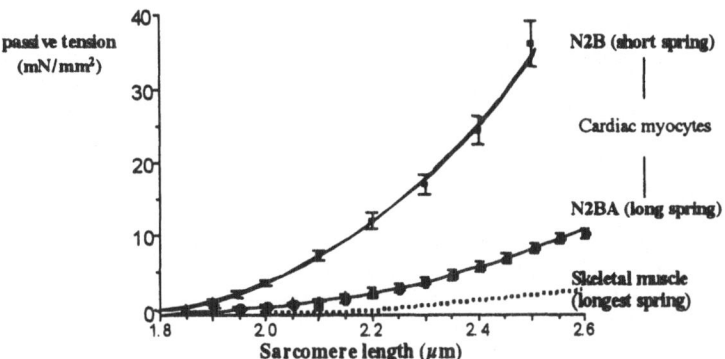

Figure 2. Passive tension – sarcomere length relations of cardiac myocytes that express predominately N2B titin, predominantly N2BA titin and of skeletal muscle fibers (soleus).

The high titin-based passive stiffness provided by N2B titin may allow rapid and stable determination of the end-diastolic volume at the high beat frequencies encountered in small mammals (where N2B titin dominates). Titin is also expected to play a role in centering the A-band within the sarcomere [46] and the high N2B-based passive stiffness may rapidly reset the A-band location during each diastole. It is also worthwhile to consider titin's contribution to restoring forces [47]. The segment of titin near the Z-line binds the thin filament and can withstand compressive forces [48-50]. When sarcomeres shorten to below the slack length, the thick filament moves into the stiff region of titin near the Z-line, and titin's extensible region extends in a direction that is opposite of that during stretch [49]. This gives rise to so-called restoring forces (pushing Z-lines away from each other) and these forces are expected to be highest in N2B expressing myocytes. Considering that restoring force may contribute to the early diastolic suction force that aids ventricular filling, expressing high levels of N2B titin may be relevant for achieving rapid diastolic filling.

Ventricular myocardium of large mammals co-expresses isoforms at the level of the half-sarcomere[51] (Figure 3), and co-expression results in passive tension levels intermediate between that of N2B and N2BA pure myocytes. Such intermediate tensions, in theory, could be achieved by varying the number of titin molecules per thick filament. However, this would also influence functions performed by titin's inextensible regions, such as thick-filament length control and construction and maintenance of Z-lines and M-lines [1]. When passive force levels are tuned via variation in the isoform expression ratio, the inextensible regions are not impacted because these regions are the same in different isoforms [10,52]. Thus co-expressing isoforms at various ratios is an effective means for tuning passive properties, and any force level intermediate between that of isoform-pure cells may be obtained.

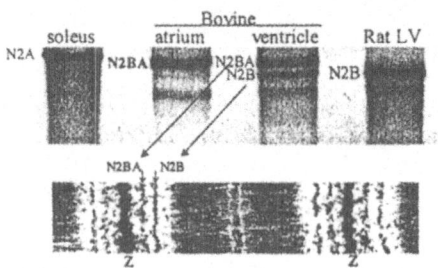

Figure 3. Top: SDS-PAGE of various muscle tissues. Bovine ventricle coexpressed N2BA and N2B titin. Bottom: immunoelectron micrograph of bovine sarcomere, revealing two I-band epitopes: one derived from N2B and another derived from N2BA titin (for detail see [51]).

6. TITIN AND HEART DISEASE

Several earlier studies reported that the amount of intact titin is reduced in patients with heart'failure [53-55], suggesting that titin is involved in altered ventricular compliance of the cardiomyopathic heart. Furthermore, titin has been implicated in familial forms of dilated cardiomyopathy (DCM) by genetic linkage studies [56]. Recent sequencing of the titin gene identified titin mutations, in and near its Z-line region as well as in the N2B exon [57-59]. A 2bp insertion in titin's 3' region in a family with DCM resulted in a truncated titin protein with missing M-line region and C-zone of A-band region of the molecule [58]. Patients that express the truncated protein suffer from heart failure of variable age of onset, in the absence of clinically detectable skeletal muscle disease. In contrast, mutations within the carboxy-terminal region of titin cause the distal type skeletal muscular dystrophy TMD without affecting the heart[60]. It is unclear why some mutations in titin cause a disease specific to the heart muscle, whereas other mutations cause a skeletal-muscle specific myopathy.

The edge of the M-line region of titin binds the novel titin-binding protein MURF-1[61] and contains a serine/threonine kinase-like domain[52], for which no physiological catalytic function has yet been shown. To investigate the physiological function(s) of the titin kinase domain, Gotthardt and colleagues have used a conditional knockout approach to selectively delete the M-line exons MEx1 (which encodes the kinase domain) and MEx2 (to maintain the reading frame) in heart and skeletal muscle[62]. When exons MEx1/MEx2 are excised early in embryonic development (using the α-MHC promoter) mice die *in utero*. In contrast, excision during late embryonic development (using the MCK promoter) allows the mice to survive but causes development of a progressive myopathy, resulting in death at 5 weeks of age. Thus, these results demonstrate an important role for MEx1 and MEx2 in early cardiac development (embryonic lethality) and postnatally when disruption of M-line titin leads to muscle weakness and death at 5 weeks of age. Myopathic changes include pale M-lines devoid of MURF-1, and gradual sarcomeric disassembly, accompanied by a normal hypertrophic gene response and upregulation of novel genes involved in myocyte signaling. Thus, this animal model indicates a critical role for the M-line region of titin in maintaining the structural integrity of the sarcomere.

To better understand titin's role in diastolic dysfunction we have also studied the canine tachycardia-induced model of DCM[63]. We investigated animals paced for four weeks, and showed that relative to controls the total amount of titin was unchanged but that the ratio of N2BA to N2B isoforms was reduced (Fig. 4). The functional consequence of this change was assessed by mechanical studies on muscle. This revealed that titin-based passive tension is significantly elevated in DCM (Fig. 4). That adjustment in cardiac isoform expression is not restricted to animal models but may also occur in human was shown by analyzing myocardium from a healthy individual and of patients with dilated cardiomypathy (DCM) or mitral regurgitation (MR)[64]. Control myocardium co-expresses N2B and N2BA isoforms at similar levels whereas in disease states the isoform expression ratios have adjusted with elevated N2B expression in DCM and elevated N2BA expression in MR. Changes in titin

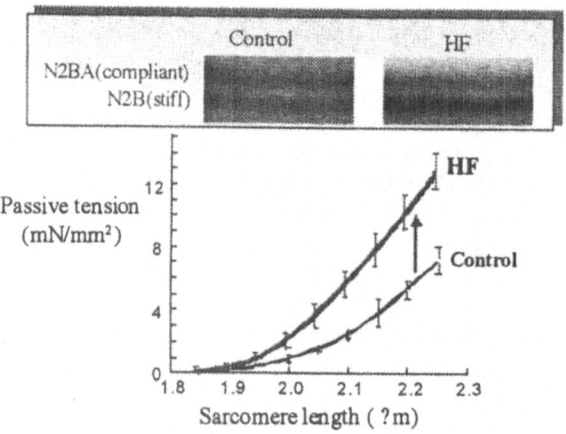

Figure 4. Top: SDS-PAGE of left ventricular myocardium of control dog and of dog with heart failure (HF) due to 4-week rapid pacing. Note that N2B titin is upregulated in myocardium of animal with HF. Bottom: Titin-based passive tension-sarcomere length relations of skinned myocardium from control and HF animals. Titin-based passive tension is significantly elevated in myocardium of HF animals. (Based on[63].)

isoform expression have also been reported recently in LV of human transplant hearts with coronary artery disease (CAD)[65]. Despite that non-transplant hearts with CAD displayed isoform expression patterns similar to those of control hearts, the changes seen in transplant hearts underscore our notion that processing of the titin pre-mRNA is subject to subtle regulatory mechanisms which control entry into either N2B or N2BA splice pathways.

It is worth highlighting that both canine and human expresses similar levels of compliant N2BA and stiff N2B titins in control myocardium. Considering that the isoforms are co-expressed at the level of the half sarcomere, each half-sarcomere in control tissue consists of an approximately equal number of stiff (N2B) and compliant (N2BA) titin molecules, organized in parallel. The titin-based stiffness will therefore be intermediate between that of sarcomeres that express solely N2B or solely N2BA titin. This intermediate stiffness allows for maximal adaptability. Thus, sarcomeres can either greatly increase compliance by increasing the N2BA/N2B expression ratio or greatly increase stiffness by reducing this ratio. We conclude that in addition to primary titin defects associated with familial forms of DCM, acquired alterations in isoform expression also play a role in diastolic dysfunction occurring in heart failure.

7. SUMMARY

The giant elastic protein titin contains an extensible segment that underlies the majority of physiological passive muscle stiffness. The extensible segment comprises mechanically distinct and serially-linked spring elements: the tandem Ig segments, the PEVK and the cardiac-specific N2B unique sequence. Under physiological conditions the tandem Ig segments are likely to largely consist of folded Ig domains whereas the N2B unique sequence and PEVK are largely unfolded and behave as wormlike chains with different persistence lengths. The mechanical characteristics of titin's extensible region may be tuned to match changing mechanical demands placed on muscle, using mechanisms that operate at different time scales and that include post-transcriptional and post-translational processes.

ACKNOWLEDGEMENTS

Supported by National Institutes of Health (HL61497 and HL62881 to HG) and Deutsche Forschungsgemeinschaft (La La668/6-2 and 7-1 to SL).

REFERENCES

1.   C. C. Gregorio, H. Granzier, H. Sorimachi, and S. Labeit, Muscle assembly: a titanic achievement?, *Curr Opin Cell Biol* **11**, 18-25. (1999).
2.   L. Tskhovrebova, and J. Trinick, Role of titin in vertebrate striated muscle, *Philos Trans R Soc Lond B Biol Sci* **357**, 199-206 (2002).
3.   K. Wang, Cytoskeletal matrix in striated muscle: the role of titin, nebulin and intermediate filaments, *Adv Exp Med Biol* **170**, 285-305 (1984).
4.   K. Maruyama, Connectin/titin, giant elastic protein of muscle, *FASEB J* **11**, 341-345. (1997).

5.   W. A. Linke, Stretching molecular springs: elasticity of titin filaments in vertebrate striated muscle, *Histol Histopathol* **15**, 799-811 (2000).

6.   H. Granzier, and S. Labeit, Cardiac titin: an adjustable multi-function spring, *Journal of Physiology (London)* **541**, 335-342 (2002).

7.   K. Wang, Titin/connectin and nebulin: giant protein rulers of muscle structure and function, *Adv Biophys* **33**, 123-134 (1996).

8.   S. Improta, A. S. Politou, and A. Pastore, Immunoglobulin-like modules from titin I-band: extensible components of muscle elasticity, *Structure* **4**, 323-337. (1996).

9.   S. Labeit, and B. Kolmerer, Titins: giant proteins in charge of muscle ultrastructure and elasticity, *Science* **270**, 293-296 (1995).

10.   A. Freiburg, K. Trombitas, W. Hell, O. Cazorla, F. Fougerousse, T. Centner, B. Kolmerer, C. Witt, J. S. Beckmann, C. C. Gregorio, H. Granzier, and S. Labeit, Series of exon-skipping events in the elastic spring region of titin as the structural basis for myofibrillar elastic diversity, *Circ Res* **86**, 1114-1121. (2000).

11.   K. Trombitas, M. Greaser, S. Labeit, J. P. Jin, M. Kellermayer, M. Helmes, and H. Granzier, Titin extensibility in situ: entropic elasticity of permanently folded and permanently unfolded molecular segments, *J Cell Biol* **140**, 853-859 (1998).

12.   W. A. Linke, and H. Granzier, A spring tale: new facts on titin elasticity, *Biophys J* **75**, 2613-2614. (1998).

13.   W. A. Linke, M. R. Stockmeier, M. Ivemeyer, H. Hosser, and P. Mundel, Characterizing titin's I-band Ig domain region as an entropic spring, *J Cell Sci* **111** (**Pt 11**), 1567-1574 (1998).

14.   W. A. Linke, M. Ivemeyer, P. Mundel, M. R. Stockmeier, and B. Kolmerer, Nature of PEVK-titin elasticity in skeletal muscle, *Proc Natl Acad Sci U S A* **95**, 8052-8057 (1998).

15.   L. Tskhovrebova, and J. Trinick, Flexibility and extensibility in the titin molecule: analysis of electron microscope data, *J Mol Biol* **310**, 755-771. (2001).

16.   A. Minajeva, M. Kulke, J. M. Fernandez, and W. A. Linke, Unfolding of titin domains explains the viscoelastic behavior of skeletal myofibrils, *Biophys J* **80**, 1442-1451. (2001).

17.   K. Trombitas, M. Greaser, S. Labeit, J. P. Jin, M. Kellermayer, M. Helmes, and H. Granzier, Titin extensibility in situ: entropic elasticity of permanently folded and permanently unfolded molecular segments, *J Cell Biol* **140**, 853-859. (1998).

18.   K. Trombitas, M. Greaser, G. French, and H. Granzier, PEVK extension of human soleus muscle titin revealed by immunolabeling with the anti-titin antibody 9D10, *J Struct Biol* **122**, 188-196 (1998).

19.   K. Ma, L. Kan, and K. Wang, Polyproline II helix is a key structural motif of the elastic PEVK segment of titin, *Biochemistry* **40**, 3427-3438. (2001).

20.   H. Li, A. F. Oberhauser, S. D. Redick, M. Carrion-Vazquez, H. P. Erickson, and J. M. Fernandez, Multiple conformations of PEVK proteins detected by single-molecule techniques, *Proc Natl Acad Sci U S A* **98**, 10682-10686. (2001).

21.   J. F. Marko, and E. D. Siggia, Fluctuations and supercoiling of DNA, *Science* **265**, 506-508 (1994).

22.   C. Bustamante, J. F. Marko, E. D. Siggia, and S. Smith, Entropic elasticity of lambda-phage DNA, *Science* **265**, 1599-1600 (1994).

23.   M. Helmes, K. Trombitas, T. Centner, M. Kellermayer, S. Labeit, W. A. Linke, and H. Granzier, Mechanically driven contour-length adjustment in rat cardiac titin's unique N2B sequence: titin is an adjustable spring, *Circ Res* **84**, 1339-1352. (1999).

24.   K. Trombitas, A. Freiburg, T. Centner, S. Labeit, and H. Granzier, Molecular dissection of N2B cardiac titin's extensibility, *Biophys J* **77**, 3189-3196. (1999).

25.   K. Watanabe, P. Nair, D. Labeit, M. S. Kellermayer, M. Greaser, S. Labeit, and H. Granzier, Molecular mechanics of cardiac titin's PEVK and N2B spring elements, *J Biol Chem* **277**, 11549-11558 (2002).

26.   M. Carrion-Vazquez, P. E. Marszalek, A. F. Oberhauser, and J. M. Fernandez, Atomic force microscopy captures length phenotypes in single proteins, *Proc Natl Acad Sci U S A* **96**, 11288-11292. (1999).

27.   M. Carrion-Vazquez, A. F. Oberhauser, S. B. Fowler, P. E. Marszalek, S. E. Broedel, J. Clarke, and J. M. Fernandez, Mechanical and chemical unfolding of a single protein: a comparison, *Proc Natl Acad Sci U S A* **96**, 3694-3699. (1999).

28.   M. Rief, M. Gautel, F. Oesterhelt, J. M. Fernandez, and H. E. Gaub, Reversible unfolding of individual titin immunoglobulin domains by AFM, *Science* **276**, 1109-1112. (1997).

29.   P. E. Marszalek, H. Lu, H. Li, M. Carrion-Vazquez, A. F. Oberhauser, K. Schulten, and J. M. Fernandez, Mechanical unfolding intermediates in titin modules, *Nature* **402**, 100-103. (1999).

30. K. Watanabe, C. Muhle-Goll, M. S. Kellermayer, S. Labeit, and H. Granzier, Different molecular mechanics displayed by titin's constitutively and differentially expressed tandem Ig segments, *J Struct Biol* **137**, 248-258 (2002).

31. M. Carrion-Vazquez, A. F. Oberhauser, T. E. Fisher, P. E. Marszalek, H. Li, and J. M. Fernandez, Mechanical design of proteins studied by single-molecule force spectroscopy and protein engineering, *Prog Biophys Mol Biol* **74**, 63-91 (2000).

32. R. Yamasaki, M. Berri, Y. Wu, K. Trombitas, M. McNabb, M. S. Kellermayer, C. Witt, D. Labeit, S. Labeit, M. Greaser, and H. Granzier, Titin-actin interaction in mouse myocardium: passive tension modulation and its regulation by calcium/S100A1, *Biophys J* **81**, 2297-2313 (2001).

33. M. Kulke, S. Fujita-Becker, E. Rostkova, C. Neagoe, D. Labeit, D. J. Manstein, M. Gautel, and W. A. Linke, Interaction between PEVK-titin and actin filaments: origin of a viscous force component in cardiac myofibrils, *Circ Res* **89**, 874-881 (2001).

34. W. Kabsch, H. G. Mannherz, D. Suck, E. F. Pai, and K. C. Holmes, Atomic structure of the actin:DNase I complex, *Nature* **347**, 37-44. (1990).

35. G. Fulgenzi, L. Graciotti, A. L. Granata, A. Corsi, P. Fucini, A. A. Noegel, H. M. Kent, and M. Stewart, Location of the binding site of the mannose-specific lectin comitin on F-actin, *J Mol Biol* **284**, 1255-1263. (1998).

36. E. Friederich, K. Vancompernolle, C. Huet, M. Goethals, J. Finidori, J. Vandekerckhove, and D. Louvard, An actin-binding site containing a conserved motif of charged amino acid residues is essential for the morphogenic effect of villin, *Cell* **70**, 81-92. (1992).

37. M. Pfuhl, S. J. Winder, and A. Pastore, Nebulin, a helical actin binding protein, *Embo J* **13**, 1782-1789. (1994).

38. R. Yamasaki, Y. Wu, M. McNabb, M. Greaser, S. Labeit, and H. Granzier, Protein kinase A phosphorylates titin's cardiac-specific N2B domain and reduces passive tension in rat cardiac myocytes, *Circ Res* **90**, 1181-1188 (2002).

39. K. T. Strang, N. K. Sweitzer, M. L. Greaser, and R. L. Moss, Beta-adrenergic receptor stimulation increases unloaded shortening velocity of skinned single ventricular myocytes from rats, *Circ Res* **74**, 542-549. (1994).

40. R. J. Solaro, A. J. Moir, and S. V. Perry, Phosphorylation of troponin I and the inotropic effect of adrenaline in the perfused rabbit heart, *Nature* **262**, 615-617 (1976).

41. H. C. Hartzell, and D. B. Glass, Phosphorylation of purified cardiac muscle C-protein by purified cAMP-dependent and endogenous Ca2+-calmodulin-dependent protein kinases, *J Biol Chem* **259**, 15587-15596 (1984).

42. K. T. Strang, and R. L. Moss, Alpha 1-adrenergic receptor stimulation decreases maximum shortening velocity of skinned single ventricular myocytes from rats, *Circ Res* **77**, 114-120 (1995).

43. M. L. Bang, T. Centner, F. Fornoff, A. J. Geach, M. Gotthardt, M. McNabb, C. C. Witt, D. Labeit, C. C. Gregorio, H. Granzier, and S. Labeit, The complete gene sequence of titin, expression of an unusual approximately 700-kDa titin isoform, and its interaction with obscurin identify a novel Z-line to I-band linking system, *Circ Res* **89**, 1065-1072 (2001).

44. M. Greaser, Identification of new repeating motifs in titin, *Proteins* **43**, 145-149. (2001).

45. O. Cazorla, A. Freiburg, M. Helmes, T. Centner, M. McNabb, Y. Wu, K. Trombitas, S. Labeit, and H. Granzier, Differential expression of cardiac titin isoforms and modulation of cellular stiffness, *Circ Res* **86**, 59-67. (2000).

46. R. Horowits, and R. J. Podolsky, The positional stability of thick filaments in activated skeletal muscle depends on sarcomere length: evidence for the role of titin filaments, *J Cell Biol* **105**, 2217-2223 (1987).

47. M. Helmes, K. Trombitas, and H. Granzier, Titin develops restoring force in rat cardiac myocytes, *Circ Res* **79**, 619-626 (1996).

48. H. Granzier, M. Kellermayer, M. Helmes, and K. Trombitas, Titin elasticity and mechanism of passive force development in rat cardiac myocytes probed by thin-filament extraction, *Biophys J* **73**, 2043-2053. (1997).

49. H. Granzier, M. Helmes, O. Cazorla, M. McNabb, D. Labeit, Y. Wu, R. Yamasaki, A. Redkar, M. Kellermayer, S. Labeit, and K. Trombitas, Mechanical properties of titin isoforms, *Adv Exp Med Biol* **481**, 283-300 (2000).

50. W. A. Linke, M. Ivemeyer, S. Labeit, H. Hinssen, J. C. Ruegg, and M. Gautel, Actin-titin interaction in cardiac myofibrils: probing a physiological role, *Biophys J* **73**, 905-919. (1997).

51.   K. Trombitas, Y. Wu, D. Labeit, S. Labeit, and H. Granzier, Cardiac titin isoforms are coexpressed in the half-sarcomere and extend independently, *Am J Physiol Heart Circ Physiol* **281**, H1793-1799 (2001).

52.   S. Labeit, and B. Kolmerer, Titins: giant proteins in charge of muscle ultrastructure and elasticity, *Science* **270**, 293-296. (1995).

53.   S. Hein, and J. Schaper, Pathogenesis of dilated cardiomyopathy and heart failure: insights from cell morphology and biology, *Curr Opin Cardiol* **11**, 293-301. (1996).

54.   S. Hein, D. Scholz, N. Fujitani, H. Rennollet, T. Brand, A. Friedl, and J. Schaper, Altered expression of titin and contractile proteins in failing human myocardium, *J Mol Cell Cardiol* **26**, 1291-1306. (1994).

55.   I. Morano, K. Hadicke, S. Grom, A. Koch, R. H. Schwinger, M. Bohm, S. Bartel, E. Erdmann, and E. G. Krause, Titin, myosin light chains and C-protein in the developing and failing human heart, *J Mol Cell Cardiol* **26**, 361-368. (1994).

56.   B. L. Siu, H. Niimura, J. A. Osborne, D. Fatkin, C. MacRae, S. Solomon, D. W. Benson, J. G. Seidman, and C. E. Seidman, Familial dilated cardiomyopathy locus maps to chromosome 2q31, *Circulation* **99**, 1022-1026. (1999).

57.   M. Itoh-Satoh, T. Hayashi, H. Nishi, Y. Koga, T. Arimura, T. Koyanagi, M. Takahashi, S. Hohda, K. Ueda, T. Nouchi, M. Hiroe, F. Marumo, T. Imaizumi, M. Yasunami, and A. Kimura, Titin Mutations as the Molecular Basis for Dilated Cardiomyopathy, *Biochem Biophys Res Commun* **291**, 385-393 (2002).

58.   B. Gerull, M. Gramlich, J. Atherton, M. McNabb, K. Trombitas, S. Sasse-Klaassen, J. G. Seidman, C. Seidman, H. Granzier, S. Labeit, M. Frenneaux, and L. Thierfelder, Mutations of TTN, encoding the giant muscle filament titin, cause familial dilated cardiomyopathy, *Nat Genet* **30**, 201-204. (2002).

59.   X. Xu, S. E. Meiler, T. P. Zhong, M. Mohideen, D. A. Crossley, W. W. Burggren, and M. C. Fishman, Cardiomyopathy in zebrafish due to mutation in an alternatively spliced exon of titin, *Nat Genet* **30**, 205-209 (2002).

60.   P. Hackman, A. Vihola, H. Haravuori, S. Marchand, J. Sarparanta, J. De Seze, S. Labeit, C. Witt, L. Peltonen, I. Richard, and B. Udd, Tibial Muscular Dystrophy Is a Titinopathy Caused by Mutations in TTN, the Gene Encoding the Giant Skeletal-Muscle Protein Titin, *Am J Hum Genet* **71**, 492-500 (2002).

61.   T. Centner, J. Yano, E. Kimura, A. S. McElhinny, K. Pelin, C. C. Witt, M. L. Bang, K. Trombitas, H. Granzier, C. C. Gregorio, H. Sorimachi, and S. Labeit, Identification of muscle specific ring finger proteins as potential regulators of the titin kinase domain, *J Mol Biol* **306**, 717-726 (2001).

62.   M. Gotthardt, R. E. Hammer, N. Hubner, J. Monti, C. C. Witt, M. McNabb, J. A. Richardson, H. Granzier, S. Labeit, and J. Herz, Conditional expression of mutant M-line titins results in cardiomyopathy with altered sarcomere structure, *J Biol Chem* (2002).

63.   Y. Wu, S. P. Bell, K. Trombitas, C. C. Witt, S. Labeit, M. M. LeWinter, and H. Granzier, Changes in titin isoform expression in pacing-induced cardiac failure give rise to increased passive muscle stiffness, *Circulation* **106**, 1384-1389 (2002).

64.   Y. Wu, S. Bell, M. M. LeWinter, S. Labeit, and H. Granzier, Titin: an endosarcomeric protein that modulates myocardial stiffness., *Journal of Cardiac Failure* **8**, S276-S286 (2002).

65.   C. Neagoe, M. Kulke, F. del Monte, J. K. Gwathmey, P. P. de Tombe, R. J. Hajjar, and W. A. Linke, Titin isoform switch in ischemic human heart disease, *Circulation* **106**, 1333-1341 (2002).

DISCUSSION

Sugi: As far as I remember, titin (connection) system did not attract attention of people when Wang and Maruyama first reported its presence. But it suddenly attracted attention of people when Podolsky's group published a paper in Nature that electromagnetic wave irradiation to the muscle fiber selectively damaged large titin molecules to produce remarkable effect on sarcomere structures. How do you think about the possibility to apply this method to locally damage titin network?

Granzier: Unfortunately, if I remember correctly, Podolsky's experiments were performed with frozen fibers and with very long periods of irradiation, and therefore this technique is not readily applicable to our study, though I will consider the use of this technique in our future work.

Pollack: Can you explain the "possible force enhancement" described by Dr. Bagni in this meeting by actin-titin binding?

Granzier: No, the binding diminishes with increase of Ca^{2+} and, thus, the effect on force can not explain force enhancement.

AN X-RAY DIFFRACTION STUDY ON CONTRACTION OF RAT PAPILLARY MUSCLE WITH DIFFERENT AFTERLOADS

Hiroshi Okuyama , Naoto Yagi**, Hiroko Toyota, Junichi Araki*, Juichiro Shimizu*, Gentaro Iribe*, Kazufumi Nakamura*, Satoshi Mohri*, Mikio Kakishita*, Katsushi Hashimoto*, Taro Morimoto*, Katsuhiko Tsujioka, Fumihiko Kajiya* and Hiroyuki Suga***

1. INTRODUCTION

In the crossbridge theory of muscle contraction (Huxley, 1957), muscle shortening is caused by crossbridges rapidly attaching and detaching from actin. At a given moment the number of crossbridges formed during shortening is smaller than in isometric contraction because they cannot keep attached to allow filament sliding. In skeletal muscles, this prediction has been tested by the x-ray diffraction technique (Podolsky et al., 1976; Huxley, 1979; Amemiya et al., 1980; Yagi & Takemori, 1995). Generally, the intensities of the equatorial (1,0) and (1,1) reflections, which are related to the number of myosin heads in the vicinity of the thin filament (Haselgrove & Huxley, 1973) are affected only by shortening with a small load (0-30%). This observation has been explained by the presence of same myosin heads that are weakly attached to actin during shortening (Yagi & Takemori, 1995). They are not actively producing force but remain in the vicinity of the thin filament, making the equatorial intensities close to those during isometric contraction.

In a physiologically working heart, the cardiac muscles have to shorten against a load to extrude blood. Thus, shortening under such a load is very important for their function. In contrast to the skeletal muscles, which contract in an all-or-none manner, the intracellular calcium concentration in cardiac muscles does not reach a level high enough to fully activate the troponin/tropomyosin regulatory system (Kurihara, 1994). Under such a condition, it is known that the crossbridges enhance the binding of calcium to troponin and that shortening causes the release of calcium from troponin (Allen &

Department of Physiology, Kawasaki Medical School, Matsushima, Kurashiki, 701-0192, Japan.
*Okayama Univ. Grad. Sch. of Medicine and Dentistry, Shikata, Okayama, 700-8558, Japan.
**SPring-8/JASRI, Sayo, Hyogo, 679-5198, Japan.
***National Cardiovascular Center Research Institute, Suita, Osaka, 565-8565, Japan.

Molecular and Cellular Aspects of Muscle Contraction
Edited by H. Sugi, Kluwer Academic/Plenum Publishers, 2003

Kurihara, 1982).

By using a third-generation synchrotron radiation facility SPring-8, it has become possible to record changes in the equatorial intensities of rat papillary muscle in a single contraction at a time resolution of 10-20 msec. We compared the equatorial intensities in contractions with different loads in the same muscle. The results are generally similar to those from skeletal muscle.

2. METHODS

2.1. Papillary muscle preparation

Male rats (Wistar, Japan Clare Co.) of 8-14 weeks of age were used. A rat was anaesthetized deeply with ether and killed by quickly removing the heart. The heart was perfused from the aorta with a Tyrode solution (136 mM NaCl, 5.4 mM KCl, 1.8 mM $CaCl_2$, 1.0 mM $MgCl_2$, 5 mM HEPES, pH 7.4 at 25°C) containing 20 mM BDM (2,3-butanedione monoxime) to remove blood and stop beating. Then a papillary muscle from the right ventricle was dissected with a chorda tendinea and tricuspid valve in the Tyrode solution with BDM. The animal experiments were conducted in accordance with the guidelines of SPring-8 for care and welfare of experimental animals.

The papillary muscle, with a small piece of right ventricular wall left, was mounted in a specimen bath using glue. One end of the muscle was introduced into the stainless specimen holder (outer diameter 1.5 mm, inner diameter 0.9 mm) and glued stiffly by a cyanoacrylate adhesive, and the other end was hooked by a valve and glued to the lever of a force-length transducer. The transducer was made of a galvanometer actuator with a length feedback so that we could measure the relative muscle length and the force simultaneously. The total compliance of the system is $0.7 \mu m/mN$ and the cutoff frequency of the length drive is 160 Hz. The size of muscle was 1.6-4.2 mm (mean ± SD, 2.67 ± 1.06) in length and 0.3-0.8 mm in width. The specimen bath was mounted on the x-ray camera. The muscle axis was horizontal. The bath had two thin glass windows (3.2 _ 5.9 mm, thickness 60-80 μm) to pass x-rays through the specimen. The oxygenated Tyrode solution (27°C) was passed through the bath (volume 2.1 ml) with a flow speed of about 10 ml/min. The bath had a pair of platinum electrodes that were aligned parallel to the muscle. The muscle was stimulated at 0.2 Hz with a supramaximal rectangular pulse of 5 msec duration. The cross-section of the muscle was determined by measuring its width at the opposite side of the chorda tendinea in two orthogonal directions using an optical microscope. The cross-section was assumed to be an ellipsoid.

2.2. X-ray diffraction technique

The experiments were made at the beamline BL45XU (Fujisawa et al., 2000) in the third-generation synchrotron radiation facility SPring-8 (Harima, Hyogo, Japan). The x-ray wavelength was 0.1 nm. The x-ray flux available at this beamline is $1 \_ 10^{12}$ photons/sec. However, in the present study, it was attenuated by using an aluminum absorber to $3 \_ 10^{11}$ photons/sec in order to avoid radiation damage. The x-ray beam (0.8 mm horizontally, 0.2 mm vertically) was smaller than the specimen. The camera length was 1.8 m. The detector system was a Be-windowed x-ray image intensifier (V5445P, Hamamatsu Photonics, Hamamatsu, Japan; Fujisawa et al., 1999) coupled with a fast CCD camera (C4880-80-12A, Hamamatsu Photonics). The frame rate was 50 frames/sec. In the present experiment, by opening a fast x-ray shutter, three and forty frames were recorded before and after a stimulus, respectively.

An equatorial intensity profile was obtained by integrating intensity in the area within 0.01 nm^{-1} from the equator. Each side of the profile was fitted with a fourth-order polynomial function, which simulates the background, and two Gaussian functions, which simulate the (1,0) and (1,1) reflection peaks. The fitting was done by using a modified Levenberg-Marquardt algorithm (subroutine UNLSF in the IMSL library, Visual Numerics, Inc., San Ramon, CA, USA). The intensity and position of a reflection were obtained from the area and center of the Gaussian function, respectively. The lattice spacing was calibrated using the 14.3-nm meridional reflection.

3. RESULTS

3.1. Conformational changes during the isotonic shortening

Figure 1 shows changes in the equatorial intensity ratio, that is the intensity of (1,0) reflection divided by that of the (1,1) reflection, during afterloaded shortening with different loads. With all loads, the ratio decreased in a twitch. The maximum change was observed slightly prior to the peak tension (Matsubara, 1980). During the high load shortening, i. e. under the load of more than 50% of the isometric active force (AL50%), the intensity ratio did not show much difference from that in the isometric contraction.

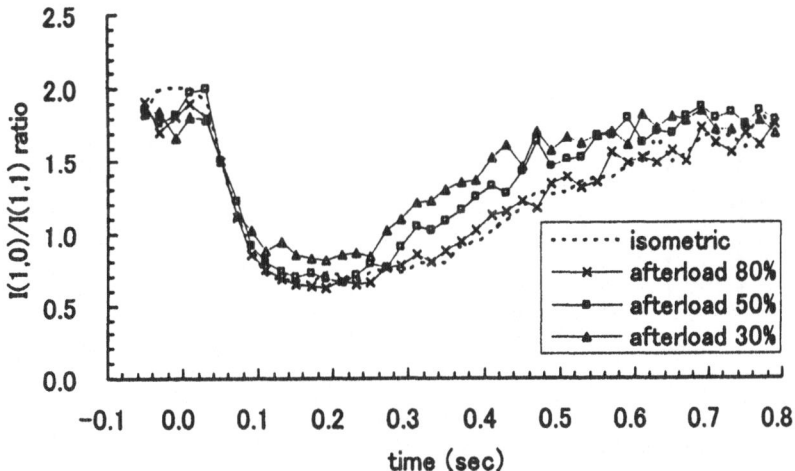

Figure 1. X-ray intensity ratio of the (1,0) and the (1,1) equatorial reflections averaged over 7 muscles, each of which contracted under isometric condition and with 80, 50, and 30 % loads.

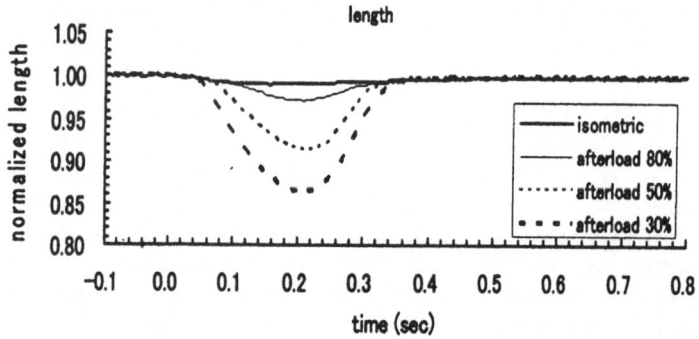

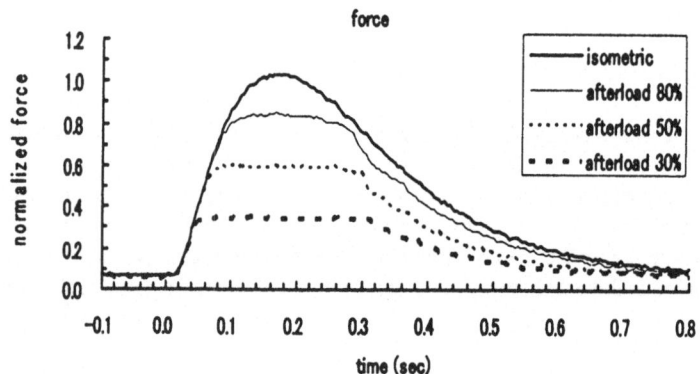

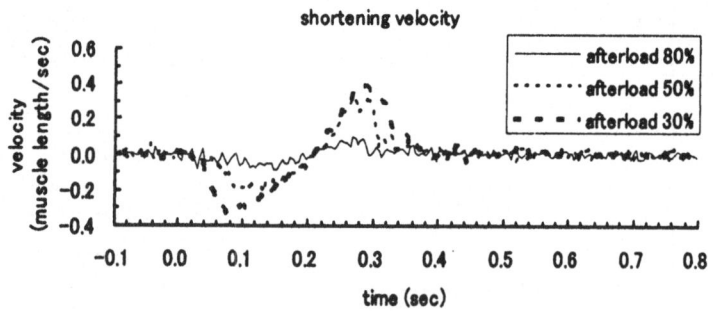

Figure 2. Length, force and shortening velocity (differentiation of the length) during contraction of rat papillary muscle with different loads. The data are averages of results obtained from 7 muscles.

During the low load shortening, i. e. under the load of about 30% of the isometric active force (AL30%), the decrease in the intensity ratio was slightly smaller. During the initial part of contraction, the intensity ratio decreased in a manner similar to that in the isometric contraction. However, after the shortening occurred, the decrease in the intensity ratio became slower. Return to the resting level after the peak change was faster than in the isometric contraction.

Figure 2 shows the mechanical data averaged over 7 muscles. In the recording of the muscle length (top), the muscle shortening in afterloaded contractions was greater with a lower load. The muscle length reached the minimum slightly after the peak time in the isometric contraction. In the recording of the force (middle), it can be seen that the force in the afterloaded contractions returned to the resting level faster than in the isometric contraction. The shortening velocity (bottom) shows that the shortening velocity (i. e. the negative velocity) reached its maximum earlier when the afterload was smaller, while the lengthening reached its maximum later when the afterload was smaller.

3.2 Changes in lattice spacing during shortening

In addition to the intensity ratio, which is related to the formation of crossbridges, the x-ray diffraction experiment also gives information on the filament spacing. The peak position of the (1,0) or (1,1) reflections gives the distance between the myosin filaments. If we assume that the volume of the muscle does not change with the muscle length, the distance between the myosin filaments, i. e. the lattice spacing, should also change. Figure 3 shows the time course of the (1,0) spacing during isometric and afterloaded contractions. As expected, the changes are approximate mirror images of those of length in Figure 2. In the isometric contraction, the spacing change reached its maximum after the peak tension is attained. The small spacing increase in isometric contraction indicates that there is a small internal shortening of sarcomere even in a fixed-end contraction.

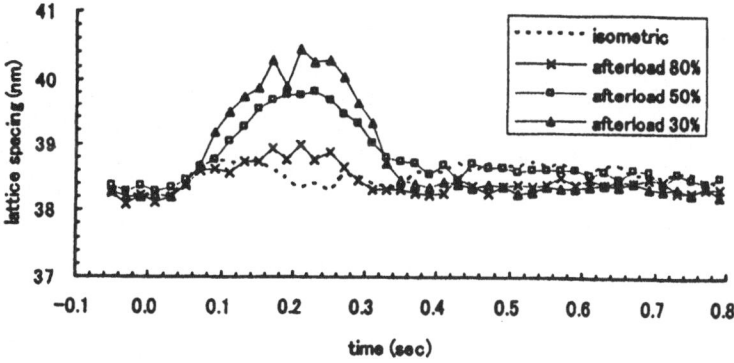

Figure 3. Lattice spacing averaged in 7 muscles. The lattice spacing is the separation of the (1,0) crystallographic planes in the hexagonal lattice of the thick filaments and is equal to $\sqrt{3}/2$ of the distance between the centers of neighboring thick filaments.

spacing showed only a small spacing increase. However, during shortening with a load of less than AL50%, the large spacing increase was observed. These results suggest that the part of the muscle that is studied by the x-rays actually shortened considerably under low loads.

4. DISCUSSION

The present results show that the equatorial intensity ratio $(I(1,0)/I(1,1))$ only begins to change appreciably when the load of the rat papillary muscle is reduced to about AL30%. These results are similar to those obtained in skeletal muscle (Podolsky et al., 1976; Huxley, 1979; Amemiya et al., 1980; Yagi & Takemori, 1995). The results on skeletal muscles were obtained in shortening during maximum tetanus, where the intracellular Ca^{2+} concentration is maintained high enough to fully activate the regulatory system of the thin filament. On the other hand, in a cardiac muscle, the calcium level during a twitch is not high enough to fully activate the contractile system (Kurihara, 1994). Thus, the similarity between the results of the present study and those obtained in skeletal muscle may indicate that the lower intracellular Ca^{2+} concentration in cardiac muscle does not affect the crossbridge kinetics when the afterload is larger than 30 %. This suggests that the weak binding of myosin heads to the thin filament takes place even when only some of the troponin molecules have bound calcium.

Since the muscle shortened in afterloaded contractions, the effect of afterload on the intensity ratio might be due to shortening of sarcomere. Generally, the the equatorial intensity ratio $(I(1,0)/I(1,1))$ during contraction is smaller at shorter sarcomere length (Haselgrove & Huxley, 1973) than longer lengths. However, we observed a larger ratio with AL30% under which the muscle shortened most. Thus, it is unlikely that this larger ratio is due to shorter sarcomere lengths during contraction.

In a skeletal muscle, Amemiya et al. (1980) measured the equatorial intensity ratio during a twitch with an AL30% in frog sartorius muscle. Their results showed that the equatorial intensity changes in the isotonic twitch were not significantly different from those in an isometric twitch. Thus, the cardiac muscle may be slightly more susceptible to the change in the load. Sugi et al. (1978) reduced the load of skeletal muscle to 2.5 % of the isometric tension and found that the equatorial intensity ratio, which decreased from a resting value of 2.0 to a minimum value of 0.5-0.6 in a preload twitch, reached only about 0.8-1.0 during shortening under the light load. This minimum value was attained before the muscle shortened maximally. The result in the present study shows that the ratio decreased from 2.0 under isometric contraction to 0.8 under AL30% in cardiac muscle. Present data would be more pronounced if we reduced the load much further in cardiac muscle.

5. ACKNOWLEDGEMENTS

This work was partly supported by Grant-in-Aid for Scientific Research (12877116) from the Ministry of Education and Research Project Grant (12-309, 13-310, 14-313) from Kawasaki Medical School. Also partly supported by Research Grants for from Kawasaki Medical School. Also partly supported by Research Grants for Cardiovascular Diseases (11C-1, 14A-1) and a 2000-2002 Health Science Research Grant for Human Genome and Regenerative Medicine from the Ministry of Health, Labor and Welfare, and Cardiac Physiome Grant from Okayama New Industry Promotion Foundation and the Science and Technology Agency. The experiments at SPring-8 were made with the approval of SPring-8 Program Review Committee.

6. REFERENCES

Allen, D. G. and Kurihara, S., 1982, The effects of muscle length on intracellular calcium transients in mammalian cardiac muscle. *J. Physiol.* **327**:79-94.

Amemiya, Y., Tameyasu, T., Tanaka, H. Hashizume, H. and Sugi, H., 1980, Time-resolved x-ray diffraction from frog skeletal muscle during shortening against an inertial load and a quick release, *Proc. Japan Acad.* B **56**:235-240.

Fujisawa, T., Inoko, Y. and Yagi, N., 1999, The use of a Hamamatsu X-ray image intensifier with a cooled CCD as a solution X-ray scattering detector, *J. Synchrotron Rad.* **6**:1106-1114.

Fujisawa, T., Inoue, K., Oka, T., Iwamoto, H., Uruga, T., Kumasaka, T., Inoko, Y., Yagi, N., Yamamoto, M. and Ueki T., 2000, Small-angle X-ray scattering station at the SPring-8 RIKEN beamline, *J. Applied Cryst.* **33**:797-800.

Haselgrove, J. C. and Huxley, H. E, 1973, X-ray evidence for radial cross-bridge movement and for the sliding filament model in actively contracting skeletal muscle, *J. Molec. Biol.* **77**:549-568.

Huxley, A. F., 1957, Muscle structure and theories of contraction, *Progr. in Biophys. and Biophys. Chem.* **7**:255-318.

Huxley H. E., 1979, Time Resolved X-Ray Diffraction Studies on Muscle, in: *Cross-bridge Mechanism in Muscle Contraction*, H. Sugi and G. H. Pollack ed., Univ. Tokyo Press, Tokyo, pp. 391-405.

Kurihara, S., 1994, Regulation of cardiac muscle contraction by intracellular Ca^{2+}, *Jpn. J. Physiol.* **44**:591-611.

Matsubara, I., 1980, X-ray diffraction studies on the heart, *Ann. Rev. Biophys. Bioeng.* **9**:81-105.

Podolsky, R. J., Onge, R. St., Yu, L. and Lymn R. W., 1976, X-Ray diffraction of actively shortening muscle, *Proc. Natl. Acad. Sci.* **73**:813-817.

Sugi, H., Amemiya, Y. and Hashizume, H., 1978, Time-resolved x-ray diffraction from frog skeletal muscle during an isotonic twitch under a small load, *Proc. Japan Acad.* B **54**:559-564.

Yagi, N. and Takemori, S., 1995, Structural changes in myosin cross-bridges during shortening of frog skeletal muscle, *J. Muscle Res. Cell Motil.* **16**:57-63.

DISCUSSION

Alpert: If fewer cross-bridge are involved at low loads, 1) is the force per cross-bridges the same at low and high loads? And 2) if that is the case, how do you get the kinetic differences (attachment time) that lead to changes in velocity at lower loads?

Okuyama: 1) At higher afterload, the cross-bridges seem to attach strongly to the actin filament, but at lower afterload they seem to attach weakly. So that, even if the generated force divided by the number of attached cross-bridges is the same, it is not directly comparable. 2) We do not have the methods to get the difference. But the diffraction patterns of (1,0) and (1,1) are broadened, which means that the fiber arrangement becomes somewhat irregular. Some cross-bridges are attached weakly and others strongly. Now what I can say is that these broadened diffractions may have some information on the attachment time.

VIII. MUSCLE FATIGUE AND ENERGETICS

ROLE OF THE T-SYSTEM AND THE Na-K PUMP ON FATIGUE DEVELOPMENT IN PHASIC SKELETAL MUSCLE

Hugo Gonzalez-Serratos, Ruzhang Chang, Monika Rozycka, Mordecai Blaustein, and Patrick DeDeyne[*]

1. INTRODUCTION

Prolonged, direct electrical stimulation of vertebrate skeletal muscles induces a state during which contractile force or the capacity to do external work declines after prolonged repetitive stimulation. The muscles then become fatigued or mechanically refractory to further stimulation. Fatigue does not involve permanent impairment of function since contractility can be restored. It is not due to neuromuscular transmission failure[1] nor is it caused by a decline in the central nervous system motor drive of muscle. Fatigue is due solely to contractile failure [2,3].

One proposal is that fatigue is caused by alterations in the intracellular concentration of ATP hydrolysis by-products (P_i, H^+ and Mg^{2+}) produced during repetitive contractions (metabolic hypothesis). According to this "metabolic hypothesis" these altered by-products may bring about a reduced concentration of phosphocreatine and ATP [4-7] accumulation of ADP, orthophosphate, Mg^{2+}, lactate and H^+ [8,9] for a review see [10-12]. After prolonged stimulation, the interaction of actin with myosin may be impaired [5], thereby decreasing force development [13-17]. Experiments done in single permeabilized muscle fibers indicated that a substantial portion of the decrease in force and the zero load velocity of shortening (V_o) [18,19,13], which also is decreased in intact muscle cells [13] during fatigue, is caused by an increase in

[*] Hugo Gonzalez-Serratos, Ruzhang Chang, Monika Rozycka, Mordecai Blaustein, and Patrick DeDeyne, University of Maryland, School of Medicine, Department of Physiology, 655 W. Baltimore Street, Baltimore, MD 21201

$H^{+18,19}$ and/or P_i [19,20]. The above experiments led to the conclusion that fatigue may be caused by these metabolic factors[7].

The metabolic hypothesis as the cause of fatigue is attractive and, in some ways, simple. But, there are other experimental results indicating that the intracellular metabolic changes may not be the main factor in fatigue development. Fatigue development follows the same time course independently of the frequency of stimulation (20 or 60 Hz) which indicates that biochemical changes may not be the primary cause of muscle fatigue [1,21]. Intracellular pH measurements have demonstrated that there is no clear correlation between pH changes, fatigue development and recovery after fatigue[22,23]. In frog-skinned fibers, a decrease in pH has no final consequence on tension development owing to the opposing effect of low pH on contractile proteins and the sarcoplasmic reticulum (SR)[24]. Fatigue cannot be explained as due to an exhaustion of the energy sources creatine phosphate and ATP (45, 89). It has also been shown that pre-fatigue tetanic tension can be recovered immediately after development of fatigue by either K depolarization or caffeine[25-27]. Furthermore, post-fatigue caffeine contractures are larger than pre-fatigue ones and develop at a subthreshold caffeine concentration[27,28]. Wavy myofibrils, i.e. non-active myofibrils, appeared in the central core of muscle fibers during fatigue development[27,16] and disappeared in the presence of caffeine at the same time that force is recovered. This indicated that the terminal cisternae (TC) of the SR Ca^{2+} content was available to be released. Furthermore, the tension-caffeine contraction relationship is shifted to the left after fatigue development suggesting a larger than normal TC or myoplasmic Ca^{2+} content. Consequently, fatigue cannot be due only to failure of the contractile machinery caused by metabolic changes. We interpreted this as either a failure in the conduction of the tubular action potential (TAP) along the T-tubule or a failure of the signal from the T-tubule to the TC to release Ca^{2+} in the regions where the T-tubules had swollen during fatigue development (43). Alterations in one or more of the events that couple excitation with contraction could also play an important role in fatigue development (e-c coupling hypothesis)[25,29]. Collectively, the experimental facts mentioned above point out the importance of the T-system in the incitation of fatigue development[29]. However, the role of the T-system in fatigue development has not as yet been fully understood. Fatigue development is a complex process that cannot be entirely explained by a single mechanism. It appears that fatigue may be caused by alterations in one or both of the above mentioned processes (metabolic or e-c coupling) depending on the degree of fatigue and/or the pattern of stimulation.

Changes in e-c coupling during fatigue development have not been well defined. We propose that T-tubular functions could be altered by the transmembrane ionic shifts caused by the cytosolic Na^+ ($[Na^+]_i$) increase with decreased K ($[K^+]_i$) concentrations that occur during fatigue[29]. We also hypothesize that during prolonged repetitive stimulations tubular Ca^{2+} ($[Ca^{2+}]_T$) and K^+ ($[K^+]_T$) increase with decreased Na^+ ($[Na^+]_T$) concentrations and swelling of the T-system occurs that lead to altered TAPs and the release of SR Ca^{2+} by T-tubular signaling. Thus, the importance of e-c coupling was investigated using two lines of reasoning. First, in newborn rat muscles the T-system and SR are not fully developed[30]. If e-c coupling plays a role in fatigue in adult skeletal muscles then, fatigue development should have different characteristics of skeletal muscles from newborn animals when compared to the same type of muscles from adult animals. Secondly, during tubular action potentials K^+ and Na^+ movements in and out of the T-system are compensated by the Na-K pump[31]. However, during repetitive

TAPs the pump may be overwhelmed causing tubular K^+ accumulation and Na^+ depletion, thereby lowering tubular membrane potential, sodium equilibrium potential and sodium ionic current. This would lead to either smaller or no TAPs causing an impairment of SR Ca^{2+} release, i.e. fatigue. Therefore, the blocking of the Na-K pump should cause a larger increase in the above ionic changes leading to an acceleration of fatigue. The present experiments were undertaken to test the e-c coupling hypothesis by investigating the above possibilities. We found that the characteristics of some of the contractility parameters in extensor digitorum longus (EDL) muscles from newborn rats are similar to slow (Type I) muscle cells without the presence of slow myosin heavy chain. Furthermore, the time it took for EDL muscles from newborn rats to fatigue was 15 times longer and decreased with age when compared with EDL muscles from adult rats. The second set of experiments were designed to study the consequences of increased $[K^+]_T$ accumulation and $[Na^+]_T$ depletion on fatigue development. To investigate this proposition it was necessary for the experiments to be performed on single isolated muscle cells to avoid ionic concentration changes that occurs in the vicinity of the sarcolemma in the spaces between muscle fibers, as it happens in whole muscle. These second set of experiments demonstrated that single fibers isolated from the tibialis anterior muscle of amphibians had a 60% faster rate of fatigue development, in the presence of ouabain, when compared to control experiments.

2. MATERIALS AND METHODS

2.1. Preparations

For the first set of experiments we used isolated EDL muscles dissected from birth up to 11 day-old rats and adult rats. The young rats were lightly anesthetized with ether followed by decapitation. The whole legs were quickly removed and placed in a modified Krebs's solution made up of (in mM): NaCl, 118; KCl, 4.7; KH_2PO_4 1.2; $MgCl_2$ 0.6; $NaHO_3$, 25; glucose, 11; and $CaCl_2$, 2.5; equilibrated with 95% O_2 - 5% CO_2 = to a pH of 7.4 at 22° C. EDL muscles were then carefully isolated and identified according to the tendon insertions. For adult rats, bundles of EDL muscles were dissected under the microscope as described previously[32] to similar diameters of young rat EDLs. In the second set of experiments single twitch skeletal muscle cells were freshly isolated from the tibialis anterior muscles of the frog *Rana temporaria* as previously described[27]. After isolation of the muscle cells, they remained resting in the dissecting dish for at least half an hour in Ringer's solution. Next, the fibers were stimulated with single low-voltage electrical shocks. If they responded with brisk twitches and there were no signs of membrane damage, the fibers were used; otherwise, they were discarded. The experiments were carried out at room temperature (20-22°C.).

2.2. Stimulating Protocols

Muscle preparations for the first set of experiments on newborn and adult animals were transferred to an experimental chamber. In the chamber one tendon of the preparation was gripped with a small clamp while the other tendon was attached to the hook of a strain gage force transducer. The stimulating electrode consisted of platinum wires placed on each side of the preparation. The muscles were then stretched 1.3 times their slack length, which corresponds to an average sarcomere length of approximately 2.6 _m. The muscles or bundles of fibers were then stimulated electrically with 0.5-ms duration pulses of x1.5 the minimum voltage necessary for activation of all the cells. The stimulating protocol consisted of a series of single twitches elicited every 3 sec until the peak twitch force was the same for 5 consecutive times. The twitches were then followed by different frequencies of tetanic stimulations with 3-min resting periods between them. Then, the preparations rested for 10 min before the fatigue protocol began. Fatigue was induced with electrical stimulation cycles of 20 Hz 3 s repeated every 6.9 s. To compare curves of fatigue development from different preparations, we used the fatigue index T_n/T_o where T_n is maximal tetanic force produced during every third tetanus and T_o is the maximal tetanic force of the first tetanus[32].

The second set of experiments (with ouabain) followed the same general procedure as described above. However, a Ringer's solution consisting of (in mM): NaCl, 115; K Cl, 2.5; CaCl$_2$, 1.8; Mg Cl$_2$, 0.2; pH adjusted with phosphates to 7.2 was used instead of Kreb's solution. The Na$^+$-K$^+$ pump was blocked with 1-1.5 mM ouabain. The fibers were attached to an Ekharts type force transducer (Senso Nor, Horten, Norway). Fatigue development was induced with electric stimulating cycles of 60 Hz 0.8 s followed by a single twitch after 2.2 s and repeated every 4.75 s.

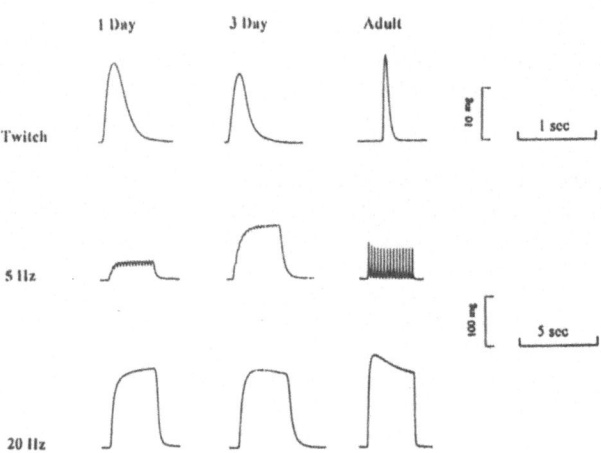

Figure 1. Tracings of force development from EDL muscles isolated from 1 and 3 days old post-partum rats and from adult rats stimulated at the frequencies indicated on the left column. The 1 s 10 mg and 5 s 100 mg scales correspond to the twitch and tetanic stimulations respectively.

2.3. Determination of Fiber Type

Muscle extracts were prepared and separated by high glycerol/acrylamide gel electrophoresis as described [33-34]. In brief, pulverized muscle was extracted in 100 mM $Na_4P_2O_7$, 5 mM ethylene glycol-bis (β-aminoethyl ether)-N, N, N', N'-tetra-acetic acid (EGTA), 0.3 M KCl, 1 mM dithiothreitol, pH 6.5, centrifuged at 12,000 G. Supernatant was collected, diluted 1:1 with glycerol and stored at -20^0C. A modified Lowry Assay (Biorad, Hercules, CA) determined the protein content. Myosin heavy chains were separated by electrophoresis in a slab gel containing 7.2 % acrylamide/bisacrylamide, 375 mM Tris-HCl, pH 8.6, 37 % glycerol, 0.1% sodium dodecyle sulfate (SDS), 0.043% ammoniumpersulfate (APS), and 0.17 % N,N,N'N'-tetramethylethylenediamine (TEMED). The stacking gel contained 4% acrylamide/bisacrylamide, 125 mM Tris-HCl, pH 6.70, 0.1% SDS, 0.09% APS and 0.15 % TEMED. The running buffer contained 26 mM Tris-HCl, pH 8.5, 204 mM glycine and 0.1% SDS. The electrophoresis was performed at a constant voltage of 65 V for 24 hours at 4^0C. MHC were visualized by using a Silver staining method.

Statistical values are given as an average ± standard error of the mean.

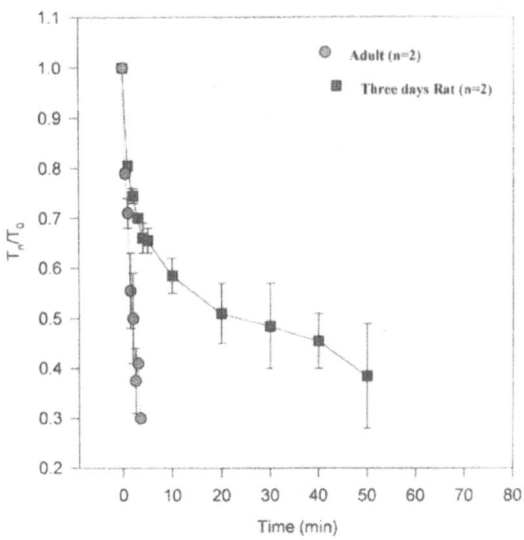

Figure 2. Relationship between fatigue index and time in EDL muscles isolated from 3 days old post-partum rats (squares) and from adult rats (circles).

3. RESULTS

3.1. Contractility Characteristics of Muscles From Newborn Rats Depend on the Age of the Rats

Reports on the contractile characteristics of developing neonatal EDL muscles are scanty and except for potassium contratures[35], they have not been well documented. Our first aim was to characterize some of the contractile properties of these muscles. We found that they changed according to the age of the rats. As shown in Fig.1, the time courses of twitches were prolonged in muscles from newborn animals compared with muscles from adult rats. The time it took for peak twitch tension to occur changed from 2.7 to 2.5, 2.6, 2.4 and 1.7 times the adult value for days 1, 2, 4, 6 and 8, respectively. Half relaxation tension time changed from 3.9 to 2.8, 3.3, 2.2 and 2 times the adult value for the same days, respectively. The time period for 80% relaxation tension followed a similar pattern. Figure 1 also shows that tetanic tension changed with age. In EDL muscles of 1-day old rats, 5 Hz produced a semifused tetanic tension that increased in magnitud by Day 3. The adult muscles each produced an individual twitch after every stimulation. However, we observed in other experiments (not illustrated) that the maximal specific tension elicited at this frequency decreased daily between Days 3 and 11. The bottom line of Fig. 1 shows that a 20 Hz frequency of stimulation produced fused tetanii of similar tensions for Days 1 and 3 and a larger tension in the EDL from an adult rat. In general, the specific tetanic tension elicited at

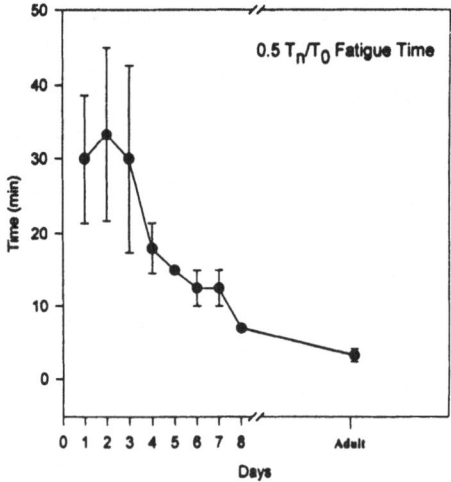

Figure 3. Relationship between the time it takes for the fatigue index to reach 0.5 and the post-partum days in which the experiments were performed

Figure 4. Myosin heavy chain (MHC) separation by electrophoresis. Adult MHC are indicated by the arrows on the left, the arrow on the right points to the neonatal MCH, which migrates between Type II_X and TypeII_B

this frequency also decreased after Day 3.The relationship between specific tension and frequency of stimulation indicated that maximal tension increased above twitch tension between 2 and 20 Hz and decreased at higher frequencies beyond 20 Hz in muscles from newborn. The tetanic force of adult rats increased between 20 and 60 Hz , and decreased at higher frequencies of stimulation.

3.2. Time Course of Fatigue Development Depends on the Age of the Rats

The time course of fatigue development was dramatically longer in EDL muscles from newborn rats compared with muscles from adult rats when stimulated with the same parameters. An example of these results is shown in Fig. 2 where the index of fatigue from 2 experiments on the EDL muscles of 3-day-old rats is compared to that of adult animals. The index of fatigue reached an average of 0.4 in 50 min in muscles of these younger rats and only 3 min in muscles from adult rats. As shown in Fig. 3, the average time for T_n/T_o to reach 0.5 deceased (except for Day 3) in general with age. The figure shows that it took 39.5 min for maximal tetanic force to decline to 50% of pre-fatigue values ($0.5T_n/T_o$). During Days 2 and 3 after birth this time increased to 33.44 ±11 and 30.5±12 min, respectively. After Day 3 the time declined to
18 ± 3.3, 15 ± 0.64, 12.5 ± 2.4, 12.5 ± 2.5 and 7 ± 0.64 min per day up to Day 11. In muscles from adult rats this time was only 3.4 ± 0.94 min. Fatigue development to $0.3T_n/T_o$ followed a similar trend. The time course of fatigue development showed three distinct slopes. However, the rates of tension decay T_n/T_o from muscles dissected from 1-day-old newborn animals were 69 and 96 as well as 32% slower than the corresponding slopes from muscles dissected from adult rats.

3.3. Fiber Type Characterization

To characterize the myosin heavy chain (MHC) profile of the muscle extracts were analyzed by gel electrophoresis (see Methods). Control samples included adult soleus muscle (SOL), which contains 90% Type slow MHC and 10 % Type IIA (fast) MHC and adult EDL muscle, which contains 42 % IIB, 24% IIX, 31 % IIA[36]. As shown in Fig. 4, muscle extracts

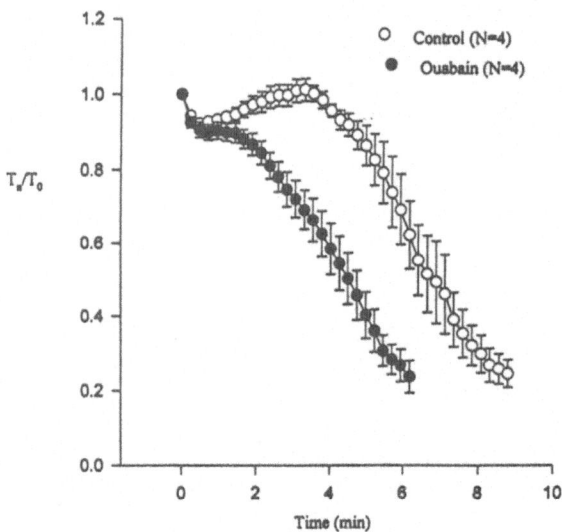

Figure 5. Time course of fatigue development in single muscle cells stimulated in Ringer's solution without ouabain (empty circles) and stimulated in the presence of 1 mM ouabain (filled circles).

from the EDL of Days 1, 2, 4, 7 and 9 post-partum rats predominantly contained neonatal MHC and Type IIA MHC or Type IIX MHC. Samples from Day 4 and from older rats contained a small amount of Type IIB MHC and the EDL from Day 9 contained an additional small amount of Type I MHC. This shift in MHC phenotype was expected as post-nataldevelopment progresses. The slow contractile properties of the EDL from newborn rats cannot be explained by the presence of slow MHC. Despite a small increase in Type IIB MHC (Day 4 and later) it is unlikely that such a small shift in protein composition would have significant impact on the contractile characteristics. On the other hand, there has been little written about the contractile properties of neonatal MHC.

3.4. Ouabain Decreases the Time of Fatigue Development

To further investigate the role of the T system, we investigated the effect of increasing $[K^+]_T$ and decreasing $[Na^+]_T$ on the time course of fatigue development. As shown in Fig. 5, we found fatigue development was faster in the presence of ouabain as compared to fatigue development without ouabain. Figure 5 also illustrates that the typical increase in maximal tetanic tension observed with this pattern of stimulation at the beginning of the repetitive stimulation disappears in the presence of ouabain. Furthermore, it takes an average of 1.9 min for maximal tetanic tension to start decreasing in the presence of ouabain after the

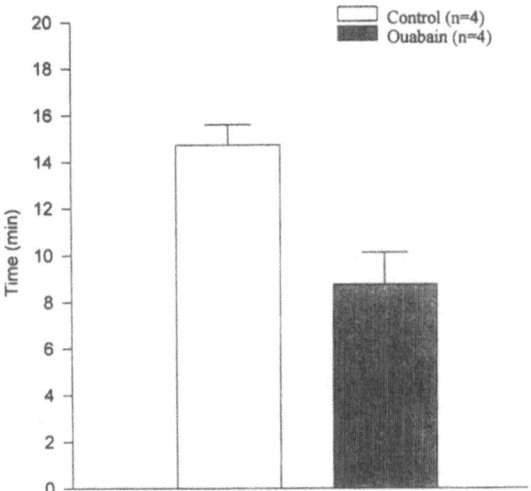

Figure 6. Bar graph illustrating the average time it takes for the fatigue index to reach 0.25 in control experiments and in experiments where 1mM of ouabaine was present in the Ringer's solution.

beginning of the repetitive stimulation, compared with 4.1 min in control experiments. We also found that, in the presence of ouabain and as fatigue developed, the sagging of tension decay that appeared normally during the plateau of individual tetanii became much faster to the point that there was no sign of a plateau as it became fused with relaxation. The average time that it took for the index of fatigue to reach 0.25 decreased by 41%, from 14.7 min in control experiments to 8.73 min in the presence of ouabain (Fig. 6). Recovery time to pre-fatigue maximal tetanic tension after fatigue development to $0.25T_n/T_o$ was 2.69 times longer in the presence of ouabain compared without ouabain (experiments not shown).

4. DISCUSSION

This study demonstrated that: 1.) The contractility characteristics of EDL muscles from rats between Days 1 and 11 after post-partum depended on the age of the rats. At birth these muscles exhibited mechanical properties that have the typical features of slow Type I muscle fibers. As the rats age, they changed to those of fast Type II muscle fibers, which is the predominant muscle fiber type found in adult EDL muscles. 2.) The time course of fatigue development depended on the age of the rats and followed the same trend as the contractile properties. For young rats, the average time that it took for maximal tetanic tension to decay by 75% during fatigue development was 15x longer than in EDL muscles from adult rats. 3.) The predominant myosin heavy chain in the EDL from newborn rats was as expected neonatal MHC and Type IIA or Type IIX. By day 11 the relative amount of neonatal MHC was decreased to the advantage of Type IIB and Type I MHC. 4.) In amphibians' freshly isolated single muscle cells, ouabain accelerated the time for fatigue development to levels similar to that of control experiments by 41%.

The classification of fiber type in muscles is based primarily on the histo-chemical and morphological characteristics of the myofibrils and the SR and T systems[37-39]. Furthermore, they also showed contractile properties that corresponded uniquely to its fiber type classification[40,41]. Predominant Type I muscles developed twitches in which the time course was approximately twice as slow as the twitches from mostly Type IIA and B muscles[40,41]. Since the time course of twitches is slower in Type I muscles, higher frequencies of stimulation fused and maximal tetanic tensions occurred at lower frequencies than from Type II A and B muscles. Skeletal muscles are composed of a mixture of different fiber types. But, they may be constituted by predominantly fast Type IIA and or Type IIB twitch fibers which have a well developed T-system[38] and fatigue faster than those of slow fiber type muscles[42]. In contrast, slow type fibers have a poorly developed T-system[39] and are fatigue resistant[42,41,43]. From the results described in this paper, it would appear that, immediately after birth, the EDL is composed of fibers containing neonatal MHV and have contratctile properties of slow muscles that continuously evolved into Type II A and B, but have not completely transformed by Day 11. We did not study the mechanical characteristics beyond Day 11. In addition, a correlation had been established between skeletal muscle fiber types and muscle fatigability. The time course of fatigue development is several times slower in predominantly Type I muscles compared to fatigue development induced with the same stimulating parameters in predominantly Type II muscles[41]. The fatigability properties described in this paper of developing muscles between Days 1 to 11 after post-partum are characteristic of Type I muscles. However, we discovered that the fiber types in these muscles were largely that of neonatal MHC and several Types II(M,B,X), which are similar to the mature EDL muscles. The above observation suggests that the regulatory Ca^{2+} mechanisms, as it exists in fully developed EDL muscles, have not yet matured during the early stages of post-partum development. There are two possible explanations. Either the myofibrilar morphological elements i.e., the SR and T-system, which are involved in the e-c coupling chain of events have not fully matured or the Ca^{2+} regulatory mechanisms are not fully functional in the early stages of post-partum development. Therefore, an alternative cytosolic Ca^{2+} regulatory mechanism must exist at this early stage of post-partum muscle development. There are two possible cytosolic Ca^{2+} regulatory mechanisms. One is the Na^+/Ca^{2+} exchanger that also regulates cytosolic Ca^{2+} concentration exists in fully developed adult skeletal muscles[44,45] and rat myotubes[46]. Furthermore, it has been proposed that this exchanger may be very prominent in postnatal EDL muscles between Days 1-5 after birth and decreases substantially after Day 6 [47]. The other Ca^{2+} regulatory mechanism is the calmodulin regulated H^+/Ca^{2+} pump located in the surface and T-system membranes[48]. This pump, that is in parallel with the Na^+/Ca^{2+} exchanger, has not been well characterized, has a lower activity[49] and lower maximal rate of Ca^{2+} efflux than the Na^+/Ca^{2+} exchanger[50,51]. At present there are no evidences of the role that may play as a cytosolic Ca^{2+} regulatory mechanism in skeletal muscle. We propose that the main regulatory cytosolic Ca^{2+} concentration mechanism between post-partum Days 1 to 11 is the Na^{+}/ Ca^{2+} exchanger rather than the SR and transverse tubular systems.

Reports using subcellular fractions and biochemical experiments have shown that the Na^{+}/K^{+} ATPase α sub-unit is primarily localized in the T-tubules[52,53]. During repetitive stimulation the K^{+} that moves into the T-tubular space could accumulate substantially because of the high surface to volume ratio of the T-system network and the restricted diffusion to the external fluid. If the Na^{+}/K^{+} ATPase α sub-unit is localized in the T-tubules, then it probably operates at its full capacity under this conditions, becomes saturated and then is not able to deal with this excessive amount of tubular K^{+}. The T-tubular functions could be altered by these T-tubular transmembrane ionic shifts and by those caused by an increase of $[Na^{+}]_i$ and a decrease in intracellular $[K^{+}]_i$ found during fatigue[29]. Increased tubular K^{+} causes tubular membrane depolarizationn which together with the depletion of $[Na^{+}]_T$[54] may lead to a decrease in the size of the tubular action potential as well as to an increased intra-tubular osmolarity, leading to swelling of the T-system. This swelling and the decreased tubular action potentials lead to an impairment of the signaling to the TC release of Ca^{2+} and, consequently, to a decline in contractile force. The exacerbation of phasic skeletal muscle fatigue of development caused by the blockade of the Na^{+}/K^{+} pump by ouabain, described in this paper, would be caused by further increases of $[K^{+}]_T$ and $[Na^{+}]_i$ gives support to the proposition described above.

Collectively, the experiments presented here support the view that alterations in e-c coupling mechanisms and in particular in the T-tubular system's physiological properties, play an important role in fatigue development as previously proposed[25,29]. During early stages of post-partum muscle development fatigue may be caused mainly by metabolic changes. In summary, fatigue development is a complex process which might not be entirely explained by a single mechanism.

We thank Dr. Hong Song from the University of Maryland School of Medicine, Department of Physiology for providing us with the newborn rats, and Mrs. Anne O. Nourse from the University of Maryland for her valuable and helpful suggestions in the editing and preparation of the manuscript. This work was supported by NIH grants to M.P. Blaustein: NS 16106 an HL45219 and NIHKOIHD01165 to P.G. DeDeyne.

5. REFERENCES

1. P. A. Merton, Voluntary strength and fatigue, *J. Physiol. (Lond)* **128**, 553-564 (1954).
2. B. Bigland-Ritchie, EMG and fatigue of human voluntary and stimulated contractions (1980), Human muscle fatigue: physiological mechanisms. *Pitman Medical, London (Ciba Foundation Symposium)* **82**, 130-156 (1981).
3. B. Bigland-Ritchie and Woods, J. J., Changes in muscle contractile properties and neural control during human muscular fatigue, *Muscle and Nerve* **7**, 691-699. (1984).
4. R. H. T. Edwards, D. K. Hille, and D. A. Jones, Metabolic changes associated with the slowing of relaxation in fatigued mouse muscle. *J. Physiol.* **251**, 303-315 (1975).
5. M. J. Dawson, D. G. Gadian, and D. R. Wilkie, Muscular fatigue investigated by phosphorus nuclear magnetic resonance, *Nature (London)*, **274**, 861-866 (1978).
6. M. M. Dawson, D. G. Gadian, and D. R. Wilkie, Mechanical relaxation rate and metabolism studied in fatiguing muscle by phosphorus nuclear magnetic resonance, *J. Physiol.* **299**, 465-484 (1980).

7. R. E. Godt and T. M. Nosek, Changes of intracellular milieu with fatigue or hypoxia depress contraction of skinned rabbit skeletal and cardiac muscle. *J. Physiol.* **412**, 155-180 (1989).
8. R. Fitts and J. Holloszy, Lactate and contractile force in frog muscle   during   development   of   fatigue   and recovery. *Am. J. Physiol.* **231**, 430-433 (1976).
9. H. Westerblad and J. Lännergren, The relation between force and intracellular pH in fatigued, single Xenopus muscle fibres. *Acta Physiol. Scand.* **133**, 83-89 (1988).
10. R. Porter and J. Whelan, Human muscle fatigue: Physiological mechanisms. *Ciba Foundation Symposium* 82. Eds. Porter, R. & Whelan, J. Pitman Medical (London) (1981).
11. N. K. Vollestad and O. M. Sejersted, Biochemical correlates of fatigue, *Eur. J. Appl. Physiol.* **57**, 336-347 (1988).
12. R. H. Fitts, Muscle fatigue: The cellular aspects; American Journal of Sports Medicine. **24**(6),S9-S13 (1006).
13. K. A. P. Edman, and A. R. Mattiazi, Effects of fatigue and altered pH on isometric force and velocity of shortening at zero load in frog muscle fibres. *J. Musc. Res. Cell Contr.* **2**, 321-334 (1981).
14. G. W. Mainwood, and J. M. Renaud, The effect of acid-base balance on fatigue of skeletal muscle. *Can. J. Physiol. Pharmacol.* **63**, 403-416 (1985).
15. K. A. P. Edman and F. Lou, Changes in force and stiffness induced by fatigue and intracellular acidification in frog muscle fibres. *J. Physiol.* **424**, 133-149 (1990).
16. K. A. P. Edman and F. Lou, Myofibrillar fatigue versus failure of activation during repetitive stimulation of frog muscle fibres. J. Physiol. **457**, 655-673 (1992).
17. N. A. Curtin and K. A. P. Edman, Force-velocity relation for frog muscle fibres: Effects of moderate fatigue and of intracellular acidification. *J. Physiol.* **475**, 483-494 (1994).
18. P. B. Chase and M. J. Kushmerick, Effects of pH on contraction of rabbit fast and slow skeletal muscle fibers. *Biophys. J.* **53**, 935-946 (1998).
19. R. Cooke, K. Franks, G. Luciani and E. Pate, The inhibition of rabbit skeletal muscle contraction by hydrogen ions and phosphate. *J. Physiol., Lond.* **395**, 77-97 (1988).
20. H. Westerblad, D. G. Allen, and J. Lannergren, Muscle fatigue: Lactic acid or inorganic phosphate the major cause? *News Physiol Sci.* **17**, 17-21 (2002).
21. P. A. Merton, C.D. Marsden, and J.C. Meadows. In Symposium on Human muscle fatigue: Physiological mechanisms, Ciba Foundation, Editors: R. Porter & J. Whelan, 287 (1980).
22. J. M. Metzger and R. H. Fitts, Role of intracellular pH in muscle fatigue, *J. Physiol.* , **62**, 1392-1397 (1987).
23. H. Westerblad and J. Lännergren, Force and membrane potential during and after fatiguing, intermittent tetanic stimulation of single Xenopus muscle fibres. *Acta Physiol. Scand.* **128**, 369-378 (1986).
24. A. Fabiato and F. Fabiato, Effects of pH on the myofilament and the sarcoplasmic reticulum of skinned cells from cardiac and skeletal muscles. *J. Physiol.* **276**, 233-255 (1978).
25. A. Berstein and A. Sandow, in: *The Effect of Use and Disuse on Neuromuscular Functions*, edited by Gutman and Hnik. (Czech. Acad. Sci., Prague, 1963), pp. 515-526.
26. W. Grabowski, E. A. Lobsiger, and H. C. Lüttgau, The effect of repetitive stimulation at low frequencies upon the electrical and mechanical activity of single muscle fibres, *Pflügers Arch. Gesamte Physiol. Menschen Tiere.* **334**, 222-239 (1972).
27. M. del C. Garcia, H. Gonzalez-Serratos, J. P. Morgan, C. L. Perreault and M. Rozycka, Differential activation of myofibrils during fatigue in phasic skeletal muscle cells. *J. Musc. Res. Cell Motil.* **12**, 412-424 (1991).
28. H. Gonzalez-Serratos and C. Garcia, Differential activation of myofibrils during fatigue in twitch skeletal muscle fibres of the frog, in: *Muscular Contraction,* edited by R. M. Simmons (University Press, Cambridge, 1982).
29. H. Gonzalez-Serratos, A. V. Somlyo, G. McClellan, H. Shuman, L. M. Borrero, and A. P. Somlyo, Composition of vacuoles and sarcoplasmic reticulum in fatigued muscle: Electron probe analysis, *Proc. Natl. Acad. Sci.* **75**, 1329-1333 (1978).
30. R. A. Bergman, Observations on the morphogenesis of rat skeletal muscle, Bull John Hopkins Hosp. **110**, 187-201 (1962)
31. T. Clausen, The Na^+, K^+ pump in skeletal muscle: Quanification, regulation, and functional significance. *Acta Physiol. Scand.* **156**, 227-235 (1996).
32. C. L. Perreault, H. Gonzalez-Serratos, S. E. Litwin,, X. Sun, C. Franzini-Armstrong, and J. P. Morgan, Alterations in contractility and intracellular Ca^{2+} transients in isolated bundles of skeletal muscle fibers from rats with chronic heart failure. *Circ. Res.* **73**, 405-412 (1993).
33. V. J. Caiozzo, M. J. Baker, S. A. McCue, and K. M. Baldwin, Single-fiber and whole muscle analyses of MHC isoform plasticity: Interaction between T3 and unloading, *Am. J. Physiol.* **273**, C944-52 (1997).
34. R. J. Talmadge and R. R. Roy, Electrophoretic separation of rat skeletal muscle myosin heavy-chain isoforms, *J. Appl. Physiol.* **75**, 2337-2340 (1993).
35. Y. Pereon, J. P. Louboutin, J. Noireaud, Contractile responses in rat *extensor digitorum longus* muscles at different times of postnatal development, *J. Comp. Physiol.* B**163**, 203-211 (1993).

36. H. H. Jung, R. L. Lieber, and A. F. Ryan, Quantification of myosin heavy chain mRNA in somatic and branchial arch muscles using competitive PCR, *Sm. J. Physiol.* **275**, C68-C74 (1998).
37. M. H. Brook and K.K. Keiser., Muscle fiber types: what kind?, Arch Neurol, **23**, 369-379 (1970)
38. L. D. Peachey, The sarcoplasmic reticulum and transverse tubules of the frogs Sartorius. *J. Cell Biology,* **25**, 209-231 (1965).
39. L. D. Peachey and A. F. Huxley, Structural identification of twitch and slow striated muscle fibers of the frog. *J. Cell Biol.* **13**, 177-180 (1962).
40. R. E. Burke and P. Tsairs, The correlation of physiological properties with histochemical characteristics in single muscle units. *Ann. N.Y. Acad. Sci.* **228**, 145-159 (1974).
41. R. I. Close, Dynamic properties of skeletal muscle. *Physiol. Rev.* **52**, 129-197 (1972).
42. J. Lännergren and R. Smith, Types of muscle fibres in toad skeletal muscle, *Acta Physiol. Scand.* **68**, 263-274 (1966).
43. J. Lännergren, Structure and function of twitch and show fibres in amphibian skeletal muscle in: *Basic mechanism of ocular motility and their clinical implications,* edited by G. Lennerstrand and P. Bach-y-rita (Perggaiman Press Oxford,1975), pp. 63-84.
44. M. P. Blaustein and W. J. Lederer, Sodium/calcium exchange: its physiological implications, *Physiol. Rev.* **79**, 763-854 (1999).
45. H. Gonzalez-Serratos, D. W. Hilgemann, M. Rozycka, A. Gauthier, and H. Rasgado-Flores, Na$^+$-Ca^{2+} exchange studies in sarcolemmal skeletal muscle, *Annals New York Academy of Sciences* **79**, C556-560 (1996).
46. R. J. Bloch, Acetylcholine receptor clustering in rat myotubes: Requirement for Ca^{2+} and effects of drugs which depolymerize microtubules, *J. Neurosci.* **3**(12), 2670-2680 (1983).
47. J-P Louboutin and J. Noireau, Sodium withdrawal contractures in Developing and regenerating rat extensor digitorum longus muscles, *Muscle Nerve* **21**, 1530-1532 (1998).
48. A. Ortega and J. R. Lepock, Use of thermal analysis to distinguish magnesium and calcium stimulated ATPase activity in isolated transverse tubules from skeletal muscle, *Biochem. Biophys. Acta* **1233**, 7-13 (1995).
49. C. Hidalgo, M.E. Gonzalez, and A. M. Garcia, Calcium transport in transverse tubules isolated from rabbit skeletal muscle, *Biochem. Biophys. Acta* **854**(2), 279-86 (1986).
50. J. R. Giolbert and G. Meisner, Sodium-calcium exchange in skeletal muscle sarcolemmal vesicles, *J. Membr. Biol.* **69**,77-84 (1982).
51. P. Donoso, and C. Hidalgo, Sodium-calcium exchange intransverse tubules isolated from frog skeletal muscle, *Biochim et Biophy Acta* **978**, 8-16 (1989).
52. H.S. Hundal, A. Marette, Y. Mitsumoto, T. Ramalal, R. Blostein, and A. Klip, Insulin induces translocation of the alpha 2 and beta 1 subunits of the Na+/K(+)-ATPase from intracellular compartments to the plasma membrane in mammalian skeletal muscle. *J. Biol. Chem.* **267**, 5040-5043 (1992).
53. L. R. Lavoie, P. Levinson, P. Martin-Vassallo, and A. Klip, The molar ratios of alpha and beta subunits of the Na+-K+-ATPase differ in distinct subcellular membranes from rat skeletal muscle, *Biochem.* **36**, 7726-7732 (1997).
54. F. Bezanilla, C. Caputo, H. Gonzalez-Serratos, and R. A. Venosa, Sodium dependence of the inward spread of activation in isolated twitch muscle fibres of the frog. *J. Physiol.* **223**, 507-523 (1972).

DISCUSSION

Westerblad: Could the difference in fatigue resistance in young and old muscle be due to lower forces in young muscle? Therefore, energy changes might be smaller.

Gonzalez-Serratos: No, the specific force varied with no clear pattern according to age of the animal for the same frequency of stimulation (tetanus) or for different frequencies of stimulation at different ages of animal at which the muscles were isolated.

Morano: Is there also different fatigue if you do the experiments at the maximal power output?

Gonzalez-Serratos: No, all the experiments were done at the same sarcomere length of 2.63μm.

MITOCHONDRIAL Ca^{2+} IN MOUSE SOLEUS SINGLE MUSCLE FIBRES IN RESPONSE TO REPEATED TETANIC CONTRACTIONS

Jan Lännergren and Joseph D. Bruton[*]

1. BACKGROUND

Mitochondrial diseases form a heterogeneous group of disorders in which mutations of the mitochondrial material frequently results in muscle dysfunction. Mammalian skeletal muscle fibres are particularly rich in mitochondria, which may make up about 10 to 15 % of a fibre's volume (Eisenberg, 1983; Chen et al., 2001). Mitochondria are differentially distributed in many rodent skeletal muscles, with a higher density found close to the sarcolemma than deep in the fibre (Eisenberg et al., 1983; Philippi & Sillau, 1994). It has long been accepted that a key function of mitochondria is to supply energy as required by the working muscle. Denton & McCormack (1990) proposed that Ca^{2+} plays a key role in this process by activating three key mitochondrial dehydrogenases. More recently, a rise in mitochondrial Ca^{2+} was suggested to directly stimulate mitochondrial oxidative phosphorylation (Kavanagh et al., 2000). While circumstantial evidence suggests that mitochondria in skeletal muscle are able to modulate their Ca^{2+} content, surprisingly little is know about Ca^{2+} movement into and out of the mitochondria in intact skeletal muscle cells during and after a bout of contractile activity. Several groups have reported that mitochondria isolated from skeletal muscle after exhaustive exercise have a higher Ca^{2+} content than those obtained from non-exercised muscle (Duan et al., 1990; Madsen et al., 1996). Other groups have reported that mitochondria are swollen or disrupted in skeletal muscle isolated from animals that were exercised to exhaustion, (Gollnick & King, 1969; McCutcheon et al., 1992; Sakai et al., 1999).

Sembrowich et al. (1985) demonstrated that mitochondria isolated from fast- and slow-twitch mammalian skeletal muscle accumulate Ca^{2+}. On the basis of the kinetics of mitochondrial Ca^{2+} uptake, they suggested that mitochondria play a significant role in lowering $[Ca^{2+}]_i$ in intact slow-twitch muscle. Later Gillis (1997) demonstrated in skinned fibres that inhibition of mitochondrial Ca^{2+} uptake slowed force relaxation. These data

---

[*] Department of Physiology and Pharmacology, Karolinska Institutet, S-171 77 Stockholm, Sweden.

Molecular and Cellular Aspects of Muscle Contraction
Edited by H. Sugi, Kluwer Academic/Plenum Publishers, 2003

suggest that mitochondria might play a role in controlling myoplasmic Ca^{2+} in situ. However, it is known that respiratory function of isolated mitochondria is different to that of mitochondria in situ (Milner et al., 2000). Similarly, it is still unknown whether or not mitochondria in intact muscle cells take up and release Ca^{2+} during physiological activity.

Recently we found that mitochondria in frog fibres took up Ca^{2+} during repeated contractions. Intriguingly, mitochondria in mouse fast-twitch toe muscle fibres did not take up Ca^{2+} even when active uptake of Ca^{2+} into the SR was inhibited (Lännergren et al., 2001). We have now investigated mitochondrial Ca^{2+} uptake and release in mouse soleus fibres. This muscle contains mainly slow-twitch fibres, which rely more heavily on mitochondrial energy production during a bout of sustained activity.

2. METHODS

Young (3-5 months) NMRI male mice were killed by rapid neck disarticulation and the soleus muscles removed. The soleus muscle in this strain of mice is composed overwhelmingly of type I and type IIA fibres (Marechal & Becker-Bleukx, 1993). Intact single fibres were isolated by mechanical dissection. Fibres were mounted at optimal length between an adjustable hook and an Akers AE801 force transducer in the perfusion channel of a muscle bath, which was placed on the stage of an inverted microscope. A control force-frequency curve was obtained with 1 to 100 Hz, 500 ms stimulus trains. Fibres were then stimulated with 70 Hz, 500 ms tetani at 2 s intervals.

Mitochondrial $[Ca^{2+}]$ was measured with rhod-2 (Babcock et al., 1997). Fibres were loaded with 5 µM rhod-2-AM for 90 –120 min at room temperature and then washed for at least 30 min. Confocal images were obtained with a BioRad MRC 1024 unit with a krypton/argon mixed gas laser (BioRad Microscopy Division, Hertfordshire., England) attached to a Nikon Diaphot 200 inverted microscope with a Nikon Plan Apo 40x oil immersion objective lens

3. RESULTS AND DISCUSSION

Fig 1 shows a typical example of the changes in mitochondrial Ca^{2+} that occurred during and after a series of 1,000 tetani. Interestingly the increase in mitochondria Ca^{2+} was predominantly observed in mitochondria lying close to the sarcolemma. The mitochondrial rhod-2 signal increased almost 10-fold over the initial 50 tetani and then declined slightly over the rest of the series (n=10). At the end of the stimulation period, tetanic force had decreased to about 60% of the starting value and relaxation speed was little affected. After the series of tetani ended, mitochondrial Ca^{2+} decreased to 50% of the value at the end of stimulation within five min and had essentially returned to its resting value within 20 min.

Four fibres showed no change in mitochondrial Ca^{2+} when stimulated with as many as 1,000 tetani. The force frequency curves of this group of fibres lay to the right of the group of fibres whose mitochondria took up Ca^{2+}. Thus, these four fibres might have been type IIA fibres, which do not take up Ca^{2+} in their mitochondria (Lännergren et al., 2001). Conversely, the other ten fibres that showed marked mitochondrial Ca^{2+} uptake would be type I fibres. The mechanism underlying this striking difference in behaviour is not clear at this point in time.

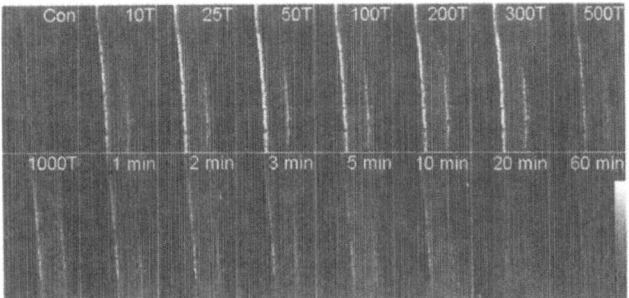

Figure 1. Mitochondrial Ca²⁺, as measured by rhod-2 intensity, increases from its resting level (Con) during a series of 1,000 tetani (T) reaching its maximum after 50T and declining slightly over the rest of the stimulation period. After the end of the series, mitochondrial Ca²⁺ falls rapidly completely to the resting value within 20 min. Fibre diameter was 35 μm. Calibration bar on the lower right shows low Ca²⁺ as black and high Ca²⁺ as white.

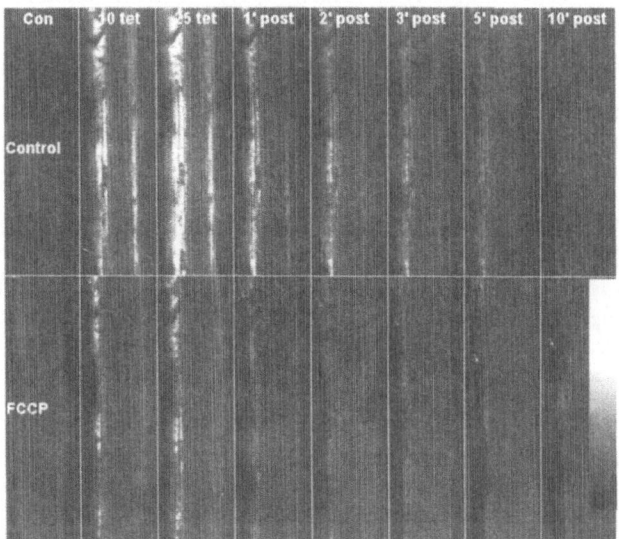

Figure 2. FCCP markedly reduces the increase in mitochondrial Ca²⁺ induced by a series of 25 tetani (10T, 25T). Top row shows the control series of 25 tetani while the bottom row shows the same fibre when exposed to 0.5 μM FCCP. Fibre diameter was 35 μm. Calibration bar on the lower right shows low Ca²⁺ as black and high Ca²⁺ as white.

In order to examine whether Ca²⁺ uptake depended on the mitochondrial potential, mitochondria were depolarised by exposing them to 0.2 - 0.5 μM carbonyl cyanide p-trifluoromethoxyphenylhydrazone (FCCP) for 5 min prior to the start of experiments. Fig 2 shows a typical example of a fibre subjected to 25 tetani first in the absence and then in the presence of FCCP. It can be seen clearly that FCCP greatly reduced but did not completely abolish the increase in the mitochondrial Ca^{2+}. Similar results were obtained in a further two fibres indicating that mitochondrial Ca uptake is largely dependent on the mitochondrial membrane potential. However, the mitochondrial membrane potential (monitored with rhodamine 123) did not change during a series of repeated tetani. This lack of change may indicate that anion movement accompanies Ca^{2+} entry into the mitochondria (Harris, 1978; Ligeti & Lukacs, 1984).

' In many cells and probably also in skeletal muscle, mitochondrial uptake of Ca^{2+} from the cytosol occurs by two pathways. The best known of these is the Ca^{2+} uniporter. In skeletal muscle, this is half-maximally activated at about 400 nM Ca^{2+} and has a maximum velocity of about 2.6 nmol(mg protein)$^{-1}$s^{-1} (Sembrowich et al., 1985). The activity of the mitochondrial Ca^{2+} uniporter can be modulated by nucleotides, inorganic phosphate and divalent cations (Litsky & Pfeiffer, 1997). Recently, a second mode of mitochondrial Ca^{2+} uptake (RaM) has been described in liver (Sparagna et al., 1995) and heart (Butinas et al., 2001) mitochondria. This mechanism was suggested to mediate very rapid uptake at the start of a Ca^{2+} pulse which takes 1 min or longer to reset after bath $[Ca^{2+}]$ is reduced to 100 nM or less, (Butinas et al., 2001). This mode of uptake has not been demonstrated in intact cells and it may simply be a special mode of operation of the mitochondrial Ca^{2+} uniporter (Litsky & Pfeiffer, 1997). Under the stimulation scheme used in our experiments, this type of uptake could operate only for the first tetanus before becoming inactivated.

Inhibition of the mitochondrial Na^+-Ca^{2+} exchanger with CGP-37157 further increased mitochondrial $[Ca^{2+}]$ signal during the series of 25 tetani and slowed the rate of decline after the series had ended (n=4). It has been suggested that the mitochondrial permeability transition pore, which can be blocked by Cyclosporin A, might play a role in the controlled release of Ca^{2+} from the mitochondria (He & Lemasters, 2002). However, the presence of Cyclosporin A did not affect the rate of mitochondrial $[Ca^{2+}]$ decline after the end of a series of 25 tetani.

Thus, removal of Ca^{2+} from the mitochondria is accomplished predominantly by the Na^+-Ca^{2+} exchanger. The speed of this exchanger is at least ten times slower than the uniporter-mediated uptake (Sembrovitch et al., 1985; Rizzuto et al., 1999). The data presented here indicate that in soleus muscle fibres, significant mitochondrial Ca^{2+} accumulation can occur during repetitive stimulation due to the high $[Ca^{2+}]_i$ (> 1µM) in each tetanus and the relatively slow extrusion of Ca^{2+} from the mitochondria.

Alterations in $[Ca^{2+}]$ in the mitochondria can play a role in at least three processes that exert profound effects on muscle performance. First, a rise in mitochondrial Ca^{2+} is believed to be important for activation of the Ca^{2+}-activated dehydrogenases (Denton & McCormack, 1990) and indeed Jouaville et al. (1999) showed that depolarisation of myotubes increased both mitochondrial $[Ca^{2+}]$ and [ATP]. However, the absence of any rise in Ca^{2+} in about 20% of the soleus fibres examined here suggests that a rise in Ca may not be a prerequisite to increase mitochondrial metabolism during a bout of activity. It may indicate that other markers of muscle activity such as ADP, Pi or phosphorylcreatine act as adequate stimulants of mitochondrial metabolism (Walsh et al., 2001). Second, an increase in mitochondrial Ca^{2+} may be important in the modulation of mitochondrial protein turnover. Mammalian mitochondria contain their own DNA, which undergoes transcription and generates the integral proteins of the respiratory chain. This process of mitochondrial protein synthesis has been shown to be Ca^{2+}-dependent (Joyal et al., 1995). Thus, it is possible that the extended contractile activity and the ensuing rise in mitochondrial Ca^{2+} are involved in mitochondrial adaptations, such as protein turnover in the mitochondria (Hood, 2001; Wu et al., 2002). Finally, mitochondrial Ca^{2+} uptake has been implicated in augmented production of reactive oxygen species (ROS) (Dykens, 1994; Grijalba et al., 1999). While excessive ROS production is believed to be deleterious to cellular performance, moderate rises in ROS production may actually be physiologically advantageous (Reid & Durham, 2002).

Acknowledgement: This work was funded by the Swedish MRC, the Swedish National Centre for Sports Research, Knut and Alice Wallenberg Foundation and funds at the Karolinska Institutet.

4. REFERENCES

Babcock, D.F., Herrington, J., Goodwin, P.C., Park, Y.B., and Hille, B., 1997, Mitochondrial participation in the intracellular Ca^{2+} network. *J. Cell Biol.* **136**: 833-844.

Butinas, L., Gunter, K.L., Sparagna, G.C., and Gunter, T.E., 2001, The rapid mode of calcium uptake into heart mitochondria (RaM): comparison to RaM in liver mitochondria. *Biochim. Biophys. Acta* **1504**: 248-261.

Chen, G., Carroll, S., Racay, P., Dick, J., Pette, D., Traub, I., Vrbova, G., Eggli, P., Celio, M., and Schwaller, B., 2001, Deficiency in parvalbumin increases fatigue resistance in fast-twitch muscle and upregulates mitochondria. *Am. J. Physiol.* **281**: C114-C122.

David, G., Barrett, J.N., and Barrett, E.F., 1998, Evidence that mitochondria buffer physiological Ca^{2+} loads in lizard motor nerve terminals. *J. Physiol.* **509**: 59-65.

Denton, R.M., and McCormack, J.G., 1990, Ca^{2+} as a second messenger within mitochondria of the heart and other tissue. *Ann. Rev. Physiol.* **52**: 451-466.

Duan, C., Delp, M.D., Hayes, D.A., Delp, P.D., and Armstrong, R.B., 1990, Rat skeletal muscle mitochondrial $[Ca^{2+}]$ and injury from downhill walking. *J. Appl. Physiol.* **68**: 1241-1251.

Duchen, M.R., Leyssens, A., and Crompton, M., 1998, Transient mitochondrial depolarizations reflect focal sarcoplasmic reticular calcium release in single rat cardiomyocytes. *J. Cell Biol.* **142**: 975-988.

Dykens, J.A., 1994, Isolated cerebral and cerebellar mitochondria produce free radicals when exposed to elevated Ca^{2+} and Na^+: implications for neurodegeneration. *J. Neurochem.* **63**: 584-591.

Eisenberg, B.A., 1983, Quantitative ultrastructure of mammalian skeletal muscle. In: *Handbook of Physiology-Skeletal Muscle*, L.D. Peachey, ed., American Physiological Society, Bethesda MD, pp 73-112.

Gillis, J.M., 1997, Inhibition of mitochondrial calcium uptake slows down relaxation in mitochondria-rich skeletal muscles. *J. Muscle Res. Cell Motil* **18**: 473-483.

Gollnick, P.D. & King, D.W., 1969, Effect of exercise and training on mitochondria of rat skeletal muscle. *Am. J. Physiol.* **216**: 1502-1509.

Grijalba, M.T., Vercesi, A.E., and Schreier, S., 1999, Ca^{2+}-induced increased lipid packing and domain formation in submitochondrial particles. A possible early step in the mechanism of Ca^{2+}-stimulated generation of reactive oxygen species by the respiratory chain. *Biochem.* **38**: 13279-13287.

Harris, E.J., 1978, Anion/calcium ion ratios and proton production in some mitochondrial calcium ion uptakes. *Biochem. J.* **176**: 983-991.

He, L., and Lemasters, J.J., 2002, Regulated and unregulated mitochondrial permeability transition pores: a new paradigm of pore structure and function? *FEBS Lett.* **512**: 1-7.

Hood, D.A., 2001, Invited Review: contractile activity-induced mitochondrial biogenesis in skeletal muscle. *J. Appl. Physiol.* **90**: 1137-1157.

Jouaville, L.S., Pinton, P., Bastianutto, C., Rutter, G.A., and Rizzuto, R., 1999, Regulation of mitochondrial ATP synthesis by calcium: evidence for a long-term metabolic priming. *Proc. Nat. Acad. Sci. USA* **96**: 13807-13812.

Joyal, J.L., Hagen, T., and Aprille, J.R., 1995, Intramitochondrial protein synthesis is regulated by matrix adenine nucleotide content and requires calcium. *Arch. Biochem. Biophys.* **319**: 322-330.

Kavanagh, N.I., Ainscow, E.K., and Brand, M.D., 2000, Calcium regulation of oxidative phosphorylation in rat skeletal muscle mitochondria. *Biochim. Biophys. Acta* **1457**: 57-70.

Lännergren, J., Westerblad, H., and Bruton, J.D., 2001, Changes in mitochondrial Ca^{2+} detected with Rhod-2 in single frog and mouse skeletal muscle fibres during and after repeated tetanic contractions. *J. Muscle Res. Cell Motil.* **22**: 265-275.

Lestienne, P., Bataille, N., and Lucas-Heron, B., 1995, Role of the mitochondrial DNA and calmitine in myopathies. *Biochim. Biophys. Acta* **1271:** 159-163.

Ligeti, E., and Lukacs, G.L., 1984, Phosphate transport, membrane potential, and movements of calcium in rat liver mitochondria. *J. Bioenerg. Biomembr.* **16:** 101-113.

Litsky, M.L., and Pfeiffer, D.R., 1997, Regulation of the mitochondrial Ca^{2+} uniporter by external adenine nucleotides: the uniporter behaves like a gated channel which is regulated by nucleotides and divalent cations. *Biochem.* **36:** 7071-7080.

Lucas-Heron, B., Le Ray, B., and Schmitt, N., 1995, Does calmitine, a protein specific for the mitochondrial matrix of skeletal muscle, play a key role in mitochondrial function? *FEBS Lett.* **374:** 309-311.

Madsen, K., Ertbjerg, P., Djurhuus, M.S., and Pedersen, P.K., 1996, Calcium content and respiratory control index of skeletal muscle mitochondria during exercise and recovery. *Am. J. Physiol.* **271:** E1044-E1050.

Marechal, G., and Beckers-Bleukx, G., 1993, Force-velocity relation and isomyosins in soleus muscles from two strains of mice (C57 and NMRI). *Pflüg. Arch.* **424:** 478-487.

Milner, D.J., Mavroidis, M., Weisleder, N., and Capetanaki Y., 2000, Desmin cytoskeleton linked to muscle mitochondrial distribution and respiratory function. *J. Cell Biol.* **150:** 1283-1298.

McCutcheon, L.J., Byrd, S.K., and Hodgson, D.R., 1992, Ultrastructural changes in skeletal muscle after fatiguing exercise. *J. Appl. Physiol.* **72:** 1111-1117.

Philippi, M., and & Sillau, A.H., 1994, Oxidative capacity distribution in skeletal muscle fibres of the rat. *J. Exp. Biol.* **189:** 1-11.

Reid, M.B., and Durham, W.J., 2002, Generation of reactive oxygen and nitrogen species in contracting skeletal muscle. *Ann. New York Acad. Sci.* **959:** 108-116.

Rizzuto, R., Pinton, P., Brini, M., Chiesa, A., Filippin, L., and Pozzan, T., 1999, Mitochondria as biosensors of calcium microdomains. *Cell Calcium* **26:** 193-199.

Sakai, Y., Iwarmura, Y., Hayashi, J.I., Yamamoto, N., Ohkoshi, N., and Nagata, H., 1999, Acute exercise causes mitochondrial DNA deletion in rat skeletal muscle. *Muscle Nerve* **22:** 268-261.

Sembrowich, W.L., Quintinskie, J.J., and Li, G., 1985, Calcium uptake in mitochondria from different skeletal muscle types. *J. Appl. Physiol.* **59:** 137-141.

Sparagna, G.C., Gunter, K.K., Sheu, S.S., and Gunter, T.E., 1995, Mitochondrial calcium uptake from physiological-type pulses of calcium: a description of the rapid uptake mode. *J. Biol. Chem.* **270:** 27510-27515.

Walsh, B., Tonkonogi, M., Soderlund, K., Hultman, E., Saks, V., and Sahlin, K., 2001, The role of phosphorylcreatine and creatine in the regulation of mitochondrial respiration in human skeletal muscle. *J. Physiol.* **537:** 971-978.

Wu, H., Kanatous, S.B., Thurmond, F.A., Gallardo, T., Isotani, E., Bassel-Duby, R., and Williams, R.S., 2002, Regulation of mitochondrial biogenesis in skeletal muscle by CaMK. *Science* **296:** 349-352.

DISCUSSION

ter Keurs: Is ATP synthesis accelerated in the mitochondria in these muscles by mitochondrial Ca^{2+} uptake?

Lännergren: I would believe so, but this is not something that we can look at in our experiments.

Rall: When mitochondrial Ca^{2+} uptake is blocked, is there any change in the mechanical fatigue time course?

Lännergren: We have not looked at this carefully in soleus fibers.

Kushmerick: Is there evidence for or against a threshold Ca^{2+} for mitochondrial uptake? Is there mitochondrial Ca^{2+} uptake with twitch?

Lännergren: We didn't study twitches. But we can observe mitochondrial Ca^{2+} uptake in a 400 ms tetanus.

CELLULAR MECHANISMS OF SKELETAL MUSCLE FATIGUE

Håkan Westerblad and David G. Allen

INTRODUCTION

High-intensity exercise leads to a rapid decline in contractile function, known as skeletal muscle fatigue. In this chapter we will review possible causes of muscle fatigue. To study mechanisms underlying fatigue we frequently use isolated muscle fibers, which are fatigued by repeated isometric tetani of short duration. The present overview focuses on results obtained in such studies as well as studies on skinned muscle fibers (i.e. muscle cells where the surface membrane has been chemically or physically removed). This is because studies on single muscle fibers provide the most direct way to address cellular mechanisms of fatigue. It may be argued that conclusions drawn from studies on single fibers are not relevant to the fatigue experienced by humans during various types of exercise. However, available data indicate that the mechanisms of fatigue are qualitatively similar in diverse experimental models, ranging from exercising humans to single fibers (Allen et al., 1995). The differences that inevitable must exist appear to be mainly of a quantitative nature.

During fatigue induced by repeated tetanic contraction, a characteristic pattern is generally observed in fast-twitch fibers: initially there is fast decline of tetanic force by 10-20% that is accompanied by an increase in tetanic free myoplasmic $[Ca^{2+}]$ ($[Ca^{2+}]_i$) (phase 1); then follows a period of rather constant tetanic force and $[Ca^{2+}]_i$ (phase 2); finally there is rapid decline of both tetanic force and $[Ca^{2+}]_i$ (phase 3) (Allen et al., 1995). Thus, the force decline in early fatigue (phase 1) would be due to impaired myofibrillar function, whereas the decline in late fatigue (phase 3) would be caused by decreased sarcoplasmic reticulum (SR) Ca^{2+} release.

* Håkan Westerblad, Department of Physiology and Pharmaclolgy, Karolinska Instituet, SE-171 77 Stockholm, Sweden. David G. Allen, Department of Physiology and Institute of Biomedical Research, University of Sydney F13, NSW 2006, Australia.

Molecular and Cellular Aspects of Muscle Contraction
Edited by H. Sugi, Kluwer Academic/Plenum Publishers, 2003

DECREASED MYOFIBRILLAR FORCE PRODUCTION

The energy consumption of skeletal muscle cells may increase up to hundred-fold between rest and high-intensity exercise. This high energy demand exceeds the aerobic capacity of the muscle cells and a large fraction of the ATP required will come from anaerobic metabolism. It therefore seems logical that there is a causal relationship between anaerobic metabolism and the decrease in force production in early fatigue; that is, some consequence(s) of anaerobic metabolism causes the decline in contractile function.

Accumulation of Lactic Acid

Anaerobic breakdown of glycogen leads to an intracellular accumulation of inorganic acids of which lactic acid is quantitatively the most important. Since lactic acid is a strong acid, it dissociates into lactate and hydrogen ions. While lactate ions would have little effect on muscle contraction (Posterino et al., 2001), the increase in hydrogen ions (i.e. reduced pH or acidosis) is the *classical* cause of skeletal muscle fatigue. During intense muscle activity the intracellular pH may fall by about 0.5 pH-units. Two major lines of evidence have been used to link this decline of pH to the contractile dysfunction in fatigue. First, studies on human muscle fatigue have often shown a good temporal correlation between the decline of muscle pH and the reduction of force or power production. Second, studies on skinned skeletal muscle fibers have shown that acidification may reduce both the isometric force and the shortening velocity.

However, in humans the temporal correlation between impaired contractile function during fatigue and reduced pH is not always present. For instance, force sometimes recovers more rapidly than pH after the end of fatiguing contractions (Sahlin and Ren, 1989). This means that if reduced pH has a direct force depressing effect in human muscles, this effect must have been counteracted by some other factor that increases force to the same extent. Such a force-potentiating factor has not been identified and hence the obvious conclusion is that there is no causal relationship between acidosis and reduced force production.

Important evidence in favour of acidosis causing reduced force production comes from studies on skinned muscle fibers that were performed at $\leq$ 15 °C (Pate et al., 1995). Recent studies have focused on the temperature dependence of the pH effects on force and the results of these studies further challenge the role of H^+ in mammalian muscle fatigue. Some early studies conducted more than ten years ago showed that acidification, if anything, resulted in an increased tetanic force at physiological temperatures (Ranatunga, 1987). More recently, Pate and colleagues (1995) studied skinned rabbit psoas fibers and observed the expected large depressive effect of lowered pH at 10 °C, but the effect of acidification on force production was small at 30 °C. Similar results have subsequently been obtained in isolated single mouse muscle fibers (Westerblad et al., 1997) and whole mouse muscles (Wiseman et al., 1996).

In summary, acidosis has little direct effect on isometric force production in mammalian muscles studied at physiological temperatures. Therefore, if acidosis is involved in skeletal muscle fatigue, the effect may be indirect. For instance, extracellular acidosis may well activate group III-IV nerve afferents in muscle and hence be involved in the sensation of discomfort in fatigue.

Accumulation of Inorganic Phosphate Ions

Creatine kinase (CK) catalyses the exchange of phosphate between ATP and phosphocreatine (PCr) via the following reaction: $PCr + ADP + H^+ \Leftrightarrow Cr$ (creatine) + ATP. During periods of high energy demand, the net result of the CK reaction is that CrP breaks down to Cr and P_i while the ATP concentration remains almost constant. Creatine has little effect on contractile function (Godt and Nosek, 1989), whereas there are several mechanisms by which increased P_i may depress contractile function. Most models of cross-bridge action propose that P_i is released in the transition from low-force, weakly attached states to high-force, strongly attached states. This implies that the transition to the high-force states is hindered by increased P_i. Therefore, fewer cross-bridges would be in high-force states and the force production would decrease as P_i increases during fatigue development. In line with this, experiments on skinned fibers consistently show a reduced maximum Ca^{2+} activated force in the presence of elevated P_i (Pate and Cooke, 1989; Millar and Homsher, 1990).

The hypothesis that increased P_i reduces maximum cross-bridge force has been difficult to test in intact muscle cells, since it has proven difficult to increase myoplasmic P_i without imposing other metabolic changes as well. Recently Steeghs et al. (1997) generated genetically modified mice, which completely lack CK in their skeletal muscles ($CK^{-/-}$ mice). Muscles from these mice provide a reasonable model to study effects of increased P_i. Fast-twitch skeletal muscle fibers of $CK^{-/-}$ mice display an increased myoplasmic P_i concentration at rest and there is no significant P_i accumulation during fatigue (Dahlstedt et al., 2000). The maximum Ca^{2+} activated force of unfatigued $CK^{-/-}$ fast-twitch fibers is markedly lower than that of wild-type fibers, which support a force depressing role of increased P_i (Dahlstedt et al., 2001). Furthermore, $CK^{-/-}$ fibers do not display the 10-20% reduction of maximum Ca^{2+} activated force observed after about ten fatiguing tetani (phase 1; see above), which has been ascribed to increased P_i (Dahlstedt et al., 2000). Even after 100 fatiguing tetani, force was not significantly affected in $CK^{-/-}$ fibers, whereas by this time force was reduced to less than 30% of the original in wild-type fibers. Additional support for a coupling between myoplasmic P_i concentration and force production in intact muscle cells comes from experiments where *reduced* myoplasmic P_i is associated with *increased* force production (Phillips et al., 1993; Bruton et al., 1997). Thus, increased myoplasmic P_i may decrease force production during fatigue by direct action on cross-bridge function. Altered cross-bridge function may also affect the force-$[Ca^{2+}]_i$ relationship via the complex interaction between cross-bridge attachment and thin (actin) filament activation (Gordon et al., 2000). In this way increased P_i may also reduce force production by causing a reduced myofibrillar Ca^{2+} sensitivity, which is a frequently observed characteristic in skeletal muscle fatigue (Allen et al., 1995).

DIFFERENT TYPES OF FATIGUE-INDUCED FAILURE OF SR CA^{2+} RELEASE

Failure of Action Potential Propagation into the T-tubular System

Failure of action potential propagation into the t-tubular system is a frequently suggested mechanism for the decreased Ca^{2+} release in fatigue (Garcia et al., 1991). This type of failure would lead to close to normal activation of myofibrils close to the surface of muscle cells whereas activation of deeper parts will be limited. We have used $[Ca^{2+}]_i$

imaging to assess t-tubular conduction failure during fatigue. With this technique t-tubular failure will show up as a lower $[Ca^{2+}]_i$ in the centre of muscle fibers as compared to the surface. A pattern consistent with t-tubular failure was observed during prolonged high-frequency stimulation both in fibers from *Xenopus* frogs (Westerblad et al., 1990) and mice (Westerblad et al., 1993). However, continuous high-frequency stimulation is unphysiological because during prolonged maximum voluntary contraction, this kind of problem is avoided by the progressive reduction of the motor unit firing frequency (Bigland-Ritchie et al., 1983). Accordingly, when the stimulation frequency was reduced, the gradient of $[Ca^{2+}]$ disappeared and force increased (Westerblad et al., 1990). With more physiological fatiguing stimulation protocols (i.e. repeated short tetani; see below), the reduction of $[Ca^{2+}]_i$ is homogeneous throughout the myoplasm, both during induction of fatigue and during the following recovery period (Westerblad et al., 1990; Westerblad et al., 1993).

Direct Effects of Metabolic Changes on SR Ca^{2+} Release Channels

Acidification has been suggested to reduce the SR Ca^{2+} release by direct action on the SR Ca^{2+} release channels (Ma et al., 1988). However, it now seems clear that while acidosis might reduce the SR Ca^{2+} leak, it has no significant inhibitory effect on the normal voltage activated release (Lamb et al., 1992; Westerblad and Allen, 1993). Furthermore, increased lactate does not affect voltage activated SR Ca^{2+} release (Posterino et al., 2001). Accumulation of lactic acid can therefore be excluded as an important cause of decreased SR Ca^{2+} release in fatigue (Bruton et al., 1998).

Inhibition of SR Ca^{2+} release channels in fatigue may be due to reduced ATP (Smith et al., 1985). A net breakdown of ATP will result in increased free myoplasmic $[Mg^{2+}]$ ($[Mg^{2+}]_i$), since ATP binds Mg^{2+} more strongly than its breakdown products (Westerblad and Allen, 1992). Increased $[Mg^{2+}]_i$ has also been shown to inhibit SR Ca^{2+} release channels (Lamb and Stephenson, 1991). Furthermore, the combination of reduced ATP and increased $[Mg^{2+}]_i$ has an additive effect (Owen et al., 1996; Blazev and Lamb, 1999). During fatigue induced by repeated short tetani, $[Mg^{2+}]_i$ starts to increase at the time when tetanic $[Ca^{2+}]_i$ starts to fall, suggesting a causal relationship (Westerblad and Allen 1992). The increase in $[Mg^{2+}]_i$ in fatigue could not on its own explain the reduction in tetanic $[Ca^{2+}]_i$ and force, because control experiments with injection of $MgCl_2$ gave a much smaller force reduction (Westerblad and Allen 1992).

In the absence of CK activity, changes of ATP and $[Mg^{2+}]_i$ might be larger and hence SR Ca^{2+} release more affected during fatigue. Accordingly, SR Ca^{2+} release decreased during the first seconds of stimulation in muscles fibers from CK deficient ($CK^{-/-}$) mice. However, this decrease was transient and tetanic $[Ca^{2+}]_i$ recovered rapidly during ongoing stimulation. In fact, $CK^{-/-}$ muscle fibers were *more* fatigue resistant than muscles from their wild-type littermates (Dahlstedt et al., 2000). This is most likely due to an increased aerobic capacity in $CK^{-/-}$ fibers. Thus, direct inhibition of SR Ca^{2+} release channels by reduced ATP and increased $[Mg^{2+}]_i$ appears to be of greatest importance at the onset of high-force contractions, whereas a role in later stages of fatigue is more uncertain.

Ca^{2+}-P_i Precipitation in the SR

A mechanism that has received increasing attention in recent years is that during fatigue Ca^{2+}-P_i might precipitate in the SR, leading to a reduced amount of free Ca^{2+} available for release. The underlying idea is that during fatigue, P_i accumulates in the myoplasm due to break-down of PCr via the CK reaction. Some P_i ions are then transported into the SR where the Ca^{2+}-P_i solubility product is exceeded, precipitation occurs and the releasable pool of Ca^{2+} is reduced (Inesi and de Meis, 1989; Fryer et al., 1995). Although no study has directly shown that this type of precipitation does occur, strong indirect evidence of its existence has been presented both in experiments on skinned fibers with intact SR exposed to high P_i solutions (Fryer et al., 1995) and intact mouse fibers microinjected with P_i (Westerblad and Allen, 1996). Recent fatigue experiments support the Ca^{2+}-P_i precipitation mechanism (Allen and Westerblad, 2001; Allen et al., 2002a): (1) The amount of Ca^{2+} in the SR that can be released by application of a high dose of caffeine or 4-choro-m-cresol is reduced in fatigued toad muscle fibers (Kabbara and Allen, 1999). Both these compounds act directly on the SR Ca^{2+} release channels and a reduced response to them would indicate that the total amount of Ca^{2+} available for release has been reduced. (2) Direct measurement of the free $[Ca^{2+}]$ in the SR ($[Ca^{2+}]_{SR}$) has recently been performed using a low affinity Ca^{2+} indicator (fluo-5N) loaded in its membrane-permeant form (Kabbara & Allen, 2001). Assuming there is esterase present in the SR, some indicator should be localised and activated within the SR and the contribution to fluorescent signal from the myoplasm will be small because of the low Ca^{2+} sensitivity of the indicator. This study showed that $[Ca^{2+}]_{SR}$ declined throughout a period of fatiguing stimulation and recovered afterwards. (3) The decline of tetanic $[Ca^{2+}]_i$ during fatiguing stimulation was markedly delayed in fibers where the CK reaction is inhibited either pharmacologically (Dahlstedt and Westerblad, 2001) or genetically (Dahlstedt et al., 2000). In this case fibers fatigue without PCr break-down and therefore no major accumulation of P_i in the myoplasm will occur, which means that P_i will not enter the SR and cause Ca^{2+}-P_i precipitation.

The SR membrane contains small conductance chloride channels, which may conduct P_i (Ahern and Laver, 1998; Laver et al., 2001). The open probability of these channels increases at low ATP, which is in accordance with the fact that P_i entry into the SR of skinned muscle fibers is inhibited by ATP (Posterino and Fryer, 1998). This dependence on ATP can explain one apparent weakness of the hypothesis that raised myoplasmic $[P_i]$ causes Ca^{2+}-P_i precipitation in the SR: myoplasmic $[P_i]$ increases relatively early during fatiguing stimulation while the decline of tetanic $[Ca^{2+}]_i$ generally occurs quite late. Moreover, in mouse fast-twitch fibers the decline of tetanic $[Ca^{2+}]_i$ temporally correlates with an increase in $[Mg^{2+}]_i$, which presumably stems from a net breakdown of ATP (Westerblad and Allen, 1992), and it is not obvious why Ca^{2+}-P_i precipitation in the SR should show a temporal correlation with ATP breakdown. The ATP-dependence of the presumed SR P_i channels can explain both why P_i enters the SR with a delay and why there is a temporal correlation between declining ATP and declining tetanic $[Ca^{2+}]_i$. Interestingly, in fibers where CK was pharmacologically inhibited, $[Mg^{2+}]_i$ increased early during fatiguing stimulation and this was not associated with a decline of tetanic $[Ca^{2+}]_i$ (Dahlstedt and Westerblad, 2001). Thus, the temporal correlation between declining tetanic $[Ca^{2+}]_i$ and increasing $[Mg^{2+}]_i$ is lost when myoplasmic P_i accumulation is prevented.

In conclusion, several different experimental approaches have shown that Ca^{2+}-P_i precipitation in the SR can occur during fatiguing stimulation and this appears to be a major cause of reduced tetanic $[Ca^{2+}]_i$ in late stages of fatigue.

Failure of SR Ca^{2+} Release Due to Glycogen Depletion

In prolonged endurance activities muscle fatigue becomes pronounced at about the time when muscle glycogen levels fall to low levels (Bergström et al., 1967). In contrast to intense short-term exercise, the changes in myoplasmic pH and [P$_i$] are relatively small at the end of endurance activities (Vøllestad et al., 1988). These observations suggest that failure of SR Ca^{2+} release might be an important factor in muscles fatigued by endurance activities and could be related to the level of glycogen within the muscle fiber.

Recent studies on single mouse fibers showed that glycogen fell to about 25 % during fatigue caused by repeated tetani and this coincided with reduced Ca^{2+} transients (Chin and Allen, 1997). If the muscle fiber was allowed to recover in the absence of glucose, glycogen did not recover and there was limited recovery of tetanic force and [Ca^{2+}]$_i$. When restimulated with repeated tetani, the fiber fatigued much more rapidly. Thus, there appears to be a correlation between the level of glycogen and the magnitude of the Ca^{2+} transients. Similar results were obtained in a recent study on isolated mouse extensor digitorum longus (EDL, fast-twitch) muscles (Helander et al., 2002). These muscles were fatigued by repeated tetani, allowed to recover for 2 hours in zero, normal or high extracellular glucose, and then fatigued again. Muscles recovering in zero glucose had lower glycogen levels (~50% of the control) at the start of the second fatigue run and fatigued more rapidly, both regarding tetanic force and [Ca^{2+}]$_i$.

Similar experiments have also been performed in cane toad muscle fibers, which have a much higher level of glycogen (Kabbara et al., 2000). These fibers fatigue quite slowly but still exhibit failure of SR Ca^{2+} release that seems to be caused by a decline in SR Ca^{2+} stores. These fibers recover well in the absence of glucose, in contrast to mouse fibers, and repeated fatigue runs can be performed in the absence of glucose though the time to fatigue gradually shortens. Interestingly, in the final brief fatigue run, the glycogen level was substantially reduced and fibers still exhibit a failure of SR Ca^{2+} release. However, the failure was no longer related to depletion of SR Ca stores. These experiments suggest two independent mechanisms of failure of SR Ca^{2+} release can contribute to fatigue induced by repeated, brief tetani. (1) Depletion of SR Ca^{2+} stores, which seems to occur particularly when glycogen levels are high and would be related to precipitation of Ca^{2+}-P$_i$ in the SR. (2) Failure of Ca^{2+} release due to depletion of glycogen, which can occur in the presence of normal SR Ca^{2+} stores.

Two further independent lines of evidence link declining glycogen levels to failure of SR Ca^{2+} release. First, electron microscopy was used to measure the distribution of glycogen at the sarcomere level in human muscle biopsies obtained under control conditions and after exhaustive exercise (Friden et al., 1989). The results showed that in fatigue glycogen particles were preferentially depleted in the region of the t-tubular-SR junction and that rapidly fatiguable fibers were more depleted than fatigue-resistant fibers. These experiments indicate that glycogen in the region of the t-tubular-SR junction is involved in SR Ca^{2+} release and that glycogen depletion in this region might contribute to the failure of SR Ca^{2+} release during fatigue. Second, the ability of skinned toad muscle fibers to respond to depolarization of the t-tubules correlated closely with the muscle glycogen content (non-soluble component) (Stephenson et al., 1999). Thus, when glycogen was depleted the muscle fibers became unresponsive to depolarization showing that SR Ca^{2+} release failed. A crucial point is that in these experiments on skinned fibers, ATP and PCr were present in the bathing solutions suggesting that the glycogen performed a structural rather than a metabolic role. In line with this, using intact mouse muscle fibers we recently showed that the premature fatigue seen during a second fatigue run after recovery in the absence of glucose (see above) was not associated with a decline in [ATP] (measured with luciferin/luciferase) (Allen et al., 2002b).

In conclusion, there is reasonable evidence that depletion of glycogen at the end of prolonged, exhausting exercise may contribute to fatigue by causing reduced SR Ca^{2+} release. However, the details of the mechanism involved remain to be established.

REFERENCES

Ahern, G. P. and Laver, D. R., 1998, ATP inhibition and rectification of a Ca^{2+}-activated anion channel in sarcoplasmic reticulum of skeletal muscle. *Biophys. J.* 74: 2335-2351.

Allen, D. G., Kabbara, A. A., and Westerblad, H., 2002a, Muscle fatigue: the role of intracellular calcium stores. *Can. J. Appl. Physiol.* 27: 83-96.

Allen, D. G., Lännergren, J., and Westerblad, H., 1995, Muscle cell function during prolonged activity: cellular mechanisms of fatigue. *Exp. Physiol.* 80: 497-527.

Allen, D. G., Lännergren, J., and Westerblad, H., 2002b, Intracellular ATP measured with luciferin/luciferase in isolated single mouse skeletal muscle fibres. *Pflügers Arch.* 443: 836-842.

Allen, D. G. and Westerblad, H., 2001, Role of phosphate and calcium stores in muscle fatigue. *J. Physiol.* 536: 657-665.

Bergström, J., Hermansen, L., Hultman, E., and Saltin, B., 1967, Diet, muscle glycogen and physical performance. *Acta Physiologica Scandinavica.* 71: 140-150.

Bigland-Ritchie, B., Johansson, R., Lippold, O. C., Smith, S., and Woods, J. J., 1983, Changes in motoneurone firing rates during sustained maximal voluntary contractions. *J. Physiol.* 340: 335-346.

Blazev, R. and Lamb, G. D., 1999, Low [ATP] and elevated $[Mg^{2+}]$ reduce depolarization-induced Ca^{2+} release in rat skinned skeletal muscle fibres. *J. Physiol.* 520: 203-215.

Bruton, J. D., Lännergren, J., and Westerblad, H., 1998, Effects of CO_2-induced acidification on the fatigue resistance of single mouse muscle fibers at 28 °C. *J. Appl. Physiol.* 85: 478-483.

Bruton, J. D., Wretman, C., Katz, A., and Westerblad, H., 1997, Increased tetanic force and reduced myoplasmic $[P_i]$ following a brief series of tetani in mouse soleus muscle. *Am. J. Physiol.* 272: C870-C874.

Chin, E. R. and Allen, D. G., 1997, Effects of reduced muscle glycogen concentration on force, Ca^{2+} release and contractile protein function in intact mouse skeletal muscle. *J. Physiol.* 498: 17-29.

Dahlstedt, A. J., Katz, A., and Westerblad, H., 2001, Role of myoplasmic phosphate in contractile function of skeletal muscle: studies on creatine kinase-deficient mice. *J. Physiol.* 533: 379-388.

Dahlstedt, A. J., Katz, A., Wieringa, B., and Westerblad, H., 2000, Is creatine kinase responsible for fatigue? Studies of skeletal muscle deficient of creatine kinase. *FASEB J.* 14: 982-990.

Dahlstedt, A. J. and Westerblad, H., 2001, Inhibition of creatine kinase reduces the rate of fatigue-induced decrease in tetanic $[Ca^{2+}]_i$ in mouse skeletal muscle. *J. Physiol.* 533: 639-649.

Friden, J., Seger, J., and Ekblom, B., 1989, Topographical localization of muscle glycogen: an ultrahistochemical study in the human vastus lateralis. *Acta Physiol. Scand.* 135: 381-391.

Fryer, M. W., Owen, V. J., Lamb, G. D., and Stephenson, D. G., 1995, Effects of creatine phosphate and P_i on Ca^{2+} movements and tension development in rat skinned skeletal muscle fibres. *J. Physiol.* 482: 123-140.

Garcia, M. C., Gonzalez-Serratos, H., Morgan, J. P., Perreault, C. L., and Rozycka, M., 1991, Differential activation of myofibrils during fatigue in phasic skeletal muscle cells. *J. Muscle Res. Cell Motil.* 12: 412-424.

Godt, R. E. and Nosek, T. M., 1989, Changes of intracellular milieu with fatigue or hypoxia depress contraction of skinned rabbit skeletal and cardiac muscle. *J. Physiol.* 412: 155-180.

Gordon, A. M., Homsher, E., and Regnier, M., 2000, Regulation of contraction in striated muscle. *Physiol. Rev.* 80: 853-924.

Helander, I., Westerblad, H., and Katz, A., 2002, Effects of glucose on contractile function, $[Ca^{2+}]_i$ and glycogen in isolated mouse skeletal muscle. *Am. J. Physiol.* 282: C1306-C1312.

Inesi, G. and de Meis, L., 1989, Regulation of steady state filling in sarcoplasmic reticulum. Roles of back-inhibition, leakage, and slippage of the calcium pump. *J. Biol. Chem.* 264: 5929-5936.

Kabbara, A. A. and Allen, D. G., 1999, The role of calcium stores in fatigue of isolated single muscle fibres from the cane toad. *J. Physiol.* 519: 169-176.

Kabbara, A. A., Nguyen, L. T., Stephenson, G. M., and Allen, D. G., 2000, Intracellular calcium during fatigue of cane toad skeletal muscle in the absence of glucose. *J. Muscle Res. Cell Motil.* **21:** 481-489.

Lamb, G. D., Recupero, E., and Stephenson, D. G., 1992, Effect of myoplasmic pH on excitation-contraction coupling in skeletal muscle fibres of the toad. *J. Physiol.* **448:** 211-224.

Lamb, G. D. and Stephenson, D. G., 1991, Effect of Mg^{2+} on the control of Ca^{2+} release in skeletal muscle fibres of the toad. *J. Physiol.* **434:** 507-528.

Laver, D. R., Lenz, G. K., and Dulhunty, A. F., 2001, Phosphate ion channels in sarcoplasmic reticulum of rabbit skeletal muscle. *J. Physiol.* **535:** 715-728.

Ma, J., Fill, M., Knudson, C. M., Campbell, K. P., and Coronado, R., 1988, Ryanodine receptor of skeletal muscle is a gap junction-type channel. *Science* **242:** 99-102.

Millar, N. C. and Homsher, E., 1990, The effect of phosphate and calcium on force generation in glycerinated rabbit skeletal muscle fibers. A steady-state and transient kinetic study. *J. Biol. Chem.* **265:** 20234-20240.

Owen, V. J., Lamb, G. D., and Stephenson, D. G., 1996, Effect of low [ATP] on depolarization-induced Ca^{2+} release in skeletal muscle fibres of the toad. *J. Physiol.* **493:** 309-315.

Pate, E., Bhimani, M., Franks-Skiba, K., and Cooke, R., 1995, Reduced effect of pH on skinned rabbit psoas muscle mechanics at high temperatures: implications for fatigue. *J. Physiol.* **486:** 689-694.

Pate, E. and Cooke, R., 1989, Addition of phosphate to active muscle fibers probes actomyosin states within the powerstroke. *Pflügers Arch.* **414:** 73-81.

Phillips, S. K., Wiseman, R. W., Woledge, R. C., and Kushmerick, M. J., 1993, The effect of metabolic fuel on force production and resting inorganic phosphate levels in mouse skeletal muscle. *J. Physiol.* **462:** 135-146.

Posterino, G. S., Dutka, T. L., and Lamb, G. D., 2001, L(+)-lactate does not affect twitch and tetanic responses in mechanically skinned mammalian muscle fibres. *Pflügers Arch.* **442:** 197-203.

Posterino, G. S. and Fryer, M. W., 1998, Mechanisms underlying phosphate-induced failure of Ca^{2+} release in single skinned skeletal muscle fibres of the rat. *J. Physiol.* **512:** 97-108.

Ranatunga, K. W., 1987, Effects of acidosis on tension development in mammalian skeletal muscle. *Muscle Nerve* **10:** 439-445.

Sahlin, K. and Ren, J. M., 1989, Relationship of contraction capacity to metabolic changes during recovery from a fatiguing contraction. *J. Appl. Physiol.* **67:** 648-654.

Smith, J. S., Coronado, R., and Meissner, G., 1985, Sarcoplasmic reticulum contains adenine nucleotide-activated calcium channels. *Nature* **316:** 446-449.

Steeghs, K., Benders, A., Oerlemans, F., de Haan, A., Heerschap, A., Ruitenbeek, W., Jost, C., van Deursen, J., Perryman, B., Pette, D., Bruckwilder, M., Koudijs, J., Jap, P., Veerkamp, J., and Wieringa, B., 1997, Altered Ca^{2+} responses in muscles with combined mitochondrial and cytosolic creatine kinase deficiencies. *Cell* **89:** 93-103.

Stephenson, D. G., Nguyen, L. T., and Stephenson, G. M., 1999, Glycogen content and excitation-contraction coupling in mechanically skinned muscle fibres of the cane toad. *J. Physiol.* **519:** 177-187.

Vøllestad, N. K., Sejersted, O. M., Bahr, R., Woods, J. J., and Bigland-Ritchie, B., 1988, Motor drive and metabolic responses during repeated submaximal contractions in humans. *J. Appl. Physiol.* **64:** 1421-1427.

Westerblad, H. and Allen, D. G., 1992, Myoplasmic free Mg^{2+} concentration during repetitive stimulation of single fibres from mouse skeletal muscle. *J. Physiol.* **453:** 413-434.

Westerblad, H. and Allen, D. G., 1993, The influence of intracellular pH on contraction, relaxation and $[Ca^{2+}]_i$ in intact single fibres from mouse muscle. *J. Physiol.* **466:** 611-628.

Westerblad, H. and Allen, D. G., 1996, The effects of intracellular injections of phosphate on intracellular calcium and force in single fibres of mouse skeletal muscle. *Pflügers Arch.* **431:** 964-970.

Westerblad, H., Bruton, J. D., and Lännergren, J., 1997, The effect of intracellular pH on contractile function of intact, single fibres of mouse muscle declines with increasing temperature. *J. Physiol.* **500:** 193-204.

Westerblad, H., Duty, S., and Allen, D. G., 1993, Intracellular calcium concentration during low-frequency fatigue in isolated single fibers of mouse skeletal muscle. *J. Appl. Physiol.* **75:** 382-388.

Westerblad, H., Lee, J. A., Lamb, A. G., Bolsover, S. R., and Allen, D. G., 1990, Spatial gradients of intracellular calcium in skeletal muscle during fatigue. *Pflügers Arch.* **415:** 734-740.

Wiseman, R. W., Beck, T. W., and Chase, P. B., 1996, Effect of intracellular pH on force development depends on temperature in intact skeletal muscle from mouse. *Am. J. Physiol.* **271:** C878-C886.

DISCUSSION

Curtin: Do the creative-kinase knock-out mice have normal levels of creatine and phosphocreatin levels?

Westerblad: Yes, they are almost normal.

Gonzalez-Serratos: Regarding your proposition of what cause skeletal muscle fatigue, i.e. precipitation of Ca^{2+}-Pi in the SR decreasing the Ca^{2+} available to be released, is difficult for me to reconcile. The following of our published findings speak against: (1) caffeine contracture after fatigue is larger than prefatigue; (2) myofibrils which are inactivated during fatigue development get activated upon caffeine application; and (3) we have presented evidences with imaging that, after fatigue development, calcium release from the SR by caffeine is substantially larger than before fatigue.

Westerblad: (1) The Ca^{2+} released from the SR by caffeine in unfatigued frog fibers is generally well above that required for full force production. Therefore, the caffeine-induced Ca^{2+} release in fatigue may be markedly reduced and still fall activation of the contractile elements occurs. The fact that the force in caffeine contractures was larger in fatigue than in prefatigue in your experiments is rather puzzling. This would mean that metabolic changes during fatigue (i.e. Pi accumulation) were absent, since otherwise the maximum cross-bridge force should be reduced. (2) Wavy myofibrils have been observed in fatigued frog fibers studied at short length. This is a strong evidence of T-tubular conduction failure under these conditions. However, at optimal length, this may not be the case, and our results indicate that other mechanisms are more important in fatigue produced by repeated tetanic stimulation. (3) In our experiments, we always see reduced SR Ca^{2+}-release in fatigue. Several different experimental approaches indicate that Ca^{2+}-Pi precipitation in the SR is an important cause of this. In the experiments you refer to, it seems like some other mechanism is dominating. Thus, depending on the experimental condition, different mechanisms may be involved.

MYOFIBRILLAR DETERMINANTS OF RATE OF RELAXATION IN SKINNED SKELETAL MUSCLE FIBERS

[1]Ye Luo*, Jonathan P. Davis*, Svetlana B. Tikunova*, Lawrence B. Smillie[†] and Jack A. Rall*

1. INTRODUCTION

Muscle relaxation occurs when Ca^{2+}, sequestrated by the sarcoplasmic reticulum (SR) Ca^{2+} pump, dissociates from troponin (Tn) to deactivate the thin filaments leading to cross-bridge detachment and force decay. It is well established that the rate of Ca^{2+} sequestration by the SR can control relaxation kinetics.[1] The aim of the present investigation is to determine the relative contribution of Ca^{2+} dissociation from TnC and cross-bridge detachment to relaxation rate induced by rapid sequestration of Ca^{2+}. Three possibilities can be envisioned. First, Ca^{2+} dissociation from TnC may be much faster than cross-bridge detachment. In this case, only cross-bridge detachment kinetics would affect relaxation rate. Second, Ca^{2+} dissociation from TnC may be similar to cross-bridge detachment. If this relationship were true, both cross-bridge and TnC kinetics would affect relaxation rate. Third, Ca^{2+} dissociation from TnC may be slower than cross-bridge detachment. If this possibility were true, then only the kinetics of Ca^{2+} dissociation from TnC would affect relaxation rate.

To directly assess the effects of myofibrillar factors on relaxation rate, skinned rabbit psoas fibers devoid of functional SR were induced to relax by rapidly lowering free $[Ca^{2+}]$ with photolysis of the caged Ca^{2+} chelator diazo-2 after: 1) increasing or decreasing Ca^{2+} dissociation rate from TnC, 2) decreasing cross-bridge kinetics, and 3) both increasing or decreasing Ca^{2+} dissociation rate from TnC and decreasing cross-bridge kinetics.[2] Mutant TnCs (NHdel and M82Q TnC) with varying Ca^{2+} affinities and Ca^{2+} dissociation rates were characterized in solution and reconstituted into fibers to change the Ca^{2+} dissociation rate

---

[1]*Ye Luo, Jonathan P. Davis, Svetlana B. Tikunova and Jack A. Rall, Department of Physiology and Cell Biology, Ohio State University, Columbus, OH 43210.

†Lawrence B. Smillie, Department of Biochemistry, University of Alberta, Edmonton, Alberta, Canada T6G 2H7.

from TnC. Effects of decreased cross-bridge kinetics on relaxation were examined under conditions of lowered intracellular $[P_i]$.[3] By comparing the kinetics of diazo-2 induced relaxation under these interventions with control values, the rate-limiting processes for skeletal muscle relaxation were determined.

2. METHODS

2.1 Skinned Fiber Preparation and Experimental Apparatus

Single skinned rabbit psoas fibers devoid of sarcolemma and SR were utilized. The single fiber experimental setup was described previously.[4] The UV flash for the photolysis of diazo-2 was provided by a frequency-doubled ruby laser which produced a 30 ns duration pulse at 347 nm with an energy of ~ 300 mJ. The energy was maintained at ~ 100 mJ at the level of the fiber by placing glass slides in the beam path to attenuate the laser output. Experiments were performed at 15 °C.

2.2 Solutions

In making the diazo-2 solutions, the published value of the Ca^{2+} binding constant for diazo-2 (k_{dCa} = 2.2 μM)[5] was adjusted for temperature, ionic strength and pH to produce solutions with anticipated free $[Ca^{2+}]$. This was done by selecting a value of k_{dCa} for diazo-2 (1.5 μM) that resulted in similar fractional force in the absence and presence of diazo-2. The Mg^{2+} binding constant (k_{dMg}) to diazo-2 was 5.5 mM.[5] The diazo-2 solutions contained (mM): ATP, 3; CP, 14.5; Imidazole, 20; HDTA, 7; glutathione, 10 and concentrations of diazo-2 from 0.8 to 16. The solutions were adjusted to 180 mM ionic strength and pH 7.0 at 15 °C. Solutions contained 1 mM free $[Mg^{2+}]$ and 2.5 mM $[Mg \cdot ATP]$. The total [diazo-2] required to induce relaxation from a desired level of pre-photolysis force, i.e., pCa, at a desired diazo-2 chelating capacity (defined as $[diazo-2]_{free}/[Ca^{2+}]_{free}$) was calculated as previously described.[2]

The low inorganic phosphate (P_i) solutions were prepared by adding an enzymatic Pi scavenger, nucleoside phosphorlyase (NP) with substrate 7-methylguanosine (MEG) into the standard bathing or caged compound solutions. Nucleoside phosphorlyase served as a P_i scavenger to catalyze the reaction 7-methylguanosine + P_i ↔ 7-methylguanine + ribose-1-phosphate strongly toward the right to reduce the level of P_i contamination.[6] The unbuffered $[P_i]$ within rabbit psoas fibers was assumed to be ~ 200 μM and with the use of this P_i scavenger, P_i contamination was reduced to less than 5 μM.[7-9]

2.3 Recombinant Troponin C: Ca^{2+} Titrations and Ca^{2+} Dissociation Rates

Recombinant TnCs of chicken fast skeletal muscle included wild-type TnC (rTnC) and mutant TnCs that exhibited increased (M82Q TnC) or decreased (NHdel TnC, deletion of N terminal residues 1-11) Ca^{2+} affinity at their regulatory sites. To introduce a spectral probe

(Trp) into the regulatory N domain of TnC, F29W was made in single mutant F29W TnC or double mutants M82Q/F29W TnC and NHdel/F29W TnC.[10, 11] The F29W mutation has minimal effect on Ca^{2+} or Mg^{2+} binding to the C domain (sites III and IV) of TnC. Neither Ca^{2+} affinity nor cooperativity between sites I and II is significantly affected with this mutation.[10] Thus, F29W served as a wild-type like intrinsic probe for monitoring Ca^{2+} exchange with the regulatory sites of TnC at 15 °C.

2.4 Diazo-2 and P_i Scavenging Protocols

Relaxation was induced in skinned fibers by photolysis of the caged Ca^{2+} chelator diazo-2. Within a few milliseconds the Ca^{2+} affinity of diazo-2 increases ~ 30-fold upon exposure to UV light.[5, 12] To induce relaxation, fibers were first soaked in a diazo-2 containing solution until force reached a plateau as the diazo-2 and Ca^{2+} diffused into the fibers. Fibers were then transferred into an empty chamber and flashed in air and the relaxation time course recorded. It was not possible to induce complete relaxation with photolysis of diazo-2 from a maximum isometric contraction, probably due to the relatively small increase in Ca^{2+} affinity upon diazo-2 photolysis. Thus in testing the effects of rTnC and TnC mutants on relaxation, the experimental conditions were adjusted so that fibers developed approximately the same sub-maximum pre-photolysis force (50 - 60% of the maximum isometric force) and a similar final post-photolysis force. Also, since relaxation rate depended on the diazo-2 chelating capacity,[2] $[diazo-2]_{free}/[Ca^{2+}]_{free}$, relaxation was induced at similar diazo-2 chelating capacities.

TnC extraction was performed at a resting sarcomere length of 3.5 μm as previously described.[2, 13] Reconstitution of TnC was accomplished by soaking TnC-extracted fibers in a solution containing 16.7 μM purified TnC at a resting sarcomere length of 2.6 μm.

The protocol to investigate effects of rTnC or mutant TnCs on relaxation was: a) determine maximum force in pCa 4.0 ($F_{4.0}$), b) determine force at pCa to be examined ($F/F_{4.0}$), c) extract TnC until force in pCa 4.0 was < 5% of $F_{4.0}$, d) reconstitute with rTnC or mutant TnC until force in pCa 4.0 was > 90% of $F_{4.0}$, e) determine force at pCa to be examined in presence of reconstituted TnC and f) add diazo-2, induce contraction and then relaxation by flash photolysis of diazo-2. The protocol was the same for fibers in the presence of endogenous TnC except that TnC was not extracted.

To scavenge P_i, fibers were first soaked in low P_i pCa 9.0 solution for 10 min and then activated in a Ca^{2+} activating solution or a diazo-2 solution containing the P_i scavenger. The protocol to study effects of low P_i on relaxation in the presence of rTnC or mutant TnCs was: a) determine maximum force in standard pCa 4.0 ($F_{4.0}$), b) determine maximum force in low P_i pCa 4.0, c) extract TnC until force in standard pCa 4.0 was < 5% of $F_{4.0}$, d) reconstitute with rTnC or mutant TnC until force in standard pCa 4.0 was > 90% of $F_{4.0}$, e) determine maximum force in low P_i pCa 4.0 in presence of reconstituted TnC and f) add diazo-2 in low P_i, induce contraction and then relaxation by photolysis of diazo-2.

3. RESULTS AND DISCUSSION

3.1 Ca^{2+} affinity and Ca^{2+} Dissociation Rate of Mutant TnC in Solution

Ca^{2+} titrations of mutant TnCs following the increase in Trp fluorescence showed that Ca^{2+} bound half-maximally (pCa_{50}) to the regulatory sites of M82Q/F29W, F29W and

NHdel/F29W TnC at pCa 6.16 ± 0.003, 5.49 ± 0.01 and 5.08 ± 0.01, respectively. Thus, compared to F29W TnC, the M82Q mutation increased Ca^{2+} affinity of TnC by ~ 4.6-fold and the NHdel mutation decreased Ca^{2+} affinity of TnC by ~ 2.6-fold. Also using F29W as a spectral probe, Ca^{2+} dissociation rates from M82Q/F29W, F29W and NHdel/F29W TnC were 76 ± 2 s^{-1}, 340 ± 5 s^{-1} and 515 ± 4 s^{-1}, respectively. Thus, consistent with the differing Ca^{2+} affinities, compared to F29W TnC, the M82Q mutation resulted in ~ 4.5-fold decrease in the rate of Ca^{2+} dissociation from the regulatory sites of TnC whereas the NHdel mutation increased the Ca^{2+} dissociation rate by ~ 1.5-fold.

3.2 Ca^{2+} Sensitivity of Isometric Force in Control and Mutant Reconstituted Skinned Fibers

The force versus pCa relationships in fibers exhibited the following order of decreasing pCa_{50} values: M82Q/F29W TnC (6.13±0.01), unextracted (endogenous TnC) (6.04±0.01), rTnC (6.01±0.01), F29W TnC (5.94±0.02) and NHdel TnC (5.69±0.02). The force versus pCa relationships for unextracted (endogenous) fibers and fibers reconstituted with either rTnC or F29W TnC were not significantly different from each other. Thus the extraction/reconstitution procedures had minimal effect on Ca^{2+} sensitivity. In contrast, compared to fibers reconstituted with F29W TnC, M82Q/F29W TnC produced an ~1.6-fold increase in the Ca^{2+} sensitivity of isometric force and NHdel TnC produced an ~1.8-fold decrease in Ca^{2+} sensitivity of force. These shifts were qualitatively the same and in the case of NHdel TnC quantitatively similar to the shifts of pCa_{50} in the Ca^{2+} titrations measured in solution. Thus the M82Q or NHdel induced increase or decrease in the Ca^{2+} affinity of TnC in solution resulted in a corresponding shift in the Ca^{2+} sensitivity of isometric force in reconstituted muscle fibers.

3.3 Effect of Altered Ca^{2+} Dissociation from TnC on Relaxation Rate

The kinetics of relaxation was measured in fibers containing endogenous TnC, rTnC or mutant TnC. In an attempt to minimize as much variation as possible, experiments were performed at nearly constant, intermediate, diazo-2 chelating capacities and in a submaximum pre-photolysis force range where neither the pre- or post-photolysis force influenced the rate of relaxation.[2] The time of half relaxation ($t_{1/2}$) increased ~ 2-fold following reconstitution with wild-type TnC (26.2 ± 1.6 ms, N = 10 to 57 ± 6 ms, N = 5). This difference between rTnC and endogenous TnC is probably not due to the extraction/reconstitution procedure because rTnC produced a similar force versus pCa relationship and similar maximum rate of contraction in response to flash photolysis of a caged Ca (Luo and Rall, unpublished observations) as observed in unextracted fibers. There is the possibility of a species difference since the rTnC is derived from chicken TnC and re-constituted into mammalian muscle.

Compared to rTnC, M82Q TnC fibers relaxed more slowly (Fig. 1A and Table 1) which resulted in an additional ~ 2-fold increase in $t_{1/2}$. This ~ 2-fold increase in $t_{1/2}$ in fibers compares with the ~ 4.5-fold decrease in the Ca^{2+} dissociation rate from M82Q TnC in solution. These results provide direct evidence that the Ca^{2+} dissociation rate from TnC can significantly influence the kinetics of relaxation. However, the kinetics of relaxation was not significantly affected in fibers reconstituted with NHdel TnC compared to those with rTnC (Table 1). This result is in contrast to the observed effect of increasing Ca^{2+} dissociation rate from TnC in solution.

3.4 Effect of Altered Cross-Bridge Kinetics on Relaxation Rate

Lowering intracellular $[P_i]$ is expected to slow the rate of relaxation by decreasing the cross-bridge kinetics via slowing the P_i association step in the cross-bridge cycle.[7, 9] Compared to the standard solution, low P_i increased both maximum isometric force (by 12%) and submaximum isometric force, i.e., increased fiber Ca^{2+} sensitivity. Thus in the presence of low P_i, relaxation from a given relative force was induced in diazo-2 solutions containing higher pCa levels than in the standard solution. This led to an increase in the diazo-2 chelating capacity in the millimolar range of total [diazo-2]. But even at higher chelating capacity (8000 – 10000 versus 2400 – 3300 for standard P_i solution), low P_i still slowed relaxation. As shown in the example in Fig. 1B, reducing $[P_i]$ produced a slow linear phase and a decreased rate of the exponential phase even though the chelating capacity was higher in low P_i (~9800) than in the standard solution (2800). On average, low P_i decreased both slow and fast rate constants of relaxation, resulting in an increased $t_{1/2}$ from 26.2 ± 1.6 ms (N=10) to 127.6 ± 16.0 ms (N=5).

3.5 Effect of Altered Ca^{2+} Dissociation from TnC and Cross-Bridge Kinetics on Relaxation Rate

Effects of Ca^{2+} dissociation from TnC on relaxation kinetics were further examined in the presence of low P_i. Relaxation was induced in fibers containing endogenous TnC, rTnC or mutant TnC under conditions of similar pre-photolysis force and diazo-2 chelating capacity in low P_i. As shown in the examples in Fig. 1C, low P_i produced two phases of relaxation: a slow linear rate k_s: 3.3 ± 0.4 (F/□F)s^{-1} and a fast exponential rate k_f: 9.6 ± 0.9 s^{-1} (N=5) in fibers with endogenous TnC. The kinetics of relaxation was not affected by reconstitution with rTnC. However, M82Q TnC further slowed relaxation with an ~ 2-fold decrease in the linear and exponential rates which resulted in an ~2-fold increase in $t_{1/2}$ compared to rTnC (Table 1). This result is almost identical to the ~2-fold increase in $t_{1/2}$ with M82Q TnC observed in standard P_i solution. Thus when decreased Ca^{2+} dissociation rate (M82Q TnC) and decreased cross-bridge kinetics (low P_i) were combined, the effect was cumulative since the average $t_{1/2}$ was increased by ~ 4-fold. In contrast NHdel TnC still did not significantly speed up relaxation after low P_i slowed relaxation. Fibers containing NHdel TnC exhibited on average an ~ 20% shorter, but not significantly different, $t_{1/2}$ for relaxation compared to fibers with rTnC. Thus, decreasing the rate of Ca^{2+} dissociation from TnC and the cross-bridge kinetics either independently or simultaneously dramatically slowed relaxation. But increasing the rate of Ca^{2+} dissociation from TnC did not speed up relaxation significantly.

Table 1 also compares the kinetics of relaxation in the presence of mutant or rTnC under standard and low [Pi] conditions. It is apparent that low Pi increased the $t_{1/2}$ of relaxation ~2-fold for each exchanged TnC when compared to its own standard Pi control. Thus, relaxation can be slowed by slower cross-bridge kinetics and/or slower Ca^{2+} dissociation from TnC.

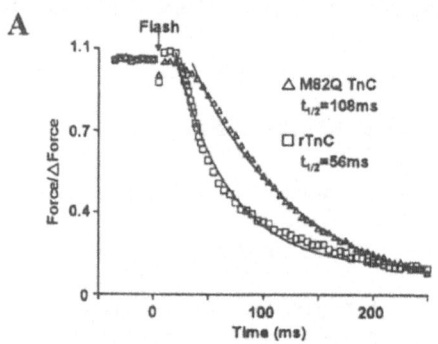

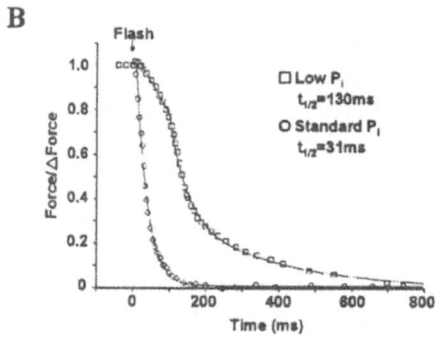

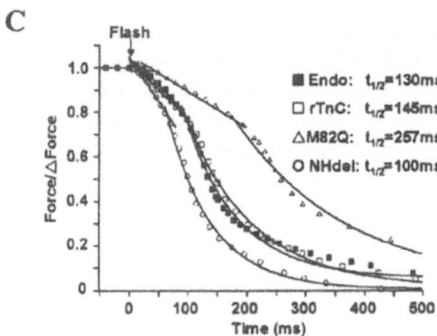

Figure 1. Diazo-2 induced relaxation at 15 °C in skinned psoas fibers with endogenous TnC, rTnC or mutant TnC in standard or low P_i solution. In order to emphasize the kinetics of relaxation, results are plotted as F/□F versus time where pre-photolysis force is set to 1.0 and post-photolysis force is set to zero and ΔF = extent of relaxation. A: M82Q TnC slows relaxation compared to rTnC in standard P_i solution. B: relaxation is slowed in fibers containing endogenous TnC in low P_i solution compared to standard P_i solution. C: M82Q TnC slows relaxation in low P_i solution compared to endogenous TnC, rTnC and NHdel TnC. The half-time of relaxation ($t_{1/2}$, the time for force to decrease to 0.5 □F) was used as a measure of relaxation kinetics. Modified figures from reference 2.

The finding that decreasing the Ca^{2+} dissociation rate from TnC with M82Q TnC slowed relaxation both under control conditions and when low P_i slowed relaxation provides the first direct evidence that Ca^{2+} dissociation from TnC can dramatically influence the kinetics of skeletal muscle relaxation. However, increasing the Ca^{2+} off-rate from TnC with NHdel did not significantly speed up relaxation in either control or after low P_i slowed relaxation. These results are consistent with the interpretation that the Ca^{2+} dissociation rate from TnC and cross-bridge detachment rate are similar in skeletal muscle fibers. In this case, the accelerating effect on relaxation of increased Ca^{2+} dissociation rate from NHdel TnC, if any, would be dampened by the rate limit of cross-bridge detachment kinetics.

3.6 Future Directions

In order to more thoroughly test the conclusions of this study, mutants of TnC with a greater than 2,000-fold range of Ca^{2+} affinities and greater than 45-fold range of Ca^{2+} dissociation rates have been constructed.[14] We are in the process of testing the ability of these mutants to support force in skinned fibers. Some of these mutants support force as well as F29W TnC but others support force poorly.[15] Those TnC mutants that support force poorly also exhibit a decreased affinity to the regulatory domain of TnI in solution. These results suggest that some TnC mutants disrupt intramolecular communication within the troponin

Table 1. Half-time of relaxation induced by photolysis of diazo-2 in skinned psoas fibers that contain rTnC or mutant TnC in standard or low P_i solution at 15 °C[a].

		Slow Cross-Bridge Detachment	
		Standard P_i (ms)	Low P_i (ms)
Fast Ca^{2+} Off TnC ↑	NHdel	62 ± 1	104 ± 12
	rTnC	57 ± 6	127 ± 11
Slow Ca^{2+} Off TnC ↓	M82Q	128 ± 13	251 ± 21

[a]Relaxation was induced at the following levels of diazo-2 chelating capacity ([diazo-2]$_{free}$/[Ca^{2+}]$_{free}$) and [diazo-2]$_{total}$: A) for standard P_i solution: rTnC (2440, 4 mM); M82Q TnC (3238, 4 mM) and NHdel TnC (3082, 15 mM) and B) for low P_i solution: rTnC (7800, 2 mM), M82Q TnC (10000, 1.25 mM) and NHdel TnC (8800, 5 mM). Pre-photolysis force ranged from 0.51 – 0.66 of $F_{4.0}$. The force in a pCa solution after extraction divided by the force in the pCa 4.0 solution before extraction averaged from 0.04 to 0.07. The force in a pCa 4.0 solution after rTnC or mutant TnC reconstitution divided by the force in a pCa 4.0 solution before extraction averaged from 0.92 to 1.05. Half-time of relaxation (ms) = time for force to fall to 50% of ☐F, the change in force after the flash. Values given as Mean ± S.E.

molecule and this leads to poor force recovery. Thus it cannot be assumed that all mutants of TnC will support force equally well. The conclusions in the present study can be further tested by examining the effects of Ca^{2+} dissociation from TnC when cross-bridge detachment is accelerated. The prediction would be that Ca^{2+} dissociation from TnC would become the rate-limiting step under these conditions. Finally the relationship of Ca^{2+} dissociation from TnC and cross-bridge detachment may be different in cardiac muscle and this relationship should be explored.

3.7 Summary

The influence of Ca^{2+} dissociation rate from TnC and decreased cross-bridge detachment rate on the time course of relaxation induced by flash photolysis of diazo-2 in rabbit skinned psoas fibers was investigated at 15 °C. A TnC mutant (M82Q TnC) that exhibited increased Ca^{2+} sensitivity caused by a decreased Ca^{2+} dissociation rate in solution also increased the Ca^{2+} sensitivity of force and decreased the rate of relaxation in fibers ~2-fold. In contrast, a TnC mutant (NHdel TnC) with decreased Ca^{2+} sensitivity caused by an increased Ca^{2+} dissociation rate in solution decreased Ca^{2+} sensitivity of force but did not accelerate relaxation. Decreasing the rate of cross-bridge kinetics by reducing $[P_i]$ slowed relaxation ~2-fold and led to two phases of relaxation, a linear phase followed by an exponential phase. In fibers, M82Q TnC further slowed relaxation in low $[P_i]$ ~2-fold whereas NHdel TnC had no significant effect on relaxation. These results are consistent with the interpretation that the Ca^{2+} dissociation rate and cross-bridge detachment rate are similar in fast twitch skeletal muscle such that decreasing either rate slows relaxation but accelerating Ca^{2+} dissociation has little effect on relaxation.

4. REFERENCES

1. J. M. Gillis, Relaxation of vertebrate skeletal muscle, *Biochim. Biophys. Acta* 811, 97-145 (1985).
2. Y. Luo, J. P. Davis, L. B. Smillie, and J. A. Rall, Determinants of relaxation rate in rabbit skinned skeletal muscle fibres. *J. Physiol.* in press (2003).
3. C. Tesi, N. Piroddi, F. Colomo, and C. Poggesi, Relaxation kinetics following sudden Ca^{2+} reduction in single myofibrils from skeletal muscle, *Biophys. J.* 83, 2142-2151 (2002).
4. P. A. Wahr, J. D. Johnson, and J. A. Rall, Determinants of relaxation rate in skinned frog skeletal muscle fibers, *Am. J. Physiol.* 274, C1608-C1615 (1998).
5. S. R. Adams, J. P. Kao, and R. Y. Tsien, Biological useful chelators that take up Ca^{2+} upon illumination, *J. Am. Chem. Soc.* 111, 7957-7968 (1989).
6. M. J. Brune, L. Hunter, E. T. Corrie, and M. R. Webb, Direct, real time measurements of rapid inorganic phosphate release using a novel fluorescent probe and its application to actomyosin subfragment 1 ATPase, *Biochem.* 33, 8262-8271 (1994).
7. N. C. Millar and E. Homsher, The effects of phosphate and calcium on force generation in glycerinated rabbit skeletal muscle fibers. *J. Biol. Chem.* 265, 20234-20240 (1990).
8. E. Pate, K. Franks-Skiba, and R. Cooke, Depletion of phosphate in active muscle fibers probes actomyosin states within the powerstroke. *Biophys. J.* 74, 369-380 (1998).
9. C. Tesi, F. Colomo, S. Nencini, N. Piroddi, and C. Poggesi, The effects of inorganic phosphate on force generation in single myofibrils from rabbit skeletal muscle. *Biophys. J.* 78, 3081-3092 (2000).
10. M. Chandra, E. F. da Silva, M. M. Sorenson, J. A. Ferro, J. R. Pearlstone, B. E. Nash, T. Borgford, C. M. Kay, and L. B. Smillie, The effects of N helix deletion and mutant F29W on the Ca^{2+} binding and functional properties of chicken skeletal muscle troponin C. *J. Biol. Chem.* 269, 14988-14994 (1994).

11. J. R. Pearlstone, T. Borgford, M. Chandra, K. Oilawa, C. M. Kay, O. Herzberg, J. Moult, J., A. Herklotz, F. C. Reinach, and L. B. Smillie, Construction and characterization of a spectral probe mutant of troponin C: application to analyses of mutants with increased Ca^{2+} affinity. *Biochem.* **31**, 6545-6553, 1992.
12. P. Mulligan, R. E. Palmer, S. Lipscomb, B. Hoskins, and C. C. Ashley, The effect of phosphate on the relaxation of frog skeletal muscle. *Pflug. Arch.* **437**, 393-399 (1999).
13. J. M. Metzger, M. L. Greaser, and R. L. Moss, Variations in cross-bridge attachment rate and tension with phosphorylation of myosin in mammalian skinned skeletal muscle fibers. *J. Gen. Physiol.* **93**, 855-883 (1989).
14. S. B. Tikunova, J. A. Rall, and J. P. Davis, Effects of hydrophobic residue substitutions with glutamine on Ca^{2+} binding and exchange with the N-domain of troponin C. *Biochem.* **41**, 6697-6705, 2002.
15. J. P. Davis, S. B. Tikunova, C. Alionte, and J. A. Rall, Conserved Hydrophobic β-Sheet and β-Sheet Interacting Residues Are Critical for Troponin C (TnC) Function in Skeletal Muscle. *Biophys. J.* in press (2003).

DISCUSSION

Westerblad: What is the relation between the off rate of Ca^{2+} from troponin C in solution and the rate of relaxation?

Rall: The off rate of Ca^{2+} from TnC in solution is much faster than the rate of relaxation, about 10 times faster. The off rate of Ca^{2+} from TnC complexed with a TnI peptide in solution is comparable to the rate of relaxation. The relative changes due to TnC mutants are the same when TnC is complexed to TnI.

Cecchi: I am sure that you know that, in intact fibers and single myofibrils, relaxation has two phases: a slow linear phase followed by a much faster exponential one. How do your results compare with those of Tesi et al. (Tesi et at. Biophys. J. 83: 2142-2151, 2002) in single myofibrils?

Rall: We only see a slow linear phase followed by an experiential phase when the concentration of Pi is lowered. Under these conditions, the relaxation time course is qualitatively similar to that observed in single myofibrils.

Stehle: Tesi et al. showed that the rate of rapid exponential force decay during relaxation is extremely sensitive to the level of final force reached at the end of relaxation. There was about 5-10 fold decrease in this rate when they increased the final force from resting levels to about 30% of the maximum tension. You do not see any dependence of the relaxation rate on the post photolysis force at all. Do you have any idea why?

Rall: I haven't shown the data, but it appears in a paper of ours in press in the Journal of Physiology. The rate of relaxation was not influenced by post-photolysis force under the conditions of these experiments. But we did not do a systematic study of the effects of post photolysis force on relaxation rate.

Maughan: Jack, I am interested in your strategy for choosing the second point mutation in the double mutants, which have an accentuated phenotype. Can you speculate whether a third mutation could make the filaments always "on" or always "off" with respect to calcium activation? How you would go about picking the site to mutate?

Rall: Our collaborator Larry Smillie engineered these mutations, and we determined that they had an increased and decreased Ca^{2+} sensitivity and Ca^{2+} dissociation rates. As such we thought that they would be useful to examine the effects of Ca^{2+} dissociation from TnC on relaxation rate in muscle fibers.

Curtin: When there is slow dissociation of Ca from troponin (as in your modified TnC's) and slow relaxation, do you think relaxation is slow because the bridges re-attach (rather than staying on for longer time)?

Rall: I don't know for certain, but we think that the relaxation rate is due to the altered Ca^{2+} dissociation from TnC. I will have to think about that.

Saeki: What is the Pi concentration under the low Pi conditions that you used? Can you fit the time course of relaxation (second phase) following the initial phase by a double exponential function?

Rall: We haven't measured the Pi concentration but have utilized the same conditions as Tesi et al. They determined that the Pi concentration was less than 10μM. The fast phase of relaxation in skeletal muscle could be adequately fit with a single exponential.

FORCE, SARCOMERE SHORTENING
VELOCITY AND ATP-ASE ACTIVITY

Henk E.D.J. ter Keurs[*], Nathan Deis, Amir Landesberg[+], The-Tin
T. Nguyen, Leonid Livshitz[+], Bruno Stuyvers and Mei Luo Zhang

ABSTRACT

We have tested the hypothesis that the transition rate (G) of the cardiac XB
from the strong force generating state to the weak state is a linear function V of the
sarcomere (V_{SL}); furthermore, we tested whether the ATPase rate of the two isoforms
of myosin can be held responsible for the difference between V_0 of rat cardiac
trabeculae containing V1 isomyosin versus those containing V3 isomyosin.
Methods: V1 isomyosin was induced by thyroid hormone treatment of the rats for 2
weeks, V3 isomyosin by PTU treatment for 1 month. Force was measured with a
strain gauge in trabeculae from the rat right ventricle in K-H solution ($[Ca]_o$=1.5 mM,
25° C). Sarcomere length (SL) was measured with laser diffraction techniques.
Twitch force at constant SL, and the force response to shortening at constant V_{SL} (0-8
μm/s; ΔSL 50-100 nm) were measured at varied time during the twitch. **Results:** The
force response to shortening consisted of a fast initial exponential decline (τ = 2 ms)
followed by a slow decrease of F. The instantaneous difference (ΔF) between
isometric force (F_M) and the declining force depended on shortening duration (Δt),
V_{SL} and instantaneous F_M: $\Delta F = G_1 \bullet F_M \bullet \Delta t \bullet V_{SL} \bullet (1 - V_{SL}/V_{MAX})$,
where V_{MAX} is the unloaded V_{SL} and G_1 was 6.15 ± 2.12 μm^{-1} (mean ± s.d.; n=6).
ΔF/F_M was independent of the time onset of shortening. G_1 of V1 and V3 trabeculae
did not differ. V_0 of V1 and V3 trabeculae differed 2-2.5 fold, as did both the
ATPase rate and the velocity of actin sliding in a motility assay of the myosin

[+]Departments of Biomedical Engineering, Technion -Israel institute of Technology,
Haifa Israel
[*]Departments Medicine, Physiology and Biophysics, Faculty of Medicine, University
of Calgary, 3330 Hospital Dr. NW T2N 4N1 Calgary CANADA

purified from V1 or V3 hearts. The temperature dependence of the ATPase rate (Q10: 4.03 and 4.33, respectively; n.s.) was similar to that of V_0 that has previously been reported for predominantly V1 trabeculae [2]. Cross-linking of actin to myosin with the short chain cross linker EDC increased the ATPase rate of the two isomyosins (200-fold and 600-fold respectively) to exactly the same final level and reduced their Q10 by 50 %. **Conclusion:** The linear interrelation between ΔF and V_{SL} is consistent with feedback, whereby XB kinetics depends on V_{SL}. This feedback provides an integrated description of cardiac muscle mechanics and energetics. The results, also, suggests that it is unlikely that the hydrolytic domain of the cross bridge determines V_0 and warrant ongoing experiments to investigate the role of the actin binding domain of the XB in cardiac sarcomere kinetics. In order to further investigate the role of the actin binding domain, we have expressed chimeric cardiac myosin, co-assembled with MLC, by mutual substitution of actin binding loop on α MHC and β MHC.

1. INTRODUCTION

Energy consumption in the cardiac muscle is characterized by two basic phenomena: 1. The well known linear relationship between energy consumption by the sarcomere and the generated mechanical energy [3-5] and 2. The ability to modulate the generated mechanical energy and energy consumption to the various loading conditions, as is manifested by the Frank-Starling Law [6] and the Fenn effect [7-9].
These phenomena are analyzed here based on the intracellular control of contraction. Rall [9] suggested that there is some internal feedback in active muscle whereby the total energy liberation is regulated by the loading conditions. What is the feedback mechanism? This question is still unresolved.
Furthermore, Barany's classical study has shown that the unloaded velocity of muscle shortening (V_0) the actomyosin ATPase rate in solution co-vary linearly [10]. This observation held true for a wide range of values of V_0 and ATPase rate in both skeletal [10] and cardiac muscle [11,12]. Also, the temperature dependence of V_0 of rat cardiac trabeculae ($Q_{10} = 4.6$) [13] is similar to that of the maximal actin-S_1 ATPase rate ($Q_{10} = 4.6$) [14]. The close correlation between V_0 and the maximal ATPase rate suggests that the same reaction step(s) in the actomyosin ATPase cycle may limit both V_0 and the maximal ATPase rate, and that this step may have a Q_{10} close to 4.6. The nature of this step is still not clear, although APD release has been put forward as a candidate [14].

1.1. Coupling Ca^{2+} binding with cross bridge cycling

The model [15-17] describes the intracellular control mechanisms of contractile filament contraction based on the structural and biochemical model of XB cycling [18-21] and couples the kinetics of Ca^{2+} binding to troponin with the regulation of XB cycling. The basic assumptions underlying the model are detailed elsewhere [15-17,22] but are summarized here:
1. The regulatory unit (regulated actin) consists of a single regulatory troponin protein complex with fourteen adjacent actin molecules [21].
2. The XB cycle is a repeated oar-like cycle between weak and strong conformations that differ in their structure [20]. The XB transitions between weak and strong

conformations are described by the biochemical rate limiting steps, which are related to nucleotide binding and release [20]. Force is produced only by the strong conformation. XB turnover from the weak to the strong conformation relates to ATP hydrolysis and phosphate release [18, 20]. Thus, energy consumption is proportional to the total XB turnover from the weak to the strong conformation.

3. The individual XB acts like a (pseudo) Newtonian viscous element: the average force-velocity relation of a single XB is linear [1].

4. Ca^{2+} binding to troponin low affinity sites regulates the activity of actomyosin ATPase [19], which is required for XB transition to the strong conformation [20]. Thus, Ca^{2+} binding to troponin regulates XB recruitment and the energy consumption by the sarcomere.

5. Ca^{2+} can dissociate from troponin before the XB turnover to the weak conformation; XBs can generate force without having bound Ca^{2+} on the adjacent troponin [23].

6. Control of sarcomere function includes a positive and a negative feedback mechanism: (i) Positive feedback, denoted as cooperativity, whereby the affinity of the troponin for Ca^{2+}, and hence the rate of XB recruitment, depends on the number of force producing XBs; (ii) Negative feedback, denoted as the mechanical feedback, whereby the filament sliding velocity, or the XB strain rate, determines the rate of XB weakening (Figure 1). Cooperativity derives from the interaction between the XBs and the sarcomere regulatory proteins; it regulates XB recruitment and plays a key role in the regulation of the force-length relationship (FLR) in cardiac muscle. Cooperativity may also exist in slow skeletal muscle, which has a similar troponin T as that found in cardiac muscle [24]. The proposed mechanism stems from analyses of skinned cardiac fiber data [16], substantiated by simulation of intact fiber function [15,17] and analysis of energy consumption in cardiac muscle [22,25].

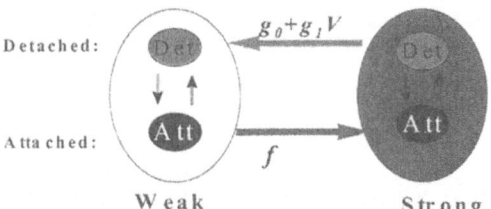

Figure 1: Proposed Cross bridge dynamics as determined by the kinetics of cross bridge cycling between the weak and strong conformations and the kinetics of cross bridge attachment-detachment. The weakening rate of the cross bridge is proportional to the velocity of sliding of actin along myosin.

The negative feedback describes the effect of V_{SL} on XB kinetics. It is based on the biochemical model of XB cycling of Eisenberg and Greene [20], who have suggested that V_{SL} affects the rate of XB weakening. The rate of XB weakening is a linear function of V_{SL}, which was substantiated [17] by the analytical derivation of the FVR in agreement with the well established, experimentally derived, Hill's equation [26].

These assumptions allow the construction of a kinetic model that corresponds to the various possible states of the regulated actin and the associated cross bridges, and allows coupling Ca^{2+} binding to Troponin-C with the regulation of XB cycling [15-17,22,25]. The combined effect of the feedback loops regulates sarcomere dynamics and explains a wide spectrum of phenomena including the Frank Starling Law [16,22], the effect of load on both free Ca^{2+} transients [17,27] and the end-systolic pressure volume relationship of the left ventricle (ESPVR) [15], and the linear relationship between energy consumption and the generated mechanical energy [28].

1.2. Cross bridge Dynamics.

Theoretical studies [17,25] have suggested that XB dynamics is determined by two kinetics, a slow and a fast kinetics, and the shortening velocity (V_{SL}) has two different effects on these kinetics and on force generation by the XBs. The first process describes the fast kinetics of the physical interactions between the myosin and the actin and are denoted as XB attachment-detachment. These kinetic properties determine the unitary force, i.e. the average force per XB, (F_{XB}). The shortening velocity decreases the unitary force as was shown by de Tombe and ter Keurs [1]. Hence the XB is described as a pseudo-viscous element:

$$F_{XB} = \overline{F} - \eta \cdot V_{SL} \qquad (1)$$

where $\overline{F}$ is the isometric unitary force and η describes the pseudo viscous property of the XBs [1]. Note that this equation applies to the steady-state, since it was measured experimentally during steady shortening [1]. However, since the kinetics of XB attachment-detachment are very fast – in the order of hundred and thousand per second [29-31] and has a time constant of several milli-seconds (see below), a steady-state is reached within a few milliseconds.

The second process describes XB cycling between weak and strong biochemical conformations. The kinetics of this process is related to the ATPase cycle and to the kinetics of nucleotides binding and release [20,31,32]. Sarcomere shortening velocity determines the rate of XB weakening (G), i.e., the rate of XB turnover from the strong to the weak state:

$$G = G_0 + G_1 \cdot V_{SL} \qquad (2)$$

where G_0 is the rate of XB weakening in the isometric regime, and G_1 is the mechanical feedback coefficient that describes the effect of the filament shortening velocity on the rate of XB weakening, and has the units of [1/m] [15,17,25]. The dominant effect of shortening, at moderate shortening velocities, on force development is attributed to its effect on the rate of XB weakening (Eq. 2). During shortening at velocities close to V_{MAX}, the viscous XB-property becomes dominant (Eq. 1).

1.3. The Force-Velocity relationship

We have shown [17,25] that the mechanical feedback provides the analytical solution, based on XB dynamics, for the experimentally established Hill's equation for the FVR:

$$V_{HILL} = b_H \frac{F_M - F}{F + a_H} = \frac{G_0(F_M - F)}{\left\{ \left(G_1 + \frac{1}{L_S}\right)F + \frac{G_0 F_M}{V_{MAX}} \right\}} \qquad (3)$$

where a_H and b_H denote Hill's coefficients, force is the isotonic force, F_M is the peak isometric force, L_S is the length of the single overlap between the actin and myosin filaments and V_{MAX} is the unloaded shortening velocity, which is determined by the pseudo-viscoelastic properties of the XBs ($V_{MAX} = \overline{F}\big/\eta$). The Hill's parameters are given by [17]:

$$a_H = \frac{G_0}{G_1 + L_S^{-1}} \cdot \frac{F_M}{V_{MAX}} = b_H \cdot \frac{F_M}{V_{MAX}} \qquad\qquad b_H = \frac{G_0}{G_1 + L_S^{-1}} \cong \frac{G_0}{G_1} \qquad (4)$$

Equation (4) provides the physiological meaning for the experimentally derived Hill's constant a_H and b_H. The dependence of V and V_{MAX} on the SL, the free Ca^{2+}, the time during the contraction, the rate of XB turnover and the internal load, are in agreement [17] with published experimental observations [1,33]. The curvature of the FVR is inversely proportional to G_1, The smaller G_1 the shallower is the curvature of the FVR. Since power is the product of force and V, the mechanical feedback regulates the FVR as well as the generated power, since a_H/F_M is inversely proportional to G_1. Moreover, the mechanical feedback describes [17] the dependence of the force deficit [34] on the instantaneous force, shortening velocity and the duration of shortening. The force-deficit is defined as the decrease in force during shortening, compared with the isometric force, at the same instant of the twitch.

1.4. Control of Energy Conversion

The exciting result of this model is that the mechanical feedback controls the FVR and power generation [15, 17]. The mechanical feedback determines how the biochemical energy, stored in the XBs as biochemical potential energy, is converted to external work [25]. Hence, the cooperativity mechanism determines the rate of ATP consumption and the mechanical feedback determines the amount that will be converted to external mechanical energy. Numerous experimental studies have convincingly shown a linear relationship between the oxygen consumption ($\dot{Vo_2}$) and mechanical energy output, both at the whole heart level [4,5] and at the isolated fiber level [3]. The observed linear correlation between the mechanical energy and energy consumption by the XBs is conveniently and convincingly explained by the mechanical feedback [15,25]. The energy consumption is calculated from the number of XBs that turn from the weak to the strong state in the XB cycle during contraction, since each cycle from the weak to the strong conformation is assumed to require the hydrolysis of one ATP. The general equation for energy conversion is given by [25]:

$$\rho \dot{E} = W + E_{PP} + Q_\eta \qquad (5)$$

where:

$$\rho = \frac{\overline{F} \cdot V_{MAX}}{E_{ATP}} \cdot \frac{1}{G_0 + G_1 V_{MAX}} \qquad E_{PP} = \frac{G_0 V_{MAX}}{\left(G_0 + G_1 V_{MAX}\right)} \int_0^T F(t) dt$$

$$W = \int_0^T F(t) V_{SL}(t) dt \qquad Q_\eta = \eta \int_0^T \frac{F(t) V_{SL}(t)^2}{\overline{F} - \eta V_{SL}} dt$$

$$(6)$$

ρ is the efficiency of biochemical to mechanical energy conversion, E_{ATP} is the free energy of ATP hydrolysis, E is the energy (ATP) consumption by the sarcomere during the twitch, W is the external work, E_{PP} is the pseudo-potential energy which is proportional to the force-time integral (as for the isometric contraction) and Q_η

represents energy dissipation due to the viscous property of XB. Q_η represents the viscous component since it is the integral over the force multiplied by the square of the velocity and by higher orders of the velocity. Note that the efficiency index of muscle contraction, which describes the utility of the muscle as a generator of external work, depends on G_1. Equations 5 and 6 provide the theoretical explanation of energy conversion, including the Fenn Effect [7,9] and the linear relationship between energy consumption and the generated mechanical energy described by Suga and Sagawa for the cardiac muscle [4,5]. The pseudo-potential energy [22], E_{pp}, is equal to Suga's potential energy for isometric contractions, and dissipates as heat. The detailed derivation of the equation based on XB dynamics is described elsewhere [15,22,25].

1.5. Current Problems

The FVR applies to the steady-state, where load and shortening velocity are "constant". Hence, the FVR depends on the experimental approach and on the time during the contraction. Figure 2 presents the force response of isolated rat trabeculae to sarcomere shortening at various, but steady velocities. As is shown in Figure 2, a steady state of constant force and shortening velocity can be defined only for one of the force responses and for only a limited time interval. Muscle generally does not work in a steady-state condition (constant activation and constant length) and cardiac muscle never works at steady state since activation by cytosolic Ca^{2+} is transient. Hence, the FVR is far from explaining the force control in vivo. The goal of the present work is to develop a general approach to describe the effect of shortening velocity on generated force, based on intracellular biochemical processes and the regulation of XB dynamics. The following questions are addressed:
• Can a biochemical model of XB cycling explain the effect of shortening on force?
• Can we quantify the regulation of force development by the sarcomere shortening velocity in isolated rat trabeculae containing fast (V1) and slow (V3) isoforms of myosin?
• What is the rate-limiting step for ATPase and can this rate-limiting step be held responsible for the difference in V_0 of V1 and V3 containing cardiac muscle?

2. METHODS

We studied the force response of isolated rat trabeculae from the right ventricle (RV) to controlled sarcomere shortening. Young adult Sprague-Dawley or Lewis Brown Norway Rats (3-4 Months) were used; the group of LBN rats was either treated with Thyroid hormone (T3) to promote expression of the α-myosin heavy chain (MHC) or with propyl thiouracil to render the animals hypothyroid and promote the expression of the β-MHC (3, 3', 5-triiodo-thyronine, Sigma, 30 µg/100g body weight, subcutaneous injection daily for 2 weeks; or 6-n-propyl-2-thiouracil (Sigma, 0.8g/l in the drinking water for 6 weeks). The presence of V1 (αα myosin) and V3 (ββ myosin) was verified using Gel electrophoresis over pyrophosphate tube gels [35].

2.1. Sarcomere Mechanics

Animals were anesthetized with diethyl ether; the heart was rapidly excised, perfused via the proximal aorta with a modified Krebs-Henseleit solution ($[Ca]_0$=1.5 mM, temp=25°C) and placed in a dissection dish beneath a binocular microscope. Spontaneous beating was stopped at $[K^+]_0$ 20 mM. The RV free wall was gently separated from the septum. Thin, unbranched, uniform trabeculae, running between the RV free wall and the A-V ring were selected. The trabeculae measured 50-200µm in width and 40-80 µm in thickness. Under microscopic control, the muscles were positioned horizontally in an experimental chamber. The stainless steel basket, which is connected to the silicon strain gauge (SenSonor, Norway) provided a stable mounting cradle in which the ventricular end of the trabeculae was positioned. The resonant frequency of the transducer is 10 KHz . The remnant of the tricuspid valve served as an attachment point for the motor arm of the servo-controlled motor (Cambridge technology) via a stainless steel hook. A fast servomotor was used to determine the muscle length and, thereby, SL which was measured by laser diffraction techniques. After mounting, the muscles were stretched to a SL of 2.1 µm, stimulated at 0.5 Hz, and left to equilibrate for 1 h at 25°C and 1.5 mM $[Ca^{2+}]_0$. The solutions were equilibrated with a 95% O_2 and 5% CO_2.

The striations of the muscle act as an optical grating to incident laser light (17-mW helium-neon laser, 632.8nm). Two types of detectors were used for measurement of the angle of the first order diffraction band, from which SL was calculated: 1). Scanning a 512-element photodiode array (Reticon) at 2 kHz; 2) photo-optical position detection (Schottky barrier detector; 25 kHz). The spatial resolution of both diffractometers is about 4nm.

To evaluate the effect of shortening, the force generated at constant SL (sarcomere isometric contraction) was measured, and used as a baseline for the evaluation of any deterioration in fiber function with time. Isometric contractions were achieved by stretching the trabecula with the servomotor. An exponential stretch of the muscle, which started ~22 ms after the stimulus generally provided constant SL with some adjustment for individual trabeculae of stretch amplitude or τ.

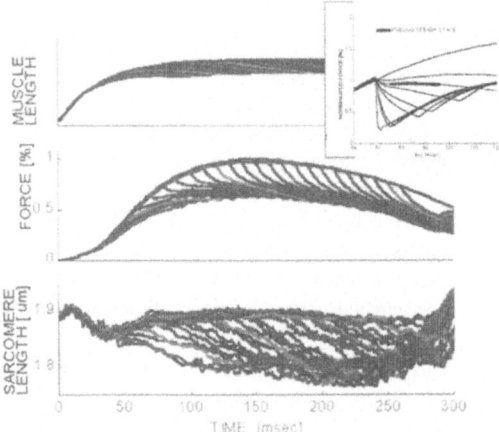

Figure 2: **Left Panel:** The effect of filament shortening at constant velocity on force. Each force response relates to a different, but constant, shortening velocity. It is clear the 'Force-Velocity relationship' is a unique example of the interaction between shortening and force. **Right Panel:** Identical sarcomere shortening at a constant velocity was imposed at various instants during the twitch. Sarcomere length was controlled by changing muscle length while the force response was observed.

The effect of shortening on force generation was quantified by imposing sarcomere shortening at constant velocity at various moments during the twitch as shown in Figure 2. This procedure was repeated with various shortening velocities. The effect of shortening velocity on force was compared to the isometric force (Figure 3).

2.2. Actin-S_1 ATPase and Cross linked Actin-S_1 ATPase

Myosin V1 and V3 were isolated from ventricles of hyperthyroid or hypothyroid rats, respectively [2,14]. Myosin subfragment-1 (S_1) was obtained from myosin by digestion with chymotrypsin [14]. S_1 was cross-linked to actin using 1-ethyl-3- [3-(dimethyl-amino)-propyl]-carbodiimide (EDC) as previously described [36]. In short, S_1 was mixed with actin, which had been activated with EDC for 2 minutes. The cross-linked mixture was incubated at 20°C for 1-60 minutes. The ATPase rate of cross-linked actin-S_1 (V_{CR}) was determined by two methods. The first method was to measure the ATPase rate of the cross-linked complex after the complex had been separated from free S_1 and actin in the mixture [37]. The drawback of this approach is that that isolation of the cross-linked complex decreases its activity by 30% [36,38]. Therefore, we also estimated the ATPase rate of cross-linked actin-S_1 from the total ATPase rate (V_T) of the cross-linked mixture using:

$$V_{CR} = [V_T - V_{NC} (1-x)]/x, \tag{7}$$

where V_{NC} was the ATPase rate of non cross-linked actin-S_1 of the same S_1 sample, measured separately. The amount of S_1 cross-linked to actin (x) in the cross-linked mixture was estimated from densitometric scanning of the SDS-polyacrylamide gels of the mixture.

All ATPase rates were assessed from the rate of inorganic phosphate release in a solution containing 135 mM KCl, 5 mM $MgCl_2$, 5 mM Na_2ATP, 1 mM DTT, 0.5 mM EGTA, 10 mM HEPES, pH 7.0. Inorganic phosphate was detected with malachite green [39]. The effect of temperature on the ATPase rate was analyzed from the Arrhenius plots in the temperature range between 15-35°C (± 0.2°C), controlled during the reaction using a thermostatically controlled water bath (model F3, Haake, Karlsrule, FRG).

2.3. Cloning of Rat Cardiac Myosin Genes

cDNA was cloned by RT-PCR of both Cardiac α- and β MHC, and the myosin light chains MLC1 and MLC2 of ventricles of LBN rats. RNA was purified from left ventricles of T3 or PTU treated rats [40] with TRIzol (Gibco BRL), and 2-5 μg was reversely transcribed to single strand DNA with Superscript II and oligo-dT primer (Gibco BRL). Myosin cDNA was amplified using primers (Gibco BRL) designed to clone full-length of two MLC and 4 or 5 overlapping segments of the two MHC [41-44]. The PCR was carried out in a total volume of 50 μl containing 20 mM Tris-HCl, 10 mM KCl, 2 mM MgSO$_4$, 10 mM (NH$_4$)$_2$SO$_4$, 0.1% Triton, 0.1 mg/ml BSA, 100 μM dNTP, 100 ng of each primer and 1.5 units of high fidelity, proofreading PfuTurbo DNA polymerase (Stratagene). PCR products were cloned into PCR Script cloning vector (Stratagene). The purified plasmid DNA was analyzed by restriction enzyme mapping and sequenced with a BigDye terminator cycle sequencing kit on an Applied Biosystems (ABI) DNA sequencer, model 377 (PERKIN-ELMER). The overlapping MHC cDNA fragments were assembled into a full length α or β MHC cDNA at the unique restriction enzyme sites of Nco I, ClaI, BstE II and Not I. A second proofreading DNA polymerase, Elongase (Gibco BRL) was used to confirm the cDNA sequences.

2.4. His-tagged MHC cDNAs with chimeric actin-binding domain

In order to produce a soluble, functional analogue of muscle heavy meromyosin (HMM), part of the C-terminal portion of the MHC was removed and a poly linker containing 6-histidine residues followed by a stop codon was added at the unique Not I site on MHC cDNAs in order to produce a HMM like myosin fragment with a His-tag at the C-terminus. The four constructs are denoted as αα-HMM, αβ-HMM, ββ-HMM and βα-HMM. This modification allowed affinity purification of myosin with nickel-bound resin. Expression of MHC and two MLCs was performed with a baculovirus expression system. Two MLC were co-purified with nickel-bound HMM indicating assembly of MHC and MLC in insect cells.
Chimeric substitution of mutual actin-binding domain on α and β MHC gene was performed with PCR technology. The actin-binding domain on the MHC is located at the highly divergent region of 50K/20K junction with amino acid sequences on α MHC: STYASADTGDSGKGKGGK and ANYAGADAPVDKGKGKAKK on β MHC gene respectively [45]. Two PCR fragments were made with the wild type α and β MHC cDNA as template. The 5'fragment was synthesized using a forward primer starting at nucleotide 1550 and a reverse mutagenic primer starting inside the region of substitution. The 3'fragment was synthesized using a reverse primer starting at nucleotide 2376 and a forward mutagenic primer starting inside the region of substitution. The two overlapping products were cleaved internally at a unique common restriction site (Age I for β MHC and Acc I for α MHC) and ligated. The resulting Chimeric PCR products were used to replace the corresponding wild type

region at the Nco I and Esp I site. DNA sequencing was performed to verify the fidelity of inserted PCR products.

2.5. Generation of Recombinant Baculovirus

Recombinant baculovirus was produced with a MaxBac 2.0 baculovirus expression system (Invitrogen). The transfer vector, pBlueBac 4.5, contained lacZ gene sequences. When co-transfected with linearized wild type Bac-N-Blue virus DNA, active β-galactosidase was produced, creating blue, recombinant plaques on agarose containing X-gal for easy selection. The cDNAs of MLC1 and MLC2 were subcloned into the pBlueBac 4.5 vector at the restriction enzyme sites of Bam HI, Xho I and Xho I, Kpn I respectively. The α HMM cDNA was subcloned into the Xho I site and β HMM cDNA was subcloned into the Kpn I and EcoR I restriction enzyme sites. The Sf9 insect cells were cultured in Grace's insect culture media supplemented with yeastolate, lactalbumin hydrolysate, 10% fetal bovine serum and antibiotics. Individual transfer vector containing myosin cDNA and linearized wild type Autographa californica nuclear polyhedrosis virus DNA were co-transfected into the cells with InsectinPlus liposomes (Invitrogen). After 4 or 5 days, well separated blue, occlusion negative plaques were selected in a plaque assay with an agarose overlay containing X-Gal. The presence of MHC and MLC cDNA in the recombinant viruses was confirmed with PCR technique using baculovirus forward and reverse PCR primers and primers within the MHC and MLC cDNAs.

2.6. Myosin Immunoblots

Sf9 cells were seeded in 6 well culture plates ($3 \cdot 10^6$ cells per plate), and infected with each of HMM and two MLC containing viruses at a multiplicity of infection of 10. The cultured cells were collected 72 hours later for SDS-PAGE and immunoblotting analysis.
A rabbit anti-pan rat cardiac myosin antibody was produced in the Hybridoma Facility at the University of Calgary. Proteins on SDS-PAGE were transferred to a nitrocellulose membrane with a semi-dry transfer system (Bio-Rad) in 25 mM Tris-HCl pH 8.5, 192 mM glycine with 20% methanol. The membranes were incubated with 3% bovine serum albumin in PBS for 30 min, blotted with rabbit anti pan-myosin serum (1:8000) for 1 hr and, after three washes, incubated with horseradish peroxidase (1:2000) conjugated to secondary antibody (New England Biolabs) for 2 hrs. Then, blots were exposed to chemiluminescence reagent (Amersham) and exposed to X-ray film (Kodak).

2.7. Purification of Myosin

Infected Sf9 cells were lysed in a high ionic strength buffer of 20 mM Tris-HCl pH 7.5, 0.5 M KCl, 10 mM $MgCl_2$, 1mM ATP, 0.5% Triton, 1% NP 40, 2 mM PMSF, 16 μg/ml benzamidine, 10 μg/ml phenanthroline and 10μg/ml leupeptin for 30 min and homogenized. Cell debris was removed by centrifugation (30 min, 10^5 g). The supernatant was incubated with Ni-NTA resin (QIAGEN) (4° C; 3 hr). The washed mixture was transferred to an open column (Phamacia); bound protein was eluted with buffer containing 0.5 M imidazole; then, separated by SDS-PAGE and Western blotting.

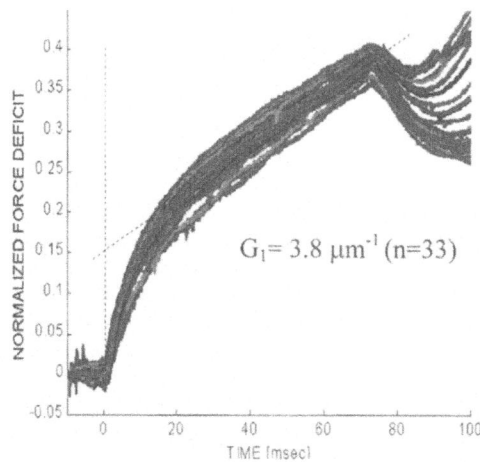

Figure 3: The dependence of the normalized force deficit, $\Delta F_N(t)$, on the duration of shortening. $\Delta F_N(t)$ was obtained during releases at constant velocity , but at different times during the twitch. The observed effect of shortening is independent of the activation level.

3. RESULTS

3.1. Force Response to Shortening

The force response to shortening was quantified by measuring the force-deficit, defined as the instantaneous force difference between the isometric force ($F_M(t)$) and the shortening force ($F_V(t)$): $\Delta F(t) = F_M(t) - F_V(t)$. The normalized force deficit $\Delta F_N(t) = (F_M(t) - F_V(t))/F_M(t)$ was calculated for all shortenings at the different times during the twitch. An identical curve was obtained for $\Delta F_N(t)$ vs. time by shifting the curves along the time axis (Figure 3) so that t=0 denotes the time of onset of shortening during the different twitches. The force response to shortening consisted two phases (Figs. 2,4): an initial fast rise of $\Delta F_N(t)$ followed by a second slow force response (Figure 3) and moderate linear increase of $\Delta F_N(t)$. The magnitude of force decline during the first phase increased with increasing shortening velocities.

As seen, $\Delta F_N(t)$, at constant shortening velocity (V_{SL}), increased in the second phase linearly with time and was independent of the time onset of shortening. Thus, the dependence of the force deficit on the length change is due to the product of velocity and time (Appendix):

$$\Delta F_N(t) = \frac{F_M(t) - F_V(t)}{F_M(t)} = G_1 \cdot V_{SI} \cdot \Delta t \cdot \left(1 - \frac{V_{SL}}{V_{MAX}}\right) + \frac{V_{SL}}{V_{MAX}} \qquad (8)$$

where V_{MAX} is the maximal unloaded velocity, ($V_{MAX} = \overline{F}/\eta$), which is determined by the viscous properties of the XB [1,17] and Δt is the time from the time onset of shortening.

The intercept of the linear approximation for the force-deficit with the force-deficit axis (Figure 3) at the time onset of shortening (Δt=0) is given by V/V_{MAX} (Eq. 8) and is determined by the decrease in the average force per XB (Eq. 1) due to shortening. The normalized force deficit was identical for all perturbations during the twitch (Figure 3); hence, the slope of the slow phase of force deficit is independent of the onset time of shortening, and the effect of shortening is independent of the activation level. Moreover, Figure-3 supported the validity of the experimental analytical methods that we have used. Although the activation level is not constant during the twitch, as is evident form the isometric contraction (Figure 2), the measurements of the normalized force deficit at various moments during the twitch is independent of the number of XBs. The analysis allows the conclusion that the slope of the slow phase of the normalized force-deficit (Figure 3) does not depend on the number of strong XBs but only on the intrinsic property of the single XB.The effect of shortening on force was tested on six trabeculae of young SD rats and the value of the mechanical feedback coefficient was: $G_1 = 6.15\pm2.12$ μm^{-1}. In experiments on trabeculae from LBN rats we observed a slightly lower value for G_1, which appeared similar for V1 (5.24 ± 0.38 μm^{-1}) and V3 (5.33 ± 0.49 μm^{-1}) containing trabeculae of hyperthyroid rat and hypothyroid rat even though the maximal velocity of sarcomere shortening differed twofold (17.6 ± 0.4 $\mu m.s^{-1}$ compared to 9.7 ± 0.4 $\mu m.s^{-1}$, respectively). This relative difference between sliding velocity of actin filaments on a myosin bed of a motility assay was confirmed when we compared V1 and V3 myosin from hyperthyroid rat and hypothyroid rat (2.45 ± 0.15 $\mu m.s^{-1}$ compared to 0.97 ± 0.1 $\mu m.s^{-1}$, respectively (20 °C).

3.2. Actin-S_1 ATPase and Cross linked Actin-S_1 ATPase

Cross-linking of S_1 to actin by EDC was shown by the formation of actin-S_1 doublets (MW ~ 150 and 170 kDa) on the SDS-polyacrylamide gel (Figure 4 A). The amount of S_1 cross-linked to actin increased linearly with incubation time (not shown). The ATPase rate of the mixture increased in proportion to the formation of cross-linked product, when about 20% of S_1 was cross-linked. The ATPase rate of cross-linked actin-S_1 was constant between 1-40 min and, then, decreased. Hence, to ensure full activation of cross-linked actin-S_1, incubation time was limited to 10 minutes.

Without cross linking, the actin-S_1 ATPase rate of V1 was about three times higher than that of V3 (Figure 4 B), but the activation energy (and Q_{10}) of ATP hydrolysis by V1 and V3 were similar (102.9 ± 4.7 (mean±s.e.m, n=4) KJ/mol versus 107.2 ± 8.6; p=0.30; Q_{10} 4.03 ± 0.25 versus 4.33 ± 0.47; p=0.30), suggesting that the rate-limiting steps of ATP hydrolysis are the same. Cross-linking S_1 to actin increased the ATPase rate of both isomyosins to the same final level (60 s^{-1}; Figure 4 B), which was substantially higher than the ATPase rate extrapolated to infinite actin concentration for both V1 and V3 (12.3 ± 0.7 s^{-1} (n=4) and 4.8 ± 0.2 s^{-1}(n=4), respectively).

Cross-linking actin to S_1 also eliminated the three-fold difference between the actin-S_1 ATPase rates of V1 and V3 and reduced their respective activation energies (71.2 ± 2.3 KJ/mol versus 77.6 ± 3.1; n.s.; (mean±s.e.m) n=4), and Q_{10} values (2.62 ± 0.11 versus 2.83 ± 0.08). The activation energy and Q_{10} of cross-linked actin-S_1

were significantly (p<0.006) lower than those of non cross-linked actin-S_1 for both isomyosins.

3.3. Chimeric Cardiac myosin expression

The isoforms of the two cardiac myosin appeared to have highly homologous amino acid sequence and the same MLCs and shared 93% amino acid identity with major differences concentrated at the loop regions [46]. [42]The MLC1 cDNA appeared to consist of 600 nucleotides while MLC2 cDNA contains 498 nucleotides. CDNA sequence alignment indicated complete match against GeneBank entries. In contrast, sequence alignment indicated complete match against random single amino acid mismatches were observed on both α (5 out of 5814 amino acids) and β MHC (8 out of 5805 amino acids) when comparing with the rat cardiac MHC sequences published previously [41,42].

The accuracy of our sequence (Labeled I) for each mismatch was confirmed by sequencing the gene products amplified with two proofreading DNA polymerases and more than two different PCR reactions with each DNA polymerase. Therefore it is unlikely the mismatches were generated by the reaction error of PCR. By comparing the MHC sequences with the MHC sequences from other species, it was found that these mismatches were compatible with the corresponding amino acid sequences in human and mice MHC molecules [47-49] (Table.1). The flanking sequences of the mismatches were mostly conserved among the different species.

Table 1. Amino acid sequence comparison of α MHC and β MHC to other cardiac MHC sequences from rat, mouse and human. (I) shows rat cardiac MHC sequences from previous published data (see text). (II) indicates the cardiac MHC sequences cloned in this lab. The amino acids which differ between our results and the published data are shown in bold and aligned with the cardiac MHC sequences from mouse (III) and human (IV). The dashes indicate missing residues. The location of the non-identical amino acids on the MHC is numbered.

α MHC

I:	DFGAA -RYLRKS	GQFIDSGKGAEK	DDVTSHMEQII	KEALIWQLTRG	DIGAKQKMHDE
II:	DFGAAAPYLRKS	GQFIDSRKGAEK	DDVTSNMEQII	KEALISQLTRG	DIGAK-KMHDE
III:	DFGAAAQYLRKS	GQFIDSRKGAEK	DDVTSNMEQII	KEALISQLTRG	DIGAK-KMHDE
IV:	DFGAAAQYLRKS	GQFIDSRKGAEK	DDVTSNMEQII	KEALISQLTRG	DIGAKQKMHDE
#	(12,13)	(740)	(1235)	(1300)	(1932)

β MHC

I:	AAFGAGAPFLR	PVYNAQVVAAY	YDYAFFSQGET	DDSEE HHMATD	EGSLDQDKKVR
II:	AAFGAAAPFLR	PVYNAEVVAAY	YDYAFISQGET	DDSEE LHMATD	EGSLEQEKKVR
IV:	AAFGAAAPFLR	PVYTPEVVAAY	YDYAFISQGET	DDSEE LHMATD	EGSLEQEKKVR
#	(12)	(137)	(313)	(329)	(1039,1041)
I:			VNDLTRQRAKL	FDKILVEWKQK	LERMKNNMEQT
II:			VNDLTSQRAKL	FDKILAEWKQK	LERMKKNMEQT
IV:			VNDLTSQRAKL	FDKILAEWKQK	LERMKKNMEQT
#			(1275)	(1454)	(1784)

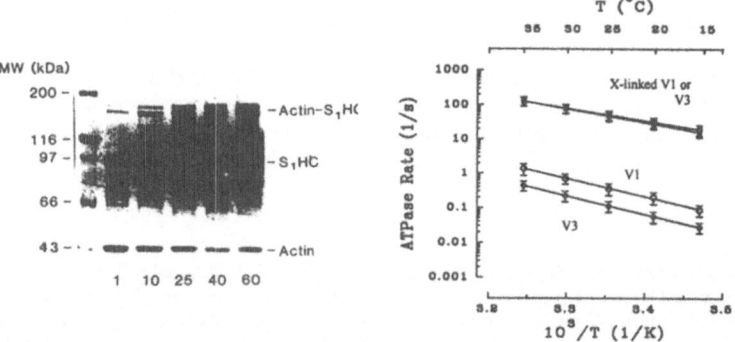

Figure 4: Cross-linking S_1 to actin (S1-HC shown in panel A at varied time in min after reaction with EDC) increased the ATPase rate of both isomyosins (shown in an Arrhenius plot in panel B). The increase in the rate of V3 was three-fold greater than that of V1, i.e. 600-fold. The cross-linked ATPase rate (60 sec^{-1} at 25°C) was substantially higher than the maximum actin-S_1 ATPase rate extrapolated to infinite actin concentration for V1 and V3. The cross-linked actin-S_1 ATPases of V1 and V3 had similar Q_{10} values (2.62±0.11 versus 2.83±0.08), which was substantially (p<0.006) lower than those of non cross-linked actin-S_1 for both isomyosins.

3.4. Expression of Recombinant Cardiac Myosin in Sf9 Cells

All 4 constructs showed distinct expression of the HMM-like MHC fragment containing the S_1 and S_2 regions of myosin, with smaller MW (144Kda) when compared with native myosin from rat ventricle. All 4 recombinant viruses expressed entire reading frame of HMM cDNA as indicated by anti-His antibody test (data not shown). Blotting with anti-pan myosin serum indicated that HMM and two MLCs were expressed in Sf9 cells co-infected with three recombinant viruses (data not shown). Affinity purification of expressed HMM with the Ni-NTA resin yielded 40-50 µg of HMM per 10^8 cells. MLC1 and MLC2 were co-purified with the resin binding HMM as indicated by immunoblot analysis (Figure 5). This result suggests that MHC and MLC are assembled in the insect cells to form a whole myosin molecule.

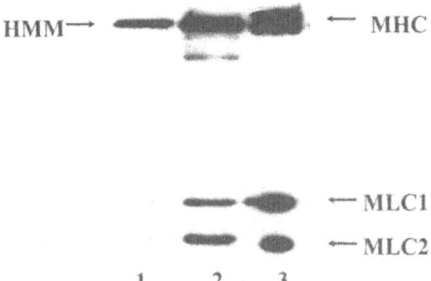

Figure 5. Immunoblots of Ni-NTA purified HMM separated on 10% SDS-PAGE gel and blotted with anti-pan myosin serum. HMM, HC and MLC positions are indicated by arrow mark. Lane 1, HMM elute; lane 2, concentrated HMM elute; lane 3, myosin from rat ventricle.

4. DISCUSSION

4.1. The effect of shortening on Cross bridge dynamics

This study provides new insights into the mechanisms affecting muscle contraction. These data suggest that XB dynamics is described by two mechanisms comprising a fast physical kinetics of XB attachment-detachment and slow biochemical XB cycling between the strong force generating conformation and the weak conformation. The fast kinetics relates to the viscoelastic property of the XB described by de Tombe and ter Keurs [1] for steady shortening. It also relates to Huxley's T_1-T_2 plots derived from quick release and to the present exponential response during the fast phase (Figure 3). These kinetic properties determine the average force per XB. The slow kinetics relates to the kinetics of nucleotides binding and release. These kinetic properties determine the number of strong XBs and the duty cycle of the strong XBs by a mechanical feedback of the shortening velocity on the biochemical rate of XB weakening. This effect of shortening on the weakening kinetics is described by the parameter G_l. Interestingly G_l appeared in this study similar for myosin of fast (V1) and slow cardiac muscle (V3), even though the maximal velocity of shortening of V1 and V3 differed twofold. The feedback loop provides an analytical solution for the force-velocity relationship and for shortening dependent power output by cardiac muscle as we have discussed elsewhere [28].

4.2. The Maximal Velocity of Shortening and Actin-S$_1$ ATPase

We also investigated the factors, which determine the maximal shortening velocity (V_o) of cardiac muscle. V_o clearly correlates with the maximal ATPase rate [11,12,38] differ two to three-fold between the fast (V1) and slow (V3) isoform of ventricular myosin. It has been shown that the amino-acid sequences of myosin S$_1$ heavy chains encoded for V1 (α) and V3 (β) are essentially similar except for the amino terminus and the binding domains for actin, ATP and the essential light chain [50]. This observation is important, because it limits the number of possible mechanisms underlying the difference in myosin kinetics to the four amino-acid domains on S$_1$, which distinguish V1 and V3.

In order to differentiate the contribution of binding and dissociation of S$_1$ to/from actin from the kinetic properties of the hydrolytic domain on S$_1$ itself, we have cross-linked actin to S$_1$ at the actin-binding site of myosin and then measured the ATPase rate of cross-linked actin-S$_1$ at varied temperature for both V1 and V3 isomyosin of rat ventricles. This type of cross-linking will eliminate the two relevant steps in the normal actin-S$_1$ ATPase cycle, i.e. binding and dissociation of S$_1$ to and from actin. Our results clearly suggest that the actin activated myosin ATPase rate of rat cardiac muscle in solution is not limited by the hydrolytic domain of myosin but is limited by the actin binding domain of myosin. Therefore, the rate-limiting step must be binding of myosin to and/or dissociation of myosin from actin.

Cross-linking S$_1$ to actin increased the ATPase rate of S$_1$ of V1 and V3 isomyosin to the same final level, i.e. 200-600 fold higher, respectively, than that observed when S$_1$ interacted reversibly with actin (2 μM) in solution (Figure 4 B). The cross-linked ATPase rate was about five and twelve times higher than the actin-S$_1$ ATPase rate at infinite actin concentration for V1 and V3. Since it has been shown that EDC does not spuriously increase the ATPase rate of isolated S$_1$ [36], the observed high ATPase rate of cross-linked actin-S$_1$ proves that the hydrolytic domain of S1 does not limit ATP hydrolysis by S1 interacting with actin in solution.

The main difference between non cross-linked and cross-linked S$_1$ in solution is the ability of S$_1$ to bind to and dissociate from actin. Since the domain of S$_1$ which is cross-linked to actin is also involved in actin binding and detaching [51], the ATPase rate must be controlled by the actin binding domain through the steps of myosin binding to and detaching from actin. So, it is reasonable to assume that the difference between the actin-S$_1$ ATPase rates of non cross-linked V1 and V3 is also due to a difference in actin binding and/or dissociation by V1 compared to V3. One would expect the difference to disappear following cross-linking if the kinetics of hydrolysis would be the same in the two isoenzymes. This is indeed shown in Figure 4. The proposal that the domain on S$_1$ which interacts with actin is responsible for the kinetics of ATP hydrolysis of actomyosin in solution is consistent with studies [52] of the chimerically altered actin-binding domain on myosin which have suggested that the actin-binding domain controls the ATPase rate of the enzyme complex in solution.

The temperature dependence of the ATPase rate of V1 was similar to that of V3 (Q_{10}= 4.0-4.3, consistent with [14]). This was also true following cross-linking, but the Q_{10} values were substantially reduced by cross-linking to 2.6-2.8 (Figure 4). The difference in temperature dependence between S_1 in solution and cross-linked S_1 is consistent with the proposal that cross-linking eliminates the rate-limiting step caused by binding and/or dissociation of myosin to actin, and that this rate-limiting step has a high activation energy (Q_{10}= 4.0-4.3). The low Q_{10} and high ATPase rate of both V1 and V3 isomyosin following cross-linking suggest that the ATPase rate by the hydrolytic domain on both V1 and V3 is rapid and has a low activation energy.

If the rate of ATP hydrolysis is limited by the binding/dissociation of myosin to/from actin, the question arises: is it possible to decide whether binding or dissociation is rate limiting. This question cannot be answered on the basis of the data in this study. Both binding and dissociation have been argued to be too fast to limit the rate of ATP hydrolysis (for review see [53]. For example the maximum rate of myosin detachment from rigor by ATP in skinned cardiac fibers [54] is four orders of magnitude higher than the ATPase rate during an isometric contraction [54,55]. However, in the presence even of low concentrations of ADP, the second order rate constant of detachment decreased 50-fold [56,57]. Similar observations have been made on actin-S_1 in solution [14,58]. Of course, the relevant ATPase rate in the sarcomere with which one would want to compare the actomyosin ATPase rate in solution, is that of the maximally activated muscle during shortening at zero load (V_o). The ATPase rate for rat cardiac muscle under these conditions is ~40/s, as can be estimated on the basis of biochemical data from Stienen *et al.* [55] and on the basis of stiffness measurements which have shown that during unloaded shortening only a small fraction of cross bridges are attached (12 % compared to the isometric contraction [1]). This value approaches the cross-linked ATPase in the present study (Figure 4) suggesting that the cross-linked ATPase may reflect the maximal ATPase rate in the sarcomere. So it is reasonable to propose that dissociation of myosin from actin at normal cytosolic [ATP] and [ADP] may be slow enough to limit the actin-S_1 ATPase rate and V_o in cardiac muscle. If dissociation of myosin from actin following binding of ATP limits the overall rate of ATP hydrolysis, it follows that this step has to have a high Q_{10}, i.e. similar to that of the overall ATPase in solution. It has been suggested indeed the dissociation step has a high Q_{10} at low temperature [59]. Whether this is true for rat cardiac muscle at higher temperatures has not been studied yet.

A.F. Huxley was the first to propose that the rate of cross-bridge detachment (g) limits the unloaded velocity of shortening (V_o) [60]; Huxley's model has been supported by many studies (for reviews see [61,62]). Studies from this laboratory have shown that the temperature dependence of V_o in cardiac muscle (Q_{10}= 4.6) is similar to the temperature dependence of actomyosin ATPase in solution as observed in the present study. This close correspondence suggests on the one hand that the rate limiting step for V_o and for ATPase in solution is the same i.e. dissociation of myosin from actin, and makes it on the other hand unlikely that steps with a low Q_{10} such as P_i or ADP release control V_o [14].

The present data provide an important reconciliation of the biochemical and mechanical data, by their support for a model in which both the myosin ATPase rate and the maximal velocity of shortening are limited by the rate of detachment of the cross bridge from actin. Therefore, the kinetic difference between fast isomyosin (V1) and slow isomyosin (V3) may well reside in only a stretch of several amino acids in the actin binding domain which is one of the four domains which distinguish V1 and V3, for which we have now successfully expressed a chimera (Figure 5).

ACKNOWLEDGEMENT

This study was financed by Grants from the Medical research council (MRC, Canada) and by the distinguished Yigal Allon Grant (to A.L., Israel). This research was also supported by the fund for promotion of research at the Technion.

REFERENCES

1. P. P. de Tombe and H. E. D. J. ter Keurs, An internal viscous element limits unloaded velocity of sarcomere shortening in rat myocardium. *J. Physiol.* **454**, 619-642 (1992).
2. P. P. de Tombe and H. E. D. J. ter Keurs, Lack of effect of isoproterenol on unloaded velocity of sarcomere shortening in rat cardiac trabeculae. *Circ. Res.* **68**, 382-391 (1991).
3. R. Hisano and G. Cooper IV, Correlation of force-length area with oxygen consumption in ferret papillary muscle. *Circ. Res.* **61**, 318-328 (1987).
4. K. Sagawa, L. Maughan, H. Suga, and K. Sunagawa, *Cardiac Contraction and The Pressure-Volume Relationship.*, Oxford University Press, Oxford (1988).
5. H. Suga. Ventricular energetics. *Physiol. Rev.* **70**, 247-277. (1990).
6. D. G. Allen and J. C. Kentish, The cellular basis of the length-tension relation in cardiac muscle. *J. Mol. Cell Cardiol.* **17**, 821-840 (1985).
7. W. O. Fenn. A quantitative comparison between the energy liberated and the work performed by the isolated sartorius muscle of the frog. *J. Physiol..Lond.* **58**, 457-497. (1923).
8. F. F. H. M. Mommaerts, I. Seraydarian, and G. Marechal, Work and mechanical change in isotonic muscular contractions. *Biochem Biophys Acta* **57**, 1-12 (1962).
9. J. A. Rall, Sense and nonsense about the Fenn effect. *Am. J. Physiol* **242**, H1-H6 (1982).
10. M. Bárány, ATPase activity of myosin correlated with speed of muscle shortening. *J. Gen. Physiol.* **50**, 197-216 (1967).
11. K. Schwartz, Y. Lecarpentier, J. L. Martin, A.-M. Lompré, J.-J. Mercadier, and B. Swynghedauw, Myosin isoenzymic distribution correlates with speed of myocardial contraction. *J. Mol. Cell Cardiol.* **13**, 1071-1075 (1981).
12. C. Delcayre and B. Swynghedauw, A comparative study of heart myosin ATPase and light chain subunits from different species. *Pflugers Arch.* **355**, 39-47 (1975).
13. P. P. de Tombe and H. E. D. J. ter Keurs, Force and velocity of sarcomere shortening in trabeculae from rat heart. Effects of temperature. *Circ. Res.* **66**, 1239-1254 (1990).
14. R. F. Siemankowski, M. O. Wiseman, and H. D. White, ADP dissociation from actomyosin subfragment 1 is sufficiently slow to limit the unloaded shortening velocity in vertebrate muscle. *Proc. Natl. Acad. Sci. USA* **82**, 658-662 (1985).
15. A. Landesberg, End-systolic pressure-volume relationship and intracellular control of contraction. *Am. J. Physiol* **270**, H338-H349 (1996).
16. A. Landesberg and S. Sideman, Coupling calcium binding to troponin-C and cross-bridge cycling in skinned cardiac cells. *Am. J. Physiol.* **266**, H1260-H1271 (1994).
17. A. Landesberg and S. Sideman, Mechanical regulation in the cardiac muscle by coupling calcium kinetics with crossbridge cycling; a dynamic model. *Am. J. Physiol.* **267**, H779-H795 (1994).
18. B. Brenner and E. Eisenberg, Rate of force generation in muscle: correlation with actomyosin ATPase activity in solution. *Proc. Natl. Acad. Sci. USA* **83**, 3542-3546 (1986).
19. J. M. Chalovich and E. Eisenberg, The effect of troponin-tropomyosin on the binding of heavy meromyosin to actin in the presence of ATP. *J. Biol. Chem.* **261**, 5088-5093 (1986).
20. E. Eisenberg and T. L. Hill, Muscle contraction and free energy transduction in biological systems. *Science* **227**, 999-1006 (1985).
21. S. S. Lehrer, The regulatory switch of the muscle thin filament: Ca2+ or myosin heads? *Journal of Muscle Research and Cell Mobility* **15**, (1994).
22. A. Landesberg and S. Sideman, Regulation of energy consumption in cardiac muscle: analysis of isometric contractions. *Am. J. Physiol* **276**, H998-H1011 (1999).
23. J. N. Peterson, W. C. Hunter, and M. R. Berman, Estimated time course of CA2+ bound to troponin c during relaxation in isolated cardiac muscle. *Am. J. Physiol.* **260**, H1013-H1024 (1991).
24. T. L. Hill, Two elementary models for the regulation of skeletal muscle contraction by calcium. *Biophys. J.* **44**, 383-396 (1983).
25. A. Landesberg and S. Sideman, Force-velocity relationship and biochemical-to-mechanical energy conversion by the sarcomere. *Am. J. Physiol Heart Circ. Physiol* **278**, H1274-H1284 (2000).
26. A. V. Hill, The effect of load on the heat of shortening. *Proc Royal Soc B* **159**, 297-318 (1964).
27. D. G. Allen and J. C. Kentish, Calcium concentration in the myoplasm of skinned ferret ventricular muscle following changes in muscle length. *J. Physiol.* **407**, 489-503 (1988).

28. A. Landesberg, L. Livshitz, and H. E. ter Keurs, The effect of sarcomere shortening velocity on force generation, analysis, and verification of models for crossbridge dynamics. *Annals of Biomedical Engineering* **28**, 968-978 (2000).

29. B. Brenner, Rapid dissociation and reassociation of actomyosin cross-bridges during force generation: a new observed facet of cross-bridge action in muscle. *Proc. Natl. Acad. Sci. USA* **88**, 10490-10494 (1991).

30. V. Lombardi, G. Piazzesi, and M. Linari, Rapid regeneration of the actin-myosin power stroke in contracting muscle. *Nature* **355**, 638-641 (1992).

·31. L. A. Stein, P. B. Chock, and E. Eisenberg, The rate-limiting step in the actomyosin adenosinetriphosphate cycle. *Biochemistry* **23**, 1555-1563 (1984).

32. L. E. Greene, D. L. Williams, and E. Eisenberg, Regulation of actomyosin ATPase activity by troponin-tropomyosin: effect of the binding of the myosin subfragment I (S-1) ATP complex. *Proc. Natl. Acad. Sci. USA* **84**, 3102-3106 (1987).

33. M. C. G. Daniels, M. I. M. Noble, H. E. D. J. ter Keurs, and B. Wohlfart, Velocity of sarcomere shortening in rat cardiac muscle: relationship to force, sarcomere length, calcium, and time. *J. Physiol.* **355**, 367-381 (1984).

34. J. K. Leach, A. J. Brady, B. J. Skipper, and D. L. Millis, Effects of active shortening on tension development of rabbit papillary muscle. *Am. J. Physiol* **238**, H8-13 (1980).

35. J. Y. Hoh, P. A. McGrath, and R. I. White, Electrophoretic analysis of multiple forms of myosin in fast-twitch and slow-twitch muscles of the chick. *Biochem J.* **157**, 87-95 (1976).

36. D. Mornet, R. Bertrand, P. Pantel, E. Audemard, and R. Kassab, Structure of the actin-myosin interface. *Nature* **292**, 301-306 (1981).

37. T. Arata, Chemical crosslinking of myosin subfragment-1 to F-actin in the presence of nucleotide. *J. Biochem.* **96**, 337-347 (1984).

38. B. Lauer, N. Thiem, and B. Swynghedauw, ATPase activity of the cross-linked complex between cardiac myosin subfragment I and actin in several models of chronic overloading. *Circ. Res.* **64**, 1106-1115 (1989).

39. P. A. Lanzetta, L. J. Alvarez, P. S. Reinach, and O. A. Candia, An improved assay for nanomole amounts of inorganic phosphate. *Anal. Biochem.* **100**, 95-97 (1979).

40. W. H. Dillmann, Hormonal influences on cardiac myosin ATPase activity and myosin isoenzyme distribution. *Molec. Cell. Endocrin.* **34**, 169-181 (1981).

41. E. M. McNally, K. M. Gianola, and L. A. Leinwand, Complete nucleotide sequence of full length cDNA for rat alpha cardiac myosin heavy chain. *Nucleic Acids Res.* **17**, 7527-7528 (1989).

42. R. Kraft, M. Bravo-Zehnder, D. A. Taylor, and L. A. Leinwand, Complete nucleotide sequence of full length cDNA for rat beta cardiac myosin heavy chain. *Nucleic Acids Res.* **17**, 7529-7530 (1989).

43. E. M. McNally, P. M. Buttrick, and L. A. Leinwand, Ventricular myosin light chain 1 is developmentally regulated and does not change in hypertension. *Nucleic Acids Res.* **17**, 2753-2767 (1989).

44. S. A. Henderson, Y. C. Xu, and K. R. Chien, Nucleotide sequence of full length cDNAs encoding rat cardiac myosin light chain-2. *Nucleic Acids Res.* **16**, 4722 (1988).

45. T. Q. Uyeda, K. M. Ruppel, and J. A. Spudich, Enzymatic activities correlate with chimaeric substitutions at the actin-binding face of myosin. *Nature* **368**, 567-569 (1994).

46. H. V. Goodson, H. M. Warrick, and J. A. Spudich, Specialized conservation of surface loops of myosin: evidence that loops are involved in determining functional characteristics. *J. Mol. Biol.* **19**, 173-185 (1999).

47. R. Matsuoka, K. W. Beisel, M. Furutani, S. Arai, and A. Takao, Complete sequence of human cardiac alpha-myosin heavy chain gene and amino acid comparison to other myosins based on structural and functional differences. *Am. J. Med. Genet.* **41**, 537-547 (1991).

48. C. C. Liew, M. J. Sole, K. Yamauchi-Takihara, B. Kellam, D. H. Anderson, L. P. Lin, and J. C. Liew, Complete sequence and organization of the human cardiac beta-myosin heavy chain gene. *Nucleic Acids Res.* **18**, 3647-3651 (1990).

49. B. K. Quinn-Laquer, J. E. Kennedy, S. J. Wei, and K. W. Beisel, Characterization of the allelic differences in the mouse cardiac alpha- myosin heavy chain coding sequence. *Genomics* **13**, 176-188 (1992).

50. E. M. McNally, R. Kraft, M. Bravo-Zehnder, D. A. Taylor, and L. A. Leinwand, Full-Length rat alpha and beta cardiac myosin heavy chain sequences. *J. Mol. Biol.* **210**, 665-671 (1989).

51. K. Sutoh, Mapping of actin-binding sites on the heavy chain of myosin subfragment 1. *Biochemistry* **22**, 1579-1585 (1983).

52. T. Q. P. Uyeda, K. M. Ruppel, and J. A. Spudich, Enzymatic activities correlate with chimeric substitutionss at the actin-binding face of myosin. *Nature* **368**, 576-579 (1994).

53. D. Applegate, A. Azarcon, and E. Reisler, Tryptic cleavage and substructure of bovine cardiac myosin subfragment 1. *Biochemistry* **23**, 6626-6630 (1984).

54. R. J. Barsotti and M. A. Ferenczi, Kinetics of ATP hydrolysis and tension production in skinned cardiac muscle of the guinea pig. *J. Biol. Chem.* **263**, 16750-16756 (1988).
55. G. J. M. Stienen, Z. Papp, and G. Elzinga, Calcium modulates the influence of length changes on the myofibrillar adenosine triphosphatase aactivity in rat skinned cardiac trabeculae. *Pflugers Arch.* **425**, 199-207 (1993).
56. H. Martin and R. J. Barsotti, Kinetics of cardiac relaxation from rigor initiated by laser photolysis of cage-ATP. *Biophys. J.* **59**, 417a (1991).
57. H. Martin, R. Iacobacci, and R. J. Barsotti, Kinetics of cardiac muscle activation from rigor initiated by laser photolysis of caged-ATP. *Biophys. J.* **61**, 19a (1992).
58. R. F. Siemankowski and H. D. White, Kinetics of the interaction between Actin, ADP, and cardiac myosin-S1. *J. Biol. Chem.* **259**, 5045-5053 (1984).
59. N. C. Millar and M. A. Geeves, The limiting rate of the ATP-mediated dissociation of actin from rabbit skeletal muscle myosin subfragment 1. *FEBS Lett.* **160**, 141-148 (1983).
60. A. F. Huxley, Muscle structure and theories of contraction. *Prog. Biophys. Biophys. Chem.* **7**, 255-318 (1957).
61. R. Cooke, H. White, and E. Pate, A Model of the release of myosin heads from actin in rapidly contrating muscle fibers. *Biophys. J.* **66**, 778-788 (1994).
62. E. Pate, H. White, and R. Cooke, Determination of the myosin step-size from mechanical and kinetic data. *Proc. Natl. Acad. Sci. USA* **90**, 2451-2455 (1993).

DISCUSSION

Sugi: It is a good idea to use the technique of actin-myosin cross-linking in comparing the ATPase activity between V1 and V3 myosins. If I understood you correctly, the different ATPase activity between V1 and V3 can be explained in terms of duty ratio. Sugiura and I have used this explanation concerning the interpretation of our motility assay experiments on V1 and V3 (Sugiura et al., Circ. Res. 82: 1029-1034, 1998).

ter Keurs: I agree with you.

EVIDENCE FOR HIGH MECHANICAL EFFICIENCY OF CROSS-BRIDGE POWERSTROKE IN SKELETAL MUSCLE

Haruo Sugi and Tsuyoshi Akimoto [*]

1. INTRODUCTION

Muscle contraction is caused by attachment-detachment cycles between the cross-bridges on the thick filament and the thin filament coupled with ATP hydrolysis (A. F. Huxley, 1957; H. E. Huxley, 1960; Bagshaw, 1994). The mechanical efficiency, with which chemical energy derived from ATP hydrolysis is converted into mechanical work, in demembranated muscle fibers can now be estimated by measuring the amount of ATP utilized for work production, using fluorescence of a phosphate-binding protein (He et al., 1997, 1999) or NADH (Reggiani et al., 1997; Sun et al., 2001). During myofilament sliding, however, the cross-bridges not only attach to the thin filament to perform their powerstroke producing positive forces, but also produce negative forces before being detached from the thin filament (A. F. Huxley, 1957). On this basis, the overall mechanical efficiency of muscle fibers may be much smaller than that of individual cross-bridges during their powerstroke, since positive forces are always opposed by negative forces due to asynchronous cross-bridge activity.

The present work was aimed at estimating the maximum mechanical efficiency of the cross-bridge powerstroke in demembranated muscle fibers. The results obtained suggest that the maximum mechanical efficiency of the cross-bridge powerstroke may be close to unity (Sugi et al., 2003).

2. MATERIALS AND METHODS

2.1. Experimental setup

---

[*] Department of Physiology, School of Medicine, Teikyo University, Itabashi-ku, Tokyo 173-0003, Japan

Single demembranated muscle fibers (diameter, 40 - 60 μm; slack length Lo, ≤ 2.5 - 3 mm) or small bundles consisting of two to three muscle fibers were prepared from glycerinated rabbit psoas muscle (Sugi et al., 1998), and mounted horizontally between a force transducer (AE801, SensoNor, Holten, Norway) and a servo-motor (G100PD, General Scanning, Watertown, MA). Fiber cross-sectional area was measured by taking photographs of laterally illuminated fibers (Blinks, 1965). Further details of the method, including the composition of experimental solutions, have been described previously (Sugi et al., 1998). The fibers were maximally activated by photolysis of DM-nitrophen (caged Ca^{2+}) with a laser light flash (duration. 8 ns; wavelength, 350 nm; intensity, 20 mJ) from a Nd: YAG laser (DCR3, Spectra Physics).

2.2. Experimental procedures

In relaxing solution, the sarcomere length of the fibers was adjusted to 2.4 μm, at which the overlap between the thick and thin filaments was just maximum (Page and Huxley, 1963). As the extent of fiber shortening was < 15 % of the initial fiber length Lo, the number of cross-bridges interacting with the thin filament was always maximum during fiber shortening. The fibers in photolysis solution containing DM-nitrophen were exposed to air to be subjected to laser flash irradiation. The ATP concentration of photolysis solution was determined to be 220 μM. Very small rigor force (≤ 1 % of Po)

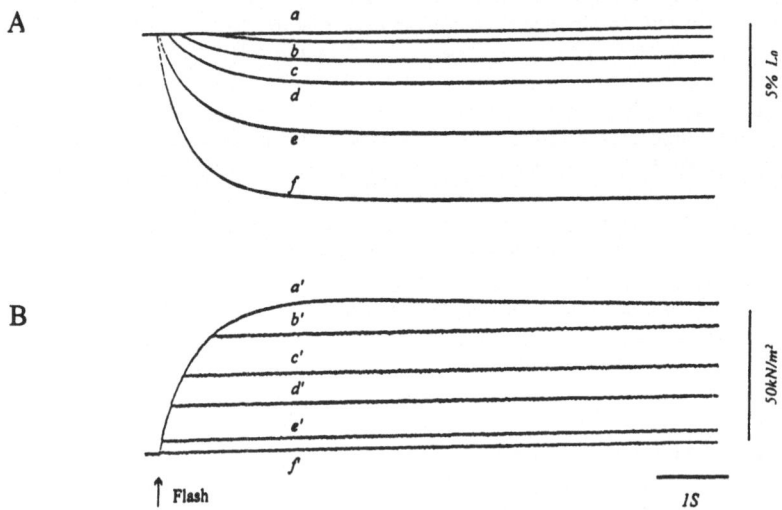

Figure 1. Laser flash-induced mechanical response. Typical fiber length (A) and force (B) changes of the preparation, contracting isometrically or to shorten shortening isotonically under five different afterloads. Length records a, b, c, d, e, and f correspond to force records a', b', c', d', e', and f', respectively. The load was Po (isometric contraction, records a - a'), 0.78 Po (records b - b'), 0.53 Po (records c - c'), 0.35 Po (records d - d'), 0.09 Po (records e - e'), and 0 (unloaded shortening, records f - f'), respectively. The length and the cross-sectional area of the fiber was 2.8 mm and 6.8 ×10⁻⁵ cm2, respectively (Sugi et al., 2003).

was always developed in the fibers immediately before flash activation, indicating that the number of ATP molecules is slightly below, but not above, that of the cross-bridges. The temperature of the space where the fibers were activated was estimated to be 4 °C (Sugi et al., 1998).

To estimate the amount of ATP (or more exactly M-ADP-Pi) utilized at 1 s after flash activation (Pu), the fibers were subjected to a quick decrease in fiber length (quick release, 1 - 2 % of Lo, complete in 1 - 2 ms) at 1 s after flash activation to drop the force to zero, and then the fiber length was clamped to allow the fibers to develop isometric force. The amount of isometric force developed (Pr, relative to the maximum isometric force Po) was taken as a measure of the amount of M-ADP-Pi remaining in the fiber at 1 s after activation. The value of Pu was obtained as $Pu = (Po - Pr)$. After a flash-induced mechanical response, the fibers were made to relax in relaxing solution. The flash activation of the fibers could be repeated 5 - 10 times at intervals of 10 min.

3. RESULTS

3.1. Laser flash-induced fiber shortening

When the fibers were activated by photoreleased Ca^{2+}, they first developed isometric force equal to the afterload P, started shortening isotonically, and then eventually stopped shortening as the fibers go into rigor state after complete exhaustion of ATP (Fig. 1). As shown in Fig. 2 A, the power output reached a peak at the early phase of fiber shortening, and then decreased with time. The higher the initial peak was, the larger

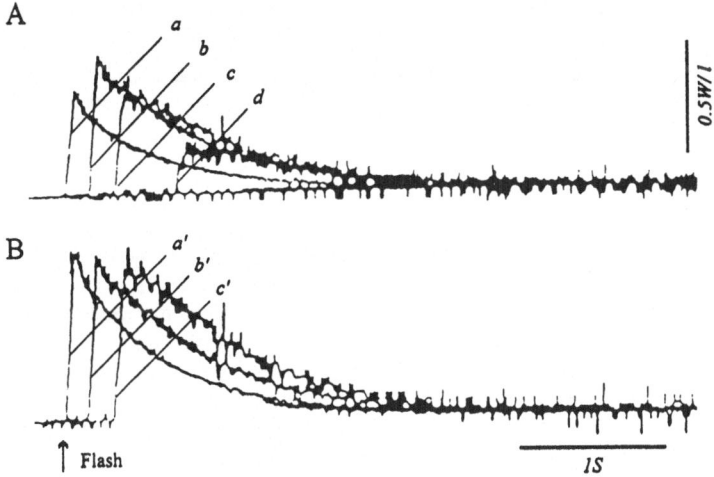

Figure 2. Power output during flash-induced fiber shortening. (A) Power output records under three different afterloads. (B) Power output records normalized relative to the peak values attained. The load was 0.09 Po (records a-a'), 0.35 Po (records b-b'), 0.53 Po (records c - c') and 0.78 Po (record d). The records were obtained from the experiment shown in Fig.1 (Sugi et al., 2003).

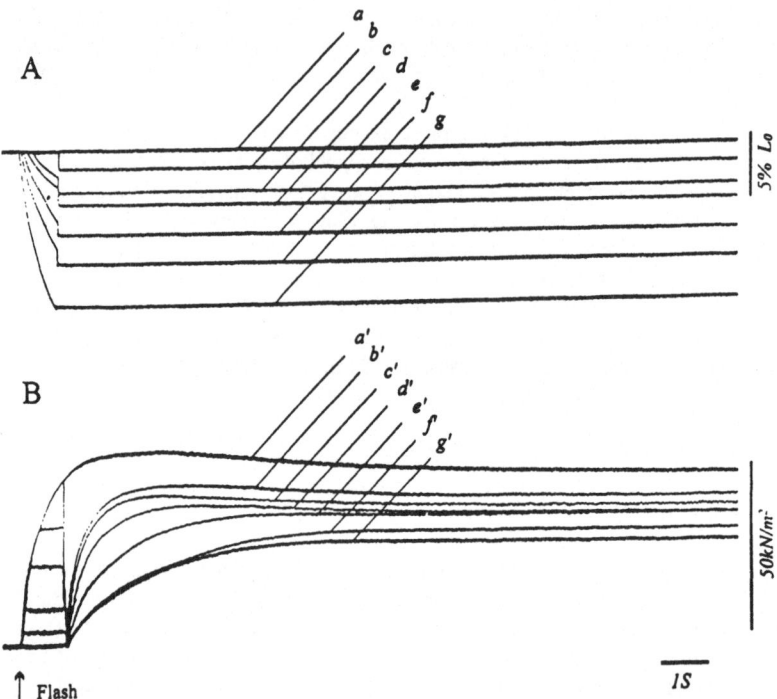

Figure 3. Fiber length (A) and force (B) changes of the preparation, which was first made to shorten isotonically under five different afterloads, and then subjected to quick releases at 1s after activation to drop the force to zero. After each release, the preparation redevelops isometric force at the decreased fiber length. Length records a, b, c, d, e, f, and g correspond to force records a', b', c', d', e', f' and g', respectively.   The load was Po (isometric condition, records b-b'), 0.63 Po (records c - c'), 0.41 Po (records d - d'), 0.20 Po (records e - e'), 0.09 Po (records f - f'), 0 Po (unloaded condition, records g - g').   Records a - a' were obtained during isometric contraction without quick release (Sugi et al., 2003).

was the area under the power output trace, i.e. the amount of work done by fiber shortening.   The power output records were almost identical when normalized with respect to their peak values (Fig. 2 B), except for the load close to Po.   The distance of fiber shortening when the power output reached a maximum did not exceed 10 nm per half sarcomere.   This suggests that, at the beginning of fiber shortening, the cross-bridges, in the form of M-ADP-Pi, start their powerstroke almost synchronously, while sensing the amount of load to determine their future energy output.

During isometric contraction, the in-phase stiffness increased approximately in parallel with isometric force, while the quadrature stiffness, reached a maximum at ∼ 0.3 s after activation, and stayed almost unchanged for the first 3 - 4 s.   This indicates no appreciable changes in number of force-generating cross-bridges during isometric force development preceding fiber shortening, since the quadrature stiffness is taken as a measure of fraction of active cross-bridges (Goldman et al., 1984).   Furthermore, during fiber shortening, no appreciable increase of internal resistance against fiber shortening

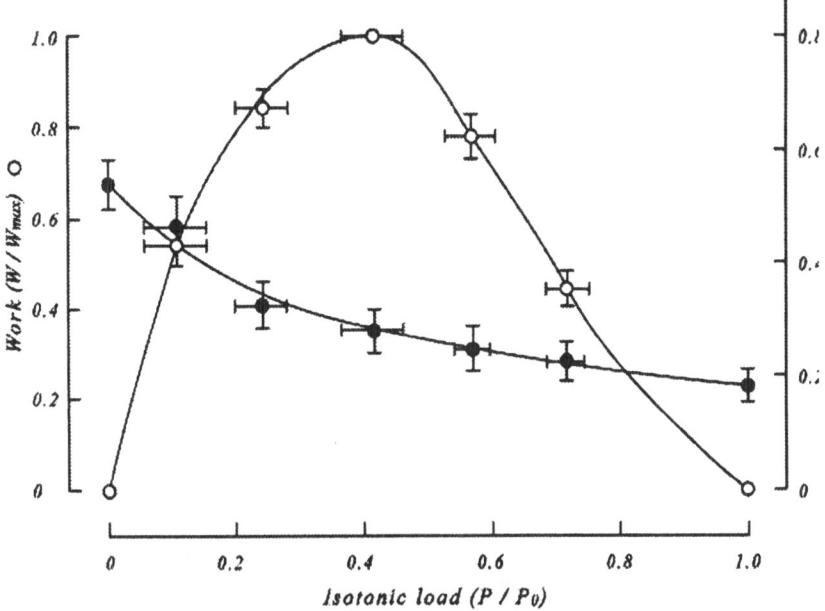

Figure 4. Dependence of the amount of ATP utilized for mechanical response (Pu, filled circles) and the amount of work done (W, open circles) on the isotonic load (P) at 1 s after flash activation. The data points were obtained from 8 different data sets (Sugi et al., 2003).

takes place at least for the first 1 - 2 s after activation (Sugi et al., 1998). It may therefore be safe to conclude that, at least for 1 - 2 s after activation, the cross-bridges may not readily form rigor links after releasing Pi and ADP.

3.2. Load-dependent amount of work done and amount of ATP utilized.

Fig. 3 shows a typical experiment, in which the fibers were activated to contract isometrically or isotonically under five different afterloads for 1 s, and then subjected to a quick release to drop the force to zero, whereon the fiber length was clamped and the fibers developed isometric force. The amount of isometric force developed after a quick release (Pr), i.e. a measure of the amount of ATP remaining in the fiber at 1 s after activation, was maximum when $P = Po$ (isometric contraction) and minimum when $P = 0$ (unloaded shortening). The amount of ATP utilized at 1 s after flash activation ($Pu = Po - Pr$) was therefore maximum during unloaded shortening ($P = 0$), and minimum during isometric contraction ($P = Po$).

On the other hand, the possibility that cross-bridges forming rigor links with the thin filaments may produce rigor force to contribute to the isometric force development after a quick release can largely be precluded by the extremely slow development of rigor force in glycerinated rabbit psoas fibers (Kobayashi et al., 1998). On this basis, the estimation of Pu value may not be influenced by rigor forces except for Pu values during isotonic shortening.   This implies that the value of Pu during isotonic shortening under small loads may be somewhat underestimated, though its extent is very small.

Fig. 4 shows the dependence of the amount of work done (W, expressed relative to the maximum value, Wmax) and the amount of ATP utilized for the whole mechanical response (Pu, expressed relative to Po) on the isotonic load (P).   The value of Pu at $P = 0$ was ~ 3 times larger than that at $P = Po$.   The value of W was maximum ($1.80 \pm 0.06 \times 10^{-8}$ J, mean $\pm$ SEM, n = 8) at ~ 0.4 Po.   The W versus P relation was bell-shaped, since W is necessarily zero at $P = 0$ and $P = Po$.

3.3. Load-dependent mechanical efficiency of individual cross-bridges.

The amount of ATP utilized for the whole mechanical response (Pu) is the sum of the amount of ATP utilized for the preceding isometric force development (Pi) and that utilized for the subsequent isotonic shortening (Ps) (see Fig. 6).   The value of Pi as a function of isotonic load were obtained by applying a quick release to isometrically contracting fibers at various times after activation and measuring the amount of force developed after each quick release (Fig. 5).   Thus, the value of Ps could be obtained by subtracting the value of Pi for a given isometric force equal to the isotonic load from Pu for the whole mechanical response.   The mechanical efficiency of individual cross-bridges (E), averaged over the period of work production, can be estimated as, $E = W / (Pu - Pi) = W / Ps$, using the results shown in Figs. 4 and 5.   The dependence of E (expressed relative to the maximum value, Emax) on the isotonic load is shown schematically in Fig. 6 together with W, Pu, Pi and Ps.   The E versus P relation was bell-shaped with a broad peak at 0.5 - 0.6 Po.

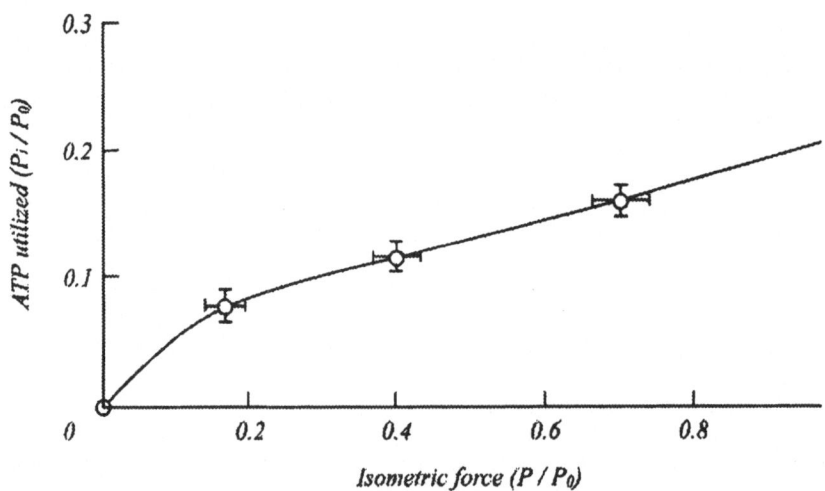

Figure 5.   Relation between the amount of ATP utilized (Pi) and the isometric force developed (P).   The data points were obtained from 5 different data sets (Sugi et al., 2003).

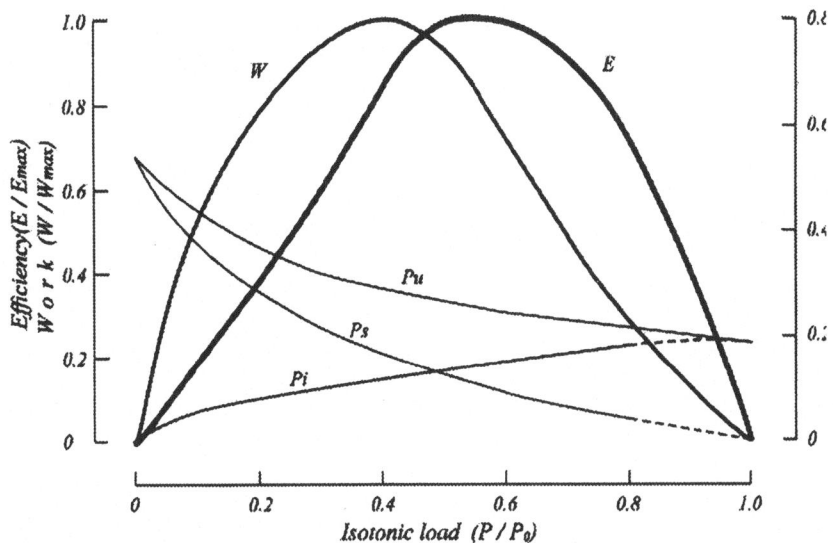

Figure 6. Dependence of the mechanical efficiency of individual cross-bridges (E) on the amount of isotonic load (P) obtained from the results shown in Figs. 4 and 5. Values are scaled to adjust the value at 1.0 P / Po to that in Fig. 4. The amount of work done (W), ATP utilized for whole mechanical response (Pu), ATP utilized for preceding isometric force development (Pi), and ATP utilized for isotonic shortening (Ps) are also shown as functions of load (Sugi et al., 2003).

3.4. Estimation of the absolute value of mechanical efficiency of individual cross-bridges

Although the mechanical efficiency of individual cross-bridges is obtained as relative values in the present study, we made a conservative estimation of its absolute value as follows. The average fiber cross-sectional area of 8 preparations, from which the data shown in Fig. 4 were obtained, was $6.1 \pm 0.1 \times 10^{-5}$ cm^2, while the fiber length was ≤ 2.5 - 3 mm. To avoid underestimation of fiber volume leading to overestimation of the efficiency, we use the maximum fiber length of 3 mm to obtain mean fiber volume of 1.8×10^{-5} cm^3. Assuming cross-bridge concentration of 200 µM (higher than widely used value of 145 or 150 µM), the amount of M-ADP-Pi immediately before flash activation is estimated to be 3.6×10^{-6} µmol $(200 \times 1.8 \times 10^{-5} \times 10^{-3}) = 3.6 \times 10^{-12}$ mol. In Fig. 7, the value of E is maximum at $P = 0.53$ Po, and the corresponding value of Ps is 0.13 Po, where Po corresponds to the initial amount of M-ADP-Pi of 3.6×10^{-12} mol. The number of ATP molecules utilized for work production is calculated to be 2.8×10^{11} $(3.6 \times 10^{-12} \times 0.13 \times 6 \times 10^{23})$. Assuming the energy released by ATP hydrolysis of 50 kJ mol^{-1} (Bagshaw, 1994; Oiwa et al., 1991), the energy available from one ATP molecule is 8.3×10^{-20} J $(50 \times 10^3 / 6 \times 10^{23})$. The energy released from ATP molecules during work production is 2.3×10^{-8} J $(2.8 \times 10^{11} \times 8.3 \times 10^{-20})$. In Fig. 6, the amount of work done at 0.53 Po is 1.6×10^{-8} J. The maximum mechanical efficiency of individual cross-bridges is therefore estimated to be $(1.6 \times 10^{-8}) / (2.3 \times 10^{-8}) = 0.7$. Since the above estimation is conservative, the actual maximum mechanical efficiency of individual cross-bridges is suggested to be 0.8 - 0.9, being close to unity.

4. DISCUSSION

4.1. Synchronized cross-bridge powerstroke on flash activation

The present work was aimed at estimating the mechanical efficiency of individual cross-bridges when they start their powerstroke synchronously. On laser flash activation, the cross-bridges, in the state of M-ADP-Pi, sequentially release Pi and ADP to build up flash-induced mechanical response, but after the product release no cross-bridges can hydrolyze ATP molecules any longer. This experimental condition may be comparable with that of quenched flow experiments, in which enzyme concentration is equal to substrate concentration to result in a single turnover for each enzyme molecule.

At the beginning of fiber shortening, the power output rose rapidly to a peak, and then decreased with time (Fig. 2). The distance of fiber shortening at the peak of power output was < 10 nm per half sarcomere. This can be taken as evidence that, at the beginning of fiber shortening, the cross-bridges start their powerstroke almost synchronously. In the present study, the period of fiber shortening was restricted to be < 1 s (Fig. 3). Since the maximum rate of ATP utilization was 0.80 s^{-1} per cross-bridge during unloaded shortening (Fig. 4), the average duration of ATP hydrolysis cycle was 1.3 s, and this value increased up to ~ 5 s with increasing load towards Po. This implies that, under large loads, a considerable fraction of cross-bridges, starting their powerstroke at the beginning of fiber shortening, would continue their ATP hydrolysis cycle over the whole period of work production. The mechanical efficiency obtained in the present study may therefore be regarded to largely reflect that of individual cross-bridges performing their powerstroke, especially under large loads.

4.2. High-mechanical efficiency of cross-bridge powerstroke

For the reasons stated above, the present results may constitute evidence that the maximum mechanical efficiency of individual cross-bridges may be very high, probably close to unity. In this connection, it is of interest to note that, in the ATP-dependent rotary motion of F_0-F_1 ATPase at the mitochondrial membrane, its mechanical efficiency has also been suggested to be close to unity (Kinosita et al., 2000).

The maximum mechanical efficiency of Ca^{2+}-activated skeletal muscle fibers has been reported to range from 0.2 to 0.46 (He et al., 1997; Reggiani et al.,1997; Sun et al., 2001), indicating that the net maximum mechanical efficiency of cross-bridges during their asynchronous activity is much smaller than the maximum mechanical efficiency of individual cross-bridges obtained in the present study. In this connection, it is of interest that, in demembranated cardiac myocytes, the maximum Ca^{2+}-activated isometric force increases by one third when the ATP concentration is reduced to 200 μM (Fabiato and Fabiato, 1975). This might result from an increased degree of synchronization of force-generating cross-bridge activity, when the ATP concentration is reduced to be nearly equal to that of cross-bridges.

The mechanism underlying the load-dependent mechanical efficiency of individual cross-bridges remains to be investigated, although it is suggested that their nucleotide affinity changes depending on the strain in the cross-bridge structure (Geeves and Holmes, 1999).

5. REFERENCES

Bagshaw, C. R., 1994, *Muscle Contraction*, Chapman & Hall, London.

Blinks, J. R., 1965, Influence of osmotic strength on cross-section and volume of isolated single muscle fibers, *J. Physiol.* **177**: 42-57..

Fabiato, A., and Fabiato, F., 1975, Effects of magnesium on contractile activation of skinned cardiac cells, *J. Physiol.* **249**: 497-517.

Geeves, M. A., and Holmes, K. C., 1999, Structural mechanism of muscle contraction, *Annual Review of Biochemistry.* **68**: 687-728.

Goldman, Y. E., Hibberd, M. G., and Trentham, D. R.,(1984, Relaxation of rabbit psoas muscle fibers from rigor by photochemical generation of adenosine-5'-triphosphate, *J. Physiol.* **354**: 577-604.

He, Z-H., Bottinelli, R., Pellegrino, M. A., Ferenczi, M. A., and Reggiani, C., 2000, ATP consumption and efficiency of human single muscle fibers with different myosin isoform composition, *Biophys. J.* **79**: 945-961.

He, Z-H., Chillingworth, R. K., Brune, M., Corrie, J. E. T., Trentham, D. R., Webb, M. R., and Ferenczi, M. A., 1997, ATPase kinetics on activation of rabbit and frog permeabilized isometric muscle fibers: a real time phosphate assay, *J. Physiol.* **501**: 125-148.

He, Z-H., Chillingworth, R. K., Brune, M., Corrie, J. E. T., Webb, M. R., and Ferenczi, M. A., 1999, The efficiency of contraction in rabbit skeletal muscle fibers, determined from the rate of release of inorganic phosphate, *J. Physiol.* **517**: 839-854.

Hill, A. V., 1964, The effect of load on the heat of shortening of muscle, *Proc. R. Soc. B.* **159**: 297-318.

Huxley, A. F., 1957, Muscle structure and theories of contraction, *Prog. Biophys. Biophys. Chem.* **7**: 255-318.

Huxley, H. E., 1960, Muscle cells, in *The cell, Biochemistry, Physiology, Morphology, vol 4*, J. Brachet, A. E. Mirsky, eds., Academic Press, New York, pp. 365-481.

Kinosita, K. Jr., Yasuda, R., Noji, H., and Adachi, K., 2000, A rotary molecular motor that can work at near 100% efficiency, *Phil. Trans. R. Soc. B.* **355**: 473-489.

Kobayashi, T., Kosuge, S., Karr, T., and Sugi, H., 1998, Evidence for bidirectional functional communication between myosin subfragments 1 and 2 in skeletal muscle fibers, *Biochem. Biophys. Res. Commun.* **246**: 539-542.

Kushmerick, M. J., and Podolsky, R. J., 1969, Ionic mobility in muscle cells, *Science* **166**: 1297-1298.

Linari, M., and Woledge, R. C., 1995 Comparison of energy output during ramp and staircase shortening in frog muscle fibers, *J. Physiol.* **487**: 699-710.

Page, S. G., and Huxley, H. E., 1963, Filament lengths in striated muscle, *J. Cell. Biol.* **19**: 369-390.

Potma, E. J., and Stienen, G. J. M., 1996, Increase in ATP consumption during shortening in skinned fibers from rabbit psoas muscle: effects of inorganic phosphate, *J. Physiol.* **496**: 1-12.

Reggiani, C., Potma, E. J., Bottinelli, R., Canepari, M., Pellegrino, M. A., and Stienen, G. J. M., 1997, Chemo-mechanical energy transduction in relation to myosin isoform composition in skeletal muscle fibers of the rat, *J. Physiol.* **502**: 449-460.

Sugi, H., Iwamoto, H., Akimoto T., and Kishi, H., 2003, High mechanical efficiency of the cross-bridge powerstroke in skeletal muscle, *J. Exp. Biol.* **206**: 1201-1206.

Sugi, H., Iwamoto, H., Akimoto T., and Ushitani, H., 1998, Evidence for the load-dependent mechanical efficiency of individual myosin heads in skeletal muscle fibers activated by laser flash photolysis of caged calcium in the presence of a limited amount of ATP, *Proc. Natl. Acad. Sci.* **95**: 2273-2278.

Sun, Y-B., Hilber K., and Irving, M., 2001, Effect of active shortening on the rate of ATP utilization by rabbit psoas muscle fibers, *J. Physiol.* **531**: 781-791.

DISCUSSION

Ranatunga: Mechanical efficiency of muscle is known to vary with temperature. At what temperature did you perform your experiment?

Sugi: At 4-5°C.

Ranatunga: The data from intact muscle show that mechanical power output increases markedly with temperature. Do you think efficiency will increase also?

Sugi: The rate of ATP utilization also increases with temperature, but it is an open question how the muscle mechanical efficiency changes with temperature.

Pollack: Do you measure heat production?

Sugi: No, ATP is known to be the only immediate source of energy for muscle contraction, and heat measurement is not necessary when the rate of ATP utilization is measured.

Woledge: We do not need to measure enthalpy changes since free energy change of ATP hydrolysis is known. You apply small quick releases to the fiber. How do you think the possibility that quick release causes ATP hydrolysis to some extent?

Sugi: It is a difficult question to answer at present. We should consider it in our future study.

RATE OF ACTOMYOSIN ATP HYDROLYSIS DIMINISHES DURING ISOMETRIC CONTRACTION

N. A. Curtin, T. G. West, M. A. Ferenczi, Z.-H. He, Y.-B. Sun, M. Irving, and R. C. Woledge[1]

INTRODUCTION

The aim of the experiments reported here was to measure the time course of ATP hydrolysis by actomyosin during isometric contraction. It is known from previous studies of energy produced as heat and work that most, but not all of this energy, is due to ATP hydrolysis stoichiometrically coupled to the creatine kinase reaction (Curtin and Woledge, 1979). Furthermore energy is produced at a much higher rate at the start of contraction than later. These facts raise the question of whether the rate of ATP hydrolysis by actomyosin, which is the major source of energy, changes during isometric contraction.

We have combined independent methods of measuring actomyosin ATP hydrolysis in our study of contracting dogfish fibres, and the results agree is showing that the rate of ATP hydrolysis by actomyosin declines substantially during contraction. This report extends our earlier accounts of the project (He et al., 1998; West et al., 2002).

[1] N.A. Curtin, T.G. West, M.A. Ferenczi, Imperial College London, Div. Biomed. Sci., BSF Section, Fleming Bldg., London SW7 2AZ, UK. Z.-H. He, National Institute for Medical Research, The Ridgeway, Mill Hill, London NW7 1AA, UK. Y.-B. Sun, M. Irving, King's College London, Sch. Biomed. Sci., New Hunt's House, Guy's Campus, London SE1 1UL, UK. R.C. Woledge, UCL Inst. Human Performance, Royal National Orthopaedic Hosp. Trust, Brockley Hill, Stanmore, Mddx. HA7 7LP, UK.

METHODS

Apparent Equilibrium Constant for Phosphate Binding to MDCC-PBP in Dogfish Fibres

One method we used was based on quantitating ATP hydrolysis from the change in fluorescence of MDCC-phosphate binding protein (MDCC-PBP) as it binds phosphate (He et al., 1997, 1999, 2000). The apparent equilibrium constant for the reaction is used to convert the fluorescence changes to amount of phosphate produced during contraction. We report here the apparent equilibrium constant measured in permeabilized white muscle fibres from the dogfish (*Scyliorhinus canicula*) under the conditions used in our experiments.

Protocol

A series of loading solutions was prepared, as described below, containing 1.2 mM active MDCC-PBP and different known concentrations of phosphate in the range 0 to 4

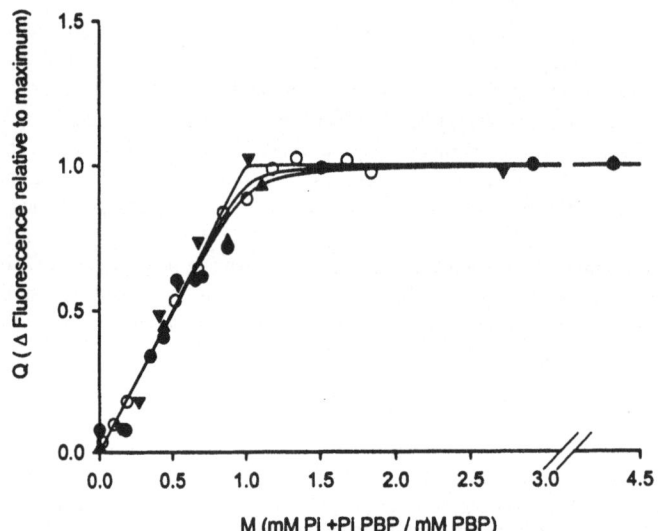

Figure 1. Fluorescence signal produced by known concentrations of phosphate and known concentrations of MDCC-PBP in permeabilized dogfish fibres (closed symbols) and rabbit fibres (open symbols). The lines show binding curves calculated for different dissociation constants and different the concentrations of MDCC-PBP. See text. Uppermost line: K_d 0.15 μM (as used by He et al. 1997) and MDCC-PBP 1.2 mM. Middle line: fitted apparent K_d 15.8 μM and MDCC-PBP 2.58 mM as used in the rabbit fibre experiments. Bottom line: fitted apparent K_d 15.8 μM and MDCC-PBP 1.2 mM as used in the dogfish fibre experiments.

mM. Permeabilized fibres were prepared and clips attached in the usual way. The sarcomere length was set to 2.1 – 2.2 μm before inducing rigor with the Ca^{2+}-free rigor solution. In each experiment the same permeabilized muscle fibre was immersed in each phosphate loading solution for 10 minutes before being transferred into silicone oil and the fluorescence signal measured.

Loading Phosphate solutions

The zero-phosphate loading solution (200 mM ionic strength) contained (in mM) TES 60.00, $MgCl_2$ 2.79, Ca^{2+}-EGTA 10.00, glutathione 10.00, K^+-propionate 138.52, and MDCC-PBP 1.25. The 4 mM phosphate solution was made from the same constituents and ionic strength was held at 200 mM by modifying K^+-proprionate concentration to 130 mM. Solutions with intermediate phosphate concentrations were prepared by mixing the 0 and 4 mM phosphate solutions in appropriate proportion.

Evaluating the Apparent K_d for the Conditions used in these Experiments

The results of experiments on 3 dogfish fibres are shown in Fig. 1 along with the results from a rabbit fibre reported by He et al. (1997). Binding curves of the following form were fitted to the observed relationship:

$$Q = 0.5 (1 + M + D - (1 + 2M + M^2 + 2DM + 2D + D^2)^{0.5})$$

Where Q is the ratio of Δfluorescence to maximum Δfluorescence (where Δfluorescence is the fluorescence minus the baseline value found by extrapolation to 0 Pi) all measured with fibre present. The value of Q is equivalent to the ratio PiPBP/C. C is the total concentration of phosphate binding protein (PiPBP + PBP, mM) used in the experiment expressed as the concentration of active phosphate binding sites. M is the ratio of the total concentration of phosphate (PiPBP + Pi, mM) in the loading solution to C (mM), and D = apparent K_d / C, and apparent K_d (units mM) is the apparent equilibrium constant for the reaction:

$$PiPBP \rightarrow Pi + PBP.$$

The best-fit value of apparent K_d was found by a least squares procedure using Excel Solver. For the experiments on dogfish and rabbit fibres, the best-fit value of apparent K_d was 15.8 μM. This value is larger than the value, 0.15 μM, used by He et al. (1997). However, both values represent relatively tight binding of phosphate by MDCC-PBP, which means that the most of the phosphate produced during contraction binds to MDCC-PBP and produces a fluorescence signal. We have corrected the records made during contraction using the apparent K_d 15.8 μM; this correction amounted to an increase of 1.86% to the maximum rate of ATP hydrolysis, and a 14.2% increase to the total ATP hydrolysis during 0.5 s of contraction. The correction does not affect our interpretation of the results for ATP hydrolysis by actomyosin.

ATP Hydrolysis by Actomyosin

All experiments were done on white myotomal muscle fibres from the dogfish. Fibres were dissected in elasmobranch saline and all experiments were conducted at 12°

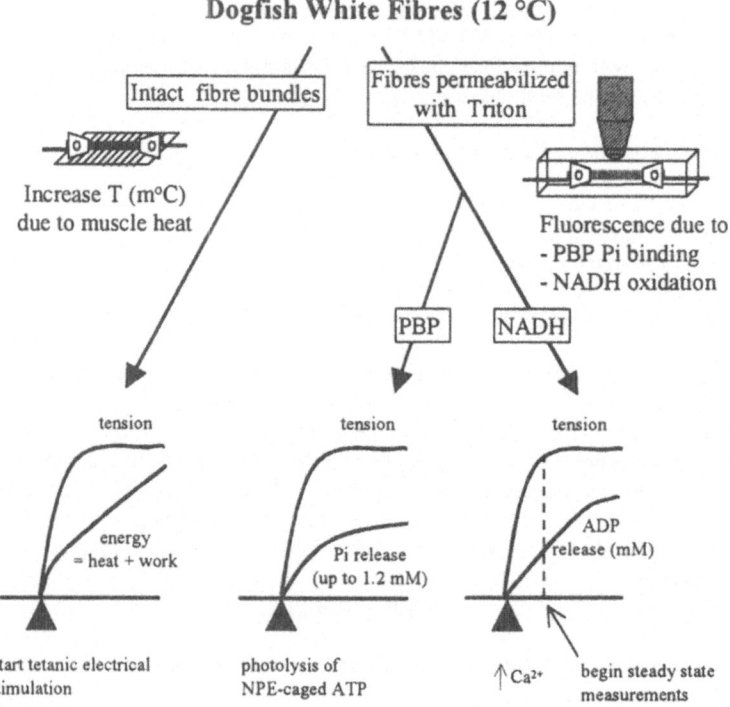

Figure 2. Diagram of the plan of the experiments to measure the energy produced and ATP hydrolysis by actomyosin in white fibres from dogfish. See text.

C, normal body temperature for these animals. The plan of the experiments is shown diagrammatically in Fig. 2. Some fibre preparations were used "intact" and force and heat were recorded during electrical stimulation under isometric conditions. Other fibres were permeabilized with Triton and ATP hydrolysis by actomyosin during isometric contraction was measured from either fluorescence signal produced by phosphate binding to MDCC-PBP or by oxidation of NADH linked to ADP production.

Intact Fibres

For experiments on intact fibres the total energy output during contraction was measured as the sum of the heat production and mechanical work. Heat was detected by an antimony-bismuth thermopile as an increase in temperature. The mechanical work done against the series elasticity in these isometric (fixed end) contractions was evaluated from the force record and the stiffness of fibre preparation as described by Curtin et al. (1998). For all fibres the optimum stimulus strength and fibre length giving maximum isometric force (Lo) was determined at the start of the experiment. Fibre bundles were stimulated for 3.5s at a frequency of 50 Hz. In some cases the first 2 pulses were given at an interval of 10 ms. Heat loss correction was made in the usual way using the time constant for heat loss measured by the Peltier method (Woledge et al., 1985). Records were corrected for stimulus heat as described by Curtin and Woledge (1993).

The method of separating the energy due to actomyosin (AM) interaction from that due to other processes is described below.

Separating AM from non-AM energy turnover by intact fibres. To separate the energy produced by intact fibres into the part due to ATP hydrolysis by AM and the part due to other processes (ATP hydrolysis by sarcoplasmic reticulum-Ca^{2+} pump, Ca^{2+}-binding reactions, etc) we made use of the linear dependence of AM interaction on filament overlap at L>Lo (see also Lou et al., 1997). In experiments on 9 fibres, force and energy production were measured during a 3.5s isometric (fixed-end) contraction at Lo and at a length greater than Lo. As expected, force "creep" occurred during the contractions at long length (Huxley and Peachey, 1961; Gordon et al., 1966a and b).

Background: what is force creep and how is it produced? Force creep is the gradual increase in force in a "fixed end" tetanus that continues after the initial rapid rise in force at the start of stimulation. It was described and characterized by Huxley and Peachey (1961) and Gordon et al. (1966a and b). Some important features are: it occurs at L>Lo and is more obvious the longer the fibre length. Huxley and Peachey (1961) presented evidence showing that when a resting fibre is stretched to length >Lo, the end sarcomeres do not stretch as much as those in the centre of the fibre (direct observations of striations). When the fibre is stimulated force rises rapidly in the usual way as bridges attach in response to Ca^{2+} release from the sarcoplasmic reticulum and Ca^{2+} binding to troponin, etc. However, the force does not remain stabilized at a plateau value, as it does in contractions at Lo. Instead the force continues to increase or "creep" up gradually. Huxley and Peachey (1961) observed that the end sarcomeres, which have more overlap than those in the centre region, shorten at the expense of those in the centre.

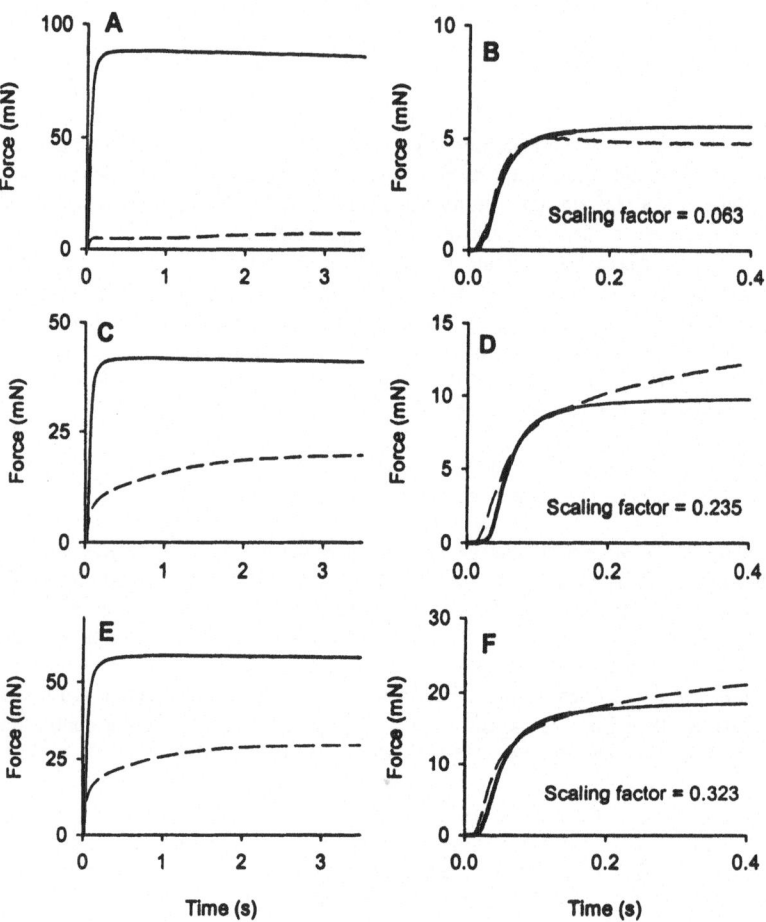

Figure 3.   Records of force during isometric (fixed end) contractions by 3 intact fibre preparations. A, C and E show superimposed records for a contraction at Lo (solid line) and at L>Lo (broken line). B, D, and F show on an expanded scale the force recorded at L>Lo (broken line) and the scaled version of the force recorded at Lo (solid line). The scaling factor was chosen to give the best match to the force recorded at L>Lo. The thick solid line shows the part of the record (initial 150 ms) that was used to calculate the scaling factor (see text).

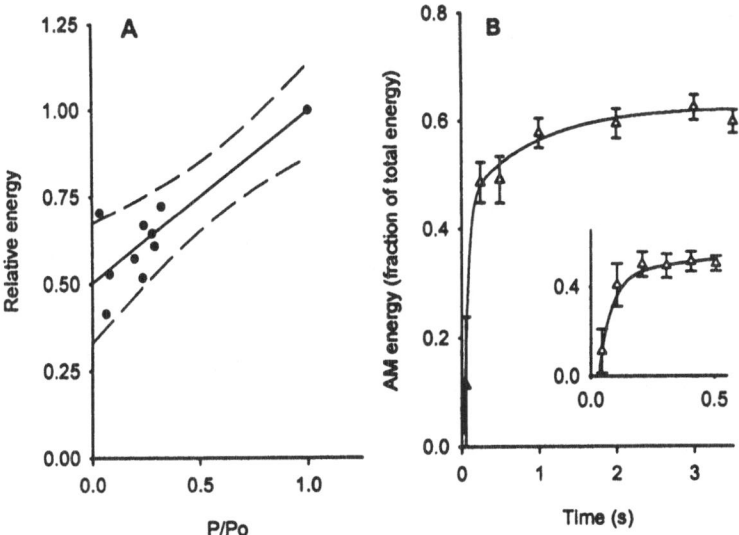

Figure 4 A. Example graph of energy output vs. force produced at different fibres lengths. Energy at each fibre length is expressed relative to that produced at Lo. Force corrected for creep (see text) is expressed relative to that produced at Lo. Energy output during 0.5 s contraction. Results for 9 fibre bundles. Solid line y = 0.497(±SEM 0.037) x + 0.501(±SEM 0.027). The slope of this regression line, 0.497(±SEM 0.037), is the fraction of the energy due to AM. Broken line 95% confidence intervals. **B.** Variation of AM energy with time during isometric contraction. AM energy expressed as a fraction of the total energy output. Symbols are means ±SEM for selected time points. Line is fitted through all time points (1point/ms). Inset shows values for early in the contraction on a larger time scale.

Separating force creep from force that is proportional to filament overlap.   In our experiments we need to separate creep force from the "normal" force because creep force represents crossbridge turnover and ATP hydrolysis that occurs **only** in the end sarcomeres, where we are not measuring energy output.

Our method of correcting force records for creep was based on the fact that creep develops gradually and progressively during a tetanus. In other words, the force at the beginning of the tetanus is least affected by it.   Thus we have used only the force produced in the initial 150 ms of the contractions at L>Lo. Fig. 3 show pairs of records for contractions at Lo and at L>Lo to illustrate the approach we used. For each fibre preparation the force recorded at Lo was scaled to fit that recorded from the same fibre preparation at L>Lo using only the initial 150 ms of each contraction. The scaling factor was found using Excel Solver and was taken as the creep-free estimate of the force expressed as a fraction of the force produced at Lo (P/Po in Fig. 4A); this fraction is also taken as the fraction of maximum filament overlap during the contraction.

Fig. 4A shows examples of the dependence of energy output on corrected force. The slope of the line is equivalent to the fraction of the energy output that is due to AM interaction.   In this example the energy is that produced in the first 0.5 s of the contraction. Similar graphs were made for all the time points in the contraction; the slopes of these graphs plotted *vs.* time show how the fraction of energy turnover due to AM changed during the course of the contraction (see Fig. 4B).

Permeabilized fibres

For experiments on permeabilized fibres Triton was used to remove all membranes and thus prevent the ATPase activity due to membrane ion pumps. Methods used here are similar to those described previously by He et al. (1997, 1999, 2000) and Hilber et al. (2001). ATP hydrolysis in the initial 0.5 s of contraction was detected as the fluorescence change due to phosphate binding to phosphate binding protein (MDCC-PBP) after activation of contraction by flash photolysis of caged-ATP (He et al., 1997, 1999, 2000). ATP hydrolysis later in contraction after isometric force had reached its plateau value was detected as the fluorescence change due to NADH oxidation coupled to ADP production by AM ATP hydrolysis; in these experiments the fibre was activated by increasing Ca^{2+} (Hilber et al., 2001).

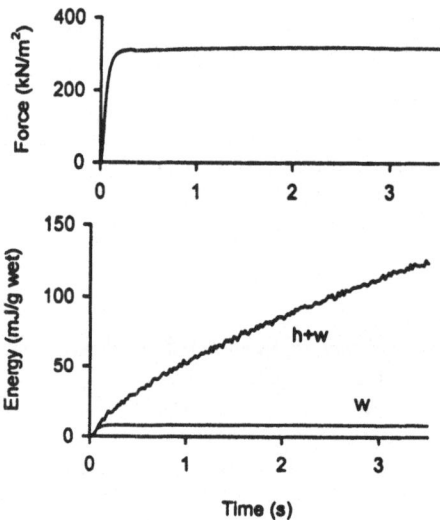

Figure 5. Example record of force (upper part) and energy (lower part) recorded during an isometric (fixed end) contraction of a bundle of intact fibres. Tetanus duration was 3.5s. The total energy is labelled h+w and the internal work against the series elasticity is labelled w.

Fixing fibre ends and attaching clips. During this project we developed a variation of the routine for fixing the ends of permeabilized fibre. In the usual routine a small bundle was dissected from the muscle slice and then permeabilized. The myosepta (tendon sheets) at the ends of the bundle were removed at this stage so that a single fibre could be drawn out from the bundle. Aluminium T-shaped clips were loosely folded over each end and used to secure the fibre during fixation of the ends with glutaraldehyde. Clips were left relatively loose so that the fixative could reach the part of the fibre inside the clip. After fixation the clip was pinched tight.

In the revised protocol an intact single fibre was dissected and the myosepta were left intact at each end of the fibre. The myosepta were used, rather than T-clips, to support the fibre during the end-fixing process. Hence, fixative flowed directly over the surface of the fibre ends. After fixation, the myosepta were removed and the T-clips were folded firmly onto the fixed portion of the fibre ends. This modification was used for 3 of the 4 fibres used in the NADH experiments. It greatly improved our success with sustaining the Ca^{2+}-activated contractions for up to 20 s, and the fibres produced higher forces than those prepared by the usual fixation and clip routine.

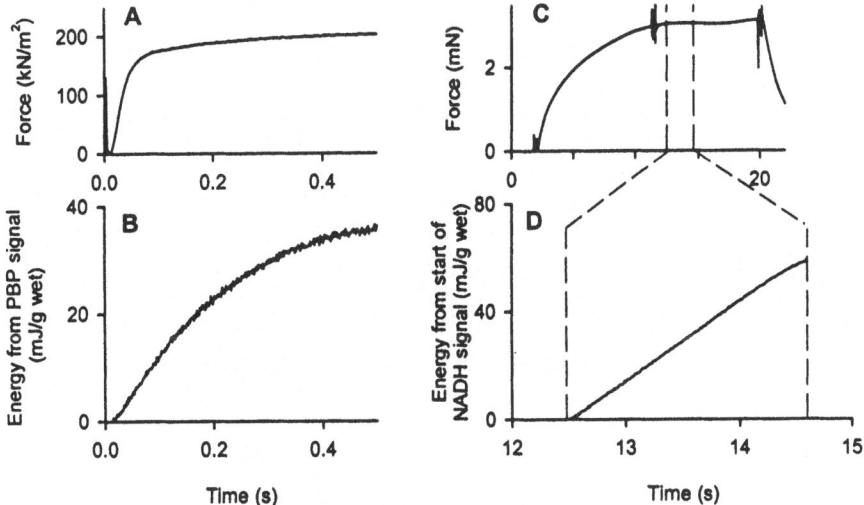

Figure 6.   Records of force and energy from ATP hydrolysis by actomyosin in permeabilized fibres.   **A** shows force and **B** shows the energy calculated from the MDCC-PBP signal during the initial 0.5 s of isometric contraction.   **C** is the force record from an experiment using the NADH assay.   The vertical broken lines mark the start and end of the period when NADH was measured.   **D** shows the energy calculated from the change in NADH from its value at the start of the measurement, which corresponded to about 10s of contraction in this experiment.   Note particularly the differences in scales in **B** and **D**.

Expressing ATP hydrolysis in energy units. Measured values of ATP hydrolysis were converted to energy units for comparison with the result from the intact fibres using the molar enthalpy for ATP hydrolysis coupled to the creatine kinase reaction, 34 kJ/mol ATP hydrolysis (Woledge et al., 1985), and assuming muscle density to be 1.06.

Normalizing for Size

Forces are expressed relative to cross sectional area to take account of variation in fibre size. Cross sectional area was measured as:

$$A = 4.9 \, w/d \, L$$

Where A is cross sectional area (mm^2), w is dry weight (mg). 4.9 is wet to dry weight ratio for white fibres from dogfish (Curtin and Woledge, 1993), d is density, which was assumed to be 1.06, and L is the length (mm) of the fibre at Lo, the length given maximum isometric force.

RESULTS AND DISCUSSION

Figure 5 shows example records of force and energy output during a 3.5 s isometric tetanus of a bundle of intact fibres. The force increases rapidly to its plateau value and remains relatively constant after the first few hundred ms of stimulation. Energy is the sum of heat production and the "internal" work done against the series elasticity. Both the total energy (heat + internal work) and "internal" work are shown in Fig. 5. The energy output is fastest at the start of the contraction and the rate declines continuously. The energy output continues to decrease even after the fibres have stopped doing "internal" work.

We evaluated the amount of energy output due to actomyosin using the relationship shown in Fig. 4B which shows how the fraction of total energy due to actomyosin varied during contraction.

$$AM_t = f_t \, E_t$$

Where AM_t is the actomyosin energy at time t, f_t is the fraction at time t (see Fig. 4B and Methods) and E_t is the total energy output at time t (Fig. 5). Results for 17 fibre preparations were analysed in this way. The rate of energy output from actomyosin was calculated at various time points in the contraction. The mean values of the rates are shown in Fig. 7.

Figure 6 shows example records from two experiments on permeabilized fibres. The panels on the left show force and energy output detected during the initial 0.5 s of contraction. In these experiments, MDCC-PBP was used to monitor ATP hydrolysis by actomyosin. Contraction was initiated by photolysis of caged ATP at time 0. Phosphate production is not reported for the times after 0.5s of contraction for two reasons: the 1.2 mM MDCC-PBP present in the fibre becomes saturated with phosphate and therefore cannot report continuing ATP hydrolysis, and ATP supply becomes depleted.

Measured values of ATP hydrolysis were expressed in the equivalent energy units (see Methods). It is clear from Fig. 6B that energy production due to actomyosin ATP hydrolysis is fastest very early in the contraction, and the rate then declines. Rates were calculated at three time points and are shown in Fig. 7 (mean values from 15 experiments). The general pattern of a sharp decline in the rate matches that found in intact fibres.

As described in the Methods, we found that the apparent equilibrium constant for phosphate interaction with MDCC-PBP was somewhat different than the K_d used by He et al. (1997, 1999, 2000). To examine the influence of the value of the apparent equilibrium constant, we calculated the rates of ATP hydrolysis by our fibres using both values, and the results are shown in Fig. 7. The small open circles are for the tighter binding constant used by He et al. and the large open circles are for the binding constant reported here as measured in the presence of the permeabilized fibres. The maximum rates early in the contraction are hardly affected by the value of the apparent equilibrium constant. While the rate at 0.5 s is dependent on which equilibrium constant is used, in both cases the rate falls dramatically during the first 0.5 s of contraction.

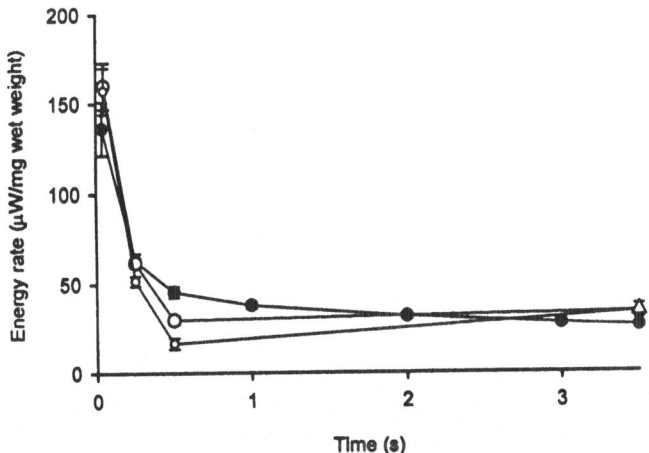

Figure 7. Energy rate during isometric contraction. Open symbols show energy rate due to ATP hydrolysis by actomyosin in permeabilized fibres detected with MDCC-PBP (n=15 fibres, apparent K_d 15.8 μM for large o, and K_d 0.15 μM for small _) and with NADH assay (n=4 fibres, Δ). Closed symbols show rate of energy release by actomyosin by intact fibres; energy release detected as heat and work (n=17 fibre preparations, •). Mean ± SEM.

Fig. 6B shows example records of force and energy output determined from the oxidation of NADH, which reports ADP formed during ATP hydrolysis by actomyosin in the skinned fibre. In these experiments the fibres were activated by increasing the Ca^{2+} and thus the rate of rise of force is less than in the other types of experiment. Once force had reached a plateau value, measurement of NADH oxidation commenced. The broken vertical lines in Fig. 6B and D mark the measurement period. The time course of energy release due to ATP hydrolysis is shown in Fig. 6D on an expanded time scale. Clearly the rate is much lower at this time late in contraction than earlier. Note the difference in the time scales for Fig. 6B and D. It is also striking that the rate of energy release due to ATP hydrolysis by actomyosin is constant during this period in contrast to that earlier in contraction (Fig. 6B). Experiments of this type were done on 4 fibres and the mean result for the rate energy due to ATP hydrolysis by actomyosin late in contraction is shown in

Fig. 7, plotted at nominal time 3.5 s. This rate is in good agreement with the rate at 0.5 s determined in the MDCC-PBP experiments.

CONCLUSIONS

The results in Fig. 7 show that the rate of energy output due to actomyosin alone as measured in intact fibres and in permeabilized fibres match well, and are in good agreement in showing that the rate is highest early in contraction and then declines dramatically. Our evidence indicates that the rate is relatively stable after the first 0.5 s of contraction.

At least part of the early high rate is due to internal work being done against the series elasticity as force rises. However, after force reaches its plateau value internal work ceases (Fig. 5). Thus it seems unlikely that internal work can completely explain the decline in ATP hydrolysis by actomyosin. As Fig. 7 shows the rate declines significantly between 0.25 and 0.5 s, a time when force is constant.

ACKNOWLEDGEMENTS

This work was supported by a grant from the Wellcome Trust.

REFERENCES

Curtin, N.A., Gardner-Medwin, A.R. and Woledge, R.C. (1998). Predictions of the time-course of force and power output by dogfish white muscle fibres during brief tetani. *J. exp. Biol.* **201**: 103-114.

Curtin, N.A. and Woledge, R.C. (1979). Chemical changes and energy production during contraction of frog muscle: How are their time courses related? *J. Physiol.* **288**: 353-366.

Curtin, N.A. and Woledge, R.C. (1993). Efficiency of energy conversion during sinusoidal movement of white muscle fibres from the dogfish, *Scyliorhinus canicula*. *J. exp. Biol.* **183**: 137-147.

Lou, F., Curtin, N.A. and Woledge, R.C. (1997). The energetic cost of activation of white muscle fibres from the dogfish, *Scyliorhinus canicula*. *J. exp. Biol.* **200**: 495-501.

Gordon, A.M., Huxley, A.F. and Julian, F.J. (1966). Tension development in highly stretched vertebrate muscle fibres. *Journal of Physiology* **184**:143-169.

Gordon, A.M., Huxley, A.F. and Julian, F.J. (1966). The variation in isometric tension with Sarcomere length in vertebrate muscle fibres. *Journal of Physiology* **184**:170-192.

He, Z-H., Chillingworth, R.K., Brune, M., Corrie, J.E.T., Trentham, D.R., Webb, M.R. and Ferenczi, M.A. (1997). ATPase kinetics on activation of rabbit and frog permeabilized isometric muscle fibres: a real time phosphate assay. *Journal of Physiology* **501**, 125-148.

He, Z-H., Chillingworth, R.K., Brune, M., Corrie, J.E.T., Webb, M.R. and Ferenczi, M.A. (1999). The efficiency of contraction in rabbit skeletal muscle fibres, determined from the rate of release of inorganic phosphate. *Journal of Physiology* **517**, 839-854.

He, Z-H., Bottinelli, R., Pellegrino, M.A., Ferenczi, M.A. and Reggiani, C. (2000). ATP consumption and efficiency of human single muscle fibers with different myosin isoform composition. *Biophysical Journal* **79**, 945-961.

He, Z-H., Ferenczi, M.A., Lou, F., Curtin N.A. and Woledge, R.C. (1998). A comparison of the energy turnover detected by heat production and by P_i assay in dogfish isolated muscle fibres. *J. Physiol.* **509**: 42P.

Hilber, K., Sun, Y.-B., and Irving, M. (2001). Effects of sarcomere length and temperature on the rate of ATP utilisation by rabbit psoas muscle fibres. *Journal of Physiology* **531**, 771-780.

Huxley, A.F. and Peachey, L.D. (1961). The maximum length for contraction in vertebrate striated muscle. *J. Physiol.* **156**: 150-165.

West, T., Curtin, N.A., Ferenczi, M.A., He, H.-Z., Sun, Y.B., Irving, M., and Woledge, R.C. (2002). Quantifying isometric force-dependent energy release in white muscle by comparing contraction energetics in intact and skinned fibres. *J. Muscle Research and Cell Motility* **23**: 26-27.

Woledge, R.C., Curtin, N.A. and Homsher, E. (1985). *Energetic Aspects of Muscle Contraction.* Monograph of the Physiological Society, No. 41, Academic Press, London.

DISCUSSION

Kushmerick: If tetani are repeated after a short interval rather than long ones, will not the high initial rate of energy release be reduced?

Curtin: We have no information yet on how the amount of the initial rate might depend on interval between tetani; but from what we know about total energy, I expect that the high rate of actomyosin ATP hydrolysis would not occur if the interval between contractions were very brief.

Sugi: Some years ago, we performed experiments with dogfish jaw muscle. During tetanic stimulation, the magnitude of isometric force was beautifully maintained to be constant over many tens of seconds. However, when I made the muscle to shorten against a constant external load, the shortening velocity gradually declined with time, indicating that the constant isometric force is not necessarily associated with the constant state of the contractile system.

Curtin: In some of our studies of dogfish muscle, the performance during shortening was very well maintained (Curtin and Woledge, J. Exp. Biol. 200: 495-501, 1997), so dogfish muscle can do as well during working contractions as in isometric contractions. However in other studies, we have seen some shortening reactivation, for reasons we do not understand.

Winegrad: What happens to the rate of energy consumption when the muscle is released and then stretched after the periods of rapid consumption?

Curtin: We have not done this experiment, but I would expect that the rate might be low and not be re-set to the high rate immediately by a release or restretch.

Rall: Have you looked at the rate of actomyosin ATP hydrolysis in slow-twitch muscle?

Curtin: No, not yet. We hope to do this in future.

Gonzalez-Serratos: From what you showed, it seemed that ion pumps made no contribution to energy. Do you think that the contribution of the energy for ion pump ATPase is negtible?

Curtin: No, it is already subtracted from the values I showed for intact fibers. The non-actomyosin reactions account for half of the total energy produced in the first 0.5sec of an isometric contraction as I showed in my presentation.

ENERGY STORAGE DURING STRETCH OF ACTIVE SINGLE FIBRES

R.C. Woledge[1], N.A. Curtin[2], and M Linari[3]

1. INTRODUCTION

One of the functions of muscles is to absorb the work done them when they are stretched while active. It has been shown by measurement of ATP hydrolysis that muscle is not a reversible energy conversion device and that work done on a muscle cannot be used to synthesise ATP (Curtin & Davies, 1973). Experiments measuring energy production as heat and work output (Hill & Howarth, 1959; Constable *et al.* 1997) have suggested that work done on an active muscle will eventually become heat but that some of it may be stored in another form before conversion to heat. One way in which this could happen is by storage of work in elastic structures in series with the muscle crossbridges, such as tendon and the muscle filaments themselves. Work would be stored as the stress in these structures rises during a stretch and would be released as the tension falls after the stretch is over. Using measurements of the heat produced by frog single muscle fibres and of the work done on them during stretch we have observed energy storage during stretch, and the release of energy after the end of a stretch. In these preparations the tendon compliance can be kept to very low values, allowing us to investigate whether stretch in series elastic structures can account for the energy stored or whether other structures within the fibre are also capable of energy storage and release.

[1] UCL Institute of Human Performance, RNOHT, Brockley Hill, Stanmore, Mddx. HA7 4LP, UK.

[2] Biological Structure and Function Section, Division of Biomedical Sciences, Faculty of Medicine, Sir Alexander Fleming Building, Imperial College, London SW7 2AZ, UK.

[3] Dipartimento di Scienze Fisiologiche, Università degli Studi di Firenze, Viale GB Morgagni 63, I-50134 Firenze, Italy.

Molecular and Cellular Aspects of Muscle Contraction
Edited by H. Sugi, Kluwer Academic/Plenum Publishers, 2003

2. METHODS

A full account of the methods used will be published elsewhere. Single fibres were dissected from the lateral head of the tibialis anterior muscles of frogs (*Rana temporaria*) The fibre tendons were held in platinum foil T-shaped clips. The sarcomere length in the resting fibre was measured by laser diffraction and set to 2.10 μm; this fibre length is defined as L_0. The fibre was mounted horizontally in a moist chamber at about 1°C. It was in contact with an antimony-bismuth thermopile made by vacuum deposition (Mulieri *et al.* 1977) on a 6 μm thick polyester film (Kapton). One end of the fibre was attached to a capacitance gauge force transducer (2.1 kHz frequency response, Model 400A, Aurora Scientific Inc., Canada) and the other to a motor which was either a Model 300B, Cambridge Technology Inc., USA or a loudspeaker motor similar to that described by Lombardi & Piazzesi (1990).

Tetanic stimulation was delivered through the platinum clips at strengths of 1.5 times threshold, with a frequency sufficient to give a fused tetanus (10-15 Hz). The fibre was stimulated for 0.5 s under isometric conditions before the start of each stretch, which was centered on L_0. The average stretch distance was $10.4 \pm$ s.e.m. 0.4 %L_0, n=55 records from 7 fibres

The cumulative work at each time point was calculated from the integral of force with respect to distance moved. Total energy output at each time point was calculated as the sum of the heat produced and the work. Work done on the fibre by the motor is treated as negative. The isometric force at L_0 (P_0), was measured in each experiment and all the energy values were expressed as the dimensionless quantity energy/($P_0 L_0$).

3. RESULTS

Figure 1 shows records of the force and energy output from three contractions with stretches at different speeds. The times at which stretch starts and end are marked by the vertical lines. It is clear that during stretch the rate of energy output falls from the value just before stretch starts. For the slowest stretch the energy output during the period of stretch is positive, but for the other two records energy is absorbed during the stretch period. When energy absorption is observed during a stretch it is generally at a higher rate in the early part of the stretch. This be seen clearly in the centre record of fig 1. Immediately after the stretch is over the rate of energy output is high, falling after about 0.2 s to a rate similar to that in the isometric period before the stretch.

Figure 2A shows, for one of these observations, the energy produced in the isometric period after the end of the stretch . The full line shows that the data can be adequately described by the function of time (t)

$$ER^* (1 - e^{-t/\tau h}) + B^*t$$

when the constants ER, τh and B are found by curve fitting. The lower record is from the corresponding period in a wholly isometric contraction at Lo; the isometric heat rate is approximately constant during this time period, and similar to the rate reached by 0.2 s after the end of the stretch.

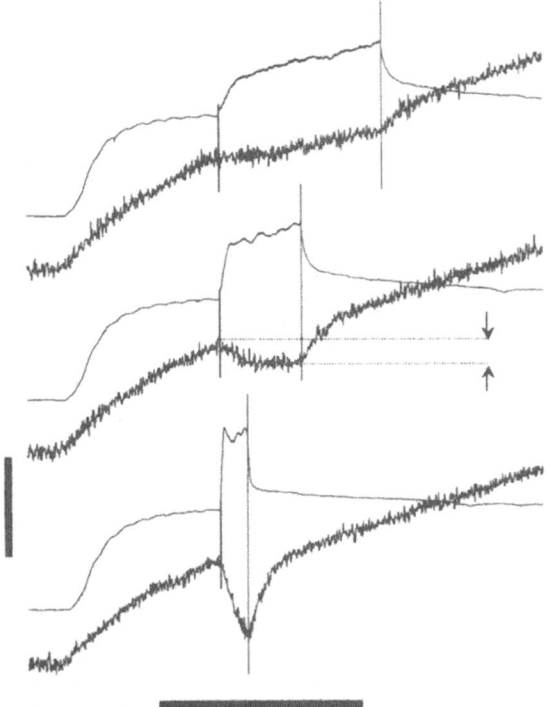

Figure 1 Records of force and energy output (heat + work) in three contractions with periods of stretching at speeds of 0.186, 0.366 and 1.091 Lo/sec In each case the distance stretched is 0.091 Lo. The horizontal bar represents 1 sec and the vertical bar represents Po for the force records and 0.125 PoLo for the energy records. The vertical lines mark the times at which stretch started and ended. The two horizontal broken lines show, for the centre record how the energy absorption, from the start to the end of the stretch was measured for each record

The decline in force after the end of the stretch in this observation is shown in Fig 2 B. The function

$$F1* (1- e^{-t/\tau f1}) + F2* (1- e^{-t/\tau f2})$$

has been fitted to the data and provides a good fit.

As is illustrated by the example in Fig 2 the time constant for the fast phase of the energy production after stretch is not similar to either of those describing the fall of force towards the isometric value. This is confirmed by the mean values of these three time constants obtained from 50 observations in 7 fibres which are shown in Fig 3. The observations have been grouped according to the speed of the stretch, and it can be seen that the time constant for the fast component of energy output after stretch is not dependent on the speed of the stretch, in contrast to the time constant of the fast component of the fall in force.

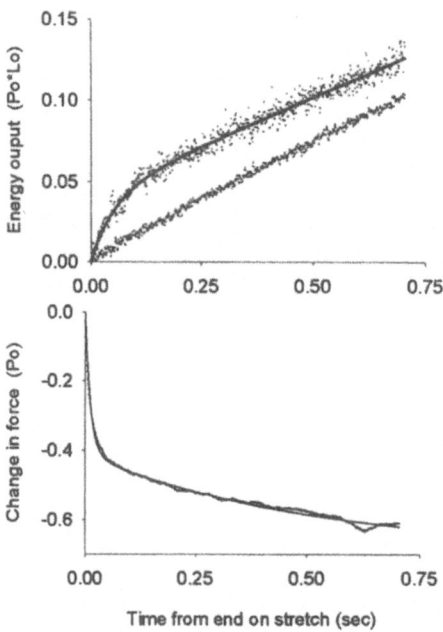

Figure 2 A Upper panel. Upper record shows the energy produced since the end of stretch (velocity 0.366 Lo/s, whole record is shown in Fig 1) The full line on this record is the fitted equation (see text) with ER = 0.041 PoLo , th = 55 ms and B = 0.121 PoLo/s. Lower record is energy produced in the corresponding period of an isometric contraction at Lo.
         B Lower panel. The thicker line shows the decline in force since the end of the stretch. The thin line is the fitted equation, (see text) with F1= - 0.279Po, τf1 = 508 ms, F2 = - 0.411 Po, τf2 = 2.4 ms.

The relation between speed of stretching and the amount of energy released in the fast phase of heat production after stretch is shown in Fig 4 (filled symbols). In this graph the energy is plotted as a fraction of the work done on the muscle during the stretch. The fraction of the work that is released as heat after the stretch rises with speed of stretch reaching 69 % at the highest velocities used. This figure also compares this energy release with the energy absorption during the stretch (open symbols). At the three lowest speeds of stretch energy is produced, not absorbed, during the stretch, and so the values in Fig 3 are negative. At the four higher speeds energy is absorbed. The amount of energy absorbed during the stretch is always less than the energy released in the fast heat production after the end of stretch. The difference between these two quantities is the amount of energy produced during the stretching period by ATP splitting and can be compared to the energy output in an isometric contraction of the same duration. A regression analysis of the relation between the isometric energy output and that during stretch gave a slope of 0.52 (s.e.m. 0.08 n=50). Thus about half the ATP splitting is suppressed by stretching

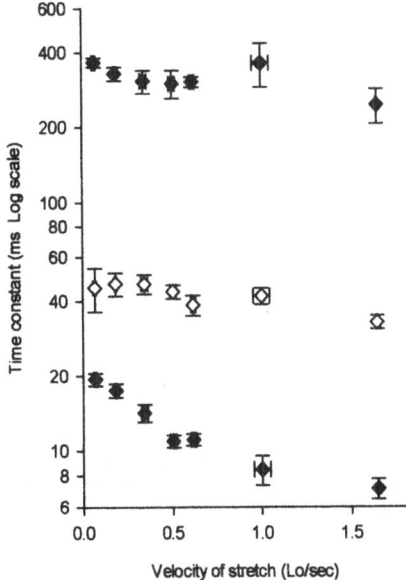

Figure 3 Measurements of the time constants of the fast phase of heat production following a period of stretch (open symbols) and of the time constants of the fast and slow phases of fall of force after the end of stretch (filled symbols). Values have been grouped by velocity. The number of observations in each group is between 5 and 11. All values are shown as means + SEM.

4. DISCUSSION

The probable energetic consequences of stretch of active muscle are (1) the reduction in the rate of ATP splitting by myofibrils during the stretch (Curtin and Davies, 1973), (2) the storage of energy in some form during the stretch (Hill & Howarth, 1959; Constable *et al.* 1997),and (3) the release as heat of the energy from this store after the stretch is over. The present observations can be interpreted in this way. The observation that the energy output during stretch is less than that in isometric contraction, could be due to either or both of (1) and (2) but the net energy absorption seen when the speed of stretch exceeds 0.3 Lo/s (Fig 4) can only be due to energy storage in some form. It is the combination of these measurements with the heat produced after a stretch that suggests energy storage is also occurring during slower stretches.

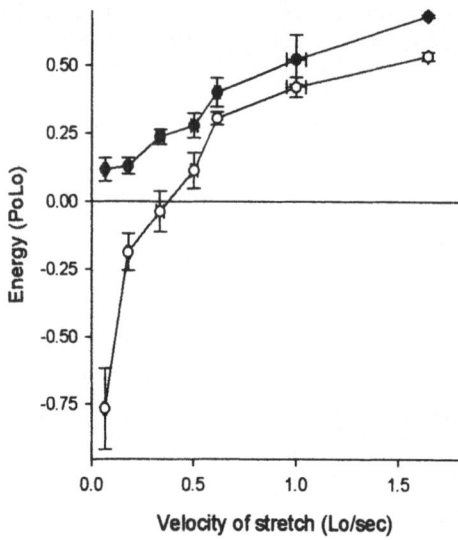

Figure 4 Filled symbols: the amount of heat released in the exponential phase of heat production after the end of a stretch. Open circles: heat absorption during the period of stretch. Values have been grouped by velocity. The number of observations in each group is between 5 and 11. All values are shown as means + SEM.

The observation that in the isometric period after the end of stretch the heat rate always exceeds that to be expected from isometric contraction is most easily interpreted as due to the dissipation as heat of the energy stored during the stretch. If we assume that after the end of stretch the energy output from ATP splitting occurs at a constant rate, as it would over this time period in a wholly isometric tetanus then it appears that the release of stored energy occurs exponentially with a time constant of about 45 msec, independent of the speed of the stretch. This simple result suggests that the nature of the storage process is the same at all speed of stretch. The amount of energy released in this process is always more that the energy absorption observed during the stretch. This is probably because some ATP splitting always continues during the stretch process, which agrees with the results of Curtin & Davies, 1973, who measured ATp splitting directly. At high speeds of stretch the energy from ATP splitting can partially mask the energy storage caused by the stretch. At low speeds the energy stored is less than the energy from the ongoing ATP splitting and the net energy output during the period of stretch is positive. Thus the measurement of the heat released after stretch may give the best measurements of the amount of energy storage, and the way this depends on the speed of stretch.

The fact that a higher proportion of the work done on a muscle is stored when the speed of stretch is greater (Fig 4) could be explained if the stored energy dissipates as heat not only after the end of stretch but also during the stretch itself. If this were so it would also provide an explanation of the time course of the energy storage during stretch. As noted in Fig 1 it is generally the case that the rate of energy absorption during stretch is greater in the early part than in the later parts of the stretch. If the stored energy could dissipate as heat during stretch this energy output would reduce the net energy absorption later in the stretch when there was an appreciable amount of stored energy present.

Is the energy absorbed during a stretch stored elastically in a structure in series with the crossbridges ? If this were the case the energy stored at each moment would be directly related to the force at that time. Energy storage and force would follow a similar time course. This is clearly not the experimental finding. For example considerable energy storage occurs during periods when the force is almost constant; an example is the bottom record in Figure 1. Although the energy release after stretch does occur at a time when the force is falling it does not follow a time course similar to that of the decline in force (Figure 3). Another argument against the energy storage being in a series elastic structure is the high proportion of the work done which can be stored. In the fastest stretches the proportion of the work stored reaches 67 %. For this to be stored in a series elastic structure, that structure would have to take up 67 % of the length change imposed which averaged 10.8 % of Lo; the length change in this element during stretch would thus have to be about 7% Lo. In contrast to this figure the length change expected from stretch of the tendons in our experiments, found from observations of their compliance, is between 0.1 and 0.2 % Lo. The compliance in the actin and myosin filaments is also much too small to provide for the energy storage we observe. We conclude that the energy storage site is not series elasticity.

Possible sites of the energy storage which cannot at present be eliminated are elastic structures in parallel with the myofilaments (Edman & Tsuchiya, 1996). and also the crossbridges themselves. Further experimental and modeling work is needed to determine whether either or both of these could be the site of the stored energy.

Acknowledgement

This research was supported by the Wellcome Trust

5. REFERENCES

Constable, J.K., Barclay, C.J. & Gibbs, C.L. (1997). Energetics of lengthening in mouse and toad skeletal muscles. *Journal of Physiology* **505**, 205-215.

Curtin, N.A. & Davies, R.E. (1973). Chemical and mechanical changes during stretching of activated frog skeletal muscle. *Cold Spring Harbor Symposium on Quantitative Biology*, **37**, 619-626.

Edman, K.A.P. & Tsuchiya, T. (1996). Strain of passive elements during force enhancement by stretch in frog muscle fibres. *Journal of Physiology* **490**, 191-205.

Hill, A.V. & Howarth, J.V. (1959). The reversal of chemical reactions in contracting muscle during an applied stretch. *Proceedings of the Royal Society, Series B* **151**, 169-193.

Lombardi, V. & Piazzesi, G. (1990). The contractile response during steady lengthening of stimulated frog muscle fibres. *Journal of Physiology* **431**, 141-191

Morgan, D.L. (1994). An explanation for residual increased tension in striated muscle after stretch during contraction. *Experimental Physiology* **79**, 831-838.

Mulieri, L.A., Luhr, G., Trefrey, J. & Alpert, N.R. (1977). Metal film thermopiles for used with rabbit right ventricular papillary muscles. *American Journal of Physiology* **233**, C146-156.

DISCUSSION

Ranatunga: Is it possible to rule out some possibilities by repeating experiments at a different temperature?

Woledge: Probably, yes. But we have not yet done such experiments.

Pollack: Have you examined whether the A-band stretches during muscle stretch?

Woledge: No, we have not measured that, although it may be possible. I hope someone else does that.

THREE DIMENSIONAL ULTRASOUND ANALYSIS OF FASCICLE ORIENTATION IN HUMAN TIBIALIS ANTERIOR MUSCLE ENABLES ANALYSIS OF MACROSCOPIC TORQUE AT THE CELLULAR LEVEL

T. Hiblar, E. L. Bolson, M. Hubka, F. H. Sheehan, and M. J. Kushmerick[*]

ABSTRACT

The purpose of this study was to test the hypothesis that the internal structure of the bipennate human tibialis anterior muscle is sufficiently homogenous throughout the muscle that the cellular stresses could be interpreted correctly from measurable anatomic properties and torque in the limb. This result is needed for facile comparison of extrinsic mechanical data and intrinsic energetic fluxes. Three-dimensional imaging of the fascicles of the human tibialis anterior muscle was made by capturing a series of ultrasound images while registering their location in space. Subsequent tracing of hundreds of structures in the ultrasound images with the use of custom software identified muscle boundaries, tendon surfaces, and fascicles as anatomic elements in 3-D space. The tendon was reconstructed as a mesh through the tracings identified as a component of the tendon. The angle of insertion of each identified fascicle at the tendon was calculated against the nearest normal in the mesh of the tendon. In three subjects the average angle of insertion of the fascicles onto the internal tendon was 11° (coefficient of variation 40%). The angle decreased along the length of the muscle from ~ 15° near the belly of the muscle to 6° near the ankle in fascicles superior and inferior to the central tendon. The angle increased by several degrees during a voluntary contraction. Despite the differences in angles of insertion that can be measured, these distinctions have little significance for the distribution of forces along cellular axes within the muscle: the angles, their distribution within the muscle and change with contraction are small. For this bipennate muscle the cosine of the angle of insertion of the cellular bundles is always

---

[*] University of Washington, Seattle, Washington 98195.

close to unity. Thus measurements of whole muscle mechanical data are simply related to mechanical stress of its cells.

1. INTRODUCTION

Analysis of experiments studying mechanics in muscle requires knowledge of fiber length, length changes during contraction and orientation of the fibers with respect to line of action of the motion and force transducer. In human muscles mechanical analyses have been complicated by the availability only of the torque measured across some joint and the unknown internal structure of the muscle unless highly invasive procedures are used. Contemporary non-invasive imaging methods can be used to measure muscle volume, the internal structure of muscle fascicles and tendons, and their anatomical relationships at rest and during activity. From this information moment arms and forces transmitted to the muscle cells in the fascicles and shortening of fascicles can be obtained allowing macroscopic measurements on intact muscle to be interpreted at the level of groups of cells within the fascicles. Human muscle is particularly suited to approach this problem because the large size of the muscles enable the use of contemporary imaging methods to measure the cellular arrangement within a muscle at a resolution of ~ 0.2 mm. These magnetic resonance imaging (MRI) and ultrasound (US) instruments have been developed for use in clinical medicine. Much work has been done in the analysis of thigh, leg and arm muscles ((Fukunaga, Roy et al. 1992; Kanehisa, Ikegawa et al. 1994; Narici, Binzoni et al. 1994; Fukunaga, Ichinose et al. 1997; Maganaris, Baltzopoulos et al. 1998; Narici 1999) and the mechanical properties of tendons (Maganaris and Paul 1999). As these authors indicated, the magnitude of the angle of insertion of the fascicles with the muscle tendons and the changes in the angle with contraction affects the cellular force by the cosine of the angle. When the angle is large these effects become very large. If there is a large distribution of angles within a muscle, the cellular forces will not be uniform despite a constant torque in the limb.

The available studies have the limitation that the results to date rely on analyses of two dimensional images. Muscles have a complex three dimensional structure (Van Leeuwen and Spoor 1992) that could benefit from a three dimensional analysis. In this report we used ultrasound to work out the three dimensional characteristics of muscle fascicles inserting onto the internal tendon of the tibialis anterior muscle, which we are extensively studying by ^{31}P NMR spectroscopy. Our goal was to use methods developed for analysis of the heart (Sheehan, Bolson et al. 1998) to sample muscle fascicles over a large fraction of the tibialis anterior muscle to measure the distribution the angles of insertion within the muscle volume and to assess the magnitude of changes in the angle during isometric contraction. The hypothesis tested is that the internal muscle structure is uniform in its anatomic properties. If so, mechanical transmission of forces from tendon to cells in the fascicles can be calculated uniformly throughout the muscle. This work reports our initial effort to measure the orientation of bundles of muscle cells (fascicles) within human muscles as part of a larger project investigating the economy and efficiency of human muscles.

2. METHODS

Three subjects (2 male, 1 female) were drawn from a population of normal volunteers. All of the subjects were recreationally active but none was involved in specific training programs involving the leg musculature at the time of the study. The experimental protocol was approved by the Human Subjects Office of the University of Washington (28-0823-A) and voluntary consent was obtained from each subject prior to their participation.

Three-dimensional imaging of the fascicles of the human tibialis anterior muscle was made by capturing a series of ultrasound images while registering the location of the probe head along fixed cartesian coordinates of the laboratory. The limb was fixed at the knee and ankle to a holder which has a constant orientation with respect to imaging apparatus. The subject kept the muscle relaxed until a voluntary contraction was requested. The holder kept the ankle at a constant angle; in some measurements the angle was adjusted to obtain images at different angles of dorsi- and plantar-flexion. The apparatus incorporated a strain gauge to measure the torque produced when the subject maintained an isometric contraction. These contractions were graded as a fraction of maximal voluntary contraction (MVC) for each subject. A static magnetic field was generated and the location of the ultrasound probe head was located within the field by a commercial apparatus (Flock of Birds®, Ascension Technology Corp, Burlington, VT). This apparatus consists of magnetic field transmitter, a receiver fixed to the ultrasound probe and electronic circuitry interfaced to a computer. The three dimensional orthogonal field is continually sensed by the receiver. The digitized image file contains the ultrasound image, the receiver's position (X, Y, and Z) and orientation vector in space, and the ultrasound imaging apparatus settings. In this way the location of the images were recorded in known 3D space accurate to ± 1 mm (Legget, Leotta et al. 1998). Imaging was performed with 10 MHz array ultrasound scan head and a commercial ultrasound apparatus (ATL HD3000, Bothell, WA). Scan depth was optimized to view the internal tendon of the tibialis anterior. Imaging was performed using freehand scanning over overlapping regions of the muscle; for each region serial images at differing orientations were obtained. The right leg was studied with cross sectional (perpendicular to long axis of the limb) and parasagittal (perpendicular to cross section) scans. The cross sectional images provided muscle boundaries and locations within the muscle down to the tendon above the retinaculum of ankle joint as a reference. The parasagittal planes started just medial of the lateral edge of the tibia and extended to the lateral edge of the tibialis anterior and adjacent extensor digitorum longus muscle. Scans were made in an overlapping fashion. Two completed sets of scans were made for each experiment. The imaged region began below the knee near the largest cross section of the tibialis anterior; the cross section of the muscle continuously decreased over the region imaged towards the ankle.

The images were suitable for analysis if the image satisfied all three of the criteria: 1. a number of fascicles were visible, 2. the central tendon was visible, and 3. fascicles were visible at the central tendon. Approximately 2/3rds of the images satisfied these criteria. Portions of the muscle close to the knee cannot be analyzed by the methods used here because the central tendon is not detectable. Vector lines were drawn on each image

by a trained observer using custom software (Sheehan, Bolson et al. 1998). Vectors were
identified as fascicles, or central tendon or muscle boundary and converted into X,Y, and
Z coordinates and vectors within the muscle. These coordinates were transformed from
the laboratory reference frame to a muscle frame of reference in which the plane of the
central tendon became the X-Y plane (X increasing towards the ankle). Z plane thus
represents distances superior (towards skin) and inferior with respect to the tendon plane.
Boundaries at the sides of the muscle represent maximum +Y and −Y co-ordinates with Y
= 0 centered on the plane of the internal tendon. Note that the actual plane of the tendon
with respect to the laboratory frame rotates approximately 40° counter-clockwise (for the
right leg) at increasing X distance toward the ankle. The software calculated each
fascicle angle with respect to the local tendon plane irrespective of its rotation in the
laboratory frame. Landmarks identified in the ultrasound images (central tendon, surfaces
of the tibia and muscle superior and inferior boundaries) were confirmed in one study of
a fresh cadaver by ultrasound and subsequent dissection and by comparison with MRI of
other volunteers.

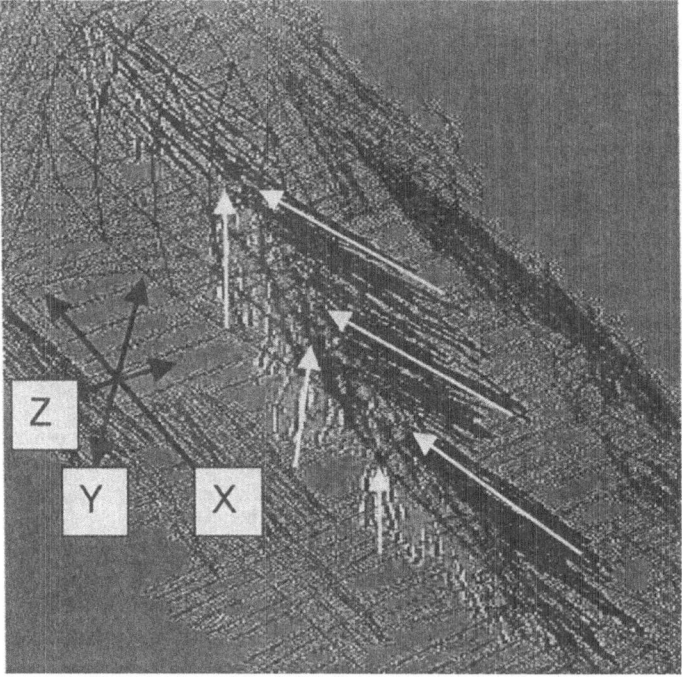

Figure 1.
Reconstruction of internal structures and boundaries of a portion of human tibialis anterior muscle. The mesh
represents the internal tendon. Vectors of tracings of fascicles are seen to touch the tendon; long white arrows
mark three fascicles superior to the tendon plane and short arrows mark three inferior ones. Tracings of the
muscle boundaries are also visible. The XYZ coordinates in the muscle are also shown with the X arrow
pointing to the ankle.

Two to four hundred fascicles were identified for each experiment. The tendon surface was reconstructed as a mesh through the tracings identified as tendon. A piecewise smooth triangular surface was fit to the 3D points labeled as tendon to define the location of the internal tendon in the anatomical reconstruction. This method accepts the marked vectors as identified and makes no assumption about the orientation of the imaging planes. In this way a well defined structure was obtained consisting of an internal tendon and fascicles inserting onto the tendon from the superior and inferior boundaries of the muscle. The angle of insertion was determined as the complement of the angle between the fascicle and the local normal of the tendon surface.

Descriptive statistics, linear regression and ANOVA tests were made with tools available in Microsoft EXCEL and JMP statistical software.

3. RESULTS

3.1. Angle of insertion of fascicles in resting muscle

The angles of insertion were acute, averaging 11° throughout the muscle and showed statistically significant differences within the muscle. The angles of insertion show a decline (i.e. become more acute) as a function of distance along the X axis towards the ankle and along the Y axis towards the boundaries of the muscle. The fascicles were also curved.

To illustrate the decrease of the fascicle angle along the length of the muscle , the fascicle angles measured at their insertion into the central tendon in one subject is displayed as a function of distance along the X axis in Fig 2.

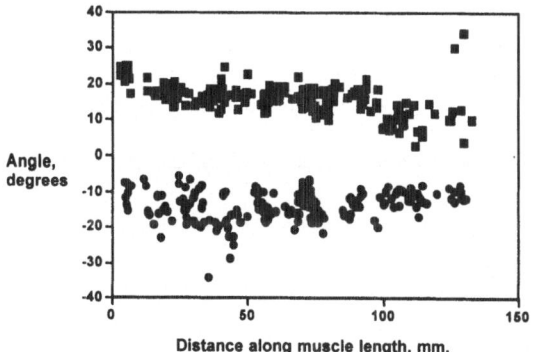

Figure 2.
Example of data from 14 cm of the muscle length spanning the thickest part of the muscle which was approximately 10 to 40 mm along the abscissa. The angle of insertion is plotted as a function of distance along the length of the muscle. Positive angles are from fascicles identified as superior to the central tendon; negative values are the fascicles inferior to the tendon. The ankle was kept at a plantar-flexion angle of 130°. Data from Subj. A r6.

The decrease in angle was tested by linear regression analysis that showed significant slopes in the fascicles superior and inferior to the central tendon (p < 0.01). Table 1 gives the results from three subjects in which the muscle was resting. The averages of the slopes of superior fascicles (0.079) and inferior fascicles (0.067) were not significantly different (one tailed p = 0.34 by ANOVA). The average angle of all the measured fascicles have relatively large coefficients of variation (100• standard deviation/mean) because the dependence of angle along X and Y planes are grouped together. The result for subject C suggests individual differences may be detectable.

Table 1

Experiment	Slope of fascicles above tendon (Degrees/mm)	P value	Slope of fascicles below tendon (Degrees/mm)	P value	Average angle of insertion, (degree)	Coefficient Of variation (%)
SubjA r9	0.050	< 0.001	0.016	0.034	13.4	41
SubjA r8	0.048	0.003	0.059	< 0.001	9.1	31
SubjA r5	0.057	< 0.001	ns	ns	9.5	38
SubjA r7	0.084	< 0.001	ns	ns	10.8	33
SubjA r6	0.060	< 0.001	0.027	0.003	14.8	43
SubjB r1	ns	ns	0.129	< 0.001	9.2	52
SubjB r4	ns	ns	0.046	<0.001	12.0	31
SubjC r1	ns	ns	0.149	< 0.001	8.1	40
SubjC r4	0.173	< 0.001	0.042	0.015	6.9	34

(ns means not significant at the level p < 0.05)

Analysis of the regression intercepts (not shown) and slopes show that the angle of insertion decreases from ~ 15° near the maximal cross section of the muscle (origin on the X axis) to ~ 7° closer to the ankle.

We also analyzed the distribution of angles of insertion of the fascicles along the sagittal plane, that is, in the medial to lateral dimension (Y plane). This analysis used the same set of images. Regression of angle of insertion against the distance along the Y axis showed for most data sets a significant relationship. The range of distances that was possible to sample with our present method was ~ 2cm. All the results from three subjects in which the muscle was resting are given in Table 2.

Table 2

Experiment	Slope superior fascicles (degrees/mm)	P value	Slope inferior fascicles (degrees/mm)	P value
Subj A r9	0.48	< 0.001	0.16	0.001
Subj A r8	0.35	0.003	0.25	0.009
Subj A r5	0.24	0.029	ns	ns
Subj A r7	ns	ns	0.21	0.003
Subj A r6	0.37	< 0.001	ns	ns
Subj B r1	ns	ns	0.34	< 0.001
Subj B r4	0.28	< 0.001	ns	ns
Subj C r1	0.12	0.004	ns	ns
Subj C r4	0.46	< 0.001	ns	ns

The averages of the slopes of superior fascicles (0.33) and inferior fascicles (0.24) were not significantly different (one tailed p = 0.09 by ANOVA). The scatter in these data is greater than in the longitudinal (X axis). In some of the cases without statistical significance there was a tendency for the angle to decrease at both extremes of the Y dimension. The results thus show that the angle of the fascicles becomes more acute towards the boundaries of the muscle as well as along the length of the muscle towards the ankle.

The fascicles were curved. This result was found by analyses of two subjects in which an additional protocol for tracing individual fascicles was used. In addition to marking fascicles seen to insert into the tendon as in all of the data reported above, segments of fascicles were also traced at various distances from the tendon plane toward the superior and inferior boundaries, over a range of 4 mm above and below the plane. The fascicle angle was determined by a linear extrapolation and projection of the traced fascicle to the central tendon. There was a reduction in the measured angle as a function of distance along the Z axis; this was tested by linear regression of fascicle angle against distance along the Z axis. Twenty of the 24 sets of images have significant (p < 0.01) slopes of the regression of angle against distance above or below the tendon plane showing that the fascicles were curved with increasing angles closer to the tendon plane.

3.2. Angle of Insertion of Fascicles in Active Muscle

Voluntary contraction of the muscle is expected to elicit some shortening ot the fascicles because they are in series with the compliance of the tendons and apparatus and

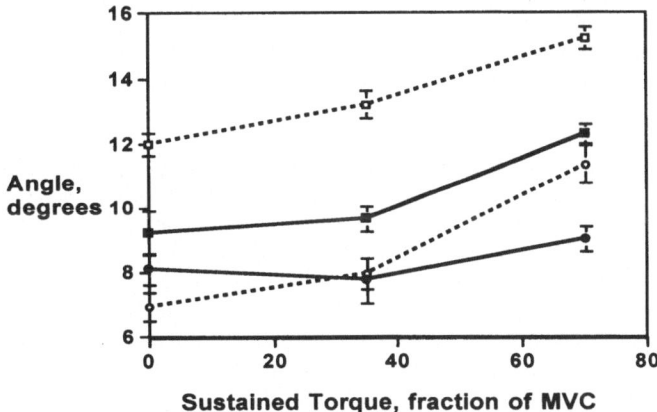

Figure 3.
Fascicle angle was averaged along the muscle length at rest during a sustained isometric contraction at 35% and at 70% of a maximal voluntary contraction for two subjects. Error bars represent ± 95% confidence limits. Two ankle positions were studied: filled symbols, ankle at neutral position (90°) and open symbols, ankle in plantar flexion (120°). Squares are from subject B and circles from subject C.

thus increase the angle of insertion. Measurements were made of the fascicle angle (averaging over all images) at rest and during sustained isometric contractions at 35% of maximal voluntary isometric contraction (MVC) and at 70% MVC with the ankle at neutral position (90°) and flexed in plantar-flexion (120°) and the results are displayed in Fig. 3. Higher activation forces could not be sustained long enough to record all the ultrasound images needed for 3D reconstruction. The results displayed in Fig 3 show that a small increase in angle was measured during the sustained contractions. All the data points at 70% MVC are significantly greater than at rest and at 35% MVC. The differences at rest are consistent with the variation noted in the different experiments in Table 1.

4. DISCUSSION

This work used an ultrasound method, originally developed for analysis of wall motion in the myocardium as an aid to clinical diagnosis (Sheehan, Bolson et al. 1998), for structural analysis of the internal structures of limb muscle. The connective tissue surrounding the bundles of muscle cells (fascicles) are highly visible in ultrasound images. Their dimensions are at or below the limit of resolution, namely, several tenths of a millimeter. Tracing the large number of fascicles gave a good statistical sampling of the muscle structure and enabled a three dimensional analysis. This three dimensional construction gave much useful and significant information on the internal arrangements of muscle fascicles and internal tendons in human muscles that is not available with the 2D images reported in the literature. The main result is that the angles of insertion of the fascicles in the tibialis anterior muscle are acute, averaging 11° over the muscle volume analyzed. The results also show systematic changes in fascicle angle along the length and cross section of the muscle. Along the long axis, the angle of insertion decreases as distance increases from the thickest cross section towards the ankle. The angle of insertion also decreases in the medio-lateral plane and towards the boundaries of the muscle. The last finding means the fascicles are curved slightly; this result was recently found in the medial gastrocnemius by normal 2D ultrasound analyses (Muramatsu, Muraoka et al. 2002). The finding of curved fascicles and variation of angles of insertion along the length and cross section of a bipennate muscle is consistent with requirements for a mechanically stable muscle structure and the generation of significant internal pressure during contraction (Van Leeuwen and Spoor 1992; Van Leeuwen and Spoor 1993; van Leeuwen and Spoor 1996). Our results show that 2-D projection images can be used to characterize the internal architecture provided the normal assumptions needed are tested by 3D analysis. Two dimensional images must be used for imaging dynamic events such as shortening of fascicles and tendons during activity, because 3 dimensional imaging requires substantially longer periods of data collection during which the structure must be constant. However the specific findings in the tibialis anterior may not be generalized to other muscles. For example the gastrocnemius has larger angles of insertion than does the tibialis anterior (Narici, Binzoni et al. 1994; Muramatsu, Muraoka et al. 2002) and thus the variations with the tibialis anterior if scaled proportionately would have a large functional significance in the gastrocnemius. Thus a three dimensional analysis should be part of a complete analysis of muscle structure and function.

Our present results have large functional significance. The angles of insertion found in the tibialis anterior are always small in magnitude; the average was 11°. These acute angles change by no more than ~6° within a resting or active muscle. This means that the stress in the muscle cells is approximately the same as in the tendon to which they insert, because the relation of the tendon stress to muscle stress is the reciprocal of the cosine of the angle of insertion. For the average angle measured (11°) the cosine equals 0.98. Thus the muscle cell stress is nearly in line with the tendon stress. Even considering the regional differences within the muscle, this quantitative conclusion remains valid. This work significantly extends prior analyses (e.g. (Fukunaga, Ichinose et al. 1997; Ito, Kawakami et al. 1998)) by giving a richer description of the range of angles of insertion throughout the muscle volume and by showing that the ratio of stress in muscle to stress in tendon is not quantitatively different given the small range of insertion angles observed. We have not evaluated the length of fascicles in this work but there is good reason to believe their lengths are also approximately constant within this muscle. The reason is because the angle of insertion decreases from the muscle belly to the ankle and so does the distance from the plane of the central tendon to the muscle surface. This measurement needs to be made but we note that this result would be similar to the analysis of force-length relationships in human cadaver specimens (Spoor, Van Leeuwen et al. 1991). Moment arm of the muscles across the ankle joint can be estimated well by ultrasound and magnetic resonance imaging techniques (Hoy, Zajac et al. 1990; Rugg, Gregor et al. 1990; Spoor, Van Leeuwen et al. 1990; Koh and Herzog 1998; Maganaris, Baltzopoulos et al. 1999). Thus all the information is available to compute average muscle cell stresses from measurements of torque at the ankle. This knowledge is critical to evaluate muscle economy as can be done by measuring total ATPase in active human muscle by ^{31}P NMR methods.

Improvements in 3 dimensional analyses is possible. We show a considerable scatter in the measurements (e.g. Fig 2). Typical coefficients of variation (percentage of standard deviation divided by mean) are 30 - 40% (Table 1). Some of the error in angle determination is due to the non-planar nature of the mesh reconstruction of the tendon and the fineness of the triangulation mesh. Our current algorithm calculated the local normal of the plane of the central tendon at the end of the traced fascicle. Neither calculation of the mesh of the tendon with different levels of fineness nor smoothing and averaging of the tendon plane was studied so the magnitude of this contribution to the error was not evaluated. Only three subjects were studied in our first use of this technique to establish validity and meaningful results. The average angle of insertion in subject C (Table 1) appears to be smaller than in the other two subjects; this subject is female and difference may reflect a sex difference. Study of a larger number of subjects will be needed to detect individual variation in their internal muscle anatomy. The central tendon is not visible in the top region of the muscle, between the knee and the thickest cross section, even though the fascicles are clearly seen. It may be possible to extrapolate the plane of the central tendon in that direction so as to enable analysis of the fascicles with respect to an internal landmark.

Acknowledgements:
This work was supported by grants from the US National Institutes of Health (AR 41928) and from the National Space Biomedical Research Institute (MA00212).

5. REFERENCES

Fukunaga, T., Y. Ichinose, et al. (1997). "Determination of fascicle length and pennation in a contracting human muscle in vivo." *J Appl Physiol* **82**(1): 354-8.

Fukunaga, T., R. R. Roy, et al. (1992). "Physiological cross-sectional area of human leg muscles based on magnetic resonance imaging." *J Orthop Res* **10**(6): 928-34.

Hoy, M. G., F. E. Zajac, et al. (1990). "A musculoskeletal model of the human lower extremity: the effect of muscle, tendon, and moment arm on the moment-angle relationship of musculotendonactuators at the hip, knee, and ankle." *J Biomech* **23**(2): 157-69.

Ito, M., Y. Kawakami, et al. (1998). "Nonisometric behavior of fascicles during isometric contractions of a human muscle." *J Appl Physiol* **85**(4): 1230-5.

Kanehisa, H., S. Ikegawa, et al. (1994). "Comparison of muscle cross-sectional area and strength between Koh, untrained women and men." *Eur J Appl Physiol* **68**(2): 148-154.

T. J. and W. Herzog (1998). "Increasing the moment arm of the tibialis anterior induces structural and Legget, functional adaptation: implications for tendon transfer." *J Biomech* **31**(7): 593-9.

M. E., D. F. Leotta, et al. (1998). "System for Quantitative Three-Dimensional Echocardiography of the Left Ventricle Based on a Magnetic-Field Position and Orientation Sensing System." *IEEE Biomed. Engineering* **45**(4): 494 - 504.

Maganaris, C. N., V. Baltzopoulos, et al. (1998). "In vivo measurements of the triceps surae complex architecture in man: implications for muscle function." *J Physiol* **512**(Pt 2): 603-14.

Maganaris, C. N., V. Baltzopoulos, et al. (1999). "Changes in the tibialis anterior tendon moment arm from rest to maximum isometric dorsiflexion: in vivo observations in man." *Clin Biomech (Bristol, Avon)* **14**(9): 661-6.

Maganaris, C. N. and J. P. Paul (1999). "*In vivo* human tendon mechanical properties." *J. Physiol.* **521**: 307 - 313.

Muramatsu, T., T. Muraoka, et al. (2002). "In vivo determination of fascicle curvature in contracting human skeletal muscles." *J Appl Physiol* **92**(1): 129-34.

Narici, M. (1999). "Human skeletal muscle architecture studied in vivo by non-invasive imaging techniques: functional significance and applications." *J Electromyogr Kinesiol* **9**(2): 97-103.

Narici, M. V., T. Binzoni, et al. (1994). "Human gastrocnemius muscle architecture from rest to the contracted state." *Journal of Physiology* **475**: 17 P.

Rugg, S. G., R. J. Gregor, et al. (1990). "In vivo moment arm calculations at the ankle using magnetic resonance imaging (MRI)." *J Biomech* **23**(5): 495-501.

Sheehan, F. H., E. L. Bolson, et al. (1998). "Three Dimensional Echocardiography System for Quantitative Analysis of the Left Ventricle." *IEEE Computers in Cardiology* **25**: 649 - 652.

Spoor, C. W., J. L. Van Leeuwen, et al. (1990). "Estimation of the instantaneous moment arms of lower-leg muscles." *J. Biomechanics* **23**(12): 1247 - 1259.

Spoor, C. W., J. L. Van Leeuwen, et al. (1991). "Active Force-Length Relationship of Human Lower-leg **29**(3): Muscles Estimated from Morphological Data: a Comparison of Geometric Muscle Models." *Europ. J. Van Morphol.* 137 - 160.

Leeuwen, J. L. and C. W. Spoor (1992). "Modelling mechanically stable muscle architectures." *Phil. Trans. R. Soc. Lond. B* **336**: 275 - 292.

Van Leeuwen, J. L. and C. W. Spoor (1993). "Modelling the pressure and force equilibrium in unipennate van muscles with in-line tendons." *Philos Trans R Soc Lond [Biol]* **342**(1302): 321-333.

Leeuwen, J. L. and C. W. Spoor (1996). "A Two Dimensional Model for the Prediction of Muscle Shape and. Intramuscular Pressure." *Europ. J. Morphol.* **34**(1): 15 – 30.

DISCUSSION

Ranatunga: Do you have measurement of fiber length?

Kushmerick: No, it is not easy to follow full length of fiber in most of our measurement.

Gonzalez-Serratos: Will you please elaborate in whether this is an efficient mechanism of force transmission?

Kushmerick: It is efficient in this muscle (tibialis anterior) because of the very low angles observed. This may not be true for other muscles, in which the angles may be larger or there is a dispersion of angles, as is likely the case in the human gastroenemius.

GENERAL DISCUSSION PART I

Chaired by G.H. Pollack, S. Ishiwata, and H. Sugi

CAN THE MECHANISM OF MUSCLE CONTRACTION BE SOLVED ONLY BY IN VITRO MOTILITY ASSAY EXPERIMENTS?

Pollack: I would like to discuss various models obtained with in vitro motility assay experiments. In general, contraction models are classified into two different types: one type of model assumes myosin head motion (e.g. lever arm mechanism) while attached to actin and therefore has clear structural constraints, while the other type of model is based on a thermal ratchet mechanism without definite structural constraints. I would like to have your comments on them.

Huxley: Such a discussion should only be made based on experimental evidences.

Sugi: I agree with Dr. Huxley's opinion. The discussion can be productive if it is based on concrete experimental evidence but not on mere speculations. And I am not happy with your manner of statement, in which you unconsciously ignore the possible involvement of myosin subfragment-2 in muscle contraction (Sugi et at., this volume, p. 319).

Frankly speaking, I have serious doubt about the effectiveness of the so-called in vitro motility assay experiments in solving muscle contraction mechanism. From the standpoint of Physiologists, the most important thing is what is taking place in living muscle during contraction, but not the movement of fluorescent actin filament and fluorescent ATP on the glass surface under conditions differing too far from those in intact muscle. For example, motility assay people observe attachment of a fluorescent ATP to an invisible myosin head fixed in position, and its detachment from it, and claim that they visualize ATP hydrolysis by a myosin head, without proving whether the detached particle is ADP or ATP. In my opinion, such experiments add almost nothing to our knowledge about actomyosin ATPase. I think the greatest contribution to this field made by biochemists is the late Dr. Tonomura's discovery of the long life intermediate M-ADP-Pi in the past 50 years.

Ishiwata: I largely agree with your comments, although I have been using techniques of in vitro motility assay frequently. Recently, I am trying to construct assay systems, in which three-dimensional myofilament lattice structure is preserved (see Ishiwata's paper in this volume, P. 107).

Solaro: I think transgenesis, in which specific changes in myofilament proteins modify function, provide a way to test whether *in vitro* biochemistry is translated to physiological function. For example, in the case of a specific replacement of cardiac troponin I with slow skeletal troponin, the blunting of the effect of acidic pH on myofilament activation can be translated to the intact *in situ* heart.

Pollack: I want to have opinions on what we have learned from single molecule work.

Granzier: We have learned that the presence of S2 is not a requirement for sliding and force development by myosin II.

Holmes: A number of *in vitro* motility assays are done on single molecule measurements, and such experiments have shown very good correlation between the number of light chain binding sites and the step size.

Sugi: The scheme of lever arm mechanism based on the structured changes of acto-S1 complex is very beautiful and interesting. However, I would like to repeat my question made after your presentation in Session II, i.e., whether the tilting of the lever arm portion of S-1 coupled with ATP hydrolysis can generate a torque strong enough to cause filament sliding under high external load or not. I made the same argument with Takeyuki Wakabayashi. He completely agreed with me in that this point should be clarified in the future. I believe that myosin S-2 is also involved in muscle contraction (this volume, p. 319).

Holmes: I am not a physiologist. Physiologists should deal with this matter.

PROBLEMS OF MYOSIN HEAD INTERACTION DISTANCE AND MUSCLE FIBER STIFFNESS

Sugi: As Dr. Huxley pointed out at the beginning of this General Discussion, arguments about possible contraction mechanisms should be done on the basis of experimental fact, but not on the basis of "arbitrary imaginations." Instead of "arbitrary postulations" about the myosin step size used by in vitro motility assay people, I would like to draw your attention on the equation widely used among muscle physiologists to obtain "interaction distance" Di of a myosin head during muscle fiber shortening. Di is the distance, at which a myosin head (cross-bridge) remains attached to the thin filament during steady myofilament sliding in muscle. Di is maximum during unloaded shortening, and minimum during isometric contraction, and expressed as,

$$Di = Ds \ Sa \ [Mo] \ / \ [ATP]_u,$$

where Ds is the total sliding distance between the thick and thin filaments determined as shortening per half sarcomere, Sa is the muscle fiber stiffness during (free loaded) shortening relative to that in rigor state, [Mo] is myosin head concentration within the fiber, and $[ATP]_u$ is the amount of ATP utilized. In the above equation, [Mo] and

[ATP]$_u$ can be measured directly, while Sa, representing the proportion of actively working myosin heads during fiber shortening, is normally estimated by applying small length perturbation in kHz region and measuring the coincident force changes.

Using the above method, Higuchi and Goldman (Nature 352: 352-354, 1991) obtained Sa to be 40% of the rigor stiffness, and got a Di value of ~40nm, a value favoring the loose coupling mechanism, in which a cross-bridge is assumed to perform multiple powerstrokes per ATP molecule.

On the other hand, however, Stehle & Brenner used length perturbations up to MHz region, and reached a conclusion that the value of Sa is only 1-5% of that in an isometrically contracting muscle fiber (Stehle and Brenner, Biophys. J. 78: 1458-1473, 2000). If one uses the above value of Sa, then the Di value is <10nm, i.e. a value within the framework of A.F. Huxley contraction model, in which one cross-bridge powerstroke is tightly coupled with hydrolysis of one ATP molecule (A.F. Huxley, Prog. Biophys. Biophys. Chem. 7: 255-318, 1957). Dr. Stehle, could you briefly explain your experiments and conclusions obtained?

Stehle: Previous studies focussing on the interaction of cross-bridges in fibers by Higuchi & Goldman were based on the assumption that the duty ratio can be directly estimated from the ratio of fiber stiffness measured during active shortening to the fiber stiffness at maximum cross-bridge attachment, i.e. the stiffness in rigor. This presumes that fiber stiffness during active shortening only results from strongly-bound cross-bridges. However, work by Brenner and co-workers (Proc. Natl. Acad. Sci. USA 79: 7288-7291, 1982) revealed that cross-bridge of the weakly-bound type also cause significant fiber stiffness, especially in the presence of high [Ca^{2+}] (Kraft et al., Proc. Natl. Acad. Sci. USA 89: 11362-11366, 1992). The aim of our recent study (Stehle & Brenner, Biophys. J. 78: 1458-1473, 2000) was therefore to determine the duty ratio not from the total fiber stiffness measured during active shortening, but from the part remaining after the contribution of weakly bound cross-bridges to fiber stiffness had been subtracted. By examining dynamic properties of stiffness in skinned rabbit psoas fibers, we found that magnitude of stiffness and the dependence of stiffness on speed of stretch (stiffness-speed relation) during active unloaded shortening were similar to those determined under the same experimental conditions ([Ca^{2+}], ionic strength, temperature and velocity of filament sliding) but with MgATP replaced by the non-hydrolyzable nucleotide analogue MgATPγS, which completely prevented active force generation, i.e., formation of strongly bound, force-generating cross-bridges (see Fig. I-1). The similarity of stiffness suggests that most of the fiber stiffness found during active unloaded shortening results from components other then strongly-bound cross-bridges, most likely weakly-bound cross-bridges. From our stiffness data, we estimated that during unloaded fiber shortening at V$_{max}$, cross-bridges spend less than 5 % (most likely 1-3 %) of their time passing an ATPase cycle in strongly-bound states. This estimate of the duty ratio, together with the ATPase rate and the magnitude of V$_{max}$, lead to an interaction distance similar to the size of a cross-bridge. Thus, there is no need to assume cross-bridge interaction with multiple actin sites while occupying strongly bound states during an ATPase cycle even at maximum speed of shortening.

In contrast to our interpretation of stiffness-speed relations observed with skinned rabbit psoas fibers under non-force generating conditions (relaxed or in the presence of

MgATP γS), i.e., those reflecting actin binding kinetics of weakly-bound cross-bridges, recent data on intact frog muscle fibers were interpreted to provide no evidence for the presence of weakly-bound cross-bridges under relaxing conditions (see following discussion by Cecchi). However, the contribution of weakly-bound cross-bridges to intact fiber stiffness at activating [Ca²⁺] is unknown, as it has not been tested experimentally by exchanging the nucleotide. However, even if viscoelasticity by non-cross-bridge components contribute to fiber stiffness, this will not change our estimate of the duty ratio. Regardless of whether actin-binding kinetics of weakly-bound cross-bridges or viscoelasticity of some non-cross-bridge components cause the significant fiber stiffness under Ca^{2+}-activated but non-force generating conditions, i.e., in the presence of ATP γS at high [Ca²⁺], the observed stiffness has to be subtracted from the fiber stiffness measured during active unloaded shortening in order to determine the fraction of strongly-bound cross-bridges which is equivalent to the duty ratio.

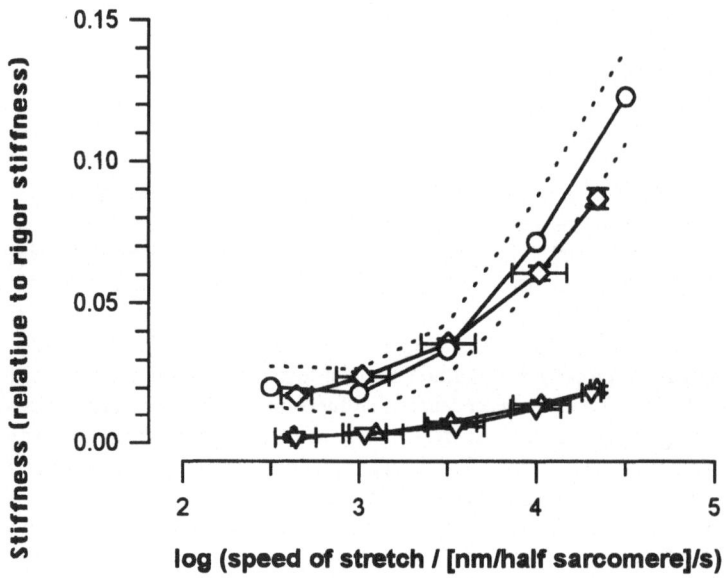

Fig. I-1: Comparison of the stiffness-speed relation derived for active unloaded shortening with those measured during passive shortening at similar velocities. Effect of nucleotide and [Ca²⁺]. Experiments were performed at $\mu = 170$ mM, at sarcomere lengths of 2.5-2.6 μm, and at -1 °C to ensure full saturation of myosin heads with ATPγS even at high [Ca²⁺]. Symbols: stiffness during passive shortening at low [Ca²⁺] (pCa < 8), in presence of 10 mM MgATP (Δ) or 10 mM MgATPγS (▽); stiffness during passive shortening at pCa 4.5 in presence of 10 mM MgATPγS (◊); stiffness derived for active unloaded shortening at pCa 4.5 in presence of 10 mM MgATP. Dashed lines represent the 95 % confidence-limits of the stiffness-speed relation for active unloaded shortening. All stiffness values are normalized to rigor stiffness. Each data point represents the mean ± SE of 5-23 stiffness measurements. Note that stiffness during active unloaded shortening is much higher than both, stiffness under relaxing conditions and stiffness in presence of MgATPγS in the absence of Ca²⁺, but very similar to the stiffness in presence of MgATPγS at high [Ca²⁺] (from Stehle & Brenner, *Biophys. J.* 78:1458-1473, 2000).

Sugi (added after the discussion): The concept of interaction distance comes from the Huxley 1957 contraction model (A.F. Huxley, 1957), in which each myosin head (cross-bridge) first interacts with actin filament producing a positive force for the myofilament sliding, and then pushed by positive forces generated by other myosin heads, so that it produces a negative force opposing sliding until it is detached from actin filament. The length of the distance along which a myosin head is kept attached to actin filament is designated as interaction distance. Since a myosin head is pear-shaped with length of ~20nm, this imposes a structural constraint ≤20nm concerning interaction distance, if each head motion is assumed to couple with hydrolysis of one ATP molecule.

PROBLEMS OF WEAKLY-BINDING CROSS-BRIDGES

Sugi: Thank you very much Dr. Stehle. I would like to point out that stiffness obtained with kHz perturbation is a very convenient value, but can also be a dangerous value leading to wrong concepts. If you extend the range of perturbation from kHz to MHz region, you get completely different features of muscle fiber stiffness. In this connection, I remember that Dr. Cecchi has applied MHz region perturbation to relaxed muscle fibers, reaching a conclusion that the "fashionable" weakly-binding myosin head concept is wrong. Dr. Cecchi, I hope you to briefly summarize your work.

Cecchi: Our experiments (Bagni et al., Biophysical J. 63: 1412-1415, 1992 and J.Physiol. 482: 391-400, 1995) were made to investigate the possibility that weakly binding bridges, demonstrated mainly in skinned rabbit muscle fibers (Brenner et al., Proc.Natl.Acad. Sci. U.S.A. 79: 7288-7391, 1982) could also be detected in intact frog muscle fibers bathed in normal Ringer solution. Schoenberg (Biophys. J. 48: 467-475, 1992) demonstrated that the force response of a simple rapid equilibrium cross-bridge model to stretches of constant velocity, is equivalent to that of a viscoelastic system having an appropriate relaxation time. Therefore we examined the force response of relaxed frog fibers to fast ramp stretches searching for a force component having a viscoelastic properties as those expected from weakly binding bridges. Results from a typical experiments are reported in Fig. I-2. It can be seen that the force response (neglecting the very small resting tension) can be split into two phases: 1) a fast initial tension increase at the beginning of the stretch (P_1) and, 2) a much slower tension rise ending at the end of the stretch (P_2).

Comparison of the force and stretching velocity traces shows that: phase 1 corresponds to the acceleration period during which sarcomere stretching velocity attains its steady state value and, 2) phase 2 coincides with the period of stretching at constant velocity. Fig. I-3 shows the dependence of P_1 amplitude from stretching speed (A) and from the reciprocal of stretch duration (B) at 3 different sarcomere lengths. It can be seen that: 1) P_1 amplitude increases linearly with stretching velocity as in a viscous response; 2) P_2 amplitude increases exponentially with the reciprocal of stretch duration to reach a plateau level as in a viscoelastic response; 3) both P_1 and P_2 responses increases progressively with sarcomere length. Analysis of the data shows that P_1 is consistent with a viscous response or with a viscoelastic response having a relaxation time much shorter than the minimum stretch duration used in our experiments (230 µs). If P_1 is assumed to be viscous, it cannot arise from weakly binding bridges since in terms of cross-bridge

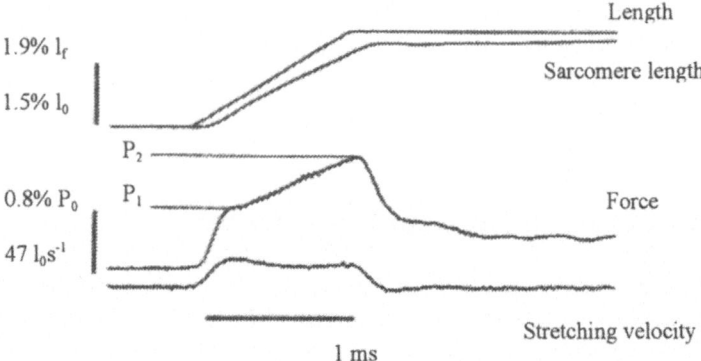

Fig. I-2. Records from a typical experiment on a resting fiber subjected to a ramp lengthening at 22 l_0s^{-1}. Initial part of the force response is clearly composed by two phases indicated as P_1 and P_2. The sharpness of the P_1-P_2 transition depends on the shape of the sarcomere length change and this in turn depends on the properties of the fiber and of the tendon attachments (Bagni et al., 1992).

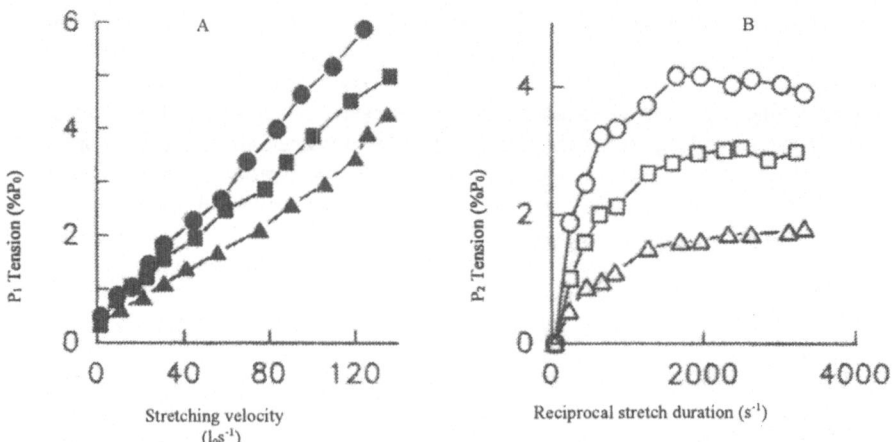

Fig. I-3. P_1 amplitude versus stretching velocity (A) and P_2 amplitude versus reciprocal time of stretch duration (B) at 2.15 μm (triangles), 2.50 μm (squares) and 2.65 μm (circles). Both P_1 and P_2 increases in spite of the decreasing of filament overlap (Bagni et al., 1995).

kinetics a viscous system means a detachment rate constant of infinite value. Alternatively, if P_1 is assumed to be viscoelastic, the force response should be distorted and delayed respect to the stretching speed by about one relaxation time. In our experiments the delay was never longer than 5 μs. In terms of cross-bridge kinetics this would mean a detachment rate constant of about $2x10^5$ s^{-1}, a values of about 20 times higher than expected.

Therefore it is unlikely that P_1 response could arise from weakly binding bridges. P_2 is a viscoelastic response and it is consistent, in principle, with the response expected from weakly binding bridges. However, the relaxation time of P_2 response is about 1ms, which would correspond to a detachment rate constant of one order of magnitude smaller than that expected from weakly binding bridges. In addition, the force records do not show any sign of cross-bridges detachment (give) even when the stretching amplitude (6% l_0) largely exceeded the elastic range of the cross-bridge (1.2-1.5% l_0) and the stretching speed was too higher to allow any cross-bridge cycling during the stretch. Another important finding contrasting with the idea that weakly binding bridges are involved in passive force response in our preparation is that both P_1 and P_2 responses increase progressively with sarcomere length in the whole range examined. Further studies from our group have shown that similar responses where obtained also in skinned preparation from frog muscle fiber both at high and low ionic strength and in a sarcomere length range up to 3.27 μm (Bagni et al., J.Physiol., 482, 391-400, 1995).

In summary, our results show that the force response of relaxed frog muscle fibres, both intact and skinned, to ramp stretches have no measurable component attributable to the presence of weakly binding bridges. These results seems to be in contrast with previous experiments showing mechanical evidences of weakly binding bridges such as those presented in the preceding discussion by Dr. Stehle. This is not necessarily true, however, since our experiments were made in frog muscle while the main experiments showing mechanical evidences of weakly binding bridges presence, including Dr. Stehle's experiments, were made on rabbit psoas muscle at low temperature. In this preparation especially at low temperature, weakly binding bridges can be effectively present in a great number as suggested by x-ray diffraction experiments.

Sugi: Thank you very much, Dr. Cecchi. Finally, I would again like to emphasize that completely unexpected, new results can be obtained by extending the frequency of perturbation to higher frequencies.

Yu: Dr. Cecchi, the most important part missing in your claim that there are no weak binding cross-bridges is that the kinetics of actin-myosin interaction of frog muscle are not known. I think it is important to remember that, unlike in rabbit skeletal muscle, there is very little biochemical information about the kinetics of the rapid equilibria in frog muscle. Furthermore, temperature dependence of the kinetics of frog muscle is not known. I assume that your experiments were performed at 4-6 °C. It could be that for frog muscle, the kinetics are indeed one order of magnitude faster than those found in rabbit muscle at these temperatures. For rabbit muscle, there has been a great deal of well established data correlating between the biochemical, mechanical and structural measurements (e.g. Brenner, 1992; Basic Res. Cardiol. 81: 1-15, 1986; Chalovich, Pharmac. Ther. 55: 95-148, 1992; Xu et al., Biophys. J. 73: 2292-2303, 1997), at various temperatures and with various ligands bound at the active site of myosin (Xu et al., Biophys. J. 77: 2665-2676,

1999). I hope that similar experiments will be carried out with the frog skeletal muscle in the future.

SPECIAL TALK: THE BEGINNING OF LOW-ANGLE X-RAY DIFFRACTION WORK ON LIVING MUSCLES

Sugi: It is almost 50 years since Dr. Hugh Huxley first discovered hexagoal double array of myofilaments in skeletal muscle. This is indeed a most remarkable example showing greatness of human wisdom.

As we have him with us on this occasion, I asked him to give us a short story about his monumental discovery. It would be helpful for young investigators to learn from him about the way by which a breakthrough is achieved.

Huxley: Professor Sugi has asked me to give a short extra talk about my early work, fifty years or more ago, on X-ray diffraction by muscle. This has been covered to some extent in my biographical article in Ann. Rev. Physiol. 58, 1-19 (1996), but it may be of interest to some readers of the published volume from this meeting if I summarize the course of events.

In 1948 I had begun work as a graduate student in a small research group, financed by the Medical Research Council, in the Cavendish Laboratory, Cambridge, U.K.

Fig.1 Skiing in Austria, 1949.

Molecular and Cellular Aspects of Muscle Contraction
Edited by H. Sugi, Kluwer Academic/Plenum Publishers, 2003

The group consisted of John Kendrew and Max Perutz, who were trying to find a way of solving the structure of proteins by X-ray diffraction. At that time, progress was very slow and the prospects of success seemed very uncertain, and in 1949, I thought that it might be more rewarding to see if the X-ray technique could yield more definite, if less detailed, information about some basic facts concerning muscle structure. Beyond what could be seen in the light microscope, this was largely unknown at that time, except that it seemed to consist of filaments of an actomyosin complex.

The basis of all the X-ray work on crystalline proteins was the discovery some fifteen years earlier, by Bernal and Crowfoot, that the secret of getting good X-ray diffraction diagrams from such crystals was to keep them in the fully hydrated state, in contact with their mother-liquor; otherwise they dried out, and, as the crystals had a large water content, became very disordered. While hydrated, they gave beautifully detailed patterns.

I wondered if the same might be true of muscle, which I knew contained about 80% of water, if it were kept in its native condition, still capable of contraction. I had estimated that muscle filaments might be several hundred angstroms apart, so I built a low-angle X-ray camera, hoping to find evidence for their arrangement and properties.

Since I suspected that the reflections might be very weak, I used a very fine focus X-ray tube developed by Ehrenberg and Spear, working in Bernal's lab in London. This produced a relatively high brightness X-ray source for my camera, which had very narrow slits (main slit 5-10 microns wide) and a very short specimen to film distance (3 or 6 cms), and was quite efficient, though it still took several hours, and in some cases days, to record the patterns.

After some teething troubles (i.e., keeping the muscle properly cooled and oxygenated), I was very excited to obtain, by 1950, some reflections on the equator of the diagram, just a pair of lines on either side of the beam stop, but indicating the presence of an hexagonal array of filaments about 440Å apart. When the muscle was stretched, the filaments moved closer together, in an approximately constant volume manner, as one would expect if the filament lattice extended across the whole cross-section of the muscle fibers. So I concluded that I must be looking at the basic contractile elements (Huxley, 1951).

Fig. 2 Taking MA degree, 1950.

More significantly, I found that

muscles in rigor also gave a pattern, with approximately the same spacings, but with the outer reflection (which had been much weaker than the inner one in the resting muscle) now much stronger. This could be accounted for if a second set of filaments, which in the resting muscle were positioned more randomly around the first set, became fixed at the trigonal positions in the lattice. (i.e., each one lying symmetrically between each group of three primary filaments, so that each primary filament had six secondary filaments spaced around it).

I knew from Szent-Gyorgyi's and H.H. Weber's work that actin and myosin formed a complex which could be dissociated by ATP, and that muscle and actomyosin fibers were inextensible in the absence of ATP, but became plasticized in its presence. So I guessed that the primary and secondary filaments that I thought were present must consist of myosin and actin, present as separate filaments, which would cross-link together in the absence of ATP. This could hold the actin filaments close to the symmetrical positions that they appeared to occupy in rigor (we now know that the "cross links" – myosin head crossbridges – also make a major contribution themselves to the change in the equatorial X-ray pattern when they attach to actin, both in rigor and in contraction). I also supposed that this crosslinking of the two sets of filaments in rigor was responsible for the greatly increased stiffness, but I did not get as far as guessing the sliding filament mechanism for contraction at that time! But I did also observe a ~ 420Å axial periodicity in relaxed muscle, several orders of which were visible, with the third one particularly strong. They appeared to include one at about 60Å (now known to come from actin). These periodicities did not change during passive stretch, showing that there was a constant subunit repeat (Huxley, 1953a). I thought that this double array of filaments was continuous through the whole sarcomere (at that time people believed that the extra density of the A-band was due to the presence of some unknown additional substance there!). So I thought that contraction might be produced by some mechanism in which actin was depolymerized as it catalyzed myosin ATPase.

I finished my Ph.D. in the summer of 1952, and in September went to M.I.T. as a post-doc, to learn about the new science of biological electron-microscopy, in Professor Frank Schmitt's lab. Alan Hodge, Dave Spiro and I collaborated in building a special microtome for thin sectioning, which we then used for our own separate projects. I used it to cut cross-sections of fixed and embedded frog and rabbit muscle, to try to see the double hexagonal lattice which I had predicted from the X-ray pattern. I was soon successful, and it was a great feeling to be able to see images, directly in the electron microscope, of the structures I had been thinking about for several years.

Then in January 1953, Jean Hanson, from the MRC Biophysics Unit in Kings College, London, also arrived in Schmitt's lab, to learn electron microscopy. She had been doing light microscope work in muscle fibrils at Kings, using the newly developed phase-contrast microscope, which gave very beautiful images of the band-pattern. There had been great controversy for over a century about what happened to the pattern when the sarcomeres shortened during contraction, and at that time there was no convincing description – or explanation – of the changes, since observations had been conducted on intact fibers, usually about 100 microns thick, with ordinary light microscopes, a system in which enormous optical artifacts could readily be produced when observing detail at the one micron level, or less. Moreover, photographic records of the observations were rarely made. Single myofibrils, about one micron thick, which give excellent contrast in phase, were relatively free of such artifact, and gave completely convincing images.   Jean and I decided to join forces, putting together her experience and expertise with light microscopy of fibrils, and my structural evidence at the submicroscopic level, from X-ray diffraction and electron-microscopy of whole muscle.

We could pass various solutions over the myofibrils very easily, between the slide and coverslip to which they were loosely attached, and observe an individual myofibril as the solution flowed past it.   We were astonished to see that the dark A-bands quickly vanished when we used solutions which extract myosin from muscle.   It was the greatest revelation of my scientific life, for we quickly realized what it must mean: that the myosin filaments were present only in the A-bands, and it was they that accounted for the much higher density there; and that they were partially overlapped by actin filaments, which extended from the Z-line to the edge of the H-zone gap (a region of lower density in the middle of the A-band) and could be clearly seen as a residual density after myosin extraction.   We described this overlapping, interdigitating filament model in a letter to *Nature* in the spring of 1953 (Hanson & Huxley, 1953).   We did not put forward a sliding filament model at that time, for we did not then have evidence which we thought would be convincing enough.   We expected to obtain such evidence from studies which we were just beginning, on the changes in band pattern during ATP-induced shortening. But we had this sliding idea very much in mind, since I had noticed that in many of Jean's earlier pictures, in which some sarcomeres had shortened during preparation, the length of the A-bands seemed to remain approximately constant.   Also, there were my X-ray results of the constant axial periodicity during stretch, which would obviously be consistent with constant filament lengths and therefore with sliding during length change. We were unsure at that time from which set of filaments the X-ray reflections came, but I mentioned the general argument in my short paper in the summer of 1953 (Huxley 1953b), where I described my electron-microscope observations on muscle cross-sections. (In fact, X-ray reflections from both sets of filaments were included in the ones I had measured, but that did not become apparent until later (Worthington, 1959).

So Jean and I then proceeded with the phase microscope studies of band-pattern changes.   These are too rapid to follow by eye with normal concentrations of ATP. Cine photography was successful up to a point, but gave somewhat low resolution images, and in the end we used extremely low concentrations of ATP to slow the shortening process down sufficiently so that we could collect excellent, sharp, phase contrast pictures of each stage of contraction, and could see clearly the H-zone closing up as the I-bands shortened, and the A-bands stayed the same length.   This showed, convincingly to us anyway – many people did not accept the conclusions for many years -- that both sets of filaments remained approximately constant in length during contraction, and that contraction worked by a sliding filament mechanism driven in some way by crossbridges. We were ready to publish early in 1954, and, knowing that A.F. Huxley and Niedergerke were doing light microscope studies on intact fibers with a specially designed interference microscopy which minimized optical artifacts, and learning that their results on A-band length were similar to ours, arranged to publish simultaneously in *Nature* in the spring of that year (H.E. Huxley and Hanson, 1954;   A.F. Huxley & Niedergerke, 1954).

Later that year we wrote a long review of all our work (Hanson & Huxley, 1955) for a conference in the summer of 1954, in which we described possible cycling crossbridge

Fig. 3   Visiting D. Podolsky, 1966.

mechanisms for producing the sliding force. One of these involved moving myosin crossbridges attaching to and detaching from actin in a step-wise fashion, and pulling the actin filament along in working stroked of 50-100Å!

But as we all know, it took many, many years, and a great deal more experimentation by ourselves and by many other people, before this basic model was generally accepted.

REFERENCES

Huxley, H. E. (1951). Low-angle X-ray diffraction studies on muscle. Disc. Faraday Soc. *11*, 148.

Huxley, H. E. (1953a). X-ray diffraction and the problem of muscle. Proc. Roy. Soc. B. *141*, 59.s

Hanson, J., and Huxley, H. E. (1953). The structural basis of the cross-striation in muscle. Nature, Lond. *172*, 530.

Huxley, H. E. (1953b). Electron-microscope studies of the organization of the filaments in striated muscle. Biochem. Biophys. Acta *12*.

Worthington, C. R. (1959). Large axial spacings in striated muscle. J. Mol. Biol. *1*, 398-401.

Huxley, H. E., and Hanson, J. (1954). Changes in the cross-striations of muscle during contraction and stretch and their structural interpretation. Nature, Lond. *173*.

Huxley, A. F., and Niedergerke, R. (1954). Structural changes in muscle during contraction. Interference microscopy of living muscle fibres. Nature (London) *173*, 971-973.

Hanson, J., and Huxley, H. E. (1955). Structural basis of contraction in striated muscle. Symp. Soc. Exp. Biol. *9*, 228-264.

GENERAL DISCUSSION PART II

Chaired by A.M. Gordon and H.E.D.J. ter Keurs

INVOLVEMENT OF TITIN IN MUSCLE MECHANICAL PROPERTIES

ter Keurs: In this period of General Discussion, I proposed to discuss on the involvement of titin in muscle mechanical properties. The topics will include muscle resting tension, extra tension produced after stretch, nature of muscle relaxation, and so on.

Sugi: First, I would like to point out that the elastic protein (titin) network within a muscle fiber can be extremely powerful in producing rapid "active relaxation." If a relaxed single muscle fiber is suspended in Ringer solution without any external load, it slowly takes vertical position under its point of support (Fig. II-1A). On tetanic stimulation, it shortens so that its shape changes to a short rod-shaped configuration (Fig. II-1B). On cessation of stimulation, the fiber is rapidly extended to take a S-shaped configuration (Fig. II-1C) by some internal force, indicating that muscle fiber relaxation is indeed an active process literature. This experiment was made by Ramsey, and was cited by Fenn (In Physical Chemistry of Cells and Tissues, ed. R. Höber, P. 509, Churchill, London, 1945).

Secondly, I would like to point out that the mechanical response of muscle fibers to applied length perturbation differs greatly depending on frequency (or speed) of perturbation, because of structural hierarchy of muscle fibers, e.g., atoms, molecules,

Fig. II-1. Diagram showing "active relaxation" in an isolated single frog muscle fiber (Fenn, 1945).

Molecular and Cellular Aspects of Muscle Contraction
Edited by H. Sugi, Kluwer Academic/Plenum Publishers, 2003

macromolecules, filaments, etc. In general, the higher the applied frequency, the smaller the structural components that are "perturbed" or "shaked."

In this connection, I am very much interested in experiments using perturbations of MHz range, which would show us new features of mechanical responses different from those obtained with conventional kHz perturbations.

Bagni: I would again like to explain the properties of the static tension examined with very rapid ramp stretches (velocity, 70×103 nm. $hs^{-1}s^{-1}$; amplitude, up to 40 nm. hs^{-1}). Fig. II-2 shows the method of measurement of the static stiffness in a BDM-treated muscle fiber (see legend).

In this particular case, the level of the static tension is about two times greater than the isometric tension. The ratio between the static tension and the sarcomere length elongation produced by the stretch represents the static stiffness of the sarcomere. To follow the time course of the static stiffness development after the activation, stretches were applied at rest and at different times after the start of stimulation. The static stiffness was independent of the active tension developed by the fiber and of stretching amplitude and stretching velocity in the whole range tested. It increased with sarcomere length in the range 2.1-2.8 µm, and decrease at longer lengths as shown in Fig. II-3.

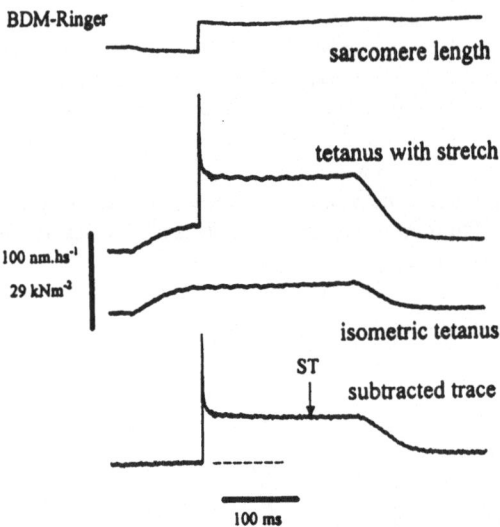

Fig. II-2. Force response to a stretch (amplitude 29.7 nm. hs^{-1}, duration 1.2 ms) applied at the tetanus plateau in a single fiber bathed in BDM-Ringer at 6 mM concentration. (*a*), sarcomere length; (*b*), tetanus with stretch; (*c*), isometric tetanus; (*d*), subtracted trace, obtained by subtracting the passive force response (not shown) and trace *c* from trace *b*. The static tension is measured on this trace after the end of the fast transient as indicated by the arrow. The sarcomere length change during the isometric tetanus is not plotted (Bagni et al., 2002).

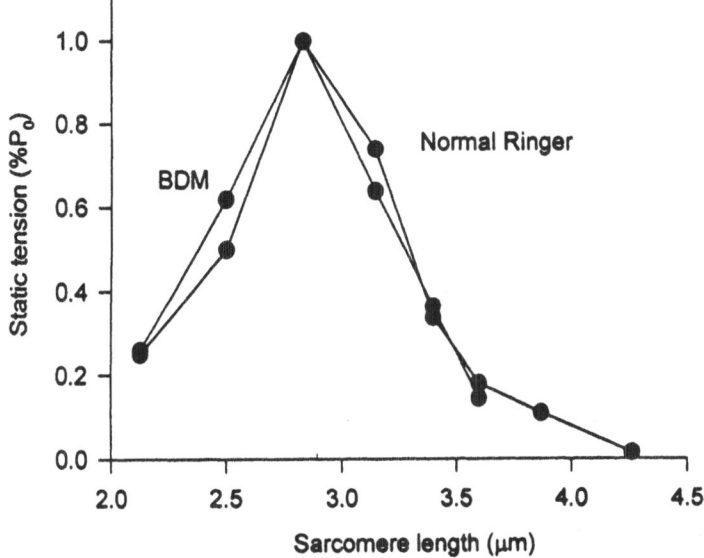

Fig. II-3. Sarcomere length dependence of the static stiffness in Normal Ringer solution (∇) and in Ringer solution containing BDM (5 mM) ($\bullet$). Stretch amplitude, 2.2 % l_0; stretching velocity, 78 l_0s^{-1}. The stretch was applied at 8 ms after the stimulus. It seems possible that the static stiffness is due to a Ca^{2+}-controlled titin-actin interaction or a direct Ca^{2+} action on titin stiffness though other possibilities can not at present be precluded.

Concerning the nature of the static stiffness, we can make the following hypothesis:
1) Static stiffness could be due to a calcium controlled titin-actin interaction. Kellemayer and Granzier (FEBS Lett. 380: 281-286, 1996) in their motility assay experiments, showed that sliding of actin is slowed down and even stopped by titin in presence of calcium.   2) Static stiffness could be due to a direct action of calcium on titin stiffness. Tatsumi et al. (J. Muscle Res. Cell Motil. 22: 149-162, 2001) have shown that calcium can bind to the elastic portion of titin. The secondary suggesting an alternation of titin stiffness. At present, these are speculative hypotheses that need to be verified. Other possible structures could be responsible for the increase in sarcomere stiffness upon activation. Actin filament stiffness, for instance, could increase upon calcium binding to troponin. However in this case a mechanism to overcome the mechanical interruption in the actin filament at the H-band should be postulated.

Huxley: A possible mechanism for the effect of calcium or titin (or whatever other long molecules or potymers are involved) might be the following. Suppose there are two (or more) classes of folded domains in the filament, which unfold to different extents as stretch proceeds, and which contribute to the filament stiffness. Suppose that in the folded state, one of these classes of domain can bind calcium and then become unable to unfold. Then stiffness will increase without changing the resting tension in the filament. The extent of the stiffness increase will depend on the stiffness of the two classes of domain, and their relative extents can probably be such as to give the observed maximum around 3.0 μm sarcomere length. If the calcium binding domains are not extended at longer initial sarcomere lengths, such binding will not affect stiffness.

Gonzalez-Serratos: Have you try to do the experiments at shorter sarcomere length than 2μm when titin probably is not contributing with tension?

Bagni: No, we have not done these experiments.

Granzier: Static stiffness could be derived from a number of sarcomeric structures, titin being one of them. I would speculate that Ca^{2+} destroys a local titin structure and that this gives rise to a persistent length reduction and hence an increase in tension / stiffness.

Lännergren: Have you tried to change the interval between the tetanus in which you measure static stiffness? Does the static stiffness become smaller if a 2nd tetanus is given very soon after the 1st ?

Bagni: I haven't tried this.

Pollack: Could actin be involved in your static stiffness?

Gonzales-Srratos: Could you exclude a contribution of the fiber membrane to the static stiffness?

Bagni: We did experiments with chemically skinned fibers, and obtained almost the same results as in the intact preparations.

Lännergren: Did you apply a second step after the first? If yes, is the static stiffness about the same in the two steps, and how long did you have to wait to obtain a similar response? Is the static stiffness also present during the fiber fatigue?

Bagni: Yes, we applied two stretches at the plateau on after the other with a delay of 100-200 ms, and the static stiffness was almost the same. For this reason, we did not make measurements at plateau of very long tetani. We did not study the static stiffness during fatigue.

Rassier: I would like to comment on the possible role of titin on the static stiffness and also extra force observed after an active stretch. As I mentioned earlier we have observed a passive force enhancement after stretch, when muscle cells are deactivated. This is true only in certain conditions, i.e., when the starting length is relatively long. This passive force enhancement has the following characteristics similar to what could be expected if titin would be the element responsible for such a phenomenon: (1) it is stretch amplitude dependant, and (2) it increases at longer fiber length. I need to reinforce that these are preliminary observations. However, we believe these findings can give important insights into the mechanisms of the long lasting force enhancement commonly observed after stretch.

Winegrad: As regards C protein being responsible for the static tension (change in stiffness or extra force) following a stretch during contraction, there are certain properties of C protein consistent with the static tension and as yet none inconsistent. However, we still know too little about C protein to attribute static tension to it.

DIFFERENCE IN REGULATORY MECHANISM OF CONTRACTION BETWEEN SKELETAL AND CARDIAC MUSCLE

Sugi (added after the discussion): During the period of General Discussion Part II, many participants asked Dr. Gordon to summarise differences in regulatory mechanism of contraction between skeletal and cardiac muscle, and Dr. Gordon kindly reviewed this topic for us.

Gordon: I have been asked to review the differences in regulation of skeletal and cardiac muscle. This discussion here will be brief because of space limitations. More extensive recent reviews of this topic may be found in Bers (Excitaion-Contraction Coupling and Cardiac Contractile Force, Kluwer, Dordvecht, 2001), Gordon *et al.* (Physiol. Rev. 80: 853-924, 2000), Solaro (in Handbook of Physiology, E. Page et al. eds., Oxford Univ. Press, New York, 2002), and Moss and Buck (in Handbook of Physiology, E. Page et al. eds., Oxford Univ. Press, New York, 2002).

Control of contraction The somatic nervous system controls skeletal muscle contraction with each motor axon innervating and controlling all muscle fibers in a motor unit. Individual muscles are composed of many such units with great variability in the number of muscle fibers in each motor unit. This provides for precise control of movement by the central nervous system since it can vary both the frequency of stimulation of each motor axon (and thus the force from each motor unit) as well as the number of active motor units. In contrast, the heart has its own pacemaker cells controlling the contraction frequency and each cardiac cell contracts on each beat. The autonomic nervous system (ANS) modulates both the heart rate and strength of contraction of each heartbeat. The strength of the each cardiac contraction is also dependent on other factors such as the end diastolic volume, the heart rate and other hormonal and paracrine factors making cardiac contraction much more variable at the cellular level than is the case for skeletal muscle. The discussion that follows will enumerate the cellular and molecular properties contributing to this much greater variability of contraction at the cellular level in cardiac muscle over skeletal muscle.

Excitation Major differences exist in excitation and excitation-contraction coupling between skeletal and cardiac muscle. Skeletal muscle is excited through neuromuscular junctional transmission whereby each muscle fiber in the motor unit is excited with each nerve impulse in the motor axon, but the number of active motor units is varied by the somatic nervous system. In cardiac muscle, activity is initiated through spontaneous diastolic depolarization in the sino-atrial nodal cells. Action potentials are then conducted by electrical coupling between cardiac muscle cells, sequentially activating atrial, atrioventricular nodal cells, rapid conduction system (Purkinje) cells, and ventricular cells so that all cells in the heart are activated. In skeletal muscle, action potentials are brief, like nerve action potentials, with brief refractory periods so that re-excitation can occur before the previous contraction subsides allowing for summation of contractions so that the steady state force depends on the frequency of stimulation. In contrast, the cardiac muscle cell action potential is long with a plateau phase involving Ca^{2+}·entry on each beat. The duration of the cardiac action potential and refractory period is about as long as the contractile response so that contractions show little summation, allowing for the relaxation of cardiac muscle so that filling can occur – a vital process for the functioning of the heart as a pump. However, because the Ca^{2+} entry and the shape of the cardiac action potential are variable, the contractile response to a single excitation is controllable in cardiac muscle.

Excitation-contraction coupling Excitation-contraction coupling occurs in skeletal muscle through direct coupling of voltage (and dihydropyridine) sensitive Ca^{2+} channels in the transverse tubule membrane with ryanodine receptor (RyR1) Ca^{2+} channels in the sarcoplasmic reticulum (SR) membrane. In this manner, depolarization of the transverse tubule surface membrane leads directly to opening of the RyR1 Ca^{2+} channels in the SR membrane allowing Ca^{2+} release from the SR into the myoplasm. The Ca^{2+} release is further enhanced by the positive feedback calcium-induced-calcium-release (CICR) property of the RyR1 SR Ca^{2+} releasing channel. In cardiac muscle there is little direct coupling. Ca^{2+} entry through the surface membrane voltage (and dihydropyridine) sensitive Ca^{2+} channels during the action potential triggers Ca^{2+} release from the SR through RyR2 ryanodine receptor Ca^{2+} channels opened by CICR. Since the Ca^{2+} entry during each action potential in cardiac muscle cells is under autonomic control, Ca^{2+} release and activation with each heart beat are also under autonomic control.

Each tissue uses specific SR Ca^{2+} pumps isoforms for Ca^{2+} uptake. In cardiac muscle the Ca^{2+} transport is modified by the protein phospholamban which inhibits uptake. However, phospholamban can be phosphorylated by PKA under adrenergic control to remove this inhibition and enhance Ca^{2+} uptake (see Bers, 2001). The enhanced uptake brings about more rapid relaxation and greater Ca^{2+} stored in the SR and available to be released on the next beat. This allows another pathway for autonomic nervous system control of the strength of each cardiac muscle contraction. In addition, the surface membrane Na^+/Ca^{2+} exchange mechanism is much more active in cardiac muscle than in skeletal muscle and can contribute to Ca^{2+} removal during relaxation. It is the major mechanism removing Ca^{2+} from the cardiac cell, thus active in controlling total cell Ca^{2+} (Bers, 2001).

Control of contractile response to Ca^{2+}: differences in protein isoforms Based on sensory input from length and force receptors, contraction in skeletal muscle can be varied precisely by the nervous system by varying the frequency of contraction of each unit and by recruitment of motor units. Contraction of all cardiac muscle cells on each beat is highly variable under autonomic control but also responds to intrinsic factors such as ventricular volume (cell length) with the maximum force and Ca^{2+} sensitivity being more sensitive to changes in muscle length than for skeletal muscle. This mechanism forms the basis for Starling's Law of the heart such that increased diastolic filling leads to increased cardiac output on the next beat. In addition to both Ca^{2+} release and uptake with each beat being modulated by the ANS, the contractile response to Ca^{2+} is also under autonomic control through phosphorylation of various regulatory proteins that change the relationship between free Ca^{2+} and force (see Solaro, 2002). These differences in contractile regulation and modulation between skeletal and cardiac muscle are the result of differences in regulatory protein isoforms since the sarcomeric structures are similar in the two tissues.

Differences in Troponin C (TnC) Cardiac TnC has only one N-terminal Ca^{2+} binding site (site II) while fast skeletal muscle TnC has two (sites I and II). (Interestingly, slow skeletal muscle also has the cardiac isoform.) Upon Ca^{2+} binding, the N-terminus of cardiac TnC does not undergo the transition to the open (E-F hand) configuration as occurs with fast skeletal TnC but also requires binding to cardiac TnI (Sia *et al.*, J. Biol. Chem. 272: 18212-18216, 1997). Furthermore, Ca^{2+} binding to cardiac TnC in the sarcomere is more enhanced by cycling cross-bridges than in skeletal muscle (Wang and Fuchs, Am. J. Physiol. 266: 1077-1082, 1994). This increases the dependence of Ca^{2+} activation of the cardiac individual regulatory unit on cross-bridge attachment.

Differences in Troponin I (TnI) and Troponin T (TnT) Significant differences are also present in TnI and TnT isoforms (see Solaro, 2002). The cardiac TnI isoform has an N-terminal extension with two PKA phosphorylation sites. Phosphorylation of both sites causes a decrease in the Ca^{2+} sensitivity (see discussions by Solaro and Ruegg in this symposium) and possibly maximum force. Phosphorylation of other PKC sites in cardiac TnI (see paper in this volume by Ruegg) may also alter Ca^{2+} sensitivity of force. None of these phosphorylation sites are available in skeletal TnI. This provides another mechanism for autonomic control of cardiac contraction.

Differences in TnT isoforms likely contribute to alterations in Ca^{2+} activation and cooperativity (Solaro, 2002). Alternative splicing of the TnT gene generates multiple isoforms, some of which give rise to differences in Ca^{2+} sensitivity and cooperative activation. Furthermore, there are numerous phosphorylation sites on skeletal and cardiac TnT isoforms, the significance of which are being investigated.

Differences in Tropomyosin (Tm) Tropomyosin in striated muscle has two major isoforms, α and β. Cardiac Tm is mostly composed of the $\alpha\alpha$ homodimer (this is species dependent with large mammals having some β Tm) whereas skeletal Tm is a mixture of the $\alpha\beta$ heterodimer and $\alpha\alpha$ homodimer. Cryo-EM data have shown that in the absence of Tn, skeletal Tm lies in a position that is equivalent to that in the presence of Tn and Ca^{2+}, a partially activated position. In contrast, in the absence of Tn cardiac, Tm occupies a position that is equivalent to that in the presence of Tn and absence of Ca^{2+}, the resting position (Lehman *et al.*, J. Mol. Biol. 302: 593-606, 2000). The strong binding of myosin to a fully activating position in both moves Tm across the surface of the thin filament to a more activating position (Vibert, J. Mol. Biol. 266: 8-14, 1997). Differences in resting Tm position suggest that activation of cardiac muscle may be more dependent on activation of the thin filament by strong cross-bridge attachment to (Gordon *et al.*, 2000).

Differences in myosin, C-protein, and titin Cardiac myosin has two heavy chain isoforms that have a lower actin-activated ATPase than fast skeletal muscle myosin.

This results in a slower contraction velocity and more efficient force maintenance. These differences in cross-bridge cycle rate could potentially affect the contribution to contractile activation by strongly attached cross-bridges.

Cardiac C-protein (also called myosin binding protein-C) isoforms provide additional phosphorylation sites (lacking in skeletal C–protein) and thus more control of cross-bridge attachment and force through C-protein modulation (see Winegrad, this volume and Solaro, 2002).

The properties of the various titin isoforms have been well characterized by Granzier and coworkers (see article in this volume by Granzier). The cardiac titin isoform a shorter resting length due to a shorter N2B region in the I band. This makes the cardiac sarcomere stiffer and less able to be extended above a sarcomere length of 2.4μm. The length restriction of the cardiac sarcomere supports Starling's Law by preventing extension into the descending limb of the length-tension curve where maximum force decreases with increasing sarcomere length, the opposite of what is required by Starling's Law.

Differences in cooperative activation These various differences in protein isoforms provide the molecular basis for the differences in cooperative activation of skeletal and cardiac muscle. As discussed in Gordon *et al.* in this volume (and see Regnier *et al.*, J. Physiol. 540: 485-497, 2002), there appears to be extensive cooperativity between adjacent regulatory units along the skeletal thin filament and less in the cardiac thin filament. This makes the skeletal functional regulatory unit substantially larger (10-12 actins) than the 7 actin structural regulatory unit (as also proposed by Maytum *et al.*, Biochemistry, 38: 1102-1110, 1999). The size of the functional regulatory unit in cardiac muscle has not been determined, but appears to be less than that for skeletal muscle. In contrast, as mentioned above, cardiac muscle demonstrates a greater sensitivity of activation by strongly attached cross-bridges. The results of these properties imply that Ca^{2+} is able to fully activate skeletal muscle. In contrast, cardiac muscle does not appear to be fully activated by Ca^{2+} binding alone, but requires strong binding cross-bridges. This makes cardiac activation more dependent on strong cross-bridge binding and more able to be varied by any factor affecting strong cross-bridge binding.

Summary The differences in properties of activation of skeletal and cardiac muscle discussed demonstrate how both muscle types are adapted to the physiological function and control. As such, one can understand how skeletal muscle is adapted to precise control by the nervous system through control of motor units and Ca^{2+} activation of individual muscle fibers. In addition, the high variability and control of cardiac muscle cells at the cellular level allow individual cardiac cells to respond over the broad range of performance required of them as they contract with each heartbeat. These molecular properties and the lack of complete activation by Ca^{2+} allow the contraction of each cardiac cell to be more easily modulated by the autonomic nervous system, hormonal regulation, and intrinsic factors such as muscle length (ventricular volume), factors that modulate cross-bridge attachment as well as Ca^{2+} activation.

I appreciate the many suggestions on the manuscript by Michael Regnier, Emilie Warner Clemmens, Alicia Moreno Gonzalez, and Kareen Kreutziger. I also appreciate the careful assistance of Martha Mathiason in preparing the manuscript for publication.

The many discussions with investigators in the laboratory and at the Fourth Fujihara Seminar were also helpful.

Pollack: You presented a nice review of the difference between cardiac and skeletal muscle, but you didn't comment on the specific molecular mechanism by which deoxy ATP gives increased force. Could you comment?

Gordon: We don't know the precise molecular mechanism. I just wanted to emphasize that cardiac muscle has many regulatory mechanisms, through which contractility can be modulated.

Gonzalez-Serratos: The mechanism of sodium/calcium exchange regulating cytosolic Ca con has not been mentioned here, but it is one of the most important mechanism regulating excitation-contraction coupling. The Na/Ca exchange mechanism is widely used as a regulatory mechanism to increase cytosolic calcium in heart failure by blocking the sodium/potassium pump with glycosides.

K. Yamada: In terms of TnC, the structural change with Ca binding observed in skeletal TnC does not occur in cardiac TnC. So, the molecular mechanism of regulation could be totally different. This is from calorimetric results.

Gordon: Yes, I agree. Sykes et al have shown that there is a small structural change in cardiac TnC with binding to Ca^{2+} compared to skeletal TnC. But the structural change is similar to cardiac TnC and skeletal TnC when cardiac TnC interacts with TnI.

K. Yamada: Yes, TnI could affect TnC when associated, of course.

ter Keurs: The presence of a collagen mesh around the cardiac cell is an important aspect of the difference between cardiac muscle and skeletal muscle. In addition, the presence of a capillary next to every myocyte of the heart allows for much tighter control of the myocyte contraction by endothelial factors than in skeletal muscle.

Kushmerick: Nature has insured that the range of metabolic and pH changes during normal function of the myocardium are small, even negligible as compared with those in skeletal muscle. In the later, Pi can rise by within 20 mM and pH decrease to 6.5 or lower. These differences in energy metabolism raise one insight. These filament regulatory processes that are identical in cardiac and skeletal muscle may function differently in the different cytosolic environment.

GENERAL DISCUSSION PART III

Chaired by R. C. Woledge, N. A. Curtin and J. A. Rall

MECHANICAL EFFICIENCY OF MUSCLE CONTRACTION

Woledge: I will start this session by showing some diagrams describing energy conversion in a manner similar to the description in Dr Tawada's presentation (this volume, p. 363). I think these diagrams can help us to understand under what conditions we can expect to see a high efficiency of energy conversion, and therefore to understand why in muscle the efficiency of energy conversion is usually not very high, in contrast to the special situation in Dr Sugi's experiments (this volume, p. 603) which I shall also refer to.

First, I want to respond to the question raised by Dr Tawada about concentration. He was asking what concentration might mean in a very small experimental system under study, perhaps even in a system in which we are following a single molecule going through the process of energy conversion. The answer that I suggest to this question is that we replace the concentration of a given state in the system by the probability of finding the system in that state, if we look at a randomly chosen moment. In fact the concentration terms in thermodynamic equations have to be dimensionless for the equations to make mathematical sense. This is often achieved by expressing all concentrations relative to a standard concentration, or rather more elegantly by using mole fractions instead of concentrations. The mole fraction of any species could be defined as the probability that a molecule chosen at random from the system will be of that species. So it is not revolutionary to replace a concentration by a probability in thermodynamic considerations. Particularly if we are interested in describing a steady state condition by studying a large number of individual events, I think the approach should work well.

I now want to illustrate the way in which the efficiency of energy conversion is dependent on the load in a tightly coupled system. I think the idea of giving a numerical value to the coupling between two processes was perhaps introduced by Prigogine (Introduction to thermodynamics of reversible processes; C.C.Thomas, U.S.A., 1955). I am familiar with it from the work of Kedem & Caplan (*Trans Faraday Soc* 61: 1897-1911, 1965). In these formulations a coupling of unity means that the two processes can occur only together, and do not occur at all as separate processes. This is sometimes called "tightly coupled", but unfortunately those words can also refer to the situation in which the coupling coefficient is not far from unity, perhaps having a value of around 0.9. So I propose to call the situation where the coupling coefficient is unity "completely coupled".

Consider a hypothetical energy conversion mechanismwhich passes through several states as it splits ATP.

$$A + ATP \longleftrightarrow B \longleftrightarrow C \longleftrightarrow A + ADP + Pi.$$

To one of the transitions (B → C) is completely coupled the performance of work by the moving of a variable load through a fixed working stroke.   This means of course that the transition between states B and C can only occur if the load is moved.

The diagrams in the upper rows of Fig 1 show the free energy changes of this system in a steady state for six different values of the load.   The thicker lines are the free energy levels of the six states with the system completely coupled, and the thin lines show the situation if the coupling constraint is momentarily lifted . But note that these lines do not show what happens if a new steady state is established without the constraint.   The relative rates of the reaction and efficiencies are also shown for each load.   As the load increases so does the efficiency and it reaches unity when the product of the load and the working stroke equals the free energy of the driving reaction.   At this point the reaction

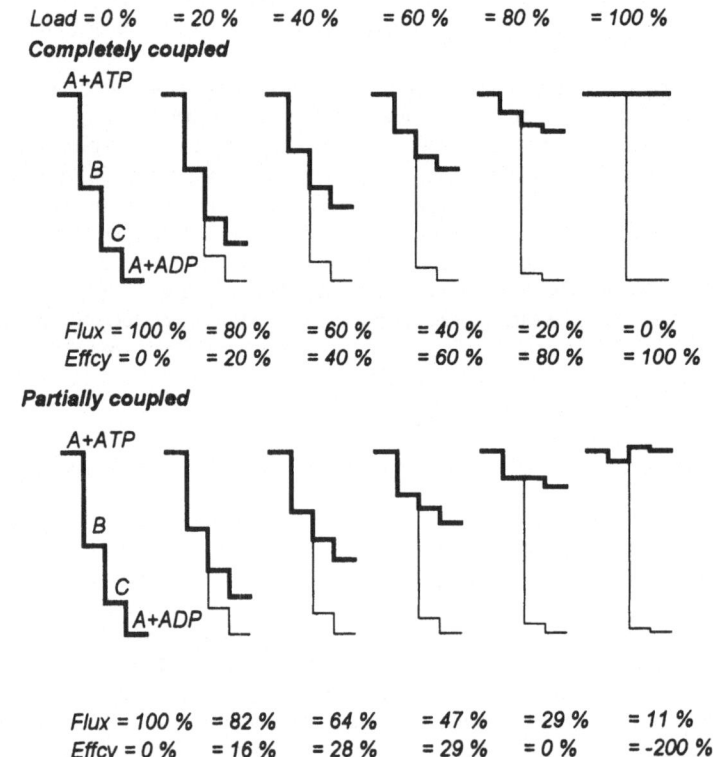

Figure III-1. Diagrams illustrating the free energy levels of the intermediate states of a hypothetical energy conversion system.   Suggestive diagrams, left to right, represent increasing loads.

stalls and would go into reverse were the load further increased. The reason for increasing efficiency is that the free energy of the coupled step increases as the load increases, at the expense of the other steps in the cycle. This happens because the reactions all slow down and so less free energy is needed to drive the uncoupled steps. Further insight into the way this happens could be obtained by considering how the concentrations of the species upstream of the coupled step increase and those downstream of the coupled step decrease as the load increases. This is what is described in the equations in Dr Tawada's paper. It follows from these considerations that complete coupling and a stalling load are sufficient and necessary conditions for 100% efficiency of energy conversion. Whenever these conditions are almost matched we should expect an efficiency close to 100%.

The bottom row of diagrams shows the comparable situation for a system which is not completely coupled. The important difference is that the efficiency falls off as the load is raised beyond 60 %, and thus shows an optimum with respect to load. This happens of course because energy is lost through the uncoupled splitting of ATP, at a rate which becomes an increasing significant part of the total ATP splitting as the reaction is slowed by raising the load In real muscles this optimum efficiency is usually found close to a load of 50 % of the maximum that can be raised (Woledge, Curtin and Homsher, Energetic Aspects of Muscle Contraction, Academic Press, New York, 1985).

We see from these diagrams that one cause of inefficiency is the free energy drop required to drive those steps in the cycle which are not coupled to the performance of work. In the incompletely coupled system these loses cannot be avoided by slowing the system down by high load, as then the uncoupled part of the ATP splitting becomes the important part of the inefficiency. In Professor Sugi's ingenious experiments (this volume, p. 603) using skinned fibers a way was found to reduce the inefficiency from this process. This was to pre-equilibrate the step of splitting of bound ATP. This process has been done before the collection of work starts and thus is not required during the work collection. If it is confirmed by the further experiments that Professor Sugi plans that under these conditions the efficiency is much higher than under normal steady state conditions this would suggest that the energy driving the process of splitting bound ATP is an important contributor to limiting the overall efficiency of muscle.

Dr. Tawada, do you wish to comment on these points?

Tawada: I do not have comments at this stage.

Maughan: Are you considering work done solely on the environment or are you including work done on structure linked to the myosin molecule in question?

Woledge: I have been considering only work done on the environment during a complete cross-bridge cycle.

Curtin: It's important to be clear about whether we are discussing complete cycles or individual steps in a cycle.

Woledge: Yes, this is very important. My diagrams refer to steady states with no change in the levels of the intermediates. Experiments are often done in a way that does allow changes in the levels of the intermediates. This is an extra source or sink of energy that must be considered in the interpretation of the results.

Kushmerick: The statement made by Dr. Woledge when the stalling force prevents advancement of ATPase flux needs amplification at the molecular level. Microscopic reversibility requires forward and reverse fluxes, which are identical at equilibrium. As techniques for visualizing molecular dynamics improve, it may be possible to observe these equilibrium phenomena.

Ranatunga: Can the marked increase of mechanical power output in mammalian muscle with temperature (~20-fold) between 10°C and 30°C be accounted for by changes in coupling? Since ATPase rate is increased, surely efficiency is also increased with heating.

Woledge: Heating is expected to increase efficiency, but exact quantities and the reasons require detailed experimental consideration.

Homes: The equilibrium constant of ATP → ADP + Pi in isolated S1 was shown by Trentham to be close to unity, i.e. there is no free energy change associated with ATP hydrolysis. The binding to actin clearly alters this. As was suggested by Dr. Kushmerick in a question to me in Session 2, the movement of switch 1 that apparently accompanies binding to actin provides a mechanism for solvating the products of hydrolysis, which could account for the larger free energy change associated with hydrolysis in the acto-myosin complex.

Sugi: We were interested in estimating the maximum mechanical efficiency, with which chemical energy derived from ATP Hydrolysis is converted into mechanical work, of

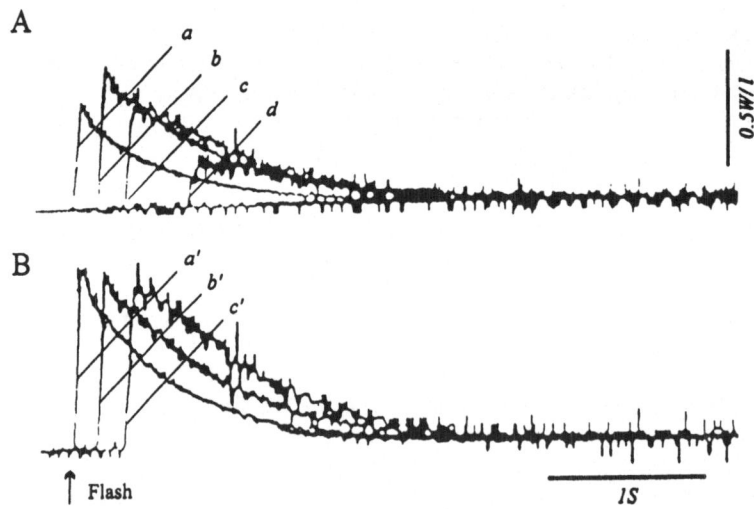

Figure III-2.   Power output during flash-induced fiber shortening.   (A) Power output recordings under four different afterloads.   (B) Power output records normalized relative to the peak values attained.   The load was 0.09 Po (a, a'), 0.35 Po (b, b'), 0.53 Po (c, c') and 0.78 Po (d).   (Sugi et al., 2003).

cross-bridge when they perform their powerstroke synchronously. For this purpose, demembranated rabbit psoas muscle fibers, containing ATP molecules almost equal in number to the cross-bridges, were activated to shorten isotonically by laser-flash photolysis of caged Ca^{2+}. In such a condition, all cross-bridge are in the state M-ADP-Pi immediately before activation, and can hydrolyzed ATP only once.

As shown in Fig III-2 power output records of the fibers following activation were almost identical if normalized with respect to their peak values. The distance of fiber shortening when the power output reached a maximum was ~10 nm per half sarcomere, suggesting that, at the beginning of fiber shortening, the cross-bridges in the form of M-ADP-Pi start their powerstroke almost synchronously. The amount of ATP remaining in the fiber at 1s after the onset of flash-induced mechanical response (Pr) was estimated from the amount of isometric force developed after interruption of fiber shortening. The amount of ATP utilized for the mechanical response (Pu) was then obtained as $Pu = (Po - Pr)$, where Po was the maximum isometric force. The amount of ATP utilized for fiber shortening i.e. for producing work, was further obtained as $Ps = (Pu - Pi)$, where Pi was the amount of ATP utilized for developing isometric force equal to the afterload P.

Fig.III-3 summarized the results. The maximum mechanical efficiency of the cross-bridges performing their powerstroke almost synchronously was maximum at 0.5 - 0.6 Po. A conservative estimate of the maximum mechanical efficiency (from fiber volume, and cross-bridge concentration in the fiber) givers the figure of 0.7, suggesting that the actual efficiency may be close to unity.

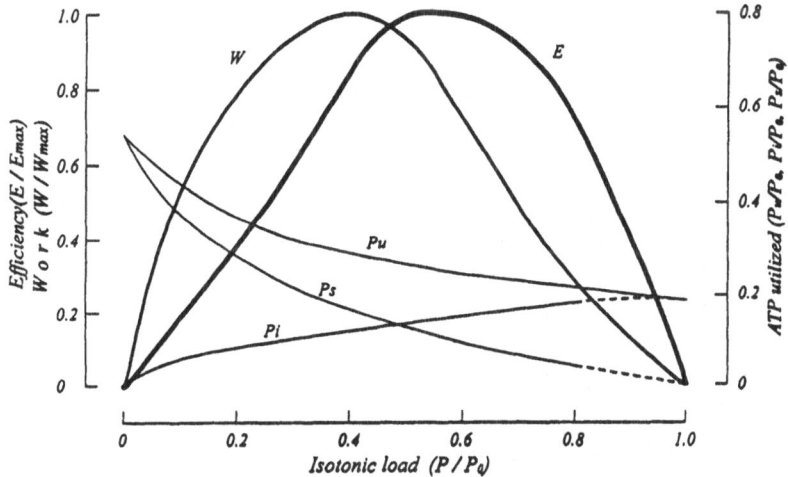

Figure III-3. Dependence of the mechanical efficiency of individual cross-bridges (E) on the amount of isotonic load (P). The amount of work done (W), ATP utilized for whole mechanical response (Pu), ATP utilized for preceding isometric force development (Pi), and ATP utilized for isotonic shortening (Ps) are also shown as functions of load (Sugi et al., 2003).

Woledge: We now want to discuss what parallels may exist between certain mechanical and energetic effects which have been discussed at this meeting and which are suggested to be not due to crossbridge action.   The effects are (1) the long lasting effects of stretch investigated by Drs. Bagni and Cecchi, and (2) under different conditions by Dr Rassier, and also (3) latency relaxation and (4) filamentary resting tension.   All except (3) are effects of stretch.   We will hear from Dr Cecchi concerning 1, from Dr Rassier concerning (2) and Dr Lännergren concerning (3)and (4), and then I will attempt a summary (Table III-1).

EFFECT OF STRETCH

Cecchi: Fig. III-4 shows the relationship between the amplitude of the applied stretch and the resulting static tension. From this it can be seen, for example for a stretch amplitude of 20nm.hs^{-1} (about 1.8% of Lo), that static tension results about 2.8 % Po. The static stiffness, calculated from the slope of the straight line fitted on the data, is 1.42×10^{-3} Po/n.mhs^{-1} which corresponds to about 7.1×10^{-3} or 0.7% of the total stiffness of a fully activated fiber at tetanus plateau in normal Ringer solution.

The above results were obtained in Ringer containing 6-10 mM BDM.   However, as

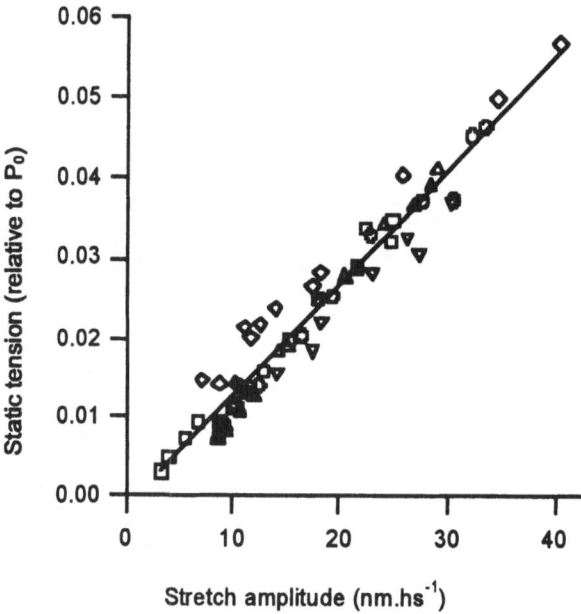

Fig. III-4. Relationships between stretch amplitude and static tension expressed relative to Po . Pooled data from 5 fibres. The regression line fitted on the pooled data (r = 0.97) is represented by the following equation: Static tension = $-1.44 \times 10^{-3} + 1.42 \times 10^{-3}$ x Stretch amplitude. The angular coefficient represents the static stiffness. Different symbols refer to different fibres.

shown by our previous data, BDM has a very small effect on static stiffness (less than 5%). The sarcomere length at which static stiffness is maximum is about 2.8 um.

Our previous results show that static tension is independent of stretching velocity, increases linearly with stretch amplitude, does not decay until the end of relaxations, occurs immediately with no delay after stretch and has a maximum values at about 2.8um sarcomere length. These properties are shared by the residual force enhancement after stretch described by Edman and collaborators (Edman, et al., *J. Gen. Physiol.* 80: 769-784, 1982) in fibers contracting in normal Ringer solution. This suggests the interesting possibility that static tension, defined and measured at very low tetanic tension in BDM Ringer, and the residual enhancement measured in normal Ringer with much lower stretching velocity, can be two different descriptions of the same phenomenon.

Rassier: The steady-state isometric force after stretch is higher than the purely isometric force observed at the corresponding final length (Edman et al., *J. Gen. Physiol.* 80: 769-784, 1982; Edman and Tsuchiya, *J. Physiol.* 490: 191-205, 1996; Herzog & Leonard, *J. Biomech.* 30: 865-872, 1997 and *J. Biomech.* 33: 531-542, 2002; Rassier et al., *J. Biomech.* In press, 2002; Sugi & Tsuchiya, *J. Physiol.* 407; 215-229, 1988). Force enhancement has the following characteristics: the amount of force enhancement increases with increasing amplitudes of stretch Rassier et al., 2002 and increasing sarcomere lengths (Edman et al., *J. Physiol.* 281: 139-155, 1978, and *J. Gen. Physiol.* 80: 769-784, 1982), but it is virtually independent of the speed of stretch (Abbott & Aubert, *J. Physiol.* 117: 77-86, 1952; Edman et al., *J. Gen. Physiol.* 80: 769-784, 1982; Rassier et al, *J. Biomech.* In press, 2002). The optimal initial sarcomere length for force enhancement is ~2.8 μm (Edman et al., *J. Gen. Physiol.* 80: 769-784, 1982). Force enhancement has not been associated with an increased stiffness in single cells (Julian & Morgan, *J. Physiol* 293: 379-392, 1979; Sugi & Tsuchiya, *J. Physiol.* 407; 215-229, 1988), but is associated with an increased stiffness in whole muscles (Herzog et al., *J. Biomech.* 33: 531-542, 2000).

It has been suggested that force enhancement may be associated with the engagement of a passive element during an active stretch (Edman & Tsuchiya, *J. Physiol.* 490: 191-205, 1996; Noble, *Exp. Physiol.* 77: 539-552, 1992) . This mechanism has been successfully incorporated into a theoretical model aimed at capturing the history-dependent effects of dynamic muscle contractions (Forcinito et al., *J. Biomech.* 31: 1093-1099, 1998). Recently, we have explored the idea of a "passive force enhancement" with a series of independent experiments, and the results of these studies produced strong evidence that the engagement of a passive element upon activation and stretch is partially responsible for force enhancement:

[1] In experiments conducted with single cells and whole muscles, we observed that, when the stretches are initiated at long lengths (when passive force starts to play a role during isometric contractions), force enhancement is accompanied by a long-lasting increase in the passive force after deactivation of the muscle/cell (Fig III-5). We refer to this phenomenon as "passive force enhancement", and it has the following characteristics, which are similar to the total force enhancement (Herzog & Leonard, *J. Exp. Biol.* 205: 1275-1283, 2002): the amount of passive force enhancement increases with increasing amplitudes of stretch, and increasing initial muscle lengths. Furthermore, passive force enhancement is independent of the speed of stretch.

[2] When muscle (cell) stretch is preceded by muscle shortening, and therefore a passive element that is engaged upon activation is fist shortened before it is   stretched,

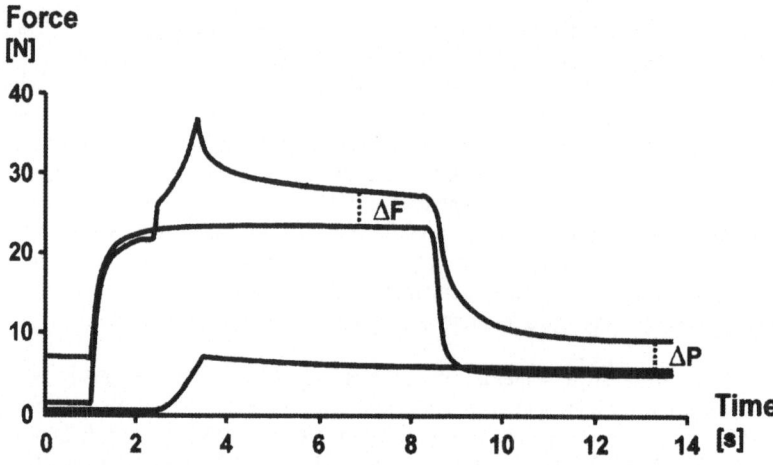

Fig. III-5. Total force enhancement (ΔF) and passive force enhancement (ΔP) during stretch of the cat soleus. The graph illustrates the time histories of an isometric contraction at the final reference length (middle, steady trace), an active stretch contraction (top trace), and a passive stretch (bottom trace). The final length is always the same. The steady-state, isometric force enhancement following the active stretch and the passive force after stretch and deactivation of the muscle are considerably greater the force produced during isometric contraction.

both total force enhancement and passive force enhancement decrease with decreasing amplitudes of shortening, i.e., the greater the amount of shortening preceding the stretch, the smaller the stretch-induced force enhancement (Herzog & Leonard, *J. Biomech.* 33: 531-542, 2000, and *J. Exp. Biol.* 205: 1275-1283, 2002) (Fig. III-6).

There are two independent observations made in other laboratories that suggest the engagement of a passive element during stretch-induced force enhancement. Bagni et al., (*Biophys. J.* 82: 3118-3127, 2002, see also this volume, p. 431) have observed an instantaneous increase in stiffness when muscle cells are activated, which is independent of cross-bridge attachment to actin. The authors refer to this phenomenon as "static stiffness". Such static stiffness increases with the magnitude of stretch, and with the initial sarcomere length.   Edman and Tsuchiya (*J. Physiol.* 490: 191-205, 1996) performed experiments in which single fibers were released to shorten against small loads after stretch and after isometric contractions. Upon release, the shortening records after stretch showed a greater and steeper initial transient than

The origin of the passive force enhancement is unknown, and it is under investigation in our laboratory. Our working hypothesis is that titin may be responsible for the passive force enhancement. It has been shown that the stiffness of titin is modulated by mechanical and chemical factors. If the stiffness of titin is changed upon activation and stretch, this protein could be responsible for the passive force enhancement, and therefore part of the total force enhancement.   Direct evidence of an increased stiffness of titin upon an active stretch is lacking, but given the characteristics of titin, it seems a prime candidate to fulfil this role.

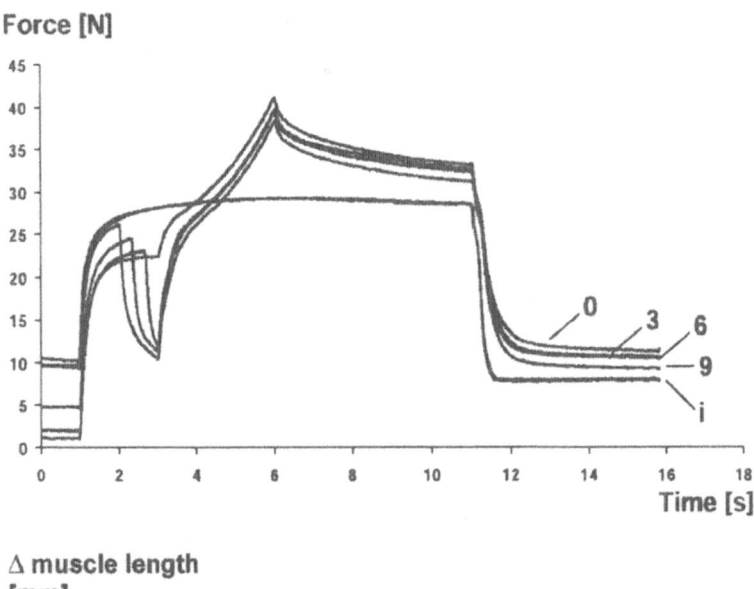

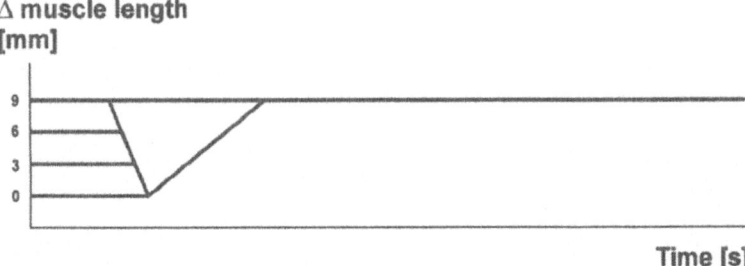

Fig. III-6. Representative force-time histories of four experimental stretch contractions (0 to +9mm at 3 mm/s) of the cat soleus that were preceded by active shortening of 0, 3, 6, and 9 mm (0, 3, 6 and 9 respectively). Also shown is an isometric reference contraction (i, +9 mm).   Increasing the amount of shortening decreases the total and the passive force enhancement to a similar degree, those obtained during the isometric contractions. This result was attributed to the elongation of a passive elastic structure during stretch, as a result of some half-sarcomeres being stretched more than others.

FILAMENTARY RESTING TENSION AND LATENCY RELAXATION

Sugi: I want to briefly describe the experiments of D.K. Hill as well as his concept of filamentary resting tension (J. Physiol. 199: 637 - 684, 1968).   He applied slow small stretches (~ 0.2%) to a resting frog sartorius, and found that, in response to stretch, tension first rose rapidly and then leveled off to give the appearance of so-called 'frictional resistance'.   He called this phenomenon as "short - range elastic component"

(SREC).   Based on the effect of hypertonic solutions on resting tension, on the other hand, Hill reached the conclusion that a small part of resting tension is due to 'active' interaction between the thick and thin filaments, and calls this 'active' component of resting tension as 'filamentary resting tension' (FRT).   According to Hill constitute a component of SREC, and may be responsible for the latency relaxation.

Lännergren: The maximum force exerted FRT is not more than 1 % of Po.   For example, Haugen & Sten-Knudsen (*Acta Physiol Scand* 112: 113-120, 1981) report a value of 0.9 % Po at a sarcomere length of 3.0, which is about the optimum sarcomere length for FRT. D. K. Hill (*J Physiol* 199: 637-684,1968) and Lännergren, (*J Gen Physiol* 58:145-162,1971) used much slower stretches and their values for FRT are only 20 to 40% as big.   The maximum stiffness reported by Haugen & Sten-Knudsen for FRT is 0.5 Po/Lo.   This is about 0.5% of stiffness during a tetanus at optimum length.

The maximum amplitude of the latency relaxation (LR) is 0.23 % Po (Haughan, *Acta Physiol Scand* 114: 487-495, 1982) Again the optimum sarcomere length is at 3.0-3.1 um and there is an almost linear decline to zero at 3.6-3.7 um.

Neither FRT nor LR has been studied in the presence of BDM.   But there is quite a bit of evidence that these effects respond to a number of other drugs in a way which is unlike the response of contractile force.   For example FRT increases with the tonicity of the bathing solution, which decreases the force developed in response to stimulation. (Lännergren & Noth, *J Gen Physiol* 61: 158-175, 1973). Substituting D_2O for H_2O greatly reduces twitch tension but the amplitude of LR is little affected (Sandow et al., *Biochim Biophys Acta* 440:733-743, 1976; Gilai & Kirsch, *J Physiol* 282: 197-202, 1978). Dantolene sodium reduced twitch tension by 40% but again LR was relatively unaffected (Gilai & Kirsch, *J Physiol* 282: 197-202, 1978).   Caffeine increases twitch tension but reduces both the amplitude and duration of LR (Herbst & Piontek, *Pflug Arch* 346: 61-76, 1974; Gilai & Kirsch, *J Physiol* 282: 197-202, 1978).

FRT cannot be measured directly but has to be inferred from the tension response to a ramp stretch (or ramp release) which consists of an initial SREC followed by a "frictional" phase. A striking thing about ramp length changes imposed on resting muscles/fibres is that the response to a ramp release is an exact mirror image of that to a ramp stretch (Hill, *J Physiol* 199: 637-684, 1968, Lännergren, *J Gen Physiol* 58:145-162, 1971). Hill never commented upon this and I'm still not sure of what it means. If one believes that a part of the resting tension is produced by a relatively small number of slowly cycling cross-bridges, then the tension loss during the initial part of a release would correspond to tension borne by these bridges. But why would those bridges be stretched and detach for exactly the same length change in the opposite direction and produce the same force change? An alternative mechanism would be that the resting tension is borne by the titin springs, with perhaps a contribution from the sarcolemma, and that the non-linear response to length changes is produced by a small number of non-force generating cross-bridges which happen to be attached when movement starts and which can detach equally easily for length changes in either direction.   Viscous forces might also contribute to these non-linear responses.

Woledge: From all these contributions the comparison between these effects can be summarised in Table III-1. It seems plausible that a common mechanism might be found for all of these effects except the last, which seems to be much larger and to exist within a different range of sarcomere lengths.

Table III-1

Effect studied	Max force % Po	Sarcomere length dependence	Stiffness Po/Lo	Resistant to acto-myosin inhibition?
Static tension	> 6 %	Optimum: 2.8 μm	1.6	Yes (BDM)
Passive force enhancement	>20 %	Optimum: about 2.8 μm	~1	Not known
FRT	<1%	Optimum: 3.0-3.1 μm	5	?Yes (see text)
LR	<1%	Optimum: 3.0-3.1 μm	Not applicable	?Yes (see text)
Energy storage during stretch	120 %	Present at: 2.0 - 2.2 μm	Not applicable	Not known

I want to finish this session by some remarks concerning D. K. Hill some of whose work we have just been discussing and who will be known personally to many of those here today. As many of you will know, Professor David Hill died earlier this year, on the 18th August and I think our discussions on his work may serve as a good obituary for him. I remember David Hill in connection with his influential work on the filamentary resting tension which has been an subject of discussion today, and also because of his early work on muscle energetics in which he measured the time course of the heat production due to oxidative recovery of frog muscle at low temperature and also the time course of the oxygen consumption itself.   These must have been extraordinary experiments made with very simple, but well designed apparatus and a great deal of skill. For further information on the many contributions to science made by Professor Hill see the obituary by Sir Andrew Huxley in Nature (419: 800, 2002).

Lännergren: Roger Woledge has already said the important things about D.K. Hill. Unfortunately I never met him personally but I remember from my time at University College that Andrew Huxley often mentioned Hill's work in very appreciative terms. And, of course my work on filamentary resting tension in single fibers was directly inspired by D.K. Hill's pioneering work on whole muscles.

CONCLUDING REMARKS : PROGRESS IN DETERMINING THE DETAILED MECHANISM OF MUSCLE CONTRACTION

H.E. Huxley

First, I would like to commend Professor Sugi for the great service he has done to the field in the last twenty five years, in organizing this series of meetings at regular intervals since the first one in 1978, covering a very critical and enormously exciting period in muscle research, and encompassing work on the force-generating mechanism and on many other important areas of muscle research as well. I will summarize, briefly, some of the key discoveries that have been made relating to the basic contractile mechanism, during this time period.

At the time of the Cold Spring Harbor Symposium (1972, published 1973), the sliding filament model and the actin and myosin filament structures had been documented in considerable detail, and the idea that force was produced by cyclic operation of myosin cross-bridges, attached to actin and tilting or changing their shape in some equivalent way (Huxley, 1969) was beginning to be accepted as the probable mechanism. There was a lack of direct experimental evidence that this was the case, but given the difficulty of obtaining such evidence, especially with the techniques that had been available up to that time, this was hardly surprising. It was generally expected that such evidence would soon be forthcoming, when the basic problem could be considered solved, though many people underestimated the difficulty of obtaining such evidence, even if the mechanism was exactly as postulated. Some even wondered if it were worth my while pursuing the problem any longer, since they thought the basic questions had been answered, and only a few details remained to be cleared up! I had a different view, and believed it was necessary to find ways of finding out in full structural detail how the crossbridges function, before we could claim that we understood the mechanism. I think my view was correct!.

After a few years, by 1978, when there had been no great further breakthroughs, the optimism of Cold Spring Harbor began to diminish a little, because of the lack of direct positive evidence for changes in structure of actin-attached crossbridges; and indeed some experiments seemed to show that that they did not change their shape or orientation at all. These experiments were misleading, because we didn't have sufficient evidence about the structure to interpret them properly, but it took many years before that became apparent. Some workers began to give credence to very poor or unsubstantiated earlier experiments, which called into question some of the basic findings which had led to the sliding filament hypothesis, such as constancy of A-band length and actin filament length. These doubts have by now evaporated, due to the force of subsequent lines of experimentation, but at the time they sowed confusion, and the mechanism was certainly not beyond dispute! So it was this fascinating period which Professor Sugi's meetings have covered, and I would like to discuss what I think were some of the important steps along the way which have brought us to our present well documented understanding of the structural mechanism.

Probably the first real steps in the right direction had come a few years earlier, with Lymn and Taylor's demonstration (1971) that the actomyosin ATPase mechanism functioned in a rather unexpected way (i.e., actin catalyzing product release rather than substrate cleavage by myosin) but one that fitted in perfectly with the attachment-detachment crossbridge cycle. Near the same time, also, A.F. Huxley and Simmons (1971) made high time-resolution tension transient measurements on single muscle fibers during and after very rapid length changes, and produced very important evidence, from rapid tension recovery processes, for a cross-bridge stroke of the order of 10 nm in size. For believers in crossbridges , this was a great step forward, but non-believers could still assume that it was just another visco-elastic response of some unknown system.

Perhaps the next significant step came as a result of the demonstration by Rosenbaum, Holmes, and Witz (1971) of the possibilities of achieving a very intense X-ray source, enormously brighter than could ever be produced using a regular X-ray tube, by making use of synchrotron radiation, emitted by circulating electrons in physicists' large accelerators. This took a number of years to implement effectively, but eventually we were able to obtain detailed X-ray patterns from contracting muscles, with high time resolution, which had seemed an impossible dream before 1971. This confirmed our predictions of the initiation of crossbridge movement as tension development began (Huxley, *et al.*, 1980, 1982), but the most crucial experiments were those concerned with the behaviour of the strong 14.5 nm meridional reflection during small, rapid applied length changes of the kind used by A.F. Huxley & Simmons (1971). With 1 msec time resolution data, these experiments showed a large decrease in the intensity of this reflection coincident with the mechanical response (Huxley *et al.*, 1981, 1983). Since the reflection comes from the axial repeat of sets of myosin crossbridges along the length of the thick filaments, this was clear evidence of a major structural change in the crossbirdges, synchronized by a sudden relative sliding of actin and myosin filaments by the order of 10 nm or less and occurring during the working stroke. The change could well be some type of crossbridge tilting (as was confirmed more recently), but that some form of crossbridge movement was involved in the crossbridge stroke could now not be in doubt.

But what opened up a totally new area of crucial experimentation was the success achieved by Kron & Spudich (1986) in implementing a system whereby fluorescently-labelled actin filaments, in the presence of ATP, could be seen sliding over surfaces to which myosin molecules were attached. This followed the earlier experiment by Sheetz & Spudich (1983) in which myosin-coated beads could be seen sliding along actin cables obtained from *Nitella*, and had the advantage that now both components of the system were purified proteins.

(Many workers, including myself, had attempted previously to construct various kinds of *in vitro* motility systems, and some even reported intial success, but had not been able to obtain a reproducible experimental results, that others could duplicate.)   After watching light-microscope videos of myriads of individual actin filaments sliding unidirectionally at speeds comparable to those in a muscle, it was difficult for anyone not to believe any longer that a similar sliding process takes place within the muscle sarcomeres.

Moreover, the Spudich group (Toyashima *et al.*, 1987) were soon able to show that actin sliding was produced in their type of assay when myosin subfragment one was used instead of intact myosin molecules. This was an extremely important finding, since it showed that the myosin filament backbone did not have to play an active role in contraction other than providing a fixed support for the myosin heads. It also showed that the myosin head units alone, in the absence of the S2 rod which connects them to the backbone, and which, it had been suggested (Harrington, 1979) might play an essential role in contraction, were perfectly capable of generating movement on their own.   This effected a considerable simplification of the range of possible mechanisms.

The *in vitro* motility approach took another major step forward with the introduction of the "3-bead" system by Finer *et al.* (1994). In this system, a single fluorescently labeled actin filament is held tight between two plastic beads suspended over a slide by two laser traps. Other optical elements were used to measure bead positions. A third bead fixed to the glass slide had been treated with very low concentrations of HMM or other myosin species, so that usually only a single molecule would attach to the bead. When the actin filament was positioned (using the laser traps), so as to lie very close to the fixed bead, and ATP supplied, unitary interactions of the HMM with the actin filament could be seen producing discrete stepwise axial movements of the filament, of the order of 10 nm, and single force transients averaging 3-4 pN. This type of arrangement had been duplicated by many other workers with basically similar results, and significant advances have been made in the techniques of data analysis. The basic experiment remains, I believe, one of the most remarkable developments in modern biology in recent years. Most people, until very recently, would have deemed it quite impossible to measure the force and movement produced by a single myosin molecule, given the problems of instrumental sensitivity, Brownian noise and optical resolution. Perhaps Millikan's famous experiment measuring the charge on a single electron in an oil drop is a fair comparison.

Certainly, these experiments confirmed beyond any possible doubt that force and movement are produced directly by a discrete interaction between a myosin head (the crossbridge) and the actin filament to which it attaches.

Undoubtedly, however, the most illuminating of all advances in that last twenty-five years was the publication in 1993 of the high resolution X-ray crystallographic structure of myosin subfragment one by Rayment and his colleagues (Rayment *et al.*, 1993a). Using the corresponding F-actin structure obtained by Holmes and his colleagues (Kabsch *et al,* 1990; Holmes *et al.*, 1990), plus electron microscope information on lower resolution structures of the actomyosin complex (Milligan *et al.*, 1990), it was now possible to see in detail the appearance of the contractile complex (Rayment *et al.*, 1993b). The most striking feature of the myosin S1 structure was the presence of a very elongated C-terminal domain, consisting of a long (~9 nm), fairly straight, continuous α-helical portion of the myosin heavy chain, with two light chains wrapped around it giving it stability, and a second more globular N-terminal domain, containing the actin binding and ATP-ase binding sites. With the S1 attached to actin, the elongated domain would point outwards, and be attached to S2 in intact myosin, and thence to the myosin filament backbone in a muscle. And so it was immediately apparent that it might function as a lever arm, to amplify smaller structural changes in the first domain (known as the catalytic subunit) into the 5-10 nm of movement which this molecule has been shown to generate.

A great deal of subsequent work on step-length size in relation to lever arm length (e.g., Uyeda *et al.*, 1996; Anson *et al.*, 1996), using optical trap and other motility measurements has produced very strong evidence that this is indeed the case. This feature of the S1 structure, with a large part of the molecule fixed on actin while the major movement occurred only in the lever arm domain, helps to explain why such changes were so difficult to detect previously; when fluorescent labels were used, they were ones which attached to the catalytic domain; and the form of the change in structure made it difficult to detect by hydrodynamic and X-ray and neutron scattering measurements.

The evidence in favor of a tilting lever-arm crossbridge mechanism has become even more overwhelming in the most recent years, from high-resolution X-ray crystallographic studies of structural changes in the S1 molecule, associated with bound nucleotide changes corresponding to the different stages in the ATPase cycle. The first sign of these was seen in truncated *Dictyostelium* S1 (lacking the lever arm but showing the structure immediately connected to it) by Smith & Rayment (1996); then in smooth muscle S1 expressed with half a lever arm, and seen in the pre-powerstroke configuration (Dominguez *et al.*, 1998); and most convincingly and dramatically of all in the case of scallop myosin subfragment one, which in its fully native form, with intact lever arm, can been seen in three different nucleotide states with three widely different lever arm angles (Houdusse *et al.*, 2000), which certainly provides more than enough movement to account for the generally accepted range of crossbridge strokes in muscle myosins.

Finally, I should refer to the interference X-ray fiber diffractions studies of contracting muscle fibers, initiated by Lombardi and his colleagues (Linari *et al.*, 2000; Piazzesi *et al.*, 2002), and the similar effects that my colleagues and I have described in interact whole muscles (Huxley *et al.*, 2000, 2001, 2002, and at the present meeting). These provide extremely powerful evidence for a tilting movement of the crossbridge lever arms during their working stroke, in a fully functioning muscle during contraction.

I think the various experiments that I have briefly summarized here have by now built up an overwhelming body of evidence in favor of the sliding-filament tilting lever arm-crossbridge mechanism for muscle contraction. The same mechanism seems to be used in many non-muscle myosins involved in other types of motility and transport, though in some cases it seems to be combined with a biased Brownian movement to extend the range of motion of the lever arm. Whether such a mechanism (e.g. Kitamura *et al*, 1999) plays any supplemental role in the muscle myosins is unclear and the evidence still controversial; in no way is it an alternative as the main force-producing mechanism.

However, we have by no means come to the end of the road as far as understanding the actual contractile mechanism. We still have to understand the detailed structural chemistry, which makes possible the conversion of the stored energy in ATP into the molecular movements inside the myosin head, and how that process is catalyzed by the binding of actin. And we have to understand exactly how the two-headed myosin molecules in a muscle attach to and detach from actin and go through their working strokes under all the conditions of muscle action. So I very much hope that Professor Sugi will be able to continue to provide such splendid opportunities for us all to discuss these and other matters, at regular intervals!

REFERENCES

Huxley, H. E. (1969). The mechanism of muscle contraction. Science *164*, 1356-1366.

Lymn, R. W., and Taylor, E. W. (1971). Mechanism of ATP hydrolysis by actomyosin. Biochemistry *10*, 4617-4624.

Huxley, A. F., and Simmons, R. M. (1971). Proposed mechanism of force generation in striated muscle. Nature *233*, 533-538.

Rosenbaum, G., Holmes, K.C., and Witz, J., (1971). Synchrotron Radiation as a Source for X-ray Diffraction. (1971). Synchrotron Radiation as a Source for X-ray Diffraction. Nature *230*, 434-437.

Huxley, H. E., Faruqi, A. R., Bordas, J., Koch, M. H. J., and Milch, J. R. (1980). The use of synchrotron radiation in time resolved X-ray diffraction studies of myosin layer-line reflections during muscle dontraction. Nature *284*, 140-143.

Huxley, H. E., Faruqi, A. R., Kress, M., Bordas, J., and Koch, M. H. J. (1982). Time resolved x-ray diffraction studies of the myosin layerline reflections during muscle contraction. J. Mol. Biol. *158*, 637-684.

Huxley, H. E., Simmons, R. M., Faruqi, A. R., Kress, M., Bordas, J., and Koch, M. H. J. (1981). Millisecond time-resolved changes in x-ray reflections from contracting muscle during rapid mechanical transients, recorded using synchrotron radiation. Proc. Nat. Acad. Sci. *78*, 2297-2301.

Huxley, H. E., and Faruqi, A. R. (1983). Time-resolved x-ray diffraction studies on vertebrate striated muscle. Ann. Rev. Biophys. Bioeng. *12*, 381-417.

Kron, S. J., and Spudich, J. A. (1986). Fluorescent actin filaments move on myosin fixed to a glass surface. Proc. Nat. Acad. Sci. *83*, 6272-6276.

Sheetz, M. P., and Spudich, J. A. (1983). Movement of myosin-coated fluorescent beads on actin cables *in vitro*. Nature *303*, 31-35.

Toyashima, Y. Y., Kron, S. J., McNally, E. M., Niebling, K. R., Toyashima, C., and Spudich, J. A. (1987). Myosin subfragment-1 is sufficient to move actin filaments *in vitro*. Nature *328*, 536-539.

Harrington, W.F. (1979). On the origin of the contractile force in skeletal muscle. Proc. Natl. Acad. Sci. U.S.A. *76*, 5066-5070.

Finer, J. T., Simmons, R. M., and Spudich, J. A. (1994). Single myosin molecule mechanics: piconewton forces and nanometre steps. Nature *368*, 113-119.

Rayment, I., Rypniewski, W., Schmidt-Base, K., Smith, R., Tomchick, D., Benning, M., Winkelmann, D., Wesenberg, G., and Holden, H. (1993). Three-dimensional structure of myosin subfragment-1: a molecular motor. Science *162*, 50-58.

Kabsch, W., Mannherz, H. G., Such, D., Pai, E. F., and Holmes, K. C. (1990). Atomic structure of the actin: DNAaseI complex. Nature *347*, 37-44.

Holmes, K. C., Popp, D., Gebhard, W., and Kabsch, W. (1990). Atomic model of the actin filament. Nature *347*, 44-49.

Milligan, R. A., Whittaker, M., and Safer, D. (1990). Localization of tropomyosin, myosin and light chain binding sites on the surface of F-actin by Cryo-EM and image analysis. Nature *348*, 217-221.

Rayment, I., Holden, H. M., Whittaker, M., Yohn, C. B., Lorenz, M., Holmes, K. C., and R.A, M. (1993b). Structure of the actin-myosin complex and its implications for muscle contraction. Science *261*, 58-65.

Uyeda, T.Q.P., Abramson, P.D. and Spudich, J. (1996) The neck region of the myosin motor domain acts as a lever arm to generate movement. Proc. Nat. Acad. Sci. *93*. 4459-4464.

Anson, M., Geeves, M.A., Kursawa, S.E., and Manstein, D.J. (1996). Myosin motors with artificial lever arms. EMBO J. *15*, 6069-74

Smith, C.A., and Rayment, I. (1996). X-ray structure of the Magnesium (II).ADP.Vanadate Complex of the Dictyostelium discoideum Myosin Motor Domain to 1.9Å resolution. Biochemistry. *33*. 3404-3417.

Dominguez, R., Freyzon, Y., Trybus, K. M., and Cohen, C. (1998). Crystal structure of vertebrate smooth muscle myosin motor domain: visualization of the pre-power stroke state. Cell *94*, 559-571.

Houdusse, A., Szent-Gyorgyi, A.G., and Cohen C. (2000). Three conformational states of scallop myosin S1. Proc. Natl. Acad. Sci. *97*, 11238-11243.

Linari, M., Piazzesi, G., Dobbie, I., Koubassova, N., Reconditi, M., Narayanan, T., Diat, O., Irving, M, and Lombardi, V. (2000). Interference fine structure and sarcomere length dependence of the axial X-ray pattern from active single muscle fibers. Proc. Natl. Acad. Sci. *97*, 7226-7231.

Piazzesi, G., Reconditi, M., Linari, M., Lucii, L., Sun, Y., Nagayanan, T., Boesecke, P., Lombardi, V., and Irving, M. (2002). Mechanism of force generation by myosin heads in skeletal muscle. *415*, 659-662.

Huxley, H. E., Reconditi, M., Stewart, A., and Irving, T. (2000). Interference changes on the 14.5nm reflection during rapid length changes. Biophys. J. *78*, 134A.

Huxley, H. E., Reconditi, M., Stewart, A., Irving, T., Fischetti, R. (2001). Use of X-ray interferometry to study crossbridge behavior during rapid mechanical transients. Biophys. J., *80*, 266A.

Huxley, H. E., Reconditi, M., Stewart, A., and Irving, T. (2002). Crossbridge and backbone contributions to interference effects on meridional X-ray reflections. Biophys. J., *82*, 5A.

Kitamura, K., Tokunaga, M., Iwane, A. H., and Yanagida, T. (1999). A single myosin head moves along an actin filament with regular steps of 5.3 nanometres. Nature *397*, 129-134.

Photo 1. Participants of the Fourth Fujihara Seminar

690

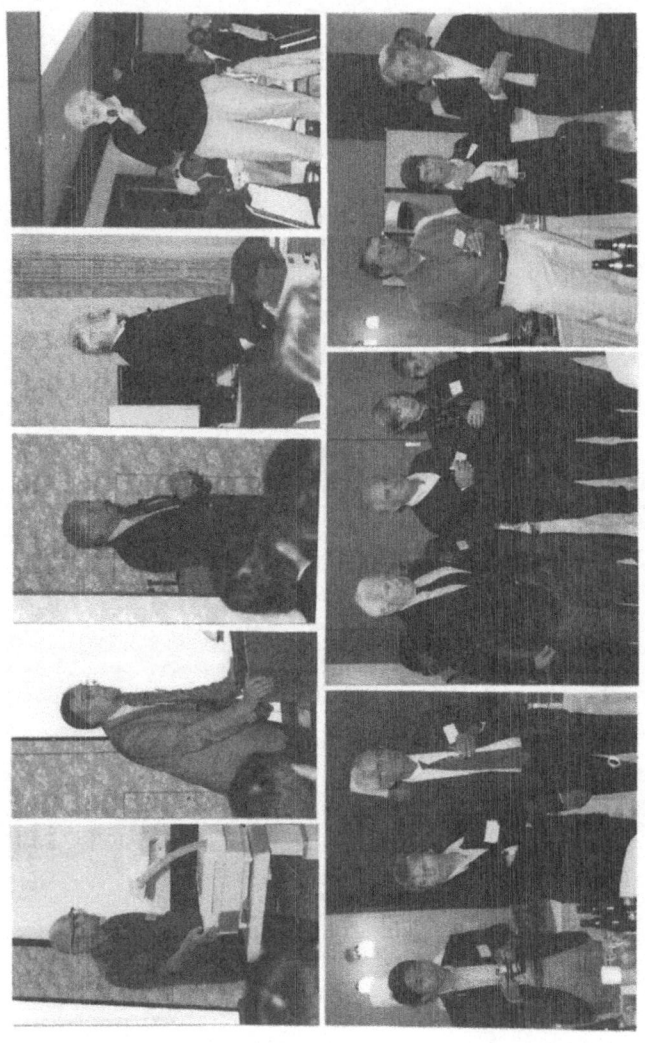

Photo 2. (First row) H. Sugi / J.M Squire / K.C. Holmes / J. Lännergren / G.H. Pollack. (Second row) S Chaen, G. Pfitzer, J.C Rüegg / H.E. Huxley, G. Cecchi, M.A. Bagni / J.A. Rall, N.A. Curtin, R.C. Woledge.

691

Photo 3. (First row) J. Lännergren, H. Sugi, H. Gonzalez-Serratos / Mrs. Holmes, S. Winegrad, H.E. Huxley, H. Sugi, Mrs. Sugi / H. Sugi, H.E.D.J. ter Keurs, R. Stehle, A.M. Gordon. (Second row) S. Winegrad, G.J.M. Stienen / H.E. Huxley, H. Sugi, J.A. Rall, M.J. Kushmerick, R.J. Solaro / H.L. Granzier, S. Nishimura, H. Sugi, G. Cecchi.

692

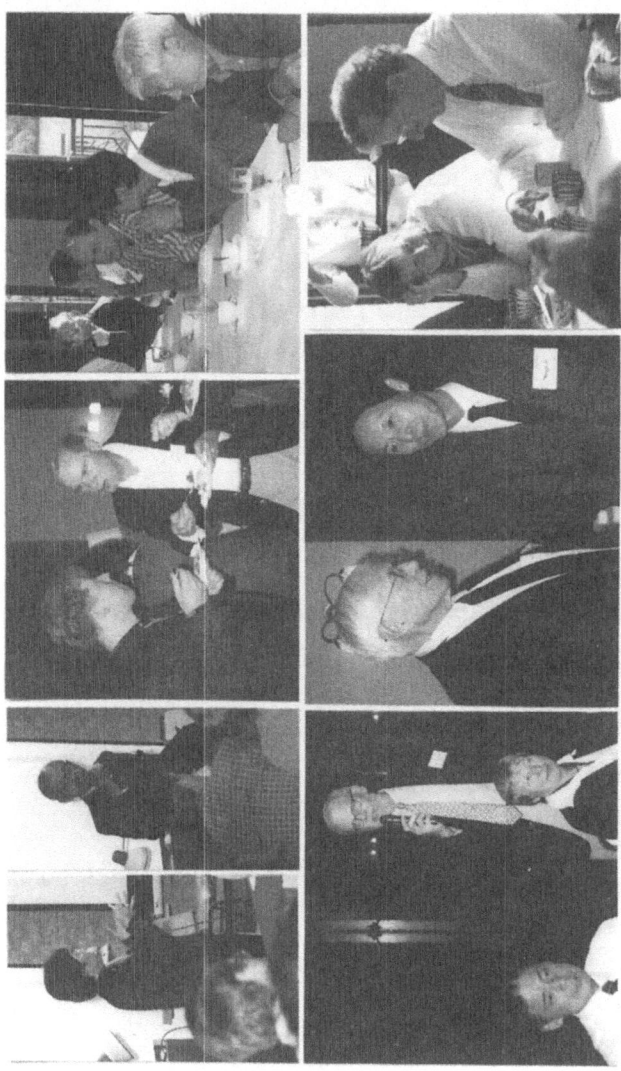

Photo 4. (First row) N.A. Curtin / H.L. Sweeney, K.W. Ranatunga / R. Stehle, D.W. Maughan / H.E.D.J. ter Keurs, D. Rassier, H.L. Granzier, H. Onishi, K. Tawada. (Second row) S. Sugiura, A.M. Gordon, G. Pfitzer / N.R. Alpert, R.Nagai / G. Pfitzer, H.E.D.J. ter Keurs.

PARTICIPANTS

Alpert, N. R.
Department of Molecular Physiology and
Biophysics
University of Vermont, College of
Medicine
Burlington, VT 05405-0075, U.S.A.

Arata, T.
Department of Biology
Graduate School of Science
Osaka University
Toyonaka, Osaka 560-0043, Japan

Bagni, M. A.
Dipartimento di Scienze Fisiologiche
Università di Firenze
I-50134 Firenze, Italy

Cecchi, G.
Dipartimento di Scienze Fisiologiche
Università di Firenze
I-50134 Firenze, Italy

Chaen, S.
Department of Applied Physics
College of Humanities and Sciences
Nihon University
Setagaya-ku, Tokyo 156-8550, Japan

Curtin, N. A.
BSF Section
Division of Biomedical Science
Imperial College
London SW7 2AZ, U.K.

Endo, M.
Department of Pharmacology
Saitama Medical School
Saitama 350-0495, Japan

Fransen, P.
Department of Pharmacology
University of Antwerp (RUCA)
B-2020 Antwerp, Belgium

Gonzalez-Serratos, H.
Department of Physiology
School of Medicine
University of Maryland
Baltimore, MD 21201-1596, U.S.A.

Gordon, A. M.
Department of Physiology and Biophysics
University of Washington
Seattle, WA 98195-7290, U.S.A.

Granzier, H. L.
VCAPP
Washington State University
Pullman, WA 99164-6520, U.S.A.

Holmes, K. C.
Max Planck Institute for Medical
Research
Abteilung Biophysik
D-69120 Heidelberg, Germany

Huxley, H. E.
Rosenstiel Center
Brandeis University

Waltham, MA 02454, U.S.A.

Ikebe, M.
Department of Physiology
University of Massachusetts
Medical School
Worcester, MA 01655, U.S.A.

Ishiwata, S.
Advanced Research Institute for
Science and Engineering
Waseda University
Shinjuku-ku, Tokyo 169-8555, Japan

Katayama, E.
Division of Biomolecular Imaging
Institute of Medical Science
University of Tokyo
Minato-ku, Tokyo 108-8639, Japan

Katoh, K.
Neuroscience Research Institute
National Institute of Advanced Industrial
Sciences and Technology
Tsukuba, Ibaraki 305-8568, Japan

Kobayashi, T.
Department of Electronic Science
Shibaura Institute of Technology
Minato-ku, Tokyo 108-8548, Japan

Kodera, N.
Department of Physics
Graduate School of Sciences
Kanazawa University
Kanazawa, Ishikawa 920-1192, Japan

Kushmerick, M. J.
Department of Radiology
University of Washington
Seattle, WA 98195-7115, U.S.A.

Lännergren, J.
Department of Physiology &
Pharmacology
Karolinska Institute
S-171 77 Stockholm, Sweden

Matsuoka, R.
Tokyo Women's Medical University
Shinjuku-ku, Tokyo 162-8666, Japan

Maughan, D. W.
Department of Molecular Physiology and
Biophysics
University of Vermont
Burlington, VT 05405, U.S.A.

Metzger, J. M.
Deprtment of Physiology
University of Michigan
Ann Arbor, MI 48109-0622, U.S.A.

Miyata, H.
Physics Department
Graduate School of Science
Tohoku University
Sendai, Miyagi 980-8578, Japan

Morano, I.
Max-Delbrück-Center for Molecular
Medicine
D-13125 Berlin, Germany

Nagai, R.
Department of Cardiovascular Medicine
University of Tokyo
Bunkyo-ku, Tokyo 113-0033, Japan

Nakayama, K.
Department of Pharmacology
Faculty of Pharmaceutical Sciences
Shizuoka Prefectural University
Shizuoka 422-8526, Japan

Nishizaka, T.
Kansai Advanced Research Center
Protein Biophysics Group
Kobe 651-2492, Japan

Ogata, M.
Institute of Health Science
Kyushu University

Fukuoka 812-8582, Japan
Ohtsuki, I.
Department of Pharmacology
Graduate School of Medicine
Kyushu University
Fukuoka 812-8582, Japan

Okamoto, Y.
Department of Applied Chemistry
Muroran Institute of Technology
Muroran, Hokkaido 050-8585, Japan

Okuyama, H.
Department of Physiology
Kawasaki Medical School
Okayama 701-0192, Japan

Onishi, H.
Department of Structural Analysis
National Cardiovascular Research Center
Suita, Osaka 565-8565, Japan

Pfitzer, G.
Department of Physiology &
Pathophysiology
University of Cologne
D-50931 Koeln, Germany

Pollack, G. H.
Department of Bioengineering
University of Washington
Seattle, WA 98195, U.S.A.

Rall, J. A.
Department of Physiology & Cell Biology
Ohio State University
Columbus, OH 43210, U.S.A.

Ranatunga, K. W.
Department of Physiology
School of Medical Sciences
University of Bristol
Bristol BS8 1TD, U.K.

Rassier, D.
Human Performance Laboratory
University of Calgary
Calgary, AB, T2N 1N4, Canada

Rüegg, J. C.
Haagackerweg 10
D-69493 Hirschberg, Germany

Saeki, Y.
Department of Physiology
Tsurumi University School of Dental
Medicine
Yokohama 230-8501, Japan

Shirakawa, I.
Department of Physiology
School of Medicine
Teikyo University
Itabashi-ku, Tokyo 173-8605, Japan

Solaro, R. J.
Department of Physiology & Biophysics
University of Illinois at Chicago
College of Medicine
Chicago, IL 60612-7342, U.S.A.

Squire, J. M.
Biological Structure & Function Section
Biomedical Sciences Division
Imperial College
London SW7 2AZ, U.K.

Stehle, R.
Institute of Physiology
University of Cologne
D-50931 Koeln, Germany

Stienen, G. J. M.
Laboratory for Physiology
Institute for Cardiovascular Research
Free University
1081 BT Amsterdam, the Netherlands

Suda, N.
Department of Physiology
The Jikei University School of Medicine
Minato-ku, Tokyo 105-8461, Japan

Sugi, H.
Department of Physiology
School of Medicine
Teikyo University
Itabashi-ku, Tokyo 173-8605, Japan

Sugiura, S.
Institute of Environmental Studies
Graduate School of Frontier Sciences
University of Tokyo
Bunkyo-ku, Tokyo 113-0033, Japan

Sweeney, H. L.
Department of Physiology
University of Pennsylvania
School of Medicine
Philadelphia, PA 19104-6085, U.S.A.

Takuwa, Y.
Department of Molecular & Cell
Physiology
Kanazawa University
Graduate School of Medicine
Kanazawa 920-8640, Japan

Tanokura, M.
Department of Applied Biological
Chemistry
Graduate School of Agricultural and
Life Sciences
University of Tokyo
Bunkyo-ku, Tokyo 113-0033, Japan

Tawada, K.
Department of Biology
Graduate School of Sciences
Kyushu University
Fukuoka 812-8581, Japan

Telley, I. A.
Laboratory for Biomechanics
Swiss Federal Institute of Technology
CH-8952 Zürich, Switzerland

ter Keurs, H. E. D. J.
Cardiovascular Research Group
Faculty of Medicine
University of Calgary
Calgary, AB T2N 4N1, Canada

Tsuchiya, T.
Department of Biology
Faculty of Science
Kobe University
Nada-ku, Kobe 657-8501, Japan

Wakabayashi, K.
Structure Biophysics Laboratory
Division of Biophysical Engineering
Graduate School of Engineering Science
Osaka University
Toyonaka, Osaka 560-8531, Japan

Westerblad, H.
Department of Physiology &
Pharmacology
Karolinska Institutet
S-171 77 Stockholm, Sweden

Winegrad, S.
Department of Physiology
University of Pennsylvania
School of Medicine
Philadelphia, PA 19104-6085, U.S.A.

Woledge, R. C.
UCL Institute of Human Performance
Royal National Orthopedic Hospital
Stanmore, Middx HA7 4LP, U.K.

Yamada, K.
Department of Physiology
Oita Medical University
Oita 879-5593, Japan

Yamada, T.
Department of Physics
Faculty of Science
Tokyo University of Science
Shinjuku-ku, Tokyo 162-8601, Japan

Yamashita, H.
Department of Cardiovascular Medicine
Graduate School of Medicine
University of Tokyo
Bunkyo-ku, Tokyo 113-0033, Japan

Yu, L. C.
Laboratory of Muscle Biology
NIAMS, National Institutes of Health
Bethesda, MD 20892-2755, U.S.A.

SUBJECT INDEX

Entries indicate pages on which descriptions on the subjects are readily accessible.